Pseudomonas syringae and related pathogens

Pseudomonas syringae and related pathogens

Biology and Genetic

Edited by

Nicola Sante Iacobellis

Alan Collmer

Steven W. Hutcheson

John W. Mansfield

Cindy E. Morris

Jesus Murillo

Norman W. Schaad

David E. Stead

Giuseppe Surico

and

Matthias S. Ullrich

Università degli Studi della Basilicata,
Dipartimento di Biologia Difesa e Biotecnologie,
Potenza, Italy

KLUWER ACADEMIC PUBLISHERS
DORDRECHT / BOSTON / LONDON

A C.I.P. Catalogue record for this book is available from the Library of Congress.

ISBN 1-4020-1227-6

Published by Kluwer Academic Publishers,
P.O. Box 17, 3300 AA Dordrecht, The Netherlands.

Sold and distributed in North, Central and South America
by Kluwer Academic Publishers,
101 Philip Drive, Norwell, MA 02061, U.S.A.

In all other countries, sold and distributed
by Kluwer Academic Publishers,
P.O. Box 322, 3300 AH Dordrecht, The Netherlands.

Printed on acid-free paper

Printed in the Netherlands.

CONTENTS

PREFACE

This volume mainly reports on new and recent advancements on different aspects of *Pseudomonas syringae,* a plant pathogenic bacterial species that include a high number of pathogens of important crops, which is an interesting model organism in plant pathology. In addition some related fluorescent *Pseudomonas* spp., responsible of new and emerging diseases, as well as some pathogens previously included in the above genus and now classified in the genera *Ralstonia*, *Acidovorax* are also considered.

The tremendous recent advancements on:
the ecology and epidemiology and, in particular, the adaptation of *P. syringae* to stresses and adverse environmental conditions;
the function and regulation of genes involved in the production of phytotoxins and on their mechanism of action in the interaction with the host cells;
the structure, function and regulation of type three secretion system (TTSS) and the transport of the effectors proteins in the host cells;
the possibility to control diseases through the induction of the systemic acquired resistance (SAR);
the development of molecular techniques for the highly specific and sensible identification and detection of pathogens;
the determination of the causal agents of new and emerging diseases as well the classification of the different pathovars of *P. syringae;*
are reported in 76 chapters cured by leading scientist in the respective fields.

This volume - a compilation of the presentations at the 6th International Conference on *Pseudomonas syringae* pathovars and related pathogens hold in Maratea, Pz, Italy, September 15-19, 2002 - has been divided in eight sections: Ecology and Epidemiology, Pathogenesis and Determinants of Pathogenicity, Genetic and Physiological Analysis of Host Pathogen Interaction, Molecular Characterisation / Genomics, Disease Management and Control, Molecular Techniques for Identification and Detection, New Emerging Pathogens and Taxonomy. All the chapter of this volume have been reviewed by the editors.

I'd gratefully acknowledge the fruitful collaboration of the scientific and organising committees, the financial support of several institutions and private companies (see later) and Dr. Pietro Lo Cantore for the precious secretariat assistance and, in particular, in the preparation of this book.

Nicola Sante Iacobellis

Sponsors and Donors

We are grateful to the following sponsors and donors:

International Society of Plant Pathology (ISPP)
Società Italiana di Patologia Vegetale (SIPaV)
Università degli Studi della Basilicata
Facoltà di Agraria - Università degli Studi della Basilicata
Regione Basilicata - Assessorato all'Agricoltura
Agenzia Lucana di Sviluppo e di Innovazione in Agricoltura (ALSIA)
Provincia di Potenza - Assessorato all'Agricoltura
Ministero dell' Istruzione, dell'Università e della Ricerca (MIUR)
Azienda di Promozione Turistica della Basilicata
Syngenta Crop Protection s.p.a.
Innotec s.r.l.
Tecnolife s.r.l.
Foss Italia s.p.a.
Cooperativa a r.l. Cantina ed Oleificio Sociale di Manduria
Chemie s.a.s.
Banca Mediterranea s.p.a.
Celbio s.r.l.
Florilab s.r.l.
Co.Vi.l.
International P.B.I.
Blackwell Publishers co.
Microtek s.r.l.
Co. Vi.Mer. a.r.l. Cooperativa Vivaistica Meridionale
Vivai Giandomenico Consalvo

Ecology and Epidemiology

Epidemiological Clues for Developing Methods of Control of Bacterial Blight of Cantaloupe Caused by *Pseudomonas syringae* pv. *aptata*

C.M.-H. RIFFAUD, C. GLAUX, C. GUILBAUD, H. DOMINGUEZ[#], P. PRIOR and C.E. MORRIS

INRA, Station de Pathologie Végétale, BP 94, Domaine St. Maurice, 85140 Montfavet, France; [#]Institut Universitaire de Technologie d'Avignon, Département de Génie Biologique, Avignon, France.

Abstract: Since the early 1990's, bacterial blight has emerged as an important disease of cantaloupe in France, particularly in the south-west but has caused epidemics in all the production regions of this country. The pathogen is a complex composed of strains identified as *Pseudomonas syringae* pv. *aptata* (70% of the strains) and the remaining resembling *P. viridiflava* (2%), *P. s.* pv. *lachrymans* (1%) or *P. syringae* in general but with no correspondence to a known pathovar. To investigate the epidemiology of this disease and to develop disease control strategies, we have attempted to identify sources of inoculum in the field, factors contributing to the explosive development of the disease, resistance of the bacterium to copper pesticides, and sources of disease resistance in *Cucumis melo*. Based on our observations and on the description of possible outbreaks of this disease elsewhere in the world, we report here a hypothetical scenario of the critical factors triggering disease development and of the potential efficiency of different control strategies.

Key words: inoculum sources, immunofluorescence colony staining, disease resistance, biofilms.

Author for correspondence: C.E. Morris; e-mail: morris@avignon.inra.fr.

1. PROLOGUE

In 1993, cantaloupe fields in the Tarn-et-Garonne district of south-western France were devastated by a bacterial blight that growers considered to be a

N.S. Iacobellis *et. al.* (eds.), *Pseudomonas syringae* and Related Pathogens, 3-15.

new disease. Since 1993, this disease has been observed in all cantaloupe regions of France when favourable conditions for the disease prevail. As of 1995, our research team has been studying the aetiology and epidemiology of the disease and the ecology of the pathogen, a complex population of *Pseudomonas syringae* pathovars and related species. This research has been conducted in the context of a national working-group composed of researchers in plant breeding and plant pathology, of staff of extension services, experimental stations, inter-professional organisations and chambers of agriculture, and representatives of grower's associations. The field observations of this working-group have shed considerable light on our research results thereby helping us to construct a hypothetical scenario of the critical factors for disease development, to predict the potential efficiency of different control strategies and to evaluate the possibility that the disease emerges elsewhere. This chapter summarizes our current view of bacterial blight of cantaloupe and the potential for control.

2. BACTERIAL BLIGHT OF CANTALOUPE CAUSED BY *PSEUDOMONAS SYRINGAE* PV. *APTATA*

2.1 Disease Symptoms and Economic Importance in France

Symptoms of bacterial blight occur on all aerial parts of cantaloupe plants: necrotic spots on leaves, cankers on stems and petioles and sunken lesions on fruits. Leaf lesions are not necessarily angular and may be encircled by water-soaked halos. Cankers are dry and light in colour. Fruit lesions may include large, dry-rotted cavities in the flesh of the fruit regardless of the size of the apparent lesions on the epidermis. Epidemics occurring at the time of early fruit development can lead to fruit drop. The similarity of these symptoms to those caused by *Cladosporium cucumerinum* and *Colletotrichum lagenarium* (Blancard and Morris, 1998), common pathogens of cantaloupe in France, may explain in part why bacterial blight has been recognised only recently.

Three principal regions of France are responsible for this country's production of 320,000 tons of cantaloupe per year: south-eastern France (42% of the annual production), the south-west (32%) and the central-west (26%). Cantaloupe production often constitutes part of a polyculture system and occupies about 2.5 ha on average for each of the 6,300 farms producing cantaloupe in France (C.T.I.F.L., 1998). Cultivars of the highly fragrant

botanical variety *cantalupensis* of *Cucumis melo* constitute nearly 100% of the French production. A mere 0.2% consists of *C. melo* var. *inodorus* that is less fragrant but is more readily stored. For the 20 tons ha^{-1} that farmers produce on average, they earned a gross income of 500 to 3,000 euro per ton in 1998 depending on the season.
The south-west has been the victim of the most important and re-occurring epidemics of bacterial blight. Since 1993, important epidemics occurred in this region in 1995, 1996 and 1997. In 1997, virtually all fields in the Lot and Tarn-et-Garonne districts were attacked and the disease was officially recognised as a natural catastrophe by the regional governments. The central-west was also heavily affected in this period. Prevailing climatic conditions (warm and windy) in the south-east are unfavourable for the disease. However, the rainy, cool spring of 2002 was conducive to the first severe epidemics witnessed in the south-east. Early plantings that were fruiting in mid-June 2002 in the south-east were completely lost.

2.2 The Causal Agent

The first comprehensive report on the aetiology of this disease was based on the physiology, genotype and host range of 23 strains of the causal agent collected from the south-west and central-west regions from 1995 to 1997 (Morris *et al.*, 2000). The results of biochemical tests of these strains were consistent with those reported for the pvs. *aptata*, *atrofaciens*, *japonica* and *syringae* of *Pseudomonas syringae*. Furthermore, nearly all strains virulent on cantaloupe were virulent on sugar beets. Similarly, reference strains of *P. s.* pv. *aptata* were also virulent on cantaloupe. Hence, we proposed that the causal agent of cantaloupe blight in France be considered as *P. s.* pv. *aptata*.
Since this initial study, we have continued to characterise an additional 225 strains collected from 1995 to 2001 from all French cantaloupe-growing regions. According to the biochemical and host range tests used previously (Morris *et al.*, 2000), all strains are *P. syringae*-like (fluoresce on King's medium-B, induce a hypersensitive reaction on tobacco, hydrolyse esculin and do not possess a terminal cytochrome oxidase nor arginine dihydrolase). Furthermore, 70% correspond to *P. s.* pv. *aptata*. However, 2% of the strains resemble *P. viridiflava* in terms of colony pigmentation, pectolytic activity and other metabolic traits. One percent correspond to *P. s.* pv. *lachrymans*. The remaining strains vary in terms of the principal diagnostic biochemical tests and do not correspond to any of the described pathovars. Interestingly, nearly half of all strains are pectolytic either on potato slices or in pectate gel. Pectolytic activity is rare among *P. syringae* pathovars according to comprehensive studies of the species (Young and Triggs, 1994), with pv. *lachrymans* being one of the few to frequently manifest this trait. Pectolytic activity is also typical of *P. viridiflava*. However, most pectolytic strains of

the cantaloupe pathogen have phenotypic profiles inconsistent with those of *P. s.* pv. *lachrymans* and *P. viridiflava.*
Our collection of strains represents a wide range of levels of aggressiveness to cantaloupe. Aggressiveness is not apparently correlated to the production of syringomycin, to pectolytic activity or to other biochemical properties of the strains. However, strains corresponding to pv. *lachrymans* are among the least aggressive. Furthermore, aggressiveness on cantaloupe is well correlated to the number of other plant species on which the strains are virulent (Morris *et al.*, 2000).

2.3 Occurrence of the Disease Elsewhere in the World

Several severe epidemics of bacterial blights of *Cucumis melo* caused by *Pseudomonas syringae* have been reported in the US (Colorado) in 1946 (Smith, 1946), in Japan in 1979 (Fujita *et al.*, 1979) and in India in 1991 (Jindal and Bhardwaj, 1991). Bacterial blights of *C. melo* caused by *P. syringae* or related species have also been observed in Argentina (Atlas de Gotuzzo, 1976), Brazil (Beriam *et al.*, 1998), Italy (Cobelli and Vicchi, 1996), Crete (Goumas and Chatzaki, 1998), Morocco (Fatmi, 2003), Turkey (Ozaktan and Bora, 1994), the Ukraine (Saleh and Korobko, 1981), China (Wang *et al*, 1992) and New Zealand (Hill, 1979). The disease generally concerns vars. *reticulatus* and *inodorus* and not var. *cantalupensis* as this latter type is a particularity of French production. Furthermore, we are not aware of recurrent, serious epidemics in any of these locations.
The causal agent for the diseases cited above is almost always referred to as *P. s.* pv. *lachrymans*, although *P. s.* pv. *syringae* (Young, 1991), *P. viridiflava* (Goumas and Chatzaki, 1998) and *P. cichorii* (Beriam *et al.*, 1998) have also been cited. Given the diversity of the pathogen population that we have observed in France, it is not surprising that these different identities have been reported. Furthermore, our identification of the pathogen as *P. s.* pv. *aptata* was based in large part on extensive host-range testing which has not been reported for the strains attacking *C. melo* elsewhere. In light of the complexity of the *P. syringae* group and related bacteria, it is likely that none of the reports to date, including ours, represents the final word about the identification of the *C. melo* pathogen.

3. SOURCES OF INOCULUM

Identifying sources of inoculum could provide partial evidence for the origin of bacterial blight of cantaloupe in France by revealing if the pathogen is widespread and apparently indigenous or if it has been introduced on seed, for example. Eradicating, reducing or avoiding these sources could also be a means of control. The pertinent potential sources to investigate for bacterial blight of cantaloupe would be those typical of *Pseudomonas syringae* in general (debris in soil, epiphytic populations on weeds or nursery plants, populations on seeds, etc.). In the case of cantaloupe production in south-western France, irrigation water may also be a possible source of inoculum as fields are irrigated with water from retention lakes typical of this hilly region. Run-off from fields on slopes could potentially transport bacteria to the lakes.

3.1 Detection

To detect *Pseudomonas syringae* pathogenic to cantaloupe, we applied immunofluorescence colony staining (IFC) (Van Vuurde, 1987) by using an appropriate semi-selective medium (Mohan and Schaad, 1987) and three of the polyclonal antisera of Saunier *et al.* (1996) as described by Riffaud and Morris (2002). Irrigation waters and washings of soil, seed and leaf samples were concentrated as much as possible to minimise the limit of detection. The ability to concentrate samples depended on the abundance of background microflora that grew in the semi-selective medium and the presence of charged particulate matter such as clay particles in soil. The limits of detection of *P. syringae* under our conditions were 4 to 200 CFU l^{-1} for irrigation water, 1 CFU g^{-1} of sandy soil, 400 CFU g^{-1} of clay soil, 100 CFU per lot of 1,000 seeds for bacteria on external parts of seeds and 2,000 CFU per 1,000 seeds for internal bacteria, and less than 1 CFU $leaf^{-1}$ for bacteria from nursery plants.

To date, we have analysed samples from 17 sources of irrigation water, 6 commercial lots of seed (2,000 seeds per lot), leaves of plants from 2 commercial nurseries and 3 “home” nursery productions, and soil from 4 fields just before transplanting of cantaloupe plants at the beginning of the season and from 3 fields at the end of the season just after cultivation. For these latter 3 fields, samples were taken at sites where debris from diseased plants was buried. *P. syringae* pathogenic to cantaloupe was not detected in any of the soil or seed samples. However, this bacterium was detected at low levels in one “home” nursery sample (0.3 CFU $leaf^{-1}$) and in water from 3 different retention lakes (10, 12 and 70 CFU l^{-1}). Irrigation and rain water are well-known means of disseminating *P. syringae* from various sources (Lindemann *et al.*, 1985; Hirano *et al.*, 1995). However, our report (Riffaud and Morris, 2002) constitutes the first report of irrigation water itself being a source of *P. syringae*.

3.2 Epidemiologically Significant Sources of Inoculum

Is it surprising that we detected *Pseudomonas syringae* of cantaloupe so infrequently in the potential sources examined? Furthermore, are the quantities detected on nursery plants and in water sufficient to incite an epidemic? Firstly, cells of *P. syringae* pathogenic to cantaloupe associated with soil, debris, seeds, water, etc. could be in a viable-but-not-culturable state and hence undetectable by IFC. At present, we have no rigorous evidence to support or reject this hypothesis. Secondly, we have followed the dynamics of *P. s.* pv. *aptata* survival by burying a mixture of soil and infected debris in nylon mesh sacks at two experimental sites (Riffaud, 2002). For the site with sandy soil, we could concentrate samples sufficiently and *P. s.* pv. *aptata* was detected at densities as low as 1 CFU g^{-1} one year after burying the sacks (study still in progress). This suggests that *P. s.* pv. *aptata* may be present in the clay soils of the south-west, but at densities below the detection limit of 400 CFU g^{-1} for these soils. Thirdly, we wondered if our detection limit is above the density of bacteria expected in soil after cultivation of diseased cantaloupe. For this latter point, we made simple calculations by taking into account the planting density of cantaloupe (9,000 plants ha^{-1}), the average number of leaves per plant at harvest (350), the fact that the first 20-cm layer of 1 ha of soil represents about 3,000 tons of soil and that fresh diseased leaves harbour total pathogen populations of ca. 10^8 CFU (the size of populations on healthy leaves is negligible for this calculation based on our observations). For an epidemic leading to an incidence of 100% diseased leaves and no bacterial mortality while leaves dry or after plant debris is buried, an average of 10^5 CFU *P. s.* pv. *aptata* would be added to each gram of soil. However, if one considers more realistic disease incidence and bacterial mortality rates (Fig. 1), the expected densities are much lower. For example, for disease incidence of 25% and mortality rates of 99%, as few as 250 bacteria g^{-1} of soil would be expected. It should be noted that these calculations do not take into account spatial heterogeneity in a field.

It is likely that bacterial mortality rates in soil are as high as 99% or greater given that cantaloupe fields are part of a 3- to 4-year rotational scheme. Likewise, the bacterial densities detected in water represent an average of 2,000 CFU $plant^{-1}$, or less than 10 CFU $leaf^{-1}$, at the time of peak irrigation (not accounting for run-off, non-homogenous distribution of bacteria in water, etc.). Are these quantities, and those detected on nursery plants, epidemiologically significant? Another simple calculation might give insight to this question.

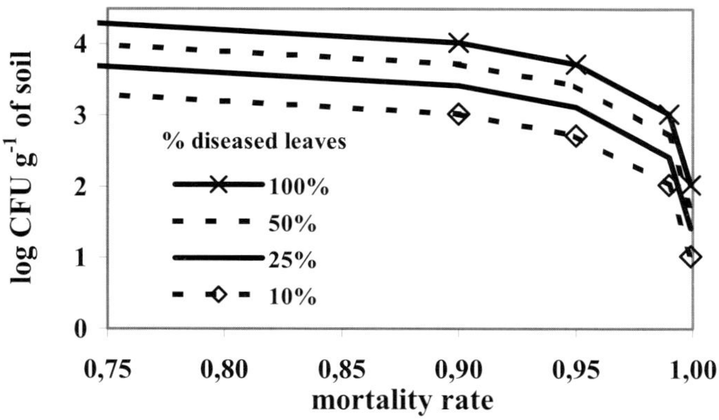

Figure 1. Estimated densities of *Pseudomonas syringae* pathogenic to cantaloupe in the top 20 cm of soil as a function of disease incidence at harvest and bacterial mortality rate after plant debris is ploughed into fields. Calculations are based on 10^8 pathogenic bacteria per diseased leaf at the end of the growing season.

The number of doublings needed to attain a certain population is expressed by the equation $2^x n=Y$, where 'x' is the number of doublings, 'n' is the initial population size and Y is the population size after multiplication. This equation represents constant exponential growth and does not take into account death rates, but gives a point of reference nevertheless. Hence, to attain a population size of 10^7 cells (a rough first estimate for the population threshold necessary for bacterial blight of cantaloupe), 23 doublings or about 5 days (5 h doubling^{-1}) would be required for an initial population of 1 cell, and 3 days would be required for an initial population of 1,000 cells. Interestingly, the forecasting model being developed for this disease has revealed that over 90% of the disease outbreaks have occurred within 3 to 5 days after the beginning of conducive conditions (mean temperature <16°C coupled to rain or high RH) (Lavigne, personal communication).

4. DISEASE RESISTANCE IN *CUCUMIS MELO*

None of the commercially available cultivars of cantaloupe adapted to the French market are resistant to bacterial blight (Guerineau and Riffaud, 2001). Sources of resistance to wild lines and cultivars grown elsewhere were explored by inoculating plants at the 2-leaf stage, detached fruits at various stages of development and plants at a range of developmental stages through maturity in greenhouses. For plants at the 2-leaf stage, resistance of 282 accessions from Europe, the Middle East, Africa, China, Japan, India,

Africa and North, South and Central America was evaluated by incising leaves and petioles with scissors dipped in inoculum. This method produced the most reliable results compared to spraying or injecting plants with inoculum (manuscript in preparation). All of the accessions tested developed symptoms after inoculation and 78% of them had mean disease severity scores greater than 3 (scale of 0-4). Results of tests in greenhouses by misting plants with inoculum at 3 different stages during their development has illustrated that the general trends in sensitivity at the 2-leaf stage correspond to those at the adult stage, with a few exceptions.

For fruits, preliminary observations from field and greenhouse studies suggested that the degree of sensitivity of fruits was independent from that of vegetative parts. Hence, to screen fruits, accessions were chosen to represent the range of different degrees of sensitivity at the 2-leaf stage. Of 13 accessions tested, one showed nearly complete resistance when fruits were dipped in bacterial suspensions of 10^8 CFU ml^{-1}. This accession from Sudan was extremely sensitive to bacterial blight when inoculated at the 2-leaf stage. The inheritance of fruit resistance from this accession is currently being studied in view of developing breeding lines.

5. PROSPECTS FOR CHEMICAL CONTROL

The only pesticides currently registered for use against bacterial blight of cantaloupe in France are copper-based. On cantaloupe, copper is applied at a dose of 800 mg l^{-1} of Cu^{+2} and often in a mixture with Mancozeb-based products. However, our working-group has regularly observed that disease incidence and severity in fields treated with copper are just as high as in untreated fields. The sensitivity of *Pseudomonas syringae* pv. *aptata* strains to copper on CYE medium (Menkissoglu and Lindow, 1991) is illustrated in Fig. 2. According to Menkissoglu and Lindow (1991) addition of 40 mg l^{-1} of Cu^{+2} to CYE corresponds to about 0.8 mg l^{-1} of the free, toxic form of Cu^{+2}. A linear decrease in copper added to the medium corresponds to an exponential decrease of toxic copper ions in the medium. Hence, this suggests that *P. s.* pv. *aptata* would be sensitive to the doses of copper applied in the field. However, most of the copper applied to leaves is likely to be bound into non-toxic forms by chemicals on the leaf surface (Menkissoglu and Lindow, 1991). Menkissoglu and Lindow (1991) estimated that only 0.1% to 1% of the copper applied to citrus and bean leaves, for example, remains in a free, toxic form. The extent to which cantaloupe leaf surface compounds bind copper has not been determined. If they bind up to 99.9% of the applied copper as for citrus, then about 2% of the strains of *P. s.* pv. *aptata* would be resistant to copper treatments.

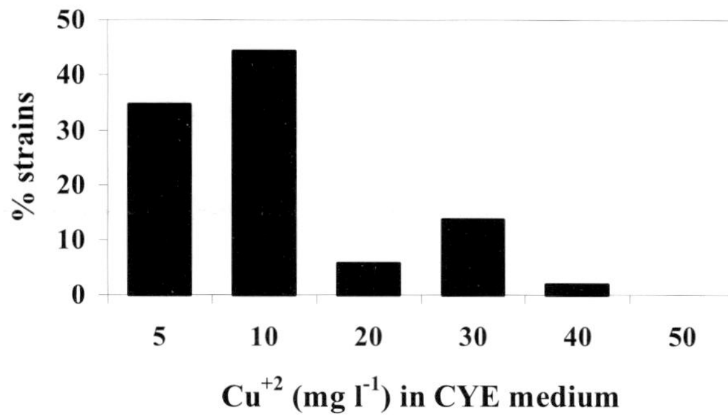

Figure 2. Copper sensitivity of *Pseudomonas syringae* pathogenic to cantaloupe as determined in CYE medium. For each concentration of copper incorporated into the medium, the figure indicates the percent of strains having this concentration as the maximum concentration tolerated.

Other factors might also influence the effectiveness of pesticides against bacteria on leaf surfaces. Features of leaf topography may interfere with contact between applied chemicals and bacteria, such as the leaf hairs and crested junctions between epidermal cells on cantaloupe leaves (Morris, 2001). Epiphytic biofilms could also protect bacteria from the action of pesticides (Morris *et al.*, 2002), and we have found that cells of *P. s.* pv. *aptata* may occasionally integrate into biofilms on leaf surfaces. For 15 symptomless leaves of cantaloupe analysed for the presence of biofilms via a quantitative filtration technique (Morris *et al.*, 1998), 4 had detectable populations of *P. s.* pv. *aptata* in the biofilm component of the epiphytic population. These biofilms harboured 30-80% of the total epiphytic populations of *P. s.* pv. *aptata* on these leaves. The physiology of *P. syringae* in its native state on cantaloupe leaves may also have a significant effect on pesticides. We are currently studying the resistance of the cantaloupe pathogen to hydrogen peroxide after the bacterium has grown on cantaloupe leaf surfaces. Hydrogen peroxide is one of the components of new bactericides (Fatmi, 2003) that may eventually be available for use on cantaloupe in France. Our results to date suggest that bacterial cells from the epiphytic phase are more resistant than those grown on nutrient medium. One of the implications of these results concerns how to conduct laboratory assays to determine the effectiveness of bactericides *in situ* as the use of assays on nutrient media are quite misleading.

6. ORIENTING RESEARCH TOWARD CRITICAL POINTS IN THE DISEASE CYCLE AND STRATEGIES OF CONTROL

At present, we view the disease cycle of bacterial blight on cantaloupe and the life cycle of the pathogenic population in the framework of what has been commonly described for diseases of annual crops caused by *Pseudomonas syringae*. From the inoculum sources where the pathogen over-winters in absence of cantaloupe, it is disseminated to aerial surfaces of cantaloupe where it multiplies epiphytically, penetrates into the intercellular spaces of leaves, stems, petioles and fruits, continues to multiply *in planta*, and then incites symptom development in the plant. Cells of the bacterium are contributed to inoculum sources primarily as cells washed from plants in run-off of irrigation or rain water that flows into sources of irrigation water or in association with debris in soil or in irrigation water. As for *P. s.* pv. *syringae* causing brown spot of bean (Hirano *et al*., 1995), the cantaloupe pathogen apparently multiplies explosively upon onset of favourable conditions. For the former pathogen, intense rain leads to rapid development of epiphytic populations (Hirano *et al*., 1995). However, for *P. syringae* of cantaloupe, prevailing temperatures below 16°C seem to be the decisive factor. However, it is not yet clear if rapid population development occurs epiphytically or only in intercellular spaces. Epiphytic populations of *P. syringae* on apparently healthy cantaloupe leaves detected to date, even those collected in disease foci, have been relatively low - below 10^6 CFU leaf^{-1} - compared to what has been observed for the brown spot pathogen (Hirano *et al*., 1995).

Throughout this chapter we have evoked potential disease control strategies. Some of these seem promising, but numerous questions remain unresolved. Firstly, sources of resistance in *Cucumis melo* have been identified for disease on fruits. However, we are concerned about the effectiveness and durability of such resistance in an environmental context of diseased leaves. Furthermore, leaves are important for fruit quality in that they contribute to fruit sugar content and to shading thereby preventing sun scalding. Screening of *C. melo* accessions by inoculation has not yet led to identification of sources of resistance of vegetative parts to cantaloupe bacterial blight. However, it may be profitable to investigate the range of physiological tolerance to temperature in *C. melo* as a strategy for developing disease tolerance. The temperatures conducive to cantaloupe blight are those for which the growth of cantaloupe is markedly slowed. At 12°C cantaloupe plants stop growing (C.T.I.F.L., 1985). This may effectively shut down the expression of plant defence mechanisms and allow bacteria to proliferate rapidly.

Reduction, eradication or avoidance of inoculum sources is also, in theory, a potential disease control strategy for cantaloupe blight and has worked effectively for other diseases caused by *P. syringae*. The calculations presented in this chapter illustrate that a 10-fold reduction of initial inoculum on leaves could delay by about one day the time necessary to attain the population size necessary for disease development. This could have a critical impact on disease incidence if weather conditions became unfavourable for disease during this delay. It would be tempting to propose that sources of irrigation water be disinfected or filtered, for example. But, even if the costs of such treatments were not prohibitive, at present it would be impossible to verify that the treated water was more pathogen-free than the untreated water given that the sizes of populations of the cantaloupe pathogen in inoculum sources are at the limit of detection with currently available techniques. Detection techniques for this bacterium need to be further refined and the question of its existence in a viable-but-not-culturable state needs to be addressed.

In light of our current understanding of bacterial blight of cantaloupe, chemical control is probably the least promising of the traditional strategies for disease control. However, development of this strategy evokes some of the most enticing questions about the biology and ecology of *P. syringae*. For example, where are the target organisms and what is their physiological state with regard to sensitivity to bactericides? Are the bacteria in biofilms protected from the action of bactericides? What amount of time occurs between ingression of bacteria into the intercellular spaces and expression of disease symptoms? Does significant bacterial multiplication occur during this time? What density of epiphytic bacteria on leaves is necessary to assure penetration into leaves?

At present, bacterial blight of cantaloupe has the stigma of being a problem specific to French cantaloupe growers and particularly those in the south-west. Our working-group is intrigued at the paucity of reports or active research about this disease elsewhere. We are currently investigating the relatedness of French strains of this pathogen to *P. syringae* and similar species attacking *C. melo* in other countries. It is not unlikely that the bacterium is widespread and emerges sporadically under conditions that are very atypical in most of the 70 different countries producing *C. melo*. The wide host range of the cantaloupe pathogen (Morris *et al.*, 2000) may allow it to persist as a disease on other crops. Furthermore, the *cantalupensis* botanical type that dominates production in France but not elsewhere might be particularly sensitive to the disease. Research on the relatedness of strains may give important insight into the potential for emergence of this disease as a significant threat to *C. melo* production and to its relationship to diseases on other hosts.

REFERENCES

Atlas de Gotuzzo, E., 1976, Angular spot of cucurbits in Argentina. *Fitopahologia* **11**: 41-46.

Beriam, L.O.S., Almeida, I.M.G., Malavolta, V.A.Jr., and Rodrigues Neto, J., 1998, Bacterial leaf blight of melon (*Cucumis melo* L.) caused by *Pseudomonas cichorii. Arq. Inst. Biol., São Paulo* **65**: 103-105.

Blancard, D., and Morris, C.E., 1998, Quel est votre diagnostic? PHM-*Revue Horticole*, **391**: 57-58.

C.T.I.F.L., 1985, Melon. Editions C.T.I.F.L., Paris, pp. 270

C.T.I.F.L., 1998, Le melon: pour un produit de qualité. Editions C.T.I.F.L., Paris, pp. 167.

Cobelli, L., and Vicchi, V., 1996, Melon and watermelon, strategies of integrated control. *Colture Protette* **25**(12): 39-42.

Fatmi, M., 2003, Use of Oxos, a Complex of Hydrogen Peroxide, Acetic Acid and Silver Ion, to Control Bacterial Speck of Tomato (*Pseudomonas syringae* pv. *tomato*) and Angular Leaf Spot of Melon (*P. s.* pv. *lachrymans*), 459-466, (this volume).

Fujita, Y., Onuma, Y., Sanada, T., Umekawa, M., and Watanabe, Y., 1979, Angular leaf spot of muskmelon I. Occurrence of the disease and infectivity of isolated bacteria. *Kita Nippon Byogaichu Kenkyukai Po.* **30**: 43-47

Goumas, D.E., and Chatzaki, A.K., 1998, Characterization and host range evaluation of *Pseudomonas viridiflava* from melon, blight, tomato, chrysanthemum and eggplant. *Euro. J. Plant Pathol.* **104**: 181-188.

Guerineau, C., and Riffaud, C., 2001, Bactériose du melon: mobilisation contre ce fléau. *Infos CTIFL* **170**: 45-48.

Hill, C.F., 1979, Angular leafspot on cucurbits. *New Zealand Commercial Grower, November,* 37-39.

Hirano, S.S., Rouse, D.I., Clayton, M.K., and Upper, C.D., 1995, *Pseudomonas syringae* pv. *syringae* and bacterial brown spot of snap bean: a study of epiphytic phytopathogenic bacteria and associated disease. *Plant Dis.* **79**: 1085-1093.

Jindal, K.K., and Bhardwaj, L.N., 1991, Occurrence of bacterial leaf spot on Persian melon (*Cucumis melo* var. *reticulatus* L.) *Plant Dis. Res.* **6**: 80.

Lindemann, J., Arny, D.C., and Upper, C.D., 1985, Aerial dispersal of epiphytic bacteria over bean plants. *Appl. Environ. Microbiol.* **50**: 1229-1232.

Menkissoglu, O., and Lindow, S.E., 1991, Relationship of free ionic copper and toxicity to bacteria in solutions of organic compounds. *Phytopathology* **81**: 1258-1263.

Menkissoglu, O., and Lindow, S.E., 1991, Chemical forms of copper on leaves in relation to the bactericidal activity of cupric hydroxide deposits on plants. *Phytopathology* **81**: 1263-1270.

Mohan, S.K., and Schaad, N.W., 1987, An improved agar plating assay for detecting *Pseudomonas syringae* pv. *syringae* and *P. s.* pv. *phaseolicola* in contaminated been seed. *Phytopathology* **77**: 1390-1395.

Morris, C.E., Monier, J.M., and Jacques, M.A, 1998, A technique to quantify the population size and composition of the biofilm component in communities of bacteria in the phyllosphere. *Appl. Environ. Microbiol.* **64**: 4789-4795.

Morris, C.E., Glaux, C., Latour, X., Gardan, L., Samson, R., and Pitrat, M., 2000, The relationship of host range, physiology and genotype to virulence on cantaloupe in *Pseudomonas syringae* from cantaloupe blight epidemics in France. *Phytopathology* **90**: 636-646.

Morris, C.E., 2001, Phyllosphere. In: Encyclopedia of Life Sciences, Nature Publishing Group, London, http://www.els.net

Morris, C.E., Barnes, M.B., and McLean, R.J.C., 2002, Biofilms on leaf surfaces: implications for the biology, ecology and management of populations of epiphytic bacteria. pp. 138-154. In: Lindow S.E., Poinar E., Elliot V. (eds.) Phyllosphere Microbiology, APS Press, Minneapolis.

Ozaktan, H., and Bora, T., 1994, Investigations on the comparison of in-vitro and in-vivo reaction tests for the determination of susceptibility of some cucurbits to *Pseudomonas syringae* pv. *lachrymans*. *J. Turk. Phytopath.* **23**: 105-111.

Riffaud, C.M.-H., and Morris, C.E., 2002, Detection of *Pseudomonas syringae* pv. *aptata* in irrigation water retention basins by immunofluorescence colony staining. *Euro. J. Plant Pathol.* **108**: 539-545.

Riffaud, C.M.-H., 2002, Thesis *La bactériose du melon: écologie et stratégies de lutte contre* Pseudomonas syringae *pv.* aptata., Univ. Claude Bernard – Lyon I, France. pp. 135.

Saleh, Y.E., and Korobko, A.P., 1981, Some aspects of bacterial diseases of melons and watermelons in Ukrainian SSR. I. *Pseudomonas lachrymans* (Smith and Bryan) Carsner. *Egypt. J. Bot.* **24**: 219-229.

Saunier, M., Maladrin, L., and Samson, R., 1996, Distribution of *Pseudomonas syringae* pathovars into twenty-three O serogroups. *Appl. Environ. Microbiol.* **62**: 2360-2374.

Smith, M. A., 1946. Bacterial spot of honeydew melon. *Phytopathology*. **36**: 943-949.

Van Vuurde, J.W.L, 1987, New approach in detecting phytopathogenic bacteria by combined immunoisolation and immunoidentification assays. *Bulletin OEPP/EPPO Bulletin* **17**: 139-148.

Wang, Z.T., Wang, J.F., Wang, M.S., and He, J., 1992, Resistance of muskmelon varieties to two diseases. *Crop Gen. Res.* **1**: 30-31.

Young, J.M., 1991. Pathogenicity and identification of the lilac pathogen, *Pseudomonas syringae* pv. *syringae* van Hall 1902. *Ann. Appl. Biol.* **118**: 283-298.

Young, J.M., and Triggs, C.M., 1994, Evaluation of determinative tests for pathovars of *Pseudomonas syringae* van Hall 1902. *J. Appl. Bacteriol.* **77**: 195-207.

Olive Knot Disease

New Insights into the Ecology, Physiology and Epidemiology of **Pseudomonas savastanoi** *pv.* **savastanoi**

G. SURICO and G. MARCHI
Dipartimento di Biotecnologie Agrarie, Sezione di Patologia vegetale, Università degli Studi, P.le delle Cascine 28, Firenze, Italy.

Abstract: The growth of *Pseudomonas savastanoi* pv. *savastanoi* on three selective media (ANS-S, PVF-1 and OKA) was studied. On ANS-S, recovery of *P. s.* pv. *savastanoi* strains was the highest and selectivity the lowest; on OKA it was the opposite (highest selectivity, lowest rate of recovery). Moreover, on OKA the colony morphology typical of the bacterium on NSA or KB, which are usually used for their growth, was lost. On PVF-1 recovery and selectivity were intermediate between ANS-S and OKA and the colonies of *P. s.* pv. *savastanoi* strains maintained their typical morphology. The spread in Tuscany of levan-positive populations of *P. s.* pv. *savastanoi* and the growth of this bacterium under adverse environmental conditions (poor nutrients and high or low agar concentration) is also reported. In this connection it was ascertained that *P. s.* pv. *savastanoi*, like other bacteria (*Bacillus subtilis*, *Proteus mirabilis*, *Escherichia coli*, etc.) adapts to hostile environmental conditions by forming complex growth patterns during colony development.

Key words: selective media, levan-positive strains, fractal growth, colony morphology.

Author for correspondence: G. Surico; e-mail: giuseppe.surico@unifi.it.

1. INTRODUCTION

Olive knot disease is common in most of the regions where olive is grown. It is caused by a bacterium, *Pseudomonas savastanoi* pv. *savastanoi* and is characterised by the formation of excrescences on young stems, branches and trunks and, more rarely, on the leaves, drupes and roots. In the last 30 years the disease has been studied in detail and from many points of view. This paper reports on the association of *P. s.* pv. *savastanoi* with an

N.S. Iacobellis *et. al.* (eds.), *Pseudomonas syringae* and Related Pathogens, 17-28.

Enterobacteriaceae in the phylloplane and in knots, on the selective isolation of *P. s.* pv. *savastanoi* from the phylloplane, on the morphology of the colonies, and on bacterial growth under limiting cultural conditions.

2. DIAGNOSIS

In his article "Les maladies de l'olivier et la tuberculose en particulier", Luigi Savastano in 1886 reported that a bacterium was the causal agent of olive knot. The Italian researcher was the first to succeed in reproducing the disease by artificially inoculating healthy plants with the bacterium he had isolated from olive knots.

Olive knot was also studied by E.F. Smith in the USA around 1904 (Smith, 1905), and at more or less the same time by L. Petri in Italy (Petri, 1907). These two researchers confirmed Savastano's work but they rejected the identity of the bacterium he had isolated. And indeed, according to Petri the bacterium isolated and described by Savastano was not in fact the causal agent of the disease but was another bacterium frequently associated with *Pseudomonas savastanoi* pv. *savastanoi* in knot tissue, *Ascobacterium luteum* Babés (= *Bacterium herbicola,* now *Pantoea agglomerans*). *A. luteum* was also isolated and erroneously identified as *P. s.* pv. *savastanoi* by Peglion in 1905 from knots on oleander. As suggested by Petri, both Savastano and Peglion probably reproduced the disease, on olive and oleander respectively, by using a culture of *A. luteum* inadvertently contaminated with *P. s.* pv. *savastanoi*, the real causal agent.

Of the two bacterial species, that of Savastano (1) and Smith (2), Petri gave the following description:

1. (*Bacillus oleae tubercolosis)* forms yellow colonies, is motile with peritrichous flagella, forms zoogloeae and capsules, liquefies gelatin, does not curdle milk, does not form spores;
2. (*Bacterium savastanoi*) consists of elongated bacteria with polar flagella, forms white colonies with entire margins, does not liquefy gelatin or curdle milk and does not produce spores.

In any case Smith recognised the priority of Savastano's discovery and named the bacterium *B. savastanoi* after the Italian researcher. Some years later the bacterium was assigned to the genus *Pseudomonas*.

3. SELECTIVE ISOLATION OF *PSEUDOMONAS SAVASTANOI*

A bacterium forming yellow colonies was thus found by earlier researchers to be frequently associated with *Pseudomonas savastanoi* pv. *savastanoi*.

Indeed, in our experience, a yellow bacterium is frequently, though not always, isolated together with *P. s.* pv. *savastanoi*, not only from olive and oleander knots, but also from the phylloplane of diseased olive and oleander plants. This yellow bacterium, which has now been tentatively identified as *Pantoea agglomerans* thus confirming the first indication by Petri, came to our attention when an attempt was made to prepare a selective medium for the isolation of *P. s.* pv. *savastanoi* from the phylloplane of olive plants. The medium we prepared, named PVF-1, was very simple, containing sodium dodecyl sulphate (SDS) as the only selective agent and appeared to be fairly useful in detecting *P. s.* pv. *savastanoi* populations on olive leaves and drupes (Surico and Lavermicocca, 1989). However two main negative aspects emerged with this medium:

1. recovery of some strains was very low: for example, the lowest values were 8.7% for strain NCPPB640 and 11.1% for strain NCPPB639. Curiously, these are the type strains for pv. *nerii* and pv. *oleae* respectively;
2. the yellow bacterium was only moderately inhibited by SDS.

Some years later another selective medium was devised by researchers of the University of California at Riverside. This medium was named OKA (<u>o</u>leander <u>k</u>not <u>a</u>gar) and the authors, on the basis of their results, stated that OKA was superior to PVF-1 for the isolation of *P. s.* pv. *savastanoi* from oleander (Azad and Cooksey, 1995). This superiority was mainly due to the fact that the growth of the yellow bacterium was completely inhibited by the selective agents (boric acid plus 6 antibiotics: vancomycin, cephalexin, bacitracin, ampicillin, novobiocin and cycloheximide) included in the OKA medium. In a recent survey on the spread of *P. s.* pv. *savastanoi* in Tuscany, we compared the two selective media PVF-1 and OKA, and two others: ANS-S, prepared by Varvaro (1983), and a modification of OKA (OKA-M, see "b" in Table 1). This survey showed that:

1. recovery of *P. s.* pv. *savastanoi* was highest on ANS-S, followed in order by PVF-1 and OKA (Table 1). On OKA some *P. s.* pv. *savastanoi* strains did not grow at all. When they did, they also usually formed two types of colonies, one small (ca. 1 mm) and another one very small (less than 1 mm), whose morphology did not presented any characteristics that could facilitate their recognition as *P. s.* pv. *savastanoi* (Fig. 1);
2. inhibition of contaminant bacteria was highest with OKA, followed in order by PVF-1 and ANS-S. The yellow bacterium (strain INQFi-1) was completely inhibited by OKA and OKA-M, moderately inhibited by ANS-S and about 90% inhibited by PVF-1 (Table 1);
3. on PVF-1 *P. s.* pv. *savastanoi* formed colonies with a morphology comparable to that on nutrient sucrose agar (NSA=ANS) and also produced fluorescent pigments.

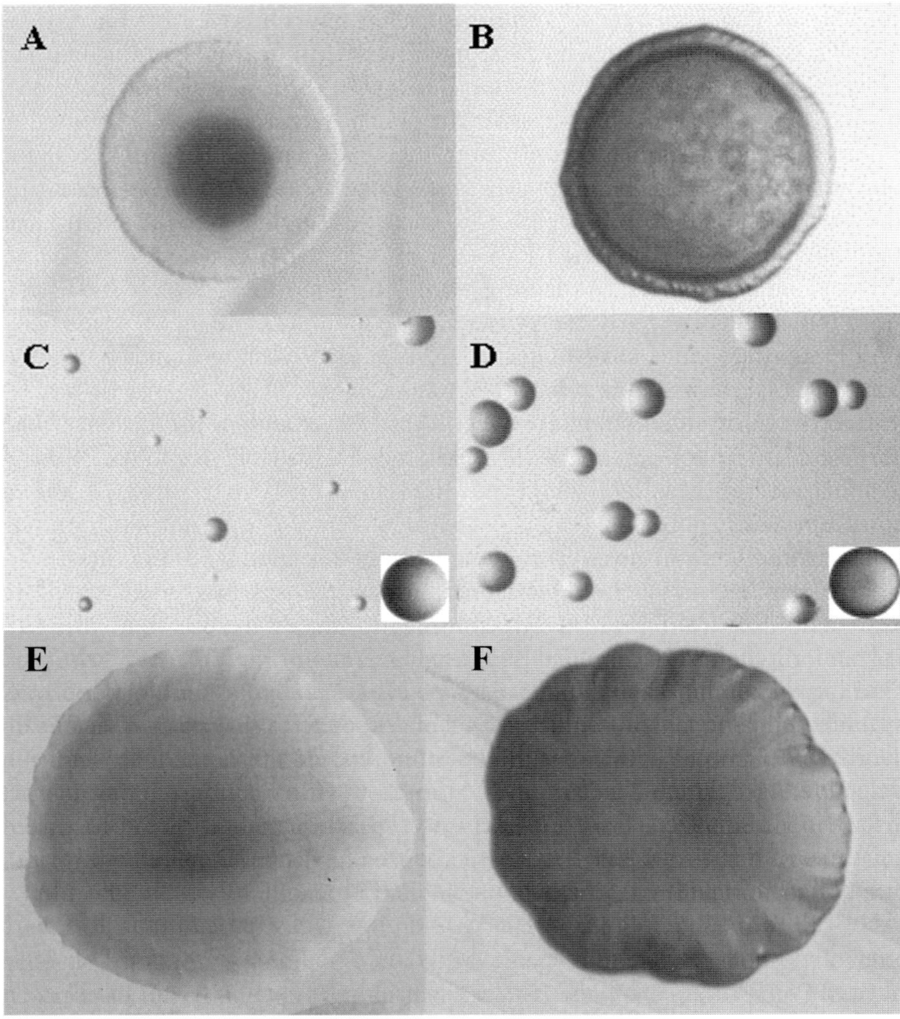

Figure 1. Morphology and size of the colonies of *Pseudomonas savastanoi* pv. *savastanoi* strains ITM519 (A, C, and E) and ITM317 (B, D, and F) on NSA (E and F) and on the selective media PVF-1 (A and B) and OKA (C and D). The diameters (in mm) are A, 3.1; B, 3.5; C, 1 and less than 1; D, 2 and less than 2; E, 3.5; F, 3.4. Insets: closer view of a colony in C and D respectively. Photographs were taken after 4 (E and F), 6 (A and B) and 9 (C and D) days of incubation at 26°C.

On the basis of these results PVF-1, despite its limitations, is a suitable substrate to isolate *P. s.* pv. *savastanoi* from knots, and from the olive or oleander phylloplane. It is true that saprophytic contaminants were less well selected on PVF-1 than on OKA, but PVF-1 maintained the morphology of *P. s.* pv. *savastanoi* colonies as well as it did that of the contaminant yellow bacterium, so both bacteria were easily recognised whenever they were present on the same plate.

Table 1. Percent recovery of six *Pseudomonas savastanoi* pv. *savastanoi* strains (NCPPB640, NCPPB639, ITM519, ITM317, PVBa229, MLLI-4) and of a saprophytic bacterial strain (INQFi-1) obtained from olive knots on four different media.

Medium	Recovery compared with NSA (%)						
	NCPPB640	NCPPB639	ITM519	ITM317	PVBa229	MLLI-4[a]	INQFi-1
ANS-S	21.4	18.6	98	96	87.4	75.8	45
PVF-1	7.9	11.1	85	91.5	92.3	0.41	10.5
OKA	0	0	15.5	22	44.7	0.13	0
OKA-M[b]	0	0	18.6	37	51.2	9.4	0

[a]Levan-positive strain of *P. s* pv. *savastanoi*. All the other strains were levan-negative.
[b]OKA modified by replacing four antibiotics (bacitracin, ampicillin, novobiocin and cycloheximide) with sodium dodecyl sulphate (Azad and Cooksey, 1995).
NCPPB639, ITM317, MLLI-4: olive strains; NCPPB640, ITM519, PVBa219: oleander strains.

4. SPREAD OF LEVAN-POSITIVE PSEUDOMONAS SAVASTANOI PV. SAVASTANOI POPULATIONS IN TUSCANY

Pseudomonas savastanoi pv. *savastanoi* typically forms levan-negative colonies on NSA. However, levan-positive (L+) strains have been isolated from ash (Janse, 1981) and olive (Iacobellis *et al.*, 1993). *P. s.* pv. *savastanoi* L+ strains were first isolated in an olive grove at Villa Bigallo, in the area of Bagno a Ripoli, near Florence. This finding prompted the question whether the presence of the strain had been a chance occurrence or whether L+ strains were well established in the grove. To answer this question the presence of L+ strains at Villa Bigallo and in groves elsewhere in Tuscany was investigated.

At Villa Bigallo, 120 knots were harvested from 6 plants randomly distributed in the orchard. Of all the isolates obtained 47 were L+, all identified as *P. s.* pv. *savastanoi* on the basis of their ARDRA profile. The same identification was obtained by analysis of some physiological characteristics, pathogenicity and plasmid profile. However, isolates varied widely. For example, plasmids of 15 sizes (from 6 to 88 kb) and varying from 1 to 6 per strain were detected. The most common plasmids were those of 6, 8 and 63 kb, while those of 65 and 70 kb, when present, were always associated with each other in the same isolate. There was also variation in the fatty acid profile. In five representative isolates (PS5, T3LAZ-2, MLLI-1, MLLI-2, MLLI-3), more than 82% of total cellular fatty acids consisted of four major components: 12:0, 16:0, 16:1 and 18:1. The hydroxy-substituted acids 10:0 3OH, 12:0 2OH and 12:0 3OH, contributed an additional 11.31%

to the total. The most important minor components, present in all the isolates, were 14:0, 17:0 ISO, 18:0, 11 methyl 18:1 w7c. Eighteen other fatty acids were minor components, found in different amounts (from 0 to 2.41%) in one or another of the five isolates. However, while the profiles (types, amounts and ratios of acids) of PS5 and T3LAZ-2 were entirely consistent with a *Pseudomonas* species (the closest match was with *P. viridiflava* and *P. cichorii*), those of the MLLI group, although characteristic of a *Pseudomonas* species, were clearly different from the profiles of the *Pseudomonas* spp. included in either the NCPPB3 or the TSBA library of the Central Science Laboratory at Sand Hutton, York, UK. If these data are confirmed, either a new *Pseudomonas* species with the same pathogenic characteristics as the pathovar *savastanoi* or some very unusual strains of *P. s.* pv. *savastanoi* will have been detected.

5. COLONY MORPHOLOGY AND FRACTAL GROWTH

Traditionally for test purposes, bacterial colonies are grown on substrates with optimal nutrient levels and an intermediate agar concentration. Under such favourable conditions, the colonies usually develop simple compact patterns with a smooth envelope. In nature, however, bacterial colonies must regularly cope with hostile environmental conditions. To study bacterial growth mechanisms under conditions of limited nutrient supplies and movement, various authors created unfavourable conditions in a Petri dish by using very low levels of nutrients, or a hard medium surface (high concentration of agar), or both. The most noticeable change in the nature of the bacterial response during colony development under such conditions was the formation of complex growth patterns (Ben-Jacob *et al.*, 1994).

In the case of *Bacillus subtilis* for example, colonies grow by a diffusion-limited aggregation process when the bacterium is inoculated on a nutrient-poor solid agar medium, but when it is inoculated on a nutrient-rich semi-solid agar medium the colonies spread out homogeneously over the dish. The morphological phase diagram of *B. subtilis* thus consists of five parts depending on the composition of the substrate: A, diffusion-limited aggregation (DLA) or DLA-like; B, Eden-like; C, intermediate; D, homogeneous spread; and E, dense branching morphology (DBM) or DBM-like (Ohgivary *et al.*, 1992).

"DLA (http://classes.yale.edu/fractals/aut02/Nov19/Nov19.html) is a type of growth in which cohesive particles follow Brownian motion paths (the "diffusion" in the name) through a medium until they encounter one another, and then they stick together, forming clusters that exhibit hierarchical branching". In region B (nutrient-rich solid agar medium) colonies have an

Eden-like, compact structure. In region D (nutrient-rich semi-solid agar medium), a colony spreads homogeneously in every direction on the medium surface. In region E (nutrient-poor semi-solid agar medium) colonies exhibit a DBM-like growth type, which consists of many thin branches all having the same width. In region C (nutrient-rich semi-solid agar) the colony grows by forming concentric rings.

In the case of a clinical isolate of *Proteus mirabilis,* colonies expand over the solid substrate by multicellular rafts of specially differentiated swarmer cells which are elongated and hyperflagellated. Raupich *et al.* (1996) distinguished three initial phases: a lag phase, a first swarming phase and a first consolidation phase, followed by repeated cycles of swarming plus consolidation. "In the lag phase there is migration of swarmer cells from the interior of the inoculation zone to the periphery, increasingly rapid back-and-forth tangential movements of individual swarmer cells, and a synchronisation of flagellar movements as groups of swarmer cells align themselves to form the migrating rafts which have been described as the basic units of swarm motility. The lag phase ends when the first swarmer rafts emerge from the colony perimeter. The first swarming phase is characterised by the relatively independent movement of swarmer cell rafts which tend to coalesce into *radially oriented tubes*. The first swarming phase ends when migration ceases and the leading edge becomes stationary. While the first swarming phase is coming to an end, a wave of cell multiplication and thickening of newly colonised areas begins adjacent to the perimeter of the inoculation spot. This wave spreads outwards until it approaches the boundary of the colonised areas (first consolidation phase), whereupon the second swarming phase begins".

When chemotactic strains of the bacterium *Escherichia coli* are inoculated on semi-solid agar containing mixtures of amino acids or sugars, the cells swarm outward in a series of concentric rings: they respond to spatial gradients of attractants generated by uptake and catabolism (Budrene and Berg, 1991). Structures similar to those produced by *E. coli* have been seen in *B. subtilis* and in *Salmonella typhimurium*.

B. subtilis, *P. mirabilis* and *E. coli* serve as examples of bacteria with a capacity for fractal growth. *Pseudomonas savastanoi* pv. *savastanoi* also when grown on a medium with low nutrients and low agar concentration formed colonies whose shapes had points of similarity with one or other of those three fractal-growth bacteria. In our experiments, several strains of *P. s.* pv. *savastanoi* with different colony shapes on NSA were used. They were grown on media containing 1, 5 and 10 g l^{-1} peptone as the carbon and nitrogen source and 5, 7.5, 10, 15, 30 and 40 g l^{-1} agar as the solidifying agent. The medium was poured into 9 cm Petri dishes (20 ml $dish^{-1}$) and inoculated by placing in the centre a 5 μl drop of bacterial suspension containing about 10^5 CFU ml^{-1}. The Petri dishes were incubated at 26°C and inspected every 2 h for the first 24 h, and then after 2, 4, 6 and 8 days. A

strain of *P. syringae* pv. *phaseolicola* and one of *P. fluorescens* were used as controls.
The levan-negative strains of *P. s.* pv. *savastanoi* formed circular colonies with all types of nutrient medium (Fig. 2).

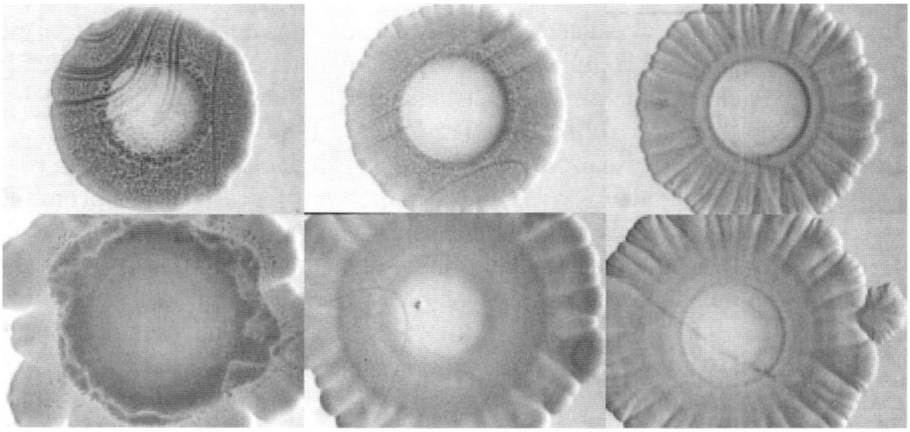

Figure 2. Colonies formed by *Pseudomonas savastanoi* pv. *savastanoi* strain PVBa229 (levan negative) on a peptone (p)-agar (a) medium (in g l^{-1}). Clockwise from upper left: p 1, a 7.5; p 1, a 10; p 1, a 30; p 5, a 30; p 5, a 10; p 5, a 7.5. Plates were inoculated by placing in the centre of the plate a 5 µl droplet of a bacterial suspension containing about 10^5 CFU ml^{-1}.

Inoculated bacteria always migrated promptly to the edge of the drop of inoculum and multiplied forming a kind of ring that was clearly visible in the Petri dish. From this ring the colony continued to spread outwards and in 3÷5 days formed colonies that were sometimes smooth and slimy (1÷5 g l^{-1} peptone and 7.5÷10 g l^{-1} agar), sometimes rough and dry (5÷10 g l^{-1} peptone and 30÷40 g l^{-1} agar), shaped with a petaloid corolla. Colony density was closely correlated with peptone concentration (low density at 1 g l^{-1} peptone and 7.5 g l^{-1} agar).
At the lower agar concentrations (7.5 and 10 g l^{-1}) aggregates of bacteria almost always formed which, depending on the bacterial strain, had the shape either of roughly spherical glomerules or of spindles. The glomerules had a regular and predetermined arrangement, but the spindles were without order in the colonies or in the nutrient medium.
In general, the shapes of the levan-negative colonies of *P. s.* pv. *savastanoi* had points of similarity with those of *P. mirabilis* in the same or nearly the same growth conditions. By contrast, *P. fluorescens, P. s.* pv. *phaseolicola,* and the four L+ *P. s.* pv. *savastanoi* strains (T3LAZ-2, MLLI-1, MLLI-2 and MLLI-4) on low-agar media (5 and 7.5 g l^{-1}) formed colonies similar in shape to that of *B. subtilis*, i.e. with DLA-like growth (Fig. 3A). However,

with *B. subtilis* the levels of peptone and agar in the nutrient medium that produced similar values in the fractal size of the colony were quite different. For example, 16 h after inoculation of *P. fluorescens* (32 hours after that of *P. s.* pv. *savastanoi*) colonies grown on media containing 5 and 10 g/l peptone and 7.5 g l^{-1} agar already had a fractal size (D_f) between 1.78 and 1.85, while colonies grown on 1 g l^{-1} peptone and 7.5 g l^{-1} agar had a D_f of about 1.92 (or about 1 if only the outer perimeter of the colony was measured) and a circular shape. Instead, a DLA-like colony was produced by *B. subtilis* at 1 g l^{-1} peptone and 9÷15 g l^{-1} agar (D_f = 1.73) and a disc-like colony at 1 g l^{-1} peptone and 5 g l^{-1} agar (Fujikawa and Matsushita, 1991; Matsuyama and Matsushita, 1992).

6. CONCLUSIONS

This paper reports on the results of some investigations on olive knot and its causal agent that have recently been started in our laboratory. As at least three semi-selective media (ANS-S, PVF-1 and OKA) are available for isolating *Pseudomonas savastanoi* pv. *savastanoi* from the phylloplane of olive and/or its knots, it seemed worthwhile to compare these substrates and determine the advantages and disadvantages of each. It seems that from a more general point of view, PVF-1 is best for monitoring the presence of *P. s.* pv. *savastanoi* on olive leaves. Nevertheless, each substrate has both positive and negative aspects, and a better understanding of the capacity of each medium will make it easier to choose one of them for particular purposes. In the course of the trials it was also found that *P. s.* pv. *savastanoi* in most cases cannot be detected on the leaves of healthy plants and that after a single application of copper-based bactericidal compounds, populations of epiphytic bacteria fell below the threshold of detection with cultural methods (washing the leaves and plating). These findings may be useful for control of the disease.

Other tests have shown that the levan positive (L+) strains of *P. s.* pv. *savastanoi* are widespread in Tuscany. To our knowledge the spread of the bacterium in nature has never been studied. It is likely however that one of the most common means of bacterial spread over shorter or longer distances is by the commercial distribution of nursery plantlets. Since Tuscan olive nurseries are among the main producers of such plantlets in Italy, of which a large proportion is also exported abroad, it is to be feared that if the L+ strains find their way into these nurseries, they will quickly become much more widespread.

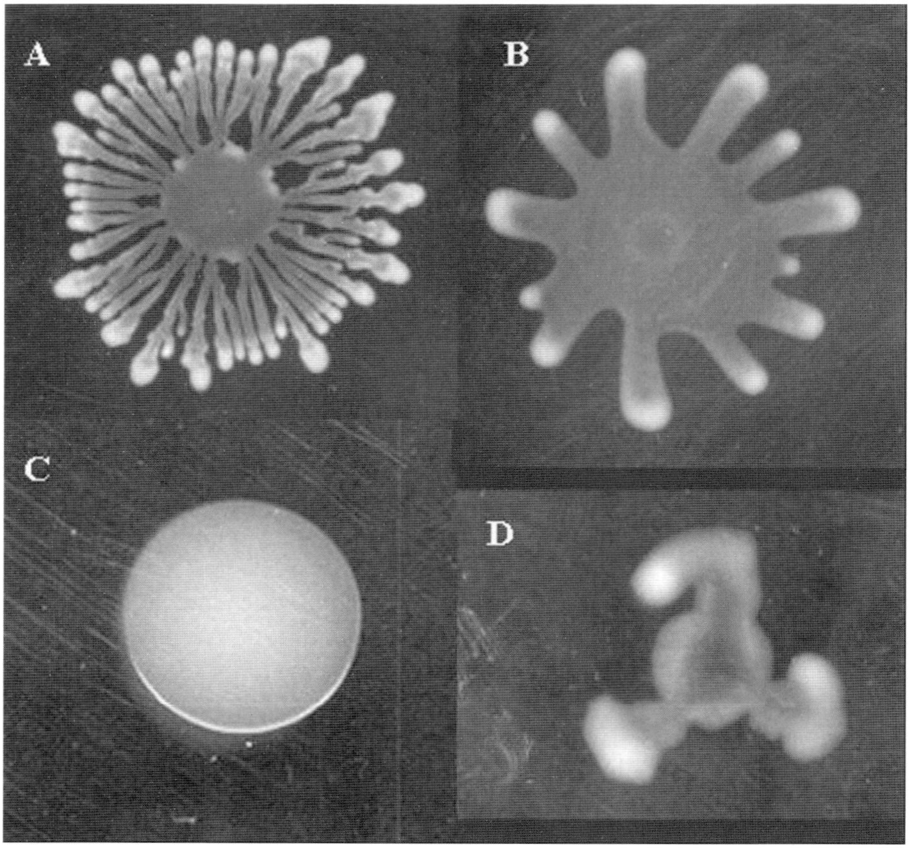

Figure 3. Colonies formed by three levan-positive bacterial strains. (A, *Pseudomonas fluorescens* Mon-1; B, *P. savastanoi* pv. *savastanoi* MLLI-2; D, *P. syringae* pv. *phaseolicola* IPV.Bo1960) and a levan negative strain (C, PVBa229) of *P. s.* pv. *savastanoi* on a medium containing 10 g l^{-1} peptone and 7.5 g l^{-1} agar. Photos (400 dpi) were taken at 16 h (A) or 36 h (B, D, C) after inoculation as reported in Fig. 2. Patterns: A, DLA-like; B, cogwheel-like; C, disc-like; D, tricorn hat-like.

There remains the question of the fractal growth of this bacterium. *P. s.* pv. *savastanoi*, unlike many other phytopathogenic bacteria, is characterised by a high diversity of its colony morphology on one and the same substrate. Most of these are levan-negative strains and produce fluorescent pigments on King's B medium, but at least in Tuscany, non-fluorescent, levan-positive populations are becoming more and more common. These populations have a capacity for fractal growth under limiting environmental conditions, a capacity that is an illustration of an evolutionarily more interesting capacity, that of adapting to adverse conditions by constitutive mechanisms. And indeed, in the opinion of some scientists the effective adaptation of the colony to adverse growth conditions requires self-organisation at all levels, and that can be achieved only by co-operation between the bacteria. In other words, these bacteria have developed sophisticated communication channels at all levels: "from cell to cell physical and chemical interactions by direct contact, through indirect physical and chemical interactions via marks left on the agar surface and chemical (chemotactic) signalling, to genetic communication via the exchange of genetic material". When bacteria are exposed to more adverse growth conditions, they become "smart" enough to develop new strategies, via genetic change, for a better adaptation to the environment. All these phenomena indicate that bacterial populations are organised as multicellular organisms. The results of our investigations, though they are still preliminary, suggest that, like other bacteria, *P. s.* pv. *savastanoi,* too, possesses some gleams of "intelligence".

ACKNOWLEDGEMENTS

The research reported here was supported, in part, by the University of Florence.

REFERENCES

Azad, Hamid, R., and Cooksey, D.A., 1995, A semiselective medium for detecting epiphytic and systemic populations of *Pseudomonas savastanoi* from oleander. *Phytopathology* **85**: 740-745.

Ben-Jacob, E., Schochet, O., Tenebaum, A., Cohen, I., Czirók, A., and Vicsek, T., 1994, Generic modelling of cooperative growth patterns in bacteria. *Nature* **368**: 46-49.

Budrene, E.O., and Berg, H.C., 1991, Complex patterns formed by motile cells of *Escherichia coli*. *Nature* **349**: 630-633.

Fujikawa, H., and Matsushita, M., 1991, Bacterial fractal growth in the concentration field of nutrient. *J. Phys. Soc. Japan* **60**: 88-94.

Iacobellis, N.S., Sisto, A., and Surico, G., 1993, Occurrence of unusual strains of *Pseudomonas syringae* subsp. *savastanoi* on olive in central Italy. *Bulletin OEPP/EPPO Bulletin* **23**: 429-435.

Janse, J.D., 1981, The bacterial disease of ash (*Fraxinus excelsior*), caused by *Pseudomonas syringae* subsp. *savastanoi* pv. *fraxini*. *Euro. J. For. Pathol.* **11**: 425-438.

Matsuyama, T., and Matsushita, M., 1992, Self similar colony morphogenesis by Gram-negative rods as the experimental model of fractal growth by a cell population. *Appl. Environ. Microbiol.* **58**: 1227-1232.

Ohgivary, M., Matsushita, M., and Matsuyama, M., 1992, Morphological changes in growth phenomena of bacterial colony patterns. *J. Phys. Soc. Japan* **61**: 816-822.

Peglion, V., 1905, La rogna o tubercolosi del *Nerium oleander*. *Rendiconti Reale Accademia Nazionale Lincei*, **14**: 462-463.

Petri, L., 1907, Untersuchungen über die Identität des Rotzbacillus des Oelbaumes. *Centralbl. f. Bakteriol. Abth. II, Bd.* **XIX**, 531-538.

Rauprich, O.,Matsushita, M., Weijer,C. J., Siegert, F., Esipov, S.E., and Shapiro, J.A., 1996, Periodic phenomena in *Proteus mirabilis* swarm colony development. *J. Bacteriol.* **178**: 6525-6538.

Surico, G., and Lavermicocca, P., 1989, A semiselective medium for the isolation of *Pseudomonas syringae* pv. *savastanoi*. *Phytopathology* **79**: 185-190.
Savastano, L., 1886, Le maladies de l'olivier et la tuberculose en particuler. *Comp Rend Acad Agr France*, 103-114.
Smith, E.F., 1905, Some observations on the biology of the olive-tubercle organism. *Centralbl. f. Bakteriol. Abth. II, Bd. XV*, 198-200.
Varvaro, L., 1983, Un substrato selettivo per *Pseudomonas syringae* pv. *savastanoi*. *Phytopath. Medit.* **22**: 78-79.

Exploring *Pseudomonas syringae* Ecology via Direct Microscopic Observations of the Leaf Surface

J.-M. MONIER and S.E. LINDOW
Department of Plant and Microbial Biology, 111 Koshland Hall, University of California, Berkeley, CA 94720-3102, USA.

Abstract: While the factors affecting the relative abundance of bacterial populations have been addressed for entire leaves or plants, little is know about how they affect the density and spatial distribution of micro-organisms at small scales corresponding to their size. The development of new approaches that combine molecular biological tools with light microscopy techniques has enabled us to extend our scale of investigation of epiphytic communities to small scales and has revealed unanticipated features of leaf surface microbial communities. We examined *Pseudomonas syringae* pv. *syringae* strain B728a inoculated onto bean leaves. Cells genetically marked with genes conferring fluorescent proteins of different colours were visualised using epifluorescence microscopy directly on leaf surfaces. Cell viability was also assessed following propidium iodide staining. Bacterial cells are not randomly distributed on the leaf surface but occur in a wide range of cluster sizes. Large cell aggregates are not frequent on a given leaf but often account for the majority of the cells present. Aggregated cells are much more capable of tolerating desiccation stress on leaves than solitary cells and the preferential survival of cells in aggregates promotes a highly clustered spatial distribution of bacteria on leaves. The fate of immigrant cells depends on both anatomical features of the leaf and the number of resident cells at their landing site. The relatively low occurrence of mixed species aggregates on leaves reveals that bacterial colonisation of leaves is characterised by a high level of spatial segregation and direct bacterial interactions between two populations is limited to only those few cells in direct contact. Description of the spatial distribution of epiphytic bacteria at small scales corresponding to their size has provided us with important biological and ecological information on processes enabling growth and survival of epiphytes and their interactions on leaf surfaces.

Key words: Epiphytic life, bacterial aggregates, epifluorescence microscopy, fluorescent proteins.

Author for correspondence: J.-M. Monier; e-mail: jm.monier@free.fr.

N.S. Iacobellis *et. al.* (eds.), *Pseudomonas syringae* and Related Pathogens, 29-40.

1. INTRODUCTION

Quantitative description of the colonisation patterns of epiphytic bacteria such as *Pseudomonas syringae* at small scales corresponding to their size is lacking. Knowledge of the spatial organisation and relationships among bacterial populations on leaf surfaces is needed in order to better understand how they survive stressful conditions encountered on leaves, to understand how disease may develop from epiphytic populations, and to determine how frequently antagonistic or syntrophic interactions occur.

Our challenge was to develop new approaches to address the spatial organisation of epiphytic populations as they appear in their natural environment, with methods that permit quantitative and detailed analysis of relevant features. A promising report revealed that fluorescent microscopy techniques in combination with appropriate fluorescent stains, could be used to visualise bacteria directly on leaf surfaces (Morris *et al.*, 1997). While this method did not allow identification of individual species, molecular tools such as use of marker genes conferring production of fluorescent proteins, should allow us to study individuals or mixtures of epiphytic bacteria at different resolutions, ranging from single cell dimensions to the entire leaf ecosystem, with minimal disruption to the leaf surface community. Development of such innovative approaches to study epiphytic bacteria directly *in situ*, should provide valuable information on spatial patterns of leaf, allow us to identify common traits or response patterns in epiphytes and thus, the processes leading to a structured organisation of bacterial populations on leaves.

Preliminary observations of leaf surfaces had revealed that colonisation of leaf surfaces by bacterial populations is characterised by a high level of spatial segregation. The presence of a majority of epiphytic bacteria in large aggregates on leaves has many implications for the behaviour of cells within such structures and raises many questions as how they form, the benefit they confer to the colonists, and how they influence the degree of interaction between bacteria on leaves. Our goal was to quantitatively describe the spatial distribution of bacteria on leaves and determine the biological significance of bacterial aggregates. While the plant pathogenic bacterium *P. s.* pv. *syringae* was the main focus of this work, the behaviour of other common epiphytic bacterial species has also been examined in order to determine the generality of our conclusions regarding the ecology of bacteria on leaf surfaces.

2. SPATIAL DISTRIBUTION OF BACTERIA ON LEAF SURFACES

2.1 Approach

2.1.1 Visualisation of Cells by Microscopy

The frequency, size and localisation of bacterial aggregates was determined at the micro-scale by analysis of digital images of micro-graphs of epiphytic bacteria over time following inoculation of *Pseudomonas syringae* strain B728a onto bean leaf surfaces. At each sampling time, leaf segments (1x1 cm) were randomly cut per leaf and individually stained with a solution of acridine orange to allow visualisation of both the bacterial cells and the anatomical structures of the leaf surface (adapted from Morris *et al.*, 1997). Samples were then observed by epifluorescence microscopy, random fields of view were selected and the corresponding images were captured.

2.1.2 Frequency and Size of Aggregates

Captured images were individually processed with Corel Photopaint software (Corel Co., Ottawa, Canada) and converted into a binary image with bacteria represented in white and the leaf surface in black. Aggregate size and frequency measurements were then conducted using IPLab software (Scanalytics Inc., Fairfax, VA) which allowed automatic measurements of the number of individual objects in an image as well as their corresponding area. In order to associate an aggregate with a particular leaf surface feature, the corresponding unprocessed images were used to define the location of a given aggregate.

2.2 Results

2.2.1 Quantification of the Spatial Distribution of Bacteria on Leaves

Several millimeter squares of leaf surfaces containing between 12,000 and 65,000 aggregates were measured at each sampling time. Bacteria covered 2.2%, 4.8% and 12.0% of the leaf surface respectively, after 0, 2 and 8 days of incubation after inoculation with a dilute *Pseudomonas syringae* B728a cell suspension (i.e. 10^4 CFU $leaf^{-1}$). After incubation for several days, cells were not randomly scattered over the leaf surface but occurred in a wide range of cluster sizes, ranging from single cells to over 10^4 cells per aggregate. The large majority of bacterial cell aggregates observed on leaves were relatively small and aggregate sizes exhibited a strong right-hand skewed frequency distribution (Fig. 1).

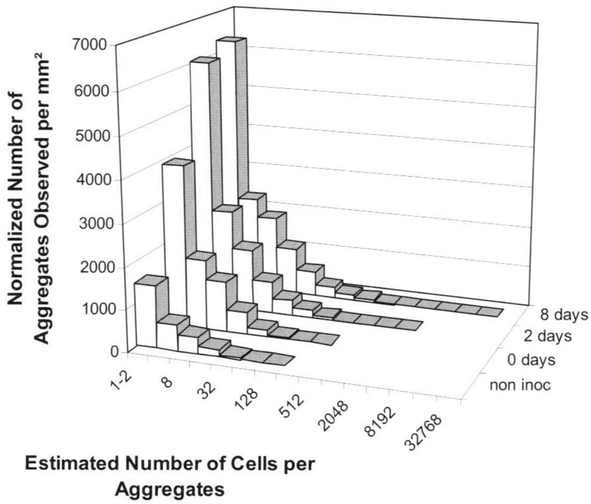

Figure 1. Frequency distribution of bacterial aggregates on leaf surfaces before inoculation and 0, 2 and 8 days following inoculation with *Pseudomonas syringae* strain B728a. The total number of aggregates was determined for 3 leaves at each sampling time and the data combined. The total surface of the leaf observed at each sampling time was slightly different and the total number of aggregates observed was normalised and expressed per mm^2.

2.2.2 Relative Percentage of Solitary and Aggregated Cells on Leaf Surfaces

While large aggregates are not frequent on a given leaf, they can account for the majority of the cells present. We also observed that the proportion of the total number of cells that were located in larger aggregates increased with incubation time (Fig. 2). Immediately after inoculation, only 2.7% of the cells are located in aggregates containing 100 cells or more. This percentage dramatically increased with incubation time and reached, on average, 22.7% and 62.2% after 2 and 8 days respectively when considered over several leaves. A great deal of variability in the degree of aggregation was observed between leaves, between leaf segments as well as within individual leaf segment. For example, the percentages of cells located in aggregates containing 100 cells or more were 32.8%, 64.1% and 71.4% respectively, for 3 leaves sampled 7 days after inoculation. In different segments taken within the same leaf, the percentage of cells located in aggregates containing 100 cells or more ranged from 1.3% to 78.9%.

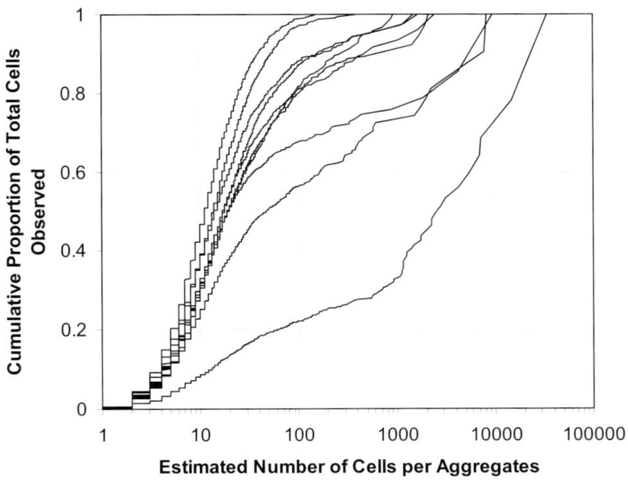

Figure 2. Cumulative proportion of the total number of cells observed as a function of the total number of cells per aggregate when considered over individual leaf segments cut randomly within the same leaf, observed 7 days following inoculation with *Pseudomonas syringae* strain B728a.

2.2.3 Localisation of Bacterial Aggregates

Aggregates were observed at the base of glandular and hooked trichomes and in the grooves between cells associated with veins and non differentiated epidermal cells. Comparison of the observed and expected numbers of aggregates associated with the different anatomical features, demonstrated however that aggregates are not randomly distributed on the leaf surface, but are preferentially associated with veins and trichomes. Aggregates observed associated with glandular trichomes were also significantly larger.

2.3 Discussion

This study provides the first quantitative information on the spatial distribution of epiphytic bacteria and on the frequency and size of bacterial aggregates, determined directly on leaf surfaces, and demonstrates that aggregated bacteria can account for the majority of the cells present on a leaf. Description of the spatial distribution of epiphytic bacteria at a scale corresponding to their size provide biological and ecological information about the processes that lead to successful colonisation. The variability in spatial aggregation clearly extends even to small regions within a given leaf segment. While the variability in population sizes observed at the plant, leaf or leaf segment levels could result from several environmental and biological factors, the variability observed within leaf segments seems to be driven by the present or absence of micro-sites conducive to bacterial growth. Cells were not randomly scattered over the leaf surface and occurred in a wide range of cluster sizes, roughly reflecting the spatial heterogeneity of nutrients availability on leaf surfaces (Leveau and Lindow, 2001). Our study revealed that *Pseudomonas syringae* B728a preferentially formed aggregates at the base of glandular trichomes and in the grooves along the veins with only smaller aggregates found in the grooves between epidermal cells and at the base of hooked trichomes. Occasionally, large aggregates were observed associated with wounded epidermal cells, reinforcing the concept that nutrient availability limits the formation of aggregates on leaf surfaces.

Observations of leaf segments sampled from non-inoculated bean plants grown under field conditions also revealed the presence of large aggregates preferentially associated with glandular trichomes or veins or occasionally observed with a wounded epidermal cell. While the spatial distribution of aggregated cells on field-grown plants followed the same colonisation pattern we were not generally able to visualise solitary cells. Samples were observed at a relatively low magnification that may not have allowed us to detect individual cells due to the presence of strongly autofluorescent soil particles or plant debris and also because the size of bacteria grown on leaves is significantly smaller than that of cultured cells (data not shown).

While *P. syringae* is a plant pathogenic bacterium with a particularly prominent epiphytic phase, our knowledge of how *P. syringae* is able to survive on leaf surfaces and how disease is initiated is still limited. The results reported in this study provide new insights into its epiphytic life, often viewed as the first step in the infection process. Formation of large aggregates might provide a different ecological niche for the majority of the cells of *P. syringae* present on a leaf than the more solitary cells which had previously been considered typical of epiphytes, and might provide protected sites in which bacteria could escape harsh conditions encountered on the leaf surface.

Aggregate formation among epiphytes has significant ecological implications that must be considered when designing strategies to control plant pathogenic bacteria. At least some cells may be protected from bactericides applied to plants especially if the cells are embedded in a mucoid matrix. In addition, a clustered colonisation pattern of the leaf surface, with a majority of the cells located in a few large aggregates, may significantly limit microbial interactions to a few sites, as well as providing "refuges" that have been proposed to account in part for the lack of efficiency of biological control agents applied to leaves (Johnson, 1994).

3. BIOLOGICAL SIGNIFICANCE OF BACTERIAL AGGREGATION

3.1 Differential Survival of Solitary and Aggregated Bacterial Cells

3.1.1 Approach

The survival of individual *Pseudomonas syringae* B728a cells in different states of aggregation was determined directly on bean leaf surfaces maintained under humid conditions or periodically exposed to desiccation stress. Cells of *P. syringae* B728a harbouring a GFP marker gene were visualised by epifluorescence microscopy, directly *in situ* or after recovery from leaves, and dead cells were identified as cells that were stained with propidium iodide in such populations.

3.1.2 Results

Under moist conducive conditions on plants, the proportion of total live cells was always high (> 87%), irrespective of whether cells were solitary or were in aggregates on leaves. In contrast, the proportion of the total cells on leaves that were periodically exposed to desiccation stress that remained alive decreased through time and was only about 15% after 5 days and 3 exposures to low relative humidity. However, the fraction of cells that were in aggregates that were alive on such plants was much higher than more solitary cells; the fraction of cells that were alive increased with increasing size of cell aggregates (Fig. 3). While immediately following inoculation, cells were randomly distributed over the leaf surface and no aggregates were observed, a very aggregated pattern of colonisation was apparent within 7 days. After 3 repeated exposures to low relative humidity, live solitary cells constituted only 0.2% of the total live cells and over 90% of the live cells were located in aggregates of 100 cells or more.

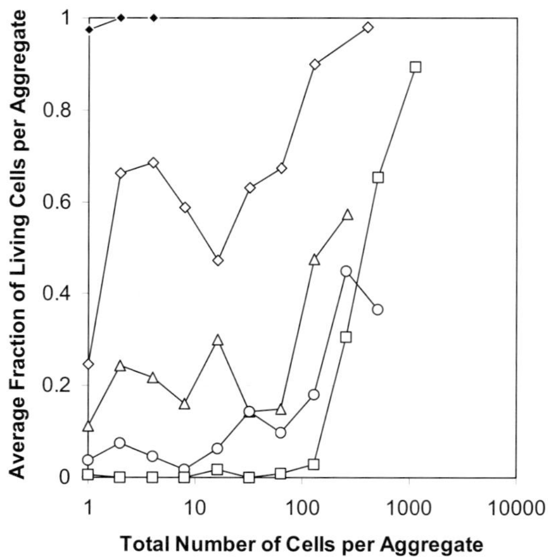

Figure 3. Average fraction of living cells in relation to the total number of cells per aggregate on plants periodically exposed to low relative humidity, 0 (black diamonds), 3 (open diamonds), 4 (triangles), 5 (circles) and 7 (squares) days after inoculation. The standard errors of the average fractions of living cells per aggregate were removed for clarity but averaged about 0.06.

3.1.3 Discussion

These results strongly suggest that, while conducive conditions favour aggregate formation, such aggregated cells are much more capable of tolerating environmental stresses on leaves and the preferential survival of cells in aggregates promotes a highly clustered spatial distribution of bacteria on leaf surfaces. While the method used in this study did not allow to visualise a matrix embedding the cells, there is evidence that bacterial aggregates on leaves are surrounded by a copious exopolymeric matrix (Gras *et al.*, 1994, Morris *et al.*, 1997). We hypothesise that the presence of an exopolymeric matrix could provide a protected micro-environment to aggregated cells against desiccation, by fostering water retention and by acting as a physical barrier. As a consequence of cell growth and differential survival of solitary and aggregated cells, the spatial distribution of epiphytic bacteria shifted from solitary cells randomly distributed on the leaf surface to a highly clustered distribution were the vast majority of live cells where found in a few large aggregates.

3.2 Spatial Organisation of Dual-species Bacterial Aggregates

3.2.1 Approach

The spatial organisation of cells within bacterial aggregates on leaf surfaces was determined for pair-wise mixtures of three different bacterial species commonly found on leaves, *Pseudomonas syringae*, *Pantoea agglomerans*, and *P. fluorescens*. Cells were co-inoculated onto bean plants, allowed to grow under moist conditions and the resulting aggregates examined *in situ* by epifluorescence microscopy. Each bacterial strain could be localised because it expressed either the green or cyan fluorescent protein constitutively and the viability of individual cells was assessed by propidium iodide staining.

3.2.2 Results

While each pair of bacterial strains co-inoculated onto leaves formed mixed aggregates, only a small fraction of the cells (less than 3%) of two given species are actually in direct contact. The degree of segregation of cells in mixed aggregates differed between the different co-inoculated pairs of strains and was higher in mixtures of *Pseudomonas fluorescens* A506 and *Pantoea agglomerans* 299R, and *P. syringae* B728a and *P. agglomerans* 299R, than for two isogenic strains of *P. agglomerans* 299R. The fraction of the total cell population that was dead in mixed and monospecific aggregates of a *gfp*-marked strain of *P. agglomerans* 299R and a *cfp*-marked strain of *P. agglomerans* 299R, or of *P. fluorescens* A506 and *P. agglomerans* 299R, was similar (*ca.* 2%). However, the proportion of dead cells in mixed aggregates of *P. syringae* B728a and *P. agglomerans* 299R was significantly higher (13.2%) than that in monospecific aggregates (1.6%) of these two strains, and increased through time. Dead cells in such mixed aggregates were preferentially found at the interface between clusters of cells of these strains, but cells of these two strains located at the interface did not exhibit an equal probability of mortality. For example, after 9 days incubation about 77% of the *P. agglomerans* 299R cells located at the interface were dead while only about 24% of the *P. syringae* B728a cells were dead.

3.2.3 Discussion

While the proportion of cells of different bacterial species that come into contact with each other on a leaf will be a function of the population sizes and respective habitat of the different species, our results suggest that only a small fraction of the cells of different species are in contact, even on heavily colonised leaves. While we observed a significant difference in the spatial organisation of mixed aggregates of the different pair-wise mixtures of strains on bean leaves, the variability in the spatial organisation of the mixed aggregates formed by such strains precludes us from identifying traits that dictate such organisation. We concluded that interactions between strains do not strongly affect the spatial organisation of cells within an aggregate. But independently of the apparent complexity of the biological and environmental factors regulating the spatial structures of epiphytic communities, our study reveals that direct bacterial interactions on leaf surfaces is limited to only a few sites, provides direct evidence of spatial segregation of bacterial populations on the leaf, which could explain in part the frequently observed incomplete biological control of disease by applied antagonistic bacteria.

3.3 Aggregates Facilitate Survival of Immigrant Bacteria

3.3.1 Approach

The fate of immigrant bacterial cells on leaves under stressful conditions was determined as a function of the anatomical features and the local spatial density of resident cells at their landing site. *Pantoea agglomerans* 299R was established on bean leaves and the survival of immigrant cells of *P. fluorescens* A506, *P. syringae* B728a, as well as *P. agglomerans* 299R itself, was determined by epifluorescence microscopy following subsequent exposure of plants to desiccation stress. Resident and immigrant bacterial strains expressed the cyan and the green fluorescent protein constitutively, respectively, and the viability of individual cells was assessed directly on leaf surfaces following propidium iodide staining.

3.3.2 Results

While only a small fraction of the immigrant cells landed on established bacterial aggregates, their fate was usually strongly influenced by the presence of indigenous bacteria at the site at which they landed. Immigrants of *Pantoea agglomerans* 299R or *Pseudomonas fluorescens* A506 that arrived as solitary cells had about double the probability of survival when landing on aggregates formed by *P. agglomerans* 299R than when landing on un-colonised areas of the leaf surface. In contrast, the survival of *P. syringae* B728a was similar irrespective of whether it landed on colonised or un-colonised parts of a leaf. The nature of plant anatomical features at which immigrant bacteria landed also strongly influenced the fate of immigrant bacteria. The fraction of immigrant cells of each species tested that landed on veins or glandular trichomes that died, was always less than when they landed on epidermal cells or at the base of hooked trichomes.

3.3.3 Discussion

When immigrant bacteria were applied to leaf surfaces and plants maintained under stressful conditions, their survival was strongly influenced by the presence of resident cells in their vicinity as well as anatomical features of the plants at sites at which they landed. Interestingly, such interaction could either enhance or inhibit survival of immigrant cells in a strain-dependent manner. The preferential survival of immigrant cells of *Pantoea agglomerans* 299R and *Pseudomonas fluorescens* A506 at sites of aggregates formed by *P. agglomerans* 299R strongly suggests that these aggregates directly enhanced the survival of immigrant cells. The absence of beneficial (and often detrimental effect) of aggregates of *P. agglomerans* 299R on the survival of immigrant *P. syringae* B728a cells was apparently a result of negative biological interaction existing between the two strains (see "*Spatial organisation of dual-species bacterial aggregates*") that overwhelmed a physical environment that might have been conducive for survival. Our observations strongly suggest that, since bacterial populations are highly aggregated across individual leaves and only a relatively small fraction of cells are in direct contact, studying interactions between bacterial populations at the whole-leaf scale may prevent the identification of such interactions.

Depending on the process by which immigrants arrive at a leaf, only a small fraction of cells may be deposited on existing bacterial aggregates. While un-colonised sites differed greatly in their ability to influence the survival of immigrant cells, the fate of an immigrant bacterium will depend on the nature of the leaf structure on which they are deposited, and indirectly on the amount of nutrients and water available at that site to support the development of a bacterial aggregate. Indeed, the preferential survival of immigrant bacteria at the base of glandular trichomes suggest that the physical and chemical environment of the leaf surface is not uniform but mediates survival. We have reported that *P. syringae* B728a cells preferentially form aggregates at the base of glandular trichomes and that such trichomes were potential sources of nutrients favouring bacterial growth under conducive conditions (see "*Spatial distribution of bacteria on leaf surfaces*"). Our data revealed that under stressful conditions, glandular trichomes may also provide protection against desiccation stress.

4. CONCLUSION

The development of new approaches that combine molecular biological tools with light microscopy techniques has enabled us to broaden our scale of investigation of epiphytic communities from the whole plant or leaf level, that has characterised most research to date, to small scale studies at single cell size dimensions. While the different approaches are complementary and some conclusions are coherent, the methods developed in this study have revealed unanticipated features of leaf surface microbial communities. Our studies strongly suggest that bacterial behaviour on leaves exhibit several density-dependent traits and reveals that colonisation of leaf surfaces by bacterial populations is characterised by a high level of spatial segregation. Observations of bacteria in their natural habitat allowed us to unveil their clustered colonisation pattern of the leaf surface, and demonstrated the quantitative importance of bacterial aggregates on leaves. Moreover, description of the spatial distribution of epiphytic bacteria at small scales corresponding to their size has provided us with important biological and ecological information leading to a better understanding of the processes fostering successful colonisation of leaf surfaces.

REFERENCES

Gras, M., H., C., Druet-Michaud, and Cerf., O., 1994, La flore bactérienne des feuilles de salade fraiche. *Science des Aliments* **14**: 173-188.

Johnson, K.B., 1994, Dose-response relationships and inundative biological control. *Phytopathology* **84**: 780-784.

Leveau, J.H.J., and Lindow, S.E., 2001, Appetite of an epiphyte: Quantitative monitoring of bacterial sugar consumption in the phyllosphere. *P. Natl. Acad. Sci. USA*. **98**: 3446-3453.

Morris, C.E., Monier, J.-M., and Jacques, M.-A., 1997, Methods for observing microbial biofilms directly on leaf surfaces and recovering them for isolation of culturable microorganism. *Appl. Environ. Microbiol.* **63**: 1570-1576.

Stress Resistance in *Pseudomonas syringae*: Mechanisms and Strategies

G.W. SUNDIN
Department of Plant Pathology, Michigan State University, East Lansing, MI 48824-1311, USA.

Abstract: The growth and survival of *Pseudomonas syringae* in association with both the external and internal leaf environment is correlated with an ability to tolerate and/or escape environmental and plant-associated stresses. These stress factors include, most notably, desiccation, nutrient deprivation, exposure to ultraviolet radiation (UVR), and oxidative stress encountered both from exposure to UV-A radiation and from the plant defence response. Information on particular physiological and genetic responses of *P. syringae* to the environmental stress associated with leaf colonisation has increased in recent years. The ability to survive on dry leaf surfaces is a critical ecological adaptation of phyllosphere microbes. Several *P. syringae* traits including exopolysaccharide production, motility, and even methionine biosynthesis, have been shown to increase survival under dry or alternating wet/dry conditions. The UV-B component of solar UVR causes direct DNA damage and cell death; thus, enzymatic mechanisms of DNA repair are essential for the survival of leaf surface-associated *P. syringae*. The *rulAB* determinant encodes a mutagenic DNA repair system that confers UVR tolerance to *P. syringae* and is required for enhanced survival on UVR-exposed leaf surfaces. *rulAB* is plasmid-encoded and widely distributed among *P. syringae* pathovars. The bacterial RpoS alternate sigma factor regulates genes in response to oxidative stress including genes encoding superoxide dismutase. Surprisingly, a *P. s.* pv. *syringae* B728a *sodAsodB* mutant exhibited increased UV-A sensitivity but no change in virulence in bean. Increasing knowledge of stress factors and the responses of *P. syringae* to stress will continue to foster a better understanding of *P. syringae* biology and the potential for disease management.

Key word: epiphytic fitness, protected sites, rulAB.

Author for correspondence: G.W. Sundin; email: sundin@msu.edu.

N.S. Iacobellis *et. al.* (eds.), *Pseudomonas syringae* and Related Pathogens, 41-49.

1. INTRODUCTION

The growth and survival of *Pseudomonas syringae* in association with both the external and internal leaf environment is correlated with an ability to tolerate and/or escape environmental and plant-associated stresses. These stress factors include, most notably, desiccation, nutrient deprivation, exposure to ultraviolet radiation (UVR), and oxidative stress encountered both from exposure to UVA radiation and from the plant defence response. Information on particular physiological and genetic responses of *P. syringae* to the environmental stress associated with leaf colonisation has increased in recent years. One aspect of current and future interest is the interrelationship between certain factors associated with pathogenicity (i.e. expression of *hrp* genes) and the ability of strains to avoid environmental stress through internal leaf colonisation. Increasing knowledge of stress factors and the responses of *P. syringae* to stress will continue to foster a better understanding of *P. syringae* biology and the potential to improve disease management.

2. GENES AND PHENOTYPES INVOLVED IN EPIPHYTIC SURVIVAL UNDER DESICCATION STRESS

Growth and survival by *Pseudomonas syringae* in response to desiccation stress appears to be conferred more through avoidance strategies than through desiccation-tolerance mechanisms. Under moist conditions, a variety of plant pathogenic and saprophytic micro-organisms, and even strains of *Escherichia coli* and *Salmonella typhimurium* can readily grow on bean leaves (O'Brien and Lindow, 1989). However, the ability to survive and increase populations on leaves under desiccation stress is a trait that seems to be unique to pathogenic strains of *P. syringae* (Wilson *et al*., 1999), because of the preferential ability of these strains to colonise internal "protected" sites within leaves. These sites were apparently unavailable for colonisation by non-pathogenic *P. syringae* strains and other leaf-associated bacteria. Indeed, pathogenicity is a trait associated with increased leaf colonisation, but linkages between pathogenicity, endophytic colonisation, preferential leaf sites for growth, and increased survival under dry conditions have not been fully established experimentally.

Motility is one of the few specific traits demonstrated to confer increased survival on leaves under dry conditions (Haefele and Lindow, 1987). Again, motility may allow strains to access protected or internal sites in the phyllosphere. Another trait of particular current interest is the production of extracellular polysaccharides (EPS) on leaves, and the role of EPS and biofilm formation in the tolerance of environmental stress. Alginate (EPS) production by *P. s.* pv. *syringae* 3525 contributed to strain survival on both bean (host) and tomato (non host) leaves under dry (40÷60% RH) conditions in growth chamber experiments (Yu *et al.*, 1999). Biofilms are of particular interest because of the distinct differences in cellular behaviour in biofilm versus planktonic growth mode, and the requirement of cell-to-cell communication to initiate biofilm formation (Davies *et al.*, 1998). Lastly, attempts to identify genes involved in epiphytic fitness of *P. s.* pv. *syringae* B728a by Tn*5*-mutageneis led to the identification of a mutant, designated MX7, that exhibited large reductions in population size on bean leaves in growth chambers at 45% RH (Andersen *et al.*, 1998). The Tn*5* insertion in mutant MX7 was in a gene designated *metX* which was part of an operon (*metXW*) involved in methionine biosynthesis (Andersen *et al.*, 1998). The MX7 mutant also grew poorly compared to the wild-type under field conditions (Beattie and Lindow, 1994), and growth under dry conditions could not be complemented through adding exogenous methionine (Andersen *et al.*, 1998). However, the contribution of methionine prototrophy to epiphytic fitness is still unclear.

3. NUTRIENT DEPRIVATION

Nutrient availability varies from leaf-to-leaf, and most likely on much smaller scales within individual leaves. Although nutrients leached onto leaf surfaces are typically available in sufficient concentrations to support microbial growth (Mercier and Lindow, 2000), smaller-scale effects on the biogeography of microbes in the phyllosphere are not well known. On a large scale, Wilson *et al* (1995) has demonstrated population increases of the non pathogenic *Pseudomonas syringae* strain Cit7 containing an opine-catabolizing plasmid on transgenic tobacco which was engineered to produce opines. A particularly promising recent approach is of the use of biosensors for sugar availability such as the fructose and sucrose-sensing *gfp* reporter utilised by Leveau and Lindow (2001) to map sugar concentrations available to *Erwinia herbicola* cells on bean leaves.

4. ULTRAVIOLET RADIATION

Solar UVR reaching the earth's surface is classified as UVA (320-400 nm) or UVB (290-320 nm) based on biological effect (Fig. 1). The UVB component of solar UVR (290 to 320 nm) causes direct DNA damage by inciting the formation of lesions such as cyclobutane pyrimidine dimers (CPDs) and pyrimidine(6-4)pyrimidinone photoproducts in cellular DNA (Pfeifer, 1997).

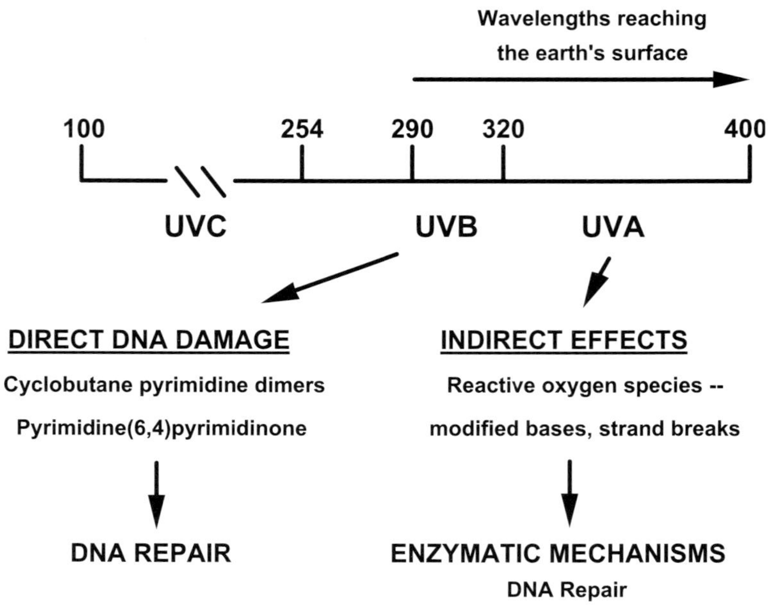

Figure 1. Wavelengths of solar ultraviolet radiation highlighting the UVA and UVB ranges, their biological effects, and the response of *Pseudomonas syringae* to exposure.

These lesions result in the blockage of DNA replication and RNA transcription; such blockages can be lethal in the absence of efficient cellular mechanisms for their removal. UVA wavelengths (320 to 400 nm) contribute more total energy (95%) than UVB (5%) in the sun's UVR spectrum and penetrate more deeply into tissue layers. Lethality due to UVA exposure is attributed to the broad spectrum effects of the intracellular generation of reactive oxygen species (ROS), including hydrogen peroxide, superoxide anion, and singlet oxygen (Pourzand and Tyrell, 1999). Solar UVR is a major stress factor for phyllosphere bacteria with much work on bacterial survival focusing on DNA repair (Sundin, 2002).

Bacterial survival following UVB-induced DNA damage is conditioned via the DNA repair response which can be separated into "light" and "dark" repair. Light repair is conferred by photolyase, an enzyme that binds CPDs and, upon activation by 375-425 nm light, efficiently reverses the lesion (Yasui and Eker, 1998). Dark repair is conferred by nucleotide excision repair (NER), a ubiquitous repair mechanism present in all cellular organisms (Sancar, 1993). We have been characterising the ecological role of photoreactivation and the contribution of NER to overall DNA repair in *Pseudomonas syringae* (Kim and Sundin, 2001). Our long-term goal is to examine the contribution of individual DNA repair mechanisms to field survival as well as assess the relative daily UV exposure stress of *P. syringae* strains on host and non host plants.

P. syringae possesses an additional plasmid-encoded dark repair mechanism termed mutagenic DNA repair (MDR), encoded by the *rulAB* operon, that confers UV tolerance (Sundin *et al.*, 1996). The *rulAB* operon is widely-distributed among *P. syringae* pathovars, and the genes are found on pPT23A-family plasmids, a conserved plasmid group that also encoded genes of importance to host-pathogen interactions (Sundin and Murillo, 1999). *rulAB* expression is rapidly induced following UVB irradiation, and, within the time-frame of colonisation, establishment, and the initiation of pathogenesis on a host plant (5 to 7 days), *rulAB* is required for the maintenance of population size and *rulAB*-mediated MDR occurs at readily-detectable levels when strains are inoculated on bean leaves that are subsequently irradiated with UV-B wavelengths (Sundin and Murillo, 1999; Kim and Sundin, 2000). *rulAB*-mediated DNA repair is mutagenic, in that the repair occurs independent of the original DNA sequence, leading to an increased chance of fixing mutations within the genome. Mutability is variable among *rulAB*-containing *P. syringae* strains and increases with increasing UV dose (Table 1).

Table 1. UVB-induced mutability to rifampicin resistance in selected *Pseudomonas syringae* pathovar *rulAB*-containing strains and one strain (*P. syringae* B728a) that does not contain *rulAB*.

	Number of Rif^R mutants per 10^8 cells recovered at each UVB dose ($J\ m^{-2}$)[a]			
Pseudomonas syringae strain	0	750	1,500	2,250
Psy B728a	nd	nd	nd	nd
Pto PT23	4.7	9.7	31.0	48.7
Psy B86-17	10.3	63.3	92.3	151.0
Pss 0886-19	1.3	26.0	33.3	42.7
Psy A2	3.3	22.0	49.7	77.0

[a] Numbers shown are the means from three replicate experiments.

[b] *Pss: Pseudomonas savastanoi* pv. *savastanoi; Psy: P. syringae* pv. *syringae; Pto, P. s.* pv. *tomato.*

Our laboratory has been investigating the potential for MDR to generate mutations with positive adaptive significance, therefore contributing positively to strain evolution. Cells with constitutive elevated mutation rates have been isolated from natural populations of animal pathogens where the phenotype is associated with an increased chance to generate adaptive variants in the changing environments encountered during animal pathogenesis (LeClerc *et al.*, 1996; Oliver *et al.*, 2000). Since *rulAB* expression is inducible, the chance of generating deleterious mutants is reduced. We have begun addressing the long-term effect of daily UVB irradiation on the accumulation of mutations within *P. s.* pv. *syringae* B86-17 and B86-17*rulB*::Km and have found no significant difference in relative fitness of lineages after approximately 1,500 generations suggesting that *rulAB*-mediated MDR does not contribute to rapid reductions in cellular fitness (Zhang and Sundin, unpublished).

The influence of UVA radiation on microbial ecology is less clear although carotenoid pigmentation, which confers UVA protection in bacteria, is a common feature of many saprophytic phyllosphere bacteria (Sundin and Jacobs, 1999 and references therein). Protection from ROS is also conferred by cellular enzymes including catalase and superoxide dismutase (SOD). My laboratory, in collaboration with the laboratory of Anne Anderson, hypothesised that because *P. syringae* is a pathogen, and would be exposed to ROS (through the oxidative burst response of plants to infection), and an epiphyte exposed to solar UVA radiation, these strains may possess enhanced UVA survival compared to other pseudomonads. Indeed, the UVA survival of *P. s.* pv. *syringae* B728a and B728a*sodAsodB* was significantly greater than that of a wild-type strain of the soil isolate *P. putida* (Fig. 2; Jacobs *et al.*, 2002). Also, the B728a *sodAsodB* mutant retained pathogenicity when inoculated to bean (Kim *et al.*, 1999) indicating that this strain possesses additional enzymatic mechanisms for the inactivation of ROS or that the organism can suppress the plant oxidative burst during infection.

5. CONCLUSION

Pseudomonas syringae strains exist as pathogens, and can grow as epiphytes, and as endophytes. The interrelationships between these various phases are important in *P. syringae* responses to environmental stress. Stress avoidance, especially to desiccation, may be a by product of preferential internal growth of pathogenic strains. Tolerance to UVB radiation encoded by the *rulAB* determinant is critical for population maintenance throughout the establishment of an infection (Kim and Sundin, 2000). Further knowledge of cellular behaviour on leaves prior to infections and mechanisms to cope with environmental stress remains a critical part of the overall quest to understand and manage *P. syringae* diseases.

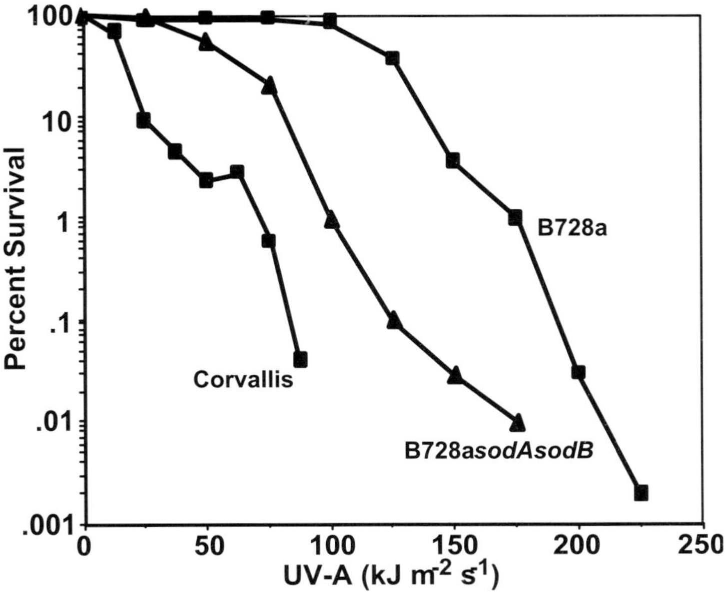

Figure 2. In vitro UVA sensitivity of *Pseudomonas syringae* pv. *syringae* B728a, B728a *sodAsodB*, and *P. putida* Corvallis.

ACKNOWLEDGEMENTS

Work in my laboratory has been supported by the United States Department of Agriculture competitive grants program and the Michigan Agricultural Experiment Station.

REFERENCES

Andersen, G.L., Beattie, G.A., and Lindow, S.E., 1998, Molecular characterisation and sequence of a methionine biosynthetic locus from *Pseudomonas syringae. J. Bacteriol.* **180**: 4497-4507.

Beattie, G.A., and Lindow, S.E., 1994, Comparison of the behaviour of epiphytic fitness mutants of *Pseudomonas syringae* under controlled and field conditions. *Appl. Environ. Microbiol.* **60**: 3799-3808.

Beattie, G.A., and Lindow, S.E., 1995, The secret life of foliar bacterial pathogens on leaves. *Annu. Rev. Phytopathol.* **33**: 145-172.

Davies, D.G., Parsek, M.R., Pearson, J.P., Iglewski, B.H., Costerton, J.W., and Greenberg, E.P., 1998, The involvement of cell-to-cell signals in the development of a bacterial biofilm. *Science* **280**: 295-298.

Haefele, D.M., and Lindow, S.E., 1987, Flagellar motility confers epiphytic fitness advantages upon *Pseudomonas syringae. Appl. Environ. Microbiol.* **53**: 2528-2533.

Hirano, S.S., Charkowski, A.O., Collmer, A., Willis, D.K., and Upper, C.D., 1999, Role of the Hrp type III protein secretion system in growth of *Pseudomonas syringae* pv. *syringae* B728a on host plants in the field. *Proc. Natl. Acad. Sci. USA* **96**: 9851-9856.

Hirano, S.S., and Upper, C.D., 2000, Bacteria in the leaf ecosystem with emphasis on *Pseudomonas syringae*: a pathogen, ice nucleus, and epiphyte. *Microbiol. Mol. Biol. Rev.* **64**: 624-653.

Jacobs, J.L., Miller, C.D., Carter, T.L., Anderson, A.J., and Sundin, G.W., 2002, Aspects of UVR sensitivity and survival strategies of relevance to phyllosphere bacteria. *Phytopathology* **92**: S38-S39.

Jacobs, J.L., and Sundin, G.W., 2001, Effect of solar UV-B radiation on a phyllosphere bacterial community. *Appl. Environ. Microbiol.* **67**: 5488-5496.

Kim, J.-J., and Sundin, G.W., 2000, Regulation of the *rulAB* mutagenic DNA repair operon of *Pseudomonas syringae* by UV-B (290 to 320 nanometers) radiation and analysis of *rulAB*-mediated mutability in vitro and in planta. *J. Bacteriol.* **182**: 6137-6144.

Kim, J.-J., and Sundin, G.W., 2001, Construction and analysis of photolyase mutants of *Pseudomonas aeruginosa* and *Pseudomonas syringae*: contribution of photoreactivation, nucleotide excision repair, and mutagenic DNA repair to cell survival and mutability following exposure to UV-B radiation. *Appl. Environ. Microbiol.* **67**: 1405-1411.

Kim, Y.C., Miller, C.D., and Anderson, A.J., 1999, Transcriptional regulation by iron and role during plant pathogenesis of genes encoding iron- and manganese-superoxide dismutases of *Pseudomonas syringae* pv. *syringae* B728a. *Physiol. Mol. Plant Pathol.* **55**: 327-339.

LeClerc, J.E., Li, B., and Cebula, T.A., 1996, High mutation frequencies among *Escherichia coli* and *Salmonella* pathogens. *Science* **274**, 1208-1211.

Leveau, J.H.J., and Lindow, S.E., 2001, Appetitie of an epiphyte: Quantitative monitoring of bacterial sugar consumption in the phyllosphere. *Proc. Natl. Acad. Sci. USA* **98**: 3446-3453.

Lindow, S.E., Andersen, G., and Beattie, G.A., 1993, Characteristics of insertional mutants of *Pseudomonas syringae* with reduced epiphytic fitness. *Appl. Environ. Microbiol.* **59**: 1593-1601.

Mercier, J., and Lindow, S.E., 2000, Role of leaf surface sugars in colonization of plants by bacterial epiphytes. *Appl. Environ. Microbiol.* **66**: 369-374.

O'Brien, R.D., and Lindow, S.E., 1989, Effect of plant species and environmental conditions on epiphytic population sizes of *Pseudomonas syringae* and other bacteria. *Phytopathology* **79**: 619-627.

Oliver, A., Canton, R., Campo, P., Baquero, F., and Blazquez, J., 2000, High frequency of hypermutable *Pseudomonas aeruginosa* in cystic fibrosis lung infection. *Science* **288**: 1251-1253.

Pfeifer, G.P., 1997, Formation and processing of UV photoproducts: effects of DNA sequence and chromatin environment. *Photochem. Photobiol.* **65**: 270-283.

Pourzand, C. and Tyrell, R.M., 1999, Apoptosis, the role of oxidative stress and the example of solar UV radiation. *Photochem. Photobiol.* **70**: 380-390.

Sundin, G.W., 2002, Ultraviolet radiation on leaves: its influence on microbial communities and their adaptations. In *Phyllosphere Microbiology* (S.E. Lindow, E.I. Hecht-Poinar, and V.J. Elliott, eds.), APS Press, St. Paul, pp. 27-41.

Sundin, G.W., and Jacobs, J.L., 1999, Ultraviolet radiation (UVR) sensitivity analysis and UVR survival strategies of a bacterial community from the phyllosphere of field-grown peanut (*Arachis hypogeae* L.). *Microb. Ecol.* **38**: 27-38.

Sundin, G.W., Kidambi, S.P., Ullrich, M., and Bender, C.L., 1996, Resistance to ultraviolet light in *Pseudomonas syringae*: sequence and functional analysis of the plasmid-encoded *rulAB* genes. *Gene* **177**:77-81.

Sundin, G.W., and Murillo, J., 1999, Functional analysis of the *Pseudomonas syringae rulAB* determinant in tolerance to ultraviolet B (290 to 320 nm) radiation and distribution of *rulAB* among *P. syringae* pathovars. *Environ. Microbiol.* **1**: 75-87.

Wilson, M., Hirano, S.S., and Lindow, S.E., 1999, Location and survival of leaf-associated bacteria in relation to pathogenicity and potential for growth within the leaf. *Appl. Environ. Microbiol.* **65**: 1435-1443.

Wilson, M., Savka, M.A., Hwang, I., Farrand, S.K., and Lindow, S.E. 1995, Altered epipytic colonisation of mannityl opine-producing transgenic tobacco plants by a mannityl opine-catabolizing strain of *Pseudomonas syringae. Appl. Environ. Microbiol.* **61**: 2151-2158.

Yasui, A., and Eker, A.P.M., 1998, DNA photolyases. In *DNA Damage and Repair, Vol. 2 DNA Damage in Higher Eukaryotes* (J.A. Nickoloff and M.F. Hoekstra, eds.), Humana Press Inc., Totowa, NJ.

Yu, J., Penaloza-Vazquez, A., Chakrabarty, A.M., and Bender, C.L., 1999, Involvement of the exopolysaccharide alginate in the virulence and epiphytic fitness of *Pseudomonas syringae* pv. *syringae. Mol. Microbiol.* **33**: 712-720.

Diversity of Epiphytic Pseudomonads on Grass and other Plant Species

D.-H. TIAN[*#], D.E. STEAD[#], and R.H.A. COUTTS[*]
[*]*Department of Biological Sciences, Imperial College London, Sir Alexander Fleming Building, London, UK;* [#]*Central Science Laboratory, Sand Hutton, York, UK.*

Abstract: Isolates of *Pseudomonas* were made from the phyllosphere of both rye grass and meadow grass at two different sites. Fatty acid profiling and rep-PCR fingerprints illustrated that most of the isolates were a genetically diverse collection of *P. syringae* strains that were possibly true epiphytes. These epiphytic strains were not themselves host-specific; genetically similar isolates were regularly isolated at different times from different host species and at different sites. The collection was largely devoid of ice nucleation activity and the ability to elicit a hypersensitive reaction on tobacco. The existence of host-unadapted, epiphytic *P. syringae* strains thus seems likely.

Key words: *Pseudomonas syringae*, epiphytes, rep-PCR.

Author for correspondence: R.H.A. Coutts; e-mail: r.coutts@imperial.ac.uk.

1. INTRODUCTION

Plant pathogenicity in bacteria has evolved independently in several genera and classification has traditionally been pathogen-orientated. *Pseudomonas syringae* comprises over 50 host-specific pathovars most of which are leaf-surface pathogens. DNA:DNA homology studies currently support the division of *P. syringae* into nine genomospecies (Gardan *et al.*, 1997) and until recently, host-specificity was the only reliable method for differentiating pathovars (Young *et al.*, 1992). The pathovar system of classification and nomenclature has been used to prevent or at least delay pathogen movement from their centres of evolution to other parts of the world where known host crops are now grown.

N.S. Iacobellis *et. al.* (eds.), *Pseudomonas syringae* and Related Pathogens, 51-60.

Recent developments in genetic fingerprinting of *P. syringae* support the concept of host-specific taxa at the pathovar level, and the tools to produce classifications, which appear to represent the major evolutionary selection pressures on the bacteria, are also available.

Many strains of *P. syringae* are also known to have epiphytic phases (Hirano and Upper, 2000). Increasingly, strains are found on a particular host species, which are known not to belong to the pathovar that causes disease on that host. It is assumed that they belong to other pathovars but by chance have been transmitted to a non-host, by for example rain-splash and survive in an epiphytic phase.

Many but not all strains of *P. syringae* cause ice-nucleation activity (INA), i.e. the creation of ice crystals at relatively high temperatures. The evolution of ice-nucleation is thought to be associated with host infection, since INA pathogens can enter frost-damaged tissues (Lindow *et al.*, 1982; Gross *et al.*, 1984). This theory holds true for many *P. syringae* pathovars because they are prevalent in cool temperate regions and recently have become more important with the advent of autumn-sown crops where young tender plants are subjected to temperate winters.

Recently, an alternative suggestion for the evolution of INA has been proposed. Here it is propounded that these bacteria are epiphytes that are spread not only by short-distance wind- and rain-splash but are spread long distances by aerosols, which move upwards into the stratosphere before eventual deposition in precipitation (Hamilton and Lenton, 1998). Such transmission would require protection from freezing damage as well as a means of precipitation supposedly provided for by INA. If this theory were correct then bacteria that we know to be host-adapted would fall to earth but would receive a suitable host plant for successful continuation of the life cycle.

Modern diagnostic methods have revealed that many non-pathogenic bacteria strains found on leaf surfaces are strains of *P. syringae.* Many of these bacterial strains do not contain the appropriate host virulence genes required to cause disease.

Host-specificity is also known to be governed by avirulence (*avr*) genes and combinations of different *avr* genes determine whether a bacterium can become pathogenic on a given host and also include the activity of other linked genes (e.g. *hrp* genes) (Mansfield *et al.*, 1997). Thus host-specificity and pathogenicity are separate systems in these bacteria. Several species (e.g. *P. viridiflava* and *P. cichorii*) thought to be non-host specific pathogens can now be classified by genetic fingerprinting into groups which may reflect host-specificity and geographic distribution.

This investigation set out to determine whether strains closely related to *P. syringae* that have no pathogenic ability, exist. If they do exist are they themselves host-specific? For instance do they possess *hrp* and *avr* genes? It is also intended to determine how common these strains are and how easily they can be isolated from the "so-called" phyllosphere. We aim to discover whether pathogenicity is simply the result of the acquisition of a few extra genes by highly evolved epiphytes and whether INA is a survival strategy for surviving low temperatures or whether it has evolved to aid the infection process.

Knowledge of the bacterial ecology of leaf surfaces is scant and is biased towards pathogenicity (Hirano and Upper, 2000). The combination of fatty acid methyl ester profiling and genetic fingerprinting for bacterial strain characterisation as described here now offers opportunities for ecological studies.

Improved understanding of the ecology of leaf surfaces will allow better evaluation of processes such as biological control of pests and diseases of crops by micro-organisms and of environmental risks of release of genetically modified (GM) micro-organisms and GM crops.

The results will also contribute to a better understanding of the evolution of pathogenicity and the epiphytic state. They should also contribute to more speculative issues relating to the importance of aerial bacteria in climatology.

2. MATERIALS AND METHODS

2.1 Sampling

Leaf sampling was originally restricted to two grass species, *Lolium perenne* and *Poa annua*, at two stable sites, one in York and the other a small site in South Kensington, London. Because of an initial paucity of *Pseudomonas* spp. several other plant species were subsequently sampled from the same habitats concentrating specifically on other grasses (36 species) but also including readily identifiable dicotyledonous weed species, shrubs (17 species) and trees (17 species) in an effort to determine if the incidence of *P. syringae* was higher. Prime sampling time was in the spring and summer months.

2.2 Cultures

Leaf washings in sterile water were plated out onto King's B (KB) medium and the cultures incubated at 28°C for 48 h. Putative *Pseudomonas* (*sensu stricto*) isolates produced a typical blue fluorescent pigment on KB media. These isolates were subjected to the LOPA(T) (levan sucrose production on 5% sucrose nutrient agar (SNA) plates, oxidase reaction, pectolysis and arginine dihydrolase production) assay to differentiate *P. syringae* from *P.*

viridiflava, *P. cichorii* and *P. fluorescens* and its allies (Lelliott *et al.*, 1966), all common saprophytes on leaf surfaces.

2.3 Fatty Acid Methyl Ester Profiles

All of the blue fluorescent, levan sucrose-producing oxidase negative isolates were subjected to fatty acid methyl ester (FAME) profiling (Stead *et al.*, 1992) and compared to a library comprising 4,000 bacterial strains, representing almost all known plant pathogens and many of their close relatives. FAMEs were quantified and identified using the peak-naming table component of the Microbial Identification System software package (MIDI; Microbial ID, Inc., Newark, Del., USA). Identification of FAMEs was based on calculation of equivalent chain lengths.

2.4 Rep-PCR

All novel isolates were compared with a panel of National Culture of Plant Pathogenic Bacteria (NCPPB) strains representing the diversity of *Pseudomonas syringae* and other *Pseudomonas* spp. by repetitive extragenic palindromic sequence PCR (rep-PCR) genomic fingerprinting using a BOX-AIR primer (Rademaker *et al.,* 1998). Template DNA solutions were prepared using the Wizard genomic DNA purification kit (Promega) and PCR-amplification and PCR-amplified products analysed as previously using DNA size markers (Marker VI, Roche Diagnostics, Lewes, UK) on each gel to allow inter-gel comparative analysis of genetic fingerprints (Richardson *et al.*, 2002). Fingerprint analysis was performed using GelCompar software Version 4.1 (Applied Maths, Kortrijk, Belgium). Eight-bit TIFF files of rep-PCR fingerprints gel images were digitised, normalised, and the similarity of banding patterns calculated by applying the Pearson product-moment correlation coefficient. Cluster analyses of similarity matrices were performed by the unweighted pair group method using arithmetic averages (UPGMA) and presented as a dendrogram.

2.5 Pathogenicity Profiling

Further investigations of individual *Pseudomonas syringae* and other *Pseudomonas* spp. concerned their potential pathogenicity and included an investigation as to their ability to elicit a hypersensitivity reaction (HR) on tobacco. Here selected isolates were purified and any HR reaction in young tobacco plants was assessed using the procedure of Lelliott and Stead (1987).

Selected *P. syringae* isolates were screened for ice nucleation activity (INA) using the method of Lindow *et al.* (1978).

3. RESULTS AND DISCUSSION

Using the selective KB medium we obtained 66 confirmed fluorescent *Pseudomonas* isolates from the two different grass species at the two different sites (Table 1). More sampling was performed at the York site as compared to the London site and larger numbers of *Pseudomonas* isolates were recovered at the former. LOPA was used to confirm the identity of the *Pseudomonas* spp., which were then subjected to FAME analysis. Using FAME analysis it was possible to differentiate *P. syringae* from other *Pseudomonas* spp. *P. syringae* strains are characterised by 3-hydroxy acids (10:030H, 12:020H and 12:030H) in the absence of or presence of minute amounts of 17:0 cyclopropane. FAME profiles for novel isolates with a similarity value of 0.85÷0.90 or above as compared to *P syringae* or other *Pseudomonas* spp. confirmed the identity of any particular isolate. All profiles had more than 50,000 units of total FAME peak area and had more than 85% of total area named. However identification by FAME profiling does represent the closest matches for some species, *e.g. P. fluorescens, P. putida* and *P. chlororaphis* and the accuracy of assignation is tentative as is also illustrated in the rep-PCR fingerprints (see later).

Table 1 Pseudomonas spp. communities on leaves of perennial rye grass (*Lolium perenne*) and meadow grass (*Poa annua*).

Grass and site Combined data 2000 and 2001	*Lolium perenne*-IC, London	*Lolium perenne*-CSL, York	*Poa annua*-IC, London	*Poa annua*-CSL, York
Pseudomonas syringae	4 (16%)	6 (60%)	2 (22%)	14 (63%)
Pseudomonas spp.	6 (24%)	2 (20%)	4 (44%)	5 (22%)
P. putida	7 (28%)	0	2 (22%)	0
P. fluorescens	0	0	0	1 (5%)
P. corrugata	0	0	0	1 (5%)
P. chlororaphis	8 (32%)	2 (20%)	1 (12%)	0
P. viridiflava	0	0	0	1 (5%)
	25	10	9	22
Pseudomonas spp.	Numbers and relative abundance (%)			

In addition to these grass *Pseudomonas* spp. isolates, which comprised 40% (26/66) *P. syringae* strains, additional isolates were obtained from dicotyledonous weed species, grasses, trees and shrubs. These included further 4 strains of *P. syringae* from *Corylus* spp., *Phyllostachys* spp., *Daucus* spp., and *Agropyron* spp. grasses.

A comparison of the rep-PCR fingerprint patterns on a dendrogram for all the novel *Pseudomonas* spp. isolates revealed that there is a deal of diversity in grass epiphytic *P. syringae* isolates (albeit a relatively small sample number) and that this diversity appears to be not related to individual grass species (Fig. 1).

Cluster analysis resulted in 5 main clusters (clusters 1-5) of similar *P. syringae* isolates being identified while several individual isolates clustered with other *Pseudomonas* spp., some closely related to *P. chlororaphis* and *P. putida*. If a similarity coefficient (*r* value) of 0.7 or more indicates strains of the same genotype then it would appear a large number of different genotypes of epiphytic *P. syringae* isolates exist in our collection. Similar observations have been reported in another study of the endemicity of fluorescent *Pseudomonas* strains (Cho and Tiedje, 2000). There is good evidence to suggest that similar *P. syringae* genotypes were repeatedly isolated from leaves of both perennial rye grass and meadow grass and indeed other plant species (e.g. *Malus* cv Ballerina and *Corylus* spp.) at different sites and at different sampling times. This is shown in detail for clusters 4 and 5 where the sampling times are shown in brackets and location of sampling site indicated. These 2 clusters and the adjacent cluster 3, which contained the *P. syringae* type strain from Lilac, were populated by the largest number of novel grass epiphytic *P. syringae* strains (16/26=62%) and a large number of apple leaf strains. The fact that the sampling times for all plants were different, i.e. different months and years for the grasses and with apple being sampled some years earlier, suggests that these clusters may well contain defined and similar genotypes. Four other grass isolates of *P. syringae* clustered with a different set of *P. syringae* pathovars in cluster 1 and the majority of the remainder of the grass isolates appeared to be related to *P. corrugata* (cluster 2), a weak opportunistic pathogen of tomato or were unrelated to any known *P. syringae* pathovar. It is likely that all of these isolates may well also represent different genotypes.

HR and INA were only detectable for a small number of the novel *P. syringae* strains *viz.* INA-TIAN120 (cluster 4), TIAN332 (cluster 2) and TIAN361 (cluster 1) and HR-TIAN336 and TIAN365 (cluster 4). Any linking with pathogenicity for our novel isolates by these analyses is not possible and tests for the presence of *hrp* genes are yet to be undertaken.

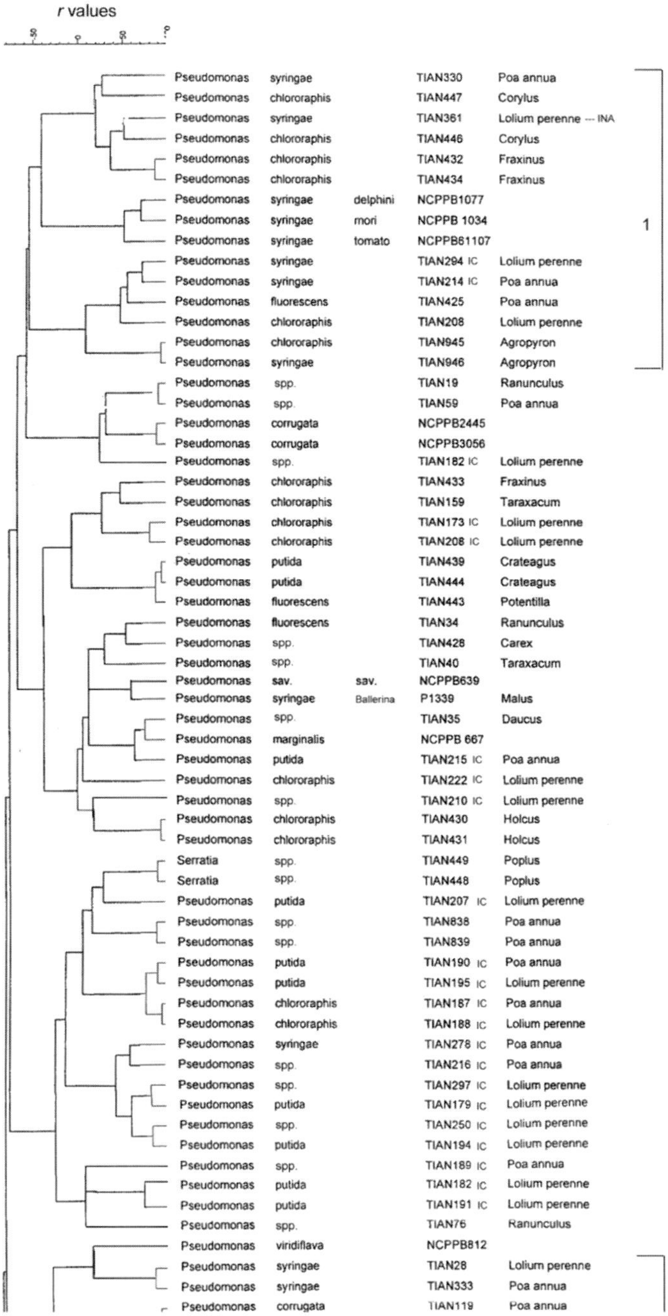
r values
-50
0
50
1.0
Pseudomonas syringae TIAN330 Poa annua
Pseudomonas chlororaphis TIAN447 Corylus
Pseudomonas syringae TIAN361 Lolium perenne --- INA
Pseudomonas chlororaphis TIAN446 Corylus
Pseudomonas chlororaphis TIAN432 Fraxinus
Pseudomonas chlororaphis TIAN434 Fraxinus
Pseudomonas syringae delphini NCPPB1077
Pseudomonas syringae mori NCPPB 1034
Pseudomonas syringae tomato NCPPB61107
Pseudomonas syringae TIAN294 IC Lolium perenne
Pseudomonas syringae TIAN214 IC Poa annua
Pseudomonas fluorescens TIAN425 Poa annua
Pseudomonas chlororaphis TIAN208 Lolium perenne
Pseudomonas chlororaphis TIAN945 Agropyron
Pseudomonas syringae TIAN946 Agropyron
1
Pseudomonas spp. TIAN19 Ranunculus
Pseudomonas spp. TIAN59 Poa annua
Pseudomonas corrugata NCPPB2445
Pseudomonas corrugata NCPPB3056
Pseudomonas spp. TIAN182 IC Lolium perenne
Pseudomonas chlororaphis TIAN433 Fraxinus
Pseudomonas chlororaphis TIAN159 Taraxacum
Pseudomonas chlororaphis TIAN173 IC Lolium perenne
Pseudomonas chlororaphis TIAN208 IC Lolium perenne
Pseudomonas putida TIAN439 Crateagus
Pseudomonas putida TIAN444 Crateagus
Pseudomonas fluorescens TIAN443 Potentilla
Pseudomonas fluorescens TIAN34 Ranunculus
Pseudomonas spp. TIAN428 Carex
Pseudomonas spp. TIAN40 Taraxacum
Pseudomonas sav. sav. NCPPB639
Pseudomonas syringae Ballerina P1339 Malus
Pseudomonas spp. TIAN35 Daucus
Pseudomonas marginalis NCPPB 667
Pseudomonas putida TIAN215 IC Poa annua
Pseudomonas chlororaphis TIAN222 IC Lolium perenne
Pseudomonas spp. TIAN210 IC Lolium perenne
Pseudomonas chlororaphis TIAN430 Holcus
Pseudomonas chlororaphis TIAN431 Holcus
Serratia spp. TIAN449 Poplus
Serratia spp. TIAN448 Poplus
Pseudomonas putida TIAN207 IC Lolium perenne
Pseudomonas spp. TIAN838 Poa annua
Pseudomonas spp. TIAN839 Poa annua
Pseudomonas putida TIAN190 IC Poa annua
Pseudomonas putida TIAN195 IC Lolium perenne
Pseudomonas chlororaphis TIAN187 IC Poa annua
Pseudomonas chlororaphis TIAN188 IC Lolium perenne
Pseudomonas syringae TIAN278 IC Poa annua
Pseudomonas spp. TIAN216 IC Poa annua
Pseudomonas spp. TIAN297 IC Lolium perenne
Pseudomonas putida TIAN179 IC Lolium perenne
Pseudomonas spp. TIAN250 IC Lolium perenne
Pseudomonas putida TIAN194 IC Lolium perenne
Pseudomonas spp. TIAN189 IC Poa annua
Pseudomonas putida TIAN182 IC Lolium perenne
Pseudomonas putida TIAN191 IC Lolium perenne
Pseudomonas spp. TIAN76 Ranunculus
Pseudomonas viridiflava NCPPB812
Pseudomonas syringae TIAN28 Lolium perenne
Pseudomonas syringae TIAN333 Poa annua
Pseudomonas corrugata TIAN119 Poa annua

continue

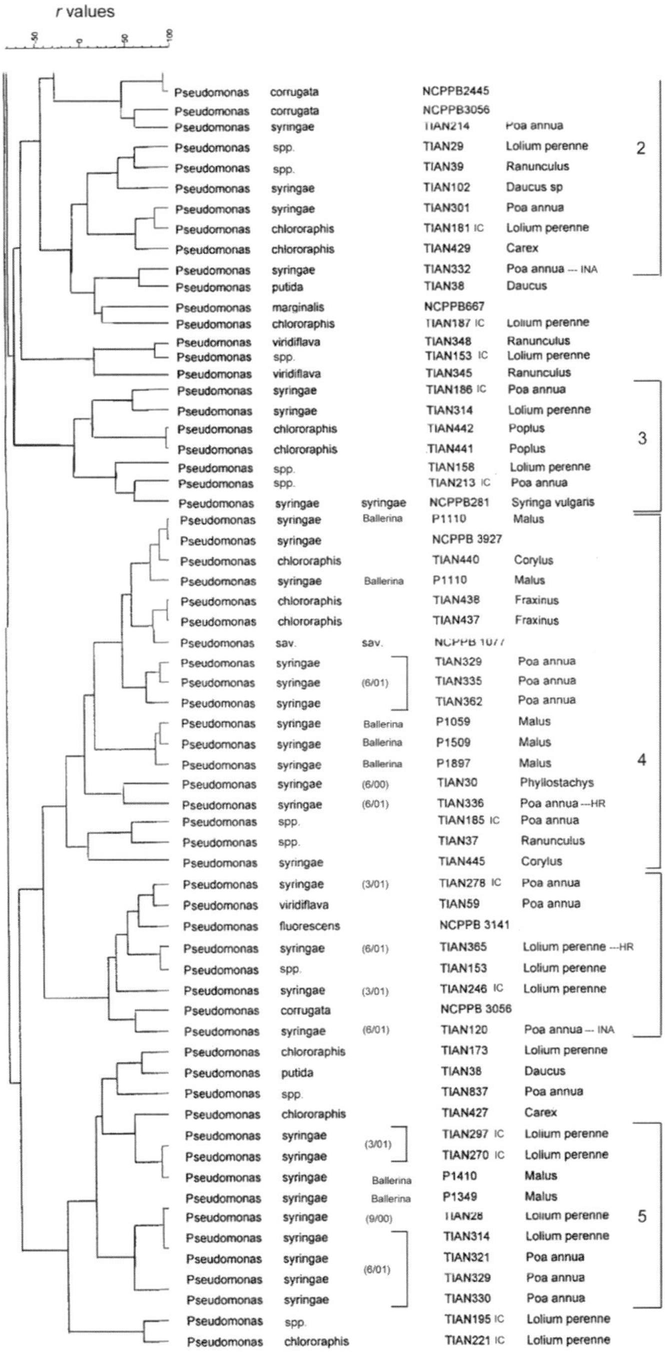

Figure 1. Product moment-UPGMA cluster analysis of BOX-PCR fingerprints for all *Pseudomonas* spp. isolates included in this study. On the scale *r* values are expressed as percentages.

Whilst we have isolated a number of novel *P. syringae* strains that are apparently not pathogenic on the plants from which they were isolated we are yet to determine whether these strains might be pathogens of other (as yet) unknown plant species. Lack of a HR for most of the novel isolates suggests they may not be pathogens but true epiphytes. While we offer evidence that perennial rye grass and meadow grass phyllosphere can support epiphytic *P. syringae* strains, genetically similar strains appear to be epiphytic on other plant species as well. More profiling of all of these isolates by RFLP of PCR-amplified 16S rRNA gene fragments is planned to understand more of their genetic fingerprints. Culture-dependent and independent methods reveal that microbial phyllosphere populations are more complex than previously realised (Yang *et al.*, 2001). Therefore it is intended that the suspected 75% of the organisms not sampled by the methods described in this study (Hirano and Upper, 2000) might by investigated further by a suitable culture-independent method. It is unresolved as to whether the strains we have isolated here are circulating in environmental water but lack of INA might preclude their long-term survival. Nevertheless the existence of host unadapted epiphytic *P. syringae* strains appears to be real.

ACKNOWLEDGEMENTS

We should like to thank The Leverhulme Trust for supporting this investigation.

REFERENCES

Cho, J-C., and Tiedje, J.M., 2000, Biogeography and degree of endemicity of fluorescent *Pseudomonas* strains in soil. *Appl. Environ. Microbiol.* **66**: 5448-5456.

Gardan, L., Shafik, H., and Grimont, P.A.D., 1997, DNA relatedness among pathovars of *Pseudomonas syringae* and related bacteria. In *Pseudomonas syringae* pathovars and related bacteria (K. Rudolph *et al,* eds.), Kluwer Academic Publishers, Dordrecht, pp. 445-558.

Gross, D.C., Cody, Y.S., Proebsting, E.L, Rademaker, G.K, and Spotts, R.A., 1984, Ecotypes and pathogenicity of ice nucleation active *Pseudomonas syringae* isolated from deciduous fruit tree orchards. *Phytopathology* **74**: 241-248.

Hamilton, W.D., and Lenton, T.M., 1998, Spora and Gaia: how microbes fly with their clouds. *Ethol. Ecol. Evol.* **10**: 1-16.

Hirano, S.S., and Upper, C.D., 2000, Bacteria in the leaf ecosystem with emphasis on *Pseudomonas syringae*-pathogen, ice nucleus and epiphyte. *Microbiol. Mol. Biol. Rev.* **64**: 624-653.

Lelliott, R.A., Billing, E., and Harward, A.C., 1966, A determinative scheme for the fluorescent plant pathogenic pseudomonads. *J. Appl. Bacteriol.* **29**: 470-489.

Lelliott, R.A., and Stead, D.E., 1987, In *Methods for the diagnosis of bacterial diseases of plants*. Blackwells Scientific Publications, Oxford.

Lindow, S.E., Arny, D.C., and Upper, C.D., 1978, *Erwinia herbicola*: A bacterial ice nucleus active in increasing frost injury to corn. *Phytopathology*, **68**, 523-527.

Lindow, S.E., Arny, D.C., and Upper, C.D., 1982, Bacterial ice nucleation - a factor in frost injury to plants. *Plant Physiol.* **70**, 1084-1089.

Mansfield, J., Tsiamis, G., Puri, N., Bennet, M., Jenner, C., Stevens, C., Teverson, D., Lyons, N., ans Taylor, J., 1997, Analysis of gene for gene interactions between *Pseudomonas syringae* pv. *phaseolicola* and *Phaseolus*. In: Pseudomonas syringae *pathovars and related bacteria* (K. Rudolph *et al.,* eds.), Kluwer Academic Publishers, Dordrecht, pp. 385-391.

Rademaker, J.L.W., Louws, F.J., and de Bruijn, F.J., 1998, Characterisation of the diversity of ecologically important microbes by rep-PCR fingerprinting. In *Molecular microbial ecology manual, supplement 3* (A.D.L. Akkermans, J.D. van Elsas and F.J. de Bruijn, eds), Kluwer Academic Publishers, Dordrecht, pp. 1-26.

Richardson, J., Stead, D.E., Elphinstone, J.G., and Coutts, R.H.A., 2002, Diversity of *Burkholderia* species from woodland rhizosphere environments. *J. Appl. Microbiol.* **93**: 1-15.

Stead, D.E., Sellwood, J.E., Wilson, J., and Viney, I., 1992, Evaluation of a commercial microbial identification system based on fatty acid profiles for rapid, accurate identification of plant pathogenic bacteria. *J. Appl. Bacteriol.* **72**: 315-321.

Yang, C-H., Crowley, D.E., Borneman, J., and Keen, N.T., 2001, Microbial phyllosphere populations are more complex than previously realized. *Proc. Natl. Acad. Sci. USA* **98**: 3889-3894.

Young, J.M., Takikawa, Y., Gardan, L., and Stead, D.E., 1992, Changing concepts in the taxonomy of plant pathogenic bacteria. *Ann. Rev. Phytopathol.* **30**: 67-105.

Survival of Two Biocontrol *Pseudomonas* Strains in Tomato Fruits After Inoculation at Flowering Through Fruit Ripening

A. PACETTO, P. BELLA, V. CATARA, R. LA ROSA and G. CIRVILLERI

Dipartimento di Scienze e Tecnologie Fitosanitarie, Sezione Patologia Vegetale, Università di Catania, Italy.

Abstract: *Pseudomonas syringae* strain 1.1S and *P. fluorescens* strain A506 were applied to tomato plants to investigate their ability to survive during plant growth and ripening of fruits. Both strains were used to spray tomato flowers and green tomato fruits. Ripe tomato fruits were harvested and subjected to microbiological analysis. Peptone wash water and homogenates of fruit pulp were plated on selective media before and after enrichment. All tomatoes produced from inoculated flowers contained *P. syringae* on the surface of fruits (100% in enriched samples) as well as in the pulp homogenates (100% before and after enrichment). All tomatoes produced from inoculated green fruits contained *P. syringae* on the surface (38% and 100% in wash water and in enriched samples, respectively) as well as in the pulp homogenates (90% and 100% before and after enrichment, respectively). Of the *P. fluorescens* surface-positive tomatoes, 37% were from fruits receiving flower inoculation and 77% were from fruits receiving green fruit inoculation. *P. fluorescens* was harboured in higher percentages in pulp homogenates of tomatoes produced from treated flowers and green fruits (62% and 100%, respectively). Results suggest that antagonistic *P. syringae* and *P. fluorescens* strains survive in and on tomato fruits from the time of inoculation at flowering or at early stage of fruit development through fruit ripening. Tomato flower and green surfaces of fruits are possible sites at which *Pseudomonas* may attach and remain viable during fruit development.

Key words: Bacterial growth, internalisation, quality, risk assessment.

Author for correspondence: G. Cirvilleri; e-mail: gcirvil@unict.it.

N.S. Iacobellis *et. al.* (eds.), *Pseudomonas syringae* and Related Pathogens, 61-68.

1. INTRODUCTION

Limitations of synthetic fungicides for the control of plant diseases, pathogen resistance to fungicides and concern for public safety have combined to increase attention in alternative methods in disease control strategies. This has led to heightened interest in the use of micro-organisms or their antimicrobial products to limit attack by phytopathogens.

Bacteria in the *Pseudomonas* group (including the genus *Burkholderia*, previously classified as *Pseudomonas)* have several features that make them attractive candidates as biological control agents against plant pathogens: their abundance in the phyllosphere and rhizosphere of plants, their aggressive colonisation ability, their ability to produce a large number of biologically active metabolites (antibiotics, antifungal, siderophores), to induce resistance, to compete aggressively with other micro-organisms for niches and to exclude phytopathogens (O'Sullivan and O'Gara, 1992).

Fluorescent pseudomonads constitute a major proportion of the indigenous microflora naturally present on the surfaces of fresh vegetables, and they are assumed to play an important role in maintaining the quality and safety of fresh-cut fruits and vegetables (Nguyen-the and Carlin, 1994).

A large body of information must be available before use of biocontrol agents in field experiments, such as ability to survive under adverse environmental conditions, antagonistic ability *in planta*, pathogenicity, absence of toxicity etc.

Little information are available on the ecology of biocontrol agents on flower and fruit surfaces and on the ability to survive outside and inside the fruits during fruit development.

Studies have been conducted on the internalisation, survival and growth of plant pathogens into fruits, vegetables and juices. Ercolani and Casolari (1966) demonstrated possible internalisation of bacteria into tomato fruits by spraying tomato blossoms with a suspension of a plant pathogen, *Xanthomonas campestris* pv. *vesicatoria*, which was reisolated from the center of fruits. Bartz (1982) found that *Erwinia carotovora* could be infiltrated in the internal part of tomato fruits when the temperature of contaminated wash water was lower than that of tomato fruits. Buchanan *et al.* (1999) also reported the internalisation of *Escherichia coli* 0157:H7 in apple fruits washed with water at temperatures lower than that of fruits. As expected, the internalised bacteria can escape the traditional surface cleaning or sanitation procedures and/or could provide the primary inocula for survival of micro-organisms in fruits.

The fate of antagonistic micro-organisms applied to tomato flowers and fruits before and after fruit set has not been described. We hypothesised that antagonistic *Pseudomonas* may internalise in tomato fruits developed from inoculated flowers. The possibility of internalisation of *Pseudomonas* strains in tomato fruits developed from inoculated green fruits was also considered. The objective of the current study was to determine if antagonistic *Pseudomonas* strains are able to internalise in tomato fruits developed from inoculated flowers and if the bacteria persist on the surface of inoculated fruits during and after fruit ripening.

2. MATERIALS AND METHODS

2.1 Bacterial Cultures

Pseudomonas syringae 1.1S and *P. fluorescens* A506 were used. *P. syringae* 1.1S has been shown to be antagonist of post-harvest pathogens *in vitro* and *in vivo* (Cirvilleri *et al.,* 2000) and is a Rif^{100} resistant strain. *P. fluorescens* A506, commercially available in USA as BlightBan for biological control of *Erwinia amylovora* and INA bacteria, has been lux-marked for risk assessment studies (Cirvilleri and Caldarera, 1998). *P. fluorescens* A506 lux53 was Rif^{100} and Tc^{10} resistant.

2.2 Inoculum Preparation

Stock cultures maintained on Nutrient Agar Glycerol (NAG) at 4°C were transferred to NAG supplemented with appropriate antibiotics and incubated at 28°C for 24 h. Cells were harvested and suspended in phosphate buffer at optical density of ca. 0.1 (OD_{600}), which is equivalent to about 10^8 CFU ml^{-1}.

2.3 Preparation of Tomato Plants for Inoculation

Young, healthy tomato plants, ca. 20 cm in height, were purchased from a local market. Each plant was transferred to potting soil in a 4-liter plastic pot on February and grown in greenhouse. The temperature was maintained at 24°C. Water was applied daily and nutrient solution was applied weekly to maintain optimum soil moisture for plant growth and flowering and fruit development.

2.4 Inoculation Procedures

Inoculation was carried out on 10 of April, on tomato plants harbouring green fruits of 2÷4 cm in diameter and fully opened flowers. Bacterial suspensions of a given strain at concentration of 10^8 CFU ml^{-1} were sprayed to runoff with an atomiser on all part of the plants. Each antagonistic strain was spray inoculated on 5 plants. Each treatment was replicated twice. Untreated tomato plants were included as control.

2.5 Microbiological Analysis

Tomatoes were harvested on 15 of May when subjectively judged to be "red ripe" and ready for consumption. Each tomato fruit was hand picked into a plastic bag and transported to the laboratory for analysis. Each tomato was placed in a Stomacher bag containing 100 ml of 0.1% sterile peptone water and incubated in a rotary shaker for 1.30 h at 100 rpm to dislodge surface population of bacteria. The peptone wash water was surface plated (50 μl in duplicate) on NAG with appropriate antibiotics. Plates were incubated at 28°C for 24÷48 h before examination for colonies of released bacteria. Pulp of each tomato was analysed for presence of antagonists. Tomatoes were removed from peptone wash water, surface sterilised with sodium hypochloride 0.1%, and the peel was removed with a sterile scalpel. The pulp of each tomato was placed in stomacher bag with 50 ml of sterile 0.1% peptone water and pummelled at medium speed for 1 min with a Stomacher 400 Instrument (Seward, London, United Kingdom). Two 100 μl aliquots of homogenate were surface plated on NAG added with appropriate antibiotics. Plates were incubated at 28°C for 24÷48 h before colonies were counted. Peptone wash water and pulp homogenates were enriched by adding 50 ml of NB (Difco) in each Stomacher bag, and the mixture was incubated at 28°C in a rotary shaker at 100 rpm for 24 h. Two 50 μl portions of enriched wash water were surface plated.

Six tomatoes from each treatment were stored at 4°C. After 20 days of incubation, fruits were subjected to the analysis previously described to determine presence of inoculated bacteria on the surface and in the pulp of fruits.

2.6 Confirmation of Antagonistic *Pseudomonas syringae* 1.1S Colonies by PCR Analysis

Approximately 10% of the colonies grown on selective medium were transferred in the same medium and examined by PCR. The analysis to detect *syr*B and *syr*D genes was conducted by using *B1/B2* and *D1/D2* primers according to the procedure previously described (Cirvilleri *et al.*, 2000).

2.7 Confirmation of Antagonistic *Pseudomonas fluorescens* A506 Colonies by Lux Expression

Approximately 10% of the colonies grown on selective medium were transferred in the same medium and analysed for bioluminescence as previously described (Cirvilleri and Caldarera, 1998).

3. RESULTS

Ripe tomato fruits were harvested and subjected to microbiological analysis. *Pseudomonas syringae* and *P. fluorescens* were detected on the surface and in the pulp homogenate of tomatoes that developed from inoculated flowers and green fruits.
Fifty-two tomatoes were harvested and 26 of them analysed for the presence of *P. syringae* 1.1S (Table 1). Five tomatoes were from plants inoculated by flower spraying and 21 were from plants inoculated by fruit spraying. *P. syringae* 1.1S was detected, even if in low concentration, on the surface of fruits developed from flower inoculation but only after the enrichment of peptone wash water with NB. The bacterium was also found on the surface of fruits developed on plants that received fruit inoculation either without (38%) or after enrichment (100%). The bacterium was detected in the pulp of tomatoes developed from flower inoculation (100%), as well in the tomatoes that received fruit inoculation (90%).

Table 1. Pseudomonas syringae 1.1S on and in tomato fruits produced by inoculated plants.

			N° of fruits colonised by *P. syringae* 1.1S			
Inoculation site	Total number of fruits	N° of fruits examined	Wash water	Enrichment wash water	Pulp	Enrichment pulp
Flower	16	5	0	5	5	5
Fruit	36	21	8	21	19	19

Sixthy-four tomatoes were harvested and 38 of them selected and analysed for the presence of *P. fluorescens* A506 (Table 2). Eight of these tomatoes were from plants inoculated by flower spraying, and 30 were from plants inoculated by fruit spraying. Compared to surface tissues (3 of 8 positive, 37%), an higher percentage of pulp (5 of 8 positive, 62%) harboured *P. fluorescens* A506 in tomatoes harvested from inoculated flowers. Similarly, a higher percentage of pulp (30 of 30 positive, 100%) harboured A506 compared to surface tissues (23 of 30 positive, 77%) in tomatoes harvested from inoculated fruits.

Table 2. Pseudomonas fluorescens A506 on and in tomato fruits produced by inoculated plants.

Inoculation site	Total number of fruits	N° of fruits examined	N° of fruits colonised by *P. fluorescens* A506			
			Wash water	Enrichment wash water	Pulp	Enrichment pulp
Flower	25	8	3	3	5	8
Fruit	39	30	23	23	30	30

After 20 days at 4°C, bacteria were recovered from the surface and the pulp of tomato fruits. No differences were observed between bacterial colonisation before and after storage at 4°C for 20 days.
PCR amplification with the *syr*B (752-bp DNA fragment) and *syr*D primers (446-bp fragment) was seen only with all *P. syringae* strains tested (Fig. 1 A and B). None of the bacteria isolated from untreated tomato plants amplified with the primers *syr*B1/B2 and *syr*D1/*syr*D2 indicating that *P. syringae* was not naturally present in the fruits.

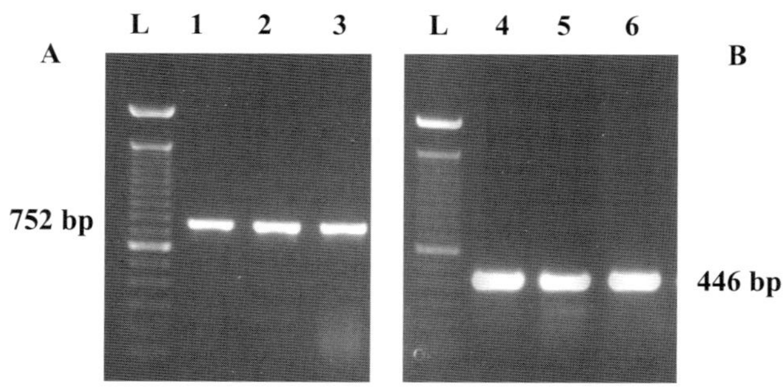

Figure 1 A. PCR products from *Pseudomonas syringae* strains amplified with *syr*B primers B1/B (lanes 1-3); B. PCR products from *P. syringae* strains amplified with *syr*D primers D1/D2 (lanes 4-6). L, 100 bp ladder.

All presumptive *P. fluorescens* A506 strains reisolated from treated tomato fruits were Rif^{100} and Tc^{10} resistant and positive for the bioluminescence activity.

4. DISCUSSION

Observations on survival of *Pseudomonas syringae* and *P. fluorescens* on and in tomato fruits (Tables 1 and 2) indicate that both epyphytic strains persisted from the time of spray inoculation at flowering through fruit ripening. A large number of *Pseudomonas* cells (ca. 10^8 CFU ml^{-1}) was spray-applied to the flowers and green tomatoes. Lower number of cells may not be realistic in terms of efficacy of biological control. A relationship between the number of *Pseudomonas* cells in the inoculum and the time required for death of all cells was not determined. The antagonistic *Pseudomonas* may not have been detectable on tomato fruits 33 days after inoculation of plants if a small number of bacterial cells had been sprayed onto flowers and fruits.

The presence of *Pseudomonas* in peptone wash water and pulp of tomatoes from inoculated plants indicates that flowers and natural opening on the surface of the fruits could be possible routes of bacterial infection of tomato fruits at different points during development and maturation. The presence of epiphytic bacteria within tissue of fruits and vegetables through various pathways was reported by Samish *et al.* (1962). By examining 8 internal locations of tomatoes, they observed that bacteria are unevenly distributed in the fruit, and entry may be from the stem scar tissue through the core and into the endocarp. This study suggested that some epiphytic bacteria might reach internal tissue of tomatoes through natural opening because of their small size and mobility. It may be that bacteria enter fruit tissue more readily in the early stages of fruit development, at a time when various channels are not yet covered by corky or waxy materials (Samish *et al.,* 1962). Broken trichomes on young fruits represent another site of entry of micro-organisms. Recently Schena *et al.* (2002) reported that *Aureobasidium pullulans,* a biocontrol agent of postharvest diseases, is able to penetrate sweet cherry fruit epidermis when applied at the bloom. Although *Pseudomonas* are ephiphityc bacteria, our study reveals their ability to survive on and in tomato fruits throughout the course of plant growth, flowering and fruit development and maturation.

REFERENCES

Bartz, J.A., 1982, Infiltartion of tomatoes immersed at different temperatures to different depths in suspensions of *Erwinia carotovora* subsp. *carotovora. Plant Dis.* **66**: 302-306.

Buchanan, R.L., Edelson, S.G., Miller, R.L., and Sapers, G.M., 1999, Contamination of intact apples after immersion in an aqueosus environment containing *Escherichia coli* 0157:H7. *J. Food Protec.* **62**: 444-450.

Cirvilleri, G., Bella, P., and Catara V., 2000, Molecular detection and biological control activity of *Pseudomonas* strains. 5th Congress of the European Foundation for Plant Pathology, Taormina, 18-22 September 2000, 534-538.

Cirvilleri, G., and Caldarera, G., 1998, Use of lux-marker genes to monitor survival of antagonistic *Pseudomonas fluorescens* in the phylloplane. *J. Pl. Dis. Protec.* **105**: 441-451.

Ercolani, G., and Casolari, A., 1966, Ricerche di microflora in pomodori sani. *Industrie Conserve Parma* **41**: 15-22.

Nguyen-the, C., and Carlin, F.C., 1994, The microbiology of minimally processed fresh fruits and vegetables. *Crit. Rev.Food Sci. Nu.* **34**: 371-401.

O'Sullivan, D.J., and O'Gara, F., 1992, Traits of fluorescent *Pseudomonas* spp. involved in suppression of plant root pathogens. *Microbiol. Rev.* **56**: 662-676.

Samish, Z., Etinger-Tulczynska, R., and Bick, M., 1962, The microflora within the tissue of fruits and vegetables. *J. Food Sci.* **28**: 259-266.

Schena, L., Sialer, M.F., and Gallitelli, D., 2002, Molecular detection of strain L47 of *Aureobasidium pullulans,* a biocontrol agent of postharvest diseases. *Pl. Dis.* **86**: 54-60.

Diversity Among *Pseudomonas syringae* Strains from Belgian Orchards

Using Secondary Metabolite-Based Tests

A. BULTREYS and I. GHEYSEN

Département Biotechnologie, Centre de Recherches Agronomiques de Gembloux, Ministère de la Région wallonne, Direction Générale de l'Agriculture, 234 chaussée de Charleroi, B-5030 Gembloux, Belgium.

Abstract: Conventional identification tests and rapid tests based on phytotoxin and pyoverdin production were used to determine the diversity in 170 fluorescent and non-fluorescent oxidase-negative strains isolated from diseased plant material in pear, sweet cherry, sour cherry and plum orchards in Belgium. Determination of phytotoxin production was based on a biological test for the detection of toxic lipodepsipeptide production and PCR tests to detect the *syrB*, *syrD* and *cfl* genes. The pyoverdin tests were visual, spectrophotometrical and isoelectric focalisation tests; they detected the Fe(III)-chelated atypical pyoverdin produced by *Pseudomonas syringae*. The *P. s.* pv. *syringae* and *P. s.* pv. *morsprunorum* were almost exclusively detected in pear and sweet cherry orchards and the phytotoxin tests proved useful in identifying these pathovars, although caution was needed regarding negative results because a few strains were unable to produce phytotoxins. In contrast, many plum and sour cherry strains could not be affiliated to a pathovar. All the non-toxin-producing strains investigated in this study were fluorescent. The pyoverdin tests classified them in the siderovar of *P. syringae*. The results therefore underscored the great variation in the *P. syringae* strains isolated from plum and sour cherry orchards. The pyoverdin tests were particularly useful in identifying the non-toxin-producers and they showed a real potential as general identification tools of fluorescent *P. syringae* strains, but they were less discriminatory than the phytotoxin tests. Also, the pyoverdin tests could not identify non-fluorescent strains comparable to 34 of the 45 *P. s.* pv. *morsprunorum* strains in this study.

Key words: coronatine, lipodepsipeptides, phytotoxin, pyoverdin.

Author for correspondence: A. Bultreys; e-mail: bultreys@cra.wallonie.be.

N.S. Iacobellis *et. al.* (eds.), *Pseudomonas syringae* and Related Pathogens, 69-77.

1. INTRODUCTION

Two pathovars of *Pseudomonas syringae* are recognised pathogens of pear, sweet cherry, sour cherry and plum orchards. *P. s.* pv. *syringae* is pathogenic on all of these crops and *P. s.* pv. *morsprunorum* is pathogenic on sweet cherry, sour cherry and plum. However, the pathovar *syringae* is heterogeneous (Little *et al.*, 1998) and recent DNA hybridisation studies underlined a possible heterogeneity within the *P. s.* pv. *morsprunorum* (Gardan *et al.*, 1999), although this pathovar is considered more homogeneous.

The identification of *P. syringae*, and the distinction between the pathovars *syringae* and *morsprunorum*, can be done using conventional, time-consuming, physiological and biochemical tests (Freigoun and Crosse, 1975). Other tests, more rapid, include genetic fingerprinting (Louws *et al.*, 1994; Weingart and Völksch, 1997), PCR detection of genes involved in phytotoxin production and tests that detect the production of phytotoxins and of pyoverdins (Bereswill *et al.*, 1994; Bultreys and Gheysen, 1999; Bultreys *et al.*, 2001; Sorensen *et al.*, 1998). Secondary metabolite-based tests also give indications of the adaptive advantages of the analysed strains. The phytotoxins can play a role in plant-pathogen interactions (Bender *et al.*, 1999; Scholz-Schroeder *et al.*, 2001) and the pyoverdin siderophores are probably important in the interactions between micro-organisms on plants (Bultreys and Gheysen, 2000). The production of toxic lipodepsipeptides is generally observed in the *P. s.* pv. *syringae* but these large-spectrum toxins are also produced by the pathovars *aptata* and *atrofaciens* within *P. syringae*, as well as by oxydase-positive *P. fuscovaginae* strains (Bultreys and Gheysen, 1999; Flamand *et al.*, 1996; Vassilev *et al.* 1996). The production of the phytotoxin coronatine is facultative in the pathovar *morsprunorum* and the molecule is also produced by strains of the pathovars *tomato*, *glycinea*, *atropurpurea* and *maculicola* within *P. syringae* (Bereswill *et al.*, 1994). Some 38 pathovars from 7 genomospecies of *P. syringae* were recently shown to produce the same atypical pyoverdin siderophore (Bultreys, 2001; Bultreys *et al.*, 2003). Several tests were recently published on identifying *P. syringae*, and the related species *P. viridiflava*, based on the detection of this atypical pyoverdin (Bultreys *et al.*, 2001; Bultreys *et al.*, 2003). However, some pathovars of *P. syringae* and strains of the pathovars *morsprunorum* and *garcae* are non-fluorescent. Also, non-fluorescent strains of *P. s.* pv. *aptata* produce a related non-fluorescent atypical dihydropyoverdin. Non-fluorescent strains of *P. s.* pv. *morsprunorum* are able to use both the atypical pyoverdin and the atypical dihydropyoverdin (Bultreys and Gheysen, 2000; Bultreys *et al.*, 2001).

In this study, conventional identification tests and secondary metabolite-based tests were combined to determine the diversity among 170 fluorescent and non-fluorescent oxidase-negative strains isolated from diseased plant material in Belgian pear, sweet cherry, sour cherry and plum orchards. The diversity among *P. syringae* strains is greater than expected.

2. MATERIALS AND METHODS

The oxidase test and the detection of fluorescence on King's medium B (King *et al.*, 1954) were performed for strains isolated from diseased plant material in pear, sweet cherry, sour cherry and plum orchards in the Gembloux area in Belgium. Only the oxidase-negative strains were further characterised.

2.1 Phytotoxin-based Tests

Phytotoxin-based tests were used to try to rapidly affiliate the isolated strains to a pathovar of *Pseudomonas syringae*. The PCR test that detects the *cfl* gene involved in coronatine production (Bereswill *et al.*, 1994) was carried out as previously described (Bultreys and Gheysen, 1999). Two recently described PCR tests that detect the *syrB* gene (Sorensen *et al.*, 1998) and the *syrD* gene (Bultreys and Gheysen, 1999), involved in toxic lipodepsipeptide production (Bender *et al.*, 1999), were used. Both PCR tests were carried out using the reagents and the program described for the *syrD* PCR test. The reliability of the biological test enabling the detection of weak production of toxic lipodepsipeptides by *P. syringae* using the PGNaClA medium and the test strain *Rhodotorula pilimanae* MUCL 30397 (Bultreys and Gheysen, 1999) was evaluated by comparison with the more complex PCR tests.

2.2 Pyoverdin-based Tests

The pyoverdin siderophore-based tests used were the recently described visual, spectrophotometrical and isoelectric focalisation (IEF) tests carried out on Fe(III)-chelated pyoverdins (Bultreys *et al.*, 2001). In these tests, pyoverdin production is carried out over 3 days in solid/liquid GASN medium cultures (Bultreys and Gheysen, 2000) at 20°C, and the culture medium is directly analysed after chelation of the pyoverdins to the Fe(III).

2.3 Conventional Identification Tests

The non-toxin-producing strains but producing the atypical pyoverdin of *Pseudomonas syringae* and the coronatine-producing strains from sweet cherry were further characterised by conventional identification tests. These tests included the API 20 NE tests (bioMérieux, France), gelatine hydrolysis, tyrosinase activity, and the utilisation by the strains of lactic acid and tartaric acid.

3. RESULTS

Among the fluorescent and non-fluorescent strains isolated from diseased plant material, 170 oxidase-negative strains were characterised in this study.

3.1 Phytotoxin-based Tests

Except for 2 strains of *Pseudomonas syringae* pv. *syringae* that were positive in the biological test but negative in both PCR tests, comparable results were obtained with the *syrB*-based and *syrD*-based PCR tests and the simpler biological test on PGNaClA medium. In the biological test, it was essential to detect a zone of total growth inhibition of the yeast (Fig. 1), even a few millimetres, to consider a strain as positive; partial inhibitions were considered as negative results.

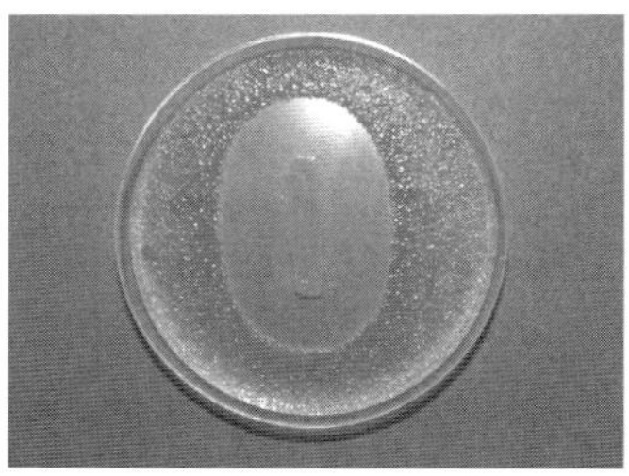

Figure 1. Total growth inhibition of *Rhodotorula pilimanae* by a *Pseudomonas syringae* pv. *syringae* strain.

The frequency of detection of phytotoxin production varied depending on the origin of the strains. Phytotoxin-producing strains were frequently encountered among the strains isolated from pear (90.4%) and sweet cherry (95.4%) orchards, but less frequently among the strains isolated from plum (54.2%) and sour cherry (21.4%) orchards. Coronatine-producing *P. s.* pv. *morsprunorum* strains were fluorescent (26.2%) or non-fluorescent (73.8%). Toxic lipodepsipeptide-producing *P. s.* pv. *syringae* strains were always fluorescent. All the non-toxin-producing strains investigated in this study were fluorescent.

3.2 Pyoverdin-based Tests

All the fluorescent oxydase-negative strains analysed produced an atypical pyoverdin that could easily be detected using the visual and the spectrophotometrical tests. In the visual test, the modification of the colour of the culture medium from orange at pH 7.0 to dark brown below pH 4.0 was easily observed; in contrast, a constant dark brown colour was observed at both pH values with saprophytic fluorescent pseudomonads (Fig. 2), which always produce typical pyoverdins (Fuchs and Budzikiewicz, 2001). Fig. 3 shows the differences of spectra obtained at pH 7.0 for a *Pseudomonas syringae* strain and a *P. fluorescens* strain. A culture medium containing typical Fe(III)-chelated pyoverdins showed a maximum near 399 nm and two broad charge transfer bands near 470 and 550 nm. A culture medium containing atypical Fe(III)-chelated pyoverdins showed a maximum between 404 and 408 nm and the broad charge transfer bands were absent.

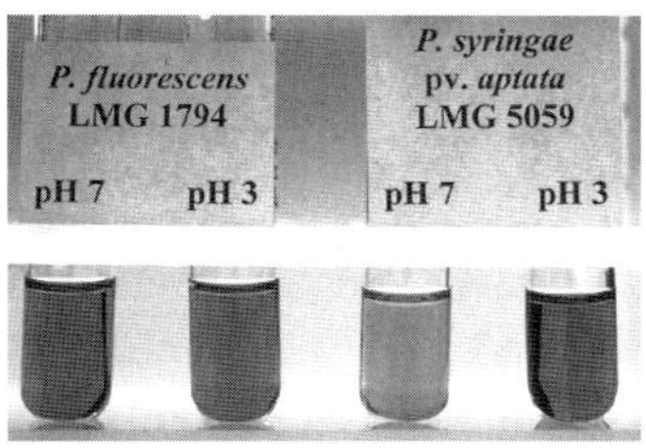

Figure 2. Pyoverdin test. Visual test for *Pseudomonas syringae* and *P. fluorescens* control strains production of pyroverdin revealed by browning of the medium.

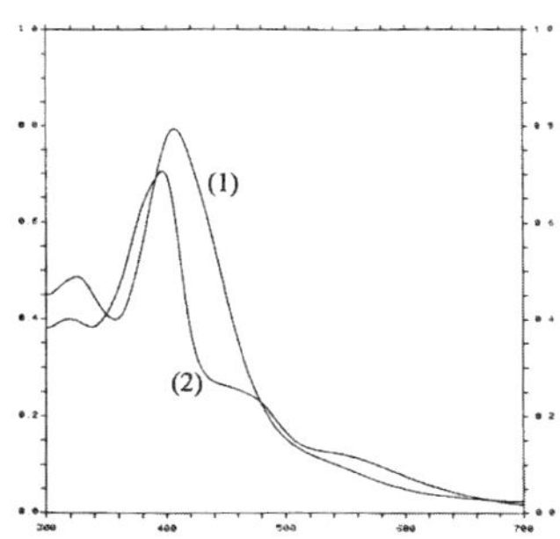

Figure 3. Pyoverdin test. Spectrophotometrical test for (1) a *Pseudomonas syringae* strain producing atypical pyoverdins and (2) a *P. fluorescens* strain producing typical pyoverdins

IEF analyses of the Fe(III)-chelated pyoverdins confirmed that all the oxydase-negative fluorescent strains belonged to the siderovar of *P. syringae* and *P. viridiflava* (Fig. 4).

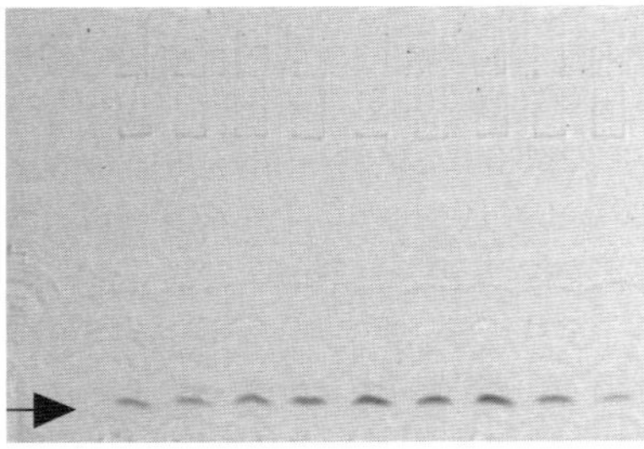

Figure 4. Pyoverdin test, IEF analyses of the culture medium of different strains containing the Fe(III)-chelated atypical pyoverdin (arrow) characteristic of the siderovar of *Pseudomonas syringae.*

3.3 Combination of Conventional and Secondary Metabolite-based Tests

Table 1 summarises the combined results. Out of 39 non-phytotoxin-producing strains affiliated with the siderovar of *Pseudomonas syringae*, only 8 strains were affiliated with either the pathovar *syringae* or the pathovar *morsprunorum* on the basis of the conventional identification tests.

Table 1. Outcome of the combined tests

Culture	Total	*P. syringae* pv. *syringae*			*P. syringae* pv. *morsprunorum*			*P. syringae* or *P. viridiflava*
		Toxin Tests	Conv. Tests	Total	Toxin test	Conv. tests	Total	
Pear	52	47	3	50	0	0	0	2
Sweet cherry	66	25	0	25	38	2	40	1
Plum	24	13	0	13	0	0	0	11
Sour cherry	28	2	2	4	4	1	5	19

Out of 52 pear strains, only 2 (3.8%) were not clearly similar to *P. s.* pv. *syringae*. However, 3 *P. s.* pv. *syringae* strains were apparently unable to produce toxic lipodepsipeptides.
Out of 66 sweet cherry strains, only 1 (1.5%) could not be assigned to a pathovar. Two strains of *P. s.* pv. *morsprunorum* out of 40 were apparently unable to produce coronatine, whereas the 25 *P. s.* pv. *syringae* strains

produced toxic lipodepsipeptides. All the coronatine-producing strains were identified as *P. s.* pv. *morsprunorum* strains by the conventional identification tests. Only 13 strains out of the 40 *P. s.* pv. *morsprunorum* strains were fluorescent.

Out of 24 plum strains, 11 (45.8%) could not be assigned to a pathovar. *P. s.* pv. *morsprunorum* was never encountered in the plum orchards near Gembloux and the 13 *P. s.* pv. *syringae* strains produced toxic lipodepsipeptides.

Out of 28 sour cherry strains, 19 (67.8%) could not be assigned to a pathovar. Out of the 4 *P. s.* pv. *syringae* strains, 2 were apparently unable to produce toxic lipodepsipeptides. Also, out of 5 *P. s.* pv. *morsprunorum* strains, 1 was apparently unable to produce coronatine and 4 strains were non-fluorescent.

4. DISCUSSION

The results of this study indicate that *Pseudomonas syringae* pv. *syringae* and *P. s* pv. *morsprunorum* are the dominant pathovars in pear and sweet cherry orchards in the Gembloux area in Belgium. The *P. s.* pv. *syringae* is encountered in both fruit crops and the *P. s.* pv. *morsprunorum* is restricted to sweet cherry. Many of the strains isolated in these orchards were able to produce phytotoxins (110 strains out of 118) and the phytotoxin-based tests were therefore particularly interesting in identifying the pathogens in these orchards. The frequent production of coronatine by *P. s.* pv. *morsprunorum* strains isolated from sweet cherry (95%) confirms similar observations made in Michigan (Liang *et al.*, 1994). However, 2 *P. s.* pv. *morsprunorum* strains from sweet cherry and 3 *P. s.* pv. *syringae* strains from pear gave negative results in the phytotoxin-based tests and caution was therefore needed concerning negative results. Few *P. s.* pv. *syringae* strains unable to produce toxic lipodepsipeptides or lacking the *syrB* and *syrD* genes have been reported previously (Bultreys and Gheysen, 1999; Sorensen *et al.*, 1998). The 5 non-toxin-producing strains were fluorescent and the pyoverdin tests were therefore particularly useful in their identification. Also, the pyoverdin tests proved useful in identifying the 3 strains that could not be affiliated to a pathovar and to reveal diversity in *P. s.* pv. *morsprunorum*.

Different conclusions can be drawn about the strains isolated from Belgian plum and sour cherry orchards. Only 13 strains out of 24 from plum, and 11 strains out of 28 from sour cherry, could be assigned to one of the investigated pathovars. These results therefore underscored the greater diversity found in the strains isolated from plum and sour cherry. The phytotoxin tests, although still useful, should therefore be taken with still greater caution in the case of a negative response. The non-toxin-producing

strains were always fluorescent in this study. Therefore, the pyoverdin-based tests were particularly useful in identifying these strains and these tests have a real potential as general identification tools for fluorescent *P. syringae* strains, whatever the pathovar, because 38 pathovars of *P. syringae* from 7 genomospecies were shown to produce the same atypical pyoverdin (Bultreys, 2001). Among the pyoverdin-based tests, a very recently described high-performance liquid chromatography-based method averred to be the most informative because it enabled distinction between slightly different atypical pyoverdins produced by *P. syringae* and *P. cichorii* (Bultreys *et al.*, 2003). However, the pyoverdin-based tests were unable to identify non-fluorescent strains comparable to 34 of the 45 *P. s.* pv. *morsprunorum* strains in this study, and they were less discriminatory than the phytotoxin-based test since they were unable to help to classify a strain in a pathovar. Also, the distinction between the species *P. syringae* and *P. viridiflava* could theoretically not be done. However, the observations of symptoms in the field give indications that can be used in identification.
The study of phenotypic characteristics is interesting because it provides information on the ecology of the investigated strains. This study underscored a greater diversity than expected concerning the production of phytotoxins and pyoverdins among the *P. syringae* strains isolated from pear, cherry and plum orchards in Belgium. Additional genetic and phenotypic studies, including pathogenicity tests, should help understand the relatedness of these strains.

ACKNOWLEDGEMENTS

This work was supported by the Ministère des Classes moyennes et de l'Agriculture de Belgique and by the Ministère de la Région wallonne, Direction Générale de l'Agriculture.

REFERENCES

Bender, C.L., Alarcón-Chaidez, F., and Gross, D.C., 1999, *Pseudomonas syringae* phytotoxins: mode of action, regulation, and biosynthesis by peptide and polyketide synthetases. *Microbiol. Mol. Biol. Rev.* **63**: 266-292.

Bereswill, S., Bugert, P., Völksch, B., Ullrich, M., Bender, C.L., and Geider, K., 1994, Identification and relatedness of coronatine-producing *Pseudomonas syringae* pathovars by PCR analysis and sequence determination of the amplification products. *Appl. Environ. Microbiol.* **60**: 2924-2930.

Bultreys, A., 2001, PhD Thesis *Utilisation de la production de métabolites secondaires de types phytotoxine et sidérophore en vue de la caractérisation et de l'identification de Pseudomonas syringae*. Université Catholique de Louvain, Louvain-la-Neuve, Belgium.

Bultreys, A., and Gheysen, I., 1999, Biological and molecular detection of toxic lipodepsipeptide-producing *Pseudomonas syringae* strains and PCR identification in plants. *Appl. Environ. Microbiol.* **65**: 1904-1909.

Bultreys, A., and Gheysen, I., 2000, Production and comparison of peptide siderophores from strains of distantly related pathovars of *Pseudomonas syringae* and *Pseudomonas viridiflava* LMG 2352. *Appl. Environ. Microbiol.* **66**: 325-331.

Bultreys, A., Gheysen, I., Maraite, H., and de Hoffmann, E., 2001, Characterisation of fluorescent and nonfluorescent peptide siderophores produced by *Pseudomonas syringae* strains and their potential use in strain identification. *Appl. Environ. Microbiol.* **67**: 1718-1727.

Bultreys, A., Gheysen, I., Wathelet, B., Maraite, H., and de Hoffmann, E., 2003, High-performance liquid chromatography analyses of pyoverdin siderophores differentiate among phytopathogenic fluorescent *Pseudomonas* species. *Appl. Environ. Microbiol.* **69**: 1143-1153.

Flamand, M.C., Pelsser, S., Ewbank, E., and Maraite, H., 1996, Production of syringotoxin and other bioactive peptides by *Pseudomonas fuscovaginae*. *Physiol. Mol. Plant Pathol.* **48**: 217-231.

Freigoun, S.O., and Crosse, J.E., 1975, Host relations and distribution of a physiological and pathological variant of *Pseudomonas mors-prunorum*. *Ann. Appl. Biol.* **81**: 317-330.

Fuchs, R., and Budzikiewicz, H., 2001, Structural studies of pyoverdins by mass spectrometry. *Curr. Org. Chem.* **5**: 265-288.

Gardan, L., Shafik, H., Belouin, S., Broch, R., Grimont, F., and Grimont, P.A.D., 1999, DNA relatedness among the pathovars of *Pseudomonas syringae* and description of *Pseudomonas tremae* sp. nov. and *Pseudomonas cannabina* sp. nov. (ex Sutic and Dowson 1959). *Int. J. Syst. Bacteriol.* **49**: 469-478.

King, E.O., Ward, M.K., and Raney, D.E., 1954, Two simple media for the demonstration of pyocyanin and fluorescein. *J. Lab. Clin. Med.* **44**: 301-307.

Liang, L.Z., Sobiczewski, P., Paterson, J.M., and Jones, A.L., 1994, Variation in virulence, plasmid content, and genes for coronatine synthesis between *Pseudomonas syringae* pv. *morsprunorum* and *P. s. syringae* from *prunus*. *Plant Dis.* **78**: 389-392.

Little, E.L., Bostock, R.M., and Kirkpatrick, B.C., 1998, Genetic characterisation of *Pseudomonas syringae* pv. *syringae* strains from stone fruits in California. *Appl. Environ. Microbiol.* **64**: 3818-3823.

Louws, F.J., Fulbright, D.W., Stephens, C.T., and de Bruijn, F.J., 1994, Specific genomic fingerprints of phytopathogenic *Xanthomonas* and *Pseudomonas* pathovars and strains generated with repetitive sequences and PCR. *Appl. Environ. Microbiol.* **60**: 2286-2295.

Scholz-Schroeder, B.K., Hutchison, M.L., Grgurina, I., and Gross, D.C., 2001, The contribution of syringopeptin and syringomycin to virulence of *Pseudomonas syringae* pv. *syringae* strain B301D on the basis of *sypA* and *syrB1* biosynthesis mutant analysis. *Mol. Plant-Microbe Interact.* **14**: 336-348.

Sorensen, K.N., Kim, K.-H., and Takemoto, J.Y., 1998, PCR detection of cyclic lipodepsinonapeptide-producing *Pseudomonas syringae* pv. *syringae* and similarity of strains. *Appl. Environ. Microbiol.* **64**: 226-230.

Vassilev, V., Lavermicocca, P., Di Giorgio, D., and Iacobellis, N.S., 1996, Production of syringomycins and syringopeptins by *Pseudomonas syringae* pv. *atrofaciens*. *Plant Pathol.* **45**: 316-322.

Weingart, H., and Völksch, B., 1997, Genetic fingerprinting of *Pseudomonas syringae* pathovars using ERIC-, REP-, and IS50-PCR. *J. Phytopathol.* **145**: 339-345.

Epiphytic Fitness of *Pseudomonas syringae* pv. *syringae* on Mango Trees is Increased by 62-Kb Plasmids

F.M. CAZORLA, E. ARREBOLA, C. ABAD, J.C. CODINA, A. PÉREZ-GARCÍA, and A. DE VICENTE
Departamento de Microbiología, Facultad de Ciencias, Universidad de Málaga. Campus Universitario de Teatinos, s/n. 29071-Málaga (Spain).

Abstract: In this work, we have studied the relationship between presence of indigenous plasmids and copper- and ultraviolet light-resistance in *Pseudomonas syringae* pv. *syringae* strains isolated from mango trees. In such strains, plasmids of 62 kb showed cross-hybridisation with specific sequences of the copper-resistance operon *cop*ABCD and the UV-resistance determinant *rul*AB. Copper-resistance was evaluated by determining the minimal inhibitory concentration (MIC) of copper sulphate, as well as analysing the increase of copper-resistant (Cu^r) bacteria in commercial field plots under copper treatments during three years. UV-resistance was evaluated by performing survival curves of *P. s.* pv. *syringae* cells exposed to doses of B+A UV-fractions, as well as the persistence of *P. s.* pv. *syringae* strains under field conditions studied on mango leaves directly exposed to solar radiation. Survival curves showed that strains harbouring 62-kb plasmids had a higher survival ability under both conditions. The role of the indigenous plasmids of 62 kb in copper- and UV-resistance was confirmed by evaluating survival of transconjugants obtained by mating assays. The most abundant 62-kb plasmid (62.1) carried copper- and UV-resistance determinants homologous to *cop*ABCD and *rul*AB genes, and was shown to be conjugative and very stable. These results strongly suggest an important contribution of 62-kb plasmids to increasing the epiphytic fitness of *P. s.* pv. *syringae* on mango plant surfaces.

Key words: epiphytic fitness, plasmid, copper resistance, ultraviolet light-resistance.

Author for correspondence: A. De Vicente; e-mail: adevicente@uma.es.

N.S. Iacobellis *et. al.* (eds.), *Pseudomonas syringae* and Related Pathogens, 79-88.

1. INTRODUCTION

Pseudomonas syringae pv. *syringae* is the causal agent of the apical necrosis of mango, a bacterial disease limiting the production of mangoes in the Mediterranean area (Cazorla *et al*., 1998). In Southern Europe, the habitat of mango plant surfaces is a difficult place for bacteria to survive. Buds and leaves are not only exposed to high levels of copper, commonly used as bactericide and fungicide, during long periods of time, but also to heavy doses of solar radiation including ultraviolet wavelengths. However, *P. s.* pv. *syringae* is generally present on mango trees all year long, even in summer, with very extreme conditions of solar radiation and temperature, suggesting that epiphytic bacterial fitness could be increased by copper- and/or ultraviolet light-resistance. In this study, we provide evidence that one contributing factor for the survival and persistence of *P. s.* pv. *syringae* on the mango surfaces could be the presence of indigenous 62-kb plasmids harbouring genetic determinants copper- and ultraviolet light-resistance.

2. MATERIALS AND METHODS

2.1 Micro-organisms and Growth Conditions

Pseudomonas syringae pv. *syringae* strains used in this study were isolated and identified in previous studies (Cazorla *et al*., 1997; 1998). Bacterial strains were routinely cultured on King's B medium (KMB) at 25°C. The strain FF5 of *P. s.* pv. *syringae* is a copper- and UV-sensitive (Cu^s and UV^s), plasmidless strain with a high efficiency of transformation by electroporation (Sundin and Bender, 1993), it was used to construct a stable kanamycin-resistant (km^r) mutant to perform the conjugation experiments (FF5-km). Biparental matings were carried out as previously described (Cazorla *et al*., 2002) to obtain 62-kb plasmid transconjugants (Cu^r-km^r) and to confirm the role in copper- and UV-tolerance of this plasmid harbouring copper and ultraviolet light resistance-genes.

2.2 Plasmid Detection and Characterisation

Plasmid DNA was isolated according to a modified alkaline lysis method (Zhou *et al*., 1990; Cazorla *et al*., 2002) and separated by electrophoresis on 0.6% agarose gels. Plasmid size was estimated by comparison with plasmids from *Pseudomonas syringae* pv. *tomato* PT23. Purified plasmid DNA digested with *Eco*RI was separated by electrophoresis on 1% agarose gels.

2.3 Southern Blot Analyses

Approximately 5 µg of plasmid DNA was digested with *Eco*RI, separated by electrophoresis in 1% agarose gels, transferred to nylon membranes, and hybridised with radioactive probes following standard procedures (Sambrook *et al.* 1989). A 4.4-kb *Pst*I fragment from pCOP2, containing the *cop*ABCD genes from *Pseudomonas syringae* pv. *tomato* PT23 (Bender and Cooksey 1987), a 0.7-kb *Pst*I-*Bam*H1 fragment from pAKC, containing part of the *rul*AB genes from *P. syringae* pv. *tomato* PT23 (Sundin *et al.* 1996), and a 0.8-kb *Eco*RI fragment from pAKC, containing the *ori*V gene from *P. s.* pv. *tomato* PT23 (Murillo and Keen, 1994), were labelled with [α-^{32}P] dCTP using the random primer method (Sambrook *et al.*, 1989) and used as hybridisation probes.

2.4 Resistance to Copper

Copper resistance was determined by using mannitol-glutamate-yeast extract medium (MGY, Bender and Cooksey, 1986) amended with $CuSO_4 \cdot 5H_2O$. Isolates were considered resistant to copper when their growth on MGY+0.8 mM $CuSO_4$ after 72 h at 27ºC was equivalent to that on MGY under the same conditions (Cazorla *et al.*, 2002).

2.5 Bacterial Populations in Field Plots Under Copper Treatments

Occurrence of total bacteria, *Pseudomonas syringae*-like populations and the proportion of Cu^r bacteria in the bacterial population on buds were determined from 1997 to 2000 in two groups of mango trees affected by apical necrosis at La Redondela (Huelva, Spain). One group of trees was sprayed monthly with Bordeaux mixture (3 g l^{-1}) during the 3 years of study, while the other group was used as a non-treated control. *P. syringae*-like and total bacterial populations were determined as previously described (Cazorla *et al.*, 1998). For each sample, the proportion of Cu^r bacteria was expressed as the percentage of bacteria growing on MGY + 0.8 mM of $CuSO_4$ relative to the total bacterial population (Cazorla *et al.*, 2002).

2.6 Resistance to Ultraviolet Light

Resistance to ultraviolet light was determined by measuring bacterial survival under exposition to ultraviolet light, fractions B and A, simulating a very sunny day in July in Málaga (Spain). Bacterial suspensions ($10^8 \div 10^9$ CFU ml^{-1}) were exposed for 4 hours, under UV-lamps into a chamber. UV irradiance values were 1.4 Wm^{-2} and 22 Wm^{-2} for B and A fractions, respectively. *Pseudomonas syringae* pv. *syringae* counts in the suspensions were carried out every hour, following protocols previously described (Cazorla *et al.*, 1998).

2.7 Survival on Mango Leaf Surfaces Under Field Conditions

Survival curves of *Pseudomonas syringae* pv. *syringae* cells exposed to direct sunlight on mango leaves were performed in two groups of two years old mango trees. One group of 15 trees was situated in a sunny area directly exposed to sunlight, while the other group was maintained in a shady area. Drops of 20 µl of bacterial suspensions ($10^8 \div 10^9$ CFU ml^{-1}) were placed on the upper surface of mango leaves, under direct sunlight in July. Leaf samples containing bacterial cells were taken every hour (from 10:00 AM to 02:00 PM) and the *P. s.* pv. *syringae* levels determined as previously described (Cazorla *et al.*, 1998).

3. RESULTS AND DISCUSSION

In phytopathogenic bacteria, several resistance determinants are commonly associated with plasmids (Cazorla *et al.*, 2002; Sundin and Bender, 1996; Sundin *et al.*, 1996). Therefore, plasmid occurrence was examined by agarose gel electrophoresis of DNA from 63 *Pseudomonas syringae* pv. *syringae* strains randomly selected from mango isolates. Results showed the presence of a 62-kb plasmid (72.9% of tested strains), either alone or together with other plasmids (88 or 45-kb) (Fig. 1; Table 1).

Different 62-kb plasmids were subjected to restriction analysis, resulting in that they could be classified in different *Eco*RI restriction patterns (Table 1; Cazorla *et al.*, 2002). Simultaneously, detection of genetic determinants that could be responsible for increased epiphytic fitness (copper- and ultraviolet resistance determinants), was performed by Southern blot experiments, showing hybridisation signals in almost all types of 62-kb plasmids. All 62-kb plasmids belonged to the family of pPT23A-like plasmids, having a replication origin homologous to *ori*V, although they presented a high degree of diversity of RFLP patterns (Cazorla *et al.*, 2002).

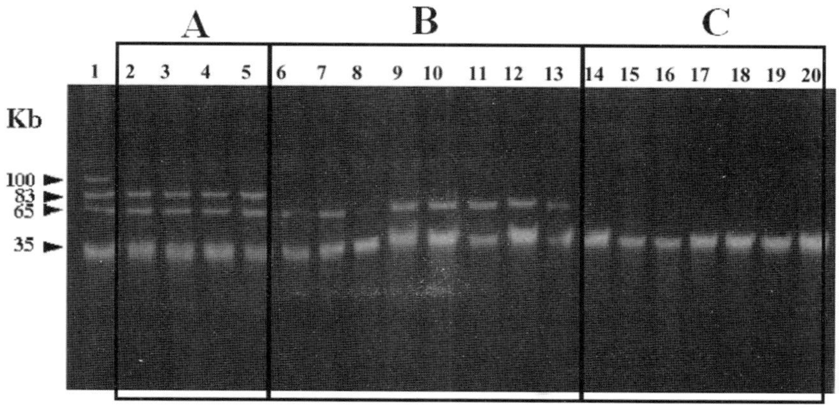

Figure 1. Representative plasmid profiles of *Pseudomonas syringae* pv. *syringae* isolated from mango trees (lanes 2-20). A) strains with two plasmids of 88 kb and 62 kb, B) strains with only one plasmid of 62 kb, C) strains in which no plasmids were detected by electrophoresis, although some of them possessed a 120 kb plasmid only detectable by Southern blot. Lane 1: *P. s.* pv. *tomato* PT23.

Table 1. Characteristics of *Pseudomonas syringae* pv. *syringae* strains harbouring different plasmid profiles, copper resistance and plasmid DNA hybridisation with different probes. The numbers 62.1 to 62.5 indicate 62-kb plasmids which were differentiated by their restriction profile with *Eco*RI.

Plasmid profile	n	Cu^r [1] (MIC>0.8 mM)	Hybridisation with[2] *cop*ABCD	*rul*AB	*ori*V
88-62 kb	6	R	+	+	+
62.1 (cured of 88-kb)	6	R	+	+	+
62-kb	44				
62.1	17	R	+	+	+
62.2	1	R	++	+	+
62.3a	1	R	+	+	+
62.3b	5	R	(+)	+	+
62.3c	11	S	-	+	+
62.4	5	S	-	+	+
62.5	4	S	-	+	+
120-kb	5	R	+	-	-
62-45 kb	1	R	+	+	-
45-kb	1	R	+	NT	NT
Plasmidless	3	R	(+)	-	-
Plasmidless	10	S	-	-	-
FF5-km	1	S	-	-	-
Transconjugants (62.1xFF5-km)	4	R	+	+	+

[1] Cu^r= resistance to copper; R: resistance; S: sensitivity.

[2] +: hybridisation; ++: strong hybridisation; (+): low stringency hybridisation; -: no hybridisation; NT: not tested.

Southern blot hybridisations showed that genetic determinants partially homologous to the *cop*ABCD operon were present in all the Cu^r strains examined, mostly associated with 62-kb plasmids (77.5% of Cu^r strains). In other Cu^r strains, the *cop*ABCD homologous signal was associated with 45 or 120-kb plasmids or chromosomal DNA. These determinants were not detected in Cu^s strains, thus, several strains harbouring 62-kb plasmids and being Cu^s did not show hybridisation to the *cop*ABCD probe (Table 1; Cazorla *et al.*, 2002).

Regarding UV resistance determinants, all 62-kb plasmids analysed showed homologous sequences to *rul*AB genes (Table 1).

The most abundant plasmid profile (62.1; Table 1) carried copper- and UV-resistance determinants homologous to *cop*ABCD and *rul*AB. Curing and mobilisation of different 62-kb plasmids were attempted in order to confirm whether the observed homologies with *cop*ABCD and *rul*AB probes were indeed identifying the genes responsible for the Cu^r and UV^r phenotypes of their harbouring strains. Curing of the plasmids was unsuccessful using conventional treatments, such as growth in presence of ethidium bromide or acridine orange, or by incompatibility with the origin of replication from pPT23A included in pAKC (data not shown). Strains containing 62.1 plasmids were successfully mated with FF5-km and *cop*ABCD and *rul*AB homologous determinants transferred to FF5-km associated with 62.1 plasmids and copper-resistance. The MIC values for copper sulphate of purified transconjugants from independent mating were 1.8 mM, very similar to donor strains.

In order to confirm the role of the copper- and UV-resistance determinants in the improvement of the epiphytic fitness of the *P. s.* pv. *syringae* strains harbouring 62-kb plasmids, survival experiments were performed under extreme conditions of copper applications and UV light exposition.

The survey of mango orchards revealed an increase in frequencies of copper resistant bacteria (mainly *P. s.* pv. *syringae*) after repeated treatments with Bordeaux mixture (Table 2). Thus, analyses of bacterial levels in commercial field plots under copper treatments showed that, both total numbers of bacteria and *P. syringae*-like counts did not differ significantly between buds from trees treated with Bordeaux mixture and those from non-treated trees, however after one year of treatments, the population of Cu^r bacteria on buds of treated trees increased sharply.

These data suggest that selection of copper-resistant strains could be a major reason for control failures following management with copper bactericides (Cazorla *et al.*, 2002). Copper resistance determinants were frequently associated to 62-kb plasmids, which could enhance the ecological fitness and persistence of their harbouring strains (Eberhard, 1989). But, additionally, it is important to note that all 62-kb plasmids harboured homologous determinants to *rul*AB genes, which could confer tolerance to UV-radiation (Sundin *et al.*, 1996).

Table 2: Populations of total bacteria, *Pseudomonas syringae*-like (log CFU ml^{-1}) and Cu^r bacteria (%) on buds of mango trees non-treated or treated with Bordeaux mixture.

	Total Bacteria			*Pseudomonas*-like bacteria			% Cu^r Bacteria*		
Year	1998	1999	2000	1998	1999	2000	1998	1999	2000
Without treatment	7.0±0.3	6.6±0.4	6.3±1.0	6.7±0.4	6.3±0.5	6.0±0.8	4.0±3.7	11.1±6.6	17.5±10.7
Bordeaux mixture treated	6.5±0.5	6.5±0.5	6.0±0.4	6.4±0.7	6.4±0.6	5.6±0.4	14.5±6.6	46.8±21.6	56.7±6.5

*Percentage of colonies on MGY+0.8 mM $CuSO_4$ referred to the total bacterial counts.

The role of the indigenous plasmids of 62-kb in UV-resistance was confirmed by evaluating UV-resistance and survival of *P. s.* pv. *syringae* strains harbouring 62-kb plasmids. All the studied strains harbouring the 62-kb plasmid hybridised to the *rul*AB probe, whereas all *P. s.* pv. *syringae* strains harbouring other plasmids or that were plasmidless did not (Table 1). UV-resistance was evaluated in these strains by performing survival curves of *P. s.* pv. *syringae* cells exposed to determined doses of B and A UV-fractions. *P. s.* pv. *syringae* strains harbouring 62-kb plasmids showed a phenotype with a significantly higher tolerance to B+A UV radiation than did plasmidless strains (Fig. 2). Furthermore, survival of transconjugants obtained by mating assays between UV-tolerant *P. s.* pv. *syringae* strains

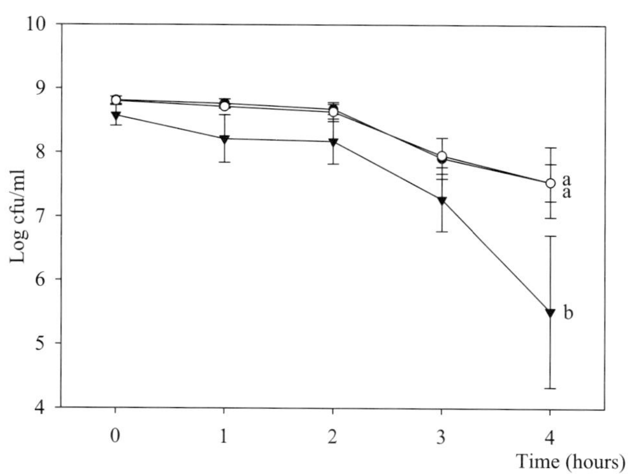

Figure 2. Survival curves of *Pseudomonas syringae* pv. *syringae* strains exposed to UV radiation by *in vitro* assays. *P. s.* pv. *syringae* strains harbouring a 62-kb plasmid (•, n=4); transconjugants (62.1 plasmid x FF5-km) harbouring the 62-kb plasmid (o, n=2); and plasmidless *Pss* strains (▼, n=3). Bacterial suspensions were exposed to B+A fractions of UV light (irradiance: 1.4 Wm^{-2} + 22 Wm^{-2}). Different letters denote significant differences according to the least significant difference test (p=0.01).

containing 62.1 plasmids and UV-sensitive *P. s.* pv. *syringae* strain FF5-km, was also similar to that of donor strains.

The persistence of *P. s.* pv. *syringae* strains harbouring 62-kb plasmids under field conditions was also evaluated on mango leaves exposed directly to solar radiation, obtaining similar results. Thus, *P. s.* pv. *syringae* strains harbouring 62-kb plasmids showed a significant higher survival ability on mango surfaces exposed to direct sunlight than did plasmidless strains (Fig. 3). Transconjugans harbouring the 62-kb conjugative plasmids showed a similar behaviour. Thus, presence of 62-kb conjugative plasmids harbouring *rul*AB homologous genetic determinants supposes an increase in epiphytic fitness for *P. s.* pv. *syringae* strains, because it results in a phenotype with a higher survival ability on plant surfaces.

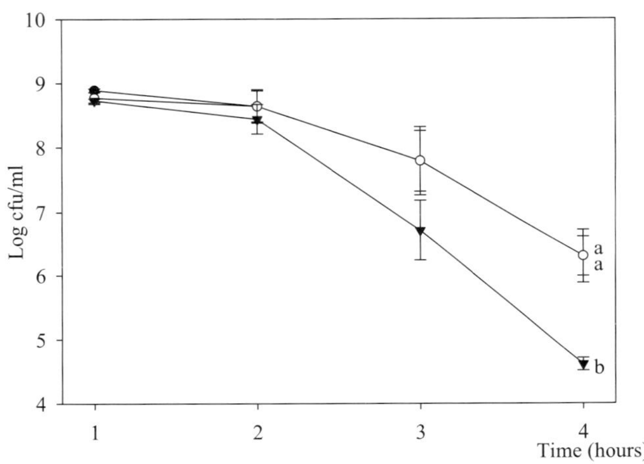

Figure 3. Survival curves of *Pseudomonas syringae* pv. *syringae* strains on upper leaf surfaces under field conditions in Málaga (Spain). *P. s.* pv. *syringae* harbouring a 62-kb plasmid (•, n=2); transconjugants (62.1 plasmid x FF5-km) harbouring the 62-kb plasmid (o, n=2); and plasmidless *P. s.* pv. *syringae* strains (▼, n=2). Different letters denote significant differences according to the least significant difference test (p=0.01).

In mango isolates of *P. s.* pv. *syringae* copper- and UV-resistance, determinants may have been incorporated into the bacterial population along with other plasmid-encoded traits that could also enhance the ecological fitness and persistence of their harbouring strains (Eberhard, 1989; Levin and Bergstrom, 2000). As previously reported, plasmids of the pPT23A-like family tend to encode determinants of importance for the ecology of diverse *P. syringae* pathovars, such as avirulence, extracellular virulence factors, and resistance to antibiotics (Jackson *et al.*, 1999; Sesma *et al.*, 2001; Sundin and Bender, 1996). In Southern Europe, mango plant surfaces are exposed to heavy solar radiation and copper levels. Therefore, it is conceivable that plasmid-encoded traits such as UV radiation- and copper-tolerances could contribute to the maintenance of the carrying plasmid and, consequently, of any other genes that might be on the plasmid. In our view, these results strongly suggest a relevant contribution of 62-kb plasmids to increased epiphytic fitness of *P. s.* pv. *syringae* on mango plant surfaces.

ACKNOWLEDGEMENTS

This work was supported by grants from Consejería de Agricultura y Pesca, Junta de Andalucia and from Plan Nacional de I+D (AGF98-0931-C02-01). We want to thank J.A. Torés, J.M. Hermoso and J.M. Farré for helpful suggestions and work facilities at E.E. La Mayora (CSIC-Spain). F.M. Cazorla thanks J. Murillo for a short stay in his laboratory. We thank C. Bender, V. de Lorenzo, J. Murillo, G.W. Sundin and A. Vivian for kindly providing us with some of the bacterial strains and DNA probes used in this study.

REFERENCES

Bender, C.L., and Cooksey, D.A., 1986, Indigenous plasmids in *Pseudomonas syringae* pv. *tomato*: conjugative transfer and the role in copper resistance. *J. Bacteriol.* **165**: 534-541.

Bender, C.L., and Cooksey, D.A., 1987, Molecular cloning of copper resistance genes from *Pseudomonas syringae* pv. *tomato*. *J. Bacteriol.* **169**: 470-474.

Cazorla, F.M., Arrebola, E., Sesma, A., Pérez-García, A., Codina, J.C:, Murillo, J., and de Vicente, A., 2002, Copper resistance in *Pseudomonas syringae* strains isolated from mango is encoded mainly by plasmids. *Phytopathology* **92**: 909-916.

Cazorla, F.M., Olalla, L., Torés, J.A., Codina, J.C., Pérez-García, A., and de Vicente, A. 1997, *Pseudomonas syringae* pv. *syringae* as microorganism involved in apical necrosis of mango: characterization of some virulence factors. In "*Pseudomonas syringae* Pathovars and Related Species". Eds. K. Rudolph, J.J. Burr, J.W. Mansfield, D. Stead, A. Vivian, and J. Von Kietzell. Kluwer Academic Publishers. Dordrecht. The Netherlands. pp. 82-87.

Cazorla, F.M., Torés, J.A., Olalla, L., Pérez-García, A., Farré, J.M., and de Vicente, A, 1998, Bacterial apical necrosis of mango in Southern Spain: a disease caused by *Pseudomonas syringae* pv. *syringae*. *Phytopathology* **88**: 614-620.

Eberhard, W.G., 1989, Why do bacterial plasmids carry some genes and not others? *Plasmid* **21**:167-174.

Jackson, R.W., Athanassopoulos, E., Tsiamis, G., Mansfield, J.W., Sesma, A., Arnold, D.L., Gibbon, M.J., Murillo, J., Taylor, J.D., and Vivian, A., 1999, Identification of a pathogenicity island, which contains genes for virulence and avirulence, on a large native plasmid in the bean pathogen *Pseudomonas syringae* pv. *phaseolicola*. *Proc. Natl. Acad. Sci. USA* **96**: 10875-10880.

Levin, B.R., and Bergstrom, C.T. 2000. Bacteria are different: observations, interpretations, speculations, and opinions about the mechanisms of adaptive evolution in prokaryotes. *Proc. Natl. Acad. Sci. USA* **97**: 6981-6985.

Murillo, J., and Keen, N.T., 1994, Two native plasmids of *Pseudomonas syringae* pathovar *tomato* strain pT23 share a large amount of repeated DNA, including replication sequences. *Mol. Microbiol.* **12**: 941-950.

Sambrook, J., Fritsch, E.F., and Maniatis, T., 1989, Molecular Cloning: A laboratory manual. 2nd ed. Cold Spring Harbor Laboratory, Cold Spring Harbor, N.Y. USA.

Sesma, A., Aizpún, M.T., Ortiz, A., Arnold, D., Vivian, A., and Murillo, J., 2001, Virulence determinants other than coronatine in *Pseudomonas syringae* pv. *tomato* PT23 are plasmid encoded. *Physiol. Mol. Plant Pathol.* **58**: 83-93.

Sundin, G.W., and Bender, C.L., 1993, Ecological and genetic analysis of copper and streptomycin resistance in *Pseudomonas syringae* pv. *syringae*. *Appl. Environ. Microbiol.* **59**: 1018-1024.

Sundin, G.W., and Bender, C.L., 1996, Molecular analysis of closely related copper- and streptomycin-resistance in plasmids in *Pseudomonas syringae* pv. *syringae*. *Plasmid* **35**: 98-107.

Sundin, G.W., Kidambi, S.P., Ullrich, M., and Bender, C.L., 1996, Resistance to ultraviolet light in *Pseudomonas syringae*: sequence and functional analysis of the plasmid-encoded *rul*AB genes. *Genes* **177**: 77-81.

Zhou, C., Young, Y., Jong, A.Y., 1990, Miniprep in ten minutes. *BioTechniques* **8**: 172-173.

Evidence that *Acidovorax valerianellae*, Bacterial Black Spot of Corn Salad (*Valerianella locusta*) Agent, is Soil Transmitted

C. GRONDEAU, V. CERCEAU, C. BUREAU[#], and R. SAMSON[#]
Comité Départemental de Développement Maraîcher; [#]Institut National de Recherche Agronomique, Unité Mixte de Recherche Pathologie Végétale 077, 42 rue Georges Morel BP 57, 49071 Beaucouzé cedex, France.

Abstract: Bacterial black spot of corn salad was first described in France in 1991. As soil was suspected to be the main primary inoculum source, field studies were undertaken under natural conditions. Soil was treated with metam sodium, a biocide with a large spectrum of action, and two successive crops were grown after the fumigation. The disease was more severe in the second crop than in the first one and high doses of metam sodium significantly reduced the number of diseased plants at the harvest of the second crop. Direct evidence of soil transmission was obtained by the isolation of *Acidovorax valerianellae* from soil samples and plant debris collected in a naturally infested field.

Key words: Soil borne bacteria, soil transmission, metam sodium, soil treatment.

Author for correspondence: C. Grondeau; e-mail: grondeau@angers.inra.fr.

1. INTRODUCTION

Bacterial black spot is the most important disease of corn salad (*Valerianella locusta*) in French production because infected crops are severely depreciated and frequently unmarketable. The causal organism is a new pathogen, *Acidovorax valerianellae* (Gardan *et al.*, to be published) and its biology is unknown. The origin of primary inoculum will determine the choice of ways of the disease control. Previous studies based on disease incidence measures (Grondeau *et al.*, 2002) tend to demonstrate that the

N.S. Iacobellis *et. al.* (eds.), *Pseudomonas syringae* and Related Pathogens, 89-91.

organism is soil borne. Indirect and direct evidence that *A. valerianellae* is soil transmitted is presented.

2. INDIRECT EVIDENCE OF SOIL TRANSMISSION BASED ON FIELD INCIDENCE

Soil in plots in a naturally infested field was treated with metam sodium, a biocide with a large spectrum of action. The fumigant was mixed with the upper 8 cm of the soil at 400, 600 or 800 l ha^{-1} and the soil was covered for 16 days with a plastic film. A first crop sown on 29/06/98 was harvested on 04/08/98 and a second one sown on 18/08/98 was harvested on 23/09/98. The number of diseased plants was determined at the harvest of each crop with an immunofluorescence test on a representative number of black spots for disease diagnosis (Grondeau *et al.*, 2000). Results indicated that the disease was more severe in the second crop than in the first one suggesting that the presence of the host plant increased the primary inoculum level. Secondly, metam sodium applied to the soil before sowing significantly acted on disease incidence in the second crop following fumigation.

3. DIRECT EVIDENCE OF SOIL TRANSMISSION BASED ON BACTERIA ISOLATION

Direct evidence was obtained with the isolation of *Acidovorax valerianellae* from soil samples and plant debris collected in two naturally infested fields, using a semi-selective medium and serological techniques to identify suspected colonies.

In a delimited area chosen in one field where the bacterium was isolated from the soil, populations of *A. valerianellae* were monitored on plants at early stages of growth. The area was divided in 25 subplots (1.4 x 5 m) in which corn salad was sampled during two successive crops (no soil fumigation between crops). Plant samples were analysed by blending and dilution-plating on the semi-selective medium and *A. valerianellae* identified by serological techniques. Results showed that fewer subplots with contaminated or diseased plants were observed in the first crop than in the next one at nearly the same stage of plant growth. At harvest, disease incidence was 12% for the first crop and 34% for the second.

4. CONCLUSION

Acidovorax valerianellae was detected in soil and contaminated root debris in naturally infested fields. Soil transmitted bacteria colonised their host in early stages of growth and disease incidence increased with the amount of inoculum in the soil and with the number of successive crops.

REFERENCES

Gardan, L., Stead, D.E., Dauga, C., and Gillis, M., *Acidovorax valerianellae* sp. nov., a new pathogen for lamb's lettuce (*Valerianella locusta* (L.) Laterr.). *Int. J. Syst. Evolution. Microbiol* (to be published).

Grondeau, C., Roos, J.R., Fouyer, L., and Hug, F., 2002, Désinfection superficielle du sol au métam sodium en culture de mâche. *PHM la revue technique des pépiniéristes, horticulteurs, maraîchers* **435**: 19-21.

Grondeau, C., Samson, R., and Gardan, L., 2000, Description and identification of the new bacterium causing bacterial black spots on corn salad (*Valerianella locusta*) as *Acidovorax valerianellae.* In *Proc. 10th Int. Conf. Plant Pathogenic Bacteria, Charlottetown, Canada, 23-27/07* (Ed. S.H. DeBoer, eds), Kluwer Academic Publisher, Dordrecht, pp.165-166.

Pseudomonas Leek Blight: Study of Seed Transmission

I. IKENE, C. GRONDEAU[#], C. BUREAU, and R. SAMSON
Institut National de Recherche Agronomique; [#]Comité Départemental de Développement Maraîcher, INRA Unité Mixte de Recherche Pathologie Végétale 077, 42 rue Georges Morel BP 57, 49071 Beaucouzé cedex, France.

Abstract: In order to monitor transmission of *Pseudomonas syringae* pv. *porri* from seed to seedlings, bacterial populations were evaluated on seed and on transplants of the same seed lots. For this work, a new semi-selective medium was defined.

Key words: *Pseudomonas syringae* pv. *porri*, seed contamination rate, epiphytic phase, semi-selective medium.

Author for correspondence: C. Grondeau; e-mail: grondeau@angers.inra.fr.

1. INTRODUCTION

Bacterial blight of leek, caused by *Pseudomonas syringae* pv. *porri* may occur at different stages of leek cultivation: seedling, transplant, edible leek, and seed production, as long as they present actively growing tissues (Samson *et al*., 1998). Leek is a biennial cultivated plant often grown by several successive farmers: seed growers working for seed industry, seedling nurserymen, and vegetable producers. These three professions may be gathered together when the same farmer produces his own seed. Seed transmission was suspected in California where bacteria were isolated from seed lots (Koike *et al*., 1999).

N.S. Iacobellis *et. al.* (eds.), *Pseudomonas syringae* and Related Pathogens, 93-95.

2. RELATIONSHIP BETWEEN EPIPHYTIC POPULATIONS ON TRANSPLANTS AND SEED CONTAMINATION RATE

We surveyed bacterial populations in leek nurseries (plug trays in greenhouses and direct-sowing in the field) by plating washes from symptomless plants on the semi-selective medium LBC (Grondeau *et al.*, 1992). *Pseudomonas syringae* pv. *porri* was able, as all epiphytic bacteria, to multiply on its host-plant without inducing symptoms. Populations colonised clandestinely and spread over the leek plots without being noticed by the grower. Thus, consequent losses could not be predicted and, when they occurred, the spatio-temporal spread of *P. s.* pv. *porri* was revealed very late after a long incubation period.

Seed analysis by isolation is difficult because of the occurrence on leek seed of an abundant bacterial flora. Industry seed coating practices increase the difficulty. We worked on the definition of selective media with two objectives: i) improve the isolation of *P. s.* pv. *porri* for commercial litigations and ii) offer a medium adapted for detection by enrichment-PCR.

For every bacterial seed-borne disease, individual seed contamination generally shows very low probability and requires analyses of large samplings. For instance, an artificially contaminated seed lot was obtained by collecting all diseased flowering stems of an inoculated leek field (located at Beaucouzé). After analysis, the seed contamination rate was not higher than 0.07%.

We surveyed epiphytic populations and disease incidence in a farmer's nursery (located at Feillens) sown with his own seed, whose contamination rate was estimated at 0.06%. Three months after sowing, 0.18% of plants (out of a total of 180,000 plants) were diseased and situated in 5 to 11 foci. Disease transmission rate from seed to seedling could therefore be estimated: in the conditions of Feillens in June-July 2001, 10 to 20 contaminated seeds were necessary to create one leek blight focus.

3. CONCLUSION

To demonstrate that *Pseudomonas syringae* pv. *porri* was seed transmitted, we faced several problems: i) seeds were heavily contaminated by a bacterial flora masking *P. s.* pv. *porri* during seed analysis, ii) *P. s.* pv. *porri* introduced with the seed thoroughly colonised leek plants without necessarily inducing blight symptoms, iii) leek blight might appear a rather long time after sowing infested seed, iv) infestation rate of a seed lot as low as 0.06% was potentially dangerous for a crop if weather conditions were favourable. Additional observations from 2001 and 2002 gave further evidence that the bacteria were seed borne. In order to help in proposing sanitary methods to propagate leek seed free of *P. s.* pv. *porri*, a seed detection test is being developed.

REFERENCES

Grondeau, C., Ladonne, F., Fourmond, A., Poutier, F., and Samson, R., 1998, Attempt to eradicate *Pseudomonas syringae* pv. *pisi* from pea seeds with heat treatments. *Seed Sci. Technol.* **20**: 515-525.

Koike, S.T., Barak, J.D., Henderson, D.M., and Gilbertson, R.L., 1999, Bacterial blight of leek: a new disease in California. *Plant Dis.* **83**: 165-170.

Samson, R., Shafik, H., Benjama, A., and Gardan, L., 1998, Description of the bacterium causing blight of leek as *Pseudomonas syringae* pv. *porri* (pv. nov.). *Phytopathology* **88**: 844-850.

Distribution and Virulence of *Pseudomonas syringae* pv. *atrofaciens*, Causal Agent of Basal Glume Rot, in Russia

I.E.V. MATVEEVA, E.SH. PEKHTEREVA, V.A. POLITYKO, A.N. IGNATOV[#], E.V. NIKOLAEVA, and N.W. SCHAAD[§]

Russian Research Institute of Phytopathology, Moscow region, 143080. Russia; [#]Centre "Bioengineering", Russian Academy of Sciences, Moscow, 117312. Russia; [§]Foreign Disease-Weed Science Research Unit, USDA, Fort Detrick, Maryland, USA.

Abstract: *Pseudomonas syringae* pv. *atrofaciens,* the causal agent of basal glume, is a major pathogen of cereals in Russia. The virulence of 23 strains isolated from infested seeds of wheat, rye, and barley plants collected from 1998 to 2001 from different regions of Russian Federation was evaluated in controlled environment chambers using plants of 13 wheat and barley cultivars. Based on differences in virulence and UPGMA cluster analysis, strains from barley and rye clustered into two separate groups whereas strains from wheat were found in both clusters. The results showed that *P. s.* pv. *atrofaciens* has a high degree of variation in virulence to the species and genotypes of cereals in Russia. The susceptibility of over 60 varieties of wheat was evaluated by several inoculation techniques against several isolates of *P. s.* pv. *atrofaciens.* Plant reaction data were analysed by Factor, k-means Cluster and ANOVA analysis. Several resistant lines were identified including, Russian winter wheat cultivars 'Moskovskaya nizkostebelnaya', 'Mironovskaya 61', 'Pamyat Fedina', 'Zhneya" and hybrid PPG-93; spring wheat 'Bezenchukskaya', 'Isheevskaya', 'Kinelskaya', and foreign cultivars 'Cooperacion Calquin', 'Pinjab 96', 'Prointa Superior', 'Prointa Imperial', 'Lhanaa', 'Bedhao' and 'Rhabour-15'.

Key words: cereals, resistance, genetic resources, pathogenicity.

Author for correspondence: A.N. Ignatov; e-mail: Ignatov@biengi.ac.ru.

N.S. Iacobellis *et. al.* (eds.), *Pseudomonas syringae* and Related Pathogens, 97-105.

1. INTRODUCTION

Basal glume rot, caused by *Pseudomonas syringae* pv. *atrofaciens* (Mc Culloch) Young, Dye and Wilke (1978), is a major bacterial disease of cereals in Russia. The pathogen has considerable physiological, genetic and virulence variability (Matveeva, 1998).The disease reduces yield of cereals by 5÷15% depending on weather conditions (Matveeva, 1998). For the first time we report on a survey of diseased plant samples from five regions of Russia and the virulence of 23 strains of *P.* s. pv. *atrofaciens* isolated from wheat, rye and barley. The purpose of the study was to determine the distribution of *P.* s. pv. *atrofaciens* in Russia and identify possible sources of resistance germplasm in cultivars of wheat.

2. MATERIALS AND METHODS

2.1 Source and Identification of Bacteria

Samples were collected and bacteria isolated from diseased leaves and grains of wheat, rye, barley and oats from different regions of Russia (Table 1). The identity of all bacteria was confirmed by classical biochemical and physiological tests (Schaad *et al.*, 2001). For pathogenicity, 4- to 5-day-old seedlings of wheat, barley and rye growing on wet filter paper in 25 cm Petri dishes (Toben *et al.*, 1989) were stabbed 2 to 3 times with a syringe and sterile 1-inch gauge needle containing approximately 1,000 CFU of *Pseudomonas syringae* pv. *atrofaciens*, as described below. Twenty-three strains from each host plant were tested for pathogenicity.

Table 1. Origin and source of strains of *Pseudomonas syringae* pv. *atrofaciens*.

Strain	Crop	Region	Year	Strain	Crop	Region	Year
B3.1	Barley, cv. Babilon	Kabardino -Balkaria	1994	BT2	cv. Zazerski-85	Moscow	1994
B3.2	-	-	-	BN	Barley, cv. Vizit	Moscow	1994
B4.1	-	-	-	R28	Rye	-	-
B50	cv. Zazerski 85	Moscow	1994	R56	cv. Vyatka	Volgo-Vatskay	1994
BT1	cv.Zazerski-85	Moscow	1994	R80	cv. Purga	Moscow	1994
B10	-	-	1994	R106	cv. Chulpan	Moscow	1994
B48	-	-	1994	R110	-	-	1994
BT36	Barley	Moscow	1995	R111	-	-	1994
BT41	Barley, cv. Zazerski-85	Moscow	1994	R200	-	Kirov	1994
BT47	cv. Zazerski-85	Moscow	1994	W7	Wheat, cv. Spatanka	Kabardini-Balkaria	1991
BD	Barley, cv. Vizit	Moscow	1994	W9	cv.Kazanskaya-84	Tatyarstan	1991
6001	Wheat, cv L-503	Saratov	1997	W6005	cv. L-503	Saratov	1997

2.2 Plant Cultivars

Sixty cultivars of wheat were collected in Russia, Argentina, Columbia, Syria and Pakistan (Table 2). Seeds of these cultivars were obtained from Vavilov's Institute of Plant Industry, St. Petersburg and collection of Russian Research Institute of Phytopathology, Moscow region.

Table 2. Wheat cultivars used in this study.

Country of origin	Type	Cultivar
Russia	winter wheat	Mironovskaya 61; Mironovskaya 27; Moskovskaya nizkostebelnaya; Pamyati Fedina; Mironovskaya 25 Mironovskaya 808; Inna; Nika Kubani
Russia	spring wheat	Besenchukskaya 139; Ishevskaya l; 503; Priokskaya; Kinelskaya; Kutulukskaya; Moskovskaya 35; Saratovskaya zolotistaya; Samsar; Allbidium; Ivolga; Kharkovskaya 46; Svetlana; Lutescens 491; Simbirka; Veresen; Vyatka; Alpha
Pakistan	spring wheat	Parwaz 94; Hybrid 9; Ob-2039; Punjab-9; Shaheen 96; Bwp-95
Columbia	spring wheat	Ica Sugamuxi; Ica Bonza; Qualmatan; Yuriya; Tota; Jacuanguer, Ica Hunza; Ica Variedad; Ica Tenza; Ica Hunza 1; Bonza; Chimborazo; Achauay
Argentina	spring wheat	Buck Chambergo; Buck Pronto; Prointa; Real Cooperation; Calquin Prointa; Federal Prointa; Imperial Pronta; Oasis; Prointa Granero
Syria	spring wheat	Fadah; Lhanaa; Bedhaa; Khabour-15; Trakk; Al-Jazurra (a); Al-Jazirra (b); Al-Jazirra (c); Tal-Alhab 8; Selkin 5; Selkin 194; Khabus-79; Ghadir; Khabor-44; Dejlee; Althaibe-6; Tal-Alshab 2; Alid 15; Ghasan Oahel; Selkin 1; Aroad; Doma 26071; Althaibe 3; Althaibe 17; Tcill 5; Hash Khabor 113; Tcill 127; Selkin 39; Khabees 71; Althabe 25

2.3 Virulence Testing

Six cultivars of wheat: "Inna", "Sarya", "Nemchnovskaya 25", "Mironovskaya 27", "Mironovskaya 808", and "Moskovskaya karlikovaya", and seven cultivars of barley "Zazerskii-85"; "Moskovskii"; "Visit", "Bios-1"; "Moskovskii-2"; "Moskovskii-3"; "Baranchik" were used for the *Pseudomonas syringae* pv. *atrofaciens* strains virulence evaluation.

A standard needle injection method was used (Braun-Kiewnick, 2000) to inoculate stems of 4÷5 day old seedlings.

The plants were grown and incubated in walk-in growth chambers set at a day/night temperature of 28/20°C, 90% relative humidity, and a 16 h day. For inoculum, 36 h old bacteria from YDC agar medium grown at 28°C were suspended in sterile tap water or physiological saline adjusted to10^8 CFU ml^{-1}. Control plants were inoculated with sterile tap water. The following disease's severity scale was used: 0, no symptoms; 1, browning near the point of inoculation; 2, browning more than 1 cm; 3, browning of stem; and 5, death of seedling.

2.4 Cultivar Reaction Tests

Three techniques were used: (i) seed contamination, (ii) needle injection and (iii) plant spraying in glasshouse and field (Toben *et al.*, 1989). For seed inoculation, surface-sterilised seeds were dipped into a bacterial suspension (10^8 CFU ml^{-1}). After 3 h, 45 seeds were removed and placed onto filter paper in Petri plates. The plates were incubated at 22÷24°C and seed germination evaluated after 5 days. Each variant was assigned to phenotype groups based on the following seed germination compared to control: 70÷100%, resistant (R); 50÷70%, partially resistant (PR); 30÷50%, partially susceptible; 10÷30%, susceptible (S); and 0÷10%, very susceptible (VS). The needle injection was made as it has been described above. For spraying, plants at the booting stage were mist-sprayed using a bacterial suspension adjusted as above. After inoculation, plants were incubated at 90÷95% relative humidity and 20÷25°C for 10 days in greenhouse. Field inoculations were made by spraying plant at tillering stage with the same bacterial suspension. After 7 days, brownish black discoloration appeared at the base of glumes of susceptible plants. All experiments were repeated 3 times. Statistical analysis of data was made by STATISTICA 5.5 package (Statsoft, USA).

3. RESULTS

Symptoms of *Pseudomonas syringae* pv. *atrofaciens* were observed on cereal seeds from all examined regions including Asian part of Russian Federation. Phenotypic properties of the *P. syringae* strains isolated from various regions of Russia corresponded to group 1a of *Pseudomonas* species including *P. s.* pv. *atrofaciens,* according to LOPAT's system. Strains of *P. s.* pv. *atrofaciens* generally grouped together according to species of origin. The different pathogen/host combinations resulted in some very different reactions. Analysis of variance showed that differences in virulence between strains and in resistance between varieties were highly significant (P=0.001). Generally, plants and seeds of winter wheat cultivars showed fewer symptoms; 2÷10% for spring wheat, 3÷20% for barley, and 5÷35% for rye. After 2÷4 days of incubation at 25°C, seedlings infected by *P. s.* pv. *atrofaciens* showed brown to black lesions occurring around the margin of the inoculation point.

The cultivars reacted very different manner when strains of *P. s.* pv. *atrofaciens* were inoculated in different cultivars. This may depend on the presence of specific plant resistance to the pathogen strains. Strains from wheat, W-7 and W-9, were most aggressive on wheat cultivars whereas strains BT-1 and B-3.1 from barley were most pathogenic for barley varieties (Table 3).

Cluster analysis of the data indicates two clearly distinguished groups of strains from rye, wheat and barley (cluster 1) and from barley only (clusters 2 and 3) (Fig. 1). Cultivars of wheat "Mironovskaya 808", "Moskovskaya karlikovaya", and barley cultivars "Zazerskii 85", were most useful to differentiate strains in clusters 1, 2 and 3.

P. s. pv. *atrofaciens* strains R-110 and R-101 from rye, W-7, W-9, W-6001, and W-6005 from wheat and B-3.1 and B-10 from barley resulted in up to 94% reduction of seed germination. Plant reactions ranged from slight resistance (R) to partially resistance (PR). Each of three inoculation techniques revealed considerable differences within the tested populations of wheat despite the fact that the wheat plants were considerably more resistant to the pathogen comparing to rye and barley. Only two of Russian spring wheat varieties resulted in a similar susceptible reaction as rye or barley. Highly susceptible accessions were found among Colombian and Syrian spring wheat varieties. On contrast, five Pakistan spring varieties were resistant. Overall 21 of 61 (33%) wheat cultivars contained some resistance or partial resistance to the basal glume rot pathogen.

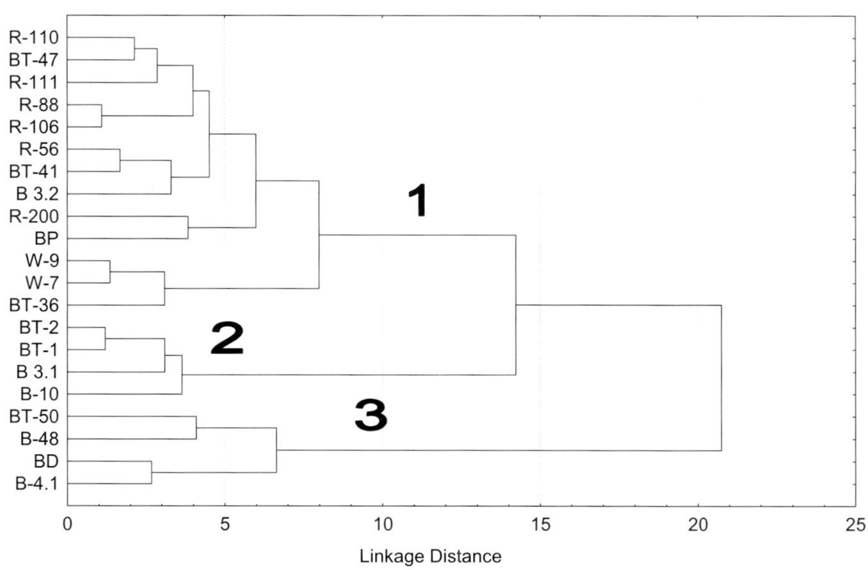

Figure 1. Cluster analysis of *Pseudomonas syringae* pv. *atrofaciens* strains virulence on wheat and barley cultivars. Cluster 1 – strains from rye and wheat; clusters 2 and 3 – strains from barley.

Table 3. Virulence of the *Pseudomonas syringae* pv. *atrofaciens* strains on wheat and barley cultivars.

Strain	Cultivar:												
	Inna	Sarya	Nemchnov-skaya 25	Mironov-skaya 27	Mironov-skaya 808	Moskovskaya karlikovaya	Zazerskii 85	Moskovskii	Visit	Bios 1	Moskovskii 2	Moskovskii 3	Baranchik
R-110	3.0b*	3.0b	4.0ab	3.0b	2.3b	3.0b	0d*	4.0ab	2.0bc	4.0a	4.0a	-	-
R-111	-	3.0b	3.0b	2.0c	1.3cd	2.5c	0d	3.0b	4.0ab	-	-	-	-
R-88	-	4.0a	3.0b	3.0b	3.0b	2.3c	0d	4.0a	3.0b	4.0a	4.0a	4.0a	-
R-106	3.0b	4.0a	2.5bc	-	2.7b	1.5cd	0d	4.0a	3.0b	4.0a	4.0a	4.0a	-
R-200	4.0a	3.5ab	2.0d	4.0a	3.0b	2.5c	0d	1.0c	-	-	4.0a	-	-
R-56	4.0a	4.0a	4.0ab	4.0a	2.5b	2.5c	0d	3.0b	3.0b	4.0a	4.0a	-	-
WT-41	4.0a	4.0a	4.0a	-	1.0cd	1.0d	-	2.0bc	3.0b	4.0a	4.0a	-	-
W-9	3.0b	3.5ab	4.0a	3.5ab	3.0b	2.7c	0d	4.0a	4.0a	3.0b	3.0b	-	4.0a
W-7	3.0b	3.5ab	4.0a	4.0a	4.0a	4.0ab	-	4.0ab	4.0a	-	3.0b	-	4.0a
WT-47	2.5c	3.0b	3.0b	3.0b	3.7ab	4.0ab	-	-	2.5c	-	4.0a	-	-
W-50	-	-	-	-	2.5bc	3.0bc	0d	3.0b	3.0bc	-	-	-	-
BT-2	-	3.5ab	-	-	0d	0d	4.0a	4.0a	4.0a	4.0a	4.0ab	4.0a	4.0a
BT-1	3.0b	3.0b	4.0a	-	2.7bc	2.7bc	4.0ab	4.0a	-	-	4.0a	4.0a	-
B-3.1	3.5ab	4.0a	4.0a	3.0b	2.5bc	2.3bc	3.0ab	4.0a	4.0a	2.0c	4.0a	-	4.0a
BN	2.0c	3.0b	3.0b	2.0c	3.0b	2.3bc	1.0c	1.0cd	2.0c	3.0b	4.0a	4.0a	2.0b
W-3.2	4.0a	4.0a	4.0a	2.5bc	2.5bc	2.0c	1.0c	4.0a	2.0c	-	4.0a	-	-
WT-36	4.0a	3.5ab	3.5ab	3.5abc	2.5bc	2.3c	0d	3.0bc	4.0a	2.0c	3.0b	-	4.0a
W-10	-	3.0b	3.0b	3.0bc	3.0b	3.0b	4.0a	4.0a	1.5bc	4.0a	4.0a	4.0a	4.0a
WD	-	-	-	-	2.0c	3.0bc	2.0bc	4.0a	4.0a	-	-	-	-
W-4.1	-	-	-	-	1.5cd	1.5cd	3.0b	4.0ab	3.0b	-	-	-	-
W-48	-	-	-	-	-	-	0	3.0b	4.0a	-	-	-	-

Three groups of cultivars for each test were distinguished by k-means cluster analysis. Three cultivars of winter wheat: "Pamyat Fedina", "Inna", "Moskovskaya nizkostebelnaya" and 18 varieties of spring wheat were most resistant. Plant reaction data treated by the Factor Analysis (Browne, 1968) revealed one major factor responsible for 49.9 to 96.5% of total variance of data. The lowest value of the 1st Factor was explained by distinctive reaction of Syrian wheat varieties against the strain W-6005 and B-10, obtained from wheat.

The scores of 1st Factor related to each cultivar were used as unambiguous characteristics of plant performance against cumulative action of distinctive bacterial strains.

The same bacterial strains were applied as mixture to inoculate seedlings and plants at booting and tillering stages, at controlled environment and at the tillering stage at field. The data obtained from this experiment showed presence of one major factor that control about 80% of resistance in environmental chamber at different stages of plant growth, and over 40% of resistance variation at field, explaining 69.9% of total variance in plant reaction in all condition. Correlation between plant resistance and susceptibility at different stages of growth in controlled environment was high (r=0.88, significant at P=0.01). There was a high correlation between both plant tests and seed inoculation tests (0.78 and 0.82, significant at P=0.01). However, the field results of inoculated plants showed a low correlation to the growth chamber tests: - 0.45 in average (Fig. 2).

4. CONCLUSION

Pseudomonas syringae pv. *atrofaciens*, the causal agent of basal glume rot of cereals, is a serious pathogen in Russia and reduces yield and quality of cereals. High genetic diversity exist in germplasm of the pathogen, although there were no clear specialisation of the bacterial strain on wheat varieties on race-like fashion. Despite the observed age-depended differences in plant reaction, laboratory and glasshouse tests give similar results, but field evaluation showed that additional factors may be important for plant ability to withstand the pathogen attack. According to the data most resistant to basal glume rot Russian wheat cultivars were: "Moskovskaya nizkostebelnaya", "Mironovskaya 61", "Pamyat Fedina", "Zhneya" and hybrid PPG-93; spring wheat "Bezenchukskaya", "Isheevskaya", "Kinelskaya", L-503.

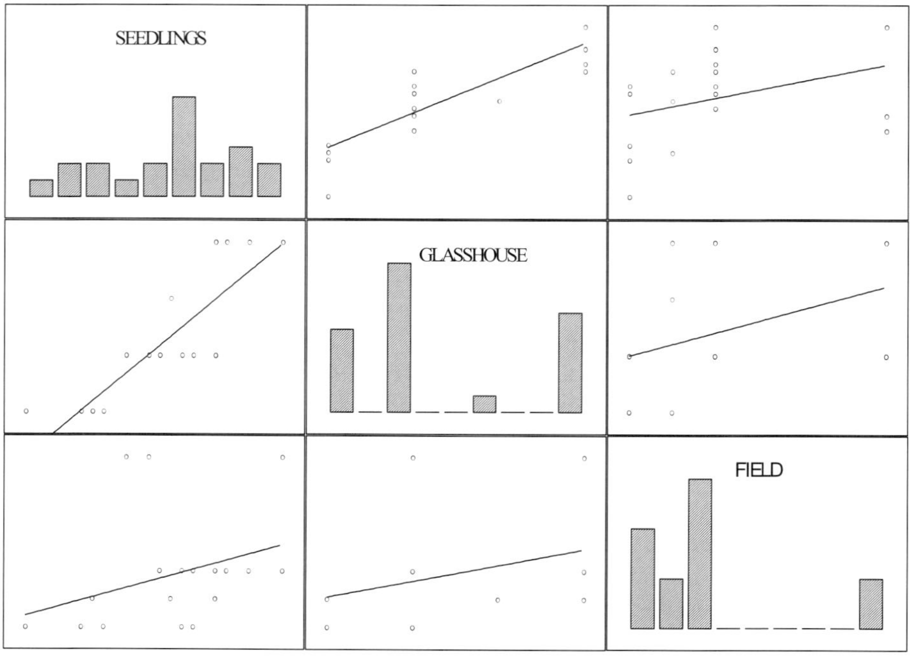

Figure 2. Analysis of Pearson's correlation between reaction of 21 wheat cultivars against mixture of *Pseudomonas syringae* pv. *atrofaciens* strains applied at different stages of plant development: at stage of 5 days-old seedlings (seedlings), at tillering stage in controlled environment (glasshouse) and in the field. Angle of the line indicates the strength of co-variation between ratings.

ACKNOWLEDGEMENTS

This work was supported from ISTC grant #1771p.

REFERENCES

Browne, M.W., 1968, A comparison of factor analytic techniques. *Psychometrika* **33**: 267-334.

Ilukhina, M.K., Schneider, Y.I., 1982, Laboratorial express method for estimating wheat resistance to bacterioses. *Reports of second national symposium of plant immunity*, Plovdiv, **1**: 189-193.

Matveeva, E.V., 1998, Bacterial diseases of cereals. *Agro 21 century* **1**: 6-7.

McCulloch, L., 1920, Basal glume rot of wheat. *J. Agric. Res.* **18**: 543-551.

Schaad, N.W., Jones, J.B., and Chun, W., (Eds), 2001, Laboratory Guide for identification of plant pathogenic bacteria. *APS Press, St. Paul, MN.*, 3rd. edition.

Toben, H., Mavridis, A., Rudolph, K.W.E., 1989, Basal glume rot (*Pseudomonas syringae pv. atrofaciens)* on wheat and barley in FRG and resistance screening of wheat. *Bulletin OEPP/EPPO Bulletin* **19**: 119-125.

Wiese, M.V., 1987, Compendium of Wheat Diseases, 2nd Ed. 1987. *APS Press, St. Paul MN.*

Survival of Biocontrol *Pseudomonas* Strains and Human Pathogens in Fruit Juices

A. PACETTO, P. BELLA, V. CATARA, R. LA ROSA and G. CIRVILLERI
Dipartimento di Scienze e Tecnologie Fitosanitarie, Sez. Patologia Vegetale, Università di Catania, Italy.

Abstract: Rifampicin resistant *Pseudomonas syringae* 1.1S and rifampicin-tetracycline resistant *P. fluorescens* A506 were analysed for survival ability in nutrient broth and fruit juices, in different incubation condition individually and in co-inoculation with *Salmonella* sp. and *Listeria* sp. strains. Persistence of bacteria was monitored by using conventional plating methods and enrichment procedures, and identity of recovered bacteria was verified through antibiotic resistance. All strains, inoculated at a level of approximately 10^6 CFU ml^{-1}, were able to growth in nutrient broth at 28°C in dynamic culture reaching population of $10^8 \div 10^{10}$ CFU ml^{-1} 2 days after inoculation. *P. syringae*, *P. fluorescens* and *Listeria* sp. strains, when individually inoculated and co-inoculated in apple, orange and tomato juices, failed to survive 9 h after incubation at 28°C, whereas *Salmonella* sp. was recovered at population densities of $10^5 \div 10^7$ CFU ml^{-1} 48 h after inoculation. When incubated at 4°C in static culture, *P. syringae* was detected in orange and peach juices 24 h after inoculation, and in apple and tomato juices, respectively, 3 and 10 days after inoculation. *P. fluorescens* was recovered in peach juice 3 days after inoculation, in orange and apple juices 7 days after inoculation, and in tomato juice 55 days after inoculation. *Salmonella* sp. and *Listeria* sp. survived longer than *Pseudomonads*. They were detectable (10^2 CFU ml^{-1}) in tomato juice 70 days after the inoculation, even if at low concentration in enrichment medium. *Salmonella* sp. and *Listeria* sp. survived 24 and 34 days respectively in orange juice, 24 days in peach juice, and 14 and 34 days respectively in apple juice. In co-inoculated juices, *P. syringae* and *P. fluorescens* populations were higher than in individually inoculated media, but their populations never influenced the growth of *Salmonella* sp. and *Listeria* sp..

Key words: bacterial growth, microbial competition, health hazard.

Author for correspondence: G. Cirvilleri; e-mail: gcirvil@unict.it.

N.S. Iacobellis *et. al.* (eds.), *Pseudomonas syringae* and Related Pathogens, 107-115.

1. INTRODUCTION

Contamination of fruits and vegetables during growth, harvest, distribution and processing with micro-organisms of different origin, and microbial survival in products and juices represents a potential health hazard to consumers.

Generally, the microflora of fresh fruits and vegetables under ambient conditions primarily consists of epiphytic strains of *Enterobacteriaceae* and *Pseudomonas* species, while lactic bacteria and fungi may be present at relatively lower numbers (Nguyen-The and Carlin, 1994). In addition, *Pseudomonas* species can occur due to use of biological control agents to prevent diseases of plants and products. Occasionally, human pathogens can occur due to use of contaminated irrigation water or organic fertiliser during cultivation, or as a consequence of poor hygiene during processing (Nguyen-The and Carlin, 1994).

Several studies have demonstrated the survival of micro-organisms in various juices. Although the low pH of some foods, like fruit juices (pH 3 to 4) does not support growth of pathogenic bacteria, some strains of *Enterobacteriaceae*, like certain strains of *Escherichia coli, Listeria* sp., *Shighella* sp. and *Salmonella* sp. species, can survive for several days or even weeks in unpasteurised apple juice and cider (Zhao *et al.,* 1993) and in orange juice (Parish, 1997; Fratamico *et al.,* 1997).

While the persistence and survival of human pathogens within fruits, vegetables and juices has received a great deal of attention, the factors affecting persistence and survival of antagonistic bacteria used in biocontrol strategies and their competitive effects with human pathogens have received much less attention.

Recently, our laboratory started to study how *Pseudomonas* species commonly used as biocontrol agents can survive in vegetables, fruits and products until they reach the consumer, in order to evaluate if they can become an hazard for plants, products and public health.

Two *Pseudomonas* strains, *P. syringae* 1.1S and *P. fluorescens* A506, able to inhibit a variety of plant pathogens *in vitro* and *in vivo* (Cirvilleri *et al.,* 2000), were selected as an example of antagonistic micro-organisms that it would be likely to use in a variety of plants in biological control strategies. *Pseudomonas* strains were evaluated, individually and in co-inoculation with *Salmonella* sp. and *Listeria* sp., for their growth in Nutrient Broth (NB) used as model system, and survival in apple, orange, tomato and peach juices stored at 28°C in dynamic culture and at 4°C in static culture.

The aim of the present work was to elucidate the influence of types of fruit juices, temperatures and growth conditions and competitive effect on the survival of artificially inoculated antagonistic *Pseudomonas* and human pathogens.

2. MATERIALS AND METHODS

2.1 Bacterial Strains

Salmonella sp. and *Listeria* sp. were obtained from the Istituto di Igiene e Profilassi (Center for Disease Control and Prevention) of Catania.
Ten rifampicin resistant (Rif^{100}) *Pseudomonas syringae* strains, able to inhibit a variety of plant pathogens *in vitro* and *in vivo* (Cirvilleri *et al.*, 2000), were initially screened for antagonistic activity *in vitro* against *Salmonella* sp. and *Listeria* sp. strains. *P. syringae* strains were individually spotted on PDA and King's B agar (King *et al.,* 1954), incubated at 28°C for 3 days, then overlaid with 3 ml of water agar (0.6%) containing 10^6 CFU ml^{-1} of *Salmonella* sp./*Listeria* sp. strains, and incubated at 28°C for an additional 3 days. Antagonistic activity was indicated by the formation of a growth inhibition zone (measured as mm of diameter) surrounding the colony. *P. syringae* 1.1S was selected on the basis of antagonistic activity.
P. fluorescens A506 resistant strain ($Rif^{100}Tc^{10}$) (Cirvilleri *et al.*, 2000) was selected as reference strain and was screened for antagonistic activity *in vitro* against *Salmonella* sp. and *Listeria* sp. as previously described.

2.2 Inoculation and Survival of Bacteria in Fruit Juices

The following packaged fruit juices were used in this study: orange juice (pH 3.8); apple juice (pH 3.9); peach juice (pH 4), tomato juice (pH 4.5).
Overnight cultures of *Pseudomonas syringae* 1.1S and *P. fluorescens* A506 grown NB at 28°C, and of *Salmonella* sp. and *Listeria* sp. grown at 37°C, were harvested by centrifugation (3,000*g)*, and the cells resuspended in phosphate buffer 0.1M pH 7.
Flasks containing 50 ml of NB and fruit juices were inoculated with 0.1 ml of appropriate dilutions of *P. syringae* 1.1S, *P. fluorescens* A506 suspensions, *Salmonella* sp. and *Listeria* sp. alone and in co-inoculation with *Salmonella* sp. and *Listeria* sp. to achieve inoculum levels of approximately 10^6 CFU ml^{-1}. Each sample was replicated three times. Uninoculated flasks of NB and fruit juices were used as control.
Flasks were kept at 28°C on rotary shaker and at 4°C in static culture. Samples were periodically removed, diluted appropriately in phosphate buffer, and surface plated in duplicate on NA+ Rif^{100} and NA+ Rif^{100} Tc^{10} at 28°C for monitoring respectively *P. syringae* and *P. fluorescen* strains, and on NA at 37°C for monitoring growth of *Salmonella* sp. and *Listeria* sp.
The enrichment of each sample was obtained by inoculating the juices in NB in the ratio of 1:1 and incubating the mixture at 28°C (*Pseudomonas)* and at 37°C (human pathogens). Aliquots (100 μl) of enrichment samples were

then plated on appropriate agar media at appropriate temperature before examining for bacterial growth.

3. RESULTS

Ten *Pseudomonas syringae* strains were screened for their ability to inhibit the growth of *Salmonella* sp. and *Listeria* sp. strains on PDA and KB. One of them, strain 1.1S, was able to inhibit the growth of *Salmonella* sp. on PDA but was not effective against *Listeria* sp.. *P. fluorescens* A506 was not effective *in vitro* on the growth of both pathogens.

Survival curves of *P. syringae* 1.1S, *P. fluorescens* A506, *Salmonella* sp. and *Listeria* sp. when cultured alone into NB, apple, orange and tomato juices, incubated in dynamic cultures for up to 2 days at 28°C are shown in Fig. 1.

P. syringae and *P. fluorescens* were able to grow in NB at 28°C reaching population of 10^8 CFU ml^{-1} 2 days after inoculation. *Pseudomonas* strains were not able to grow and survive in orange, apple and tomato juices at 28°C. At this temperature cells of *P. syringae* were not detectable after 3, 6, and 9 h from the inoculation in orange, apple and tomato juices, respectively. *P. fluorescens* reached non detectable levels after 9, 24 and 24 h from inoculation respectively in orange, apple and tomato juices. *Salmonella* sp. and *Listeria* sp. were able to grow in NB at 28°C reaching population of 10^{10} CFU ml^{-1} 2 days after inoculation. *Salmonella* sp. survived in orange, apple and tomato juices for 2 days, with population densities of $10^5 \div 10^6$ CFU ml^{-1} 48 h after inoculation, whereas *Listeria* sp. disappeared after 9 h from inoculation.

When incubated at 4°C in static culture, *P. syringae* and *P. fluorescens* strains survived longer then at 28°C. *P. syringae* was detectable in peach and orange juices only 1 day after the inoculation. In apple and tomato juices *P. syringae* survived, respectively, for up to 3 and 10 days from the inoculation (Fig. 2).

P. fluorescens was recovered in peach juice 3 days after inoculation, in apple and orange juices 7 days after the inoculation, and in tomato juice 55 days after the inoculation.

Salmonella sp. and *Listeria* sp. survived longer then *Pseudomonas* at 4°C. *Salmonella* sp. were still detectable in orange and peach juices after 24 days from inoculation, and in apple juice after 34 days. In tomato juice *Salmonella* sp. was still detectable (below 10^2 CFU ml^{-1} in enriched samples) after 70 days from the inoculation. *Listeria* sp. reached not detectable levels after 24, 34, 55 and 70 days from inoculation respectively in apple, peach, orange and tomato juices.

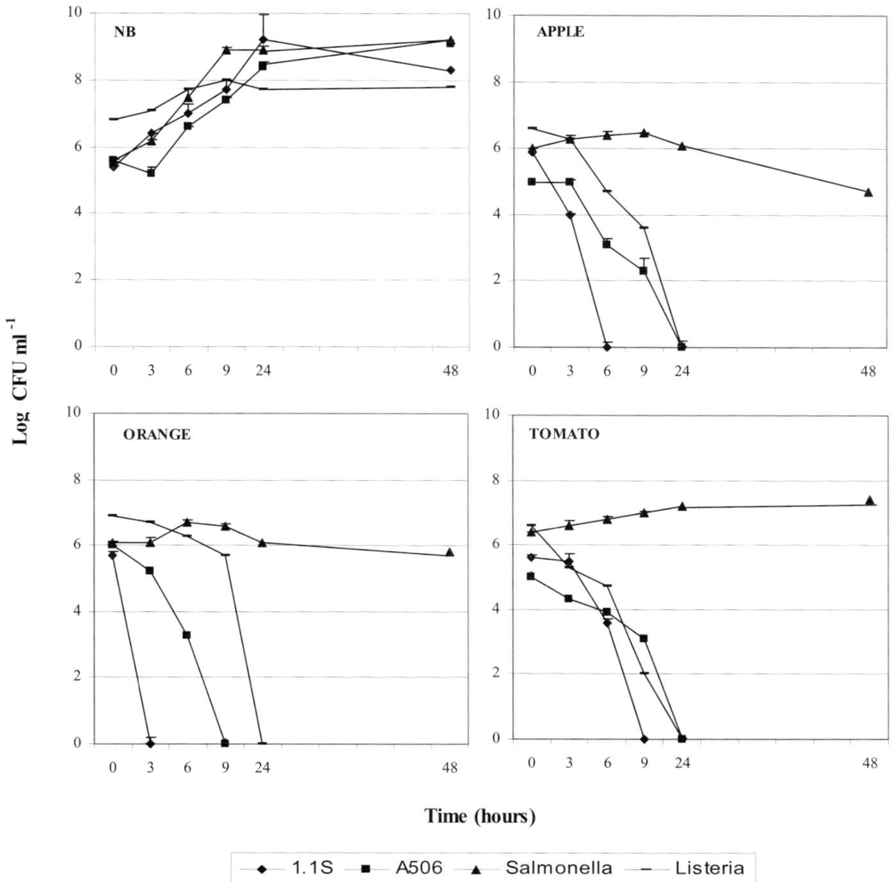

Figure 1. Survival of *Pseudomonas syringae* 1.1S, *P. fluorescens* A506, *Salmonella* sp. and *Listeria* sp. strains in Nutrient Broth (NB), apple, orange and tomato juices incubated in dynamic culture at 28°C for two days. The initial population in each case was between 10^5 and 10^6 CFU ml^{-1}. Presence of cells, when population lowered below the detection limit of 20 CFU ml^{-1}, was assessed by enrichment of samples.

The behaviour of the antagonistic *P. syringae* 1.1S and *P. fluorescens* A506 strains was also monitored in co-inoculation experiments with *Salmonella* sp. and *Listeria* sp. strains, in which *P. syringae, P. fluorescens, Salmonella* sp., and *Listeria* sp. were simultaneously inoculated at the same population densities (10^6 CFU ml^{-1}).

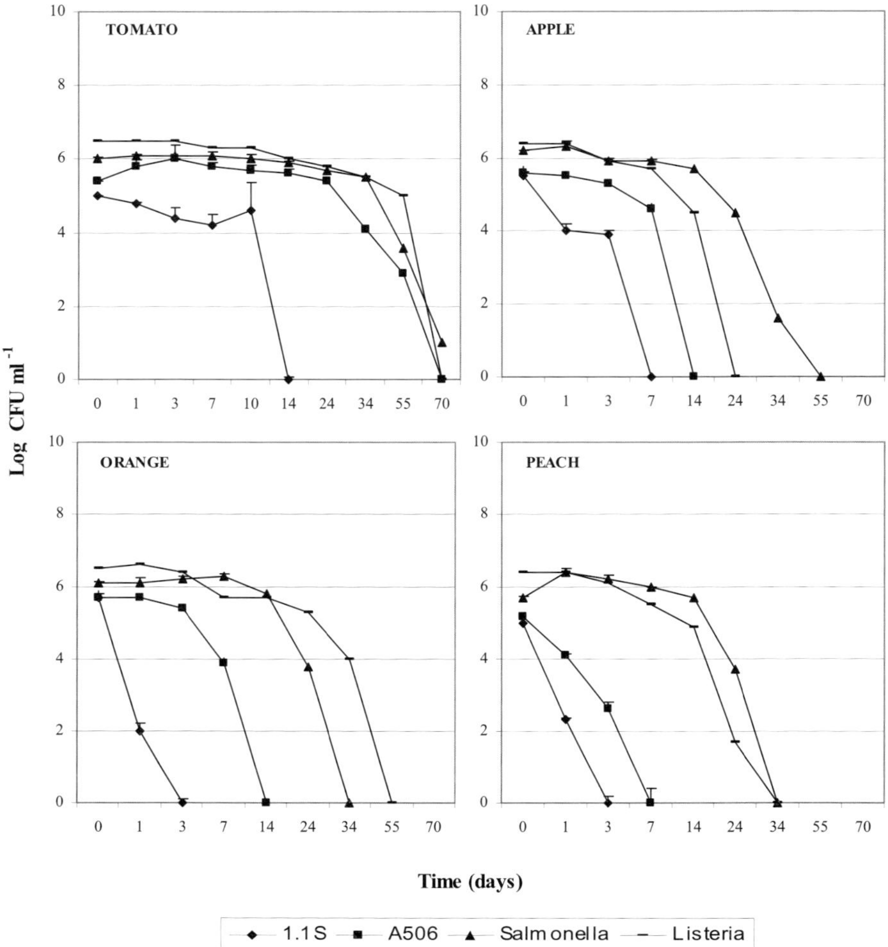

Figure 2. Survival of *Pseudomonas syringae* 1.1S, *P. fluorescens* A506, *Salmonella* sp. and *Listeria* sp. strains in tomato, apple, orange and peach juices incubated in static culture at 4°C. The initial population in each case was between 10^5 and 10^6 CFU ml^{-1}. Presence of cells, when population lowered below the detection limit of 20 CFU ml^{-1}, was assessed by enrichment of samples.

In co-inoculated NB, *P. syringae, P. fluorescens, Salmonella* sp. and *Listeria* sp. population increased to about log 9 cells per ml and was maintained at high level 2 days after inoculation, without differences with individual inoculated survival of both strains.

In co-inoculated fruit juices at 28°C and 4°C *P. syringae* and *P. fluorescens* populations were higher than in individually inoculated ones, but their populations never influenced the survival of *Salmonella* sp. and *Listeria* sp. Examples of growth curves where *P. syringae* and *P. fluorescens* did not affect growth of *Salmonella* sp. and *Listeria* sp. respectively in tomato and apple juices at 28 and 4°C are shown in Fig. 3.

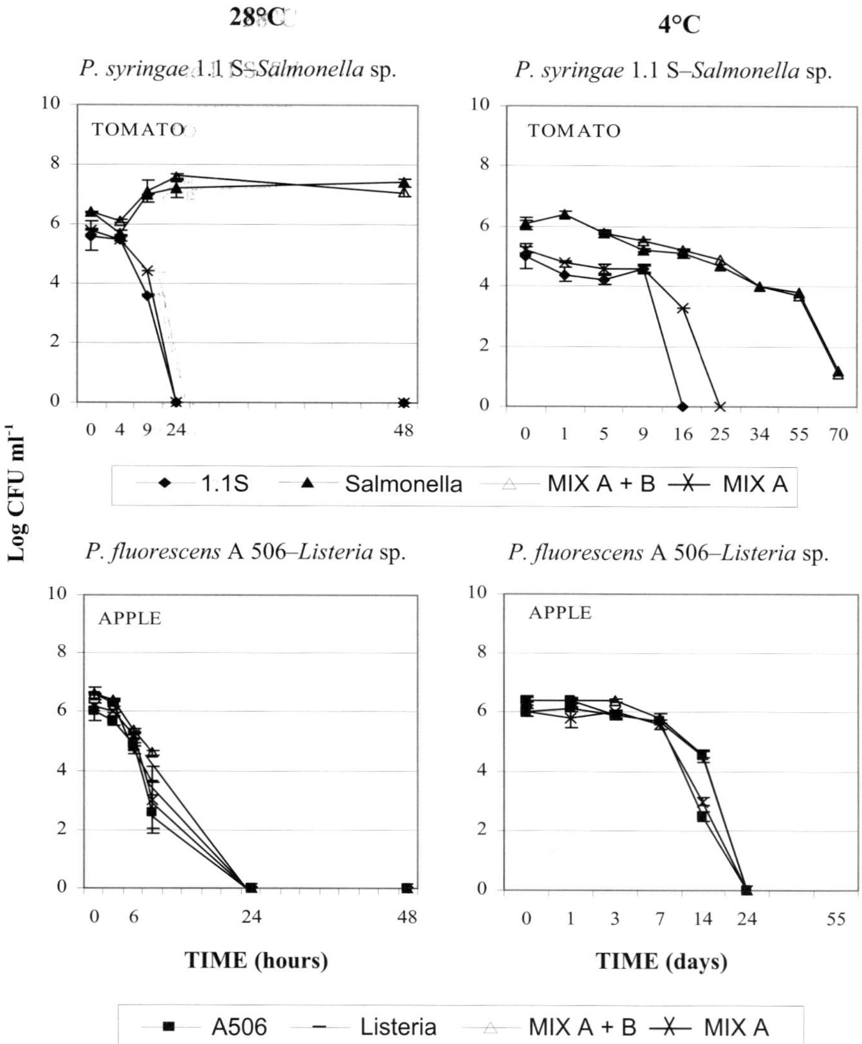

Figure 3. Growth curves of *Pseudomonas syringae* 1.1S, *P. fluorescens* A506, *Salmonella* sp. and *Listeria* sp. in tomato and apple juices at 28 and 4°C. (Mix A+B = *P. syringae* or *P. fluorescens* + *Salmonella* sp. or *Listeria* sp. in co-inoculated juices; Mix A = *P. syringae* or *P. fluorescens* in co-inoculated juices).

In co-inoculation experiments the biological control agents had no significant effect on the population size of pathogens in juices at 28 and 4°C.

4. DISCUSSION

Pseudomonas syringae 1.1S and *P. fluorescens* A506 strains were selected because of the differences in their antagonistic activity *in vitro* against *Salmonella* sp. and *Listeria* sp. strains. *P. syringae* 1.1S was able to inhibit *in vitro* the growth of *Salmonella* sp. whereas had no effect on the *Listeria* sp. growth. *P. fluorescens* A506 had no effect *in vitro* on the growth of both human pathogens. In this study survival of *P. syringae* 1.1S, *P. fluorescens* A506, *Salmonella* sp. and *Listeria* sp. varied with type of juice and with storage temperatures.

In dynamic culture at 28 °C survival was found to be similar for all strains in NB. On apple, citrus and tomato juices *P. fluorescens* A506, *P. syringae* 1.1S and *Listeria* sp. disappeared between 3÷9 h from the inoculation whereas *Salmonella* sp. survived well maintaining the same population densities (about 10^6 CFU ml^{-1}) having when inoculated.

All strains, when stored at 4°C in static culture, survived longer than at 28°C in dynamic culture. In addition, *Salmonella* sp. and *Listeria* sp. survived longer than *Pseudomonas.* A 10^4-10^5-fold reduction in numbers of organisms (*P. syringae, P. fluorescens, Listeria* sp.*)* took place during the first 9 h of storage at 28°C in apple, citrus and tomato juices, whereas no reduction of *Salmonella* sp. took place at 28°C during the first 2 days of storage. On the opposite, no reduction of *P. fluorescens* A506, *Listeria* sp., *Salmonella* sp. took place during the first 7÷14 days of storage at 4°C, whereas *P. syringae* 1.1S disappeared after 1÷10 days. *P. syringae* 1.1S, *P. fluorescens* A506, *Salmonella* sp. and *Listeria* sp. were capable of surviving the environment of tomato juices longer than that of orange, apple, and peach juices at 4°C.

In co-inoculated juices both at 28°C and 4°C *P. syringae* 1.1S and *P. fluorescens* A506 populations were higher than in individual inoculated media, but their populations never influenced the growth of *Salmonella* sp. and *Listeria* sp.

Others have shown that refrigeration enhances survival of bacteria such as *E. coli* O157:H7 (Fisher and Golden, 1998) in ground apples used in cider production, and in apple, orange and mango juices (Graells *et al.,*1998), and this was hypothized to be due to a reduced permeability of the cell membrane to protons and/or a reduced metabolic activity (Garcia-Graells *et al.,* 1998). Zhao *et al.* (1993) found that *Escherichia coli* O157:H7 inoculated at level of 10^5 CFU ml^{-1} survived in different lots of apple cider for 10 to 31 days at 8°C and for 2 to 3 days when the cider was kept at 25°C. Fratamico *et al.* (1997) found that using a recombinant strains of *E. coli* expressing green fluorescent protein at an initial inoculum level of 10^4 CFU ml^{-1}, bacteria survived for at least 21 days in the cider at refrigeration temperatures.

Survival of bacteria over a period of several days or weeks on different juices at different storage temperature will be evaluated also with lower, more realistic levels of initial contamination.

REFERENCES

Cirvilleri, G., Bella, P., and Catara, V., 2000, Molecular detection and biological control activity of *Pseudomonas* strains. 5^{th}Congress of the European Foundation for Plant Pathology, Taormina, 18-22 September 2000, 534-538.

Fisher, T.L., and Golden, D.A., 1998, Fate of *Escherichia coli* 157:H7 in round apples used in cider production. *J. Food Protec.* **61**(10): 1372-1374.

Fratamico, P.M., Deng, M.Y., Strobaugh, T.P., and Palumbo, S.A., 1997, Construction and characterisation of *Escherichia coli* = 197:H7 strains expressing firefly luciferase and green fluorescent protein and their use in survival studies. *J. Food Protec.* **60** (10): 1167-1173.

Garcia-Graells, C., Hauben, K.J.A., and Michiels, C.W., 1998, High-pressure inactivation and sublethal injury of pressure-resistant *Escherichia coli* mutants in fruit juices. *Appl. Environ. Microbiol.* **64**: 1566-1568.

King, E.O., Ward, M.K., and Raney, D.E., 1954, Two simple media for demonstration of pyocyanin and fluorescein. *J. Lab. Clin. Med.* **44**: 301-307.

Nguyen-The, C., and Carlin, F., 1994, The microbiology of minimally processed fresh fruits and vegetables. *Crit. Rev. Food Sci. Nu.* **34**: 371-401.

Parish, M.E., 1997, Public health and nonpasteurized fruit juices. *Crit. Rev. Microbiol.* **23**: 109-119.

Zhao, T., Doyle M.P., and Besser, R.E., 1993, Fate of enterohemorrhagic *Escherichia coli* 157:H7 in apple cider with and without preservatives. *Appl. Environ. Microbiol.* **59**: 2526-2530.

Antagonistic Activity of *Pseudomonas syringae* subsp. *savastanoi*: Preliminary Results on the Identification of a Plasmid-located Genetic Determinant

A. SISTO, M.G. CIPRIANI, M. MOREA, S.L. LONIGRO, and P. LAVERMICOCCA

Istituto di Scienze delle Produzioni Alimentari, C.N.R., Viale Einaudi 51, 70125 Bari, Italy.

Abstract: Strain ITM317 of *Pseudomonas syringae* subsp. *savastanoi,* which causes hyperplastic symptoms (knots) on olive plants, shows antagonistic activity against other strains of this plant pathogenic bacterium because of the production of a bacteriocin-like substance (BLIS). BLIS was sensitive to heat and proteolytic enzymes, slightly affected by lipolytic and glycolytic enzymes and active between pH 5.5 and 7.5, and showed a molecular mass, based on ultrafiltration, ranging between 100 and 300 kDa. In order to identify the genetic determinants involved in the antagonistic activity, a collection of Tn*5*-induced mutants from strain ITM317 was screened and prototrophic mutant ITM317-636, which had lost its ability to inhibit the growth of other *P. s.* subsp. *savastanoi* strains, was identified. Southern blot analysis demonstrated that the genome of mutant ITM317-636 contained a single copy of the Tn*5* element and that Tn*5* insertion occurred on a plasmid. The responsibility of Tn*5* insertion for the above-mentioned mutated phenotype (BLIS$^-$ phenotype) was demonstrated by marker-exchange mutagenesis carried out in parental strain ITM317. The *Eco*RI Tn*5*-containing fragment from mutant ITM317-636 was cloned and the DNA regions flanking Tn*5* insertion were in part sequenced. Preliminary results on sequence similarity searches revealed that Tn*5* insertion occurred within a DNA region encoding a predicted product 37% identical and 52% similar to a putative RHS-related protein of *Xanthomonas campestris* pv. *campestris*.

Key words: olive knot disease, plasmid-encoded bacteriocin production, bacteriocin from plant pathogens, *rhs*-element.

Author for correspondence: A. Sisto; e-mail: angelo.sisto@ispa.cnr.it.

N.S. Iacobellis *et. al.* (eds.), *Pseudomonas syringae* and Related Pathogens, 117-124.

1. INTRODUCTION

Pseudomonas syringae subsp. *savastanoi* [= *Pseudomonas savastanoi* pv. *savastanoi*] (Young *et al*., 1996) is the causal agent of olive knot disease; it also causes hyperplastic symptoms on oleander (*Nerium oleander* L.) and ash (*Fraxinus excelsior* L.) as well as on other minor host plants. Strains isolated from different host species may be differentiated on the basis of host range and other physiological and pathological characteristics (Surico *et al*., 1985; Mugnai *et al*., 1994; Caponero *et al*., 1995; Iacobellis *et al*., 1998). Among the other features, the ability to produce a bacteriocin-like substance (BLIS) has been reported as a trait which may distinguish the different strains of the pathogen (Iacobellis *et al*., 1995). It has been suggested that strains isolated from oleander are unable to infect olive trees under natural conditions in part because of their sensitivity to putative bacteriocin-like substances produced by the olive strains which generally survive on the surface of olive plants (Caponero *et al*., 1995; Iacobellis *et al*., 1995). Bacteriocins are antimicrobial compounds whose activity is usually restricted to bacterial species or strains that are closely related to the producing bacterium and those produced by *P. syringae* pathovars are generally high-molecular-weight proteinaceous substances (Gross and Vidaver, 1990). Because of their high specificity, bacteriocins are considered to be safe for both users and the environment and they appear to be excellent candidates for agricultural use in controlling bacterial plant pathogens (Vidaver, 1983; Lavermicocca *et al*., 2002). Genes related to bacteriocin production are frequently plasmid-located and they are usually organised in multigene operons including the structural gene, a gene involved in translocation to the exterior of the cell and a gene dedicated to the self protection of the producer strain (Jack *et al*., 1995). Strain ITM317 of *P. s.* subsp. *savastanoi* isolated from olive shows antagonistic activity against other strains of this plant pathogenic bacterium because of the production of a BLIS (Iacobellis *et al*., 1995). The aim of this work is to identify and study the genetic determinants involved in the production of BLIS by strain ITM317. To this purpose a collection of Tn*5*-induced mutants from strain ITM317 was screened for antagonistic activity, and prototrophic mutant ITM317-636, which had lost the ability to inhibit the growth of other sensitive *P. s.* subsp. *savastanoi* strains (BLIS$^-$), was identified. Here we report on the molecular characterisation of mutant ITM317-636 and on the preliminary results regarding the identification of a genetic determinant involved in bacteriocin production by the parental strain; in addition, preliminary information on the characterisation of the bacteriocin produced by *P. s.* subsp. *savastanoi* are also reported.

2. MATERIALS AND METHODS

2.1 Production and Characterisation of BLIS

The following media (Lavermicocca *et al.*, 1999) were compared for BLIS production: Woolley (W), Woolley-Peptone (WP) [Woolley's medium supplemented with Bacto-Peptone (Difco), 15 g l^{-1}], NBY, Potato-dextrose broth supplemented with Casamino acids (PDB-CA), KB (King *et al.*, 1954). Cultures were inoculated with *Pseudomonas syringae* subsp. *savastanoi* ITM317 and grown under shaking (100 rpm) at 26°C; BLIS titer was determined by a quantitative serial dilution test in an agar-spot assay against the *P. s.* subsp. *savastanoi* indicator strain PVBa204. After 24 h incubation at 26°C, the growth inhibition of the indicator strain was expressed in arbitrary units (AU) of activity.

A culture of *P. s.* subsp. *savastanoi* ITM317 grown in WP for 4 days (100 rpm, 26°C) was centrifuged (9,000 *g,* 10 min, 4°C) and the culture filtrate (1,800 ml) was added with ammonium sulphate (66% w/v saturation) and stored overnight at 4°C. The precipitate obtained after centrifugation was resuspended in Tris-phosphate buffer (0.01 mol l^{-1}, pH 6.5) and dialysed against the same buffer. The precipitate was then sequentially ultrafiltrated through a series of Spectra/Por membranes of different pore sizes (300, 100, 50 and 20 kDa).

One-ml aliquots of BLIS (3,200 AU ml^{-1}) were treated with the following enzymes: protease, α-chymotrypsin, lipase and α-amylase. The assays were performed at a final concentration of 1 mg ml^{-1} at pH 6.5. Samples with and without enzymes were kept at the appropriate temperature (depending on the enzymes) for 1 h.

To test the stability at various temperatures and pH values, (i) BLIS aliquots (1 ml, 3,200 AU ml^{-1}, pH 6.5) were adjusted to pH values ranging from 2 to 10 (with 1 M HCl or NaOH), incubated for 15 min at 26°C and then brought back to initial pH value; (ii) aliquots were kept for 1 h at 25 and 50°C and for 15 min at 60 and 100°C, then rapidly cooled. After each treatment, samples were assayed for antimicrobial activity against the indicator strain *P. s.* subsp. *savastanoi* PVBa204.

2.2 Detection of Antagonistic Activity of Tn*5*-induced Mutants

Production of the BLIS was detected by deferred antagonism (Vidaver *et al.*, 1972). The producer strains were toothpicked onto the surface of KB plates and allowed to form colonies at 26°C for 24 h. The colonies were killed with chloroform vapours and the plate was covered with 3 ml of 0.7% agar seeded with the indicator strain. Bacteriocin production was detected on account of the formation of a clear inhibition zone in the indicator lawn after further 24 h incubation.

2.3 Genetic Techniques and DNA Manipulations

Genomic-DNA of *Pseudomonas syringae* subsp. *savastanoi* strains was extracted using the AquaPure Genomic-DNA isolation kit (Biorad, CA, USA). Preparations of plasmid-DNA from *P. s.* subsp. *savastanoi* and *Escherichia coli* strains were made using the Quantum Prep Plasmid Miniprep kit (Biorad, CA, USA). Standard procedures were followed for restriction endonuclease digestions, ligations and agarose gel electrophoresis (Sambrook *et al.*, 1989). Plasmid-DNA preparations from *P.* s. subsp. *savastanoi* were also obtained using the method of Hansen and Olsen (1978) with some modifications; then the undigested plasmid preparations were analysed by pulsed field gel electroforesis (PFGE) using a FIGE Mapper apparatus (Biorad, CA, USA). DNA labelling, hybridisation and detection were carried out according to the supplier's instructions, using a non-radioactive digoxigenin-labeling and detection kit (Roche Diagnostics, Monza, Italy). The *Eco*RI Tn*5*-containing fragment from mutant ITM317-636 was cloned into the plasmid vector pBluescript SK II (Stratagene, La Jolla, CA) and the recombinant plasmid was introduced into *E. coli* TG1 by electroporation, as previously described (Sisto *et al.*, 1999). After gel purification, two fragments containing DNA regions flanking the Tn*5* element in mutant ITM317-636 were individually subcloned from the above recombinant plasmid into the same plasmid vector. The *Eco*RI Tn*5*-containing fragment was also cloned into the plasmid vector pRK7813 and the recombinant plasmid was used for marker exchange mutagenesis, as previously described (Sisto *et al.*, 1999).

2.4 DNA Sequencing and Analysis

Regions flanking the Tn*5* element in mutant ITM317-636 were sequenced by using as a primer a single oligonucleotide complementary to and extending outward from the ends of the inverted repeat of transposon (Rich and Willis, 1990) and by primer walking. DNA sequencing was performed with BigDye^{TM} Terminator Chemistry from Applied Biosystems on an ABI 310 sequencer (Applied Biosystems, Foster City, CA, USA). Sequenced DNA fragments were examined for similarity with deposited sequences using the BLASTn or the BLASTx program (Altschul *et al.*, 1997), both available from the National Centre for Biotechnology Information (http://www.ncbi.nlm.nih.gov/).

3. RESULTS AND DISCUSSION

3.1 Production and Characterisation of BLIS Produced by *Pseudomonas syringae* subsp. *savastanoi* ITM317

Among the liquid media tested for the production of BLIS, the highest titer (3,200 AU ml^{-1} referring to the sensitive test strain *Pseudomonas syringae* subsp. *savastanoi* PVBa204) was obtained in the medium WP after 4 days' growth at 26°C.
Ultrafiltration of the dialysed ammonium sulphate precipitate through a series of Spectra/Por membranes with different pore sizes (300, 100, 50 and 20 kDa) indicated that BLIS activity was completely recovered in the retentate of 100 kDa.
The amount of antimicrobial activity recovered after treating BLIS with several enzymes indicated that activity was completely lost after treatment with proteolytic enzymes protease and α-chymotrypsin and was reduced to half by treatment with lipolytic or glycolytic enzymes. Assays performed on the effects of pH and temperature on BLIS activity indicated that incubation of aliquots for 15 min at a temperature of 50°C resulted in a complete loss of antimicrobial activity. Moreover, activity was stable between pH 5.5 and 7.5. Preliminary characterisation indicated that the BLIS produced by *P. s.* subsp. *savastanoi* ITM317 belong to the group of high molecular weight thermolabile, proteolytic enzyme-sensitive bacteriocins (Vidaver, 1976; Lavermicocca *et al.*, 1999). Its molecular mass based on ultrafiltration was estimated to be in the range of 100-300 kDa.

3.2 Identification and Molecular Characterisation of the BLIS - Mutant ITM317-636

About 1,400 Tn*5*-induced mutants from *Pseudomonas syringae* subsp. *savastanoi* olive-strain ITM317 were evaluated for the antagonistic activity against other strains of *P. s.* subsp. *savastanoi*. Among them, prototrophic mutant ITM317-636 failed to inhibit the growth of four *P. s.* subsp. *savastanoi* strains isolated, respectively, from olive (strains PVBa229, PVBa230 and ITM304) or oleander (strain PVBa204) known for their sensitivity to the BLIS produced by parental strain ITM317.
To determine if a single Tn*5* insertion had occurred in the genome of mutant ITM317-636, total DNA preparations were digested to completion with *Eco*RI, *Hin*dIII or *Eco*RI/*Hin*dIII restriction endonucleases. The DNA fragments were separated by agarose gel electrophoresis and hybridised with plasmid pGS9 (carrying Tn*5*). The results of the hybridisation demonstrated

that a single Tn*5* insertion had occurred in the genome of mutant ITM317-636. When a *Hin*dIII digested plasmid-DNA preparation of mutant ITM317-636 was hybridised with the above probe, the hybridising bands were the same as when total DNA was used, thus indicating that transposon insertion occurred in a plasmid. In addition, PFGE of undigested plasmid preparations revealed a plasmid larger in mutant ITM317-636 than in parental strain ITM317; Southern blot hybridisation confirmed that the difference in size was due to Tn*5* insertion.

The *Eco*RI Tn*5*-containing fragment from the BLIS$^-$ mutant ITM317-636 was cloned into the plasmid vector pBluescript SK II and the resulting recombinant plasmid was named pITM-636. *Eco*RI and *Eco*RI/*Hin*dIII digestions of the above recombinant plasmid and Southern blot analysis using pGS9 as a probe confirmed that the clone contained the expected *Eco*RI fragment. The approximate size of this *Eco*RI fragment was estimated to be 16 kb. Two fragments from plasmid pITM-636, containing part of the Tn*5* transposon (1.2 kb) and the flanking DNAs from mutant ITM317-636, were individually subcloned; they were designated as 636-FrB (*Hin*dIII-*Hin*dIII) and 636-FrS (*Eco*RI-*Hin*dIII) and had an approximate size of 7.5 and 3.8 kb, respectively. To ascertain whether Tn*5* insertion was responsible for the above-mentioned mutated phenotype (BLIS$^-$ phenotype), marker-exchange mutagenesis was carried out. For the purpose, the *Eco*RI Tn*5*-containing fragment from mutant ITM317-636 was cloned into the plasmid vector pRK7813 and the recombinant plasmid was introduced in parental strain ITM317 by electroporation. Putative marker-exchange mutants were obtained and it was demonstrated that, when Tn*5* insertion occurred in their genome in the same locus as in mutant ITM317-636, they exhibited the same BLIS$^-$ phenotype.

DNAs flanking Tn*5* insertion within 636-FrB and 636-FrS fragments were in part sequenced. Sequence similarity searches, using the BLASTn program which compares nucleotide sequences, did not reveal any significant homology. However, using the BLASTx software, Tn*5* insertion was found to occur within a DNA region, spanning over 1,517 bp, encoding a predicted product 37% identical and 52% similar to a putative RHS-related protein of *Xanthomonas campestris* pv. *campestris*. The function of *Rhs* elements is unknown; however, their predicted products are hydrophilic substances, characterised by a regularly repeated peptide motif. These features are sometimes related to ligand-binding proteins either associated with the bacterial cell surface or secreted (Wang *et al.*, 1998). Further studies need to ascertain whether this putative *Rhs*-related element in *P. s.* subsp. *savastanoi* plays a direct role in bacteriocin production or whether Tn*5* insertion occurred within an operon and thus whether the BLIS$^-$ phenotype of mutant ITM317-636 results from a polar effect on hypothetical downstream gene/s.

ACKNOWLEDGEMENTS

Work supported by a grant from the Italian Ministry of Scientific and Technological Research (MURST), Plan for the Development of Research Networks, Law 488/92, Cluster C06 + 07, Project 1.1: Micro-organisms and Microbial Metabolites in Plant Protection. We wish to thank Dr. D. K. Willis, University of Wisconsin, Madison, WI, for generously providing plasmid pRK7813. We thank Mr. G. Stea for his technical assistance in DNA sequencing.

REFERENCES

Altschul, S.F., Madden, T.L., Schäffer, A.A., Zhang, J., Zhang, Z., Miller, W., and Lipman, D.J., 1997, Gapped BLAST and PSI-BLAST: a new generation of protein database search programs. *Nucleic Acids Res.* **25**: 3389-3402.

Caponero, A., Contesini, A.M., and Iacobellis, N.S., 1995, Population diversity of *Pseudomonas syringae* subsp. *savastanoi* on olive and oleander. *Plant Pathol.* **44**: 848-855.

Gross, D.C., and Vidaver, A.K., 1990, Bacteriocins. In *Methods in Phytobacteriology* (Z. Klement, K. Rudolph and D.C. Sands, eds.), Akademiai Kiado, Budapest, pp. 245-249.

Hansen, J.B., and Olsen, R.H., 1978, Isolation of large bacterial plasmids and characterization of the P2 incompatibility group plasmids pMG1 and pMG5. *J. Bacteriol.* **135**: 227-238.

Iacobellis, N.S., Caponero, A., and Evidente, A., 1998, Characterization of *Pseudomonas syringae* ssp. *savastanoi* strains isolated from ash. *Plant Pathol.* **47**: 73-83.

Iacobellis, N.S., Contesini, A.M., and Surico, G., 1995, Bacteriocin production by *Pseudomonas syringae* subsp. *savastanoi*. *Phytopath. Medit.* **34**: 15-22.

Jack, R.W., Tagg, J.R., and Ray B., 1995, Bacteriocins of gram-positive bacteria. *Microbiol. Rev.* **59**: 171-200.

King, E.O., Ward, M.K., and Raney, D.E., 1954, Two simple media for the demonstration of pyocyanin and fluorescin. *J. Lab. Clin. Med.* **44**: 301-307.

Lavermicocca, P., Lonigro, S.L., Evidente, A., and Andolfi, A., 1999, Bacteriocin production by *Pseudomonas syringae* pv. *ciccaronei* NCPPB2355. Isolation and partial characterization of the antimicrobial compound. *J. Appl. Microbiol.* **86**: 257-265.

Lavermicocca, P., Lonigro, S.L., Valerio, F., Evidente, A., and Visconti, A., 2002, Reduction of olive knot disease by a bacteriocin from *Pseudomonas syringae* pv. *ciccaronei*. *Appl. Environ. Microbiol.* **68**: 1403-1407.

Mugnai, L., Giovanetti, L., Ventura, S., and Surico, G., 1994, The grouping of strains of *Pseudomonas syringae* subsp. *savastanoi* by DNA restriction fingerprinting. *J. Phytopathology* **142**: 209-218.

Rich, J.J., and Willis, D.K., 1990, A single oligonucleotide can be used to rapidly isolate DNA sequences flanking a transposon Tn*5* insertion by the polymerase chain reaction. *Nucleic Acids Res.* **18**: 6673-6676.

Sambrook, J., Fritsch, E.F., and Maniatis, T., 1989, *Molecular Cloning: A Laboratory Manual. 2nd edition*. Cold Spring Harbor Laboratory Press, New York.

Sisto, A., Morea, M., Zaccaro, F., Palumbo, G., and Iacobellis, N.S., 1999, Isolation and characterisation of *Pseudomonas syringae* subsp. *savastanoi* mutants defective in hypersensitive response elicitation and pathogenicity. *J. Phytopathology* **147**: 321-330.

Surico, G., Iacobellis, N.S., and Sisto, A., 1985, Studies on the role of indole-3-acetic acid and cytokinins in the formation of knots on olive and oleander plants by *Pseudomonas syringae* pv. *savastanoi*. *Physiol. Plant Pathol.* **26**: 309-320.

Vidaver, A.K., 1976, Prospects for control of phytopathogenic bacteria by bacteriophages and bacteriocins. *Annu. Rev. Phytopathol.* **14**: 451-465.

Vidaver, A.K., 1983, Bacteriocins: The lure and the reality. *Plant Dis*. **67**: 471-475.

Vidaver, A.K., Mathys, M.L., Thomas, M.E., and Schuster, M.L., 1972, Bacteriocins of the phytopathogens *Pseudomonas syringae*, *P. glycinea*, and *P. phaseolicola. Can. J. Microbiol.* **18**: 705-713.

Wang, Y.D., Zhao, S., and Hill, C.W, 1998, *Rhs* elements comprise three subfamilies which diverged prior to acquisition by *Escherichia coli*. *J. Bacteriol.* **16**: 4102-4110.

Young, J.M., Saddler, G.S., Takikawa, Y., De Boer, S.H., Vauterin, L., Gardan, L., Gvozdyak, R.I., and Stead, D.E., 1996, Names of plant pathogenic bacteria 1864-1995. *Rev. Plant Pathol.* **75**: 721-763.

Pathogenesis and Determinants of Pathogenicity

Methods for the Identification of Virulence Genes in *Pseudomonas syringae*

J. MURILLO, D. BUTCHER[#], R. JACKSON[§], G.W. SUNDIN[ϕ], A. VIVIAN[#] and D. ARNOLD[#]

Departamento de Producción Agraria, Universidad Pública de Navarra, 31006 Pamplona, Spain; [#]Centre for Research in Plant Science, UWE-Bristol, UK; [§]Department of Plant Sciences, University of Oxford, Oxford, UK; [ϕ]Michigan State University, Department of Plant Pathology, East Lansing, MI, USA.

Abstract: The molecular analysis of pathogenicity and virulence in *Pseudomonas syringae* is hampered by functional gene redundancy and by the usually small contribution of each individual virulence gene. In consequence, classical techniques, such as transposon mutagenesis, have proved of limited use for the identification of virulence genes. Curing of native plasmids, however, has allowed the identification of diverse virulence genes and pathogenicity islands in several *P. syringae* pathovars. In this paper, we summarise the strategies and techniques that we have used successfully for the eviction of native plasmids and the analysis of the resulting strains.

Key words: native plasmids, pPT23A-like family, replication, curing.

Author for correspondence: J. Murillo; e-mail: jesus@unavarra.es.

1. INTRODUCTION

The ability of *Pseudomonas syringae*[1] to produce disease in its plant hosts, and its virulence, are based upon the action of two extensive groups of molecules: 1) a set of unrelated molecules collectively know as effectors, and 2) a diverse collection of extracellular virulence factors, of which the most relevant are enzymes, phytotoxins and phytohormones. Effectors are delivered into the plant cell by a specialised Type III secretion system, and

[1] Under this name we are also referring to strains of *Pseudomonas savastanoi*.

N.S. Iacobellis *et. al.* (eds.), *Pseudomonas syringae* and Related Pathogens, 127-135.

their specific interaction with plant molecules can result in the production of disease or in the triggering of the hypersensitive reaction. The accumulated evidence suggests that effectors are the main virulence determinants in *P. syringae*, and that the capacity to produce disease results from the combined action of the collection of effectors that the bacterium possesses. However, the majority of the effectors are not essential for pathogenicity in a given host and can be mutated without any appreciable loss of virulence, probably because their effect is usually small and redundant (e. g., Lorang *et al*., 1994; Jackson *et al*., 1999; Tsiamis *et al*., 2000). Extracellular virulence factors are also not essential for pathogenicity, although they generally contribute to virulence. Sometimes, as happens with effectors, their contribution is small or it is modified by their interaction with other bacterial molecule(s). A good example is the phytotoxin coronatine, whose genes are included in plasmid pPT23A in *P. s.* pv. *tomato* PT23; the contribution of this toxin to virulence is small and difficult to measure, although it is significantly increased in the presence of a DNA region included in another native plasmid, pPT23B present in strain PT23 (Sesma *et al*., 2001).

The additive and functional redundancy of the virulence factors in *P. syringae* makes it difficult to investigate the molecular basis of pathogenicity and host-range determination, because it is frequently the case that quantifiable loss of virulence can be observed only as the result of cumulative virulence loss resulting from inactivation of several virulence genes. A clear consequence is that classical techniques, such as transposon mutagenesis or functional complementation, have been of limited help in the identification of virulence factors. Plasmid curing is a realistic strategy to uncover the existence of virulence genes in *P. syringae* because virulence genes tend to be encoded in plasmids in many bacterial species (Eberhard, 1989), including *P. syringae* (Vivian *et al*., 2001), and to cluster in "pathogenicity islands" that are also often plasmid-borne (Hacker and Kaper, 2000). The removal of a large amount of DNA, that can be of several hundreds kilobases, from the cell offers the possibility of detecting multiple gene effects and evaluating their possible interactions.

The majority of the native plasmids in *P. syringae* strains belong to the pPT23A-like family (Murillo and Keen, 1994; Sesma *et al*., 1998; Gibbon *et al*., 1999), and most strains contain from two to six native plasmids of this family. These plasmids are characterised as containing cross-hybridising replication genes, although they often also harbour other (large) regions of homology. The origins of replication of the pPT23A-like plasmids are highly conserved (Gibbon *et al*., 1999; Sesma *et al*., 2000), which has greatly facilitated their cloning and use in curing experiments. These origins are included in fragments of about 1.6 kb that contain the gene *repA*, coding for a protein essential for replication. In this paper, we present methods to cure native plasmids and their applications for the study of gene function, and especially of virulence genes, in *P. syringae* and related bacteria.

2. MATERIALS AND METHODS

Minipreparations of plasmid DNA from *Escherichia coli* and *Pseudomonas syringae* were done following the protocol of Zhou *et al.* (1990). For *P. syringae*, cells were grown in King's medium B (KMB) (King *et al.*, 1954) and several steps were taken to minimise the isolation of chromosomal DNA: cultures were collected during exponential growth; TENS was freshly prepared using freshly prepared 2N NaOH; mixing of solutions was done gently; lysis was allowed to proceed for 5 min in ice, and tubes were maintained for at least 5 min in ice after adding ice-cold 3 M sodium acetate. Plasmid DNA of adequate quality could not be isolated from certain *P. syringae* strains using this protocol, and in these cases the protocol of Moulton *et al.* (1993) was used. Undigested plasmid DNA was separated on 0.6% agarose gels in 1×TAE (Sambrook *et al.*, 1989) for 3÷4 h at around 4 V cm^{-1} at room temperature, and gels were stained (after electrophoresis) with ethidium bromide. Care was taken not to boil the agarose excessively, which was usually the cause of poor resolution. Other standard molecular techniques were as described (Sambrook *et al.*, 1989). Inoculation of bean, pea, soybean and tomato plants; evaluation of symptoms, and estimation of *in planta* bacterial populations were done as previously described (Moulton *et al.*, 1993; Murillo *et al.*, 1994; Jackson *et al.*, 1999).

3. RESULTS AND DISCUSSION

Eviction of native plasmids can be attempted by using several methods of variable efficacy. In our experience, however, curing by incompatibility is in general very efficient and has the advantage that we can select *a priori* which plasmid would be cured. The phenomenon of incompatibility determines that two plasmids that share replication regions cannot be stably maintained in the same cell and, therefore, they are randomly segregated in such a way that the progeny will contain either one or the other (Novick, 1987). In consequence, the strategy for curing a native plasmid from a *Pseudomonas syringae* strain is based in the cloning of replication regions from this plasmid into an appropriate selectable vector, and the introduction of the resulting construct into the *P. syringae* host. The antibiotic selection for this construct determines that the transformants recovered will contain it and, due to incompatibility, they will have lost the native plasmid by segregation. An outline of this strategy is presented in Fig. 1.

3.1 Curing of Plasmids by Incompatibility

The origin of replication from a given plasmid of *Pseudomonas syringae* can be cloned using a functional approach, in which the origin is identified by its ability to support autonomous replication in this bacterium. To do so, purified plasmid DNA is treated with restriction endonucleases and cloned in a vector that does not replicate in *P. syringae* and that can be easily selected. Among others, we have successfully used plasmid pK184 (2.4 kb; Jobling and Holmes, 1990), which confers resistance to kanamycin and has a medium-copy number in *Escherichia coli*, thereby decreasing the occurrence of possible deleterious side-effects. To minimise the chances of cloning a truncated origin of replication, restriction enzymes with different frequencies of cleavage, such as *Sau*3AI, *Kpn*I or *Pst*I, can be used. An example of this type of functional cloning was employed by Sesma *et al.* (1998) to clone four origins of replication from a strain of *P. s.* pv. *glycinea*.

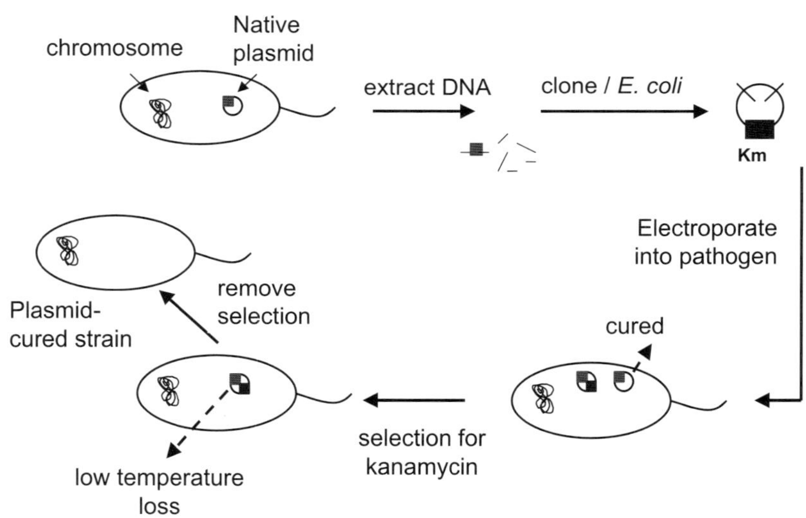

Figure 1. Strategy for curing a native plasmid by incompatibility. Appropriate DNA fragments are cloned in a vector that do not replicate in the bacterial pathogen. The constructs containing the origin of replication of the native plasmid are selected by functional analysis, DNA hybridisation or PCR, and introduced into the pathogen. Due to incompatibility, segregation of the plasmids occurs during cell division and cells arise that contain either the native plasmid or the construct; maintenance of antibiotic selection ensures the survival of cells that contain the construct and are, in consequence, free of the native plasmid. Further propagation of the resulting strain without selection results in the loss of the construct.

Cloning of origins of replication is greatly facilitated by the fact that most native plasmids in *P. syringae* belong to the pPT23A-like family and, hence, contain highly conserved origins that can be rapidly identified (and cloned) by DNA hybridisation or PCR. The starting material for cloning can be genomic DNA, a genomic library, or a plasmid library, constructed by cloning purified plasmid DNA using a cosmid vector. A plasmid or genomic library can be screened by hybridisation using a probe specific for *repA*, such as the 0.8 kb *Eco*RI fragment from pAKC containing the 5'-end of *repA* (Murillo and Keen, 1994; Gibbon *et al.*, 1999), and the cosmid clone, or subclones thereof, can be used for eviction experiments. When the *P. syringae* strain we are working with contains two or more pPT23A-like plasmids, it would be necessary to identify which native plasmid contains the cloned origin of replication. This can be done by DNA hybridisation of plasmid profiles using as a probe a fragment that contains the less conserved 3'-end of *repA* or DNA sequences surrounding the origin of replication (Sesma *et al.*, 1998); however, this can be a difficult task due to the repetition of other sequences among pPT23A-like plasmids.

We were also able to clone the origin of replication from a diverse collection of pPT23A-like plasmids by PCR amplification (Sesma *et al.*, 2000). The *repA* genes were amplified from genomic DNA using primers 532 (5′-GAACGGTGGACTTATGG-3′) and 1588 (5′-CTCCAGCTTGCGGCCCC-3′) which flank a conserved fragment of 1399 bp containing 1279 bp of the *repA* coding region plus 120 bp upstream of the putative start codon (Gibbon *et al.*, 1999). The resulting amplicons were cloned in pCR2.1 (Km^r, Invitrogen) and, in spite of lacking the last 35 nt of the *repA* coding sequence, most were shown to be functional *in vivo*.

Once a putative origin is cloned we can test its functionality by examining its capacity to replicate in a plasmidless strain; although probably any plasmidless *P. syringae* strain will be convenient, we had good results with *P. s.* pv. *syringae* FF5 (Sundin and Bender, 1994), which is able to accept incoming plasmid DNA at a high frequency. Transfer of DNA can be done by electroporation (Keen *et al.*, 1992) or, if possible, by conjugation. After we have made sure that the clone is functional, it is transferred to the parental *P. syringae* strain and appropriate antibiotic selection is applied. If incompatibility is strong, the native plasmid is already absent in the cells growing in the transformation plate; otherwise, it would be necessary to serially transfer the cells a few times (maintaining the antibiotic selection for the cloned origin) before the plasmid is evicted. The loss of a native plasmid can be easily checked by the examination of the plasmid profile and later confirmed by DNA hybridisation. The construct can subsequently be lost from the recipient cell in most cases by growth at low temperature (5-8 °C) in the absence of selection (Gibbon *et al.*, 1999).

Using the strategies summarised above, we have cloned several origins of replication from different pathovars of *P. syringae* (Table 1) that were used to demonstrate plasmid involvement in virulence in *P. syringae* pathovars *glycinea*, *phaseolicola*, *pisi*, and *tomato* (Jackson *et al.*, 1999; Sesma *et al.*, 2001, J. Murillo and A. Vivian, unpublished results).

Table 1. Origins of replication cloned from *Pseudomonas syringae* native plasmids.[a]

P. syringae host	Plasmid	Size	Construct[b]	Auto Inc[c]	Reference[d]
pv. *glycinea*					
race 6	pR6C	80	pORI601	NF	1
	pR6D	75	pORI603	NF	1
pv. *phaseolicola*					
1302A	pAV505	140	pPPY51	NF	2
1449B	pAV511	154	pAV539	yes	3
	pAV512	25	pAV536	yes	3
pv. *pisi*					
870A	pAV232	106	pAV606	yes	3
	pAV230	60	pAV621	yes	3
	pAV231	50	pAV622	yes	3
pv. *tomato*					
PT23	pPT23A	100	pAori1	yes	4
	pPT23B	83	pBKE	yes (weak)	4
	pPT23D	36	pDBE	yes	4
DC3000	pDC3000A	75	pREP614	NT	5
pv. *syringae*					
B86-17	pB8617A	75	pREP202	NT	5
7B12	pPRS14	70	pREP241	NT	5

[a] All the constructs were able to support autonomous replication in *Pseudomonas syringae*.

[b] Constructs are based in pK184, except pPPY51, based in pBluescript, and pREP plasmids, based on pCR2.1.

[c] Auto Inc, incompatibility with parental native plasmid; NF, nontransformable strain; NT, not tested.

[d] References are: 1, Sesma *et al.*, 1998; 2, Jackson *et al.*, 1999; 3, D. Butcher and A. Vivian, unpublished results; 4, Murillo and Keen, 1994; 5, Sesma *et al.*, 2000.

3.2 Complementation of Cured Strains

The possible loss of virulence of a plasmid-cured strain should be examined in different hosts in comparison with the wild type strain. Sometimes, differences in virulence are not obvious to the naked eye and it is necessary to carry out a careful evaluation of different parameters, such as lesion number and size, bacterial populations, etc. The identification of the putative virulence genes is most easily done by complementation of the plasmid-cured strain with a plasmid gene library: after transferring the different cosmids to the plasmid-cured strain, the virulence of the resulting transconjugants is tested as above. However, it is necessary to take into account that sometimes the virulence region could be too large to fit into a cosmid or may be divided between two or more native plasmids (e.g., Sesma *et al.*, 2001); also, many strains of *Pseudomonas syringae* are prone to suffer genomic reorganisations in response to their genetic manipulation. In these cases, we could evaluate the complementation by transferring the cured plasmids back to the plasmid-cured strain and testing for virulence of the resulting strain.

3.3 Identification of Virulence Genes by Plasmid Curing

The strategies outlined above have allowed us to identify different genes that contribute to virulence in several pathovars of *Pseudomonas syringae* (Table 2). In some cases, as it happens with *virPphA* (present in pAV511; Table 2), the gene was essential for pathogenicity, although in other cases the contribution to virulence was additive and difficult to demonstrate by transposon mutagenesis.

Table 2. Involvement of native plasmids from *Pseudomonas syringae* pathovars in virulence.[a]

P. syringae strain	Plasmid	Size	Plant host	Effect in virulence[c]	Remarks (Reference)[b]
pv. *glycinea*					
49a/90	p49a90C		Soybean	No necrotic lesions produced	(1)
pv. *phaseolicola*					
1449B	pAV511	154	Bean	Loss of pathogenicity	pAV511 contains a pathogenicity island with several effector genes; (2)
			Soybean	No necrotic lesions produced	Effect due to *avrPphF*; (2)
pv. *pisi*					
299A	pAV212	54	Pea	Wider host range	Effect due to loss of plasmid carrying *avrPpiB*; (3)
299A	pAV213	50	Pea	Reduced host range	Loss of virulence/suppressor gene; (3)
870A	pAV232	106	Pea	Loss of pathogenicity	Uncharacterised pathogenicity determinants located on this plasmid; (3)
870A	pAV231	42	Pea	Wider host range	Effect due to loss of plasmid carrying *avrPpiB*; (3)
pv. *tomato*					
PT23	pPT23A	100	Tomato	Reduction in lesion size and in planta populations	Due to coronatine; its effect is modulated by pPT23B; (4, 5)
	pPT23B	83	Tomato	Reduction in lesion size	The effect is only seen in strains producing coronafacic acid due to pPT23A; (5)

[a] Evidence obtained by plasmid curing;

[b] References are: 1, J. Murillo, unpublished; 2, Jackson *et al.*, 1999; 3, D. Butcher and A. Vivian, unpublished; 4, Bender *et al.*, 1987; 5, Sesma *et al.*, 2001.

ACKNOWLEDGEMENTS

This work was supported by grants AGL2001-1948-CO2-01, from the Ministerio de Ciencia y Tecnología (to JM), and from BBSRC (to AV and DA).

REFERENCES

Bender, C.L., Stone, H.E., Sims, J.J., and Cooksey, D.A., 1987, Reduced pathogen fitness of *Pseudomonas syringae* pv. *tomato* Tn*5* mutants defective in coronatine production. *Physiol. Mol. Plant Pathol.* **30**: 273-283.

Eberhard, W.G., 1989, Why do bacterial plasmids carry some genes and not others? *Plasmid* **21**: 167-174.

Gibbon, M.J., Sesma, A., Canal, A., Wood, J.R., Hidalgo, E., Brown, J., Vivian, A., and Murillo, J., 1999, Replication regions from plant-pathogenic *Pseudomonas syringae* plasmids are similar to ColE2-related replicons. *Microbiology* **145**: 325-334.

Hacker, J., and Kaper, J.B., 2000, Pathogenicity islands and the evolution of microbes. *Annu. Rev. Microbiol.* **54**: 641-679.

Jackson, R.W., Athanassopoulos, E., Tsiamis, G., Mansfield, J.W., Sesma, A., Arnold, D.L., Gibbon, M.J., Murillo, J., Taylor, J.D., and Vivian, A., 1999, Identification of a pathogenicity island, which contains genes for virulence and avirulence, on a large native plasmid in the bean pathogen *Pseudomonas syringae* pathovar phaseolicola. *Proc. Natl. Acad. Sci. USA* **96**: 10875-10880.

Jobling, M.G., and Holmes, R.K., 1990, Construction of vectors with the p15a replicon, kanamycin resistance, inducible *lacZ*α and pUC18 or pUC19 multiple cloning sites. *Nucleic Acids Res.* **18**: 5315-5316.

Keen, N.T., Shen, H., and Cooksey, D.A., 1992, Introduction of DNA into plant pathogenic bacteria. In *Molecular Plant Pathology. A Practical Approach* (S.J. Gurr, M.J. McPherson and D.J. Bowles, eds.), IRL Press, Oxford, Vol. I, pp. 45-50.

King, E.O., Ward, N.K., and Raney, D.E., 1954, Two simple media for the demonstration of pyocyanin and fluorescein. *J. Lab. Clin. Med.* **44**: 301-307.

Lorang, J.M., Shen, H., Kobayashi, D., Cooksey, D., and Keen, N.T., 1994, *AvrA* and *avrE* in *Pseudomonas syringae* pv. *tomato* PT23 play a role in virulence on tomato plants. *Mol. Plant-Microbe Interact.* **7**: 508-515.

Moulton, P.J., Vivian, A., Hunter, P.J., and Taylor, J.D., 1993, Changes in cultivar-specificity toward pea can result from transfer of plasmid RP4 and other incompatibility group P1 replicons to *Pseudomonas syringae* pv. *pisi*. *J. Gen. Microbiol.* **139**: 3149-3155.

Murillo, J., and Keen, N.T., 1994, Two native plasmids of *Pseudomonas syringae* pathovar tomato strain PT23 share a large amount of repeated DNA, including replication sequences. *Mol. Microbiol.* **12**: 941-950.

Murillo, J., Shen, H., Gerhold, D., Sharma, A.K., Cooksey, D.A., and Keen, N.T., 1994, Characterisation of pPT23B, the plasmid involved in syringolide production by *Pseudomonas syringae* pv. *tomato*. *Plasmid* **31**: 275-287.

Novick, R.P., 1987, Plasmid incompatibility. *Microbiol. Rev.* **51**: 381-395.

Sambrook, J., Fritsch, E.F., and Maniatis, T., 1989, *Molecular Cloning: a Laboratory Manual*. Cold Spring Harbor Laboratory, Cold Spring Harbor, N Y.

Sesma, A., Aizpún, M.T., Ortiz, A., Arnold, D., Vivian, A., and Murillo, J., 2001, Virulence determinants other than coronatine in *Pseudomonas syringae* pv. *tomato* PT23 are plasmid-encoded. *Physiol. Mol. Plant Pathol.* **58**: 83-93.

Sesma, A., Sundin, G., and Murillo, J., 1998, Closely related replicons coexisting in the phytopathogen *Pseudomonas syringae* show a mosaic organisation of the replication region and altered incompatibility behaviour. *Appl. Environ. Microbiol.* **64**: 3948-3953.

Sesma, A., Sundin, G.W., and Murillo, J., 2000, Phylogeny of the replication regions of pPT23A-like plasmids from *Pseudomonas syringae*. *Microbiology* **146**: 2375-2384.

Sundin, G.W., and Bender, C.L., 1994, Relative fitness in vitro and in planta of *Pseudomonas syringae* strains containing copper and streptomycin resistance plasmids. *Can. J. Microbiol.* **40**: 279-285.

Tsiamis, G., Mansfield, J.W., Hockenhull, R., Jackson, R.W., Sesma, A., Athanassopoulos, E., Bennett, M.A., Stevens, C., Vivian, A., Taylor, J.D., and Murillo, J., 2000, Cultivar specific avirulence and virulence functions assigned to *avrPphF* in *Pseudomonas syringae* pv. *phaseolicola*, the cause of bean halo-blight disease. *EMBO J.* **19**: 3204-3214.

Vivian, A., Murillo, J., and Jackson, R., 2001, The role of plasmids in phytopathogenic bacteria: mobile arsenals? *Microbiology* **147**: 763-780.

Zhou, C., Yang, Y., and Jong, A.Y., 1990, Miniprep in ten minutes. *Biotechniques* **8**: 172-173.

Characteristics of the *syr-syp* Genomic Island of *Pseudomonas syringae* pv. *syringae* Strain B301D

D.C. GROSS, I. GRGURINA[#], B.K. SCHOLZ-SCHROEDER[§], and S.-E. LU

Department of Plant Pathology and Microbiology, Texas A&M University, College Station, TX, USA, [#]Dipartimento di Scienze Biochimiche 'A. Rossi Fanelli,' Università 'La Sapienza' di Roma, Rome, Italy, and [§]Department of Plant Pathology, Washington State University, Pullman, WA, USA.

Abstract: *Pseudomonas syringae* pv. *syringae* produces two classes of pore-forming lipodepsipeptide phytotoxins that target host plasma membranes. Both syringomycin and syringopeptin are synthesised by modular nonribosomal peptide synthetases. The syringomycin (*syr*) and syringopeptin (*syp*) gene clusters are located adjacent to one another on the chromosome and are estimated to be 55 kb and 90 kb in size, respectively. Here we describe the functional organisation of the clusters, along with flanking DNA regions, which altogether cover an approximately 155-kb region of the genome. The predominant feature of the toxin clusters is the occurrence of peptide synthetase genes for syringomycin (i.e., *syrB1* and *syrE*) and syringopeptin (i.e., *sypA*, *sypB*, and *sypC*). Associated with the toxin clusters are genes that are predicted to function in toxin secretion and regulation of biosynthesis genes. Progress in characterising functions of specific genes of the *syr-syp* cluster dedicated to syringomycin, syringopeptin, or both toxins is summarised.

Key words: phytotoxins, lipodepsipeptides, peptide synthetases, nonribosomal peptide synthesis.

Author for correspondence: D.C. Gross; e-mail: dgross@ppserver.tamu.edu.

N.S. Iacobellis *et. al.* (eds.), *Pseudomonas syringae* and Related Pathogens, 137-145.

1. INTRODUCTION

A distinctive feature of *Pseudomonas syringae* pv. *syringae* is the production of two classes of pore-forming lipodepsipeptide phytotoxins that target host plasma membranes. Recent genetic analysis of the biosynthetic mechanisms responsible for syringomycin and syringopeptin production by strain B301D of *P. s.* pv. *syringae* showed that both toxins are assembled on modular nonribosomal peptide synthetases. The syringomycin (*syr*) and syringopeptin (*syp*) gene clusters are located adjacent to one another on the B301D chromosome and are estimated to be 55 kb and 90 kb in size, respectively. The nucleotide sequence of the *syr* and *syp* gene clusters was completed. Here we describe the functional organisation of the clusters, along with flanking DNA regions, which altogether cover an approximately 155-kb region of the genome.

2. BIOSYNTHESIS OF SYRINGOMYCIN AND SYRINGOPEPTIN

The predominant feature of the toxin clusters (Fig. 1) is the occurrence of peptide synthetase genes for syringomycin (i.e., *syrB1* and *syrE*) and syringopeptin (i.e., *sypA*, *sypB*, and *sypC*). Each peptide synthetase contains one or more modules required for the addition of a specific amino acid to the peptide chain (Bender *et al.*, 1999; Guenzi *et al.*, 1998; Scholz-Schroeder *et al.*, 2001a; 2001b; 2003). As predicted from the toxin structures, a total of nine syringomycin and 22 syringopeptin modules were identified and shown to contain characteristic domains for condensation, aminoacyl adenylation, and thiolation of substrate amino acids. Associated with the toxin clusters are genes that are predicted to function in toxin secretion and regulation of biosynthesis genes. Three regulatory loci that encode members of the LuxR regulatory family were identified at the right border region of the *syr* gene cluster (Lu *et al.*, 2002).

2.1 Nonribosomal Peptide Synthetases

A thiotemplate mechanism of peptide synthesis is responsible for production of both syringomycin and syringopeptin (Bender *et al.* 1999; Grgurina and Mariotti, 1999; Guenzi *et al.* 1998). The nonribosomal mechanism occurs on multifunctional synthetases that catalyse the activation and addition of amino acids to form the specific peptide chain. Synthetases are composed of one or more amino acid activation modules that carry domains for condensation, aminoacyl adenylation, and thiolation.

The dominant feature of the *syr* gene cluster is *syrE*, which encodes a synthetase containing eight amino acid activation modules (Fig. 1). The ninth module is encoded by the *syrB1* synthetase gene and is located upstream of *syrE*. In ATP-PP_i exchange reactions SyrB1 was shown to recognise L-threonine as substrate, which indicated that SyrB1 is responsible for the addition of the last amino acid in syringomycin synthesis (Guenzi *et al.*, 1998). The *syrB2* gene encodes a protein required for syringomycin biosynthesis. Finally, the *syrC* gene, located between the *syrB1* and *syrE* synthetase genes, encodes an enzyme with a thioesterase motif (Bender *et al.*, 1999).

The major feature of the *syp* gene cluster is the occurrence of a DNA region (Scholz-Schroeder *et al.*, 2001a; 2001b; 2003) encoding the *sypA*, *sypB* and *sypC* synthetase genes that are 16.1, 19.3 and 40.6-kb in size, respectively (Fig. 1). The syringopeptin synthetases are predicted to encode 22 amino acid activation modules, with SypC carrying a total of 12 modules and representing the largest linear prokaryotic synthetase described to date. Syringopeptin production is completely disrupted by mutation of any one of the 22 modules. At the left border region of the *syp* cluster, ORF14 is predicted to encode a protein with significant homology to aminotransferases involved in the formation of the modified amino acid, 2,4-diaminobutyric acid. Accordingly, mutation of ORF14 leads to reduction of both syringomycin and syringopeptin production.

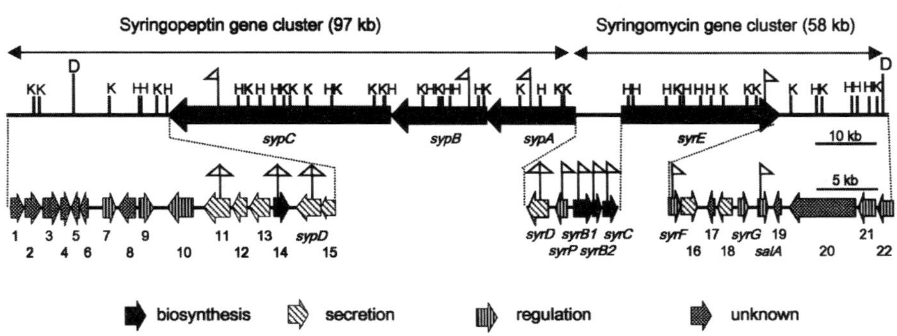

Figure 1. A physical map of a 155 kb genomic DNA region of *Pseudomonas syringae* pv. *syringae* strain B301D containing both syringomycin and syringopeptin gene clusters. The black solid, diagonally lined, and vertically lined arrows represent genes that are predicted to be involved in the biosynthesis, secretion, and regulation, respectively, of syringomycin and/or syringopeptin production. The arrows filled with diagonally hatched lines represent those genes for which the functions remain unknown. The numbers identify the potential opening reading frames (ORFs) that are being characterised. The phenotypic results of disruption of genes or ORFs are designated with a flag pointed right for loss of syringomycin production; a flag pointed left for loss of syringopeptin production; and a double-sided flag for loss of both syringomycin and syringopeptin production. The restriction enzyme sites for *Hin*dIII (H), *Kpn*I (K), and *Dra*I (D) are identified.

3. TYPE I SECRETION SYSTEMS ASSOCIATED WITH TOXIN PRODUCTION

Three prospective toxin efflux systems were identified in the *syr-syp* genomic island (Fig. 1), and phenotypic analysis of mutants showed a strong effect on production of both lipopeptide toxins. All three efflux systems exhibit characteristic components of Type I secretion that directly secrete the substrate into the surrounding environment of the bacterium. The efflux systems are representative of two secretion families, namely the ATP-binding cassette (ABC) superfamily and the Resistance-Nodulation-Division (RND) superfamily. Both secretion superfamilies have been associated with antibiotic efflux in Gram-negative bacteria (Borges-Walmsley and Walmsley, 2001), and are associated with secretion of the lipodepsipeptide toxins synthesised by enzymes encoded within the *syr-syp* genomic island.

3.1 ABC Transporters

Two genes encoding ABC transporters (i.e., *syrD* and *sypD*) were identified along with outer membrane and periplasmic elements of an efflux system. It appears that both SyrD (Quigley *et al.*, 1993) and SypD function as ATP-driven efflux pumps for toxin production across the cytoplasmic membrane. ORF15, which encodes a periplasmic efflux protein, and the *sypD* gene are located immediately downstream of the *sypC* synthetase gene (Fig. 1). A second periplasmic efflux protein is found at the right border region of the *syr* cluster together with an outer membrane efflux protein (Fig. 1; ORFs 18 and 16, respectively). The outer membrane protein shows significant homology to the OprM protein associated with multidrug efflux systems in *Pseudomonas aeruginosa* (Paulsen *et al.*, 1996), and likely functions as the secretion channel across the outer membrane of *P. syringae* pv. *syringae* cells. Although SyrD and SypD are predicted to serve as the inner efflux proteins in a tripartite transport complex, it is unknown if they interact with distinct periplasmic efflux and outer membrane efflux proteins. Nevertheless, strains carrying mutations in either the *syrD* or *sypD* genes are defective in production of both syringomycin and syringopeptin.

3.2 RND Transporters

At the left border region of the *syp* cluster, an efflux pump belonging to the RND superfamily of transporters was identified (Fig. 1; ORFs 11, 12, and 13). Components of the RND efflux pump are organised within an operon that encodes an outer membrane efflux protein (ORF13), a membrane fusion protein (ORF12), and finally an inner membrane efflux protein (ORF11) with 12 membrane-spanning structures characteristic of the RND family of transporters (Paulsen *et al.*, 1996). Furthermore, the protein encoded by ORF11 is homologous to multidrug efflux proteins of *Pseudomonas aeruginosa*, including MexD, which exhibit relatively broad substrate specificity. An RND mutant generated by insertion of the *nptII* cassette in ORF11 resulted in reduction of both lipopeptide toxins (H. Kang and D.C. Gross, unpublished).

4. REGULATORY GENES ASSOCIATED WITH THE *SYR-SYP* GENOMIC ISLAND

Analysis revealed nine loci in the *syr-syp* genomic island that were predicted to encode proteins that share significant homology to bacterial regulatory proteins or proteins involved in signal transduction (Fig. 1). Five of the nine loci were localised at the right border region of the *syr* gene cluster (*syrF*, *syrG*, *salA*, ORF21, and ORF22). One (*syrP*) was identified at the adjunct border between the *syr* and *syp* gene clusters. The other three (ORF7, ORF9, and ORF10) were localised at the right border of the *syp* gene cluster.

4.1 The *salA*, *syrF*, and *syrG* Genes at the Right Border Region of the *syr* Cluster

Three regulatory genes were identified at the border region of the *syr* gene cluster that code for members of the LuxR regulatory protein family. Mutations in the *syrF, syrG*, and *salA* genes were significantly reduced in virulence relative to the wild-type strain B301D, as indicated by lesion sizes in immature sweat cherry fruits (Lu *et al.*, 2002). A *salA* mutant failed to produce syringomycin, whereas *syrF* and *syrG* mutants produced 12 and 50%, respectively, of syringomycin relative to the wild-type strain (Fig. 2).

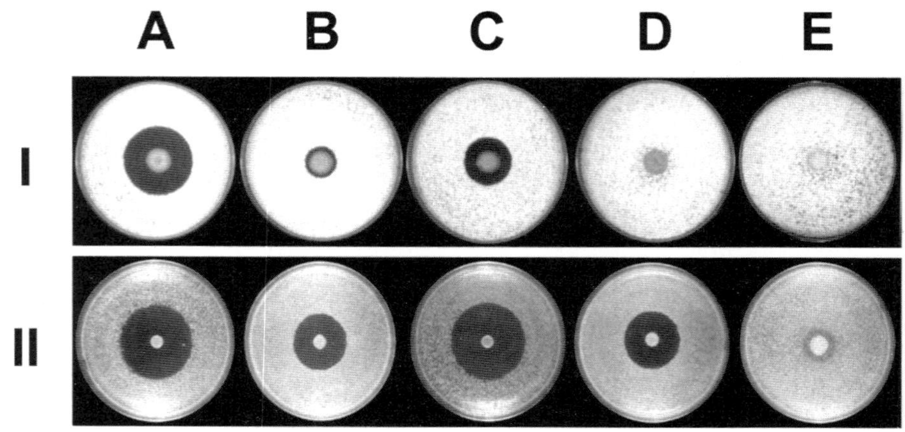

Figure 2. Plate bioassays for syringomycin and syringopeptin production by *Pseudomonas syringae* pv. *syringae* strains. (I): Strains of *P. s.* pv. *syringae* were incubated on PDA plates for 4 days and oversprayed with *Geotrichum candidum*. AI, B301D (parental strain); BI, B301DSL1 (*syrF* mutant); CI, B301DSL6 (*syrG* mutant); DI, B301DSL7 (*salA* mutant); EI, BR132 (*syrB1* mutant). (II): The *syrB1* mutant derivatives of *P. s.* pv. *syringae* were incubated on PGNA plates for 4 days and oversprayed with *Bacillus megaterium*. AII, B301DSL8 (*syrB1* mutant); BII, B301DSL10 (*syrF-syrB1* double mutant); CII, B301DSL11 (*syrG-syrB1* double mutant); DII, B301DSL12 (*salA-syrB1* double mutant); EII, BR-DBL1 (*sypA-syrB1* double mutant).

Investigations using the *uidA* reporter gene revealed that both *salA* and *syrF* are essential for expression of *syrB1*, a biosynthesis gene for syringomycin. Furthermore, it was demonstrated that the *salA* gene is located downstream of the *gacS/gacA* regulators, but upstream of the *syrF* gene in the regulatory hierarchy for syringomycin. Mutations in the *syrF* and *salA* genes resulted in only a slight reduction of syringopeptin production. No effect on syringopeptin production was observed for a *syrG* mutant on PGNA medium. These results suggest that regulation of syringopeptin biosynthesis might be distinct from that of syringomycin.

4.2 SyrP

The *syrP* gene, which was predicted to encode a histidine kinase, was identified between *syrB1* and *syrD* (Fig. 1). A *syrP* mutant exhibited an unusual pleiotropic phenotype and was substantially less virulent than the wild-type strain B301D in cherry fruits (Zhang *et al.*, 1997). In particular, syringomycin production by the *syrP* mutant is relatively insensitive to high inorganic phosphate concentrations in agar plates. These results indicated that SyrP might be involved in a phosphorylation cascade controlling syringomycin production. Additionally, no effect of the *syrP* mutation was observed on expression of a *sypA-uidA* fusion.

4.3 Other Potential Regulatory Genes in the Genomic Island

An additional five ORFs were predicted to encode proteins that might be involved in a signal transduction cascade controlling toxin production (Fig. 1). At the right border of the *syr* gene cluster, ORF21 and ORF22 encode two proteins that share 37 and 27% identity to a member of the GGDEF family and a McpA chemotaxis protein (Lu *et al*., 2002), respectively. At the left border of the *syp* gene cluster, ORF7, ORF9, and ORF10 were predicted to encode three proteins that show significant similarity to a chemotaxis protein, sensor protein, and serine acyltransferase, respectively. The functions of the unknown loci are under investigation.

5. CONSERVATION OF THE *SYR-SYP* REGION IN RELATED STRAINS OF *PSEUDOMONAS SYRINGAE*

The *syr-syp* genomic island is conserved in *Pseudomonas syringae* pv. *syringae* strains B301D and B728a, which were originally isolated from pear and bean, respectively. A draft of the genomic sequence of strain B728a was released by the Joint Genome Institute and is available for BLAST analysis (http://www.jgi.doe.gov). Preliminary sequence analysis revealed that the genomic island from ORF7 to the left end of the *syr* gene cluster of strain B301D contains the same genes and ORFs in the same organisation as strain B728a (Fig. 1). The genes and ORFs identified from the genomic islands of the two strains exhibited approximately 98% identity at the nucleotide level. An especially interesting observation was a DNA region (approximately 3 kb in size) at the left border of the genomic island of strain B301D that was not identified in the genomic sequence data available for strain B728a. The unique region of strain B301D was predicted to contain six ORFs of unknown function (N. Wang, S.-E. Lu, and D.C. Gross, unpublished). Investigations of the effect of the unique region on toxin production are in progress.

Production of syringomycin and syringopeptin was not detected for representative strains of *P. s.* pv. *tomato* (Quigley and Gross, 1994). BLAST analysis against the genomic sequence of strain DC3000 (http//www.tigr.org) demonstrated that, as expected, the entire 155-kb *syr-syp* region was not present in the genome of this strain.

6. CONCLUSION

Characterisation of the *syr-syp* genomic island provides a valuable foundation for understanding the nonribosomal peptide synthetase mechanisms dedicated to syringomycin and syringopeptin biosynthesis, efflux systems required for toxin secretion, and the regulatory network responsible for expression of toxigenesis in the plant-pathogen interaction. The facts that the *syr-syp* region constitutes more than 2% of the predicted genome of strain B301D and that the genomic island is highly conserved in strains of *Pseudomonas syringae* pv. *syringae* underscore the importance of toxigenesis to pathogenicity.

ACKNOWLEDGEMENTS

This work was supported by grant 2001-35319-10400 from the National Research Initiative Competitive Grants Program of the U.S. Department of Agriculture, Science and Education Administration to D. Gross. This work was also supported, in part, by grants of the University of Rome "La Sapienza" to I. Grgurina.

REFERENCES

Bender, C.L., Alarcón-Chaidez, F., and Gross, D.C., 1999, *Pseudomonas syringae* phytotoxins: Mode of action, regulation and biosynthesis by peptide and polyketide synthetases. *Microbiol. Mol. Biol. Rev.* **63**: 266-292.

Borges-Wamsley, M.I., and Walmsley, A.R., 2001, The structure and function of drug pumps. *Trends Microbiol.* **9**: 71-79.

Grgurina, I., and Mariotti, F., 1999, Biosynthetic origin of syringomycin and syringopeptin 22, toxic secondary metabolites of the phytopathogenic bacterium *Pseudomonas syringae* pv. *syringae*. *FEBS Lett.* **462**: 151-154.

Guenzi, E., Galli, G., Grgurina, I., Gross, D.C., and Grandi, G, 1998, Characterization of the syringomycin synthetase gene cluster: A link between prokaryotic and eukaryotic peptide synthetases. *J. Biol. Chem.* **273**: 32857-32863.

Lu, S.-E., Scholz-Schroeder, B.K., and Gross, D.C., 2002, Characterization of the *salA*, *syrF*, and *syrG* regulatory genes located at the right border of the syringomycin gene cluster of *Pseudomonas syringae* pv. *syringae*. *Mol. Plant-Microbe Interact.* **15**: 43-53.

Paulsen, I.T., Brown, M.H., and Skurray, R.A., 1996, Proton-dependent multidrug efflux systems. *Microbiol. Rev.* **60**: 575-608.

Quigley, N.B., and Gross, D.C., 1994, Syringomycin production among strains of *Pseudomonas syringae* pv. *syringae*: Conservation of the *syrB* and *syrD* genes and activation of phytotoxins production by plant signal molecules. *Mol. Plant-Microbe Interact.* **7**: 78-90.

Quigley, N.B., Mo, Y.-Y., and Gross, D.C., 1993, SyrD is required for syringomycin production by *Pseudomonas syringae* pathovar *syringae* and is related to a family of ATP-binding secretion proteins. *Mol. Microbiol.* **9**: 787-801.

Scholz-Schroeder, B.K., Hutchison, M.L., Grgurina, I., and Gross, D.C., 2001a, The contribution of syringopeptin and syringomycin to virulence of *Pseudomonas syringae* pv. *syringae* strain B301D on the basis of *sypA* and *syrB1* biosynthesis mutant analysis. *Mol. Plant-Microbe Interact.* **14**: 336-348.

Scholz-Schroeder, B.K., Soule, J.D., Lu, S.-E., Grgurina, I., and Gross, D.C, 2001b, A physical map of the syringomycin and syringopeptin gene clusters localised to an approximately 145-kb DNA region of *Pseudomonas syringae* pv. *syringae* strain B301D. *Mol. Plant-Microbe Interact.* **14**: 1426-1435.

Scholz-Schroeder, B.K., Soule, J.D., and Gross, D.C., 2003, The *sypA*, *sypB*, and *sypC* synthetase genes encode twenty-two modules involved in the nonribosomal peptide synthesis of syringopeptin by *Pseudomonas syringae* pv. *syringae* B301D. *Mol. Plant-Microbe Interact.* **16**: 271-280.

Zhang, J.-H., Quigley, N.B., and Gross, D.C, 1997, Analysis of the *syrP* gene, which regulates syringomycin synthesis by *Pseudomonas syringae* pv. *syringae*. *Appl. Environ. Microbiol.* **63**: 2771-2778.

Regulation and Detection of Effectors Translocated by *Pseudomonas syringae*

S.W. HUTCHESON, J.R. BRETZ, J.C. CHARITY, L. LOSADA and T. SUSSAN
Dept. Cell Biology and Molecular Genetics, Microbiology Bldg, University of Maryland, College Park, MD 20742, USA.

Abstract: *Pseudomonas syringae* strains express a *hrp*-encoded type III protein secretion system (TTSS) during pathogenesis that injects effector proteins into the cytoplasm of host cells. A complex regulatory system controlling *hrp* regulon expression has recently been characterised that involves positive and negative regulatory factors. The interaction of these regulatory factors and their roles are discussed. Several genetic screens for effectors that employ components of the regulatory system are presented.

Key words: Type III protein secretion, pathogenesis, gene regulation, genomics.

Author for correspondence: S.W. Hutcheson; e-mail: sh53@umail.umd.edu.

1. INTRODUCTION

While many virulence factors contribute to the ability of *Pseudomonas syringae* strains to cause disease in host plants, an underlying property is their ability to colonise tissue of the host. Colonisation of the host requires the ability to obtain the nutrients and water necessary for growth and to evade the cellular defences of the host. The ability of *P. syringae* strains to colonise tissue can now be attributed in part to the activities of a type III protein secretion system (TTSS) encoded by the *hrp* pathogenicity island (PAI) (Alfano *et al.*, 2000; Hutcheson, 1999). Type III protein secretion systems are complex systems that essentially inject effector proteins into host cells (Galan and Collmer, 1999). These effectors serve two known purposes during compatible interactions with susceptible hosts. They are

N.S. Iacobellis *et. al.* (eds.), *Pseudomonas syringae* and Related Pathogens, 147-156.

thought to alter the membrane permeability of the host cell to favour leakage of nutrients (Lee *et al.,* 2001) and they appear to disrupt cellular signalling pathways involved in defence responses (Chen *et al.,* 2000). In essence, the pathogen engineers its host to be a host. Resistant plants, on the other hand, are able to recognise one or more of these same effectors to initiate defence responses to prevent further colonisation of the tissue (Dangl and Jones, 2001; Pirhonen *et al.*, 1996). One of the most surprising advances in recent years has been the realisation that the recognition process leading to the initiation of a defence response in resistant plants occurs in the cytoplasm of host cells for many bacterium/host interactions. Thus the effectors translocated into host cells appear to define the pathogenicity and host range of that strain (Hutcheson, 2001).
The assembly and activity of the TTSS in *P. syringae* requires the co-ordinated expression of 4 operons that encode components of the TTSS (Hutcheson, 1999), such as the secretion complex and the *hrp* pilus (Jin and He, 2002), together with many independently transcribed genes for effectors or accessory proteins (Kim *et al.,* 1998). Analysis of the DC3000 (Fouts *et al.,* 2002) and B728A genomes (Greenberg *et al.*, 2002) suggest that as many as sixty genes for effectors may exist in each genome. The genes encoding the TTSS and effectors form the *hrp* regulon that is environmentally regulated. Expression of the *hrp* regulon is low during growth in rich media but is induced during pathogenesis or by acidic minimal salts media (Rahme *et al.,* 1992; Xiao *et al.,* 1992). Below we describe recent progress elucidating the mechanisms involved in the regulation of TTSS in *P. syringae* strains and how we have utilised this information to develop novel methods to identify effectors.

2. THE REGULATORY SYSTEM CONTROLLING EXPRESSION OF THE *HRP* REGULON

A complex regulatory system regulates expression of the *hrp* regulon in *Pseudomonas syringae* strains through the activity of positive and negative regulatory factors. Early experiments showed that the activity of a *hrp* regulon promoter fusion was dependent upon three genes, *hrpR*, *hrpS* and *hrpL* (Xiao *et al.,* 1994). Sequence analysis revealed that *hrpR* and *hrpS* encode unusual enhancer binding proteins and *hrpL* produces an alternative sigma factor. HrpR and HrpS were unusual because the modulatory receiver domains found in most other members of the protein family appeared to be absent from both proteins (Osuna *et al.,* 1997). Such truncated proteins, when constructed in the laboratory, are constitutively active (Lee *et al.,* 1994). Subsequent experiments showed that HrpR and HrpS specifically activated the *hrpL* promoter in *Escherichia coli* and *P. syringae* backgrounds

(Xiao *et al.*, 1994). No other accessory proteins, like a protein kinase, were required for this activity. The two proteins acted synergistically with each other. Partial activation of the sigma 54-dependent *hrpL* promoter was induced by HrpS (Grimm *et al.,* 1995; Hutcheson *et al.,* 2001) but maximal activation required both HrpR and HrpS (Hutcheson *et al.*, 2001; Xiao *et al.*, 1994). A physical interaction between HrpR and HrpS could be detected (Hutcheson *et al.*, 2001).

These observations suggested that a regulatory cascade controls the expression of the *hrp* regulon in which expression of the *hrpRS* operon would be predicted to regulate the levels of HrpR and HrpS in the cell. HrpR and HrpS, in turn, would form a complex to activate the *hrpL* promoter. HrpL then would functions as a sigma factor to direct transcription of the genes for the TTSS and the various effectors produced by these strains. If this were the case, expression of the *hrpRS* operon would control the expression of the remaining regulon. Analysis of *hrpRS* promoter activity under non-inductive and inductive conditions instead revealed that the *hrpRS* operon was constitutively expressed. Since the *hrpL* promoter was environmentally regulated (Bretz *et al.,* 2002) and HrpR and HrpS were constitutively active once expressed, some form of negative regulation must be affecting the ability of HrpR and/or HrpS to activate the *hrpL* promoter. By isolating and characterising constitutive expression mutants, a strong homolog of Lon, a well known regulatory ATP-dependent protease of many eubacteria (Gottesman, 1996), was identified as a negative regulator of *hrpL* expression (Bretz *et al.*, 2002). Since Lon is a protease, it likely functioned to degrade one or more of the transcriptional factors required for activity of the *hrpL* promoter. Levels of HrpR, but not HrpS, were found to be unstable in Lon^{+} cells but were stable in Lon^{-} cells (Bretz *et al.*, 2002).

We next examined whether the Lon-associated degradation of HrpR was affected by the environmental conditions known to affect expression of the *hrp* regulon. HrpR was unstable when cells were grown under non-inductive conditions (any rich medium) but was stable when cells were grown under *hrp* inductive conditions. Since Lon protease is constitutively expressed (Bretz and Hutcheson, unpublished results), the observed changes in Lon activity must have occurred post-transcriptionally. One possibility is that the activity of Lon protease is environmentally regulated by physiological conditions of the cell, like reported recently for *E. coli* (Kuroda *et al.,* 2001). Alternatively, the apparent Lon-mediated degradation of HrpR is reduced through stabilisation of HrpR under these conditions. Irrespective of the mechanism, the Lon-associated degradation of HrpR provides a mechanism through which the physiological state of the cell can influence expression of the *hrp* regulon.

HrpV, a product of the HrpL-dependent *hrp*C operon, was shown to be another negative regulator of the *hrp* regulon (Preston *et al.,* 1998). HrpV is a small acidic polypeptide that lacks obvious homologs or known domains.

Insertional inactivation of HrpV enhanced the expression of the *hrp* regulon whereas ectopic expression of HrpV reduced expression. The repressive activity of HrpV could be suppressed by directed expression of the *hrpRS* operon, thereby suggesting that HrpV regulates *hrpRS* expression (Preston *et al.*, 1998). This, however, was inconsistent with the observed constitutive expression of the *hrpRS* operon described above. When the effects of ectopic expression of HrpV and *hrpV::kan* on specific promoters were examined directly, HrpV, like Lon, specifically affected the activity of the *hrpL* promoter. An alternative explanation for the suppressive effect of *hrpRS* expression on HrpV activity, however, could be the physical interaction of HrpV with HrpR and/or HrpS. Using a yeast two-hybrid assay, HrpV was shown to specifically interact with HrpS (Sussan and Hutcheson, unpublished results). This observation was confirmed by co-purification experiments. Interestingly, HrpV mutants also exhibit altered pathogenicity (Preston *et al.*, 1998), suggesting that, in agreement with the observations of Grimm *et al* (1995), HrpV could function in the HrpS-dependent transduction of one or more signals generated during the interaction with the host.

Sequence analysis of the *hrpRS* promoter from several strains suggests that a third level of control for the *hrpL* promoter may, in fact, occur at the level of the *hrpRS* promoter as originally postulated. Significant differences in the sequence of upstream regulatory regions for the *hrpRS* promoter have been detected in different strains (Bretz, Sussan and Hutcheson, unpublished results). This opens the possibility that another regulatory system influences the expression of this promoter that, in turn, would affect the level of HrpR and HrpS in cells and *hrpL* expression.

We conclude that expression of the *hrp* regulon is controlled through three distinct mechanisms acting to modulate the activity of the *hrpL* promoter (Fig. 1). Differences in *hrpRS* promoter sequences open the possibility of strain-specific regulation of *hrpRS* expression. The Lon-associated turnover of HrpR provides a mechanism to link the physiology of the cell to the expression of the *hrp* regulon. Lon activity would provide an adaptable threshold for level of *hrpRS* expression required before *hrpL* promoter activity is induced. Under repressive conditions, HrpR turnover would be comparatively rapid, thereby preventing the formation of the HrpR/HrpS regulatory complex. This restriction, however, could be overcome by elevated expression of *hrpRS*. Under inductive conditions, Lon-associated degradation of HrpR is minimal, and much lower levels of *hrpRS* expression would be required to activate the *hrpL* promoter. HrpL would then drive the expression of the remaining *hrp* regulon and allow the accumulation of HrpV. The HrpV-mediated regulation of HrpS activity could provide the means to regulate expression of effectors, and possibly, activity of the TTSS, during pathogenesis. A mechanism to inactivate HrpV would be required during pathogenesis to allow further expression of the *hrp* regulon.

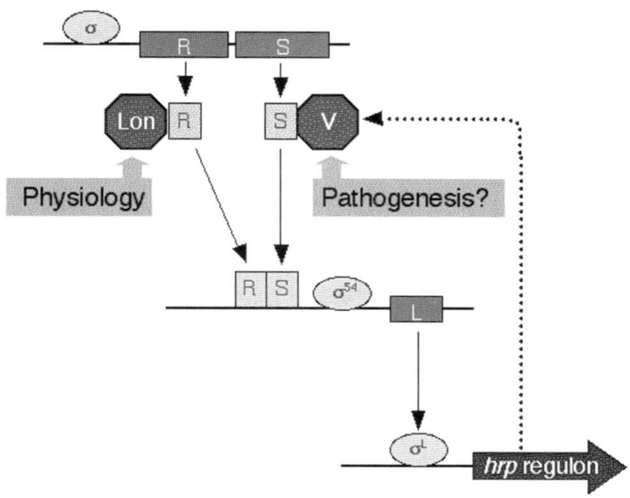

Figure 1. The regulatory system controlling *hrp* regulon expression in *Pseudomonas syringae* strains. Function of system components is described in the text.

3. UTILIZATION OF THE *HRP* REGULATORY SYSTEM TO IDENTIFY TRANSLOCATED EFFECTORS

3.1 HrpL-Dependent Promoter Trap Screen for Effector Genes

The elucidation of the *hrp* regulatory system has provided a number of useful tools that can be used to identify translocated effectors. For example, one unifying feature of all known *Pseudomonas syringae* genes for TTSS-dependent effectors is the requirement of HrpL for expression. This opens the possibility of screening genomes of *P. syringae* strains for translocated effectors by identifying the genes transcriptionally dependent upon HrpL.

To screen *P. syringae* genomes for HrpL-dependent promoters, we first developed a system for conditional expression of *hrpL*. A tightly regulated *hrpL* expression system was constructed using a modified AraC-linked expression system that restricted expression of *hrpL* in the absence of arabinose. When known HrpL-dependent promoters fused to '*lacZYA* were transformed into an *Escherichia coli* strain carrying this regulated *hrpL* construct, strains exhibited a Lac^+ phenotype only in the presence of arabinose. Thus, the Lac phenotype of the cells could be used to identify transformants carrying HrpL-dependent promoters. Plated *E. coli* cells are then scored for arabinose-dependent Lac phenotypes on indicator media after transformation with a genomic library of the strain of interest. Candidate *hrpL*-dependent promoters are amplified by PCR and nucleotide sequence is obtained and screened for homologs (manuscript in review).

We have applied this promoter trap screen to two strains thus far. A genomic library of Pss61, carrying 1-2 kb fragments, was screened for HrpL-dependent promoters in *E. coli* transformants using the aforementioned HrpL-dependent promoter trap screen. Each of the known HrpL-dependent promoters of the *hrp* PAI were detected in this genomic library except for the *hrpU* operon promoter (Bretz *et al*., 2002). The promoters for homologs of *hrpW* and *avrE*, which had been identified in the conserved effector locus of other strains (Alfano *et al*., 2000), were also present in this library. In addition, the promoters for 9 candidate effectors were detected. One of these genes has been shown to encode a 20 kDa protein that reduced the virulence of DC3000. The promoter of this gene carried a HrpL-dependent promoter consensus sequence and required HrpL for activity in *P. syringae*. Fusions to *avrRpt2* elicited *hrp*-dependent programmed cell death in $RPS2^+$ *Arabidopsis thaliana.* These results are consistent with the translocation of this candidate effector into host cells. As such, this effector has been designated as HopPsyL (manuscript in review).

A similar screen of the DC3000 genome led to the identification of two unusual alleles of *avrPphD*. A fragment was isolated from a genomic library of DC3000 that carried a strong candidate HrpL-dependent promoter, and subsequently, was found to encode an allele of *avrPphD* (Bretz *et al*., 2002). Further sequence analysis showed that this allele carried an IS*52* element internal to the coding sequence, indicating that this allele was probably inactive. To determine if the DC3000 genome carried a wild-type allele, the amino terminal region of the *avrPphD*::IS*52* allele was used as a probe and two additional alleles of *avrPphD* were identified in the DC3000 genome. One allele was a strong homolog to the wild-type *avrPphD* and exhibited greater than 90% identity with the *P. syringae* pv. *phaseolicola avrPphD* over its entire length. The other allele showed strong sequence similarity to *avrPphD* at the amino terminal 174 aa that carries the TTSS secretion signals, but exhibited little similarity to *avrPphD* in the carboxy terminal 298 aa. This carboxy terminal region, however, exhibited similarity to tyrosine phosphatases that have been identified as translocated effectors of several mammalian pathogens. This candidate effector was shown to be translocated into host cells by using '*avrRpt2* fusions and to encode tyrosine phosphatase activity (manuscript in press).

3.2 Detection of Effectors in the Exchangeable Effector Locus

During the promoter trap screen of the Pss61 genome, we also isolated the promoter for the *shcA-hopPsyA* operon together with upstream sequences. These sequences carried *queA* and the gene for $tRNA_{leu}$ as had been reported for DC3000 and B728A (Alfano *et al.*, 2000), that lie outside of the *hrp* PAI. This allowed us to use conserved sequences internal to *queA* and *hrpK*, a gene of the *hrp* PAI conserved central region, to design primers for amplification of the exchangeable effector loci (EEL) from a variety of strains. A distinct EEL had been identified in each strain characterised thus far (Alfano *et al.*, 2000; Hutcheson, 1999), suggesting that the EEL was a likely source for novel effectors. Using these primers, we were able to amplify and characterise the EEL from more than 30 strains (Charity *et al.,* 2003). Ten unique types of EEL were identified, 7 of which had not been described previously. These newly characterised EEL included the genes for 6 candidate effectors, and 2 potential accessory genes. These candidate effectors were expressed as part of the *hrp*K operon and retained most features of TTSS-dependent effectors. Partial homologs of HopPsyA, AvrPphE, AvrPphC, HopPtoB, and AvrXct were identified that differed significantly in amino acid sequence and/or length of the gene product.

Several gene cassettes were identified that were conserved between different types of EEL, but were found in different combinations to form the various types of EEL observed. The presence of these conserved gene cassettes suggests that the EEL formed by integron-like assembly of transposed gene cassettes. The DC3000 genome was found to carry as many as 5 alleles of a tRNA-dependent integrase (Hutcheson, unpublished results) and tRNA-linked excision of gene cassettes has been reported in *Pseudomonas syringae* strains (Jackson *et al.,* 2000), suggesting that these integrases could be active in transposition of the cassette that form the EEL.

3.3 Identification of Hypersecretion Mutants of *Pseudomonas syringae*

When our *Pseudomonas syringae* ΔLon mutants were screened for plant responses, we noticed that several plant responses occurred more rapidly (Bretz *et al*., 2002). Culture filtrates of Pss61 *lon*::Tz mutants were found to contain at least 12 new proteins that were not present in culture filtrates of the wild-type bacteria nor *hrp* secretion mutants (Lisboa and Hutcheson, unpublished results). To determine if any of these apparently novel proteins could be secreted effectors, we examined the secretion of a known effector, AvrPto, from the Lon mutants. AvrPto could be detected in culture filtrates of the Lon mutants but was essentially undetectable in culture filtrates of the wild-type bacterium (Bretz *et al*., 2002). The enhanced secretion of AvrPto was subsequently found to be linked to the enhanced expression of the *hrp* TTSS and to stabilisation of AvrPto prior to secretion. Preliminary results suggest that *P. syringae* Lon mutants are hyper-piliated consistent with enhanced expression of the *hrp* TTSS (Lisboa and Hutcheson, unpublished results). The half-life of AvrPto was increased markedly in the Lon mutants (Losada and Hutcheson, unpublished results). This suggests that Lon modulates expression of the *hrp* TTSS by its effect on HrpR stability and affects the activity of the TTSS by controlling the level of translocated effectors in bacterial cells prior to secretion.

4. CONCLUDING REMARKS

The regulation of the *Pseudomonas syringae hrp* regulon is obviously complex and responds to physiological signals linked to the growth conditions as well as to postulated signals generated during pathogenesis. The regulatory system described above provides a mechanism to transduce these disparate signals to regulate *hrp* expression. This process, however, appears to involves more components than reported here. For example, we have isolated several mutants that carry wild-type copies of all of the regulatory genes described above but exhibit altered expression of the *hrp* regulon. Identification of these components should help elucidate the signal transduction cascades functioning in *P. syringae* strains, and ultimately, the signals initiating *hrp* expression during pathogenesis.

Although our knowledge of the regulatory system controlling *hrp* expression is still incomplete, several useful tools have been identified that can be exploited to improve our understanding of *P. syringae* interactions with its hosts. Identification of HrpL-dependent promoters can be used to identify the effectors produced by a strain. Constitutive expression mutants appear to be derepressed for effector secretion and proteomics can now be applied to the identification of the effectors produced by these mutants. Manipulation of the regulatory system can facilitate the identification of structural elements of the TTSS and assist the elucidation of early events in host responses to effectors.

ACKNOWLEDGEMENTS

This work was supported by grants MCB9729524 and MCB0215417 from the National Science Foundation. We thank X. Wei for her assistance with our studies on *hrpV,* K. Lisboa for her isolation of *hrp* constitutive mutants and I. Rozenbaum, K. Pak and S. Zeyad for their assistance with the promoter trap screen.

REFERENCES

Alfano, J.R., Charkowski, A.O., Deng, W.-L., Badel, J.L., Petnicki-Ocwieja, T., van Dijk, K., and Collmer, A., 2000, The *Pseudomonas syringae* Hrp pathogenicity island has a tripartite mosaic structure composed of a cluster of type III secretion genes bound by exchangeable effector and conserved effector loci that contribute to parasitic fitness and pathogenicity in plants. *Proc. Natl. Acad. Sci. USA* **97**: 4856-4861.

Bretz, J., Lisboa, K., Thareja, N., Losada, L., and Hutcheson, S.W., 2002, Lon protease functions as a negative regulator of type III protein secretion in *Pseudomonas syringae. Mol. Microbiol.* **45**: 397-409.

Charity, J., Pak, K.C., Delwiche, and Hutcheson, S., 2003, Novel exchangeable effector loci associated with the *hrp* pathogenicity island of *P. syringae*: evidence for integron-like assembly from transposed gene cassettes. *Mol. Plant-Microbe Interact.* **16**: (in press).

Chen, Z., Kloek, A.P., Boch, J., Katagiri, F., and Kunkel, B.N., 2000, The *Pseudomonas syringae avrRpt2* gene product promotes pathogen virulence from inside plant cells. *Mol. Plant-Microbe Interact.* **13**: 1312-1321.

Dangl, J., and Jones, J., 2001, Plant pathogens and integrated defence responses to infection. *Nature* (London) **411**: 826-833.

Fouts, D., Abramovitch, R., Alfano, J., Baldo, A., Buell, C., Cartinhour, S., Chatterjee, A., D'Ascenzo, M., Gwinn, M., Lazarowitz, S., Lin, N.-C., Martin, G., Rehm, A., Schneider, D., vanDijk, K., Tang, X., and Collmer, A., 2002, Genomewide identification of *Pseudomonas syringae* pv *tomato* DC3000 promoters controlled by the HrpL alternative sigma factor. *Proc. Natl. Acad. Sci. USA* **99**: 2275-2280.

Galan, J.E., and Collmer, A., 1999, Type III secretion machines: bacterial devices for protein delivery into host cells. *Science* **284**: 1322-1328.

Gottesman, S., 1996, Proteases and their targets in *Escherichia coli. Annu. Rev. Genet.* **30**: 465-506.

Greenberg, J.,T., Vinatzer B.N., and Jelenska J., 2002, Genetic and genomic analysis of *Pseudomonas syringae* effectors. Book of Abstracts, 6th International Conference on *Pseudomonas syringae* Pathovars and Related Pathogens, September 15 – 19, 2002, Acquafredda di Maratea (PZ), Italy, p. 23.

Grimm, C., Aufsatz, W., and Panopoulos, N.J., 1995, The *hrpRS* locus of *Pseudomonas syringae* pv. *phaseolicola* constitutes a complex regulatory unit. *Mol. Microbiol.* **15**: 155-165.

Hutcheson, S.W., 1999, The *hrp* cluster of *Pseudomonas syringae*: a pathogenicity island encoding a type III protein translocation complex?, p. 309-329. *In:* J. B. Kaper and J. Harker (ed.), Pathogenicity islands and other mobile virulence elements. Amer. Soc. Microbiol. Washington, D.C.

Hutcheson, S.W., 2001, The molecular biology of hypersensitivity to plant pathogenic bacteria. *J. Plant Pathol.* **83**: 151-172.

Hutcheson, S.W., Bretz, J., Sussan, T., Jin, S., and Pak, K., 2001, The enhancer binding protein HrpR and HrpS interact to regulate *hrp*-encoded type III protein secretion in *Pseudomonas syringae* strains. *J. Bacteriol.* **183**: 5589-5598.

Jackson, R., Mansfield, J., Arnold, D., Sesma, A., Paynter, C., Murillo, J., Taylor, J., and Vivian, A., 2000, Excision from tRNA genes of a large chromosomal region, carrying *avrPphB*, associated with race change in the bean pathogen, *Pseudomonas syringae* pv. *phaseolicola. Mol. Microbiol.* **38**: 186-197.

Jin, Q., and He, S., 2002, Role of the Hrp pilus in type III protein secretion in *Pseudomonas syringae. Science* **294**: 2556-2558.

Kim, J.F., Charkowski, A.O., Alfano, J.R., Collmer, A., and Beer, S.V., 1998, Sequences related to transposable elements and bacteriophages flank avirulence genes of *Pseudomonas syringae*. *Mol. Plant-Microbe Interact.* **11**: 1247-1252.

Kuroda, A., Nomura, K., Ohtomo, R., Kato, J., Ideda, T., Takiguchi, N., Ohtake, H., and Kornberg, A., 2001, Role of inorganic polyphosphate in promoting ribosomal protein degradation by the Lon protease in *E. coli*. *Science* **293**: 705-708.

Lee, J., Klusener, B., Tsiamis, G., Stevens, C., Neyt, C., Tampakaki, A.P., Panopoulos, N.J., Noller, J., Weiler, E., Cornelis, G.R., Mansfield, J.W. and Nurnberger, T., 2001, $HrpZ_{Psph}$ from the plant pathogen *Pseudomonas syringae* pv. *phaseolicola* binds to lipid bilayers and forms an ion-conducting pore *in vitro*. *Proc. Natl. Acad. Sci. USA* **98**: 289-294.

Lee, J.L., Scholl, D., Nixon, B.T., and Hoover, T., 1994, Constitutive ATP hydrolysis and transcription activation by a truncated form of *Rhizobium meliloti* DctD, a σ^{54}-dependent transcriptional activator. *J. Biol. Chem.* **269**: 20401-20409.

Osuna, J., Soberon, X., and Morett, E., 1997, A proposed architecture for the central domain of the bacterial enhancer-binding proteins based on secondary structure prediction and fold recognition. *Prot. Sci.* **6**: 543-555.

Pirhonen, M.U., Lidell, M.C., Rowley, D., Lee, S.W., Silverstone, S., Liang, Y., Keen, N.T. and Hutcheson, S.W., 1996, Phenotypic expression of *Pseudomonas syringae avr* genes in *E. coli* is linked to the activities of the *hrp*-encoded secretion system. *Mol. Plant-Microbe Interact.* **9**: 252-260.

Preston, G., Deng, W.-L., Huang, H.-C. and Collmer, A., 1998, Negative regulation of *hrp* genes in *Pseudomonas syringae* by HrpV. *J. Bacteriol.* **180**: 4532-4537.

Rahme, L.G., Mindronos, M.N., and Panopoulos, N.J., 1992, Plant and environmental sensory signals control the expression of *hrp* genes in *Pseudomonas syringae* pv. *phaseolicola*. *J. Bacteriol.* **174**: 3499-3507.

Xiao, Y., Lu, Y., Heu, S., and Hutcheson, S.W., 1992, Organisation and environmental regulation of the *Pseudomonas syringae* pv. *syringae* 61 *hrp* cluster. *J. Bacteriol.* **174**: 1734-1741.

Xiao, Y., Heu, S., Yi, J., Lu, Y. and Hutcheson, S.W., 1994, Identification of a putative alternate sigma factor and characterisation of a multicomponent regulatory cascade controlling the expression of *Pseudomonas syringae* pv. *syringae* Pss61 *hrp* and *hrmA* genes. *J. Bacteriol.* **176**: 1025-1036.

Genetic Organisation and Proposed Function of the *hrpM* Locus of *Pseudomonas syringae* pv. *syringae*

R. SHIMIZU, B.W. RUSSELL[#] and D. MILLS[§]

Department Plant Pathology, Cornell University, Ithaca, New York, USA; [#]Chemica Synthesis, LLC, 325 SW Cyber Drive, Bend, Oregon, USA; [§]Department of Botany and Plant Pathology, Oregon State Univ., Corvallis, Oregon, USA.

Abstract: A fully functional *hrpM* gene of the *hrpM* locus of *Pseudomonas syringae* pv. *syringae* is required for expression of brown spot disease symptoms in susceptible bean cultivars. *orf1* and *hrpM* genes of the *hrpM* locus have considerable homology with *mdoG* and *mdoH*, respectively, of the *mdoA* locus of *Escherichia coli.* The *mdoA* locus is essential for the synthesis of membrane-derived oligosaccharides found in the periplasmic space. A new, revised genetic organisation of *orf1* and *hrpM* is presented based on additional nucleotide sequencing and comparative homologies with the *mdoA* locus. The function of *hrpM* is inferred based on genetic complementation of the *hrpM* mutation with the *mdoA* locus.

Key words: *hrpM*, pathogenicity, glucosyl transferase.

Author for correspondence: D. Mills; e-mail: millsd@science.oregonstate.edu.

1. INTRODUCTION

Brown spot disease of bean (*Phaseolus vulgaris* L) is characterised by small necrotic lesions on the leaves and pods that result from infection by *Pseudomonas syringae* pv. *syringae*. Genes affecting pathogenicity of this pathovar (Anderson and Mills, 1985) and other pathovars of *P. syringae* have been identified by insertion mutagenesis (Lindgren *et al*., 1986; Willis *et al*., 1991; Huang, *et al*., 1995). One class of mutant was termed *hrp* (Lindgren *et al*., 1986), because mutations eliminated the typical hypersensitive reaction (HR) on non host plants and pathogenicity to

N.S. Iacobellis *et. al.* (eds.), *Pseudomonas syringae* and Related Pathogens, 157-164.

susceptible host plants (Path⁻). A cluster of at least 28 genes comprises one class of *hrp* genes that are essential for the components of the Type III secretion apparatus (for a review, see Alfano and Collmer, 1997; Lindgren, 1997; Collmer, 1998; Hueck, 1998; He, 1998). A second region containing the *hrpM* locus is comprised of two genes (*orf1* and *hrpM*), and in addition to the *hrp* phenotype, mutants also exhibit a mucoid, colony morphology (Niepold *et al.*, 1985; Mukhopadhyay *et al.*, 1988).

1.1 Comparative Studies of the *hrpM* and *mdoA* Loci

Mutations in *hrpM* were shown to be complemented by a 3.9 kb *Hin*dIII fragment and it was assumed that the entire *hrpM* was contained within that fragment. Two ORF's were identified in this fragment that appeared to compose the locus (Mukhopadhyay *et al.*, 1988). Initial searches of the genomic data bases revealed no obvious sequence similarities to either gene, but subsequently, similarities to the *Escherichia coli mdoA* locus were detected (Loubens *et al.*, 1993). The *mdoA* locus is composed of two genes, *mdoG* and *mdoH*, that are essential for the synthesis of membrane-derived oligosaccharides (MDO) or periplasmic glucans. The striking similarity of the *hrpM* and *mdoA* loci provided clues about the possible structure and function of *hrpM*. Periplasmic glucan synthesis by *mdoG* and *mdoH* genes is under osmotic control (Lacroix *et al.*, 1989; 1991). Whereas *mdoG* encodes a 56-kDa periplasmic protein of unknown function, *mdoH* encodes a 97-kDa membrane spanning protein required for glucosyl transferase activity (Loubens *et al.* 1993). DNA sequence comparisons of the *mdoA* and *hrpM* loci strongly inferred that the promoter for the *hrpM* locus and upstream sequences of *orf1* were not present on the 3.9 kb *Hin*dIII fragment. A weak promoter identified within *orf1*, and initially believed to be the promoter for the *hrpM* locus, apparently provided for expression of the downstream *hrpM* gene that was cloned on the *Hin*dIII fragment.

Genetic complementation of a *mdoH* mutant with the 3.9 kb *Hin*dIII fragment carrying *hrpM* was the first demonstration that a *hrp* gene would function in another species, and it suggested that the *hrpM* gene would encode a glucosyl transferase activity (Loubens *et al.*, 1994). In this study, we provide a new, revised genetic organisation of the two genes that comprise the *hrpM* locus based on additional nucleotide sequencing and speculate about the function of *hrpM*.

2. MATERIAL AND METHODS

2.1 Bacterial Strains, Plasmids, and Culture Conditions

Pseudomonas syringae pv. *syringae* wild-type strain PS9020 and a *hrpM* mutant PS9021::Tn*5* (Niepold *et al.*, 1985) used in this study were cultured in MaNY broth medium (Curiale and Mills, 1982) on a rotary shaker at 23°C. *Escherichia coli* was grown on LB medium at 37°C. Plasmid pOSU226 was described previously (Zhou, 1991) and pBlueScript and pUC vectors were used to clone fragments for sequencing**.** Seed of the bean cultivar *Phaseolus vulgaris* "Eagle", a gift from Asgrow Seed Company, Twin Falls, ID, were sown in a 1:1:1 mixture of peat moss, perlite, and vermiculite and plants were grown in a growth chamber at 22°C. Plants were inoculated on the under side of the leaves with a diluted suspension of bacteria (ca. 0.25 ml) and disease symptom development was monitored daily.

2.2 DNA Manipulations, Transcription and Sequencing

Restriction enzyme digestions, agarose gel electrophoresis, DNA, RNA and plasmid extractions, DNA ligations, radioisotope labeling and Southern blot hybridisation were performed as described by Sambrook *et al.* (1989). The *hrpM* locus was cloned from strain PS9020 as a 5.5 kb fragment by long PCR according to instructions provided with the DyNAzyme EXT DNA Polymerase (MJ Research, Inc., Waltham, MA.). Primer extension was performed using instructions with the RETROscript™ Kit (Ambion, Inc., Austin, TX, USA). DNA sequence was obtained through the Central Service Laboratory, Oregon State Univ. using an ABI 373A automated sequencing system (Applied Biosystems, Foster City, CA, USA), and analysed using the GCG 9 software package of the University of Wisconsin Genetic Computer Group. Database searches were performed using BLAST at NCBI. The primers used for sequencing were purchased from Oligos Etc. (Wilsonville, OR, USA). A revised sequence for the *hrpM* locus can be found under accession number M23555.

3. RESULTS

3.1 Further Sequence of the *hrpM* Locus

The nucleotide sequence of the promoter region and a portion of the *orf1* and *hrpM* genes were obtained from cloned overlapping fragments extending upstream from within the left *Hin*dIII site in pOSU3105 (Fig. 1). The *hrpM* locus was then cloned by long PCR using primers designed to be complementary to the ends of a 5,490 bp segment of DNA and engineered to contain *Bam*HI sites upstream of the *Pst1* site and downstream of the right-most *Hin*dIII site to facilitate cloning the PCR product into pOSU226. The resulting plasmid, pOSU55 (Fig. 1), was used to confirm the presence of restriction sites in the upstream region and for genetic complementation of a *hrpM* mutant.

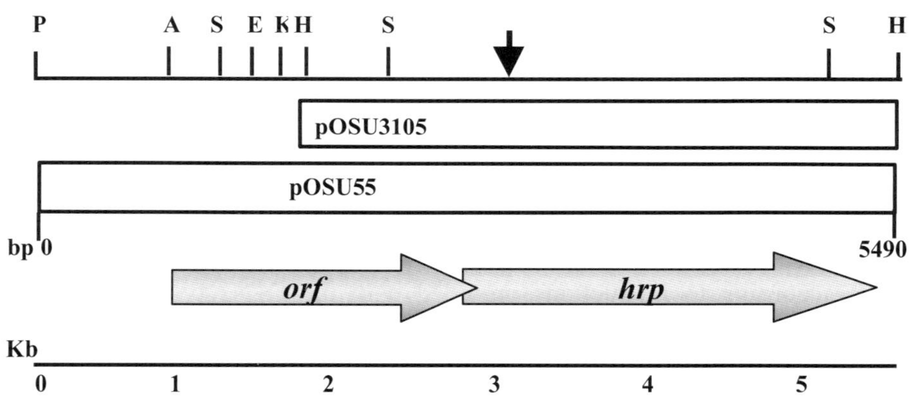

Figure 1. The proposed organisation of the *hrpM* locus of *Pseudomonas syringae* pv. *syringae*. The restriction sites within the genome of strain PS9020 are indicated at the top. Plasmids described in the text are indicated by open rectangles. The coding regions of *orf1* and *hrpM* slightly overlap as indicated by the arrows. The approximate site of Tn*5* insertion in PS9021::Tn*5* is indicated by the vertical arrow. Restriction enzyme symbols: P, *Pst*I; A, *Alu*I; S, *Sal*I; E, *Eco*RI; K, *Kpn*I; H, *Hin*dIII.

The nucleotide sequence extending upstream of the 3.9 Kb *Hin*dIII fragment of pOSU3105 had high sequence similarity with *orf1* of the *mdoA* locus until a putative promoter region was detected. Translation alternative codon GTG (Val) (Preston *et al.*, 2000) based on the occurrence of a ribosome-binding site (RBS) located immediately upstream. *orf1* terminates at a TAA, resulting in a putative polypeptide of 642 amino acids with a molecular weight of 71,489 daltons (Fig. 2). Automated sequencing resolved specific G+C-rich regions upstream and within the *hrpM* gene that were unresolved in the initial sequence of this locus. This has extended the size of *hrpM*, and the putative start codon is now proposed to be an ATG that resides 5 nucleotides inside the terminus of *orf1* (Fig. 1). In frame with this ATG is Met^{74} with its consensus RBS (tggag) 14 nucleotides upstream, which was previously presumed to be the initiation codon for *hrpM* (Mukhopadhyay *et al.*, 1988). It is unclear whether both ATG codons may be used in translation of a functional HrpM polypeptide. Using the longer open reading frame, the predicted HrpM polypeptide of 860 amino acids has a molecular weight of 97,084 daltons.

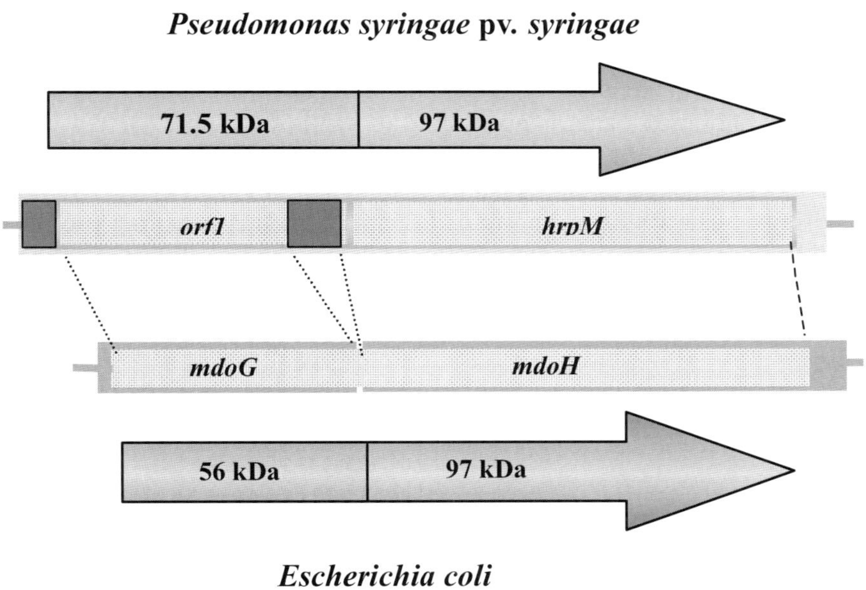

Figure 2. Genetic organisation of the *hrpM* and the *mdoA* loci. The dark, shaded regions near the 5 prime and 3 prime ends of *orf1* refer to 42 and 348 nucleotides, respectively, that do not correspond to sequences in *mdoG*. Regions identified by small squares encode polypeptides that have approximately 64 percent identical and 70 percent similar amino acids.

3.2 Sequence Similarity of the *hrpM* and *mdoA* Loci

The *mdoA* locus of *Escherichia coli* and the *hrpM* locus of *Pseudomonas syringae* pv. *syringae* are both composed of two slightly overlapping genes that have extensive similarity (Fig. 2). ORF1 (642 aa, 71 kDa) is larger than MdoG (511 aa, 56 kDa) primarily because of a 116 aa near the carboxyl end and a 14 aa region near the amino terminus that are not found in MdoG. These polypeptides have 64% identity and 70% similarity in other regions after introduction of a single gap of one amino acid. MdoH and HrpM have

64% identity and 71% similarity with the introduction of 4 gaps of a single amino acid and one gap of two amino acids. HrpM (860 amino acids, 97,084 daltons) and MdoH (847 amino acids, 96,939 daltons) are very similar in size, and the proposed ATG initiation codons for these polypeptides are located 5 and 6 nucleotides inside the termini of *orf1* and *mdoG*, respectively.

3.3 *hrpM* and *mdoH* are Isofunctional in *Escherichia coli* and *Pseudomonas syringae*

Loubens *et al.*, (1994) showed that the 3.9 kb *Hin*dIII fragment of pOSU3105 (Fig. 1) containing a truncated copy of *orf1* and a full-length *hrpM* gene could genetically complement an *mdoH* mutant resulting in complete restoration of periplasmic glucan synthesis (Table 1). We cloned the *mdoA* operon on an 8 kb fragment in pOSU226, a vector that autonomously replicates and is stably maintained in *Pseudomonas syringae* pv. syringae. The PS9021::Tn*5* transformants with this construct (pOSU55, Fig. 1) appeared to be restored for glucosyl transferase activity resulting in pathogenicity and the ability to cause disease symptoms in the susceptible bean cultivar "Eagle" (Table 1).

Table 1. Phenotypes lost through mutation of *mdoH* and *hrpM* of *Escherichia coli* and *Pseudomonas syringae* pv *syringae* and restored following transformation with the corresponding wild type gene from either species.

	Phenotypes Restored		
Transformant/ Wild-type Gene	Glucosyl Transferase Activity	Symptom Development *In Planta*	Mucoid Colony Morphology
E. coli			
mdoH	+[a]	NA	NA
hrpM	+[a]	NA	NA
P. syringae pv. *syringae*			
mdoH	ND	+	+
hrpM	+[b]	+	+

[a]Data obtained by Loubens *et al.*, 1994. [b]Data obtained by Talaga, *et al.*, 1994;
+= Restored phenotype; NA= Not Applicable; ND= Not Determined.

4. CONCLUSIONS

Through nucleotide sequence comparisons of the *mdoA* and *hrpM* loci we have learned that only a portion of the *hrpM* locus was located on the 3.9 kb *Hin*dIII fragment that complements the *hrpM* mutation. A weak promoter located within *orf1* likely drives expression of *hrpM* on that fragment, rather than the native promoter for the *hrpM* locus. This promoter corresponds to a similar, weak promoter located within *mdoG* (Loubens *et al.*, 1993). Additional nucleotide sequence upstream adjacent to that fragment revealed continued high homology with *mdoG* until the region of the putative promoter. Revaluation of previous sequence data upstream of *hrpM* now suggests that the two genes of the *hrpM* locus slightly overlap as is the case for *mdoG* and *mdoH*. The occurrence of G+C-rich regions has led to some uncertainty concerning the size of the coding region of the *hrpM* gene. Two potential translation start codons are found within the open reading frame of *hrpM*. Isolation and determination of the size of the HrpM polypeptide would alleviate that uncertainty.

Membrane-derived oligosaccharides (MDO) or periplasmic glucans appear to be essential for growth of *Pseudomonas syringae* in hypo-osmotic environments, as would be characteristic of the intercellular spaces of plant tissue (Grignon and Sentenac, 1991). The loss of virulence associated with MDO mutants in both *Escherichia coli* and *P. syringae* could adversely affect function of the Type III secretion apparatus, which is encoded by genes of the highly conserved, large *hrp* gene cluster in *P. syringae*. Of interest in future studies will be ascertaining why the osmotically regulated synthesis of glucans in the periplasm of the plant pathogen *P. s.* pv. *syringae*, as well as in the animal pathogen *E. coli*, are essential for growth and expression of disease symptoms in their respective hosts.

ACKNOWLEDGEMENTS

This work was supported in part by the Oregon Agricultural Experiment Station from which this is Technical paper 11867. We thank Pascale Williams for assistance in sequencing regions of the *hrpM* locus. R. Shimizu was an Exchange Visitor from Okayama University, Okayama, Japan.

REFERENCES

Alfano, J.R., and Collmer, A., 1997, The type III (Hrp) secretion pathway of plant pathogenic bacteria: Trafficking harpins, avr proteins and death. *J. Bacteriol.* **179**: 5655-5662.

Anderson, D.M., and Mills, D., 1985, The use of transposon mutagenesis in the isolation of nutritional and virulence mutants in two pathovars of *Pseudomonas syringae*.

Phytopathology **77**: 832-835.

Collmer, A., 1998, Determinants of pathogenicity and avirulence in plant pathogenic bacteria. *Curr. Opin. Plant Biol.* **1**: 329-335.

Curiale, M.S., and Mills, D., 1982, Integration and partial excision of a cryptic plasmid in *Pseudomonas syringae* pv. *phaseolicola*. *J. Bacteriol.* **152**: 797-802.

Grignon, C., and Sentenac, H., 1991, pH and ionic conditions in the apoplast. *Annu. Rev. Plant Physiol.* **42**: 103-28.

He, S.Y., 1998, Type III Protein secretion systems in plant and animal pathogenic bacteria. *Annu. Rev. Phytopathol.* **36**: 363-392.

Huang, H.C., Lin, R.H., Chang, C.J., Collmer, A., and Deng, W.L., 1995, The complete hrp gene cluster of *Pseudomonas syringae* pv. syringae 61 includes two blocks of genes required for harpin$_{Pss}$ secretion that are arranged colinearly with Yersinia ysc homology. *Mol. Plant-Microbe Interact.* **8**: 733-746.

Hueck, C.J., 1998, Type III protein secretion systems in bacterial pathogens of animals and plants. *Microbiol. Mol. Biol Rev.* **62**: 379-433.

Lacroix, J.-M., Loubens, 1., Tempête, M., Menichi, B., and Bohin, J.-P., 1991, The *mdoA* locus of *Escherichia coli* consists of an operon under osmotic control. *Mol. Microbiol* **5**: 1745-1753.

Lacroix, J.-M., Tempête, M., Menichi, B., and Bohin, J.-P., 1989, Molecular cloning and expression of a locus *(mdoA)* implicated in the biosynthesis of membrane-derived oligosaccharides in *Escherichia coli. Mol. Microbiol* **3**: 1173-1182.

Lindgren, P.B., 1997, The role of *hrp* genes during plant-bacterial interactions. *Annu. Rev. Phytopathol.* **35**: 129-152.

Lindgren, P.B., Peet, R.C., and Panopoulos, N.J., 1986, Gene cluster of *Pseudomonas syringae* pv. phaseolicola controls pathogenicity on bean plants and hypersensitivity on nonhost plants. *J. Bacteriol.* **168**: 512-522.

Loubens, I., Debarbieux, L., Bohin, A., Lacroix, J.-M., and Bohin, J.-P., 1993, Homology between a genetic locus (*mdoA*) involved in the osmoregulated biosynthesis of periplasmic glucans in *Escherichia coli* and a genetic locus (*hrpM*) controlling the pathogenicity of *Pseudomonas syringae*. *Mol. Microb.* **10**: 329-340.

Loubens, I., Richter, G, Mills, D., and Bohin, J-P., 1994, A pathogenicity gene of *Pseudomonas syringae* pv. *syringae* complements a defect in periplasmic glucan biosynthesis in *Escherichia coli* K-12. *Plant Pathogenic Bacteria 8th International Conference* (L. Lemattre, S. Freigoun, K. Rudolph, and J.G. Swings, eds.), INRA Editions,Versailles, France. pp. 491-496.

Mukhopadhyay, P., Williams, J., and Mills, D., 1988, Molecular analysis of a pathogenicity locus in *Pseudomonas syringae* pv. *syringae*. *J Bacteriol* **170**: 5479-5488.

Niepold, F., Anderson, D., and Mills, D., 1985, Cloning determinants of pathogenesis from *Pseudomonas syringae* pathovar *syringae*. *Proc. Natl. Acad. Sci. USA* **82**: 406-410.

Preston, L.A., Wong, T.Y., Bender C.L., and Schiller, N.L., 2000, Characterization of alginate lyase from *Pseudomonas syringae* pv. *syringae*. *J. Bacteriol.* **182**: 6268-6271.

Sambrook, J., Fritsch, E.F., and Maniatis, T., 1989, *Molecular Cloning. A Laboratory Manual* (Second edition). Cold Spring Harbor Laboratory Press, Cold Spring Harbor, New York:

Talaga, P., Fournet, B., and Bohin, J-P., 1994, Periplasmic glucans of *Pseudomonas syringae* pv. *syringae*. *J. Bacteriol.* **176**: 6538-6544.

Zhou, Y., 1991, Ph.D. Thesis *Molecular genetic characterization of a pathogenicity attenuated mutant of Pseudomonas syringae* pv. *syringae.* Oregon State University, Corvallis, OR. USA.

Is Pore Formation Activity of HrpZ Required for Defence Activation in Plant Cells?

J. LEE and T. NÜRNBERGER
Institute of Plant Biochemistry, Weinberg , D-06120, Halle, Germany.

Abstract: The *HrpZ* gene product, harpin, is an export substrate of the type III secretion system of phytopathogenic *Pseudomonas syringae*. The role of this protein in pathogenesis is largely unknown. We previously determined that HrpZ binds to lipids and can form cation pores in synthetic lipid bilayers. Such pore-forming activity may allow nutrient release during bacterial colonisation of host plants. In addition, HrpZ is known to trigger plant defence responses in a variety of plants, such as tobacco. We have previously also characterised a binding site in tobacco plasma membranes that likely mediates HrpZ-induced defence responses. In order to reconcile these findings, we pose the question as to whether the activation of plant defence responses by HrpZ is mediated through a "classical" receptor perception mode or if plant membrane perturbation through the inherent pore-forming activity of HrpZ may induce defence responses. As defence in parsley cells can be induced both in a receptor-mediated manner or through ionophores these cells served as an ideal system for our analysis. We first performed ligand binding studies to characterise the presence of a binding site/receptor. We further digested HrpZ with endopeptidases and used subfragments of HrpZ to assess the elicitor-active domain of HrpZ. A C-terminal region of HrpZ appears to be sufficient to elicit plant defence responses. A novel assay involving dye-loaded liposomes was developed to validate previous electrophysiological findings on HrpZ-mediated cation pore formation. More importantly, this assay was used to establish if the elicitor-active C-terminal fragment of HrpZ could form pores. Our findings suggest that the structural requirements for ion pore formation and activation of plant defence responses by HrpZ are different. Thus, ion pore formation alone may not explain the activation of plant defence by HrpZ.

Key words: receptor, ionophore, harpin, lipid bilayer.

Author for correspondence: J. Lee; e-mail: jlee@ipb-halle.de.

N.S. Iacobellis *et. al.* (eds.), *Pseudomonas syringae* and Related Pathogens, 165-173.

1. INTRODUCTION

Gram-negative phytopathogenic bacteria possess a gene cluster (*hrp*, for hypersensitive reaction and pathogenicity) that is required for pathogenicity in susceptible plants and to elicit the hypersensitive reaction (HR) in resistant cultivars of host plants or in non host plants (Galan and Collmer, 1999). Several of the *hrp* genes encode an ubiquitous secretion system of Gram-negative pathogenic bacteria that exports virulence factors and delivers them into the cytosol of host cells (Bogdanove *et al.*, 1996). One of the first export substrates identified to be secreted by the type III secretion system of phytopathogenic *Pseudomonas syringae* is the *hrpZ* gene product, harpin (He *et al.*, 1993). The role of this protein in pathogenesis is largely unknown, especially since *Pseudomonas* spp. mutated in *hrpZ* alone are still virulent (Alfano *et al.*, 1996; Charkowski *et al.*, 1998; Lee *et al.*, 2001b). This may be explained by the presence of proteins with redundant function (e.g. Harpin-like proteins, such as HrpW) (Charkowski *et al.*, 1998). Infiltration of high levels of HrpZ can cause HR-like symptoms in non host plants, such as tobacco (Preston *et al.*, 1995). We previously characterised a binding site in tobacco plasma membranes which may mediate the activation of defence responses in this plant (Lee *et al.*, 2001a). In parallel, we also determined that HrpZ could bind to lipid bilayers and shows pore-forming activity in synthetic planar lipid bilayer assays (Lee *et al.*, 2001b). The question therefore arises as to whether the activation of defence responses to HrpZ in plants is mediated through a "classical" receptor perception mode (Nürnberger *et al.*, 1994; Nürnberger *et al.*, 1995) or if plant membrane perturbation through inherent pore activity of HrpZ may be the cause for the activation of defence responses by this protein (Klüsener and Weiler, 1999).

2. MATERIALS AND METHODS

2.1 Recombinant HrpZ Elicitor

Recombinant HrpZ was expressed in *Escherichia coli* and purified as described (Lee *et al.*, 2001a). Full length HrpZ was digested with either of the endopeptidases trypsin, GluC, LysC or AspN, respectively, according to the manufacturer's instructions (Roche Applied Science, Mannheim, Germany).

2.2 Cell Cultures, Microsomal Preparations and Binding Studies

Parsley (*Petroselinum crispum*) cell suspension cultures were maintained as described (Nürnberger *et al.*, 1994). Experiments were performed on cells 6 days after subculturing. To ensure uniform conditions for all experiments, cells were filtered, washed and resuspended in fresh medium at a density of 10 g 100 ml^{-1}. The cells were allowed to recover from the transfer for about 1 hour prior to elicitation. Elicitors were added at 500 nM (HrpZ) or 100 nM (pep13). Phytoalexins secreted into the culture medium were quantified by spectrophotometry (Ex. 360/Em. 440)

Microsomes from parsley cells were prepared as described in Nürnberger *et al.* (1994). The final microsomal pellet was resuspended in 20 mM sodium phosphate (pH 6.0), 100 mM KCl, 2 mM $MgCl_2$ and 1 mM $CaCl_2$. Microsomes containing 200 µg of protein was used for binding assays in a total volume of 200 µl. Rapid filtration and washing steps were as described (Lee *et al.*, 2001a).

2.3 Encapsulated Liposomes and Pore Formation Assays

Encapsulated liposomes were kindly prepared by Novosom AG (Halle, Germany). Details to lipid composition and dyes are available on request. The encapsulated liposomes were desalted on PD-10 columns prior to use. Ten microliters of the liposomes were placed into wells of microtiter plates in a final volume of 100 µl. Proteins were added in the presence or absence of NaCl. Changes in fluorescence were monitored for the indicated duration with a Cytofluor Microplate Fluorescence reader (Millipore, Eschborn, Germany) using the filter combinations (Ex. 485/Em. 530).

3. RESULTS

Parsley cell cultures have a long history as a robust model system in studying defence signal transduction pathways, and offer numerous experimental advantages. For example, in contrast to the often variable HR assay in tobacco, the response to elicitors, such as HrpZ, in terms of phytoalexin accumulation is quantitative and is useful for comparison to other response kinetics values. In addition, defence responses in parsley can also be induced by ionophore treatment (Jabs *et al.*, 1997). Hence, in addition to previous data collected for tobacco cells, we decided to use parsley cells to investigate the two conceivable modes of HrpZ perception that may mediate activation of plant defence responses: pore-formation vs. receptor binding.

Furanocoumarin phytoalexins accumulate in the culture medium of parsley cells after treatment with elicitors, such as the oomycete-derived pep13 elicitor (Nürnberger *et al.*, 1994). Addition of purified HrpZ also led to phytoalexin accumulation in parsley cells 12÷24 hours after treatment (Fig. 1). A dose-response curve showed that the effective concentration to obtain 50% of the maximum response (EC_{50}) is 235 nM (Table 1). If this elicitation by HrpZ is receptor-mediated, binding site(s) should be present in membranes of parsley. Binding studies were used to ascertain such binding sites in parsley membranes. A tyrosine residue was added to the carboxyl-terminus of HrpZ by recombinant technology and iodinated with ^{125}I (Lee *et al.*, 2001a).

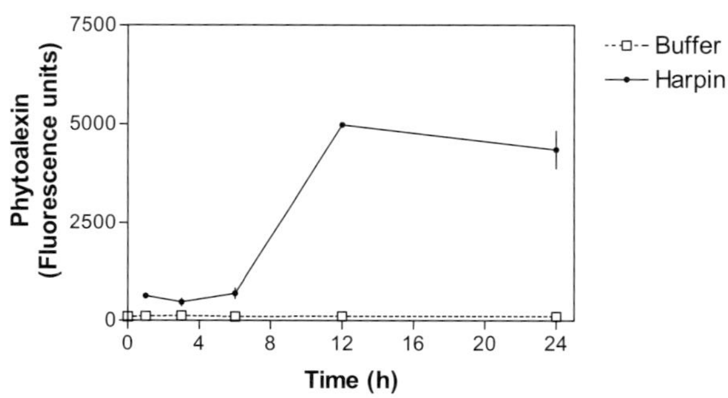

Figure 1. Time course of phytoalexin accumulation in parsley cells after elicitation. (Values = means with standard errors).

Table 1. Comparison of values determined from binding and response studies of HrpZ in tobacco and parsley cells.

	Tobacco	**Parsley**
EC_{50}	120 nM[a]/82 nM[b]	235 nM[c]
K_d	425 nM	133 nM
No. of binding sites	1	1
IC_{50}	550 nM	250 nM

[a,b]: EC_{50} for *Hin1* and *Hsr203* transcript accumulation, respectively (Lee *et al.*, 2001a).
[c]: EC_{50} for phytoalexin induction.

The binding conditions for interaction of HrpZ with parsley microsomes were optimised, but were essentially very similar to those for tobacco microsomes. In both systems, binding was not affected significantly by various salt concentrations. However, the pH values for maximal binding differed by one pH unit between the two systems (pH 6.0 for parsley and pH 7.0 for tobacco microsomes). Table 1 compares some of the data determined from binding studies of $[^{125}I]$-HrpZ to tobacco and parsley microsomal membrane preparations. In particular, the ligand affinity (K_d) is in the same range as the EC_{50} value for phytoalexin production as well as the concentration of non-radioactive ligand needed to reduce radioligand binding by 50% (IC_{50}) (Table 1). The Scatchard analysis of data obtained from saturation binding assays revealed a linear graph (data not shown), indicating the presence of a single HrpZ binding site in parsley membranes similar to that found in tobacco. Since the tobacco binding site was shown to mediate activation of HrpZ-induced defence responses in this plant, it is tempting to speculate that the parsley binding site might serve a similar role.

Surprisingly, protease or heat treatment of microsomal membranes did not abolish the ability of $[^{125}I]$-HrpZ to bind to the parsley binding site. This is in contrast to binding studies performed with the oomycete-derived elicitor, pep13, which revealed a parsley 100 kDa proteinaceous receptor to mediate elicitor perception and initiation of an intracellular signalling cascade (Nürnberger *et al.*, 1994; Nürnberger *et al.*, 1995). Binding of radiolabelled pep13 to parsley membranes was specifically blocked by non-radioactive pep13 competitor, but not by HrpZ (unpublished data, T.N.), suggesting further that perception of the two elicitors is not mediated by the same binding site.

In view of the previously determined lipid binding and cation pore-forming capacity of HrpZ, the non-proteinaceous binding site in microsomes might reflect direct insertion of HrpZ into the lipid bilayer. Hence, the resulting pore activity could also be the cause for the observed defence responses, and, thus, represent an alternative to receptor-mediated activation of plant defence. Indeed, non-ionspecific ionophores, such as amphotericin B, have been shown to trigger phytoalexin production in parsley cells (Jabs *et al.*, 1997). To test if the same fragments of HrpZ that retained the ability to induce phytoalexin production in parsley cells would still facilitate ion fluxes across lipid bilayer membranes and/or bind to the observed binding site, we digested purified HrpZ with several endopeptidases and tested them as elicitors of phytoalexin production in parsley cells.

With the exception of endopeptidase LysC, all endopeptidases tested destroyed the elicitor activity of HrpZ (Fig. 2). However, analyses of purified recombinant fragments of HrpZ revealed that a C-terminal fragment, F3 (amino acid residues 201-345), was sufficient to elicit phytoalexin accumulation in parsley (Fig. 2). Since a C-terminal fragment (amino acid residues 174-304) remained in the LysC digest, it is likely that the region spanning amino acid residues 201-304 of HrpZ is sufficient for elicitor activity.

When the F3 fragment was used as competitor in the binding assays, it was equally effective as the full length HrpZ to reduce binding of $[^{125}I]$-HrpZ to parsley microsomal membranes (data not shown). Binding of HrpZ is thus determined by this region of the protein. However, since this part of HrpZ is also the most hydrophobic section of the protein and since it was found to bind to lipid bilayer-coated silica beads (Lee *et al*., 2001b), it is conceivable that the ability of HrpZ not only to bind, but also to insert into bilayers could be determined by this domain as well. Pore formation and subsequent activation of plant defence responses in an ionophore-like fashion is not at all an unprecedented phenomenon (Jabs *et al*., 1997; Klüsener and Weiler, 1999).

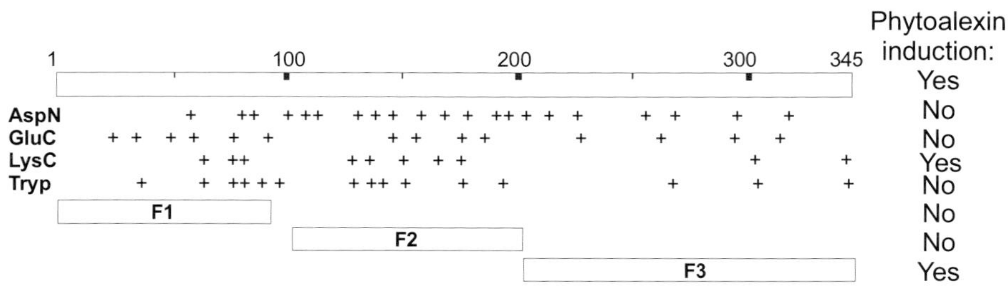

Figure 2. Ability of endopeptidase-digested HrpZ and subfragments of HrpZ to induce phytoalexin accumulation in parsley cells. Positions of the respective protease sites are indicated by plus symbols.

Pore formation assays were thus conducted to compare the activities of HrpZ and truncated subfragments to evoke ion fluxes across lipid bilayer membranes. A novel assay involving dye-loaded liposomes was developed for this purpose. Previous electrophysiological experiments had revealed a specificity of the HrpZ pore for cations, such as Na^+, K^+ or Ca^{2+} (Lee *et al*., 2001b). Therefore, encapsulated liposomes containing a sodium-sensitive fluorescent dye were prepared in a sodium-free buffer. Permeabilization of the lipid bilayer through pore-forming molecules (such as HrpZ) would facilitate diffusion of sodium ions from the external medium into the liposomes and thus evoke fluorescence emission from the activated dye. Fluorescence spectrometry performed in real time allowed us to visualise and quantify the permeabilization of the lipid bilayer.

As shown in Fig. 3, addition of HrpZ led to increased fluorescence. When HrpZ was added in the absence of sodium ions, no change in fluorescence was observed (data not shown), indicating that the flux of sodium into the liposomes, but not the presence of HrpZ, was the cause of the observed increase in fluorescence. Moreover, this assay was used to establish if elicitor-active and inactive fragments of HrpZ could form pores. Importantly, all subfragments of HrpZ were proved inactive in this assay. Hence, the ion pore-forming activity of HrpZ may not form the molecular basis for defence activation in parsley and tobacco (Lee *et al*., 2001a).

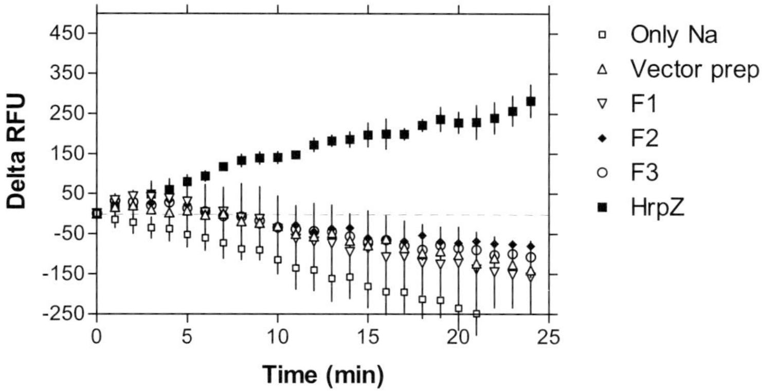

Figure 3. Pore formation assays using encapsulated liposomes. The indicated substances were added and the changes in relative fluorescence units (Delta RFU) were plotted against time.

4. DISCUSSION

In summary, we show that a C-terminal region of HrpZ is sufficient to elicit plant defence responses in parsley and tobacco cells (Lee *et al*., 2001a). This fragment does not possess pore formation activity (Fig. 3), but does compete for binding of [^{125}I]-HrpZ to parsley microsomal membranes. These findings, together with the fact that similar concentrations of either HrpZ fragment were found to trigger phytoalexin production in parsley and to compete for binding of [^{125}I]-HrpZ to parsley microsomal membranes, strongly suggest that this (apparently) non-proteinaceous binding site may mediate activation of plant defence responses in plants. In addition, it was clearly shown that ion pore formation activity of HrpZ alone may not explain the activation of plant defence.

Since HrpZ is well conserved and abundant in most *Pseudomonas syringae* pathovars, it is conceivable that plants have co-evolved the capacity to recognise *P. syringae* by HrpZ receptors, which may ‘guard’ the perturbance of the plasma membrane integrity by HrpZ-mediated ion pore formation. The C-terminus of HrpZ may thus betray the pathogenic bacterium to the plant surveillance system. In this respect, it is interesting that the *HrpZ* gene of *P. s.* pv. *tabaci* (isolate 6605), a tobacco pathogen, encodes a truncated HrpZ with only the N-terminal 140 amino acid residues (GeneBank Acc. No. BAB40655).

Our findings, however, do not exclude the possibility that ionophore-like activation of plant defence responses could occur in parallel to receptor-mediated activation of plant defence. Moreover, it is possible that the C-terminal HrpZ fragment may bind to a specific target site (with the kinetic properties described above), which might be the virulence target of HrpZ in plant-pathogen encounters, and which may facilitate the release of nutrients into the apoplasts for nourishment of the bacteria. This docking site might then facilitate insertion into the membrane and subsequent activation of plant defence responses through ion pore formation. The reason why this HrpZ fragment did not cause pore formation in our liposome-based assay could be that this specific docking site was missing in the liposome membranes used in this assay.

We are yet unable to answer the question how HrpZ recognition and activation of plant defence responses functions at the molecular level. However, our findings show that direct and mere insertion of HrpZ into plant membranes may not explain the plant defence-inducing capacity of HrpZ.

ACKNOWLEDGEMENT

We would like to thank Dr. Steffen Panzner (Novosom AG, Halle, Germany) for preparation of the dye-loaded liposomes. Helpful discussions from members of our lab, especially Dierk Scheel and Jason Rudd, is gratefully acknowledged.

REFERENCES

Alfano, J.R., Bauer, D.W., Milos, T.M., and Collmer, A., 1996, Analysis of the role of the *Pseudomonas syringae* pv. syringae HrpZ harpin in elicitation of the hypersensitive response in tobacco using functionally non-polar *hrpZ* deletion mutations, truncated HrpZ fragments, and *hrmA* mutations. *Mol. Microbiol.* **19**: 715-728.

Bogdanove, A.J., Beer, S.V., Bonas, U., Boucher, C.A., Collmer, A., Coplin, D.L., Cornelis, G.R., Huang, H.C., Hutcheson, S.W., Panopoulos, N.J., and Van Gijsegem, F., 1996, Unified nomenclature for broadly conserved *hrp* genes of phytopathogenic bacteria. *Mol. Microbiol.* **20**: 681-683.

Charkowski, A.O., Alfano, J.R., Preston, G., Yuan, J., He, S.Y., and Collmer, A., 1998, The *Pseudomonas syringae* pv. *tomato* HrpW protein has domains similar to harpins and pectate lyases and can elicit the plant hypersensitive response and bind to pectate. *J. Bacteriol.* **180**: 5211-5217.

Galan, J.E., and Collmer, A., 1999, Type III secretion machines: Bacterial devices for protein delivery into host cells. *Science* 284: 1322-1328.

He, S.Y., Huang, H.-C., and Collmer, A., 1993, *Pseudomonas syringae* pv. *Syringae* $harpin_{Pss}$: a protein that is secreted by the hrp pathway and elicits the hypersensitive response in plants. *Cell* **73**: 1255-1266.

Jabs, T., Tschöpe, M., Colling, C., Hahlbrock, K., and Scheel, D., 1997, Elicitor-stimulated ion fluxes and O_2^- from the oxidative burst are essential components in triggering defense gene activation and phytoalexin synthesis in parsley. *Proc. Natl. Acad. Sci. USA* **94**: 4800-4805.

Klüsener, B., and Weiler, E.W., 1999, Pore-forming properties of elicitors of plant defense reactions and cellulolytic enzymes. *FEBS Lett.* **459**: 263-269.

Lee, J., Klessig, D.F., and Nürnberger, T., 2001a, A harpin binding site in tobacco plasma membranes mediates activation of the pathogenesis-related gene *HIN1* independent of extracellular calcium but dependent on mitogen-activated protein kinase activity. *Plant Cell* **13**: 1079-1093.

Lee, J., Klüsener, B., Tsiamis, G., Stevens, C., Neyt, C., Tampakaki, A.P., Panopoulos, N.J., Noller, J., Weiler, E.W., Cornelis, G.R., Mansfield, J.W., and Nürnberger, T., 2001b, $HrpZ_{Psph}$ from the plant pathogen *Pseudomonas syringae* pv. *phaseolicola* binds to lipid bilayers and forms an ion-conducting pore in vitro. *Proc. Natl. Acad. Sci. USA* **98**: 289-294.

Nürnberger, T., Nennstiel, D., Hahlbrock, K., and Scheel, D., 1995, Covalent cross-linking of the *Phytophthora megasperma* oligopeptide elicitor to its receptor in parsley membranes. *Proc. Natl. Acad. Sci. USA* **92**: 2338-2342.

Nürnberger, T., Nennstiel, D., Jabs, T., Sacks, W.R., Hahlbrock, K., and Scheel, D., 1994, High affinity binding of a fungal oligopeptide elicitor to parsley plasma membranes triggers multiple defense responses. *Cell* **78**: 449-460.

Preston, G., Huang, H.C., He, S.Y., and Collmer, A., 1995, The HrpZ proteins of *Pseudomonas syringae* pvs. *syringae*, *glycinea*, and *tomato* are encoded by an operon containing *Yersinia ysc* homologs and elicit the hypersensitive response in tomato but not soybean. *Mol. Plant-Microbe Interact.* **8**: 717-732.

An Antimetabolite Toxin (Mangotoxin) is Produced by *Pseudomonas syringae* pv. *syringae* Isolated from Mango

F.M. CAZORLA, E. ARREBOLA, E. DEL MORAL, E. RIVERA, F. OLEA, A. PÉREZ-GARCÍA and A. DE VICENTE
Departamento de Microbiología, Facultad de Ciencias, Universidad de Málaga. Campus Universitario de Teatinos, s/n. 29071-Málaga (Spain).

Abstract: In this study, we reported the production of an antimetabolite toxin (named mangotoxin), mainly by strains of *Pseudomonas syringae* pv. *syringae* pathogenic to mango. The target of mangotoxin is the enzymatic step where ornithine is produced from *N*-acetyl ornithine in the biosynthetic pathway of arginine. The inhibition of ornithine acetyltransferase (OAT) activity by mangotoxin was confirmed by exposing tomato leaf protein extracts to filtrates from cultures of *P. s.* pv. *syringae*. Mutants impaired in the production of mangotoxin were constructed. Sequencing of Tn5 flanking regions of 9 non/low-producers mutants revealed that several genes codifying for putative acetyltransferase, non-ribosomal peptide synthetase and the global regulators proteins GacA and LemA could be involved in mangotoxin production. Mangotoxin also showed antimicrobial activity against some micro-organisms, usually found on plant tissues as epiphytic or pathogenic micro-organisms. To conclude, the possible role of the mangotoxin in pathogenesis is discussed.

Key words: ornithine, arginine, ornithine *N*-acetyltransferase, antimicrobial activity.

Author for correspondence: A. De Vicente; e-mail: adevicente@uma.es.

1. INTRODUCTION

Some pathovars of the species *Pseudomonas syringae* produce non host-specific toxins, but it has been difficult to define the role of toxins in the interactions between these bacteria and their hosts. Structurally, they belong

N.S. Iacobellis *et. al.* (eds.), *Pseudomonas syringae* and Related Pathogens, 175-183.

to diverse chemical families including small peptides, polyketides or complex carbohydrates, and genetically, they are mainly clustered in the chromosome, but plasmidic exceptions have been reported (Bender *et al.*, 1999). These toxins could function as virulence factors in disease development by inhibiting specific enzymes of the host plant primary metabolism, by disrupting membrane functions or by subjecting the host to increased amounts of phytohormones, thereby upsetting its normal regime and leading the typical disease symptoms (Durbin, 1991). Besides, some of these toxins also have antibiotic effects against other micro-organisms (Volksch and Weingart, 1998). Therefore, toxins could play a role in the parasitic phase of the producer to harm the host plant as well as in the saprophytic phase to inhibit other invaders of the plant habitat. Hence, production of toxins may be widespread.

P. s. pv. *syringae* is the causal agent of the apical necrosis of mango, a bacterial disease limiting the production of mangoes in the Mediterranean area (Cazorla *et al.*, 1998). Some strains of this bacterium have been described as producers of different virulence factors, including toxins (Cazorla *et al.*, 1997). *P. s.* pv. *syringae* is known as a producer of different lipodepsipeptides, as syringomycins, acting at cellular membrane level and causing necrotic symptoms on plants (Sinden *et al.*, 1971; Iacobellis *et al.*, 1992). On the other hand, different pathovars of *P. syringae* produce antimetabolite toxins, as tabtoxin or phaseolotoxin, that inhibit enzymes involved in the biosynthetic pathways of aminoacids (Bender *et al.*, 1999). In this study, the presence and characterisation of an antimetabolite toxin (mangotoxin) produced by strains of *P. s.* pv. *syringae* is reported and its potential role in pathogenesis and epiphytic survival discussed.

2. MATERIAL AND METHODS

2.1 Micro-organisms and Growth Conditions

Most of *Pseudomonas syringae* strains used in this study were isolated and identified in previous studies (Cazorla *et al.*, 1997; 1998). Bacterial strains were routinely cultured on King's B medium at 25°C. To obtain toxic filtrates, *P. s.* pv. *syringae* strains were inoculated on PMS liquid medium (Gasson, 1980) and incubated for 5÷7 days at 22°C. Culture supernatants were then filtered through 0.2 μm filters.

2.2 Detection of Syringomycin

To detect the production of lipodepsinonapeptides (mainly syringomycins) by the strains of *Pseudomonas syringae* pv. *syringae*, agar diffusion assays on PDA against *Geotrichum candidum* (Gross and de Vay, 1977) and *Rhodotorula pilimanae* (Iacobellis *et al.*, 1992) were used.

2.3 Bioassay of Antimetabolite Toxins

The production of antimetabolite toxins was assayed by the indicator technique, involving the inhibition zone production on a lawn of *Escherichia coli* (Cazorla *et al.*, 1997; Gasson, 1980). Detection of the putative target enzyme of mangotoxin was assayed by the same technique, but adding different aminoacids.

2.4 Ornithine Acetyltransferase Enzymatic Assay

Ornithine acetyltransferase (OAT) activity was measured from crude protein extracts of tomato leaves as the increase of ornithine amounts, detected by the ninhydrin procedure (Dénes, 1970), after addition of cell-free culture filtrates from mangotoxin-producer and -non producer strains. The compound p-chloromercuribenzoic acid (PCMB) (Denés, 1970) and sterilised minimal medium (PMS) were used as controls.

2.5 Partial Characterisation of Mangotoxin

To characterise the toxic molecules, toxic cell-free filtrates from mangotoxin-producing *Pseudomonas syringae* pv. *syringae* strains were treated at different temperatures (-20÷121°C), pH values (2÷12), protease treatments, filtration (3÷10 kDa) and extractions with different organic solvents.

2.6 Genetic Basis of Mangotoxin Production

In order to get information about the genetic basis involved in the production and/or regulation of mangotoxin, mini-Tn5km2 mutants (de Lorenzo and Timmis, 1994) impaired in the production of mangotoxin were constructed from the mangotoxin-producing wild type *Pseudomonas syringae* pv. *syringae* strain UMAF0158. After checking the absence or a very weak production of mangotoxin, 9 mutants were selected for sequencing of the flanking regions of the Tn5 insertion, using the transposon sequence as a primer. The sequences obtained were compared with those from data bases with the NCBI Genbank BLAST program (Altschul *et al*., 1997).

2.7 Role of Mangotoxin in Pathogenesis

Because these *Pseudomonas syringae* pv. *syringae* strains are also pathogenic to tomato, mangotoxin-producer and -non producer strains were inoculated onto tomato leaves, using a detached leaf method (Pérez-García *et al*., 1995). The necrotic and chlorotic areas on the leaf surface were recorded at different times. In addition, chlorophyll levels and bacterial counts were determined at different times after inoculation.

2.8 Bioassay of Antimicrobial Activities

The inhibitory activities of mangotoxin against typical epiphytic micro-organisms were evaluated by an agar diffusion assay, using a modified procedure previously reported (Gasson, 1980). Briefly, after the inoculation of the indicator micro-organisms (bacteria, fungi or yeast) as a double layer onto PMS medium, mangotoxin producing strains were stabbed and incubated for 48 to 96 h in presence or absence of ornithine or *N*-acetyl ornithine (NAO). The strains showing growth inhibition haloes in presence of NAO, but reverting the effect in presence of ornithine, were considered sensitive to mangotoxin.

3. RESULTS AND DISCUSSION

Toxin detection analyses showed that most of the strains of *Pseudomonas syringae* pv. *syringae* produce syringomycin (Table 1), as it has been previously described for this pathovar (Gross and de Vay, 1977). Interestingly, the production of an antimetabolite toxin was detected in many of the *P. s.* pv. *syringae* strains tested (76.4%) and in the most of the *P. s.* pv. *syringae* strains isolated from mango (87.6%). The toxic effect of this antimetabolite toxin was characterised by the reversion of the growth inhibition of *Escherichia coli* in presence of ornithine (Orn), but not in presence of *N*-acetyl ornithine (NAO) (Table 1). These results suggest that ornithine *N*-acetyltransferase (OAT), an enzyme involved in the biosynthetic pathway of ornithine and arginine, could be the target of the antimetabolite toxin produced by *P. s.* pv. *syringae* strains (Fig. 1). This toxin was named mangotoxin, in reference to the plant source (mango) of most of producer strains. OAT is a critical enzyme of the urea cycle and its disruption could leads to an accumulation of *N*-acetyl ornithine and a decrease of the levels of ornithine in plant tissues. In fact, the target of mangotoxin is the reaction that

Table 1. Toxin production by *Pseudomonas syringae* pv. *syringae* strains isolated from different sources. Mangotoxin production by *P. s.* pv. *syringae* strains was revealed by growth inhibition of *Escherichia coli* after addition of *N*-acetyl ornithine, but not in presence of ornithine.

P. s. pv. *syringae* strains			Syringomycin production	Antimetabolite toxin activity in:			
Host Plant	Location (Country)	n		PMS	PMS +NAO	PMS +Orn	Toxin produced
Mango	Málaga (Spain)	69	+	+	+	-	Mangotoxin
Mango	Málaga (Spain)	5	-	+	+	-	Mangotoxin
Mango	Málaga (Spain)	3	+	-			
Mango	Huelva (Spain)	13	+	+	+	-	Mangotoxin
Mango	Huelva (Spain)	5	nd	+	+	-	Mangotoxin
Mango	Huelva (Spain)	1	-	+	+	-	Mangotoxin
Mango	Huelva (Spain)	3	nd	-			
Mango	Granada (Spain)	5	nd	-			
Mango	Canary Isl. (Spain)	4	nd	+	+	-	Mangotoxin
Mango	Algarve (Portugal)	13	+	+	+	-	Mangotoxin
Mango	Algarve (Portugal)	3	-	-			
Mango	Bet Dagan (Israel)	3	nd	+	+	-	Mangotoxin
Mango	Bet Dagan (Israel)	2	nd	-			
Tomato	Málaga (Spain)	1	+	+	+	-	Mangotoxin
Chestnut	Málaga (Spain)	1	+	+	+	-	Mangotoxin
Phlomis sp	Málaga (Spain)	2	+	+	+	-	Mangotoxin
Peach	Málaga (Spain)	1	+	+	+	-	Mangotoxin
Orange	Málaga (Spain)	1	+	-			
Almond	Málaga (Spain)	2	-	-			
Croccus sp	Málaga (Spain)	1	+	-			
Pear	Gerona (Spain)	1	+	+	+	-	Mangotoxin
Pear	Gerona (Spain)	5	+	-			
Pear	Gerona (Spain)	1	-	-			
Laurel	Madrid (Spain)	1	+	-			
Crataegus sp	Valencia (Spain)	1	+	+	+	-	Mangotoxin
Pea	France	1	+	-			
Bean	Central Africa	5	+	-			
Bean	Central Africa	4	-	-			
Vetch	France	1	+	+	+	+	Phaseolotoxin
Pss CECT-4429 (NCPPB-281)			+	+	+	-	Mangotoxin
Pss CECT-127			+	+	+	-	Mangotoxin
Pss NCPPB-1239			+	+	+	-	Mangotoxin
Psc CECT-4389			-	+	+	+	Tabtoxin
Psp CECT-4390			-	+	+	+	Phaseolotoxin
Pst CECT-4393			-	+	-	-	Unknown
Pst UMAF-4007			-	+	-	-	Unknown

nd: no determined.; CECT:Spanish Type Culture Collection; NCPPB: National Collection Plant Pathogenic Bacteria-UK; PMS: minimal medium (Gasson, 1980); *NAO:* *N*-acetyl ornithine *Orn:* ornithine; *Pss*: *P. s.* pv. *syringae*; *Psc*: *P. s.* pv. *coronafaciens*; *Psp*: *P. s.* pv. *phaseolicola*; *Pst*: *P. s.* pv. *tomato*.

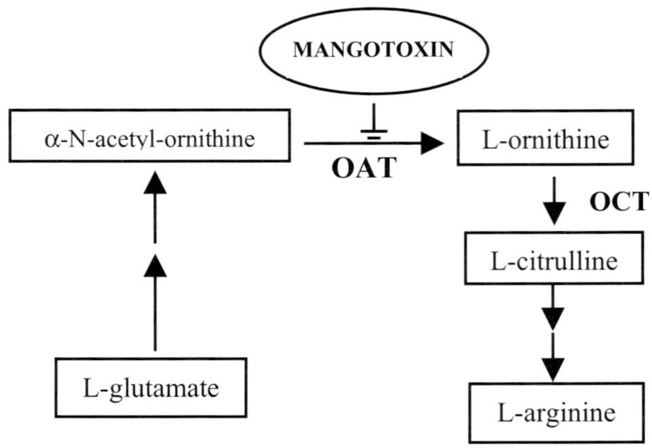

Figure 1. Schematic representation of the step affected by mangotoxin. The enzyme inhibited is the ornithine acetyltransferase (OAT), while phaseolotoxin inhibits ornithine carbamoyl transferase (OCT).

produces ornithine from N-acetyl ornithine; this activity is catalysed by ornithine acetyltransferase (OAT) in most of the organisms (bacteria, fungi, plants), but in *Enterobacteriaceae*, ornithine is produced by simple deacetylation by acetylornithinase, an enzyme apparently also inhibited by mangotoxin. The confirmation of OAT as a target enzyme for mangotoxin was obtained by measuring the enzymatic activity of OAT from plant extracts. Cell-free filtrates from *P. s.* pv. *syringae* strains producers of mangotoxin strongly decrease OAT activity of tomato leaf protein extracts, at similar levels than the specific chemical inhibitor PCMB (positive inhibition control, Table 2). Conversely, filtrates from mangotoxin-non producing strains showed levels of OAT activity close to the controls with PMS medium (negative inhibition control).

The target enzyme affected by mangotoxin is clearly different from those affected by other toxins which act in the biosynthetic pathway of glutamine and arginine, as tabtoxin, inhibiting glutamine sinthetase, or phaseolotoxin, inhibiting the ornithine carbamoyl transferase (OCT) (Bender *et al.*, 1999; Durbin, 1991).

A preliminar characterisation of mangotoxin showed that it is secreted to the media as an hydrophilic molecule, with small size (< 3 kDa) and peptidic nature (sensitive to proteases), and resisting extreme levels of pH and high temperatures (100°C - 1 h), as happens with other antimetabolite toxins (Iacobellis and Latorraca, 1999).

Table 2. Inhibition of ornithine acetyltransferase (OAT) activity in crude tomato leaves extracts by cell-free filtrates of different *Pseudomonas syringae* pv. *syringae* strains. As controls, minimal medium for growing the bacteria (PMS), the specific inhibitor of OAT (PCMB), and bacterial filtrates containing phaseolotoxin (*P. s.* pv. *coronafaciens* CECT4389) and tabtoxin (*P. s.* pv. *phaseolicola* CECT4390), were used.

Filtrates from	*Characteristics*	*Inhibition of* Escherichia coli *on*			*Decrease of OAT activity (%)*
		PMS	*PMS+NAO*	*PMS+Orn*	
PMS	Negative control	-	-	-	0
PCMB	Positive control: inhibitor of OAT	+	+	-	83.1
Pss (n=5)	Mangotoxin-producing	+	+	-	77.8±5.3
Pss (n=3)	Mangotoxin-non producing	-			9.7±4
Psc CECT4389	Tabtoxin-producing	+	-	-	13.6%
Psp CECT4390	Phaseolotoxin-producing	+	+	+	17.5%

PMS: minimal medium (Gasson, 1980); NAO: *N*-acetyl ornithine; Orn: Ornithine; *Pss: P. s.* pv. *syringae*; *Psc*: *P. s.* pv. *coronafaciens*; *Psp*: *P. s.* pv. *phaseolicola.*

The production of mangotoxin appears to be independent of the production of syringomycin, as it shows in Table 1, but the analysis of toxin-defective mutants suggests that the regulation of these two groups of toxins could be related. Disruption of some genes provokes lack of production of mangotoxin and an increase of the production of syringomycin (Table 3), suggesting a co-ordinated regulation of these different toxins, possibly by the global regulatory proteins GacA and LemA. Likewise, mutants affected in putative acetyltransferase and non-ribosomal peptide synthetase are being characterised.

Table 3. Analysis of the DNA sequences of flanking regions of the Tn5 insertions from 9 selected mutants impaired in the production of mangotoxin, obtained by mini-Tn5 mutagenesis of the wild type mangotoxin-producing *Pseudomonas syringae* pv. *syringae* strain UMAF0158.

Tn5 mutants	*Escherichia coli growth inhibition*	*Syringomycin production*	*Sequence (pb)*	*Putative homology to*
3αE10	-	++	410	Two-component regulatory protein LemA
5αC5	(-)	+	935	Acetyltransferase
8αE1	-	++	974	Hrp pathogenicity island
2βB7	-	++	408	Two-component regulatory protein GacA
2βH4	-	++	941	Hrp pathogenicity island
4βA2	(-)	++	988	No homology
3γH1	(-)	+	200	No homology
6γF6	-	+	256	Non-ribosomal peptide synthetase
6γD7	-	++	930	Regulatory protein LemA

-/(-): no/weak inhibition; +/++: production/high production.

Preliminary results about the role of mangotoxin in pathogenesis did not show significant differences on disease symptoms induction between mangotoxin-producing and non-producing strains, including Tn5 mutants and naturally occurring defective strains, when inoculations were performed on detached tomato leaves. Determinations of chlorotic and necrotic areas, chlorophyll levels and bacterial growth did not reveal a primary role of mangotoxin in bacterial pathogenesis.

In relation to the role of mangotoxin in epiphytic interactions, preliminary results showed no effects of mangotoxin against the different fungi (n=9) or yeast (n=3) assayed. However, mangotoxin producing strains showed antimicrobial activity against *Bacillus* and *Erwinia amylovora* (Table 4), which was not observed when Orn was added. In this sense, mangotoxin could provide a competitive advantage for bacteria to colonise various habitats. It could explain that *P. s.* pv. *syringae* has a broader host range compared with other pathovars and that it occurs on various habitat in its saprophytic phase, as discussed previously (Volksch and Weingart, 1998). In addition, it might also minimise the growth of secondary invaders in infected tissues to ensure a high population of the pathogen for overwintering in diseased plants or seeds (Cazorla *et al.*, 1998). On the other hand, during the parasitic phase, mangotoxin could contribute to increase disease severity as a virulence factor, by disturbing plant cell ornithine biosynthesis.

Table 4. Antagonistic activity of mangotoxin-producers and non-producers *Pseudomonas syringae* pv. *syringae* strains against epiphytic bacteria. Assays were carried out on minimal medium (PMS) and PMS added with ornithine (PMS+Orn). Data (inhibition halo diameter in mm) represent the mean values and standard deviation from three replicate experiments and "–" denote no detection of inhibition halo.

Target bacteria	*Pss mangotox + (n=18)*	*Pss mangotox + (n=18)*	*Pss mangotox - (n=8)*	*Pss Tn5-mutants mangotox - (n=2)*	*Psp CECT-4390*
	PMS	*PMS+Orn*	*PMS*	*PMS*	*PMS*
Bacillus megaterium	11.0 ± 2.5	-	-	-	-
Bacillus subtilis	9.0 ± 2.4	-	-	-	-
Clavibacter michiganensis	-	-	-	-	-
Escherichia coli	16.4 ± 1.0	-	-	-	10.0 ± 0.9
Erwinia amylovora	4.5 ± 1.8	-	-	-	-
Pantoea agglomerans	-	-	-	-	-
Pseudomonas fluorescens	-	-	-	-	-

Pss: *P. s.* pv. *syringae* strains; *Psp*: *P. s.* pv. *phaseolicola.*

ACKNOWLEDGEMENTS

This work was supported by grants from Junta de Andalucía (AGR-169) and from Plan Nacional de I+D (AGF98-0931-C02-01). We want to thank J.A. Torés. J.M.

Hermoso and J.M. Farré for helpful suggestions and work facilities at E.E. La Mayora (CSIC-Spain). We thank A. Litcher, M. López, C. Manceau, E. Montesinos and A. Vivian for kindly providing us some of the bacterial strains.

REFERENCES

Alstchul, S.E., Madden, T.L., Schaffer, A.A., Zhang, J., Zhang, Z., Miller, W., and Lipman, D.J., 1997, Gapped BLAST and PSI-BLAST: a new generation of protein database search programs. *Nucleic Acids Res.* **25**: 3389-3402.

Bender, C.L., Alarcón-Chaidez, F., and Gross, D.C., 1999, *Pseudomonas syringae* phytotoxins: Mode of action, regulation, and biosynthesis by peptide and polyketide synthetases. *Microbiol. Mol. Biol. Rev.* **63**: 266-292.

Cazorla, F.M., Olalla, L., Torés, J.A., Codina, J.C., Pérez-García, A., and de Vicente, A., 1997, *Pseudomonas syringae* pv. *syringae* as microorganism involved in apical necrosis of mango: characterization of some virulence factors. In "*Pseudomonas syringae* Pathovars and Related Species". Ed. K. Rudolph, J.J. Burr, J.W. Mansfield, D. Stead, A. Vivian, and J. Von Kietzell. Kluwer Academic Publishers. Dordrecht. The Netherlands. pp. 82-87.

Cazorla, F.M., Torés, J.A., Olalla, L., Pérez-García, A., Farré, J.M., and de Vicente, A., 1998, Bacterial apical necrosis of mango in Southern Spain: a disease caused by *Pseudomonas syringae* pv. *syringae*. *Phytopathology* **88**: 614-620.

De Lorenzo, V., and Timmis, K.N., 1994, Analysis and construction of stable phenotypes in gram-negative bacteria with Tn5 and Tn10-derived mini-transposons. *Methods Enzymol.* **235**: 386-406.

Denés, G., 1970, Ornithine acetyltransferase (*Chlamydomonas reinhardti*). *Methods Enzymol.* **17A**: 273-277.

Durbin, R.D., 1991, Bacterial phytotoxins: mechanisms of action. *Experientia* **47**: 776-783.

Gasson, M.J., 1980, Indicator technique for antimetabolite toxins production by phytopathogenic species of *Pseudomonas*. *Appl. Environ. Microbiol.* **39**: 25-29.

Gross, D.C., and De Vay, J.E., 1977, Production and purification of syringomycin, a phytotoxin produced by *Pseudomonas syringae*. *Physiol. Plant Pathol.* **11**: 13-28.

Iacobellis, N.S., and Latorraca, M. 1999, Production of antimetabolites by *Pseudomonas syringae* pathovars. Molecular Plant-Microbe Interactions, 9th International Congress. Amsterdam (The Netherlands). pp.180.

Iacobellis, N.S., Lavermicocca, P., Grgurina, I., Simmaco, M., and Ballio, A., 1992, Phytotoxic properties of *Pseudomonas syringae* pv. *syringae* toxins. *Physiol. Mol. Plant Pathol.* **40**: 107-116.

Pérez-García, A., Cánovas, F.M., Gallardo, F., Hirel, B., and de Vicente, A., 1995, Differential expression of glutamine synthetase isoforms in tomato detached leaflets infected with *Pseudomonas syringae* pv. *tomato*. *Mol. Plant-Microbe Interact.* **8**: 96-103.

Sinden, S.L., de Vay, J.E., and Backman, P.A., 1971, Properties of syringomycin, a wide spectrum antibiotic and phytotoxin produced by *Pseudomonas syringae*, and its role in the bacterial canker disease of peach trees. *Physiol. Plant Pathol.* **1**: 199-213.

Volksch, B., and Weingart, H., 1998, Toxin production by pathovars of *Pseudomonas syringae* and their antagonistic activities against epiphytic microorganisms. *J. Basic Microbiol.* **38**: 135-145.

Antimicrobial Lipodepsipeptides from *Pseudomonas* spp: a Comparison of Their Activity on Model Membranes

G. MENESTRINA, M. CORAIOLA, V. FOGLIANO[#], A. FIORE[#], I. GRGURINA[§], A. CARPANETO[‡], F. GAMBALE[‡] and M. DALLA SERRA
CNR-ITC Istituto di Biofisica, Sezione di Trento, Povo (Trento), Italy; [#]Dipartimento di Scienza degli Alimenti, Università di Napoli "Federico II", Napoli, Italy; [§]Dipartimento di Scienze Biochimiche "A. Rossi Fanelli", Università di Roma "La Sapienza", Roma, Italy; [‡]CNR Istituto di Biofisica, Sezione di Genova, Genova, Italy.

Abstract: Lipodepsipeptides (LPDs) are a group of cyclic, acylated peptides produced by several *Pseudomonas* species. They are usually divided in two groups, mycins and peptins, on the basis of the size of the amino acidic part of the molecule. Mycins have a ring of 9 amino acids closed between the first and the last residue, peptins contain a more complex peptide moiety of up to 25 amino acids, partially cyclized. Both mycins and peptins attack the plasma membrane, but may have different target organisms. Comparing the mode of action of these two classes of LDPs on natural and model membranes we observed that all peptides induced red blood cell haemolysis and leakage of tonoplasts and liposomes by the formation of pores. The haemolytic activity of the smaller mycins was higher than that of the bigger peptins and proportional to the amphipathic index of the molecule. The extent of permeabilization was dependent also on the composition of the lipid membrane. In particular, mycins show a preference for sterols, whereas peptins are more active on phospholipids, especially sphingomyelin. These differences may have physiological implications. The formation of discrete ion channels, with anionic selectivity, was directly demonstrated by electrophysiological experiments performed on planar lipid bilayers or sugar beet vacuoles. The channels show sub-states and their properties in vacuoles and in planar lipid membranes were remarkably similar.

Key words: Haemolysis, toxin pores, surface activity, lipid bilayer.

Author for correspondence: G. Menestrina; e-mail: menes@itc.it.

N.S. Iacobellis *et. al.* (eds.), *Pseudomonas syringae* and Related Pathogens, 185-198.

1. INTRODUCTION

Lipodepsipeptides (LDPs) are a group of cyclic secondary metabolites produced by many *Pseudomonas* species including some pathovars of *P. syringae* (Bender *et al.*, 1999). On the basis of their amino acid chain length, they are usually divided into two groups: the mycins (Takemoto *et al.*, 2003) and the peptins (Dalla Serra *et al.*, 2003). Syringomycins, syringotoxin and pseudomycins, belonging to the first group, consist of acylated peptides of 9 amino acids closed in a cyclic structure between the first and the last residue (Segre *et al.*, 1989; Ballio *et al.*, 1990, 1994). Syringopeptins 22 and 25, together with fuscopeptins (19 amino acids, Ballio *et al.*, 1996) and corpeptins (22 amino acids, Emanuele *et al.*, 1998) instead, belong to the second group (Ballio *et al.*, 1991). They are bigger and more complex than the former, being composed of 19 to 25 amino acids, again cyclized (usually in an 8 or in a 5 residues ring) and acylated. Both kind of peptides, mycins and peptins, are expressed *in vivo* (Gallo *et al.*, 2000), they significantly contribute to bacterial pathogenesis and have the plasma membrane as their primary site of action, but may have different target organisms. We studied and compared the activity of the members of these two classes of LDPs on red blood cells (RBC) from different origin and on model membranes (unilamellar vesicles and planar lipid bilayers).

2. MATERIALS AND METHODS

2.1 Chemicals

Phospholipids used were all obtained from Avanti Polar Lipids (Alabaster, AL, USA) at the highest possible grade (usually more than 99% pure). Cholesterol was from Fluka, (Buchs, Switzerland), ergosterol and stigmasterol from Sigma (St. Louis, MO, USA).
Different combinations were used as specified in the text. Calcein, EDTA, and Sephadex were from Sigma.

2.2 Peptides and Their Preparation

Syringomycins E, G (SRE, SRG), syringotoxin (ST), syringopeptins 22A, 22Phv, 25A and 25Lau ($SP_{22}A$, $SP_{22}Phv$, $SP_{25}A$ and $SP_{25}Lau$), pseudomycin A, B, C, C' (PSA, PSB, PSC, PSC') were prepared and purified from cultures of *Pseudomonas syringae* pv. *syringae* strains as previously described (Segre *et al.*, 1989; Ballio *et al.*, 1990, 1991, 1994; Scaloni *et al.*,

1997; Grgurina *et al.*, 2002). Primary structures of these LDPs are reported in Fig.1.

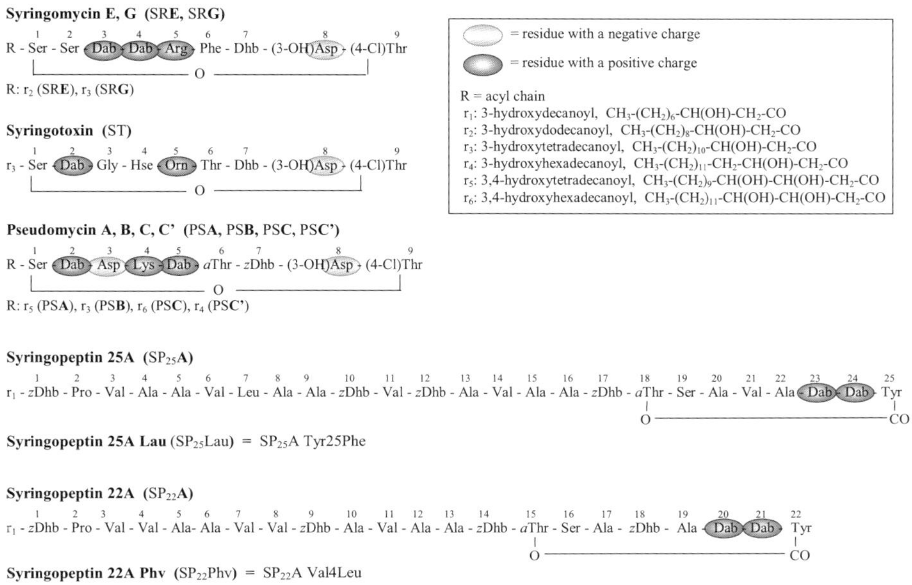

Figure 1. Primary structure of the *Pseudomonas syringae* lipodepsipeptides investigated. Non standard amino acids are: Dab, 2,4-diaminobutanoic acid; Dhb, 2,3-dehydro-2-aminobutyric acid; Hse, homoserine; Orn, ornithine.

2.3 Red Blood Cell Haemolysis

Haemolytic activity of LDPs was determined turbidimetrically with a 96-well microplate reader as described earlier (Dalla Serra *et al.*, 1999a). Briefly: human red blood cells (HRBC) were prepared from fresh heparinized blood from healthy volunteers, by extensively washing in NaCl 0.85%. LDPs were plated in a 96-well plate with 2-fold serial dilutions in a buffer containing NaCl 120 mM, MES 10 mM, EDTA 1 mM, pH 6.0. Each well was supplemented with 0.02% HRBC and continuously stirred. Percent haemolysis was calculated as $(A_i-A_f)/(A_i-A_o) \times 100$, where A_i, A_f and A_o are the initial, final and completely lysed absorbances at 650 nm. A_i was about 0.13 OD and A_0 (as obtained by hypotonical lysis with pure water) was around 0.03 OD.

2.4 Membrane Damaging Action on Plant Protoplasts

Root protoplasts were obtained from maize seedlings (DK 502 DeKalb) by a slight modification of a described procedure (Bregante *et al.*, 1997). Maize seeds were grown in the dark at 26°C from 4 to 7 days on wet filter-paper. The tip of each root was removed and about 30 root segments where chopped in the enzyme solution: $CaCl_2$ 1 mM, polyvinylpyrrolidone 0.5% (w v^{-1}), BSA 0.5%, cellulose 0.8% (Onozuka RS, Yakult Honsha Co. LTD., Tokyo), pectolyase 0.08% (Sigma), sorbitol 500 mM, MES 5 mM and KOH at pH 5.5. The solution was stirred at 30°C for one hour. The digest was filtered through 100 μm muslin and centrifuged at 60 g, 10°C, for 5 min. The supernatant was discarded and the pellet resuspended in 0.5÷1 ml solution P1 (KCl 30 mM, $CaCl_2$ 1 mM, $MgCl_2$ 2 mM, sorbitol 530 mM, Mes 10 mM and Tris at pH 5.6) and stored on ice for a maximum of two days. To assay LDPs effects, the protoplasts were incubated in solution P2 (neutral red 5 mg ml^{-1}, KCl 30 mM, $CaCl_2$ 1 mM, $MgCl_2$ 2 mM, sorbitol 530 mM, Hepes 10 mM and Mes at pH 7.4) for 5÷10 min at room temperature. Neutral red is an acidotropic dye that permeates through the plasma membrane and the tonoplast at neutral pH. In the acidic environment of the vacuole the dye becomes protonated, trapped and accumulated as a cation. The protoplasts with the stained vacuoles were resuspended in solution P1 or in solution P3 ($CaCl_2$ 1 mM, sorbitol 575 mM, Mes 10 mM and KOH 1.5 mM, pH 5.6) with the addition of variable amounts of LDPs (e.g. 1.3÷15 μM $SP_{25}A$ or 0.13÷15 μM SRE). The acidic bath solution allows the protonated neutral red to escape from the vacuole only after permeabilisation by the LDPs.

2.5 Permeabilization of Artificial Lipid Membranes

Membrane permeabilization was determined fluorimetrically with a 96-well fluorescence microplate reader (Fluostar, SLT, Groeding, Austria) as already described (Dalla Serra *et al.*, 1999a). Briefly: large unilamellar vesicles (LUVs), were prepared by extrusion through polycarbonate filters with 100 nm pores, loaded with calcein at a self-quenching concentration and washed by gel-filtration on sephadex G50 medium (Amersham-Biosciences, Uppsala, Sweden). Various lipid compositions were used, as detailed in the text. LDPs were 2-fold serially diluted in a buffer containing NaCl 100 mM, MES 10 mM, EDTA 1 mM, pH 6.0, supplemented with washed LUVs (at a final lipid concentration of 6 μM in 200 μl) and continuously shaken. The percent of permeabilization, $R_{\%}$, was calculated as $(F_{fin} - F_{in})/(F_{max} - F_{in})$ x 100, where F_{in}, F_{fin} are the initial and final fluorescence, and F_{max} corresponds to maximal dye release, which was induced by 1 mM Triton X-100.

2.6 Patch-Clamp Experiments

Ion channel formation by LDPs in plant tonoplasts was measured by patch-clamp as described in Carpaneto *et al.* (2002). Sugar beet vacuoles were extruded directly into the recording chamber by slicing the sugar beet root in their standard bath solution (KCl 150 mM, EDTA 1 mM, Mes 10 mM, pH 6.0, adjusted by KOH ~5 mM) as in Gambale *et al.* (1996). Currents were recorded in the whole-vacuole or in the cytoplasmic-side-out excised-patch configurations. Applied voltages and currents were controlled and recorded with a List EPC7 patch-clamp amplifier interfaced with an Instrutech AD/DA board (Instrutech, Elmont, N.Y., USA). Data analysis were done using home-made programs based on IgorPro (Wavemetrics Inc., Lake Oswego, OR, USA). Upon reproduction, single channel data were filtered with a digital low-pass filter at a frequency of 5 or 10 Hz.

2.7 Planar Lipid Bilayer Experiments

Ion channel formation by LDPs in lipid bilayers was measured in planar lipid membranes (PLM) prepared by the apposition technique (Montal and Mueller, 1972) as already described (Dalla Serra *et al.*, 1999b; Carpaneto *et al.*, 2002). Briefly: two monolayers were spread from a 6 mg ml^{-1} lipid solution in n-hexane and attached on both sides of a n-hexadecane pre-treated hole (0.1 mm diameter) in a 12 μm Teflon foil. Lipid used were either asolectin (re-purified from Sigma) or diphytanoylphosphatidylcholine (DPhPC). Four ml of NaCl 100 mM, Mes 10 mM, EDTA 1 mM at pH 6.0 (adjusted by NaOH) were present on both sides of the membrane. LDPs were added only on one side of a stable PLM. Untreated membranes had a typical capacity of 100 pF and conductance of 10 pS at 100 mV. Voltage clamp experiments were performed using a Dagan 3900A integrated amplifier (Dagan Corporation, Minneapolis, MN, USA), with conversion factor 1 mV pA^{-1}, low-pass filter at 1 kHz, and were digitally stored via a PCM recorder (PCM-701ES from Sony, Tokyo, Japan). For display and analysis the traces were low-pass filtered at 100 Hz, acquired on a PC and analysed with the software package IgorPro.

3. RESULTS

3.1 Red Blood Cell Lysis

LDPs of *Pseudomonas syringae* presumably exert their antibiotic activity via a relatively non-specific attack of the plasma membrane (Hutchison *et al.*,

1995). An example of this non-specific action is the fact that they induce haemolysis of erythrocytes (Dalla Serra *et al.*, 1999a; Agner *et al.*, 2000), which are not supposed to be a biological target cell. All the LDPs we tested proved to be haemolytic on HRBC, albeit to a different extent (Table 1). Results with rabbit or sheep RBC were similar (not shown). An estimate of the haemolytic activity is given by C_{50}, the concentration necessary to produce haemolysis of 50% of cells. This was clearly different among the LDPs tested. In general, the C_{50} of mycins is smaller than that of the bigger peptins, indicating a higher activity. Interestingly, within a certain group of peptides, the activity increased with the length of their acyl chain of the fatty acid moiety. For example the activity of PSC and PSC', carrying an hydroxyhexadecanoyl chain, was about one order of magnitude above that of PSA and PSB, carrying an hydroxytetradecanoyl chain. Similarly, SRG, carrying an hydroxytetradecanoyl chain, was about ten times more active than SRE, carrying an hydroxydodecanoyl chain. When different LDP groups were compared, instead, it appears that within molecules carrying an acyl chain of the same length, the activity is higher for those that are more basic. In fact, SRG (with a net balance of two positive charges) is more active than PSB and ST, that have both a net balance of one positive charge, whereby all three these LDPs have a 3-hydroxytetradecanoyl chain. Furthermore, when PSA and ST are directly compared it appears that, at the same chain length and net charge, a higher polarity (5 charged residues for PSA instead of three for ST) gives slightly more activity.

Table 1. Lytic activity of *Pseudomonas syringae* lipodepsipeptides on human red blood cell.

	Peptide	C_{50}[(a)] ± SEM (µM)	h[(b)]	n[(c)]
Mycins				
	SRE	0.64 ± 0.09	6 ± 1	10
	SRG	0.05 ± 0.02	6 ± 1	5
	ST	9.5	5 ± 1	2
	PSA	7.8 ± 1.4	6 ± 1	3
	PSB	5.9 ± 0.2	6 ± 1	3
	PSC	0.4	6 ± 1	2
	PSC'	0.6	6 ± 1	2
Peptins				
	SP_{25} A	3.7 ± 0.8	5 ± 1	14
	SP_{25} Lau	12.5	2 ± 1	2
	SP_{22} A	11.7 ±1.7	2 ± 1	3
	SP_{22} Phv	19.4	2 ± 1	1

[a] C_{50}, is the concentration required for half of the total haemolysis.
[b] h, is the Hill coefficient of co-operativity obtained when the dose dependence of haemolysis is fit to the Hill equation.
[c] n, is the number of experiments.

The dose dependence of haemolysis was, in all cases, sigmoidal and quite steep, suggesting a highly co-operative behaviour compatible with an oligomerisation mechanism (Dalla Serra *et al.*, 1999a). All the curves were fitted to the Hill equation and the resulting Hill coefficients are reported in Table 1. In general Hill coefficients are bigger for mycins than for peptins suggesting the formation of larger oligomeric lesions by the first class of molecules.

3.2 Action on Maize Roots Protoplast

The action of some LDPs was next examined on a more natural cellular model, i.e. the plant protoplast. The effects of $SP_{25}A$ on isolated maize roots protoplast, stained with neutral red, are shown in Fig. 2. At 30°C, after 1.5

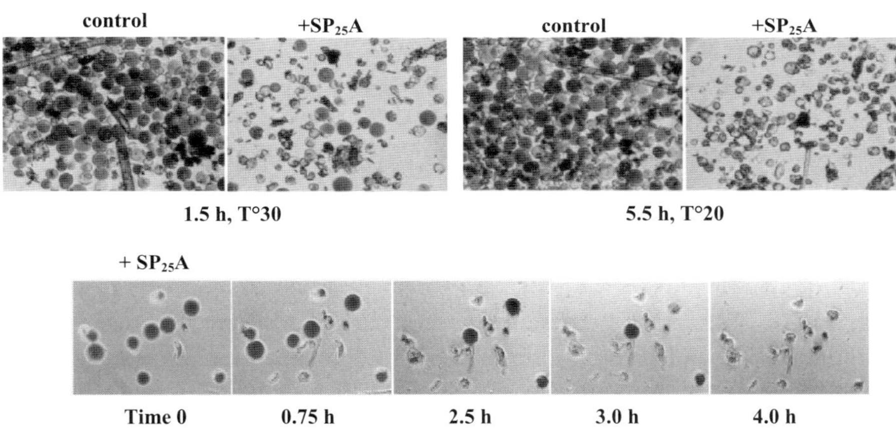

Figure 2. Effects of $SP_{25}A$ on the tonoplast of isolated maize roots protoplasts. Upper panels: Protoplasts and vacuoles, previously stained with neutral red (dark colour), were incubated in solution P1 without (control) or with (+$SP_{25}A$) 15 μM $SP_{25}A$, either for 1.5 hours at 30°C (two left panels), or for 5.5 hours at 20°C (two right panels). The average diameter of the protoplasts was 40÷50 μm. Bottom panels: The permeabilisation of the tonoplast induced by $SP_{25}A$ in nine isolated protoplasts was followed in time. Bath solution was P3, $SP_{25}A$ was 10 μM, T was 30°C.

hours exposure to 15 μM $SP_{25}A$, most of the protoplasts had lost their vacuolar content, even if it was still possible to clearly recognise the plasma membrane barrier (upper left panels). At 20°C, instead, around 5.5 hours were required to observe the same effect. (upper right panels). A time course of vacuolar content release is reported in the lower panels. Here, a group of protoplast is shown at different times after the addition of 15 μM $SP_{25}A$ at 30°C. Interestingly, the release of the dye appears to be an all-or-nothing

mechanism. Although the frames we show are only a selection the whole kinetic was recorded at a rate of 20 frames $second^{-1}$ and it was not possible to observe a gradual decrease on dye in any of the protoplasts. The rate limiting event is probably the loss of integrity of the plasma membrane, that allows the access of the LDPs to the vacuolar membrane where they quickly form channels with radius bigger than the dye and allow an almost immediate release. Control fields, not exposed to any toxin, clearly demonstrated that spontaneous release of the dye was virtually impossible at the acidic pH that we used. SRE had a similar effect, although at a concentration around ten times smaller than that of $SP_{25}A$ (not shown).
Therefore it appears that also in this case the smaller mycins are more active than the bigger peptins.

3.3 Action on Unilamellar Lipid Vesicles

All peptides were also active on very simple model membranes comprised only of purified lipid components. In fact they all induced leakage of calcein encapsulated into liposomes, in an assay that we already described (Dalla Serra *et al.*, 1999a). The extent of permeabilisation was dependent, besides on the concentration and the nature of the peptide, also on the composition of the lipid vesicles. The C_{50} for vesicle permeabilisation, derived each from a dose dependence experiment, are reported in Table 2 for the different LDPs and LUV compositions. We observed that mycins are poorly active on LUV comprised only of PC (the main phospholipid component of natural membranes), but these are sensitive to peptins. When the other main components of eukaryotic animal cell membranes are introduced, i.e. cholesterol and sphingomyelin, it appears that mycins have a strong preference for liposomes containing cholesterol, whereas peptins are more active if sphingomyelin is present. If cholesterol is replaced by other sterols, e.g. ergosterol and stigmasterol (which are typical for fungal and plant membranes, respectively) the effect of peptins are enhanced but those of mycins are reduced. The introduction into LUVs containing a fixed amount of cholesterol and sphingomyelin, of variable amounts of a negatively charged phospholipid, PS, in place of PC, enhanced the activity of $SP_{25}A$, but reduced that of SRE and SRG.
Similarly to the haemolytic activity, also LUVs permeabilisation increased with the length of the acyl chain of the fatty acid moiety of different LDPs. In fact, PSC and PSC', were in general around one order of magnitude more active than PSA and PSB, and SRG was more active than SRE. It increased also with the net positive charge of the peptide part (syringomycins, with two charges, were much more active than pseudomycins, with only one charge). Within the peptins, activity was inversely related with the size of the molecule, being higher for $SP_{22}A$ and SP_{22} Phv than for $SP_{25}A$.

Table 2. Permeabilising action of the different LDPs on LUVs of different composition.

Peptide	Lipid composition[a]								
	PC 100	Chol 50 PC 50	Chol 33 SM 17 PC 50	Chol 17 SM 33 PC 50	SM50 PC50	Erg 50 PC 50	Stig 50 PC 50	Chol 33 SM 17 PC 45 PS 5	Chol 33 SM 17 PC 10 PS 40
mycins									
SRE	7.1 ± 1.2[b]	0.05 ± 0.02	0.3 ± 0.1	1.3 ± 0.1	0.9 ± 0.1	0.4 ± 0.1	24 ± 10	0.53	1
SRG	15	0.16	0.22	0.20	0.15	4.0	16.5	0.32	0.4
PSA	_	21.6	56.3	>54	>54	_	_	74.2	>54
PSB	_	21.6	61.5	>32.4	125.0	_	_	>32.4	>32.4
PSC	_	2.34	7.2	22.0	44.7	_	_	5.9	>12.6
PSC'	_	2.02	6.2	9.4	44.7	_	_	8.4	>5.4
peptins									
$SP_{25}A$	2.7	8.10	2.3	0.8	0.40 ± 0.04	3.7	4.35	3.6	0.5
$SP_{22}A$	0.26 ± 0.01	2.1 ± 0.2	0.7 ± 0.1	0.29 ± 0.01	0.24 ± 0.04	0.6 ± 0.1	0.4	_	_
SP_{22} Phv	0.7 ± 0.2	5.1 ± 1.8	2.2 ± 0.3	0.64 ± 0.04	0.46 ± 0.12	1.8 ± 0.4	_	_	_

[a] Lipid compositions are expressed in percentage of each component on a molar basis. Abbreviations are: PC, phosphatidycholine; Chol, cholesterol; SM, sphyngomielin; Erg, ergosterol; Stig, stigmasterol; PS, phosphatidylserine.
[b] Experimental values (expressed in μM) are C_{50}, i.e. LDP concentrations causing 50% of calcein release, measured as in (Dalla Serra *et al.*, 1999a). Values are either mean ± SEM of three experiments; or single determinations.

3.4 Ion Channel Formation in Planar Lipid and Vacuolar Membranes

The formation of LDPs induced ion channels was directly demonstrated with electrophysiological experiments performed on planar lipid bilayers or on sugar beet vacuoles. In both systems, $SP_{25}A$ and SRE increase membrane permeability by forming discrete ion channels (Fig. 3). These channels show anionic selectivity and intrinsic rectification properties that presumably derive from an asymmetric distribution of fixed charges on the pore (Dalla Serra *et al.*, 1999b; Carpaneto *et al.*, 2002).
The properties of $SP_{25}A$ channels in vacuoles and in planar lipid membranes were remarkably similar. In both systems channels were stable on a time scale of seconds, but a slow deactivation was detected with time. This effect, which was noted also by others (Ziegler *et al.*, 1984; Dalla Serra *et al.*, 1999a) in different model membranes, could be related to a desorbtion of the toxin from the lipid membrane. In both systems we observed the presence of channel sub-states with lower conductance (Fig. 3) (Kaulin *et al.*, 1998; Dalla Serra *et al.*, 1999b).

Similar sub-states were observed also with SRE and are evident also in the trace in Fig. 3B were they cause regular splitting in the conductance histograms.

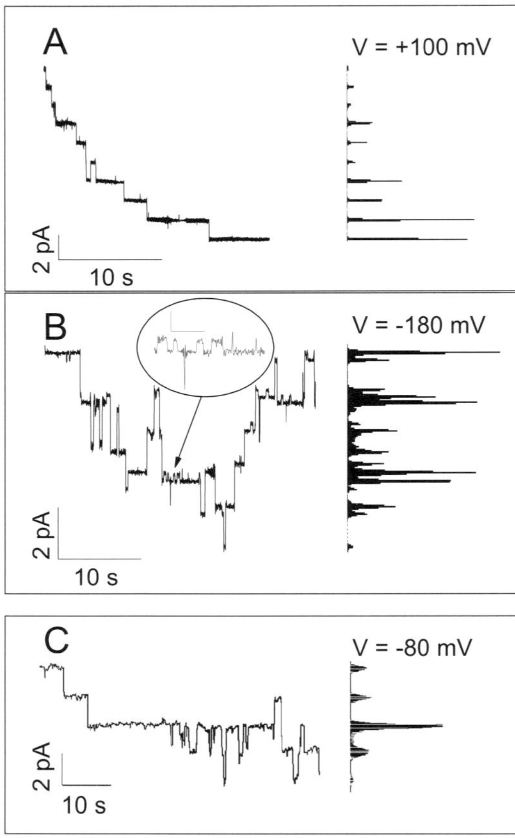

Figure 3. Electrophysiological demonstration of ion channel formation by LDPs in lipid membranes. Upper panels: discrete ionic current fluctuations in a planar lipid membrane after addition of 0.1 μM $SP_{25}A$ (panel A) or 10 μM SRE (panel B). The membrane was comprised of asolectin (a plant lipid mixture) in A and DPhPC in B. The conductance of the $SP_{25}A$ pore was 15 pS; that of SRE 8.5 pS. SRE sub-state levels of ~1.7 pS were also observed which correspond to 1/5 of the large channel. They are resolved in the amplified section of the trace in panel B (scale bars are 1s and 0.5pA). Panel C: single channel transitions in sugar beet vacuolar membranes perfused with 1 μM $SP_{25}A$. The conductance is here around 25 pS.

4. DISCUSSION

Being poorly specific permeabilisers of the cell membrane, all LDPs induce also haemolysis. In general, mycins are haemolytically more potent than peptins (Table 1). In addition, it appears that the haemolytic activity is proportional to the amphipathic index of the molecule. i.e. it increases either when the hydrophobic character of the acyl chain is increased (longer chains), or when the hydrophilic character of the peptide moiety is increased (more basic or more polar peptidic portion). Besides more active, mycins show also a higher degree of cooperativity, suggesting they may form larger oligomers. This is consistent with the results of osmotic-protection experiments (Dalla Serra *et al.*, 1999a), in which the presence of a sugar added in the external solution retarded the kinetics of haemolysis to an extent that increased with the size of the osmoticant. This is due to the time necessary for the sugar to diffuse into the cells through the toxin-induced lesions, and indicated that mycins formed pores with larger radius than peptins. Furthermore, mycins pore radius increased with peptide concentration (from 0.7 nm to 1.7 nm for SRE Dalla Serra *et al.*, 1999a), suggesting that the channels could be formed by a variable number of monomers, with larger aggregates favoured at higher LDP concentration. Apparently this was not the case with peptins that exhibited a radius (0.98 ± 0.03 nm) independent of the LDP dose.

LDPs permeabilise also plant cells. The experiments we report here demonstrate that the tonoplast can be a true target for the toxin. In fact, by using pre-stained plant protoplasts, we could show that both peptins and mycins (e.g. $SP_{25}A$ and SRE) were able to empty the tonoplasts in a time, dose and temperature dependent way. Evidently, to do this, the toxin had first to surmount the protoplast plasma-membrane barrier and then diffuse to the tonoplast membrane. This may result from the surfactant activity of the LDPs, i.e. their capacity to disorganise the lipid membrane if they are present at concentrations higher then their critical micelle concentration (CMC, Che *et al.*, 1992; Hutchison *et al.*, 1995; Hutchison and Gross, 1997; Dalla Serra *et al.*, 1999a). It should be noted that in an *in vivo* situation, getting access to the protoplast itself would require the presence of a cell-wall degrading enzyme (like those we have used in our protoplast preparation). Such enzymes are likely those produced by the bacteria itself, and indeed a strong synergism between $SP_{25}A$ activity and bacterial cell-wall degrading enzymes has been recently demonstrated (Fogliano *et al.*, 2002). The action of SP_{22} on plant protoplasts has been already studied with similar results (Di Giorgio *et al.*, 1994).

SRE appears to be more potent than these peptins, possibly simply because of its higher surfactant activity, expressed by a lower CMC (Dalla Serra *et al.*, 1999a), or because of its smaller size which allows a faster diffusion through the damaged plasma membrane to reach the tonoplast. The fact that LDPs can cross the protoplast membrane and permeabilize the internal vacuoles may be physiologically important for the bacteria, since it would allow full exploitation of the nutrients stored by the plant into these compartments.

The differences observed in the membrane permeabilising activity, depending on LUVs composition, may also have physiological implications. It appears that peptins are most effective on LUVs reproducing a lipid composition typical for plant cell membranes. In fact, they are more active on membranes containing phospholipids, in particular those negatively charged, and sphingomyelin. Even among sterols, they prefer stigmasterol, which is again a plant lipid. The mycins, instead, have a stronger preference for sterol-containing LUVs, whereby the best sterols are cholesterol and ergosterol (Julmanop *et al.*, 1993; Wangspa and Takemoto, 1998). This may reflect a better activity of mycins on fungi, and explains also their stronger haemolytic action, since RBC contain high levels of cholesterol, besides PC. Electrophysiological experiments are paramount to demonstrate the channel-forming properties of LDPs (Feigin *et al.*, 1996; Hutchison and Gross, 1997; Kaulin *et al.*, 1998; Dalla Serra *et al.*, 1999b; Carpaneto *et al.*, 2002). Both peptins and mycins form discrete ion channels in model systems, either as simple as an artificial planar lipid membrane, or more complex as the plant vacuole membrane. Channels in both systems have similar properties: conductance, rectification and anion-selectivity that can be reconducted to the presence of positive charges in the peptide and their asymmetric transmembrane distribution (Dalla Serra *et al.*, 1999b). Even in this case, the lipid composition of the target membrane influences the activity of LDPs. In particular, the voltage dependence of channel opening is different and is large in membranes composed of neutral phospholipids, e.g. pure PC, but much less in membranes composed of plant lipids like asolectin or the vacuole membrane (Dalla Serra *et al.*, 1999b; Carpaneto *et al.*, 2002).

ACKNOWLEDGEMENTS

This work was financially supported by the Italian Consiglio Nazionale delle Ricerche (CNR), by the Istituto Trentino di Cultura (ITC) and by a special grant from the Provincia Autonoma di Trento (PAT, project AGRIBIO). We want to thank Prof. Alessandro Ballio for constant stimulation and advice.

REFERENCES

Agner, G., Kaulin, Y.A., Gurnev, P.A., Szabo, Z., Schagina, L.V., Takemoto, J.Y., and Blasko, K., 2000, Membrane-permeabilizing activities of cyclic lipodepsipeptides, syringopeptin 22A and syringomycin E from *Pseudomonas syringae* pv. *syringae* in human red blood cells and in bilayer lipid membranes. *Bioelectrochemistry* **52**: 161-167.

Ballio, A., Bossa, F., Collina, A., Gallo, M., Iacobellis, N.S., Paci, M., Pucci, P., Scaloni, A., Segre, A., and Simmaco, M., 1990, Structure of syringotoxin, a bioactive metabolite of *Pseudomonas syringae* pv. *syringae*. *FEBS Lett.* **269**: 377-380.

Ballio, A., Barra, D., Bossa, F., Collina, A., Grgurina, I., Marino, G., Moneti, G., Paci, M., Pucci, P., Segre, A., and Simmaco, M., 1991, Syringopeptins, new phytotoxic lipodepsipeptides of *Pseudomonas syringae* pv. *syringae*. *FEBS Lett.* **291**: 109-112.

Ballio, A., Bossa, F., Di Giorgio, D., Ferranti, P., Paci, M., Pucci, P., Scaloni, A., Segre, A., and Strobel, G.A., 1994, Novel bioactive lipodepsipeptides from *Pseudomonas syringae*: the pseudomycins. *FEBS Lett.* **355**: 96-100.

Ballio, A., Bossa, F., Camoni, L., Di Giorgio, D., Flamand, M.C., Marcite, H., Nitti, G., Pucci, P., and Scaloni, A., 1996, Structure of fuscopeptins, phytotoxic metabolites from *Pseudomonas fuscovaginae*. *FEBS Lett.* **381**: 213-216.

Bregante, M., Carpaneto, A., Pastorino, F., and Gambale, F., 1997, Effects of mono- and multivalent cations on the inward-rectifying potassium channel in isolated protoplasts from maize roots. *Euro. Biophys. J. Biophys. Lett.* **26**: 381-391.

Carpaneto, A., Dalla Serra, M., Menestrina, G., Fogliano, V., and Gambale, F., 2002, The phytotoxic lipodepsipeptide syringopeptin 25A from *Pseudomonas syringae* pv *syringae* forms ion channels in sugar beet vacuoles. *J. Membrane Biol.* **188**: 237-248.

Che, F.S., Kasamo, K., Fukuchi, N., Isogai, A., and Suzuki, A., 1992, Bacterial phytotoxins, syringomycin, syringostatin and syringotoxin, exert their effect on the plasma membrane H^+-ATPase partly by detergent-like action and partly by inhibition of the enzyme. *Physiol. Plant.* **86**: 518-524.

Dalla Serra, M., Fagiuoli, G., Nordera, P., Bernhart, I., Della Volpe, C., Di Giorgio, D., Ballio, A., and Menestrina, G., 1999a, The interaction of lipodepsipeptide toxins from *Pseudomonas syringae* pv. *syringae* with biological and model membranes: a comparison of syringotoxin, syringomycin and syringopeptins. *Mol. Plant-Microbe Interact.* **12**: 391-400.

Dalla Serra, M., Nordera, P., Bernhart, I., Di Giorgio, D., Ballio, A., and Menestrina, G., 1999b, Conductive properties and gating of channels formed by syringopeptin 25-A, a bioactive lipodepsipeptide from *Pseudomonas syringae* pv. *syringae*, in planar lipid membranes. *Mol. Plant-Microbe Interact.* **12**: 401-409.

Dalla Serra, M., Menestrina, G., Carpaneto, A., Gambale, F., Fogliano, V., and Ballio, A., 2003, Molecular mechanism of action of syringopeptins, antifungal peptides from *Pseudomonas syringae* pv. *syringae*. In *Pore-forming peptides and protein toxins* (G. Menestrina, M. Dalla Serra and P. Lazarovici, eds.), Taylor & Francis Group, London, UK, pp. 272-295.

Di Giorgio, D., Camoni, L., and Ballio, A., 1994, Toxins of *Pseudomonas syringae* pv. *syringae* affect H^+-transport across the plasma membrane of maize. *Physiol. Plant.* **91**: 741-746.

Emanuele, M.C., Scaloni, A., Lavermicocca, P., Iacobellis, N.S., Camoni, L., DiGiorgio, D., Pucci, P., Paci, M., Segre, A., and Ballio, A., 1998, Corpeptins, new bioactive lipodepsipeptides from cultures of *Pseudomonas corrugata*. *FEBS Lett.* **433**: 317-320.

Feigin, A.M., Takemoto, J.Y., Wangspa, R., Teeter, J.H., and Brand, J.G., 1996, Properties of voltage-gated ion channels formed by syringomycin E in planar lipid bilayers. *J. Membrane Biol.* **149**: 41-47.

Fogliano, V., Ballio, A., Gallo, M., Woo, S., Scala, F., and Lorito, M., 2002, *Pseudomonas* lipodepsipeptides and fungal cell wall degrading enzymes act synergistically in biocontrol. *Mol. Plant-Microbe Interact.*.

Gallo, M., Fogliano, V., Ritieni, A., Peluso, L., Greco, M. L., Lops, R., and Graniti, A., 2000, Immuno-assessment of *Pseudomonas syringae* lipodepsipeptides (syringomycins and syringopeptins). *Phytopath. Medit.* **39**: 410-416.

Gambale, F., Bregante, M., Stragapede, F., and Cantù, A. M., 1996, Ionic channels of the sugar beet tonoplast are regulated by a multi-ion single-file permeation mechanism. *J. Membrane Biol.* **154**: 69-79.

Grgurina, I., Mariotti, F., Fogliano, V., Gallo, M., Scaloni, A., Iacobellis, N. S., Lo Cantore, P., Mannina, L., van Axel Castelli, V., Greco, M. L., and Graniti, A., 2002, A new syringopeptin produced by bean strains of *Pseudomonas syringae* pv. *syringae*. *Biochim. Biophys. Acta* **1597**: 81-89.

Hutchison, M.L., and Gross, D.C., 1997, Lipopeptide phytotoxins produced by *Pseudomonas syringae* pv. *syringae*: Comparison of the biosurfactant and ion channel-forming activities of syringopeptin and syringomycin. *Mol. Plant-Microbe Interact.* **10**: 347-354.

Hutchison, M.L., Tester, M.A., and Gross, D.C., 1995, Role of biosurfactant and ion channel-forming activities of syringomycin in transmembrane ion flux: a model for the mechanism of action in the plant-pathogen interaction. *Mol. Plant-Microbe Interact.* **8**: 610-620.

Julmanop, C., Takano, Y., Takemoto, J.Y., and Miyakawa, T., 1993, Protection by sterols against the cytotoxicity of syringomicin in the yeast *Saccharomyces cerevisiae*. *J. Gen. Microbiol.* **139**: 2323-2327.

Kaulin, Y.A., Schagina, L.V., Bezrukov, S.M., Malev, V.V., Feigin, A.M., Takemoto, J.Y., Teeter, J.H., and Brand, J.G., 1998, Cluster organization of ion channels formed by the antibiotic syringomycin E in bilayer lipid membranes. *Biophys. J.* **74**: 2918-2925.

Montal, M., and Mueller, P., 1972, Formation of bimolecular membranes from lipid monolayers and a study of their electrical properties. *Proc. Natl. Acad. Sci. U S A.* **69**: 3561-3566.

Scaloni, A., Camoni, L., DiGiorgio, D., Scortichini, M., Cozzolino, R., and Ballio, A., 1997, A new syringopeptin produced by a *Pseudomonas syringae* pv. *syringae* strain isolated from diseased twigs of laurel. *Physiol. Mol. Plant Pathol.* **51**: 259-264.

Segre, A., Bachmann, R.C., Ballio, A., Bossa, F., Grgurina, I., Iacobellis, N.S., Marino, G., Pucci, P., Simmaco, M., and Takemoto, J.Y., 1989, The structure of syringomycins A1, E and G. *FEBS Lett.* **255**: 27-31.

Takemoto, J.Y., Brand, J.G., Kaulin, Y.A., Malev, V.V., Schagina, L.V., and Blasko, K., 2003, The syringomycins: lipodepsipeptide pore formers from plant bacterium,*Pseudomonas syringae*. In *Pore-forming peptides and protein toxins* (G. Menestrina, M. Dalla Serra and P. Lazarovici, eds.), Taylor & Francis Group, London, UK, pp. 260-271.

Wangspa, R., and Takemoto, J.Y., 1998, Role of ergosterol in growth inhibition of *Saccharomyces cerevisiae* by syringomycin E. *FEMS Microbiol. Lett.* **167**: 215-220.

Ziegler, W., Pavlovkin, J., and Pokornj, J., 1984, Effect of syringotoxin on the permeability of bilayer lipid membranes. *Biologia* **39**: 693-699.

Temperature-Regulated Biosynthesis of Coronatine by *Pseudomonas syringae in vitro* and *in planta*

H. WEINGART, S. STUBNER[#] and M.S. ULLRICH
School of Engineering and Sciences, International University Bremen, Bremen, Germany; [#]*Max-Planck-Institute for terrestrial Microbiology, Marburg, Germany.*

Abstract: Coronatine (COR) is a chlorosis-inducing phytotoxin produced by the plant pathogenic bacterium *Pseudomonas syringae*. A variety of environmental factors affect COR production. Previously, it was shown *in vitro* that temperature had a highly significant effect on COR synthesis in *P. s.* pv. *glycinea* PG4180, with maximal production rates at 18°C and low activity at 28°C. However, little was known about the actual expression of biosynthetic genes required for production of COR *in planta*. Confocal laser scanning microscopy was used to investigate *in vitro* and *in planta* expression of COR genes by two model organisms, *P. s.* pv. *glycinea* PG4180, a pathogen of soybean, and *P. s.* pv. *tomato* DC3000, a pathogen of tomato. The results indicate that the signals for induction of COR biosynthesis differ in these strains. COR biosynthesis by *P. s.* pv. *glycinea* PG4180 was temperature-dependent in minimal medium as well as inside the plant tissue. However, transcription of COR biosynthetic genes was not significantly affected by temperature in *P. s.* pv. *tomato* DC3000. In contrast, DC3000 showed higher levels of COR gene expression when inoculated into host plants as compared to DC3000 grown in minimal medium.

Key words: phytotoxin, gene expression, egfp.

Author for correspondence: H. Weingart; e-mail: h.weingart@iu-bremen.de.

1. INTRODUCTION

Coronatine (COR) is a non-host-specific phytotoxin produced by five pathovars of *Pseudomonas syringae*, i.e. pv. *atropurpurea*, *glycinea*, *maculicola*, *morsprunorum*, and *tomato*, which infect ryegrass, soybean,

N.S. Iacobellis *et. al.* (eds.), *Pseudomonas syringae* and Related Pathogens, 199-205.

crucifers, *Prunus* spp., and tomato, respectively (Bender *et al*., 1999; Mitchell, 1982; Völksch and Weingart, 1998). The structure of COR is unusual and consists of two distinct components; coronafacic acid (CFA) is of polyketide origin, whereas coronamic acid (CMA) is an ethylcyclopropyl amino acid derived from isoleucine. CFA and CMA are fused to each other by an amide bond. Both CFA and CMA function as discrete intermediates and are secreted by COR-producing strains at low levels (Bender *et al*., 1999; Ichihara *et al*., 1977; Mitchell *et al*., 1994).

The primary symptom observed in leaf tissue treated with COR is an intense spreading chlorosis that can be induced on a wide variety of plant species (Gnanamanickam *et al*., 1982). COR is also known to induce hypertrophy of storage tissue, compression of thyllakoids, thickening of plant cell walls, accumulation of protease inhibitors, inhibition of root elongation, and stimulation of ethylene production in some but not all plant species (Sakai *et al*., 1979; Kenyon and Turner, 1992; Palmer and Bender, 1995). The structure of COR revealed a high degree of homology to methyl jasmonate (MeJA), a plant growth regulator of the octadecanoid signalling pathway involved in plant defence mechanisms and global cell differentiation processes. COR and MeJA induce analogous biological responses in plants, which led researchers to suggest that COR functions as a molecular mimic of octadecanoid signaling molecules produced by higher plants (Koda *et al*., 1996; Weiler *et al*., 1994). Furthermore, Feys *et al*. (1994) generated a COR-insensitive mutant (coi1) of *Arabidopsis* that was insensitive to the effects of both COR and MeJA, suggesting a similar mode of action.

COR production plays an important role in virulence of toxin-producing *P. syringae* strains. Studies with COR-defective mutants have shown that COR synthesis contributes significantly to lesion expansion, the development of chlorosis, and bacterial multiplication in infected leaves (Bender *et al*., 1987; Mittal and Davis, 1995; Tamura *et al*., 1998). These findings were confirmed by Budde and Ullrich (2000), who investigated the impact of COR production by *P. s.* pv. *glycinea* PG4180 during infection of soybean plants. Interestingly, they observed a significant delay of the hypersensitive response (HR) on tobacco plants treated with a CFA-overproducing derivative of PG4180 indicating that COR/CFA play also a role during incompatible plant-bacteria interactions.

2. REGULATION OF CORONATINE PRODUCTION

A variety of environmental and nutritional factors, such as pH, osmolarity, carbon sources, nutrient levels, amino acid supplements, and complex carbon/nitrogen sources, affect COR production in *Pseudomonas syringae* (Palmer and Bender, 1995). Temperature had a highly significant effect on

COR biosynthesis in *P. s.* pv. *glycinea* PG4180, with maximal production rates at 18°C and negligible yields at 28°C, the optimal growth temperature (Fig. 1).

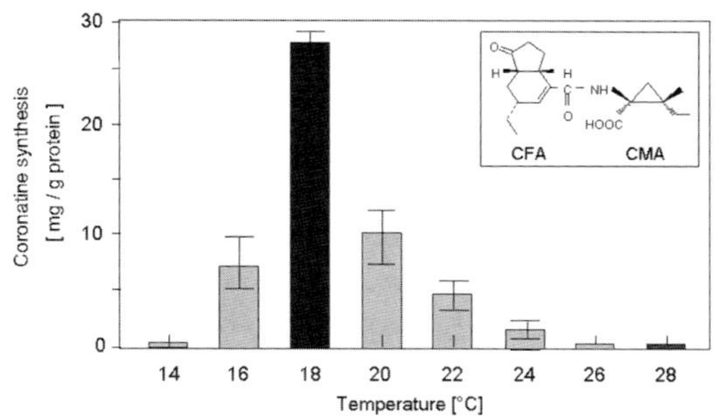

Figure 1. Effect of temperature on coronatine production in *Pseudomonas syringae* pv. *glycinea* PG4180.

The production of both intermediates, CFA and CMA, in PG4180 is regulated at the transcriptional level by temperature. Characterisation of the 32-kb COR gene cluster revealed two regions containing structural genes for CFA and CMA biosynthesis, respectively. These were separated by a 3,4-kb regulatory region. Transcriptional fusions of the CFA and CMA promoter region to a promoterless glucuronidase gene showed maximal transcriptional activities at 18°C and significantly lower activities at 28°C (Bender *et al.*, 1999).

The regulatory region which controls both CFA and CMA production encodes a modified two-component system consisting of two response regulators, CorR and CorP, and the histidine protein kinase, CorS (Ullrich *et al.*, 1995). CorR contains a DNA-binding, helix-turn-helix motif and acts as a positive regulator of COR gene expression. CorP is highly conserved to CorR but lacks the DNA-binding domain. CorP is required for the activation of CorR by histidine protein kinase CorS at low temperature. This modified two-component system was able to achieve temperature-dependent activation of an introduced COR biosynthetic promoter in a *P. syringae* strain lacking the COR gene cluster, indicating that this triad of genes has the primary control over COR biosynthesis and is responsible for thermoregulation (Ullrich, 2002).

In phytopathogenic bacteria, several classes of genes important for plant-bacteria interactions are induced by plant-derived signals. Well-known examples are the *vir* genes of *Agrobacterium*, encoding the T-DNA transfer machinery, the *hrp* genes of gram-negative phytopathogenic bacteria, encoding a type III secretion machinery, and the *syrB* gene of *P. s.* pv. *syringae*, involved in production of the phytotoxin syringomycin (van Gijsegem, 1997).

However, little was known about the signals influencing COR production in infected plant tissue. Palmer and Bender (1993) amended the growth medium of *P. s.* pv. *glycinea* PG4180 with extracts of soybean and with plant-derived secondary metabolites but found no influence on COR production. In contrast, Ma *et al.* (1991) demonstrated that growth on tomato plants increased COR gene expression in *P. s.* pv. *tomato* DC3000 over that observed *in vitro*. Shikimic and quinic acid were identified by Li *et al.* (1998) as active components in tomato leaf extracts inducing expression of COR genes in DC3000.

One important question currently addressed in our laboratory is whether the *in vitro* thermoregulation of COR gene expression also plays a role *in planta*. Therefore, plasmid pHW01 containing a transcriptional fusion of the temperature-regulated *cmaABT* promoter region to a promoterless *egfp* gene was constructed. Confocal laser scanning microscopy was used to investigate *in vitro* and *in planta* expression of COR genes by *P. s.* pv. *glycinea* PG4180, a pathogen of soybean, and *P. s.* pv. *tomato* DC3000, a pathogen of tomato.

Transcription of *cmaABT*::*egfp* in *P. s.* pv. *glycinea* PG4180 (pHW01) and *P. s.* pv. *tomato* DC3000 (pHW01) was first examined in HSC-medium, which is optimised for COR production. PG4180 (pHW01) showed a 10-fold higher fluorescence at 18°C as compared to 28°C (Fig. 2). In contrast, expression of EGFP from the *cmaABT* promoter in DC3000 (pHW01) was not affected by temperature (Fig. 2).

To analyse the *cmaABT* promoter activity *in planta*, bacteria containing pHW01 were inoculated into leaves of their host plants by infiltration. Subsequently, plants were kept in growth chambers at 18°C or 28°C. Three days after inoculation bacteria were isolated by macerating infected leaf areas. Fluorescence analysis of bacteria with confocal laser scanning microscopy revealed that COR gene expression in *P. s.* pv. *glycinea* PG4180 (pHW01) is also temperature-dependent *in planta*. PG4180 (pHW01) isolated from soybean plants kept at 18°C showed a strong fluorescence, whereas fluorescence of bacteria from plants incubated at 28°C was difficult to detect (Fig. 3). Similar levels of fluorescence were detected in PG4180 (pHW01) isolated from soybean plants incubated at 18°C and in PG4180 (pHW01) grown in HSC-medium at 18°C, indicating that the *cmaABT* promoter is temperature-sensitive but not plant-inducible in this strain.

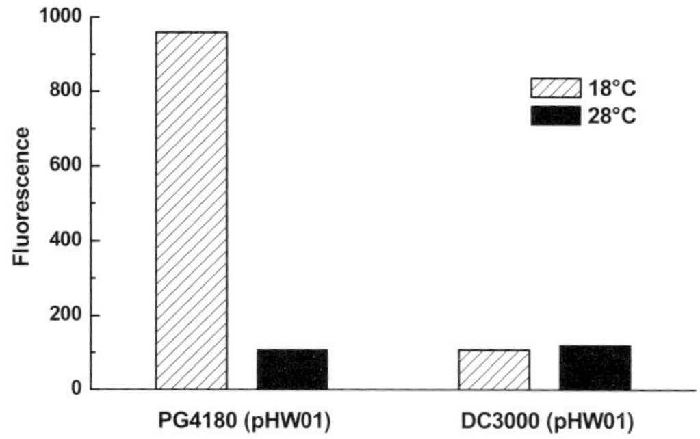

Figure 2. Fluorescence of *Pseudomonas syringae* pv. *glycinea* PG4180 (pHW01) and *P. s.* pv. *tomato* DC3000 (pHW01) in HSC-medium at 18°C and 28°C. To evaluate fluorescence, aliquots of the cultures were harvested at an OD_{600} value of 5, centrifuged, resuspended in PBS and adjusted to an OD_{600} value of 0.5. Fluorescence was quantified using a Fluorolite microplate reader set to an excitation wavelength of 470 nm with emission detection at 516 nm.

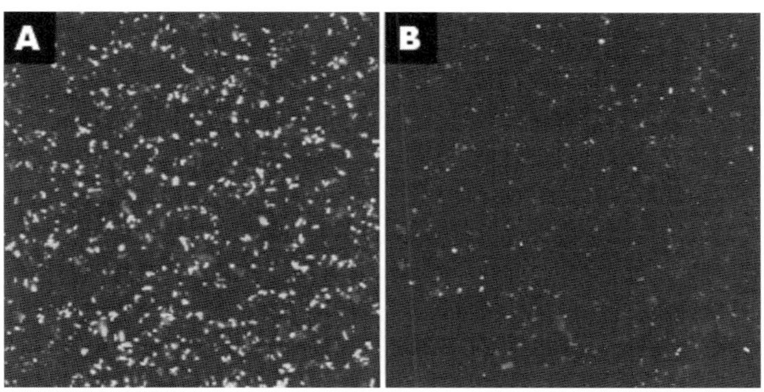

Figure 3. Fluorescence of *Pseudomonas syringae* pv. *glycinea* PG4180 (pHW01) isolated from soybean plants, which were incubated at 18°C (**A**) and 28°C (**B**). To assay fluorescence of bacterial cells, infected plant tissue was macerated in isotonic NaCl and imaged by confocal laser scanning microscopy.

Interestingly, the opposite was observed for strain *P. s.* pv. *tomato* DC3000. DC3000 (pHW01) isolated from *Arabidopsis* showed higher levels of fluorescence than DC3000 (pHW01) grown in HSC-medium and no significant effect of temperature could be detected.

Recently, Fouts *et al.* (2002) used computational and gene expression techniques to identify promoters controlled by the HrpL alternative sigma factor in the complete genome sequence of DC3000. Interestingly, they found a Hrp promoter-like sequence upstream of the *corR* gene. It is tempting to speculate that COR production in DC3000 is plant-inducible due to the activation by the Hrp regulon.

3. CONCLUSIONS

Our results show that COR biosynthesis is differently regulated in *Pseudomonas syringae* pv. *glycinea* PG4180, a pathogen of soybean, and *P. s.* pv. *tomato* DC3000, a pathogen of tomato and *Arabidopsis*. *P. s.* pv. *glycinea* PG4180 showed a temperature-regulated COR production *in vitro* and *in planta*. In contrast, COR gene expression in *P. s.* pv. *tomato* DC3000 is plant-inducible but not temperature-dependent. The differences in the regulation of phytotoxin production possibly reflect the adaptation of these strains to different host plants.

REFERENCES

Bender, C.L., Stone, H.E., Sims, J.J., and Cooksey, D.A., 1987, Reduced pathogen fitness of *Pseudomonas syringae* pv. *tomato* Tn*5* mutants defective in coronatine production. *Physiol. Mol. Plant Pathol.* **30**: 273-283.

Bender, C.L., Alarcón-Chaidez, F., and Gross, D.C., 1999, *Pseudomonas syringae* phytotoxins: Mode of action, regulation, and biosynthesis by peptide and polyketide synthetases. *Microbiol. Mol. Biol. Rev.* **63**: 266-292.

Budde, I.P. and Ullrich, M.S., 2000, Interactions of *Pseudomonas syringae* pv. *glycinea* with host and nonhost plants in relation to temperature and phytotoxin synthesis. *Mol. Plant-Microbe Interact.* **13**: 951-961.

Feys, B.J.F., Benedetti, C.E., Penfold, C.N., and Turner, J.G., 1994, Arabidopsis mutants selected for resistance to the phytotoxin coronatine are male, sterile, insensitive to methyl jasmonate, and resistant to a bacterial pathogen. *Plant Cell* **6**: 751-759.

Fouts, D.E., Abramovitch, R.B., Alfano, J.R., Baldo, A.M., Buell, C.R., Cartinhour, S., Chatterjee, A.K., D'Ascenzo, M., Gwinn, M.L., Lazarowitz, S.G., Lin, N.-C., Martin, G.B., Rehm, A.H., Schneider, D.J., van Dijk, K., Tang, X., and Collmer, A., 2002, Genomewide identification of *Pseudomonas syringae* pv. *tomato* DC3000 promoters controlled by the HrpL alternative sigma factor. *Proc. Natl. Acad. Sci. USA* **99**: 2275-2280.

Gnanamanickam, S.S., Starratt, A.N., and Ward, E.W.B., 1982, Coronatine production *in vitro* and *in vivo* and its relation to symptom development in bacterial blight of soybean. *Can. J. Bot.* **60**: 645-650.

Ichihara, A., Shiraishi, K., Sato, H., Sakamura, S., Nishiyama, K., Sakai, R., Furusaki, A., and Matsumoto, T., 1977, The structure of coronatine. *J. Am. Chem. Soc.* **99**: 636-637.

Kenyon, J.S. and Turner, J.G., 1992, The stimulation of ethylene production in *Nicotiana tabacum* leaves by the phytotoxin coronatine. *Plant Physiol.* **100**: 219-224.

Koda, Y., Takahashi, K., Kikuta, Y., Greulich, F., Toshima, H., and Ichihara, A., 1996, Similarities of the biological activities of coronatine and coronafacic acid to those of jasmonic acid. *Phytochemistry* **41**: 93-96.

Li, X.-Z., Starratt, A.N., and Cuppels, D.A., 1998, Identification of tomato leaf factors that activate toxin gene expression in *Pseudomonas syringae* pv. *tomato* DC3000. *Phytopathology* **88**: 1094-1100.

Ma, S.-W., Morris, V.L., Cuppels, D.A., 1991, Characterization of a DNA region required for production of the phytotoxin coronatine by *Pseudomonas syringae* pv. *tomato. Mol. Plant-Microbe Interact.* **4**: 69-74.

Mitchell, R.E., 1982, Coronatine production by some phytopathogenic pseudomonads. *Physiol. Plant Pathol.* **20**: 83-89.

Mitchell, R.E., Young, S.A., and Bender, C.L., 1994, Coronamic acid, an intermediate in coronatine biosynthesis by *Pseudomonas syringae*. *Phytochemistry* **35**: 343-348.

Mittal, S. and Davis, K.R., 1995, Role of the phytotoxin coronatine in the infection of *Arabidopsis thaliana* by *Pseudomonas syringae* pv. *tomato. Mol. Plant-Microbe Interact.* **8**: 165-171.

Palmer, D.A. and Bender, C.L., 1993, Effects of environmental and nutritional factors on production of the polyketide phytotoxin coronatine by *Pseudomonas syringae* pv. *glycinea. Appl. Environ. Microbiol.* **59**: 1619-1626.

Palmer, D.A. and Bender, C.L., 1995, Ultrastructure of tomato leaf tissue treated with the pseudomonad phytotoxin coronatine and comparison with methyl jasmonate. *Mol. Plant-Microbe Interact.* **8**: 683-692.

Sakai, R., Nishiyama, K., Ichihara, A., Shiraishi, K., and Sakamura, S., 1979, Studies on the mechanism of physiological activity of coronatine: Effect of coronatine on cell wall extensibility and expansion of potato tuber tissue. *Ann. Phytopathol. Soc. Japan* **45**: 645-653.

Tamura, K., Zhu, Y., Sato, M., Teraoka, T., Hosokawa, D., and Watanabe, M., 1998, Role of coronatine production by *Pseudomonas syringae* pv. *maculicola* for pathogenicity. *Ann. Phytopathol. Soc. Japan* **64**: 299-302.

Ullrich, M.S., Penaloza-Vázquez, A., Bailey, A.-M., and Bender, C.L., 1995, A modified two-component system is involved in temperature-dependent biosynthesis of the *Pseudomonas syringae* phytotoxin coronatine. *J. Bacteriol.* **177**: 6160-6169.

Ullrich, M.S., 2002, Thermoresponsive phytotoxin production in plant pathogenic bacteria. In Advances in Microbial Toxin Research and its Biotechnological Exploitation. (R.K. Upadhyay, ed.), Kluwer Academic/ Plenum Publishers, New York, pp. 81-91.

Van Gijsegem, F., 1997, In planta regulation of phytopathogenic bacteria virulence genes: relevance of plant-derived signals. *Euro. J. Plant Pathol.* **103**: 291-301.

Völksch, B. and Weingart, H., 1998, Toxin production by pathovars of *Pseudomonas syringae* and their antagonistic activities against epiphytic microorganisms. *J. Basic Microbiol.* **38**: 135-145.

Weiler, E.W., Kutchan, T.M., Gorba, T., Brodschelm, W., Niesel, U., and Bublitz, F., 1994, The *Pseudomonas* phytotoxin coronatine mimics octadecanoid signalling molecules of higher plants. *FEBS Letters* **345**: 9-13.

Interaction of Syringomycin E Structural Analogues with Biological and Model Membranes

M. DALLA SERRA, G. MENESTRINA, M. CORAIOLA and I. GRGURINA[§]

CNR-ITC Istituto di Biofisica, Sezione di Trento, Povo (Trento), Italy; [§]Dipartimento di Scienze Biochimiche "A. Rossi Fanelli", Università di Roma "La Sapienza", P.le A. Moro 5, 00185 Roma, Italy.

Abstract: Syringomycin E (SRE), a lipodepsinonapeptide produced by many *Pseudomonas syringae* pv. *syringae* strains, displays a prominent antifungal and haemolytic activity. The former is already exploited for post-harvest biocontrol of fruit and appears interesting in the perspective of medical applications; the latter hampers the use of this compound as a systemic drug. The ability to interact with biological membranes and form pores is at the basis of its biological activity. To gain insight into the structural features which are important for this process, we produced structural analogues of the metabolite and evaluated their activity on red blood cells and on artificial membranes of different compositions. The substitution of chlorine with hydrogen in the C-terminal amino acid caused a marked decrease in the pore-forming activity, in accordance with the previously observed trend in the antifungal activity assay. The cooperativity of the process is not affected, but the activity of the deschloro-SRE analogue (SREH) at difference from SRE, is not influenced by the presence of sterols in the artificial lipid bilayers. The residual activity of SREH is abolished when the lactone ring is hydrolysed. Thus, both the presence of chlorine in the C-terminal residue and the integrity of the macrocycle appear to be important for the pore-forming activity of SRE.

Key words: syringomycin, liposome, membranes, structure-activity relationship studies.

Author for correspondence: I. Grgurina; e-mail: Ingeborg.Grgurina@uniroma1.it.

N.S. Iacobellis *et. al.* (eds.), *Pseudomonas syringae* and Related Pathogens, 207-215.

1. INTRODUCTION

A characteristic trait of the secondary metabolism of the phytopathogenic bacterium *Pseudomonas syringae* pv. *syringae* is the production of structurally related toxic lipodepsipeptides (LDPs). The first discovered and the best studied member of the group, syringomycin is produced in three homologous forms differing in the length of the lipid chain, among which syringomycin E (SRE) is the most abundant. It has been detected in infected plants (Grgurina *et al.*, 1997; Fogliano *et al.*, 1999), where its biosynthesis is triggered by plant secondary metabolites (Mo and Gross, 1991a; 1991b; Mo *et al.,*1995). Purified SRE induces necrotic lesions on plant tissues, similar to those observed in diseased plants (Iacobellis *et al.,*1992). Investigations on the role of this phytotoxin in the development of plant disease showed, however, that it is not indispensable for pathogenesis, but contributes to bacterial virulence (Scholz-Schroeder *et al.,*2001).

Syringomycin displays a prominent antifungal activity, a property which is at the basis of the application of some *P. syringae* strains in post-harvest biocontrol of fruit (Bull *et al.,*1998; Janisieiwicz *et al.,*1995) and which also appears interesting in the perspective of biomedical applications (Sorensen *et al.,*1996; De Lucca *et al.,*1999). In addition, the toxin is strongly haemolytic, indicating damage of the erythrocyte membrane (Lavermicocca *et al.,*1997; Menestrina *et al.,* 2003).

The mechanism at the basis of the toxic activities of SRE was widely investigated on natural and artificial membranes (Zhang and Takemoto, 1989; Hutchison et al, 1995; Hutchison and Gross, 1997; Dalla Serra *et al.,*1999, and ref. cited therein). One of the advantages in the use of artificial membranes, is the possibility to evaluate how the changes in the membrane composition influences the interaction with the toxin. The metabolite causes the formation of ion channels which result in ion leakage and in the impairment of membrane functions.

The interaction with biological membranes is a mechanism shared by many toxic peptides, produced by microbes and higher organisms (Hancock and Chapple, 1999). The attractive therapeutic applications of the antimicrobial activity of most of these compounds are often hampered by their haemolytic activity. Because both properties are based on the same type of mechanism, it is difficult to obtain structural analogues with enhanced antimicrobial and decreased haemolytic activity. However some of the structure-activity relationship studies aimed at obtaining peptide analogues with improved haemolysis/antibiosis profile were successful (Hancock and Chapple, 1999).

The structure of syringomycin (Segre *et al.*, 1989; Fukuchi *et al.,*1992) is shown in Fig. 1a. The presence of a hydrophylic nine amino acids containing lactone macrocycle and a hydrophobic acyl chain confers an amphipathic character to this molecule, which facilitates its interaction with membranes. Fig.1b shows the structure of deschlorosyringomycin (SRH), differing from

SRE only in the C-terminal residue (4-chlorothreonine in SRE and threonine in SREH) can be obtained by growing the bacteria in a chloride-free medium (Grgurina *et al.,*1994). This compound was shown to be 75% less active than SRE in the antifungal activity test against *Rhodotorula pilimanae* (Grgurina *et al.,* 1994).

```
a)      R – Ser Ser Dab Dab Arg Phe Dhb Asp (3-OH) Thr(4-Cl)
            |                                      |
            O______________________________________CO

b)      R – Ser Ser Dab Dab Arg Phe Dhb Asp (3-OH) Thr
            |                                 |
            O_________________________________CO
```

Figure 1. Structures of SRE (a) and SREH (b); R: 3-hydroxydodecanoyl, Dab: 2,4 diaminobutyric acid; Dhb: dehydroaminobutyric acid; Asp(3-OH): 3-hydroxyaspartic acid; Thr(4-Cl): 4-chlorothreonine.

We were interested to evaluate the relevance of this modification at the C-terminal residue in the pore forming activity of SRE. Here we report the results of a comparative study of membrane permeabilization by SRE and SREH using artificial membranes of different compositions. In addition, the importance of the integrity of the macrocyclic lactone was also evaluated.

2. MATERIALS AND METHODS

2.1 Chemicals

Phospholipids used were obtained from Avanti Polar Lipids (Pelham, Al, USA) at the highest possible grade (usually more than 99% pure). Cholesterol was from Fluka (Buchs, Switzerland).

2.2 Preparation of Lipopeptides

SRE, SREH and its linear analogue (SREH-L), were prepared and purified as previously described (Grgurina *et al.,* 1994; Segre *et al.,* 1989).

2.3 Red Blood Cell Haemolysis

Haemolytic activity of LDPs was determined turbidimetrically with a 96-well microplate reader. Human red blood cells were prepared from fresh heparinized blood from healthy volunteers, by extensively washing in NaCl 0.85%. LDPs were plated in a 96-well plate with 2-fold serial dilutions in a buffer containing 120 mM NaCl, 10 mM MES, 1 mM EDTA, pH 6.0. Each well was supplemented with 0.02% HRBC and continuously stirred. Percent haemolysis was calculated as $(A_i-A_f)/(A_i-A_o)$ x 100, where A_i, A_f and A_o are the initial, final and completely lysed apparent absorbances at 650 nm. A_i was about 0.13 OD and A_0 (as obtained by hypotonical lysis with pure water) was around 0.03 OD.

2.4 Permeation of Artificial Lipid Membranes

Membrane permeation was determined fluorimetrically with a 96-well kinetic fluorescence microplate reader (Fluostar, SLT, Groeding, Austria). Large unilamellar vesicles (LUVs), were prepared by extrusion through polycarbonate filters with 100 nm pores, loaded with calcein at a self-quenching concentration and washed by gel-filtration on sephadex G50 medium (Sigma). Two lipid compositions were used, as detailed in the text. LDPs were 2-fold serially diluted in a buffer containing 100 mM NaCl, 10 mM MES, 1 mM EDTA, pH 6.0, supplemented with washed LUVs (at a final lipid concentration of 6 μM) and continuously stirred. The percent of permeation, $R_{\%}$, was calculated as $(F_{fin} - F_{in})/(F_{max} - F_{in})$ x 100, where F_{in}, F_{fin} are the initial and final (steady state) value of fluorescence, before and after the toxin addition. F_{max} corresponds to maximal dye release, which was induced by 1 mM Triton X-100.

3. RESULTS

To understand the relevance of the chlorine at the C-terminal residue for the membrane activity of SRE and its deschlorinated analogue, SREH, were compared in an haemolysis assay which provides a rapid indication of pore-forming activity. A difference of one order of magnitude between SRE and SREH, was observed at saturation levels, SREH being the less active (data not shown). This parallels the decrease observed for the antifungal activity. Additional studies were carried out on liposomes composed either of pure phosphatidylcholine (PC) or PC/cholesterol=1:1molar (PC/Chol). The pore-forming activity was measured as the percentage of calcein release. As shown in Fig. 2., 30% calcein release was obtained with approximately 3 μM of SRE and 14 μM of SREH on PC liposomes (upper panel) and with 0.05 μM SRE and 1,9 μM SREH on cholesterol-containing liposomes (bottom panel). Analogous trend was observed at 50% release of calcein (0.07 μM SRE and 2.9 μM SRH). At saturation levels, when 90% release of calcein is observed, the difference in the activity between the two toxins is less pronounced (0.4 μM SRE and 15 μM SREH). The concentration-dependence of calcein release shows a co-operative process for both SRE and SREH. The presence of the sterol in the liposome membrane enhances strongly the activity of SRE, as previously reported (Dalla Serra *et al.*, 1999), but does not affect significantly the activity of SREH. The difference in the activity between the two syringomycin forms is most pronounced on cholesterol-containing membranes and at the mid-point of the process where the co-operativity is more pronounced.

The results of our structure-activity relationship studies on the relevance of the C-terminal residue show the importance of chlorine in both antifungal and pore-forming activity. The residual activity of the deschlorinated analogue, however, points to the contribution of other structural elements. The lactone ring was taken into consideration on the basis of previous results on lipodepsipeptides (Asselineau, 1991). A linear analogue of SRE, which would be necessary to study this particular aspect, is not accessible by chemical transformations because under the mild basic conditions used for opening the lactone ring, the chloromethylene group of the C-terminal residue is also hydrolysed. This hampers the selective evaluation of the relevance of each structural element of SRE for its biological activity. Because this reaction clearly cannot take place in SREH, the only modification occurring during the hydrolysis of this compound affects its macrocyclic structure. The linear analogue, SREH-L, was indeed completely inactive both in cholesterol free and cholesterol containing liposomes, (Fig. 2). This demonstrated that also the cyclic structure is, in fact, important for the membrane activity of syringomycins.

4. DISCUSSION

Our results show that the modifications of the C-terminal residue and in the lactone ring are both reflected in the changes on the antifungal (haemolytic and pore-forming activities of SRE. At the onset of this investigation, it was hypothesised that the toxic activities of SRE could be due, at least in part, to a covalent interaction with some intracellular target, considering that it contains two reactive moieties: chloromethylene group and dehydrobutyrine. The decrease of the antifungal activity observed upon substitution of chlorine with hydrogen, could be interpreted as the inability of SRH to undergo nucleophilic substitution (peculiar of 4-chlorothreonine) in a reaction with a target molecule. The α,β-unsaturated residue could be involved in a Michael-type addition, as was shown for the cyanobacterial lipopeptide microcystin (MacKintosh *et al.*,1995).

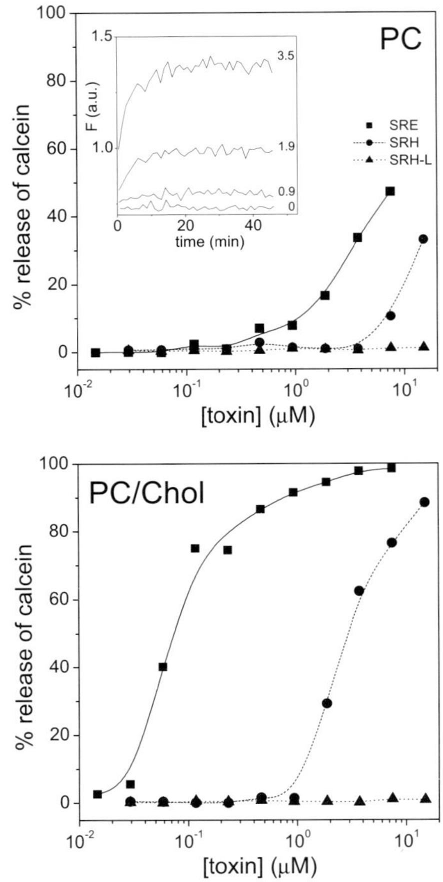

Figure 2. Pore-forming effect of SRE (4), SREH (•) and SREH-L (▲) on liposomes composed of pure phosphatidylcholine (PC) (upper panel) and of a mixture phosphatidylcholine/cholesterol (1:1 mol) (PC/Chol) (bottom panel). **Inset**: time-course of calcein release from PC liposome after addition of different amount of SRE, whose final micromolar concentration is reported next to each trace.

The presence of chlorine in the C-terminal residue strongly affects the pore forming activity of SRE and is also important for the enhancement of this activity by a sterol. The relevance of sterol membrane components was shown both by biophysical (Feigin *et al.,* 1997; Dalla Serra *et al*., 1999) and by a genetic approach (Taguchi *et al.,* 1994; Takemoto *et al.,*1993; Hemmi *et al.,* 1995). The deschloro analogue, SRH, is less active than SRE and it is also insensitive to cholesterol, but does not differ from SRE in the degree of co-operativity.

Presently available data are not enough to unambiguously explain the difference in the activity between SRE and SRH. The pore forming process requires both their interaction with some membrane component/s and the aggregation of several toxin molecules to form an ion channel (Feigin *et al.*, 1996; Kaulin *et al.*, 1998). The structural change might conceivably affect either one or both of these processes. Further studies at molecular level would be needed to clarify the mode of interaction of syringomycin with the membrane structures.

ACKNOWLEDGEMENTS

We would like to acknowledge the financial support offered: by the Italian Consiglio Nazionale delle Ricerche (CNR), the Istituto Trentino di Cultura (ITC) and by a special grant from the Provincia Autonoma di Trento (PAT, project AGRIBIO), the Italian Ministero dell'Istruzione, dell'Università e della Ricerca (MIUR) and the Fondazione Pasteur – Cenci Bolognetti (grant to I.G.). We also thank Prof. A. Ballio for encouragement.

REFERENCES

Asselineau, J., 1991, "Bacterial Lipids containing amino acid or peptides linked by amide bonds" *Progress in the Chemistry of Organic Natural Products* **56**: 1-86.

Bull, C.T., Wadsworth, M.L., Sorensen, K.M., Takemoto, J.Y., Austin, R.K., and Smilanick, J.L., 1998, Syringomycin E produced by biological control agents controls green mold on lemons. *Biol. Control* **12**: 89-95.

Dalla Serra, M., Fagiuoli, G., Nordera, P., Bernhart, I., Della Volpe , C., Di Giorgio, D., Ballio, A., and Menestrina, G., 1999, The interaction of lipodepsipeptide toxins from *Pseudomonas syringae* pv. *syringae* with biological membranes: a comparison of syringotoxin, syringomycin and two syringopeptins. *Mol. Plant-Microbe Interact.* **12**: 391-400.

De Lucca, A.J., Jacks, T.J., Takemoto, J.Y., Vinyard, B., Peter, J., Navarro, E., and Walsh, T.J., 1999, Fungal lethality, binding and cytotoxicity of syringomycin E. *Antimicrob. Agents Chemother.* **43**: 371-373.

Feigin, A.M., Schagina, L.V., Takemoto, J.Y., Teeter, J.H., and Brand, J.G., 1997, The effect of sterol on the sensitivity of membranes to the channel-forming antifungal antibiotic, syringomycin E. *Biochim. Biophys. Acta* **1324**: 102-110.

Feigin, A.M., Takemoto, J.Y., Wangspa, R., Teeter, J.H., and Brand, J.G., 1996, Properties of voltage-gated ion channels formed by syringomycin E in planar lipid bilayers. *J. Membr. Biol.* **149**: 41-47.

Fogliano, V., Gallo, M., Vinale, F., Ritieni, A., Randazzo, G., Greco, M.L., Lops, R., and Graniti, A., 1999, Immunological detection of syringopeptins produced by *Pseudomonas syringae* pv. *lachrymans*. *Physiol. Mol. Plant Pathol.* **55**: 255-261.

Fukuchi, N., Isogai, A., Nakayama, J., Takayama, S., Yamashita, S., Suyama, K., Takemoto, J.Y., and Suzuki, A., 1992, Structure and stereochemistry of three phytotoxins, syringomycin, syringotoxin and syringostatin, produced by *Pseudomonas syringae* pv. *syringae*. *J. Chem. Soc. Perkin Trans.* **1**: 1149-1157.

Grgurina, I., Barca, A., Cervigni, S., Gallo, M., Scaloni, A., and Pucci, P., 1994, Relevance of chlorine-substituent for the antifugal activity of syringomycin and syringotoxin, metabolites of the phytopathogenic bacterium *Pseudomonas syringae* pv. *syringae*. *Experientia* **50**: 130-133.

Grgurina, I., Iacobellis, N.S., Camele, I., and Curci, R., 1997, Detection of Syringomycin in plant tissues infected with *Pseudomonas syringae* pv. *syringae*. In: Pseudomonas syringae *Pathovars and Related Pathogens*. (K. Rudolph, T.J. Burr, J.W. Mansfield, D. Stead, A. Vivian and J. von Kietzell, Eds.) Kluwer Academic Publishers, Dordrecht, The Netherlands, 188-191.

Hancock, R.E.W., and Chapple, D.S., 1999, Peptide antibiotics, *Antimicrob. Agents Chemother.* **43**: 1371-1323.

Hemmi, K., Julmanop, C., Hirata, D., Tsuchiya, E., Takemoto, J.Y., and Miyakawa, T., 1995, The physiological roles of membrane ergosterol as revealed by the phenotypes of *syr1/erg3* null mutant of *Saccharomyces cerevisiae*. *Biosci. Biotech. Biochem.* **59**: 482-486.

Hutchison, M.L., and Gross, D.C., 1997, Lipopeptide phytotoxin produced by *Pseudomonas syringae* pv. *syringae*: comparison of the biosurfactant and ion channel-forming activities of syringopeptin and syringomycin. *Mol. Plant-Microbe Interact.* **10**: 347-354.

Hutchison, M.L., Tester, M.A. and Gross, D.C., 1995, Role of biosurfactant and ion channel-forming activities of syringomycin in transmembrane ion flux: a model for the mechanism of action in the plant-pathogen interaction. *Mol. Plant-Microbe Interact.* **8**: 1-10.

Iacobellis, N.S., Lavermicocca, P., Grgurina, I., Simmaco, M., and Ballio, A., 1992, Phytotoxic properties of *Pseudomonas syringae* pv. *syringae* toxins. *Physiol. Mol. Plant Pathol.* **40**: 107-116.

Janisiewicz, W.J., and Bors, B., 1995, Development of a microbial community of bacterial and yeast antagonists to control wound-invading postharvest pathogens of fruits. *Appl. Environ. Microbiol.*, **9**: 3261-3267.

Kaulin, Y.A., Schagina, L.V., Bezrukov, S.M., Malev, V.V., Feigin, A.M., Takemoto, J.Y., Teeter, J.H., and Brand, J.G., 1998, Cluster organisation of ion channel formed by the antibiotic syringomycin E in bilayer lipid membranes. *Biophys. J.* **74**: 2918-2925.

Lavermicocca, P., Iacobellis, N.S., Simmaco, M., and Graniti, A., 1997, Biological properties and spectrum of activity of *Pseudomonas syringae* pv. *syringae* toxins. *Physiol. Mol. Plant Pathol.* **50**: 129-140.

MacKintosh, R.W., Dalby, K.N., Campbell, D.G., Cohen, P.T.W., Cohen, P., and MacKintosh, C., 1995, The cyanobacterial toxin microcystin binds covalently to cysteine-273 on protein phosphatase 1. *FEBS Lett.* **371**: 236-240.

Menestrina, G., Coraiola, M., Fogliano, V., Fiore, A., Grgurina, I., Carpaneto A., Gambale, F., and Dalla Serra, M., 2003, Antimicrobial Lipodepsipeptides from *Pseudomonas* spp: a Comparison of their Activity on Model Membranes, 185-198, (this volume).

Mo, Y.Y., and Gross, D.C., 1991a, Expression in vitro and during plant pathogenesis of the *syrB* gene required for syringomycin production by *Pseudomonas syringae* pv. *syringae*. *Mol. Plant-Microbe Interact.* **4**: 28-36.

Mo, Y.Y., and Gross, D.C., 1991b, Plant signal molecules activate the *syr*B gene, wich is required for syringomycin production by *Pseudomonas syringae* pv. *syringae*. *J. Bacteriol.*, **18**: 5784-5792.

Mo, Y.Y., Geibel, M., Bonsall, R.F., and Gross, D.C., 1995, Analysis of sweet cherry (*Prunus avium* L.) leaves for plant signal molecules that activate the *syrB* gene required for the synthesis of the phytotoxin, syringomycin, by *Pseudomonas syringae* pv. *syringae*. *Plant Physiol.* **107**: 603-612.

Scholz-Schroeder, B.K., Hutchison, M.L., Grgurina, I., and Gross, D.C., 2001, The contribution of syringopeptin and syringomycin to virulence of *Pseudomonas syringae* pv. *syringae* strain B 301 D on the basis of *sypA* and *syrB1* biosynthesis mutant analysis. *Mol. Plant-Microbe Interact.* **14**: 336-348.

Segre, A., Bachman, R.C., Ballio, A., Bossa, F., Grgurina, I., Iacobellis, N.S., Marino, G., Pucci, P., Simmaco, M., and Takemoto, J.Y., 1989, The structure of syringomycin A_1, E and G. *FEBS Lett.* **255**: 27-31.

Sorensen, K.N., Kim, K-H., and Takemoto, J.Y., 1996, In vitro antifungal and fungicidal activities and erythrocyte toxicities of cyclic lipodepsinonapeptides produced by *Pseudomonas syringae* pv. *syringae*. *Antimicrob. Agents Chemother.* **40**: 2710-2713.

Taguchi, N., Takano, Y., Julmanop, C., Wang, Y., Stock, S., Takemoto, J., and Miyakawa, T., 1994, Identification and analysis of the *Saccharomyces cerevisiae SYR1* gene reveals that ergosterol is involved in the action of syringomycin. *Microbiology* **140**: 353-359.

Takemoto, J.Y., Yu, Y., Stock, S.D., and Miyakawa, T., 1993, Yeast genes involved in growth inhibition by *Pseudomonas syringae* pv. *syringae* syringomycin family lipodepsipeptides. *FEMS Microbiol. Lett.* **114**: 339-342.

Zhang, L., and Takemoto; J.Y., 1989, Syringomycin stimulation of potassium efflux by yeast cells. *Biochim. Biophys. Acta* **987**: 171-175.

Substrate Specificity of Syringomycin Synthetase Adenylation Domains

S. PASCARELLA, P. GIOVANNINI and I. GRGURINA
Dipartimento di Scienze Biochimiche "A. Rossi Fanelli", Università di Roma "La Sapienza", P.le A. Moro 5, 00185 Roma, Italy.

Abstract: The three-dimensional structures of the adenylation domains (A) of eight modules of SyrE, a peptide synthetase involved in the biosynthesis of syringomycin, was investigated by homology modeling using as a template the adenylation domain of gramicidin synthetase 1. Multiple sequence alignment of the adenylation domains of each of the eight modules of SyrE, allowed to identify the residues which delimitate the active site pockets. The active sites of the A domains of SyrE1 and SyrE2, involved in the insertion of the two serine residues in syringomycin, are essentially equivalent. The docking of these models with L-Ser and D-Ser showed that L-Ser is preferred as a substrate on the basis of stabilizing interactions. This is in accordance with previously shown stereo-specificity of these modules. Analogous investigation carried out on the modules SyrE3 and SyrE4 showed the preference for the L-isomer of their substrate, 2,4-diaminobutyric acid. In addition, the model of SyrE8, whose substrate amino acid is arginine, allowed to evidence specificity – conferring residues for this residue, which had not been previously reported

Key words: non-ribosomal code, enantiospecificity, molecular modelling.

Author for correspondence: I. Grgurina; e-mail: Ingeborg.Grgurina@uniroma1.it.

1. INTRODUCTION

Syringomycin (SR) is a toxic, virulence conferring secondary metabolite produced by many strains of the widespread phytopathogenic bacterium *Pseudomonas syringae* pv. *syringae*. As shown in Fig. 1, it is a lipodepsipeptide containing nine amino acid residues arranged in a lactone macrocycle. The N-terminal residue, serine, is N-acylated by a medium

N.S. Iacobellis *et. al.* (eds.), *Pseudomonas syringae* and Related Pathogens, 217-223.

chain hydroxylated fatty acid. It can be observed that this compound is particularly rich in modified amino acid residues, as shown by the C-terminal tripeptide, which is composed of dehydroaminobutyric acid (Dhb), 3-hydroxyaspartic acid, Asp(3-OH) and 4-chlorothreonine, Thr(4-Cl) and by the presence of two residues having the D-configuration (Fukuchi *et al.,*1992; Segre *et al.,*1989; Scaloni *et al.,*1994).

R – Ser Ser Dab Dab Arg Phe Dhb Asp (3-OH) Thr(4-Cl)
| |
O ———————————————————— CO

Figure 1. Structure of syringomycin; R=3-hydroxydodecanoyl.

SR is synthesised by a Non Ribosomal Peptide Synthesis (NRPS) mechanism (Grgurina and Benincasa, 1994; Zhang *et al.,* 1995), in which the formation of the peptide is catalysed by multifunctional enzymes called peptide synthetases (Döhren, *et al.,* 1997). The functional units of these enzymes, modules, are involved in the processing of single amino acids to be inserted into the peptide metabolite which requires the ATP-dependent activation, thiolation on the phosphopantetheine residue and the formation of the peptide bond. These reactions are catalysed by an adenylation domain (A), a peptidyl carrier protein (P), and a condensation domain (C), respectively (Beside these basic activities, some synthetases contain domains involved in the modifications of canonical amino acid residues, among which that involved in the epimerization at the α-carbon of an amino acid is the most frequently encountered (Döhren *et al.,* 1997).

In the past years, considerable progress has been made in the investigations of the structural bases of the catalytic mechanisms of peptide synthetases and their domains. Among the most relevant achievements are the determination of the crystallographic structure of the A domain of gramicidin synthetase 1 (GS1) (Conti *et al.,* 1997) co-crystallised with its substrates, and the discovery of the non-ribosomal code (Stachelhaus *et al.,* 1999). The identification of the region of the enzyme active site involved in the binding of the substrate was at the basis of subsequent *in silico* studies on the alignment of 160 sequences of the A domains of peptide synthetases with known substrate specificity. The set of ten amino acids involved in the substrate recognition and binding of most amino acids frequently occurring in microbial peptides was thus identified providing the first data on the structural bases of molecular recognition between an adenylation domain of a synthetase module and its substrate amino acid. The "non ribosomal code" is a long awaited explanation of the substrate selectivity of peptide synthetases.

In the light of this remarkable progress in the understanding of the functioning of peptide synthetases, we undertook an investigation on the syringomycin biosynthetic cluster, which had been sequenced (Guenzi *et al.*,

1998) with the objective to clarify some of the less understood aspects of this system. The biosynthesis of syringomycin is catalysed by four proteins, SyrB1, SyrB2, SyrC and SyrE which comprises eight modules (Guenzi *et al*., 1998). The adenylation domains of the modules SyrE1 and SyrE2 relative to the activation of the first two Ser residues, showed stereospecificity for the L-isomers of the substrate; however, Ser-2 has the D-configuration in the peptide and no epimerization domains were found in the modules of SR cluster. These findings prompted us to undertake an investigation on the structural bases of the stereospecificity of syringomycin synthetase domains. Our approach relied on the homology modelling of the adenylation domains in syringomycin synthetase using as a template the A domain of gramicidin synthetase 1 (PheA), which activates phenylalanine.

2. MATERIALS AND METHODS

2.1 Sequence Analysis

Protein sequences and structures were retrieved from the databanks through the ENTREZ interface (http://www.ncbi.nlm.nih.gov). Multiple sequence alignments were performed with the program CLUSTALW (Thompson *et al.,* 1994) and were manually manipulated to optimise the matching of several characteristics, including the observed and predicted secondary structural elements, the hydrophobic regions in the three-dimensional structures, the structurally and functionally conserved residues, and *indel* regions in the structures. The HOMOLOGY package in INSIGHTII (Accelerys, 2000, San Diego, CA, U.S.A.) was used for the manipulation of structures and alignments.

2.2 Model Building

The three-dimensional crystal structure of gramicidin synthetase I (PDB code 1amu) was the template for the homology-derived construction of the SyrE adenylation domains. Models were calculated with the MODELLER-4 package (Šali *et al.,*1995). Ten different models were built for each target protein with the highest refinement procedure and the one displaying the lowest objective function (which measures the extent of violation of constraints from the templates) was taken as the representative model. All the models were built with the inclusion of the adenosine-5'-monophosphate (AMP) and the water molecule found in the template active site in order to maintain the proper side chain conformation at the corresponding model site. Superimposition and RMSd calculation of C_α traces of the ten models indicated the most variable and therefore less reliable regions. However, the subsequent analyses were restricted to the active site region of the models. PROCHECK (Laskowski *et al.,* 1993) was used to monitor the stereochemical quality of the final models, whereas PROSAII (Sippl, 1993) to measure the overall protein quality in packing and solvent exposure. Secondary structures were determined using the program DSSP (Kabsch and Sander, 1983). Amino acid substrates were manually docked into the model active sites according to the position of phenylalanine (Phe) at the gramicidin synthetase 1 active site. The complexes were subsequently energy-minimised with the module discover INSIGHT II. All the programs were run under IRIX 6.5 on O_2 Silicon Graphics stations.

3. RESULTS

Multiple sequence alignment of the A-domains of syringomycin synthetases and PheA (the A domain of GS1), carried out with CLUSTALW and subsequently refined manually to achieve greater precision allowed to identify, within the primary structures of the A domains of SR synthetase, the residues that delimitate the active site pocket. The less conserved region of the domains comprised between residues 235 and 331, is involved in the selection and binding of the amino acid substrate to be activated.

Table 1 shows the non ribosomal code found for SR-synthetase adenylation domains compared to those found by Stachelhaus *et al.* (1999). The residues Asp 235 and Lys 517, involved in the binding of the amino group and a carboxyl group of the substrate amino acid, respectively, are common to all the models.

The percentage of identity between the A domains of SR synthetase and PheA between 29 and 39%, was considered sufficiently high to allow the construction of spatial models of syringomycin synthetase A domains. In fact, the three-dimensional structures of PheA and of the adenylate forming enzyme from firefly luciferase were remarkably similar, although the identity of the primary structures between the two proteins was not greater than 16%.

The homology modelling of the adenylation domains in syringomycin synthetase provided the spatial models of A domains of SR synthetase, which were used for simulations of substrate docking into active sites of the binding pockets.

The active sites in the models of the A domains of SyrE1 and SyrE2, are essentially equivalent. The only difference, which does not appear significant, is in the position 236 Leu of which is occupied by Leu and Val in SyrE1 and SyrE2, respectively. Both models were docked with L-Ser and with D-Ser in order to evaluate the compatibility and appeared spatially adequate to host both isomers of Ser. However, the insertion of the L-Ser could be preferred on the basis of a stabilising hydrogen-bond interaction of the hydroxymethylene group of Ser side chain with Ser 301 in the active site, which cannot be formed with the differently oriented side chain of the D-enantiomer.

SyrE3 and SyrE4 (corresponding to the insertion of the two 2,4-diaminobutyric acid (Dab) residues possess equivalent active sites, exactly coincident with that reported by Stachelhaus *et al.* (1999) for this residue. Docking of L-Dab suggests that this substrate is sterically compatible with the insertion both into SyrE3 and into SyrE4. The amino group of the side chain forms a salt bridge with the carboxyl group of the Glu 239, one of the residues of the NRPS code, responsible for the substrate recognition, which stabilises the substrate in the enzyme active site. Analogous procedure carried out with D-Dab evidenced the steric hindrance in the active sites of the A domains of the modules SyrE3 and SyrE4, due to the collision with Phe 234 and Thr 292. Furthermore, the salt bridge between the side-chain amino group and Glu 239, observed with L-Dab, cannot be formed because of the different orientation of the side chain. The energy minimisation of the complex formed between D-Dab and the Syr3 A domain and D-Dab and the Syr4 A domain shows that the substrate can adapt to the two active sites only assuming an energetically unfavourable conformation.

The code for the residue 5 in SR, Arg, shown in Table 1, is the first report on the set of amino acids responsible for the binding of Arg in an A domain. Two residues at the active site Asp331 and Asp278 are particularly important in determining the substrate specificity. They are involved in the key interactions between the substrate and the enzyme, namely in stabilising arginine in the active site by forming hydrogen bonds with the guanidine group.

The A domains of SyrB1 and SyrE7 are essentially equivalent. SyrE7 is coincident with the reported Thr code (Stachelhaus *et al.*, 1999). The exchange Ser278 into Asn278 between SyrE7 and SyrB1 does not appear to be particularly significant. Threonine was inserted into both models. The substrate is stabilised in the active site of SyrE7 by a hydrogen bond with the His331 residue and by insertion of the methyl group of the side chain into the hydrophobic pocket formed by Phe234 and Phe236.

The A domain binding pocket in SyrE8 was docked both by Asp. and by Asp(3-OH). In the model the side chain of Asp can interact with Lys239. Similar interaction had been hypothesised by Stachelhaus *et al.* (1999) for the A-domain specific for glutamate. This gives rise to a strongly stabilising interaction of the negatively charged substrate with a residue of the opposite charge in the enzyme binding pocket.

Table 1. Comparison between the non ribosomal code found in SR-synthetase A-domains and the code (italics) reported by Stachelhaus *et al* (1999).

A domain substrate	**Sequence position**									
	235	**236**	**239**	**278**	**299**	**301**	**322**	**330**	**331**	**517**
E1 (L-Ser)	D	L	W	H	L	S	L	I	D	H
E2 (D-Ser)	D	V	W	H	L	S	L	I	D	H
Ser	*D*	*V*	*W*	*H*	*L*	*S*	*L*	*I*	*D*	*H*
E3 (L-Dab)	D	L	E	H	N	T	T	V	S	K
E4 (D-Dab)	D	L	E	H	N	T	T	V	S	K
Dab	*D*	*L*	*E*	*H*	*N*	*T*	*T*	*V*	*S*	*K*
E5 (Arg)	D	V	A	D	V	G	A	I	D	K
E6 (Phe)	D	A	P	I	M	G	G	T	C	K
Phe	*D*	*A*	*W*	*T*	*I*	*A*	*A*	*I*	*C*	*K*
E7 (Dhb)	D	F	W	N	V	G	M	V	H	K
B (4-Cl-Thr)	D	F	W	S	V	G	M	V	H	K
Thr	*D*	*F*	*W*	*N*	*I*	*G*	*M*	*V*	*H*	*K*
E8 (3-OH-Asp)	D	M	K	D	L	G	M	V	D	K
Asp	*D*	*L*	*T*	*K*	*V*	*G*	*H*	*I*	*G*	*K*
Glu	*D*	*A*	*K*	*D*	*L*	*G*	*V*	*V*	*D*	*K*

4. DISCUSSION

The results of this work provide new information on the non ribosomal code of the SR synthetase A domains and, in particular, on the degree of degeneracy. In addition the set of residues conferring the specificity for Arg is reported here for the first time.

The inspection of the specificity conferring residues of syringomycin synthetase A domains allows to gain some insight into the timing of the amino acid modifications in the biosynthesis of syringomycin. SyrB1 and SyrE7, are equivalent and coincident with the Thr code. It can therefore be inferred that their common substrate is modified subsequently to the activation as the residues introduced by these two domains appear in the peptide as Thr (4-Thr) and dehydroaminobutyric acid (Dhb), respectively. Conversely, the low similarity between the putative specificity conferring residues in SyrE8, corresponding to Asp(3-OH), the NRPS codon reported for Asp, suggests the direct introduction of the pre-formed hydroxyaspartic acid residue. The clarification of this point warrants further studies.

The most important and innovative result of this work, however, is the demonstration of the structural bases of the enantioselectivity of the adenylation modules of peptide synthetases. The timing and the mechanism of the epimerization of the L-residues, remains to be determined.

ACKNOWLEDGEMENTS

This work was funded by the Italian Ministero dell'Istruzione, dell'Università e della Ricerca (MIUR) and by the Fondazione Pasteur – Cenci Bolognetti.
The authors thank Prof. A. Ballio and Prof. F. Bossa for encouragement.

REFERENCES

Conti, E., Stachelhaus, T., Marahiel, M.A., and Brick, P., 1997, Structural basis for the activation of phenylalanine in the non-ribosomal biosynthesis of gramicidin S. *EMBO J.* **16**: 4174-4183.

Döhren, von H., Keller, V., Vater, J., and Zocher, R., 1997, Multifunctional peptide synthetases. *Chem. Rev.* **97**: 2675-2705.

Fukuchi, N., Isogai, A., Nakayama, J., Takayama, S., Yamashita, S., Suyama, K., Takemoto, J.Y. and Suzuki, A., 1992, Structure and stereochemistry of three phytotoxins, syringomycin, syringotoxin and syringostatin, produced by *Pseudomonas syringae* pv. *syringae*. *J. Chem. Soc. Perkin Trans.* **1**: 1149-1157.

Grgurina, I., and Benincasa, M., 1994, Evidence of the non ribosomal biosynthetic mechanism in the formation of syringomycin and syringopeptin, bioactive lipodepsipeptides of the phytopathogenic bacterium *Pseudomonas syringae* pv. *syringae*. *It. Biochem. Soc. Trans.* **5**: 143.

Guenzi, E., Galli, G., Grgurina, I., Gross, D.C., and Grandi, G., 1998, Characterisation of the syringomycin synthetase gene cluster. *J. Biol. Chem.* **273**: 32857-32863.

Kabsch, W., and Sander, C., 1983, Dictionary of protein secondary structure: pattern recognition of hydrogen-bonded and geometrical features. *Biopolymers* **22**: 2577-2637.

Laskowski, R.A., MacArthur, M.W., Moss, D.S., and Thornton, J.M., 1993, PROCHECK: a program to check the stereochemical quality of protein structures. *J. Appl. Cryst.* **26**: 283-291.

Scaloni, A., Bachmann, R.C., Takemoto, J.Y., Barra, D., Simmaco, M., and Ballio, A., 1994, Stereochemical structure of syringomycin, a phytotoxic metabolite of *Pseudomonas syringae* pv. *syringae*. *Nat. Prod. Letters* **4:** 159-164.

Segre, A., Bachman, R.C., Ballio, A., Bossa, F., Grgurina, I., Iacobellis, N.S., Marino, G., Pucci, P., Simmaco, M., and Takemoto, J.Y., 1989, The structure of syringomycin A_1, E and G. *FEBS Lett.* **255**: 27-31.

Stachelhaus, T., Mootz, H.D., and Marahiel, M.A., 1999, The specificity-conferring code of adenylation domains in non ribosomal peptide synthetase. *Chem. Biol.* **6**: 493-505.

Šali, A., Potterton, L., Yuan, F., Van Vlijmen, H., and Karplus, M., 1995, Evaluation of comparative protein modeling by MODELLER. *Proteins* **23**: 318-326.

Sippl, M.J., 1993, Recognition of errors in three-dimensional structures of proteins. *Proteins* **17**: 355-362.

Thompson, J.D., Higgins, D.G., and Gibson, T.J., 1994, CLUSTALW: improving the sensitivity of progressive multiple alignment through sequence weighting, position-specific gap penalties and weight matrix choice. *Nucleic Acids Res.* **22**: 4673-4680.

Zhang, J.H., Quigley, N.B., and Gross, D.C., 1995, Analysis of the *syrB* and *syrC* gene of *Pseudomonas syringae* pv. *syringae* indicates that syringomycin is synthesised by a thiotemplate mechanism. *J. Bacteriol.* **177**: 4009-4020.

Identification of Virulence Factors from *Pseudomonas syringae*

J. BOCH, A. LANDGRAF, J. KRAUSE, B.N. KUNKEL[#], and U. BONAS
Institute of Genetics, Martin-Luther-University Halle-Wittenberg, Weinbergweg 10, D-06120 Halle, Germany; [#]Department of Biology, Washington University, St. Louis, MO 63130, USA.

Abstract: Plant pathogenic bacteria carry a large set of virulence genes that enables them to exploit plant hosts to support their growth. A genetic screen (IVET) was established to identify bacterial genes that are expressed during the infection process. About one third of the *in planta* expressed (*ipx*) genes were shown to be coregulated with the Hrp-type III secretion system. The *hrp*-coregulated *ipx* genes were compared to putative virulence genes that were identified by other genomic screens. A subset of the *hrp*-coregulated *ipx* genes encode known effectors that are translocated by the Hrp-system into the plant cell as well as putative effectors. Using a reporter system for translocation we were able to identify two novel effectors. Further studies will be aimed at characterizing the effect of these virulence factors in the plant cell.

Key words: IVET, HrpL, genomic, effector.

Author for correspondence: J. Boch; e-mail: boch@genetik.uni-halle.de.

1. INTRODUCTION

Gram-negative bacterial plant pathogens have been intensively investigated because they are amenable to a wide range of molecular biological analyses. To date, the best studied model organism is *Pseudomonas syringae* which is further strengthened by the recent completion of its genome sequence analysis (The Institute for Genomic Research, www.tigr.org). The identification of virulence factors and analysis of their function is one of the main goals persecuted today. A core component of virulence is the Hrp-type III secretion system (TTSS) which facilitates the direct transfer of bacterial

N.S. Iacobellis *et. al.* (eds.), *Pseudomonas syringae* and Related Pathogens, 225-231.

virulence proteins (effectors) into the plant cell. 36 of these effectors have been identified in the model strain *P. s.* pv. *tomato* DC3000 (Collmer *et al.*, 2002) and the molecular analysis of their function will be one of the major challenges for the future. The availability of the genome sequence of one of its host plants, *Arabidopsis thaliana* (The Arabidopsis Genome Initiative, 2000), will greatly enhance the methods to further elucidate the molecular functions of these virulence factors.

2. IDENTIFICATION OF GENES INDUCED DURING PLANT PATHOGEN INTERACTION

Plant pathogenic bacteria have to adapt to a specialised habitat during infection of a host plant. The expression of a large set of virulence genes is essential for a successful invasion and bacterial multiplication within the plant. Many of these genes are not constitutively expressed but induced upon infection (Francis *et al.*, 2002). We thus reasoned, that by identifying *Pseudomonas syringae* genes induced upon infection of *Arabidopsis thaliana* we could determine components of the pathogen which are involved in the adaptation to the plant and the establishment of pathogenicity. We used a promoter-trap strategy similar to IVET (*in vivo* expression technology) (Mahan *et al.*, 1993). In contrast to most IVET studies which were based on the complementation of a general growth deficiency or antibiotic sensitivity, we developed a *P. s.* pv. *tomato* strain for our IVET screen that was only deficient in pathogenicity. This strain carried a deletion of *hrcC*, a core component of the TTSS which is essential for pathogenicity of *P. syringae* (Staskawicz *et al.*, 2001). Genomic fragments were cloned upstream of a promoterless *hrcC* gene and screened for a promoter activity that directs expression of this *hrcC* gene *in planta* thus complementing the endogenous *hrcC* mutation of the strain. A promoterless β-glucuronidase (GUS) reporter gene downstream of the *hrcC* gene was used to subsequently identify those promoters that are not expressed on rich laboratory media but active *in planta* (Boch *et al.*, 2002). Several of these promoters should direct the expression of genes which are involved in the adaptation of the bacteria to the *in planta* growth conditions. 79 *ipx* (*in planta* expressed) fusions were sequenced and identified (1) several genes known to be involved in *P. syringae* pathogenicity, (2) novel genes predicted to be virulence determinants, and (3) genes with unknown function in virulence or no homology to known sequences. The identified virulence factors included TTSS-structural components and effectors, as well as enzymes directing phytotoxin (coronatine) production, exopolysaccharide (alginate) biosynthesis, detoxification of reactive compounds (catalase), and a pectin lyase predicted to degrade plant material (Boch *et al.*, 2002).

3. IDENTIFICATION OF IPX GENES THAT ARE CONTROLLED BY HRPL

The identification of genes encoding components and substrates of the TTSS in the IVET analysis is not surprising, because they are controlled by a key regulator of virulence in *Pseudomonas syringae*, the alternative sigma factor HrpL (Xiao *et al.*, 1994). The importance of the Hrp regulon for virulence of *P. syringae* prompted us to analyse which *ipx* fusions are controlled by HrpL. This was accomplished by comparing reporter gene activity (GUS) of the *ipx* fusions in the presence and absence of HrpL in *Escherichia coli* (Boch *et al.*, 2002). 24 different HrpL-dependent *ipx* promoters were found (Table 1), including 5 within the *hrp* gene cluster, 3 within the conserved effector locus (CEL) of the *hrp*-flanking region, 3 connected to known effector genes, the pectin lyase promoter, and 11 directing expression of uncharacterised open reading frames (ORFs). The large majority of HrpL-dependent genes in *P. s.* pv. *tomato* DC3000 seem to be part of the Hrp secretome, but a few exceptions are known (e.g. *corR*, the regulator of coronatine biosynthesis genes) (Fouts *et al.*, 2002). Accordingly, the 11 uncharacterised HrpL-regulated ORFs presumably encode novel virulence genes including TTSS-secreted factors. Several groups have recently identified genes from *P. s.* pv. *tomato* DC3000 that are related to the TTSS, either because their expression is controlled by HrpL, or because they are secreted by the TTSS (Guttman *et al.*, 2002; Fouts *et al.*, 2002; Petnicki-Ocwieja *et al.*, 2002; Zwiesler-Vollick *et al.*, 2002). The methods used to identify HrpL-dependent genes were based on identification of conserved DNA elements (*hrp* boxes) in the genome sequence of *P. s.* pv. *tomato* DC3000. HrpL exerts its effect on expression through a *hrp* box that is present in the promoter region of a HrpL-regulated gene (Xiao *et al.*, 1994). Putative effectors were postulated based on additional criteria including amino acid biases, which are believed to be part of a Hrp-secretion motif. Unfortunately, the use of different nomenclatures within each publication left the reader unable to compare the analyses without a lengthy exploration of supplementary data. Table 1 displays a list of the HrpL-dependent *ipx* genes and indicates which genes have also been identified in the genomic approaches.

4. TRANSLOCATION ANALYSIS OF PUTATIVE EFFECTORS

The proteins secreted by the TTSS either affect the outer barriers (plasma membrane and plant cell wall) or they are translocated into the interior of the plant cell where they can target different subcellular compartments (Nimchuk *et al.*, 2000; van den Ackerveken, 1996). The pivotal role of the TTSS in the establishment of pathogenicity implies that the concerted action of these TTSS-secreted factors is the trigger for disease development. Nevertheless, the molecular function of most TTSS-secreted proteins is not understood.

To prove the nature of a new effector gene, translocation of the corresponding protein into the plant cell has to be shown. This has been difficult due to the fact that fusion of an effector protein to a reporter often interferes with secretion. The preferred method to date uses the effector AvrRpt2 as a reporter for translocation (Mudgett *et al.*, 2000; Guttman and Greenberg, 2001). Hrp-secretion signals are believed to reside in the N-terminus of effector proteins and accordingly, an N-terminal deletion of AvrRpt2 is not secreted anymore. The N-termini of putative effectors are analysed for Hrp-secretion signals by fusing them to the C-terminal part of AvrRpt2. Translocation of a fusion protein into the plant cell leads to a recognition of the AvrRpt2-moiety by the cytoplasmatically localised resistance gene product in *RPS2* but not in *rps2 Arabidopsis thaliana* plants and induces a visible plant tissue collaps (hypersensitive response). This method was used to analyse several *ipx* genes and other HrpL coregulated genes (Table 1) (Guttman *et al.*, 2002; Petnicki-Ocwieja *et al.*, 2002; Zwiesler-Vollick *et al.*, 2002; J. Boch unpublished data). Ipx7 and Ipx68 were translocated as AvrRpt2-fusion proteins and shown to be effectors of *Pseudomonas syringae* pv. *tomato* DC3000. Further studies will clarify, which HrpL-regulated *ipx* fusions describe effector genes and which define additional virulence factors that are not secreted, but co-regulated with the TTSS.

5. CONCLUSION

The identification of virulence factors from *Pseudomonas syringae* using an IVET analysis has enabled the identification of several genes controlled by HrpL (Boch *et al.*, 2002). For two of these, translocation into the plant cell could be shown, which indicates that these genes encode novel effectors.

Table 1. Pseudomonas syringae pv. *tomato* DC3000 *ipx* genes preceded by a *hrp* box or controlled by HrpL. The nomenclature of corresponding genes identified in genomic screens is shown[a].

ipx[b]	Similarity[c]	Putative effectors[d]	miniTn*5gus*[e]	Secretion signal[f]	*hrp* co-regulated[g]	Effector[h]
ipx1-6	*hrp*-genes					n.d.
ipx7, ipx66[i]	*virPphA*	*VirPphA$_{Pto}$*	*avrPtoB*	*AvrPtoB$_{DC}$*	*virPphA*	+
ipx8[j]	*avrPpiB*	*AvrPpiB$_{Pto}$* *AvrPpiB2$_{Pto}$*	*avrPpiB11$_{Pto}$*	*AvrPpiB1$_{DC}$*	*avrPpiB*	n.d.
ipx9[k]	*avrPphD*					n.d.
ipx10, ipx11	CEL ORF1		CEL ORF1			n.d.
ipx12, ipx13	CEL ORF3	*HolPtoX*				n.d.
ipx14	CEL ORF7	*HolPtoT*	CEL ORF7			n.d.
ipx15	CEL ORF8		CEL ORF8			n.d.
ipx31	Pectin lyase	*HopPmaH$_{Pto}$*	Pectin lyase	ORF12		(+)
ipx32[l]	CEL ORF5	*HopPtoA2*				(+)
ipx35	Phospho-transferase					n.d.
ipx41	Carboxypep			ORF27		n.d.
ipx45	Thiamine biosynthesis		ORF07856		*chp7*	n.d.
ipx52	Rhs family					n.d.
ipx53	hp					n.d.
ipx54[m]	hp		*iaaL*	*IaaL*		n.d.
ipx67	no homology	*HopPmaI$_{Pto}$*	ORF04349	ORF10	*chp2*	(+)
ipx68[i]	no homology	*HolPtoV*		ORF3	*chp1*	+
ipx69, ipx70[i]	no homology	*ChpPtoZ/ holPtoZ*	ORF05707	ORF25/ ORF26	*chp4*	n.d.
ipx71	no homology					n.d.

[a] A full comparative list of *ipx* fusions and effector genes isolated in different labs based on their position in the DC3000 genome can be found at www.biologie.uni-halle.de/genet/plant/research/supplementary/DC3000effectors.html

[b] *ipx* gene designation according to Boch *et al.* (2002).

[c] Similar gene found by BLAST analysis of GenBank. hp, hypothetical protein.

[d] Putative effector genes identified by *hrp* box and putative secretion signal (Guttman *et al.*, 2002).

[e] HrpL-regulated genes identified by Tn*5*-mutagenesis or hidden Markov model-based identification of *hrp* boxes (Fouts *et al.*, 2002).

[f] Putative effector genes identified by *hrp* box and N-terminal motives that are believed to define a secretion signal (Petnicki-Ocwieja *et al.*, 2002).

[g] *hrp* co-regulated genes identified by *hrp* box, microarray analysis and northern blot (Zwiesler-Vollick *et al.*, 2002).

[h] Translocation of the protein into the plant cell has been analysed as AvrRpt2 fusion. +: translocated; (+): homologous to translocated protein; n.d.: not determined.

[i] The *ipx*-fusions *ipx66, ipx68,* and *ipx70* are 491 bp, 462 bp, and 55 bp downstream of *virPphA$_{Pto}$*, *holPtoV*, and *holPtoZ*, respectively, but their transcription is probably dependent on these promoters.

[j] Two homologous copies of *avrPpiB* exist in the DC3000 genome, but it is unknown which one *ipx8* is linked to.

[k] *ipx9* corresponds to a fragmented homolog of *avrPphD*. Two other genes homologous to *avrPphD* exist in the genome of DC3000 which are described by the genomic analyses.

[l] *ipx32* corresponds to a gene homologous to CEL ORF5, but located elsewhere in the genome

[m] The *ipx54*-fusion is located in the first ORF of a putative operon that includes the *iaaL* gene identified by other screens. *ipx54* is preceded by a *hrp* box.

Recent genomic approaches have led to the identification of additional genes encoding putative effectors and yield a comprehensive list of 36 putative TTSS-secreted proteins (Collmer *et al.*, 2002). Future analysis will aim to elucidate the molecular function of these proteins, their distribution within different plant pathogenic bacteria, and their interplay which orchestrates the development of plant disease.

ACKNOWLEDGEMENTS

Parts of this work were supported by the grant BO 1496/4-1 from the DFG to J.B. We are grateful to Daniela Büttner for critical comments on the manuscript. Sequencing of the *Pseudomonas syringae* pv. *tomato* DC3000 genome by The Institute for Genomic Research (TIGR) was accomplished with support from the National Science Foundation.

REFERENCES

Boch, J., Joardar, V., Gao, L., Robertson, T.L., Lim, M., and Kunkel, B.N., 2002, Identification of *Pseudomonas syringae* pv. *tomato* genes induced during infection of *Arabidopsis thaliana*. *Mol. Microbiol.* **43**: 73-88.

Collmer, A., Lindeberg, M., Petnicki-Ocwieja, T., Schneider, D.J., and Alfano, J.R., 2002, Genomic mining type III secretion system effectors on *Pseudomonas syringae* yields new picks for all TTSS prospectors. *Trends Microbiol.* **10**: 462-469.

Fouts, D.E., Abramovich, R.B., Alfano, J.R., Baldo, A.M., Buell, C.R., Cartinhour, S., Chatterjee, A.K., D´Ascenzo, M., Gwinn, M.L., Lazarowitz, S.G., Lin, N.-C., Martin, G. B., Rehm, A.H., Schneider, D.J., van Dijk, K., Tang, X., and Collmer, A., 2002, Genomewide identification of *Pseudomonas syringae* pv. *tomato* DC3000 promoters controlled by the HrpL alternative sigma factor. *Proc. Natl. Acad. Sci. USA* **99**: 2275-2280.

Francis, M.S., Wolf-Watz, H., and Forsberg, Å., 2002, Regulation of type III secretion systems. *Curr. Opin. Microbiol.* **5**: 166-172.

Guttman, D.S., and Greenberg, J.T., 2001, Functional analysis of the type III effectors AvrRpt2 and AvrRpm1 of *Pseudomonas syringae* with the use of a single-copy genomic integration system. *Mol. Plant-Microbe Interact.* **14**: 145-155.

Guttman, D.S., Vinatzer, B.A., Sarkar, S.F., Ranall, M.V., Kettler, G., and Greenberg, J.T., 2002, A functional screen for the type III (Hrp) secretome of the plant pathogen *Pseudomonas syringae*. *Science* **295**: 1722-1726.

Mahan, M.J., Slauch, J.M., and Mekalanos, J.J., 1993, Selection of bacterial virulence genes that are specifically induced in the host tissues. *Science* **259**: 686-688.

Mudgett, M.B., Chesnokova, O., Dahlbeck, D., Clark, E.T., Rossier, O., Bonas, U., and Staskawicz, B.J., 2000, Molecular signals required for type III secretion and translocation of the *Xanthomonas campestris* AvrBs2 protein to pepper plants. *Proc. Natl. Acad. Sci. USA* **97**: 13324-13329.

Nimchuk, Z., Marois, E., Kjemtrup, S., Leister, R.T., Katagiri, F., and Dangl, J.L., 2000, Eukaryotic fatty acylation drives plasma membrane targeting and enhances function of several type III effector proteins from *Pseudomonas syringae. Cell* **101**: 353-363.

Noël, L., Thieme, F., Nennstiel, D., and Bonas, U., 2001, cDNA-AFLP analysis unravels a genome-wide *hrpG*-regulon in the plant pathogen *Xanthomonas campestris* pv. *vesicatoria. Mol. Microbiol.* **41**: 1271-1281.

Petnicki-Ocwieja, T., Schneider, D.J., Tam, V.C., Chancey, S.T., Shan, L., Jamir, Y., Schechter, L.M., Janes, M.D., Buell, C.R., Tang, X., Collmer, A., and Alfano, J.R., 2002, Genomewide identification of proteins secreted by the Hrp type III protein secretion system of *Pseudomonas syringae* pv. *tomato.* DC3000. *Proc. Natl. Acad. Sci. USA* **99**: 7652-7657.

Staskawicz, B.J., Mudgett, M.B., Dangl, J.L., and Galan, J.E., 2001, Common and contrasting themes of plant and animal diseases. *Science* **292**: 2285-2289.

The Arabidopsis Genome Initiative, 2000, Analysis of the genome sequence of the flowering plant *Arabidopsis thaliana. Nature* **408**: 796-815.

Van den Ackerveken, G., Marois, E., and Bonas, U., 1996, Recognition of the bacterial avirulence protein AvrBs3 occurs inside the host plant cell. *Cell* **87**: 1307-1316.

Xiao, Y., Heu, S., Yi, J., Lu, Y., and Hutcheson, S.W., 1994, Identification of a putative alternate sigma factor and characterization of a multicomponent regulatory cascade controlling the expression of *Pseudomonas syringae* pv. *syringae* Pss61 *hrp* and *hrmA* genes. *J. Bacteriol.* **176**: 1025-1036.

Zwiesler-Vollick, J., Plovanich-Jones, A.E., Nomura, K., Bandyopadhyay, S., Joardar, V., Kunkel, B.N., and He, S.Y., 2002, Identification of novel *hrp*-regulated genes through functional genomic analysis of the *Pseudomonas syringae* pv. *tomato* DC3000 genome. *Mol. Microbiol.* **45**: 1207-1218.

Toxic Metabolites and Lipopolysaccharides from *Pseudomonas cichorii*

S. LAZZARONI, A. EVIDENTE and G. SURICO[#]

Dipartmento di Scienze del Suolo, della Pianta e dell'Ambiente, Università di Napoli Federico II, Viale Università, 100, Portici (NA), Italy; [#]Dipartimento di Biotecnologie Agrarie, Sezione di Patologia Vegetale, Università di Firenze, P.le delle Cascine, 28, Firenze, Italy.

Abstract: *Pseudomonas cichorii* is the causal agent of "varnish spot", a serious disease that damages field-grown lettuce. This paper reports preliminary results on the chemical and biological characterisation of the lipophylic phytotoxic and the hydrophilic antibiotic metabolites produced by a strain of *P. cichorii* (ICMP5707), isolated in Germany from infected tissues of endive (*Cichorium endivia* L.). These features, together with the results obtained on the chemical characterisation of some components of the lipopolysaccharides (LPS), contained in the outer membrane, were compared with those of a Canadian isolate (ICMP9417) from the *cultivar* Park of spring wheat (*Triticum aestivum* L.) and preliminarily classified as *P. cichorii.*

The initial results of this comparison indicated that the pattern of phytotoxic metabolites was similar for both strains, while strain ICMP5707 differs from strain ICMP9417 of *P. cichorii* for the production of the antibiotic metabolites. In fact, the culture filtrates of strain ICMP9417, grown under identical conditions, do not show significant antibiotic activity.

Moreover, preliminary investigations of the LPS of *P. cichorii* ICMP9417 showed that the repeating unit of O-specific polysaccharide chain was certainly different from the one of *P. cichorii* ICMP5707. Conversely the LPS lipid moiety is the same in the fatty acid composition.

Key words: lettuce, varnish spot, phytotoxins; antibiotic.

Author for correspondence: A. Evidente; e-mail: evidente@unina.it.

N.S. Iacobellis *et. al.* (eds.), *Pseudomonas syringae* and Related Pathogens, 233-243.

1. INTRODUCTION

Pseudomonas cichorii (Swingle) Stapp is a bacterial pathogen of several plants of agricultural and ornamental interest such as celery, cabbage, tomato, chrysanthemum, gerbera, etc. In particular, *P. cichorii* is known to be the causal agent of "varnish spot", a serious disease that damages field-grown lettuce. Infected plants show dark-brown, shiny, firm, necrotic spots on the blades and petioles of leaves (Grogan *et al.*, 1977). This bacterium is also known as the causal agent of "stem melanosis" on the cultivar Park of spring wheat (*Triticum aestivum* L.), a disease characterised by bleached empty heads and blackening of the rachids, peduncle and stem immediately beneath the nodes (Piening *et al.*, 1985). Most of these and of the above symptoms may be caused by phytotoxic substances produced by the bacterium. Although previous studies (Shirata *et al.*, 1983; Hu *et al.*, 1998) had ascertained a biocidal activity in several strains of *P. cichorii*, no bioactive metabolite had been so far isolated and characterised. These results prompted further investigations on the phytotoxic and antibiotic metabolites produced by *P. cichorii* in order to first isolate and biologically/chemically characterise these molecules and to then determine their role in the ecology and pathogenesis of the bacterium.

Knowledge of the lipopolysaccharide (LPS) structure, which may have an important role in host-pathogen interaction, may be fundamental to obtain more information about the mechanisms of pathogenicity and the correct taxonomic classification of bacteria. Therefore, bacterial cells of *P. cichorii* strain ICMP9417, preliminarily classified as *P. cichorii*, were used for the extraction, purification and chemical characterisation of the LPS. The results were then compared with those recently reported on the characterisation of the lipid A (Molinaro *et al.,* 2002) and the O-chain polysaccharide (Jimenez-Barbero *et al.*, 2002) of *P. cichorii* ICMP5707 LPS.

This paper reports on preliminary results obtained on the production and chemical and biological characterisation of antibiotic and phytotoxic metabolites produced in culture by two strains of *P. cichorii*.

1.1 Initial Comments

A strain of *Pseudomonas cichorii* (ICMP5707), isolated in Germany from infected tissues of endive (*Cichorium endivia* L.), was grown in still culture on IMM medium (Surico *et al.*, 1988) for the production of phytotoxins and metabolites with antibiotic activity. Bacterial culture filtrates caused necrosis when assayed on tobacco leaves and significantly inhibited the growth of *Geotrichum candidum* and *Rhodotorula pilimanae*. Therefore the attention was directed toward the isolation and the chemical and biological characterisation of the antibiotic and phytotoxic metabolites. The same experiments were then carried out on the culture filtrates of *P. cichorii* strain ICMP9417 grown under identical conditions. However, in this case only a significant phytotoxic activity was observed. Studies on the chemical characterisation of phytotoxic metabolites, as well as on that of the LPS, were performed to contribute to the correct classification of this strain, in addition to its virulence and morphological and biomolecular properties.

2. MATERIALS AND METHODS

2.1 *Pseudomonas cichorii* and Bacterial Cultivation

Pseudomonas cichorii ICMP5707 was grown in IMM medium (Surico *et al.*, 1988). Bacterial suspension (8 ml, A_{600} = 0.35) was added to 1 l Erlenmeyer flasks containing 400 ml of medium and incubated at 25°C. The bacterial suspension was centrifuged (9,000 *g* X 20 min) and the harvested cells were washed three times with 0.85% NaCl and lyophilised.
P. cichorii ICMP9417 was grown in IMM under identical conditions and the bacterial suspension were worked up as above described.

2.2 Biological Assays

2.2.1 Phytotoxic Activity

Phytotoxic activity was assayed on tobacco leaves using the leaf mesophyll infiltration method. The assays were performed with 5-day-old culture filtrates, before and after lyophilisation, the corresponding ethyl acetate extract and fractions obtained by its preparative TLC purification. The samples were dissolved in a minute amount of MeOH and brought up to the required concentration with distilled water. If needed the pH of the solutions was adjusted to about 6.

2.2.2 Antibiotic Activity

Bacteria were grown as separate colonies on potato-dextrose agar (PDA) in petri dishes and incubated at 25 °C for 5 days. The plates were then seeded with the test organism (*Rhodotorula pilimanae* and/or *Geotrichum candidum*) by spraying a suspension of cells and/or mycelial fragments on the surface of the agar. The test organisms were grown as previously reported in shake cultures in potato-dextrose broth (Surico *et al.*, 1988). Antibiotic activity was also determined on the solutions prepared for the phytotoxic activity assay. 100 μl aliquot from each preparation was pipetted into a hole (0.6 cm in diameter) made in the PDA in Petri dishes. After absorption of the solvent (a minute amount of MeOH and distilled water) the plates were seeded with the test organisms as above. Plates were scored after 24 h for the presence of zones of inhibition around bacterial colonies or holes.

2.3 Experimental on *Pseudomonas cichorii* ICMP 5707 Strain

2.3.1 Phytotoxins Purification

Bacterial culture filtrates (1,830 ml, pH 6.5) were extracted four times with 400 ml of ethyl acetate, and the combined organic extracts were dried (Na_2SO_4) and evaporated under reduced pressure to give a crude residue of ca. 32 mg. Its thin layer chromatography (TLC) analysis was performed on reversed-phase plates (Whatman, Statocrom KC 18 F_{254}, 0.20 mm). The spots were visualised by exposing the plates to UV radiation and to iodine vapour. The residue (29 mg) was partially purified by reversed-phase preparative TLC (eluent: ethanol:water 1:1) to give thirteen fractions. Antibiotic Purification

The residual aqueous phase from the phytotoxins extraction was concentrated under reduced pressure and lyophilised to give a solid residue (23.4 g). An aliquot of this residue (282 mg) was dissolved with 20 ml of water, precipitated with ammonium sulphate (11.4 g) and stored at 4 °C. After six days the formation of a precipitate was observed. This was removed by centrifugation (10,000 rpm, 1 h, 4 °C) and resuspended in water to give, after lyophilisation, a crude residue of 222 mg.

TLC analysis of this residue was performed on silica gel plates (Merck, Kieselgel, 60 F_{254}, 0.25 mm). The spots were visualised by exposing the plates to UV radiation and spraying with 0.5 % ninhydrin in Me_2CO, followed by heating at 110 °C for 10 minutes.

2.4 Experimental on *Pseudomonas cichorii* ICMP 9417 Strain

2.4.1 Isolation and Purification of Phytotoxic Metabolites

Bacterial culture filtrate (100 ml, pH 6.2) was extracted four times with 100 ml of ethyl acetate, and the combined organic extract of culture filtrate was

dried (Na_2SO_4) and evaporated under reduced pressure to give a crude residue of ca. 11.6 mg.
TLC analysis of this residue was performed as described (2.3.1).

2.4.2 Isolation and Purification of the LPS Fraction

Dried cells (0.620 g) were extracted by the phenol/water method (Westphal *et al.*, 1965). Both phases were separately worked up and freeze-dried to yield 40 and 37 mg of LPS fractions in the phenol and water phases, respectively. Both fractions were analysed by discontinuous SDS-PAGE electrophoresis (Laemmli, 1970) on a 12% gel with a Bio-Rad miniprotean gel system; the samples (4 μg) were run at constant voltage (150 V) and stained according to the procedure of Kittelberger *et al.*, 1993. LPS fraction from the phenol phase was further purified on Sephacryl HR 400 (Pharmacia, 1.5 x 50 cm, eluent NH_4HCO_3 50 mM, flow 0.4 ml min^{-1}); eluate was monitored with an R.I. refractometer (R410 Waters), and the collected peaks were analysed for the LPS content on SDS-PAGE.

2.4.3 Sugar and Lipid Analysis of the LPS Fraction

LPS (1 mg) was dried in a desiccator over P_2O_5 for 1 h *under vacuum* and then treated with 1 M methanolic HCl at 80 °C for 18 h. After cooling, the solution was directly extracted twice with equal volumes of *n*-hexane; the top phase (*n*-hexane) was dried and the fatty acid methyl esters were analysed by GC-MS. The methanolic phase was dried and the methyl glycosides were acetylated as follows: first with Ac_2O (200 μl) at 80 °C for 15 min, followed by drying in a stream of air, and then by treatment with dry pyridine (200 μl) and Ac_2O (100 μl) at the same temperature for 30 min. The reagents were removed by evaporation and the mixture of peracetylated methyl-glycosides was analysed by GC-MS. GS-MS analysis conditions for both fatty acids and methyl glycoside derivatives were the same. The analysis were run with a Hewlett-Packard 5890 instrument, on an SPB-5 capillary column (Supelco, 30 m × 0.25 i.d. flow rate, 0.8 ml min^{-1}; He as carrier gas), with the temperature program: 150 °C for 5 min, 150 → 300 °C at 5.0 °C min^{-1}, 300 °C for 15 min. The mass spectra were recorded with an ionisation energy of 70 eV and an ionising current of 0.2 mA.

2.4.4 Isolation of the Specific Polysaccharide Fraction of the LPS

LPS fraction (8 mg) was dissolved in 1 ml of 1% acetic acid solution and kept at 100 °C for 2 h. After cooling, the solution was centrifuged (6,000 rpm, 20 min) and the clear supernatant was freeze-dried.

Further purification of this sample was performed by Size Exclusion Chromatography (SEC) on Sephacryl HR 300 (Pharmacia, 1.5 × 70 cm, NH_4HCO_3 50 mM, flow 0.4 ml min^{-1}) and the eluate was monitored by refractive index as mentioned above. 2 mg of the O-chain was isolated from purified LPS.

2.5 NMR Spectroscopic Analysis

NMR experiments were carried out with a Bruker DRX 400 and 500 MHz instruments equipped with reverse multinuclear probe at 45 °C or at 30 °C. Chemical shifts of spectra recorded in D_2O are expressed in δ relative to internal acetone (δ = 2.225 and 31.4 ppm). Two-dimensional spectra (DQF-COSY, phase-sensitive TOCSY and ROESY, and phase-sensitive gradient HSQC) were measured with standard Bruker software.

3. RESULTS AND DISCUSSION

3.1 Biological Assays

Phytotoxicity of the bacterial culture filtrates and of their extracts was highlighted by the comparison of necrosis in the treated area of tobacco leaves. Within 3÷4 h the infiltrated tissues appeared hydropic, and then necrotized and assumed a papery appearance and a brown colour within 24÷48 h. Of the two *Pseudomonas cichorii* strains assayed both gave positive results in this assay.

With regard to antibiotic activity, positive effects were indicated by the comparison of an inhibition halo of the target micro-organism growth. The inhibition halos were observed around colonies of strain ICMP5707 of *P. cichorii* but not those of strain ICMP9417. Similar antibiotic activity was shown by the culture filtrates and the residual aqueous phase of ethyl acetate extraction of strain ICMP5707. Of the two test organisms used in this assay, *Rhodotorula pilimanae* resulted to be the most sensitive to the antibiotic substance/s produced by ICMP5707 strain of *P. cichorii.*

3.2 Results on *Pseudomonas cichorii* ICMP5707 Strain

3.2.1 Isolation, Purification and Preliminary Characterisation of Phytotoxic and Antibiotic Metabolites

Preliminary dialysis (cut-off 3,500 and 12÷14,000 Da) and extraction experiments with butanol (at pH 2.0) and ethyl acetate (at different pH: 2.0, 6.5 and 10.0) were carried out on the culture filtrates. The phytotoxic metabolites proved to be low molecular weight compounds and were exhaustively extracted from the bacterial culture filtrates by using ethyl acetate at pH 6.5. This organic extract showed phytotoxic activity when assayed on tobacco leaves and the corresponding aqueous phase retained the antibiotic activity against *Rhodotorula pilimanae*. The crude extract was then fractionated by reversed-phase TLC, and thirteen fractions (A, B, C, D, E, F, G, H, 1, 2A, 2B, 3 and 4) were recovered. The crude extract [1 mg (corresponding to 112 ml culture filtrate) ml^{-1}] and all fractions (from 0.5 to 1.7 mg ml^{-1}) were assayed on tobacco leaves. Only the crude extract and fraction 4 [1.5 mg corresponding to 630 ml culture filtrate], containing the less polar metabolites, showed phytotoxic activity. However, the fraction 4 is still a complex mixture of metabolites. Preliminary dialysis (cut-off 3,500 and 12÷14,000 Da) demonstrated that antimicrobial substances both diffused through the membrane and were retained in the tubes, respectively. The bioactive metabolites with antibiotic properties, probably of peptidic nature, were precipitated from the residual aqueous phase of ethyl acetate extraction by addition of ammonium sulphate. The pellet, collected by centrifugation (10,000 rpm, 1 h, 4 °C) and containing salts, was resuspended in water and analysed by TLC. It resulted to be constituted by essentially two compounds, one of which successively identified as histidin, the amino-acid present in the culture medium.

Attempts performed to further purify the antibiotic metabolites by gel filtration (Biogel P-2 and Sephadex G-10), cationic exchange (Dowex-H^+) or reversed-phase columns were unsuccessful.

3.3 Results on *Pseudomonas cichorii* ICMP9417 Strain

3.3.1 Isolation and Purification of Phytotoxic Metabolites Isolation and Purification of the LPS Fraction

Also for strain ICMP9417 of *Pseudomonas cichorii*, bacterial culture filtrate was exhaustively extracted by using ethyl acetate at pH 6.5 and the organic extract retained all the phytotoxic activity initially observed in the culture filtrates assayed on tobacco leaves at 3.4 mg (corresponding to 29 ml culture filtrate) ml^{-1}. This crude extract was analysed by reversed-phase TLC and showed a similar but not identical pattern of metabolites to that of the corresponding organic extract obtained from the culture filtrate of ICMP5707 of *P. cichorii* when analysed under identical conditions.

3.3.2 Isolation and Purification of the LPS Fraction

The extraction of dried bacterial cells using phenol/water method yielded LPS in the phenol phase, as revealed by SDS-PAGE analysis. The LPS was obtained after centrifugation of the phenol suspension and successive extensive dialysis of solution against distilled water and lyophilisation. The crude LPS was purified by SEC on Sephacryl HR 400. Under the conditions used, the LPS molecules were eluted in the void volume, due to their high apparent molecular weight; the other contaminants, such as low molecular weight carbohydrates (mainly glucans) and proteins, were strongly retained in these conditions.

3.3.3 Sugar and Lipid Analysis of the LPS Fraction

Methanolysis was performed on the purified fraction in order to determine both the lipid and the monosaccharide compositions. The GC-MS analysis of the fatty acid methyl esters revealed the presence of the following components: lauryl-, 2-hydroxylauryl-, 3-hydroxylauryl- and 3-hydroxycaprinic methyl esters, characteristic for *Pseudomonas* sp. (Zahringer *et al.*, 1999). Unsaturated fatty acid methyl esters, probably due to contamination by glycolipids, were present in small amounts. This fatty acid composition is the same as the one shown by *P. cichorii* ICMP5707 strain. More interesting was the monosaccharide composition of the polysaccharide moiety as determined by GC-MS. Besides the acetylated methyl glycosides of rhamnose, glucose and 2-acetamido-2-deoxyglucose (glucosamine), the presence of 3-deoxy-D-*manno*-oct-2-ulosonic acid (Kdo) was detected, supporting the LPS nature of the fraction analysed. A peak with a similar fragmentation pattern, consistent with the structure of a acetamidodideoxy sugar and identified as 2-N-acetyl-quinovosamine, was also detected. Two main monosaccharides, still not identified, were also present. These latter were not present in the LPS fraction of ICMP5707 strain.

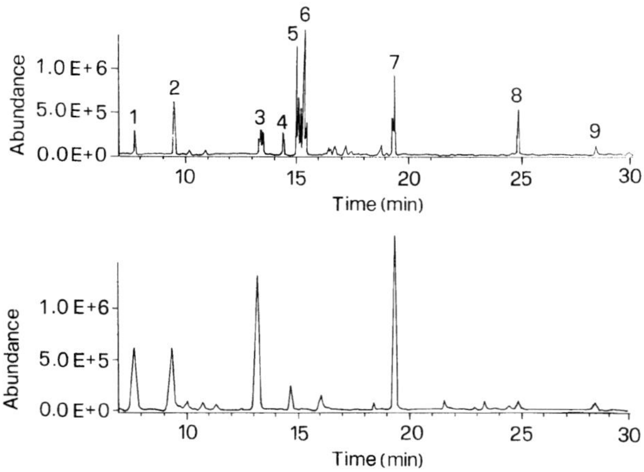

Figure 1. GC-MS profiles of monosaccharide components of *Pseudomonas cichorii* ICMP5707 LPS (above) and *P. cichorii* ICMP9417 LPS (below). (1) Not identified; (2) Rhamnose; (3) Not identified; (4) 3-N-Acetyl-Quinovosamine; (5) 2-N-Acetyl-Quinovosamine; (6) 2-N-Acetyl-Fucosamine; (7) Glucose; (8) Glucosamine; (9) Kdo.

3.3.4 Isolation of the Specific Polysaccharide Fraction of the LPS and Preliminary NMR Spectroscopic Analysis

The O-specific polysaccharide part associated with this LPS, isolated after mild acid hydrolysis and gel chromatography on a Sephacryl HR 300 column, was preliminary studied by NMR spectroscopy. Its ^{1}H and HSQC NMR spectra (Fig. 2) suggested a regular structure built up of a repeating unit. In particular, this polymer showed different signal systems in the anomeric regions both of ^{1}H and ^{13}C spectra. In agreement with the sugar GC-MS analysis, signals in the region of methyl groups of N-acetyl residues, as well as signals in the secondary methyl regions of deoxysugars, were also present, suggesting the presence of amino or acetamido deoxysugars; in the proton spectrum the presence of an AB system at δ 2.80 and 2.49 correlated with a methylenic carbon signal at δ 43.8 in the ^{13}C-NMR spectrum, assigned to a deoxysugar, is also significative.

4. CONCLUSION

This study expands previous researches (Hu *et al.*, 1998) highlighting that several strains of *Pseudomonas cichorii*, including the two used in this study (ICMP5707 and ICMP9417), were able to accumulate in culture low molecular weight diffusible toxins as well as non-dialysable syringomycin-

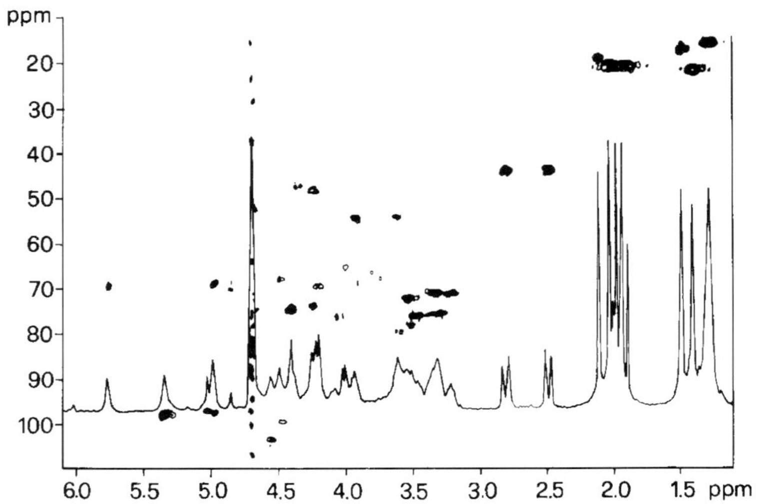

Figure 2. Superimposition of proton and HSQC spectra of LPS O-chain fraction of *Pseudomonas cichorii* ICMP9417 strain.

like toxins. Their conclusion was mainly based on the results of extensive antifungal assays. In the present study the ability of strain ICMP5707, but not of strain ICMP9417, to produce antifungal substances of probable peptidic nature was confirmed; it was also ascertained, throughout phytotoxic assays on tobacco leaves, that both strains produce lipophylic necrosis-inducing toxins, exhaustively extractable with ethyl acetate. However, it is not possible yet to determine the chemical nature of both the antimicrobial and phytotoxic substances. In fact, the phytotoxic fraction 4, obtained by preparative reversed-phase TLC of the organic extracts from culture filtrates, is still a complex mixture of metabolites, and therefore further purification and subsequent chemical and biological characterisation are needed to fully characterise the phytotoxin/s produced by *P. cichorii* as well as the antibiotic substance/s.

The preliminary spectroscopical analyses (NMR) conducted on the LPS O-specific polysaccharide part of ICMP9417 of *P. cichorii*, together with those of GC-MS monosaccharide composition, demonstrated that the repeating unit of this molecule was certainly different from that of the recently described strain ICMP5707 of *P. cichorii* (Jimenez-Barbero *et al.*, 2002). Conversely the lipid moiety is the same in the fatty acid composition. Further studies by NMR spectroscopy are now in progress to determine the complete O-chain chemical structure.

ACKNOWLEDGEMENTS

The work was supported by grants from the Italian National Research Council (CNR). The authors would like to thank Prof. M. Parrilli and his research group for critical discussion on the LPS investigations. Contribution DiSSPA 37.

REFERENCES

Grogan, R.G., Misaghi, I.J., Kimble, K.A., Grethead, A.S., Ririe, D., and Bardin, R., 1977, Varnish spot, destructive disease of lettuce in California caused by *Pseudomonas cichorii*. *Phytopathology* **67**: 957-960.

Jimenez-Barbero, J., C. de Castro, C., Evidente, A., Molinaro, A., Parrilli, M., and Surico, G., 2002, Structural determination of the O-specific chain of the lipopolysaccharide from *Pseudomonas cichorii*. *Eur. J. Org. Chem.* 1770-1775.

Hu, F.P., Young, J.M., and Fletcher, M.J., 1998, Preliminary description of biocidal (syringomycin) activity in fluorescent plant pathogenic *Pseudomonas* species. *J. Appl. Microbiol.* **85**: 365-371.

Kittelberger, R., and Hilbink, F., 1993, Sensitive silver-staining detection of bacterial lipopolysaccharides in polyacrylamide gels. *J. Biochem. Biophys. Methods* **26**: 81-86.

Laemmli, U.K., 1970, Laemmli, U.K., 1970, Cleavage of structural proteins during the assembly of the head of bacteriophage T4. *Nature*. **227**: 680-685.

Molinaro, A., Silipo, A., Lanzetta, R., Parrilli, M., Malvagna, P., Evidente, A., and Surico, G., 2002, Determination of the structure of the Lipid A fraction from the lipopolysaccharide of *Pseudomonas cichorii* by means of NMR and MALDI-TOF spectrometry. *Eur. J. Org. Chem.* 3119-3125.

Piening, L.J., and MacPherson, D.J., 1985, Stem melanosis, a disease of spring wheat caused by *Pseudomonas cichorii*. *Can. J. Plant Pathol.* **7**: 168-172.

Shirata, A., Tsuchiya, Y., and Ohata, K., 1983, Antibacterial activity of *Pseudomonas cichorii* against the same species and other phytopathogenic bacteria. *Bulletin of the National Institute of Agricultural Science, Japan.* **38C**: 139-148.

Surico, G., Lavermicocca, P., and Iacobellis, N.S., 1988, Produzione di siringomicina e siringotossina in colture di *Pseudomonas syringae* pv. *syringae*. *Phytopathol. Medit.* **27**: 163-168.

Zahringer, U., Lindner, B., and Rietschel, E.T., 1999, Chemical structure of lipid A: recent advances in structural analysis of biologically active molecules. In *Endotoxin in Health and Disease* (Morrison, D.C., Brade, H., Opal, S. and Vogel, S., eds), M. Dekker Inc., New York, pp. 93-114.

Chemical and Biological Characterisation of Tolaasins A-E: New Lipodepsipeptides Produced by *Pseudomonas tolaasii*

S. LAZZARONI, C. BASSARELLO[#], G. BIFULCO[#], P. LO CANTORE[§], A. EVIDENTE, N.S. IACOBELLIS[§], R. RICCIO[#] and L. GOMEZ-PALOMA[#]

Dipartimento di Scienze del Suolo, della Pianta e dell'Ambiente, Università degli Studi di Napoli Federico II, Viale dell'Università, 100, Portici (Napoli), Italy; [#]Dipartimento di Scienze Farmaceutiche, Università di Salerno, Fisciano (Salerno), Italy; [§]Dipartimento di Biologia, Difesa e Biotecnologie Agro-Forestali, Università degli Studi della Basilicata, Potenza, Italy.

Abstract: *Pseudomonas tolaasii*, the causal agent of brown blotch disease of *Agaricus bisporus* and of the yellowing of *Pleurotus ostreatus,* produces in culture an extracellular toxin, called tolaasin I, which has been shown to have an important role in the development of the disease symptoms. The toxin, previously characterised as a lipodepsipeptide with a long and hydrophobic peptide chain, was isolated from the culture filtrates together with tolaasin II, a minor component structurally related to tolaasin I. This paper reports the preliminary structural elucidation data (based on NMR studies and MS spectra) and biological activity of four minor analogs, named tolaasins A, B, D and E, as well as the complete characterisation of the tolaasin C.

Key words: cultivated mushrooms, *Agaricus bisporus* brown blotch, *Pleurotus ostreatus* yellowing, toxins.

Author for correspondence: A. Evidente; e-mail: evidente@unina.it.

1. INTRODUCTION

Several phytopathogenic *Pseudomonas* spp. produce *in vitro* phytotoxic and antimicrobial lipodepsipeptides, which frequently contain non-proteinogenic

N.S. Iacobellis *et. al.* (eds.), *Pseudomonas syringae* and Related Pathogens, 245-254.

amino acids or residues with the uncommon D-stereochemistry, classified in two groups according to their primary structures. The first group includes nonapeptides such as syringomycins, syringotoxins, syringostatins and pseudomycins. The second group comprises molecules such as syringopeptins, tolaasins, fuscopeptins and corpeptins containing 18 to 25 amino acid residues most of which having a D-stereochemistry (Bender *et al.,* 1999). In the latter compounds the C-terminal region forms a lactone ring of 5 (tolaasins, fuscopeptins and corpeptins) to 8 (syringopeptins) amino acids.

Tolaasin I is a lipodepsipeptide produced in culture by virulent strains of *Pseudomonas tolaasii* (Rainey *et al.*, 1991; 1992) the causal agent of brown blotch disease of *Agaricus bisporus* (Lange) Imbach (Tolaas, 1915) and of the yellowing of *Pleurotus ostreatus* (Jacq. Ex Fr.) (Ferri, 1985; Iacobellis and Lo Cantore, 1997; Iacobellis and Lo Cantore, 1998; Lo Cantore, 2001; Lo Cantore e Iacobellis, 2002). The primary structure of tolaasin I, elucidated by Nutkins *et al.* (1991) (Fig. 1) bears a β-OH octanoic acid blocking group at the N-terminus, a sequence of seven successive D-amino acids in the N-terminal region of the peptide (Pro 2 - Val 8), with a Ser-Leu-Val repeat, and then alternate L- and D-amino acids. It also contains a 2,3-dehydro-2-aminobutyric acid (ΔBut) residue at positions 1 and 13, a D-homoserine (Hse) and a D-2,4-diaminobutyric acid (D-Dab) at positions 16 and 17, respectively. Finally, a lactone ring is formed between the hydroxyl of D-Thr14 and the C-terminal L-Lys18. Previous studies showed that this molecule is active toward microbial and/or animal cells through the formation of transmembrane pores (Rainey *et al.*, 1991; Brodey *et al.*, 1991; Hutchison and Johnstone, 1993; Lo Cantore, 2001; Lo Cantore *et al.*, 2003a; 2003b). The strong resistance to enzymatic degradation and inactivation as well as an increased antigenicity of tolaasin suggests a potential use of this lipodepsipeptide as a therapeutic peptide.

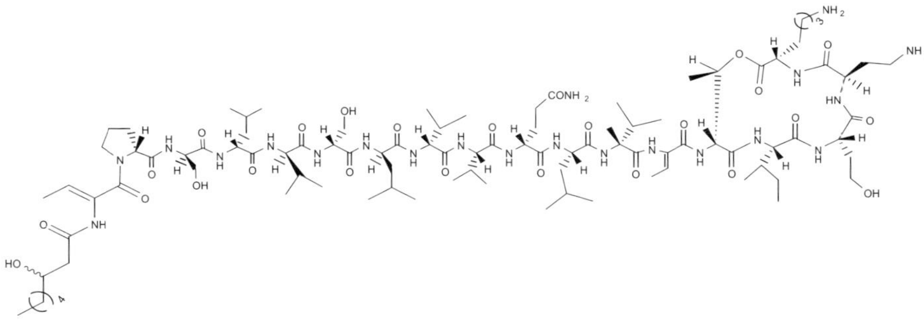

Figure 1. Tolaasin I primary structure.

In the frame of this project, five new analogs have recently been isolated from *P. tolaasii* culture filtrates by HPLC methods. Here we reported preliminary results on the biological activity and the structural elucidation (based on NMR studies and MS spectra) of the four minor tolaasins A, B, D and E as well as the complete structural characterisation of tolaasin C.

2. MATERIALS AND METHODS

2.1 Growth of Bacteria

Strain NCPPB2192 of *Pseudomonas tolaasii* was grown as reported by Lo Cantore *et al.* 2003b. The antimicrobial activity of cultures against *Bacillus megaterium* was determined following an already established procedure (Lavermicocca *et al.,* 1997). The active supernatants were lyophilised and stored at -20°C.

2.2 Isolation and Purification of Tolaasins

Tolaasins were isolated and purified from cell-free culture filtrate according to a modification of the method described by Peng (1986). Lyophilised cell-free culture was taken up in distilled water, filtered through Whatman n° 42 paper, acidified to pH 3 with concentrated HCl, left overnight at room temperature and then the precipitate was removed by centrifugation. $CaCl_2$ was added to the supernatant and the centrifugation-derived pellet was dried at 50°C and treated three times with small amounts of methanol. The mixture was finally dried *in vacuo* and then dissolved in water. The suspension was centrifuged and the residue was desalted by a gel-filtration column (Sephadex G-10). The collected fractions were monitored by TLC, and the fractions containing tolaasins were combined and lyophilised to give a white solid residue. TLC analyses were carried out on silica gel plates (Merck, Kieselgel, 60 F 254, 0.25 mm), and the fractionated substances were visualised by exposing the plates to UV radiation and spraying with 0.5% ninhydrin in Me_2CO, followed by heating at 110°C for 10 minutes.

2.3 Reversed-Phase HPLC

Reversed-phase high performance chromatography (HPLC) of tolaasin preparations was performed on a Shimadzu instrument, equipped with HPLC Pump LC-10 ADVP and UV-VIS Detector SPD-10AV. Analytical (25 cm X 4.6 mm) and semipreparative (25 cm X 10 mm) Brownlee Aquapore RP-300 (C8) columns were used for the analysis and purification of the above

mixture, respectively. A solvent system of water (0.1 % TFA, v v^{-1}) and CH_3CN with flow rates of 1.0 ml min^{-1} for analytical column and 4.7 ml min^{-1} for semipreparative column were used for elution, respectively. All solvents were degassed before use.
The fractions obtained from the semipreparative column were collected, partially dried under reduced pressure and finally lyophilised.

2.4 ES-MS and FAB-MS

The ES and FAB MS spectra of tolaasins mixture and HPLC-grade tolaasins I and II as well as tolaasins A, B, C, D, and E were recorded on a Perkin-Elmer API 100 LC-MS and a VG ZAB 2SE (Manchester, UK) spectrometers, by direct injection in glycerol/thioglycerol matrix, respectively. The analysis of the mass fragmentation pattern of tolaasin C was performed recording the mass spectra on a LCQ-DECA Finnigan spectrometer equipped with an electrospray source.

2.5 NMR Spectroscopy

NMR measurements were performed at 300 K on a Bruker DRX-600 and Bruker Avance-300 spectrometers operating at 600 and 300 MHz, respectively. All spectra were acquired in the phase-sensitive mode and TPPI method (Marion *et al.*, 1983) was used for quadrature detection in the ω1 dimension.
Tolaasin C sample was prepared dissolving 2.0 mg in 0.5 ml of DMSO-d_6 (Carlo Erba, 99.95% D). All 2D-NMR were acquired setting the transmitter at the centre of the spectral width, presaturating at the residual HOD resonance (3.323 ppm) in DMSO. Samples of tolaasins A, B, D and E were prepared following the same procedures.
The 2Q spectrum (Braunschweiler *et al.*, 1983) was executed on the Avance 300 spectrometer recording a total 96 scans t_1^{-1} value with a t_{1max} of 53.3 ms. The TOCSY spectrum (Braunschweiler and Ernst, 1983) was acquired using a DIPSI-2 sequence (Rance, 1987) for spin-locking with t_{mix} = 70 ms, 256 scans t_1^{-1} value and t_{1max} = 29.7 ms. The NOESY spectrum (Jeener *et al.*, 1979) was recorded with (256 scans t_1^{-1} and t_{1max} = 41.6 ms) and without (128 scans t_1^{-1} and t_{1max} = 24.1 ms) the saturation of the residual HOD resonance during the preparation and mixing periods. The mixing time was fixed to 200 ms.

2.6 Biological Assays

The antimicrobial activity of two-fold serial dilutions of stock solutions of tolaasins in distilled water has been evaluated according to the previously reported soft agar procedure (Iacobellis *et al.*, 1992; Lavermicocca *et al.*, 1997). After 24÷48 hr the last dilution determining the growth inhibition of the test micro-organisms was recorded. The assays were performed twice with three replicates.

3. RESULTS AND DISCUSSION

3.1 Isolation, Purification and Characterisation of Tolaasins

Tolaasins A-E, together with tolaasins I and II, were isolated from the cell-free culture filtrate of *Pseudomonas tolaasii* essentially by precipitation with calcium chloride. After partial purification by fractionated precipitation and desalting by gel-filtration chromatography, a crude preparation which retained full biological activity on *Bacillus megaterium* was obtained. HPLC analysis of the above preparation showed that besides the two main metabolites, namely tolaasin I and II, also five minor peaks, named tolaasins A, B, C, D, and E, were present (Fig. 2).

Through multistep reversed-phase HPLC on a C8 semipreparative column, these minor substances were purified in very low amounts together with tolaasin I and tolaasin II. The identity of tolaasin I and II was ascertained by comparing their NMR (1D and 2D) and MS data with those reported by Nutkins *et al.* (1991). In particular, positive-ion fast atom bombardment mass spectrometry (FAB-MS) and electrospray mass spectrometry (ES-MS) showed molecular weights of 1985 and 1941, respectively. ES-MS analysis of tolaasins A-E indicated for these molecules the molecular weights 1956, 1971, 2003, 1985, and 1941, respectively.

These results, in agreement with the preliminary ^{1}H NMR data, indicated that these metabolites are structurally related to either tolaasin I or tolaasin II.

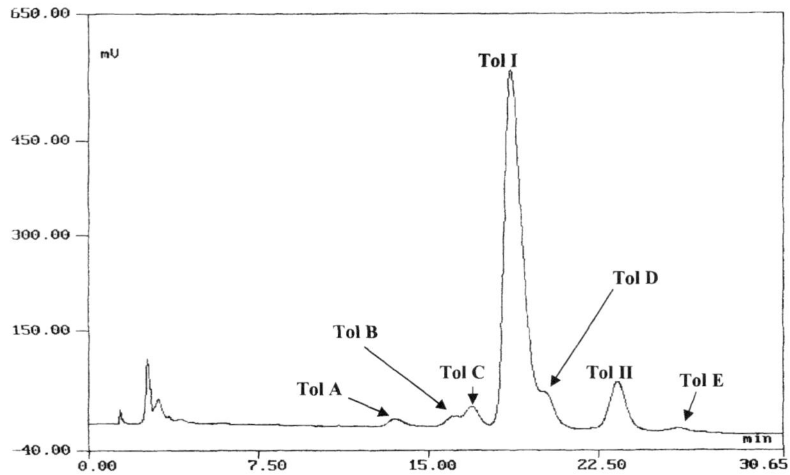

Figure 2. HPLC analysis of tolaasins mixture.

Tolaasin C showed a pseudomolecular ion peak at $m\ z^{-1}$ 2005.1 $(M+H)^+$ corresponding to a molecular weight of 2003 a.m.u., 18 mass units greater than that of tolaasin I, suggesting that tolaasin C may be the open-chain version of tolaasin I deriving from the hydrolysis of its lactone ring. The ES-MS spectrum also contained peptide fragment ions generated by a stepwise spontaneous loss of amino acid residues from the N-terminus. In particular, seven ions at $m\ z^{-1}$ 1781.1, 1682.9, 1595.7, 1482.8, 1385.5, 1296.6 and 1183.1 could be identified and interpreted as fragments originated by removal of the so called ϕ chain and ΔBut1, Pro2, Ser3, Leu4, Val5 and Leu7 residues, respectively. Comparison of the fragmentation patterns of tolaasins C and I showed that all tolaasin C fragment ions were found 18 a.m.u greater than the corresponding ones of tolaasin, supporting the hypothesis that tolaasin C is a linear peptide deriving from the hydrolysis of the lactone ring. NMR data along with the above ES-MS data allowed to derive for tolaasin C a molecular formula $C_{94}H_{165}N_{21}O_{26}$.

The NMR study of tolaasin C was complicated by the low availability of the compound. Hence the carbon chemical shifts could not be assigned. However, a full characterisation of the proton spin systems, including the identification of the N-terminal side chain, could be achieved through extensive use of 2D NMR techniques. In particular, the assignment of the proton spin systems of each amino acid residue (Table 1) was obtained from TOCSY and 2Q experiments. The 2Q spectrum was acquired in order to confirm the proton-proton connectivities preliminarily examined in the TOCSY spectrum. A sequence-specific assignment for the residues of the molecule was then made by studying the pattern of sequential dipolar couplings contained in a 2D NOESY spectrum (Wüthrich *et al.*, 1986). A comparative analysis of tolaasin I and tolaasin C 2D NMR spectra provided additional support to the hypothesis of tolaasin C being a derivative of tolaasin I with the opened lactone ring. Further evidence of the linear nature of the peptide was provided by the analysis of the proton chemical shifts of the dehydroaminobutyric acid residue at position-13 (ΔBut_{13}) and the threonine at position-14. In fact, NMR data of tolaasin C showed a moderate low-field shift for the ΔBut_{13} NH (δ 9.55) and H β (δ 6.34) protons and a moderate up-field shift for the threonine NH resonance (δ 7.88). A large up-field shift was observed for the Hα (δ 4.18) and Hβ (δ 3.84) of the same residue, in agreement with the expected random coil value of threonine α (4.35 ppm) and β protons (4.22 ppm), suggesting that the hydroxyl group of the side chain of this residue should not be involved in an ester functionality.

Table 1. Experimental Chemical Shifts of Tolaasin C in DMSO-d_6 at 300 K.

	NH	Hα	Hβ	Hγ	Hδ	others
φ[a]		2.48 2.36	3.84	1.39[b]	1.27[b]	nd[c] OH 4.73
ΔBut1	9.81	-	5.62	1.70		
Pro2	-	4.24	2.21 1.79	1.87 1.79	3.58 3.45	
Ser3	7.84	4.20	3.72[b]			OH 4.89
Leu4	7.58	4.16	1.58 1.44	1.58	0.78[b]	
Val5	7.37	4.13	2.02	0.85[b]		
Ser6	7.83	4.34	3.60[b]			OH 4.95
Leu7	7.95	4.32	1.63 1.49	nd[c]	0.86 0.82	
Val8	7.71	4.20	2.02	0.84[b]		
Val9	7.82	4.19	2.06	0.85[b]		
Gln10	8.08	4.20	1.88 1.78	2.08[b]	7.26 6.79	
Leu11	8.03	4.40	1.52[b]	nd[c]	0.88 0.83	
Val12	8.09	4.13	2.09	0.87[b]		
ΔBut13	9.55	-	6.34	1.60		
Thr14	7.88	4.18	3.84	1.14		
Ile15	7.82	3.78	1.65	1.56 1.16 γCH_3 0.89	0.89	
Hse16	9.30	3.84	2.27 1.87	3.54 3.40		OH 4.54
Dab17	8.03	4.30	2.22 1.99	2.90 2.82		δNH_2 nd[c]
Lys18	7.59	4.48	1.82 1.58	1.25[b]	1.54 1.50	Hε 2.72[b] ζNH_2 nd[c]

[a] φ, β-hydroxyoctanoic acid; [b] degenerate; [c] not determined.

3.2 Biological Assays

The antimicrobial activity of HPLC grade tolaasins A-E, assayed in comparison with tolaasin I and II, against strain ATCC26423 of the yeast *Rhodotorula pilimanae*, the strain ITM100 of the Gram-positive bacterium *Bacillus megaterium* and the strain ITM103 of the Gram-negative bacterium *Escherichia coli* K12, is reported in Table 2. *B. megaterium* resulted the most sensible test micro-organisms used. In fact, all the analogs, but tolaasin C, inhibited the grown of this Gram positive bacterium though differences among their specific activities were observed. The most active analog appeared tolaasin D followed by tolaasin I and II with a minimal inhibitory quantity of 0.16, 0.32 and 0.64 μg, respectively. Tolaasins A and B, and E were less active, with a minimal inhibitory quantity of 1.28 and 2.56, respectively. None of the analogs inhibited the growth of the Gram-negative bacteria *E. coli.* Finally, the growth of the yeast *R. pilimanae* was inhibited only by tolaasins I, II and D but the sensibility of the yeast was reduced in comparison the one shown by *B. megaterium* (Table 2).

Table 2. Antimicrobial activity of tolaasin analogs.

Micro-organism	**Minimal Inhibitory Quantity (μg)**						
	I	**II**	**A**	**B**	**C**	**D**	**E**
Rhodotorula pilimanae ATCC26423	2.56	5.12	> 5.12	> 5.12	> 5.12	2.56	> 5.12
Bacillus megaterium ITM100	0.32	0.64	1.28	2.56	>5.12	0.16	2.56
Escherichia coli K12 ITM103	> 5.12	> 5.12	> 5.12	> 5.12	> 5.12	> 5.12	> 5.12

4. CONCLUSIONS

In conclusion in this work we confirmed previous evidences (Shirata *et al.*, 1995) that *Pseudomonas tolaasii* besides tolaasin I and II produces other not yet characterised analogs. The preliminary spectroscopic investigations reported here highlight that the five new tolaasins isolated from the culture filtrates of *P. tolaasii*, seem to be structurally related to both tolaasin I and II (Nutkins *et al.*, 1991). They showed chemical modifications either in the fatty acid residue, the peptide side chain and in the lactone macrocycle as well. Tolaasin C, showing the opening of the macrocyclic lactone ring deriving from the hydrolysis of tolaasin I, appear of interest. It may be an artefact due to the extraction and purification procedures, however, the possible corresponding linear peptides originated from the other analogs, such as tolaasin II, were not found. In this regard, however, the low level in the culture filtrates of the latter compounds must be taken in consideration.
The antimicrobial activity of tolaasins appeared to be modulated by the structural modifications present in the different natural analogs of both tolaasin I and II. In fact, the opening of the lactone ring occurring in the case of tolaasin C, as reported for other lipodepsipeptides (Grgurina *et al.*, 1994; Bender *et al.*, 1999), eliminate the antibacterial activity.

ACKNOWLEDGEMENTS

This work was supported in part by grants from the Italian Ministry of University and Research (MIUR), Italian National Research Council (CNR) and by "Progetto Giovani Ricercatori", Biological characterisation of the tolaasin and WLIP, lipodepsipeptides of *Pseudomonas tolaasii* and *P. "reactans"* of the Università degli Studi della Basilicata. Contribution DiSSPA 38.

REFERENCES

Bender, C.L., Alarcón-Chaidez, F., and Gross, D.C., 1999, *Pseudomonas syringae* phytotoxins: mode of action, regulation, and biosynthesis by peptide and polyketide synthetases, *Microbiol. Mol. Biol. Rev.* **63**: 266-292.

Braunschweiler, L., Bodenhausen, G., and Ernst, R.R., 1983, Analysis of networks of coupled spins by multiple quantum NMR. *Mol. Phys.* **48**: 535-560.

Braunschweiler, L., and Ernst, R.R., 1983, Coherence transfer by isotropic mixing: application to proton correlation spectroscopy. *J. Magn. Reson.* **53**: 521-528.

Brodey, C.L., Rainy, P.B., Tester, M., and Johnstone, K., 1991, Bacterial blotch disease of the cultivated mushroom is caused by an ion channel forming lipodepsipeptide toxin. *Mol. Plant-Microbe Inter.* **4**: 407-411.

Ferri, F., 1985, *I funghi. Micologia, isolamento coltivazione*. Edagricole, Bologna, 398 pp.

Grgurina, I., Barca, A., Cervigni, S., Gallo, M., Scaloni, A., and Pucci, P., 1994, Relevance of chlorine substituent for the antifungal activity of syringomycin and syringotoxin, metabolites of the phytopathogenic bacterium *Pseudomonas syringae* pv. *syringae. Experientia* **50**: 130-133.

Hutchison, M.L., and Johnstone, K., 1993, Evidence for the involvement of the surface active properties of the extracellular toxin tolaasin in the manifestation of brown blotch disease symptoms by *Pseudomonas tolaasii* on *Agaricus bisporus*. *Physiol. Mol. Plant Pathol.* 42: 373-384.

Iacobellis, N.S., Lavermicocca, P., Grgurina, I., Simmaco, M., and Ballio, A., 1992, Phytotoxic properties of *Pseudomonas syringae* pv. *syringae* toxins. *Physiol. Mol. Plant. Pathol.*, **40**: 107-116.

Iacobellis, N.S., and Lo Cantore, P., 1997, Bacterial diseases of cultivated mushrooms in southern Italy. In: *Proceedings of X Congress of Mediterranean Phytopathological Union*, Montpellier, 2-6 june, 33-37.

Iacobellis, N.S., e Lo Cantore, P., 1998, Studi sull'eziologia dell'ingiallimento dell'ostricone (*Pleurotus ostreatus*). *Agricoltura Ricerca*, **176**: 55-60.

Jeneer, J., Meier, B.H., Bachmann, P., and Ernst, R.R., 1979, Investigation of exchange processes by two-dimensional NMR spectroscopy. *J. Chem. Phys.* **71**: 4546-4553.

Lo Cantore, P., 2001, Dottorato Thesis *Aspetti patogenetici, fisiologici e molecolari di* Pseudomonas tolaasii *e* P. "reactans". Università degli Studi della Basilicata, Potenza.

Lo Cantore, P., Coraiola, M., Dalla Serra, M., Menestrina, G., Lazzaroni, S., Evidente, A., and Iacobellis, N.S., 2003a, Interaction of Tolaasin I and WLIP, Lipodepsipeptides of *Pseudomonas tolaasii* and *P. "reactans"*, with Biological and Model Membranes, 263-273, (this volume).

Lo Cantore, P., e Iacobellis, N.S., 2002, Recenti acquisizioni sulla eziologia delle malattie batteriche dei funghi del genere *Agaricus* e *Pleurotus. Micologia Italiana* **2**: 18-27.

Lo Cantore, P., Lazzaroni, S., Evidente, A., Cafarchia, C., and Iacobellis, N.S., 2003b, Production of Tolaasin I and WLIP by *Pseudomonas tolaasii* and *P.* "*reactans*", their Antimicrobial Activity and Possible Role in the Virulence of the Pathogens, 255-262, (this volume).

Lavermicocca, P., Iacobellis, N.S., Simmaco, M., and Graniti, A., 1997, Biological properties and spectrum of activity of *Pseudomonas syringae* pv. *syringae* toxins. *Physiol. Mol. Plant Pathol.*, **50**: 129-140.

Marion, D., and Wüthrich, K., 1983, Application of phase sensitive two-dimensional correlated spectroscopy (COSY) for measurements of ^{1}H-^{1}H spin-spin coupling constants in proteins. *Biochem. Biophys. Res. Commun.* **113**: 967-974.

Nutkins, J.C., Mortishire-Smith, R.J., Packman, L.C., Brodey, C.L., Rainey, P.B., Johnstone K., and William, D.H., 1991, Structure determination of tolaasin, an extracellular lipodepsipeptide produced by the mushroom pathogen *Pseudomonas tolaasii* Paine. *J. Am. Chem. Soc.* **113**: 2621-2627.

Peng, J.T., 1986, PhD Thesis *Resistance to disease in* Agaricus bisporus *(Lange) Imbach.* University of Leeds.

Rainey, P.B., Brodey, C.L., and Johnstone, K., 1991, Biological properties and spectrum of activity of tolaasin, a lipodepsipeptide toxin produced by the mushroom pathogen *Pseudomonas tolaasii. Phisiol. Mol. Plant Pathol.*, **39**: 57-70.

Rainey, P.B., Brodey, C.L., and Johnstone, K., 1992, Biology of *Pseudomonas tolaasii*, cause of brown blotch disease of the cultivated mushroom. *Plant Pathol.* **8**: 95-117.

Rance, M., 1987, Improved techniques for homonuclear rotating-frame and isotropic mixing experiments. *J. Magn. Reson.* **74**: 557-564.

Shirata, A., Sugaya, K., Takasugi, M., and Monde, K., 1995, Isolation and biological activity of toxins produced by a Japanese strain of *Pseudomonas tolaasii*, the pathogen of bacterial rot of cultivated oyster mushroom. *Ann. Phytopathol Soc. Japan*, **61**, 493-502.

Tolaas, A.G., 1915, A bacterial disease of cultivated mushrooms. *Phytopathology* **5**: 51-54.

Wüthrich, K., 1986, NMR of Proteins and Acid Nucleic, J. Wiley & Sons, New York, 1-292.

Production of Tolaasin I and WLIP by *Pseudomonas tolaasii* and *P.* "*reactans*", their Antimicrobial Activity and Possible Role in the Virulence of the Pathogens

P. LO CANTORE, S. LAZZARONI[#], A. EVIDENTE[#], C. CAFARCHIA[§] and N.S. IACOBELLIS

Dipartimento di Biologia, Difesa e Biotecnologie Agro Forestali, Università degli Studi della Basilicata, C.da Macchia Romana, 85100 Potenza, Italy; [#]Dipartimento di Scienze del Suolo, della Pianta e dell'Ambiente, Università di Napoli "Federico II", Portici (Napoli), Italy. [§]Dipartimento di Sanità e Benessere Animale, Università di Bari, Italy.

Abstract: Shaken culture in KB of strain NCPPB2192 of *Pseudomonas tolaasii* showed an activity of 3,200 and 800 units ml^{-1} toward *Bacillus megaterium* and *Rhodotorula pilimanae*, respectively. On the contrary, an activity of 400 units ml^{-1} toward *B. megaterium* was observed with the cultures of strain NCPPB1311 of *P.* "*reactans*". The purification of lipodepsipeptides from the cultures of *P. tolaasii* and *P.* "*reactans*" yielded about 13 mg l^{-1} of tolaasin and 169 mg l^{-1} of White Line Inducing Principle (WLIP), respectively. Antimicrobial assays showed that HPLC grade tolaasin I inhibited the growth of filamentous fungi and yeasts and Gram positive bacteria. Only a limited activity was observed toward some Gram negative bacteria. Of particular interest is the activity of tolaasin I toward phytopathogenic fungi, *Pleurotus* and *Agaricus* spp., as well as yeasts and filamentous fungi responsible for mycoses of mammals. WLIP inhibited the growth of fungi and Gram positive bacteria, though at higher concentration when compared to tolaasin I, but not that of Gram negative bacteria. Assays of the two toxins on blocks of *A. bisporus* confirmed the antifungal activity of the two lipodepsipeptides and suggested their possible role in the symptoms caused by the mushroom bacterial pathogens.

Key words: cultivated mushrooms, *Pleurotus* spp., *Agaricus bisporus*, lipodepsipeptides.

Author for correspondence: N.S. Iacobellis; e-mail: iacobellis@unibas.it.

N.S. Iacobellis *et. al.* (eds.), *Pseudomonas syringae* and Related Pathogens, 255-262.

1. INTRODUCTION

Brown blotch of *Agaricus bisporus* and yellowing of *Pleurotus ostreatus*, diseases of cultivated mushrooms known to be caused by *Pseudomonas tolaasii*, have been recently shown to be complex diseases. In fact, besides *P. tolaasii* also other bacteria including *P. "reactans"* are responsible, though in different manner, of the symptom development on the hosts (Iacobellis and Lo Cantore, 1997; Iacobellis and Lo Cantore, 1998b; Lo Cantore, 2001; Lo Cantore and Iacobellis, 2002). Furthermore, *P. "reactans"*, a bacterial entity not yet classified and known in bibliography for its use in the "white line" assay for the specific identification of *P. tolaasii* (Wong and Preece, 1979), has been shown to be the casual agent of the yellowing of *P. eryngii* (Iacobellis and Lo Cantore, 1998a). Indications on the pathogenicity of some strains of *P. "reactans"* have been also obtained in independent studies (Kim *et al.*, 1995; Wells *et al.*, 1996).
Virulent strains of *P. tolaasii* produce in culture an extracellular lipodepsipeptide called tolaasin I (Nutkins *et al.*, 1991) which is considered the main virulence factor of the pathogen (Rainey *et al.*, 1991; 1992). In fact, it has been reported that tolaasin I has antifungal activity and this is apparently due to the ability of this molecule to disrupt cell membrane function by transmembrane pore formation (Brodey *et al.*, 1991; Rainey *et al.*, 1991; Hutchinson and Johnstone, 1993; Lo Cantore *et al.*, 2003). Also *P. "reactans"* produces in culture a lipodepsipeptide called the "White Line Inducing Principle" (WLIP) (Mortishire-Smith *et al.*, 1991), known for its ability to interact with tolaasin and form a white precipitate in the "white line" assay (Wong and Preece, 1979). The biological activities of WLIP are as yet unknown. The fact that non-pathogenic morphological variants of *P. "reactans"* do not produce WLIP (Lo Cantore and Iacobellis, 2000), suggests the possible importance of this lipodepsipeptide in pathogenicity, and prompted us to characterise it and compare it with tolaasin I. Here we report some of the results obtained.

2. MATERIALS AND METHODS

2.1 Growth of Bacteria

Strains NCPPB2192 and NCPPB1311, respectively, of *Pseudomonas tolaasii* and *P. "reactans"* were grown at 25°C under shaking (180 rpm) in 500 ml Erlenmeyer flasks filled with 150 ml of liquid King's B medium (King *et al.*, 1954) inoculated with 1.5 ml of a bacterial suspension containing 10^8 CFU ml^{-1}. After 48 h incubation cultures were centrifuged

(20,000 *g* for 15 min.) and the resulting supernatants were evaluated for the antimicrobial activity against *Bacillus megaterium* following an already established procedure (Lavermicocca *et al.,* 1997), lyophilised and stored at -20°C before further processing.

2.2 Isolation and Purification of Lipodepsipeptides

Tolaasins were isolated and purified from cell-free culture filtrate according to a modification of the method described by Peng (1986). The resulting tolaasin preparations were further purified by reversed-phase high performance chromatography (HPLC) as reported by Lazzaroni *et al.* (2003). The isolation and purification of crude WLIP was performed according to the procedure reported by Mortishire-Smith *et al.* (1991).
The identity of both tolaasin I and WLIP was ascertained by comparing NMR and MS data with those reported in the literature (Nutkins *et al.*, 1991; Mortishire-Smith *et al.*, 1991).

2.3 Antimicrobial Assays

The antimicrobial activity of cultures of strains NCPPB2192 and NCPPB1311 of *Pseudomonas tolaasii* and *P.* "*reactans*", respectively, as well as of pure lipodepsipeptides have been evaluated on Gram positive, Gram negative bacteria, yeasts, and fungi according to the procedure reported by Lavermicocca *et al.* (1997).

2.4 Assay on Blocks of *Agaricus bisporus*

Blocks of *Agaricus bisporus*, prepared according to the procedure reported by Gandy (1968), were treated with 5 μl drops of two-fold serial dilutions of pure lipodepsipeptide solutions, incubated at 25°C and observed after 16 and 40 hours.

3. RESULTS

3.1 Isolation and Purification of the Lipodepsipeptides

The purification of lipodepsipeptides from cultures of strains NCPPB2192 and NCPPB1311 of *Pseudomonas tolaasii* and *P.* "*reactans*", respectively, yielded about 13 mg l^{-1} of tolaasins and 169 mg l^{-1} of crystalline WLIP, respectively.

The further purification of the above tolaasin preparations by HPLC showed that they were not homogeneous and beside tolaasin I, also tolaasin II and other minor compounds were present in the mixture (Lazzaroni *et al*., 2003).

3.2 Antimicrobial Assays

Forty-eight hours-old cultures of strain NCPPB2192 of *Pseudomonas tolaasii,* assayed for the antimicrobial activity on KB medium, showed an activity of 3,200 and 800 units ml^{-1} toward *Bacillus megaterium* and *Rhodotorula pilimanae,* respectively. On the contrary, cultures of strain NCPPB1311 of *P.* "*reactans*" showed an activity of 400 units ml^{-1} toward *B. megaterium* but they did not inhibit the growth of *R. pilimanae.* In the above assay conditions no activity toward strain ITM103 of *Escherichia coli* was observed.

HPLC grade tolaasin I and crystalline WLIP showed a different capability to inhibit the target micro-organisms. In particular, the minimal inhibitory quantity (M.I.Q.) of tolaasin I toward strains ATCC26423 of *Rhodotorula pilimanae,* ITM103 of *Escherichia coli* K12, and ITM100 of *Bacillus megaterium* was 0.64 μg, and 0.04 μg. The M.I.Q. toward fungi ranged between 0.08 and 0.32 μg with strains of *Agaricus bisporus* and *Rhizoctonia solani* the most sensitive among fungi used in this study (Table 1). Besides *B. megaterium* also other phytopathogenic Gram positive bacteria were quite sensitive to the toxin with a M.I.Q. ranging from 0.04 to 0.16 μg. Among Gram negative bacteria, as expected, strains of *P. tolaasii* appeared resistant to the tolaasin I whereas strains of *E. coli, Agrobacterium tumefaciens* and *P.* "*reactans*" showed a M.I.Q. ranging from 0.64 to 1.28 μg (Table 2). Tolaasin I inhibited the growth of *Candida albicans* and *Cryptococcus neoformans,* yeast-like fungi responsible of systemic and cutaneous mycoses in animals and humans but the sensitivity was lower when compared to the one shown by phytopathogenic fungi. Furthermore, in the assay condition no activity on the growth of *C. parapsilosis* or *Malassezia pachydermatis* was observed.

Crystalline WLIP inhibited the growth of fungi and Gram positive bacteria, though the M.I.Q. were higher as compared to tolaasin I, but it did not inhibit the growth of Gram negative bacteria including *E. coli* and *P.* "*reactans*" (Tables 1 and 2). The micro-organisms highly sensitive to WLIP were Gram positive bacteria. In particular, WLIP inhibited the growth of strain ITM100 of *B. megaterium* with a M.I.Q. of 0.32 μg (Table 2).

No activity either toward the yeast, *R. pilimanae*, or the above yeast-like fungi responsible of systemic and cutaneous mycoses in animals and humans was observed also when 10.24 μg of the substance was used.

Table 1. Antimicrobial activity of tolaasin I and WLIP towards yeasts, soil inhabiting phytopathogenic fungi and *Pleurotus* and *Agaricus* spp. cultivated mushrooms.

Micro-organism	Minimal Inhibitory Quantity (μg)	
	Tolaasin I	WLIP
Agaricus bisporus	0.08	1.28
Armillaria mellea	0.32	2.56
Pleurotus eryngii	0.32	2.56
P. ostreatus	0.16	1.28
Rhizoctonia solani	0.08	0.64
Rhodotorula pilimanae	0.64	> 10.24
Sclerotinia sclerotiorum	0.32	1.28
Verticillium dahliae	0.16	1.28

Table 2. Antimicrobial activity of tolaasin I and WLIP towards bacteria.

Micro-organism	Minimal Inhibitory Quantity (μg)	
	Tolaasin I	WLIP
Bacillus megaterium	0.04	0.32
Clavibacter michiganensis subsp. *michiganensis*	0.32	0.64
Curtobacterium flaccumfaciens	0.16	0.64
Agrobacterium tumefaciens	0.64	> 10.24
Escherichia coli	0.64	> 10.24
Pseudomonas tolaasii	> 1.28	> 10.24
P. "reactans"	1.28	> 10.24

3.3 Assay on Blocks of *Agaricus bisporus*

Assays of HPLC grade tolaasin I and crystalline WLIP on tissue blocks of *Agaricus bisporus* (Gandy, 1968) showed that the deposition on their surface of a drop of solution containing 0.08 μg of tolaasin I caused brown sunken lesions. In the same assay condition a similar effect was observed when drops of solutions containing 1.28 μg of WLIP were used.

4. DISCUSSION

The purification of lipodepsipeptides from cultures of strains NCPPB2192 and NCPPB1311 of *Pseudomonas tolaasii* and *P. "reactans"*, respectively, yielded about 13 mg l^{-1} of tolaasins and 169 mg l^{-1} of crystalline WLIP whose chemical nature was confirmed by spectroscopic analysis (Nutkins *et al.*, 1991; Mortishire-Smith *et al.*, 1991). Further purification of the above tolaasin I preparation by HPLC confirmed previous indications that *P. tolaasii,* besides the already known tolaasin I and II (Nutkins *et al.*, 1991), produced *in vitro* other apparently chemical correlated metabolites whose chemical structure is not yet elucidated (Shirata *et al.*, 1995). Preliminary results on the chemical and biological characterisation of the new analogs of tolaasiin I and II will be reported elsewhere (Lazzaroni *et al.*, 2003).

The antimicrobial assays of HPLC grade tolaasin I and crystalline WLIP confirmed the antimicrobial activity of tolaasin I (Rainey *et al.*, 1991; 1992) and showed that also WLIP is a lipodepsipeptide active in the growth inhibition of bacteria and fungi. In general tolaasin I showed higher antimicrobial activity than WLIP. The Gram positive bacterium *Bacillus megaterium* resulted the most sensitive micro-organism to both tolaasin I and WLIP with a M.I.Q of 0.04 μg and 0.32 μg, respectively. In this case, WLIP was about 8-fold less active than tolaasin I. Also the other Gram positive bacteria used in this study were inhibited by both lipodepsipeptides with M.I.Q. not significantly different. On the contrary the Gram negative bacteria, including *Escherichia coli*, were inhibited by tolaasin I but not by WLIP.

In general filamentous fungi were inhibited by both lipodepsipeptides but on average tolaasin I was eight/ten-fold more active than WLIP.

The assays of the two toxins on tissue blocks of *Agaricus bisporus* confirmed the antifungal activity of both tolaasin I and WLIP which, in fact, both caused brown and sunken lesions on the surface of the above blocks. However, in this assay WLIP was about 15-fold less active than tolaasin I.

The results of this study confirmed previous findings on the toxicity of tolaasin I on fungi (Rainey *et al.*, 1991; 1992) and clearly demonstrated that WLIP, as other lipodepsipeptides of bacterial origin (Bender *et al.*, 1999), is also toxic on fungi, though, its activity was in general lower than that of tolaasin I. The above findings strongly suggest that also WLIP, as already reported for tolaasin I (Rainey *et al.*, 1992), is important in the interaction with mushrooms. In this respect, when considering the observed lower antifungal activity of WLIP it is also necessary to remind that strain NCPPB1311 of *P. "reactans"*, at least *in vitro*, produced a quantity of WLIP (on average 169 mg l^{-1} of broth) which is more than ten-fold higher then the quantity of tolaasin I produced by the type strain NCPPB2192 *P. tolaasii* (on average 13 mg l^{-1} of broth). The lower antifungal activity of WLIP appear to be compensated by the higher quantity produced. Furthermore, the observation that morphological variants of *P. "reactans"* strains are avirulent and do not produce WLIP (Lo Cantore and Iacobellis, 2000; unpublished results) further support the possibility that the latter compound may play an important role in the interaction of *P. "reactans"* with the cultivated mushrooms.

ACKNOWLEDGEMENTS

The financial supports of the "Progetto Giovani Ricercatori", Biological characterisation of the tolaasin and WLIP, lipodepsipeptides of *Pseudomonas tolaasii* and *P. "reactans"* of the "Università degli Studi della Basilicata", MURST ex 60% and Regione Basilicata are acknowledged.

REFERENCES

Bender, C.L., Alarcón-Chaidez, F., and Gross, D.C., 1999, *Pseudomonas syringae* phytotoxins: mode of action, regulation, and biosynthesis by peptide and polyketide synthetases. *Microbiol. Mol. Biol. Rev.* **63**: 266-292.

Brodey, C.L., Rainy, P.B., Tester, M., and Johnstone, K., 1991, Bacterial blotch disease of the cultivated mushroom is caused by an ion channel forming lipodepsipeptide toxin. *Mol. Plant-Microbe Interact.* **4**: 407-411.

Gandy, D.G., 1968, A technique for screening bacteria causing brown blotch of cultivated mushrooms. In: *Rep. Glasshouse Crops Res. Insti.*, pp. 150-154.

Hutchison, M.L., and Johnstone, K., 1993, Evidence for the involvement of the surface active properties of the extracellular toxin tolaasin in the manifestation of brown blotch disease symptoms by *Pseudomonas tolaasii* on *Agaricus bisporus*. *Physiol. Mol. Plant Pathol.* **42**: 373-384.

Iacobellis, N.S., and Lo Cantore, P., 1997, Bacterial diseases of cultivated mushrooms in southern Italy. In: *Proceedings of X Congress of Mediterranean Phytopathological Union*, Montpellier, 2-6 june, 33-37.

Iacobellis, N.S., e Lo Cantore, P., 1998a, Recenti acquisizioni sul determinismo della batteriosi del cardoncello (*Pleurotus eryngii*), *Agricoltura Ricerca* **176**: 51-54.

Iacobellis, N.S., e Lo Cantore, P., 1998b, Studi sull'eziologia dell'ingiallimento dell'ostricone (*Pleurotus ostreatus*), *Agricoltura Ricerca* **176**: 55-60.

Kim, J.W., Kwon, S.I., and Kang, H.J., 1995, Studies on the pathogenic *Pseudomonas* causing bacterial diseases of cultivated mushrooms in Korea. 2. Bacteriological characteristics of *P. tolaasii* causing mushroom brown blotch and white line reacting organisms. *Korean J. Plant Pathol.* **11**: 353-360.

King, E.O., Ward, M.K., and Raney, D.E., 1954, Two simple media for the demonstration of pyocyanin and fluorescein. *J. Lab. Clin. Med.* **44**: 301-307.

Lavermicocca, P., Iacobellis, N.S., Simmaco, M., and Graniti, A., 1997, Biological properties and spectrum of activity of *Pseudomonas syringae* pv. *syringae* toxins. *Physiol. Mol. Plant Pathol.* **50**: 129-140.

Lazzaroni, S., Bassarello, C., Bifulco, G., Lo Cantore, P., Evidente, A., Iacobellis N.S., Riccio, R., and Gomez-Paloma, L., 2003, Chemical and Biological Characterisation of Tolaasins A-E: New Lipodepsipeptides Produced by *Pseudomonas tolaasii*, 245-354, (this volume).

Lo Cantore, P., 2001, Dottorato Thesis *Aspetti patogenetici, fisiologici e molecolari di* Pseudomonas tolaasii *e* P. "reactans". Università degli Studi della Basilicata, Potenza.

Lo Cantore, P., and Iacobellis, N.S., 2000, Phenotypic variability in *Pseudomonas "reactans"* and *P. tolaasii*. 5th Congress of the European Foundation for Plant Pathology, Taormina (Ct) Italy, 18-22 September 2000, p. 46.

Lo Cantore, P., e Iacobellis, N.S., 2002, Recenti acquisizioni sulla eziologia delle malattie batteriche dei funghi del genere *Agaricus* e *Pleurotus. Micologia Italiana* **2**: 18-27.

Lo Cantore, P., Coraiola, M., Dalla Serra, M., Menestrina, G., Lazzaroni, S., Evidente, A., and Iacobellis, N.S., 2003, Interaction of Tolaasin I and WLIP, Lipodepsipeptides of *Pseudomonas tolaasii* and *P. "reactans"*, with Biological and Model Membranes, 263-273, (this volume).

Mortishire-Smith, R.J., Nutkins, J.C., Packman, L.C., Brodey, C.L., Rainey, P.B., Johnstone, K., and Williams, D.H., 1991, Determination of the structure of an extracellular peptide produced by the mushroom saprotroph Pseudomonas "reactans". Tetrahedron*, 47: 3645-3654.*

Nutkins, J.C., Mortishire-Smith, R.J., Packman, L.C., Brodey, C.L., Rainey, P.B., Johnstone, K., and William, D.H., 1991, Structure determination of tolaasin, an extracellular lipodepsipeptide produced by the mushroom pathogen *Pseudomonas tolaasii* Paine, *J. Am. Chem. Soc.* **113**: 2621-2627.

Peng, J.T., 1986, PhD Thesis *Resistance to disease in* Agaricus bisporus *(Lange) Imbach.* University of Leeds, Department of Plant Science.

Rainey, P.B., Brodey, C.L., and Johnstone, K., 1991, Biological properties and spectrum of activity of tolaasin, a lipodepsipeptide toxin produced by the mushroom pathogen *Pseudomonas tolaasii. Physiol. Mol. Plant Pathol.* **39**: 57-70.

Rainey, P.B., Brodey, C.L., and Johnstone, K., 1992, Biology of *Pseudomonas tolaasii*, cause of brown blotch disease of the cultivated mushroom. *Plant Pathol.* **8**: 95-117.

Shirata, A., Sugaya K., Takasugi M., and Monde, K., 1995, Isolation and biological activity of toxins produced by a Japanese strain of *Pseudomonas tolaasii*, the pathogen of bacterial rot of cultivated oyster mushroom. *Ann. Phytopathol Soc. Japan*, **61**, 493-502.

Wells, J.M., Sapers, G.M., Fett, W.F., Butterfield, J.E., Jones, J.B., Bouzar, H., and Miller, F.C., 1996, Postharvest discoloration of the cultivated mushrooms *Agaricus bisporus* caused by *Pseudomonas tolaasii*, *P. "reactans"*, and *P. "gingeri"*. *Phytopathology* **86**: 1098-1104.

Wong, W.C., and Preece, T.F., 1979, Identification of *Pseudomonas tolaasii*: the white line in agar and mushroom tissue block rapid pitting tests. *J. Appl. Bacteriol.* **47**: 401-407.

Interaction of Tolaasin I and WLIP, Lipodepsipeptides of *Pseudomonas tolaasii* and *P. "reactans"*, with Biological and Model Membranes

P. LO CANTORE, M. CORAIOLA[#], M. DALLA SERRA[#], G. MENESTRINA[#], S. LAZZARONI[§], A. EVIDENTE[§] and N.S. IACOBELLIS

Dipartimento di Biologia, Difesa e Biotecnologie Agro Forestali, Università degli Studi della Basilicata, C.da Macchia Romana, 85100 Potenza, Italy; [#]CNR- ITC Istituto di Biofisica - Sezione di Trento, Povo (Trento), Italy; [§]Dipartimento di Scienze del Suolo, della Pianta e dell'Ambiente, Università di Napoli "Federico II", Portici (Napoli), Italy.

Abstract: The activity of tolaasin I and White Line Inducing Principle (WLIP), produced *in vitro* by virulent strains of *Pseudomonas tolaasii* and *P. "reactans"*, respectively, was comparatively evaluated on red blood cells and large unilamellar vesicles. Tolaasin I and WLIP caused red blood cell haemolysis through a colloid-osmotic shock apparently caused by transmembrane pores formation. Both lipodepsipeptides were able to form channels in lipid vesicles. Their activity was dependent on the toxin concentration and liposomes composition, and independent on the pH. Studies of dynamic light scattering suggested a detergent-like activity for WLIP at high concentration (higher than 27 μM). This effect was not detected for tolaasin I at the concentrations tested (up to 28.2 μM).

Key words: lipodepsipeptides, toxin pores, liposomes, transmembrane pores formation.

Author for correspondence: N.S. Iacobellis; e-mail: iacobellis@unibas.it.

1. INTRODUCTION

Tolaasin I and the White Line Inducing Principle (WLIP) are two lipodepsipeptides (LDPs) produced by virulent strains of *Pseudomonas tolaasii* and *P. "reactans"*, respectively. It has been recently demonstrated

N.S. Iacobellis *et. al.* (eds.), *Pseudomonas syringae* and Related Pathogens, 263-273.

that brown blotch disease of *Agaricus bisporus* and yellowing of *Pleurotus ostreatus* are complex diseases (Wells *et al.*, 1996; Iacobellis and Lo Cantore, 1997; Iacobellis and Lo Cantore, 1998b; Lo Cantore, 2001; Lo Cantore and Iacobellis, 2002; Lo Cantore and Iacobellis, unpublished results) and both bacteria participate to the development of the disease symptoms tough in a different way. Furthermore, it has been shown that *P. "reactans"*, which is a bacterial entity not yet classified and known as a saprophytic bacterium associated to the cultivated mushrooms, is the causal agent of the yellowing of *P. eryngii* (Iacobellis and Lo Cantore, 1998a; Lo Cantore and Iacobellis, 2002; Lo Cantore and Iacobellis, unpublished results).

Tolaasin I, considered to be the main virulence factor of *P. tolaasii*, appeared to cause the mushroom tissue alteration by forming transmembrane pores (Rainey *et al.*, 1991; Brodey *et al.*, 1991; Rainey *et al.*, 1992; Hutchinson and Johnstone, 1993). On the contrary, neither the biological properties of the WLIP nor its possible role in the symptom induction on mushrooms are yet known. In fact, this lipodepsipeptide is known mainly for its ability to form a white precipitate when interact with tolaasin I and this feature is used for the specific identification of *P. tolaasii* through the "white line" assay (Wong and Preece, 1979). The novelty of the pathogenetic capability of *P. "reactans"* and, in particular, the fact that avirulent variants of the pathogen have also lost the ability to produce WLIP (Lo Cantore, 2001; Lo Cantore and Iacobellis, unpublished results) prompted us to evaluate the action of WLIP directly on cell membranes, also in comparison to tolaasin I.

2. MATERIALS AND METHODS

2.1 Growth of Bacteria

Strains NCPPB2192 and NCPPB1311 of *Pseudomonas tolaasii* and *P. "reactans"*, respectively, were grown as reported by Lo Cantore *et al.* (2003). The antimicrobial activity against *Bacillus megaterium* was determined following an already established procedure (Lavermicocca *et al.*, 1997). The active preparations were finally lyophilised and stored at -20°C.

2.2 Isolation and Purification of the Lipodepsipeptides

Tolaasin I was isolated and purified from cell-free culture filtrate according to a modification of the method described by Peng (1986). Tolaasin preparations active in the inhibition of *Bacillus megaterium* were further

purified by reversed-phase high performance chromatography (HPLC) as reported by Lazzaroni *et al.* (2003).
Isolation and purification of WLIP was performed according to the procedure reported by Mortishire-Smith *et al.* (1991).
The identity of both tolaasin I and WLIP was ascertained by comparing NMR and MS data with those reported in literature (Nutkins *et al.*, 1991; Mortishire-Smith *et al.*, 1991).

2.3 Red Blood Cell Lysis and Protection

Haemolytic activity of tolaasin I and WLIP on human and rabbit red blood cells (RBC) has been determined by measuring the turbidity at 650 nm using a 96 well microplate Uvmax reader (Molecular Devices, Sunnyvale, CA, USA) as described by Dalla Serra *et al.* (1999). Erythrocytes have been prepared from fresh heparinized blood by washing it three times in 0.85% NaCl. In the haemolytic test, lipodepsipeptides have been two-fold serially diluted from well to well with a buffer containing 30 mM Tris-HCl, 100 mM NaCl, 1 mM EDTA, pH 7. The initial absorbance of intact RBC was about $OD_{650} = 0.1$. The percentage of haemolysis was calculated as follows:

$$\% \text{ haemolysis} = 100\,(A_i - A_f)/(A_i - A_w)$$

where A_i and A_f are the absorbance at the beginning and at the end of the reaction, and A_w the value obtained after hypotonical lysis with pure water.
The determination of the pore radius was performed by using osmolite protectants with different hydrodynamic radii following a procedure already described by Ginsburg and Stein (1987) and Tejuca *et al.* (2001). 30 millimoles of the following osmolites were supplemented to the haemolysis buffer: sucrose (0.54), raffinose (0.66), maltohexaose (0.90) (Sigma), stachiose (0.75) (Acros), and polyethylene glycols 1.000 (0.92), 1.500 (1.1), 2.000 (1.2), 3.000 (1.6), 4.000 (2.0) and 6.000 (2.5) (Fluka). The hydrodynamic radius in nm of each osmolite is reported in brackets after the compound names. The optical density data were analysed by using the equation of Renkin (Renkin, 1954) as described by Ginsburg and Stein (1987).

2.4 Calcein Release Assay

Large unilamellar vesicles (LUV100) loaded with calcein at a self-quenching concentration (80 mM) have been used to measure membrane permeabilization as reported by Kayalar and Duzgunes (1986). LUV100 were prepared using pure lipids ($4 \div 8$ mg ml^{-1}) by extrusion through stacked polycarbonate filters with 100 nm pores (MacDonald *et al.*, 1991). Lipid used were egg phosphatidylcholine (EggPC) and sphingomyelin (SM) from Avanti Polar Lipids (Alabaster, AL, USA), ergosterol (Erg) from Sigma, cholesterol (Chol) from Fluka. The assay was performed with the microplate fluorimeter (Fluostar, SLT, Groeding, Austria) as described by Tejuca *et al.* (1996). Before using, the external non encapsulated calcein was removed by washing the vesicle suspension through Sephadex G-50 minicolumns (Alvarez *et al.*, 2001). Aliquots of washed LUV100 were introduced into each well to a final lipid concentration of 3÷6 μM in 200 μl of 20 mM Tris-HCl, 100 mM NaCl, 1 mM EDTA, pH 7.0. Interferometric filter at 485 nm in excitation and at 538 nm in emission were used. Maximum calcein release (F_{max}) was determined by adding 1 mM Triton X-100 and used as 100%. The extent of permeabilization, $R\%$, was then calculated as follows:

$$R\% = 100\,(F_{fin} - F_{in})/(F_{max} - F_{in})$$

where F_{in} and F_{fin} represent the initial and final value of fluorescence before and after toxin addition. Unspecific binding of protein and liposomes to plastic microplate walls were strongly reduced by pre-treating the microplate with 0.1 mg ml^{-1} of Prionex (Pentapharm, CH).

2.5 Dynamic Light Scattering

The dimension of vesicles was measured by dynamic light scattering (DLS) at a fixed angle (90°) following procedures previously reported in Alvarez *et al.* (2001). The light scattering apparatus (Zetasizer3, Malvern, UK) was equipped with a 30 mW laser diode light source (675 nm), 64 channels autocorrelator and a quartz cell with a Peltier temperature control. The quartz cell has been filled with 0.5 ml buffer (20 mM Tris-HCl, 100 mM NaCl, 1 mM EDTA, pH 7.0) and 40 μl of phosphatidylcholine LUV100. The final lipid concentration was 80 μM.
The calculated autocorrelation functions were submitted to computer analysis (with the Malvern Application Software) using the method of cumulants. From the first two cumulants the mean intensity-weighted diffusion coefficient (D) and the variance of the sample were obtained. The D value is used to provide information on the size applying the Stokes-Einstein relation to obtain the mean hydrodynamic radius R:

$$R=kT/(6\pi\eta D)$$

where kT is the thermal energy (4.12×10^{-21} J at 25°C) and η is the viscosity of the solvent (0.89×10^{-3} Kg m^{-1} s^{-1} for water at 25°C).

3. RESULTS

3.1 Red Blood Cell Lysis

Both, tolaasin I and WLIP, caused red blood cell lysis. The haemolytic activity appeared to be higher for WLIP than for tolaasin I (Fig. 1). In fact, tolaasin I and WLIP showed the minimal haemolytic concentration of 12.6 and 2.7 μM, respectively. Furthermore, the kinetics of haemolysis caused by WLIP were more rapid than those of tolaasin I (not shown). The haemolytic activity of WLIP is apparently due to a colloid-osmotic shock caused by the formation of transmembrane pores which, in fact, can be prevented just adding to the external medium osmoticants of adequate dimensions. Preliminary results suggest that WLIP and tolaasin I formed pores with a diameter of 1.1±0.1 and 2.0±0.4 nm, respectively.

3.2 Calcein Release Assay

Tolaasin I and WLIP were both able to form channels in vesicles of different lipid composition but, the permeabilising activity, expressed as the reciprocal of the concentration of toxin causing the release of 50% calcein (C_{50}^{-1}), of tolaasin I was higher than that of WLIP (Table 1).
The release of calcein caused by tolaasin I increased with the increase of the sphingomyelin content in the membrane and, on the contrary, it decreased with the increase of sterols. Similar tendency has already been observed for other LDPs (Dalla Serra *et al.*, 1999a; Dalla Serra *et al.*, 2003). However, differences in the above behaviour were observed between cholesterol and ergosterol. In fact, tolaasin I was more active on ergosterol containing-membranes. This effect was more pronounced when sterols and phosphatidylcholine were present in equal amount. Similar effects of lipid composition on the permeabilising activity were also observed for WLIP, though in general WLIP appeared less active than tolaasin I (Table 1).

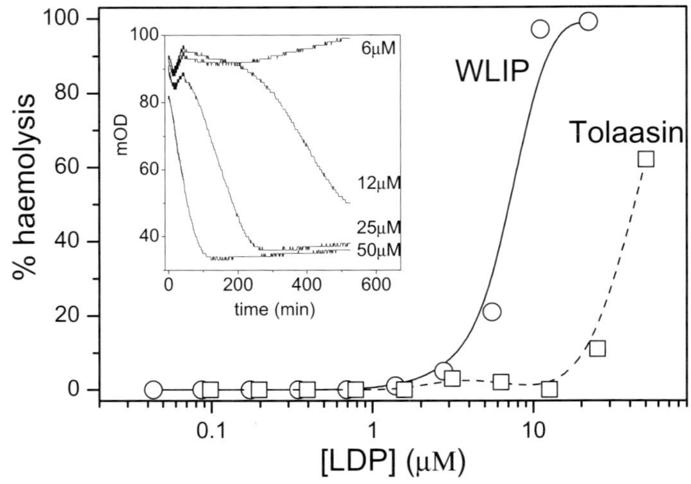

Figure 1. Haemolytic activity of tolaasin I and WLIP on human red blood cells. Extent of haemolysis induced after 45 min by WLIP (circles) and tolaasin I (squares). **Inset**: Kinetics of haemolysis of human RBC exposed to different amounts of tolaasin I. The decrease of turbidity indicated haemolysis. Both extent and rate of haemolysis depend on peptide concentration.

Table 1. Permeabilizing activity of tolaasin I and WLIP on vesicles of different lipid composition.

lipid composition (mol %)			**C_{50}^{-1b} (μM^{-1})**	
EggPC[a]	**SM[a]**	**Sterol[c]**	**WLIP**	**Tolaasin I**
100	0	0	0.59	2.84
50	50	0	1.41	4.96
50	33	16.5*	0.70	2.84
50	16.5	33*	0.59	1.17
50	0	50*	0.45	0.38
50	33	16.5**	0.70	2.48
50	16.5	33**	0.59	1.32
50	0	50**	0.54	0.99

[a] EggPC = egg phosphatidylcholine; SM = sphyngomyelin; [b] activity is expressed as $\mathbf{C_{50}^{-1}}$, where C_{50} is the concentration of LDP causing 50% release of calcein; [c] the sterol included in the vesicles was cholesterol* or ergosterol**.

3.3 Dynamic Light Scattering

Studies of "dynamic light scattering" (DLS) showed that treatment of phosphatidylcholine vesicles (80 μM) with tolaasin I in the range 0.3÷4.0 μM caused a slight increase of vesicle average size (Table 2) which was about 20% for tolaasin I and 9% for WLIP.
At these toxin concentrations, which correspond to a toxin to lipid ratio ($T\ L^{-1}$) lower than 0.3, the release of calcein was lower than 70%. Therefore, at such doses both toxins are able to form lesions allowing calcein escape without causing any destruction of the vesicles. A further increase of WLIP (more than 27 μM) probably induces vesicles micellization, through a detergent-like activity as suggested by the decrease of the scattered light intensity (Table 3). This effect was not observed in the case of tolaasin I, at least at concentrations up to 28.2 μM.

Table 2. Effect of tolaasin I on phosphatidylcholine vesicles size as determined by dynamic light scattering.

Tolaasin I (μM)	**Light scattering intensity (x1000 counts s^{-1})**	**Diameter (nm)**	**SD $Mean^{-1}$* %**
0.0	45.0	151	32
0.3	34.6	153	32
0.5	31.5	152	32
0.8	29.5	150	28
4.0	29.7	151	32
8.1	31.8	161	30
12.1	33.0	162	35
20.2	36.9	168	51
28.2	41.0	180	33
Tx100 (1 mM)	1.3	ND	ND

* width of the distribution of vesicle diamenter. SD is obtained from polydispersity of the cumulant fit as reported in Santos and Castanho (1999).
ND = not determinable.

From the static light scattering signal it is evident that both toxins caused a decrease of the intensity of the diffused light. The increase of the dimensions and decrease of the intensity caused by the toxin addition well correlated with vesicle aggregation, which increased the measured diameters, and precipitation of vesicle macro-aggregates which caused a decrease of signal intensity through a decrease of the concentration of scattering centres.
This effect on the intensity of the diffused light is much more marked with WLIP which caused up to 67% intensity diminution with monotonous course. On the contrary, tolaasin I showed only a 34% decrease.

Table 3. Effect of WLIP on phosphatidylcholine vesicles size as determined by dynamic light scattering measures.

WLIP (μM)	**Light scattering intensity (x1000 counts s^{-1})**	**Diameter (nm)**	**SD Mean^{-1}* %**
0.0	51.9	155	20
4.4	43.4	165	24
8.9	41.7	173	33
13.3	36.2	179	33
17.8	37.3	178	32
22.2	17.0	169	37
26.7	35.4	189	32
31.1	1.1	ND	ND
Tx100 (1mM)	1.5	ND	ND

* width of the distribution of vesicle diamenter. SD is obtained from polydispersity of the cumulant fit as reported in Santos and Castanho (1999); ND = not determinable.

4. DISCUSSION

In this paper the permeabilizing activity of tolaasin I and WLIP, two lipodepsideptides both able to damage biological and artificial membranes through the formation of transmembrane pores, have been compared. The pore forming ability of tolaasin is well known (Brodey *et al.*, 1991; Rainey *et al.*, 1991), however, this is the first report showing the effect of WLIP on membranes. These findings suggest that plasma membrane is the target also for this LDP. WLIP showed a higher haemolytic activity than tolaasin I. In fact, the haemolytic rate was higher for WLIP and the minimum haemolytic concentrations were 12.6 and 2.8 μM for tolaasin I and WLIP, respectively. In similar assays, Rainey *et al.* (1991) reported that 0.25 μM solution of tolaasin I determined an haemolytic effect. However, the latter experiments were performed on horse erythrocytes at 37 °C and measuring the rate of lysis.

Theoretical analysis of the protection experiments with the Renkin equation permits an estimation of the functional pore radius which was 1.1±0.1 and 2.0±0.4 nm, for WLIP and tolaasin I, respectively. These radius values are similar to those found for other LDPs (Dalla Serra *et al.*, 1999a). The small molecular size of LDPs suggests that such large pores could be formed by an aggregate of monomers, with a molecularity, and therefore a pore diameter, possibly dependent on the toxin concentration. For the smaller WLIP this

dependence is more pronounced and, in fact, at higher toxin dose a bigger osmoticant is necessary for fully protection of RBC from lysis.
The previous smaller estimation of the functional pore radius of tolaasin I, reported by Rainey *et al.* (1991) to be 0.9÷1.0 nm, could depend on the lower tolaasin concentration used in those experiments by the authors.
Tolaasin was more active on SM-containing membranes and, on the contrary, it was less active when cholesterol was present in the artificial membranes. This feature was also observed for other peptin-like LDPs but not by the smaller LDPs such as mycines whose activity is related to the membrane cholesterol content (Julmanop *et al.*, 1993; Taguchi *et al.*, 1994; Feigin *et al.*, 1997; Baginski *et al.*, 1997; Silberstein, 1998; Bender *et al.* 1999; Dalla Serra *et al.,* 1999a; 1999b; Dalla Serra *et al.,* 2003; Menestrina *et al.*, 2003). Also WLIP, on the contrary of the mycin-like LDPs, showed a similar feature although the inhibitory effect of sterols was less effective.

ACKNOWLEDGEMENTS

This work has been supported by the "Progetto Giovani Ricercatori", Biological characterisation of the tolaasin and WLIP, lipodepsipeptides of *Pseudomonas tolaasii* and *P. "reactans"* of the Università degli Studi della Basilicata, and by Provincia Autonoma di Trento, Fondo Progetti (Project AgriBio).

REFERENCES

Alvarez Valcarcel, C., Dalla Serra, M., Potrich, C., Bernhart, I., Tejuca M., Martinez, D., Pazos, F., Lanio, M.E., and Menestrina, G., 2001, Effects of lipid composition on membrane permeabilization by Sticholysin I and II, two cytolysins of the sea anemone Stichodactyla helianthus. *Biophys. J.* **80**: 2761-2774.

Baginski, M., Resat, H., McCammon, J.A., 1997, Molecular properties of amphotericin B membrane channel: A molecular dynamics simulation. *Mol. Pharmacol.* **52**(4): 560-570.

Brodey, C.L., Rainy, P.B., Tester, M., and Johnstone, K., 1991, Bacterial blotch disease of the cultivated mushroom is caused by an ion channel forming lipodepsipeptide toxin. *Mol. Plant-Microbe Inter.* **4**: 407-411.

Cauci, S., Monte, R., Ropele, M., Missero, C., Not, T., Quadrifoglio, F., and Menestrina G., 1993, Pore-forming and haemolytic properties of the *Gardnella vaginalis* cytolysin. *Mol. Microbiol.* **9**: 1143-1155.

Dalla Serra, M., Fagiuoli, G., Nordera, P., Bernhart, I., Della Volpe, C., Di Giorgio, D., Ballio, A., and Menestrina, G., 1999a, The interaction of lipodepsipeptide toxins from *Pseudomonas syringae* pv. *syringae* with biological and model membranes: a comparison of syringotoxin, syringomycin, and two syringopeptins. *Mol. Plant-Microbe Inter.*, **12**: 391-400.

Dalla Serra, M., Bernhart, I., Nordera, P., Di Giorgio, D., Ballio, A., and Menestrina, G., 1999b, Conductive Properties and Gating of Channels Formed by Syringopeptin 25A, a Bioactive Lipodepsipeptide from *Pseudomonas syringae* pv. *syringae*, in Planar Lipid *Membranes. Mol. Plant-Microbe Inter.* **12**: 401-409.

Dalla Serra, M., Menestrina, G., Carpaneto, A., Gambale, F., Fogliano, V., and Ballio, A., 2003, Molecular mechanism of action of syringopeptins, antifungal peptides from *Pseudomonas syringae* pv. *syringae*. In: *Pore-forming peptides and protein toxins* (G. Menestrina, M. Dalla Serra and P. Lazarovici, eds.), Taylor and Francis Group, London, UK, pp. 272-295.

Feigin, A.M., Schagina, L.V., Takemoto, J.Y., Teeter, J.H., and Brand, J.G., 1997, The effect of sterols on the sensitivity of membranes to the channelforming antifungal antibiotic, syringomycin E. *Biochim. Biophys. Acta* **1324**: 102-110.

Ginsburg, H., and Stein, W.D., 1987, Biophysical analysis of novel transport pathways induced in red blood cell membranes. *J. Membrane Biol.* **96**: 1-10.

Hutchison, M.L., and Johnstone, K., 1993, Evidence for the involvement of the surface active properties of the extracellular toxin tolaasin in the manifestation of brown blotch disease symptoms by *Pseudomonas tolaasii* on *Agaricus bisporus*. *Physiol. Mol. Plant Pathol.* **42**: 373-384.

Iacobellis, N.S., and Lo Cantore, P., 1997, Bacterial diseases of cultivated mushrooms in southern Italy. In: *Proceedings of X Congress of Mediterranean Phytopathological Union*, Montpellier, 2-6 June, 33-37.

Iacobellis, N.S., e Lo Cantore, P., 1998a, Recenti acquisizioni sul determinismo della batteriosi del cardoncello (*Pleurotus eryngii*), *Agricoltura Ricerca* **176**: 51-54.

Iacobellis, N.S., e Lo Cantore, P., 1998b, Studi sull'eziologia dell'ingiallimento dell'ostricone (*Pleurotus ostreatus*), *Agricoltura Ricerca* **176**: 55-60.

Julmanop, C., Takano, Y., Takemoto, J.Y., and Miyakawa, T., 1993, Protection by sterols against the cytotoxicity of syringomycin in yeast *Saccharomyces cerevisiae*. *J. Gen. Microbiol.* **139**: 2323-2327.

Kayalar, C., and Duzgunes, N., 1986, Membrane action of colicin E1: Detection of the release of carboxyfluorescein and calcein from liposomes. *Biochim. Biophys. Acta* **869**: 51-56

Lavermicocca, P., Iacobellis, N.S., Simmaco, M., and Graniti, A., 1997, Biological properties and spectrum of activity of *Pseudomonas syringae* pv. *syringae* toxins. *Physiol. Mol. Plant Pathol.* **50**: 129-140.

Lazzaroni, S., Bassarello, C., Bifulco, G., Lo Cantore, P., Evidente, A., Iacobellis, N.S., Riccio, R., and Gomez-Paloma, L., 2003, Chemical and Biological Characterisation of Tolaasins A-E: New Lipodepsipeptides Produced by *Pseudomonas tolaasii*, 245-254, (this volume).

Lo Cantore, P., 2001, Dottorato Thesis *Aspetti patogenetici, fisiologici e molecolari di Pseudomonas tolaasii e P. "reactans"*. Università degli Studi della Basilicata, Potenza.

Lo Cantore, P., e Iacobellis, N.S., 2002, Recenti acquisizioni sulla eziologia delle malattie batteriche dei funghi del genere *Agaricus* e *Pleurotus*. *Micologia Italiana* **2**: 18-27.

Lo Cantore, P., Lazzaroni, S., Evidente, A., Cafarchia, C., and Iacobellis, N.S., 2003, Production of Tolaasin I and WLIP by *Pseudomonas tolaasii* and *P. "reactans"*, their Antimicrobial Activity and Possible Role in the Virulence of the Pathogens, 255-262, (this volume).

MacDonald, R.C., MacDonald, R.I., Menco, B.P.M., Takeshita, K., Subbarao, N.K., and Hu, L., 1991, Small-volume extrusion apparatus for preparation of large, unilamellar vesicles. *Biochim. Biophys. Acta* **1061**: 297-303.

Menestrina, G., Coraiola, M., Fogliano, V., Fiore, A., Grgurina, I., Carpaneto, A., Gambale, F., and Dalla Serra, M., 2003, Antimicrobial Lipodepsipeptides from *Pseudomonas* spp: a Comparison of Their Activity on Model Membranes, 185-198, (this volume).

Mortishire-Smith, R.J., Nutkins, J.C., Packman, L.C., Brodey, C.L., Rainey, P.B., Johnstone, K., and Williams, D.H., 1991, Determination of the structure of an extracellular peptide produced by the mushroom saprotroph *Pseudomonas "reactans"*. *Tetrahedron* **47**: 3645-3654.

Nutkins, J.C., Mortishire-Smith, R.J., Packman, L.C., Brodey, C.L., Rainey, P.B., Johnstone K., and William, D.H., 1991, Structure determination of tolaasin, an extracellular lipodepsipeptide produced by the mushroom pathogen *Pseudomonas tolaasii* Paine. *J. Am. Chem. Soc.* **113**: 2621-2627.

Peng, J.T., 1986, PhD Thesis *Resistance to disease in* Agaricus bisporus *(Lange) Imbach.* University of Leeds.

Rainey, P.B., Brodey, C.L., and Johnstone, K., 1991, Biological properties and spectrum of activity of tolaasin, a lipodepsipeptide toxin produced by the mushroom pathogen *Pseudomonas tolaasii. Physiol. Mol. Plant Pathol.* **39**: 57-70.

Rainey, P.B., Brodey, C.L., and Johnstone, K., 1992, Biology of *Pseudomonas tolaasii*, cause of brown blotch disease of the cultivated mushroom. *Plant Pathol.* **8**: 95-117.

Renkin, E.M., 1954, Filtration, diffusion, and molecular sieving through porous cellulose membranes. *J. Gen. Physiol.* **38**: 225-243.

Santos, N.C., and Castanho, M.A.R.B., 1999, Teaching Light Scattering Spectroscopy: The Dimension and Shape of Tobacco Mosaic Virus. *Biophys. J.* **71**: 1641-1650.

Taguchi, N., Takano, Y., Julmanop, C., Wang, Y., Stock, S., Takemoto, J., and Miyakawa, T., 1994, Identification and analysis of the *Saccharomyces cerevisiae* SYR1 gene reveals thatergosterol is involved in the action of syringomycin. *Microbiology* **140**: 353-359.

Tejuca, M., Dalla Serra, M., Ferreras, M., Lanio, M.E., and Menestrina, G., 1996, The mechanism of membrane permeabilisation by sticholysin I, a cytolysin isolated from the venom of the sea anemone *Stichodactyla helianthus. Biochemistry* **35**: 14947-14957.

Wells, J.M., Sapers, G.M., Fett, W.F., Butterfield, J.E., Jones, J.B., Bouzar, H., and Miller, F.C., 1996, Postharvest discoloration of the cultivated mushrooms *Agaricus bisporus* caused by *Pseudomonas tolaasii, P. "reactans"*, and *P. "gingeri". Phytopathology* **86**: 1098-1104.

Wong, W.C., and Preece, T.F., 1979, Identification of *Pseudomonas tolaasii*: the white line in agar and mushroom tissue block rapid pitting tests. *J. Appl. Bacteriol.* **47**: 401-407.

In *Pseudomonas syringae* pv. *phaseolicola* the Synthesis of Phaseolotoxin and the Concurrent Expression of the *argK* Gene Coding for the Phaseolotoxin-Resistant Ornithyl-Carbamoyl Transferase Occur Independent of the Global Arginine Regulator ArgR

K. LÓPEZ-LÓPEZ, J.L. HERNÁNDEZ-FLORES, R. GARCIDUEÑAS-PIÑA and A. ALVAREZ-MORALES

Center for Research and Advanced Studies (CINVESTAV - Irapuato), Irapuato, Guanajuato, México. Apdo. Postal 629, Irapuato, Gto. Mexico CP 36500.

Abstract: *Pseudomonas syringae* pv. *phaseolicola*, the causal agent of halo blight in beans produces a non-host specific toxin capable of inhibiting the enzyme ornithine carbamoyltransferase (OCTase) of plant, bacterial and mammalian origin. When *P. s.* pv. *phaseolicola* grown under conditions leading to the synthesis of phaseolotoxin, synthesises a phaseolotoxin-resistant OCTase, the product of the *argK* gene. Under any other growth conditions, the synthesis of arginine is mediated by a phaseolotoxin-sensitive OCTase encoded by the *argF* gene. In *P. aeruginosa* genes involved in the catabolism of arginine, the synthesis of carbamoylphosphate and the expression of *argF* are regulated by arginine and the ArgR protein. We isolated and characterised the *argR* gene of *P. s.* pv *phaseolicola*, and constructed a null *argR* mutant. Northern blot analysis using this mutant showed that in this bacterium regulated expression of *argF* is dependent on the ArgR product, but *argK* expression and phaseolotoxin synthesis occur independent of this regulatory protein. Also, we have shown that exogenous arginine has an effect on the expression of *argF* but not on *argK*. These results strongly indicate a lack of metabolic co-ordination between the synthesis of phaseolotoxin and the system regulating arginine metabolism mediated by ArgR

Key words: Arginine regulation, *argF*, phaseolotoxin, ArgR.

Author for correspondence: A. Alvarez-Morales; e-mail: aalvarez@ira.cinvestav.mx.

N.S. Iacobellis *et. al.* (eds.), *Pseudomonas syringae* and Related Pathogens, 275-282.

1. INTRODUCTION

Pseudomonas syringae pv. *phaseolicola*, the causal agent of halo blight in beans produces a toxin, known as phaseolotoxin. It consists of a tripeptide, ornithinyl-alanyl-homoarginine and an inorganic moiety, N′-sulpho-diaminophosphinyl (Mitchell, 1976; Moore *et al.*, 1984). In beans phaseolotoxin inhibits the enzyme ornithyl carbamoyltransferase (OCTase) involved in the biosynthesis of L- citrulline from L- ornithine and carbamoylphosphate in the pathway leading to the biosynthesis of arginine (Ferguson and Johnston, 1980). This toxin inhibits not only plant OCTases but also bacterial and mammalian enzymes. Therefore, in order to protect itself from its own toxin, *P. s.* pv. *phaseolicola* synthesises a phaseolotoxin-resistant OCTase, the product of *argK* (Mosqueda *et al.*, 1990; Hatziloukas and Panopoulos, 1992). This gene is only expressed under conditions leading to the synthesis of phaseolotoxin, such as growth in minimal medium at 16-18°C. Under any other growth condition, the synthesis of arginine in *P. s.* pv. *phaseolicola* is mediated by a phaseolotoxin-sensitive OCTase encoded by *argF*. In *P. aeruginosa* genes involved in the catabolism of arginine, the synthesis of carbamoylphosphate and the expression of the *argF* are regulated by arginine and the ArgR protein (Park *et al.*, 1997).
Arginine is a key amino acid for the biosynthesis of phaseolotoxin because it is the substrate of the enzyme amidinotransferase involved in the synthesis of homoarginine, an unusual amino acid present in the phaseolotoxin molecule. This amidinotransferase is encoded by the *amtA* gene (Hernández-Guzmán and Alvarez-Morales, 2001), which is expressed in coordination with the synthesis of phaseolotoxin and expression of *argK*.
There is evidence suggesting that the genes involved in the biosynthesis of phaseolotoxin were acquired horizontally by *P. s.* pv. *phaseolicola* from another micro-organism at some point during evolution (Sawada *et al.*, 1999). We wanted to investigate whether the *argK* gene was being regulated by the ArgR protein to ensure co-ordination of its function with the overall metabolism of the cell.

2. MATERIALS AND METHODS

Pseudomonas syringae pv. *phaseolicola* strain NPS3121 was grown in KB medium or M9 minimal medium at 28°C. For phaseolotoxin production *P. s.* pv. *phaseolicola* was grown in M9 medium at 18°C for 36 h. *Escherichia coli* strains were grown at 37°C in LB medium. Routine techniques of molecular biology were performed as described previously (Sambrook *et al.*, 1989), and when using "kits" for DNA or plasmid purification, labelling, etc., and restriction or modification enzymes, the manufacturer's instructions were followed.

3. RESULTS AND DISCUSSION

3.1 Identification and Cloning of the *argR* Gene from *Pseudomonas syringae* pv. *phaseolicola*

Through Southern blot hybridisation of a genomic library from *Pseudomonas syringae* pv. *phaseolicola* and using the *argR* gene from *P. aeruginosa* PAO1 as a probe, we identified a genomic clone carrying the homologous *argR* gene. The 1.6 kbp clone obtained was sequenced and then analysed to look for potential open reading frames. One ORF was selected and its deduced amino acid sequence was aligned with the global arginine regulator ArgR from *P. aeruginosa* using MegAlign software (DNASTAR, Madison WI, U.S.A.). The alignments of the ArgR protein from *P. s.* pv. *phaseolicola* showed 81.5% similarity with the ArgR protein from *P. aeruginosa* (Fig. 1). The sequence was submitted to GenBank (accession No. AF443787).

```
P.phaseolicola   AHRIGFLVWPGTKALTLALAEEALRVAQRVHPEVVYELSFLQAEAGEPTAVAGAWQLPGEPWTGR
P. aeruginosa    PQRIGFLLWP ATRALTLSLAEEALRAARRLHPEA LYEPLFL LAEA - - PAEEEG-WR LPGTAWNGR

Consensus        : : RIGFL : WP : T : ALTL :LAEEALR .A : R :HPE .  :YE   FL  AEA    P : .  . G W  : LPG .  :W .GR

P. phaseolicola  LKQLVRAGCSIGGLSAGVYPLAQLGLLDGYRAAVHWRWQDDFAERFPKVIATSHLFDWDRDRLTA
P. aeruginosa    LKQLARSGAA IGALSAG IYPLAQLGLLDGYRAAVHWRWHDDFTERFPKVIATNHLFEWDRDRMTA

Consensus        LKQL . R :G   : IG : LSAG :YPLAQLGLLDGYRAAVHWRW :DDF : ERFPKVIAT :HLF : WDRDR :TA

P. phaseolicola  VSEELVVERIREGGERQRIPLQNRLGSSHPKLTQAVLLMEANIEEPLTTDEIAQHVCVSRRQLER
P. aeruginosa    VSEELVVERIREGNERQRIPL KNRLGSSHPKLTQAVLLMEANIEEPLTTDEIAQHVCVSRRQLER

Consensus        VSEELVVERIREG . ERQRIPL : NRLGSSHPKLTQAVLLMEANIEEPLTTDEIAQHVCVSRRQLER

P. phaseolicola  LMQTSKSIIQIGLSCGFSSGPHFSSAYRNFFGATPREDRNQ
P. aeruginosa    LMQTSKSIIQIGLSCGFSSGPHFSSAYRNFFGVTPREDRNQ

Consensus        LMQTSKSIIQIGLSCGFSSGPHFSSAYRNFFG . TPREDRNQ
```

Figure 1. Alignment of the ArgR protein from *Pseudomonas syringae* pv. *phaseolicola* with the ArgR protein from *P. aeruginosa*. Identical residues are bold. The conserved domain of HTH ARAC (helix_ turn_ helix similar to the arabinose operon control protein) is underlined.

3.2 Characterisation of the *argR* Gene from *Pseudomonas syringae* pv. *phaseolicola*

To confirm the regulatory role of the putative *argR* gene we had cloned, we constructed a mutant carrying an interrupted *argR* gen. To this end, the *argR* gene was interrupted with a tetracycline resistance cassette and was introduced via electroporation into *Pseudomonas syringae* pv. *phaseolicola.* The double recombination event was selected using tetracycline and confirmed through Southern blot hybridisation. As expected for an *argR*- (Park *et al.*, 1997), the mutants were unable to use arginine as sole carbon and/or nitrogen source. Furthermore, the *argR*- mutant showed similar levels of OCTase activity in both, rich and minimal medium when grown at 28°C, whereas the wild type strain showed low levels when grown in rich medium (Fig. 2).

This result was expected assuming the role of the ArgR regulatory protein in *P. s.* pv. *phaseolicola* to be similar to that described for *P. aeruginosa,* that is, it mediates repression of *argF* in the presence of arginine (rich medium or M9 plus 10 mM arginine) (Park *et al.*, 1997). This result was confirmed through Northern blot analysis using the *argR*- mutant grown in rich medium which shows constitutive expression of the *argF* gene (Fig. 3).

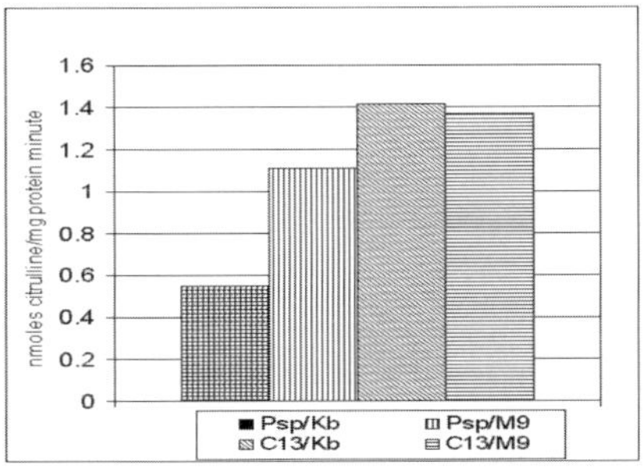

Figure 2. In *Pseudomonas syringae* pv. *phaseolicola* the expression of *argF* is dependent of ArgR protein. OCTase activity was assayed in cell-free extracts of wild type *P. s.* pv. *phaseolicola* grown at 28°C in rich medium (Psp/Kb), in M9 minimal medium (Psp/M9) and the C13 *argR*- mutant grown in rich medium (C13/Kb) and in minimal medium (C13/M9). The values are means of least two repeats.

To test whether the ArgR protein could play a role in the synthesis of phaseolotoxin, perhaps through the regulation of *argK*, both the wild type strain and *argR*- mutant of *P. s.* pv. *phaseolicola* were grown under conditions permissive (18°C) and non-permissive (28°C) for phaseolotoxin synthesis. In both cases the wild type phenotype was observed, contrary to the constitutive expression of *argF* in the *argR*- background (Fig. 4). These results have shown that the expression of *argF* is under control of the global arginine regulator ArgR encoded by the *argR* gene cloned and analysed in this work and that phaseolotoxin synthesis is independent of this factor.

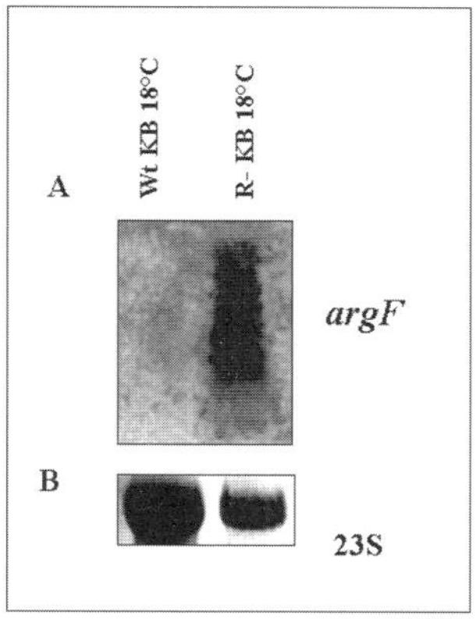

Figure 3. Constitutive expression of the *argF* gene in the *argR-* mutant. (A) Northern blot of total RNA from wild type *Pseudomonas syringae* pv. *phaseolicola* (lane 1) and from the *argR-* mutant C13 (lane 2) grown at 28°C in rich medium. The blots were hybridised with a ^{32}P-labeled probe internal to the *argF* gene. (B) 23S rRNA staining with ethidium bromide.

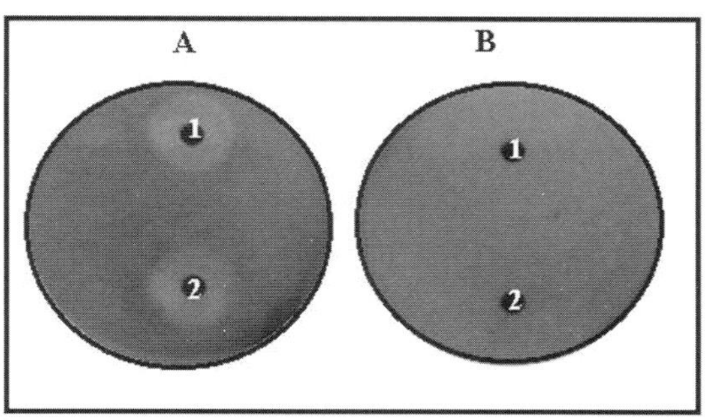

Figure 4. Growth inhibition assays of *Escherichia coli* JM103 by phaseolotoxin. Assays were made with supernatants from Wt *Pseudomonas syringae* pv. *phaseolicola* (1) and the *argR-* mutant (2) grown in M9 medium at: A) 18°C; and B) 28°C.

3.3 The *argF* Gene from *Pseudomonas syringae* pv. *phaseolicola* is Repressed Under Conditions of Phaseolotoxin Synthesis

To investigate whether there was a mechanism that would repress *argF* when *Pseudomonas syringae* pv. *phaseolicola* was growing under conditions of phaseolotoxin synthesis, we constructed an *argK-* mutant by inserting a chloramphenicol resistant cassette into the *argK* coding sequence, and

obtaining a homogenate through a double recombination event, which was further confirmed by Southern hybridisation.
If the *argK-* mutant is grown at 18°C, it could be expected that the mutation would be lethal since there would be no phaseolotoxin-resistant OCTase, and perhaps the phaseolotoxin-sensitive OCTase from *argF* would be repressed under this conditions. However, we observed that the *argK-* mutation was not lethal under these conditions; the strain grow at a very low rate (Fig. 5). To our surprise this mutant still produced a small amount of phaseolotoxin (Fig. 6).

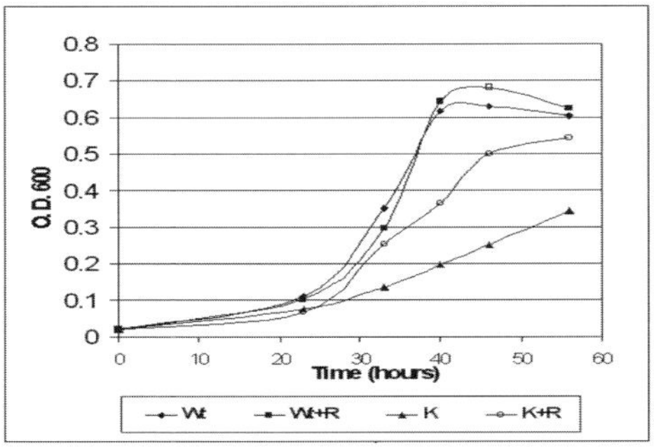

Figure 5. Growth of the *argK-* mutant at 18°C in minimal medium (K), minimal medium supplemented with 20 mM arginine (K+R); wild type *Pseudomonas syringae* pv. *phaseolicola* grown at 18°C in minimal medium (Wt) and minimal medium supplemented with 20 mM arginine (Wt+R).

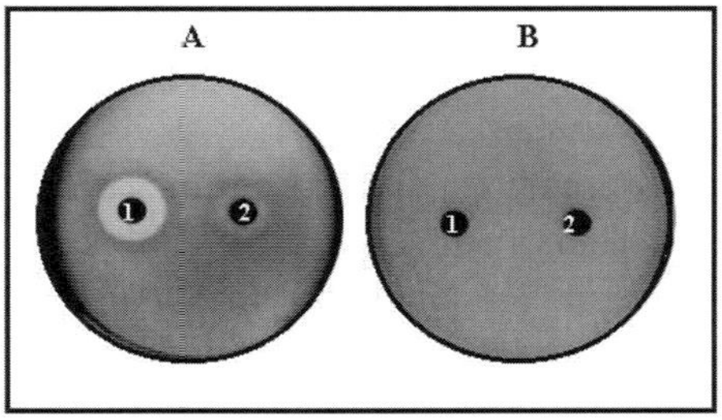

Figure 6. Growth inhibition assays of *Escherichia coli JM103* by phaseolotoxin. Assays were made with supernatants from *Pseudomonas syringae* pv. *phaseolicola* Wt (1) and the *argK-* mutant (2) grown at 18°C in (A) minimal medium and (B) minimal medium supplemented with 10 mM arginine.

The *argK-* mutant cultured at 18°C showed a reduced rate of growth that could be increased to almost normal levels when exogenous arginine was added (Fig. 5, **K+R**). This suggests that in the *argK-* mutant limited growth is caused by deficiencies in the arginine pools since the phaseolotoxin-sensitive OCTase produced from *argF* is being almost completely inhibited by phaseolotoxin. This would imply that the *argF* gene is expressed wastefully, and presumably at high level if the arginine pools are maintained low. A Northern blot assay to detect the *argF* transcript in the *argK*- mutant grown in M9 at 18°C confirmed these predictions (Fig. 7). High expression of *argF* under these conditions is apparently the only way to sustain a low rate of growth and low levels of phaseolotoxin synthesis.

Taken together these results strongly indicate a lack of metabolic co-ordination between the synthesis of phaseolotoxin and the system regulating arginine metabolism mediated by ArgR protein.

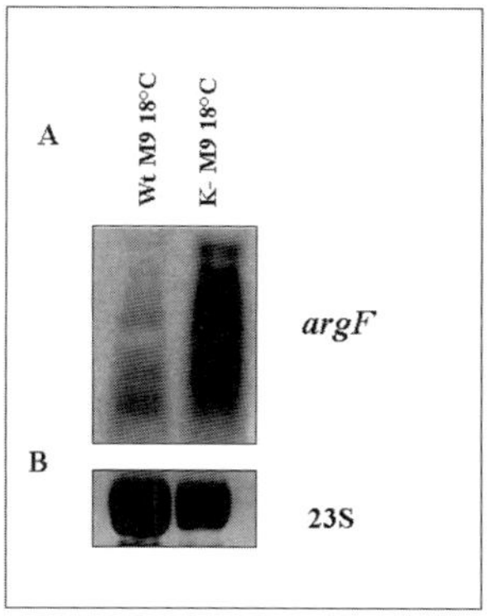

Figure 7. The *argF* gene is expressed at high level in the *argK*- mutant grown at 18°C. (A) Northern blot of total RNA from *Pseudomonas syringae* pv. *phaseolicola* Wt (lane 1) and from the *argK*- mutant (lane 2) grown at 18°C in minimal medium. The blots were hybridised with a ^{32}P-labeled probe corresponding to the *P. s.* pv. *phaseolicola argF* gene. (B) 23S rRNA staining with ethidium bromide.

REFERENCES

Ferguson, A.R., and Johnston, J.S., 1980, Phaseolotoxin: chlorosis, ornithine accumulation and inhibition of ornithine carbamoyltransferase in different plants. *Physiol. Plant Pathol.* **16**: 269-275.

Hatziloukas, E., and Panopoulos, N.J., 1992, Origin, structure and regulation of *argK*, encoding the phaseolotoxin-resistant carbamoyltransferase in *Pseudomonas syringae* pv. *phaseolicola,* and functional expression of *argK* in transgenic tobacco. *J. Bacteriol.* **174**: 5895-5909.

Hernández-Guzmán, G., and Alvarez-Morales, A., 2001, Isolation and characterisation of the gene coding for the amidinotransferase involved in the biosynthesis of phaseolotoxin in *Pseudomonas syringae* pv. *phaseolicola. Mol. Plant Microbe Interact.* **14**: 545-554.

Mitchell, R.E., 1976, Isolation and structure of chlorosis inducing toxin of *Pseudomonas phaseolicola. Phytochemistry* **15**: 1941-1947.

Moore, R.E., Niemczura, W.P., Kwok, O.C.H., and Patil, S.S., 1984, Inhibitions of ornithine carbamoyltransferase from *Pseudomonas syringae* pv. *phaseolicola.* Revised structure of phaseolotoxin. *Tetrahedron lett.* **25**: 3931-3934.

Mosqueda, G., Den Broeck, G.V., Saucedo, O., Bailey, A.M., Alvarez-Morales, A., and Herrera-Estrella, L., 1990, Isolation and characterisation of the gene from *Pseudomonas syringae* pv. *phaseolicola* encoding the phaseolotoxin-insensitive ornithine carbamoyl-transferase. *Mol. Gen. Genet.* **222**: 461-466.

Park, S., Lu, C., and Abdelal, A.T., 1997, Cloning and characterisation of *argR*, a gene that participates in regulation of arginine biosynthesis and catabolism in *Pseudomonas aeruginosa* PA01. *J. Bacteriol.* **179**: 5300-5308.

Sambrook, J., Fritsch E.F., and Maniatis, T., 1989, Molecular cloning: a laboratory manual, 2nd ed. Cold Spring Harbor Laboratory, Cold Spring Harbor, NY, U.S.A.

Sawada, H., Suzuki, F., Matsuda, I., and Saitou, N., 1999, Phylogenetic analysis of *Pseudomonas syringae* pathovars suggests the horizontal gene transfer of *argK* and the evolutionary stability of *hrp* gene cluster. *J. Mol. Evol.* **49**: 627-644.

Genetic and Physiological Analysis of Host Pathogen Interaction

The HRP Pilus of *Pseudomonas syringae*

C.M. LI, I. BROWN[#], C. STEVENS[#], J.W. MANSFIELD[#], T. BOUREAU, E. HIENONEN, M. ROMANTSCHUK and S. TAIRA
University of Helsinki, Department of Biosciences, Division of General Microbiology. P.O. Box 56, FIN-00014, University of Helsinki, Finland; [#]Imperial College at Wye, University of London, Ashford, Kent, TN25 5AH, UK.

Abstract: The Hrp pilus is an indispensable component of type III secretion system, a common virulence factor shared by a number of Gram negative plant pathogens. The type III secretion system enables the bacteria to translocate virulence proteins through the plant cell wall into the plant cytoplasm. The Hrp pilus is a long, up to 5 μm, filamentous structure that penetrates the plant cell wall. Expression of the Hrp pili are induced rapidly after transferring bacteria to Hrp inducing minimal medium and also inside the plant tissues. The molecular mechanism of pilus function in the type III secretion system-dependent translocation of proteins remains poorly understood, although it has been suggested that the pilus somehow aids the protein delivery from bacteria into plants. We have studied the expression and assembly of the Hrp pilus, and have shown that the pilus grows by addition of new pilin subunits at the distal end of the filament. Additionally, we have shown that HrpZ protein is secreted from the tip of the pilus. These observations strongly indicate that the Hrp pilus acts as a conduit for secreted proteins, and directs their delivery through the plant cell wall.

Key words: Type III secretion, Hrp pilus, HrpZ, virulence.

Author for correspondence: S. Taira; e-mail: suvi.taira@helsinki.fi.

1. INTRODUCTION

The type III secretion system (TTSS) is an essential virulence determinant in the different pathovars of *Pseudomonas syringae.* In addition to allowing the secretion of harpin proteins to the extracellular millieu, TTSS is able to translocate effector proteins directly into the plant cell (Casper-Lindley *et*

N.S. Iacobellis *et. al.* (eds.), *Pseudomonas syringae* and Related Pathogens, 285-292.

al., 2002). Effector proteins are usually termed as Avirulence (Avr) or Virulence (Vir) proteins depending on the phenotype that was first associated with the protein in question. The effectors in turn interact with host proteins, and the interaction can lead to successful colonisation by the bacterium and disease, or the hypersensitive reaction (HR) and restriction of bacterial multiplication (Leach and White, 1996).

The TTSS is found in a number of pathogenic and symbiotic bacteria that interact with phylogenetically distant hosts, including plants, insects and mammals (Hueck, 1998, Dale *et al.*, 2002). The most striking morphological difference between the TTSS of plant and animal pathogens is the extracellular appendage associated with the core secreton. With the exception of *Escherichia coli*, all animal pathogens possess a needle-like extension shorter than 100nm on the bacterial surface (Knutton *et al.*, 1998, Kubori *et al.*, 1998, Blocker *et al.*, 1999, Hoiczyk and Blobel, 2001). Plant pathogens possess a pilus-like extension, which can grow to several microns in length (Brown *et al.*, 2001). The differences between the extracellular appendages in animal and plant pathogens have been proposed to reflect the need for the latter to deliver proteins across the plant cell wall.

The Hrp pilus of *P. syringae* is the best-characterised TTSS-associated surface appendage. Its main structural subunit is a small, about 10 kDa α-helical protein encoded by *hrpA* gene (Roine *et al.*, 1997a). The Hrp pili can be efficiently induced in Hrp inducing minimal medium, and bacteria expressing them appear densely decorated with long autoaggregative filaments, 6÷8 nm in diameter, in electron microscopy (EM), (Roine *et al.*, 1997a). The Hrp pili of *P. s.* pv. *tomato* DC3000 have been purified, and the purification protocol partly relies on the fact that monomeric pilin subunits are able to auto-assemble into intact filaments under proper conditions (Roine *et al.*, 1997b). Although the Hrp pilin subunits share limited homology between different pathovars of *P. syringae,* and even less homology compared with other TTSS-harbouring plant pathogens, Hrp pili appear morphologically very similar (Jin *et al.*, 2001, our unpublished results).

Hrp-pilus negative mutants of *P. syringae* lose their pathogenicity towards the host, fail to elicit HR in resistant plants and are not able to secrete harpins or effector proteins (Roine *et al.*, 1997, Wei *et al.*, 2000). In other words, the Hrp pilus is an indispensable component of the TTSS. The role of the pilus in the secretion process has, however not been clear. In most cases, bacterial pili and fimbriae have been associated with attachment and adhesion. However, for the Hrp pilus, a more intriguing alternative is that it has a more active role in the secretion process itself. It has been shown that the pilus forms a physical link between the bacterium and the plant (Brown *et al.*, 2001), and it appears to penetrate the plant cell wall. The pilus might therefore function as a sensor of contact with the plant cell, and regulate the start of the secretion process and/or it could function as a device for directing the translocated proteins into plant cells.

We studied the possibility that the Hrp pilus forms a conduit, through which the translocated proteins move across the plant cell wall. First we addressed the question of the polarity of pilus elongation. The flagellar apparatus is considered to be a specialised TTSS, and the flagellum has been demonstrated to grow by adding flagellar subunits at the distal end of the structure (Emerson *et al.*, 1970). The flagellin monomers hence travel through the forming filament, a process similar to the conduit model of Hrp pilus function. Secondly we asked if secreted harpin proteins can be detected at the tip or at the base of Hrp pilus, after pulse-induction of the harpin production. Our results clearly support the conduit model of Hrp pilus function.

2. RESULTS AND DISCUSSION

2.1 FLAG-tagging of the HrpA Pilin

In order to distinguish newly translated and incorporated pilin subunits from the previously assembled ones, we tagged the HrpA pilin with the immunologically detectable FLAG epitope. The site of the FLAG tag was chosen based on our earlier results defining a rough domain map of the pilin gene (Fig. 1; Taira *et al.*, 1999). The mid-amino terminal region of HrpA tolerates short insertions, hence the FLAG insertions were placed there. We tested four different locations (Fig. 1) for the FLAG tag in their ability to form functional pili and to display the FLAG epitope on the pilus surface. FLAG insertion at all four locations allowed the formation of normal pili, which could complement the chromosomal *hrpA* mutation in pathogenesis and HR assays. The two insertions at the edges of the insertion-tolerating region did not display the FLAG epitope in such a way that it could be visualised by immuno-EM using monoclonal anti-FLAG antiserum. However, the two FLAG insertions situated at the middle of the region were detectable by immunocytochemistry (Fig. 2). The HrpA pilin with FLAG epitope at codon 23 of the HrpA gene was chosen for further studies and it is referred as FLAG-HrpA in the following text.

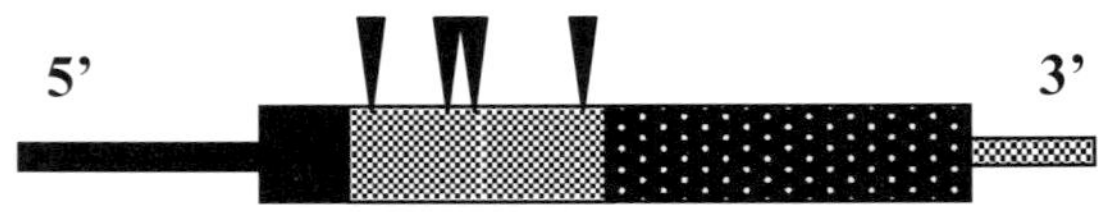

Figure 1. Schematic representation of the HrpA domain structure and the location of FLAG tags. Narrow rectangles are non-translated region, thick rectangles are the open reading frame. Black region: secretion signals and mRNA stability; grey: regions that allow insertions without affecting pilus formation and function; dotted: region needed for pilus assembly. Black triangles show the locations of FLAG insertions.

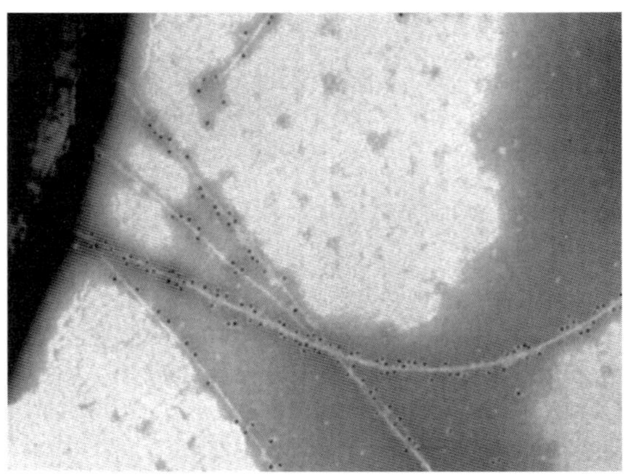

Figure 2. FLAG-tagged Hrp pili detected with monoclonal anti-FLAG antiserum. Individual pili are separated when the bacteria are grown on electron microscopy grids. (Brown *et al.*, 2001).

2.2 Placing the FLAG-hrpA Under an Inducible Promoter

We next wanted to express the FLAG-HrpA independently from the rest of the *hrp* genes. The inducible promoter we chose was the mercury-inducible *mer* promoter from Tn*21* (Misra *et al.*, 1984). The transcript of HrpA is extremely stable (our unpublished results), and the region encoding the amino-terminal secretion signal, as well as the 5' non-coding region are important for transcript stability (Taira *et al.*, 1999, Hienonen *et al.*, 2002). Our earlier studies have shown that fusion of *hrpA* with the *lacZ* promoter results in a low expression level if the fusion site is the translational start (ATG-codon), but significantly higher when the fusion site is the transcriptional start site of *hrpA,* -42 bp from the translation start (Hienonen *et al.*, 2002). We therefore included the 5' non-coding region in the *mer-FLAG-hrpA* construct.

2.3 Demonstration of the Polarity of Elongation by Immuno-EM

The plasmid harbouring the mercury-inducible *FLAG-hrpA* was then introduced to wild type DC3000 strain. The resulting strain produced FLAG-tagged HrpA in the presence 100 nM $HgCl_2$ both in rich and in Hrp-inducing minimal media, and non-tagged wild type HrpA only in Hrp-inducing medium regardless of the presence of mercury. Hence the question of the Hrp pilus assembly could be approached with the experimental set up described in Fig. 3. The bacteria were grown on EM-grids under Hrp-inducing conditions, which allowed the expression of the chromosomally encoded wild type pilus growth. Mercury was then added on the grid, which induced the expression of the FLAG-tagged pilin. The appearance of the FLAG tags at the tip of the pilus was followed by immuno-EM using anti-FLAG antiserum. Quantitative analysis and time course analysis after mercury induction showed the appearance of the growing FLAG-tagged distal end of the pilus, while the basal non-tagged portion remained unchanged in length.

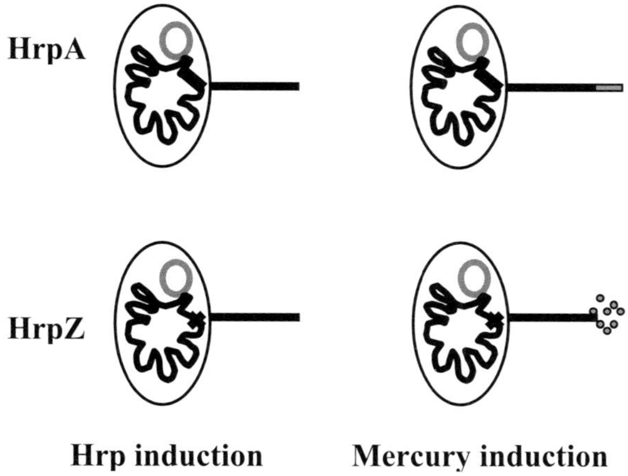

Figure 3. The appearance of newly translated HrpA and HrpZ proteins at the tip of the Hrp pilus. Genes for mercury-inducible FLAG-HrpA or HrpZ are in a plasmid (grey). Chromosomally encoded wild type Hrp pilus is expressed in Hrp-inducing medium and the FLAG-tagged pilin or HrpZ is expressed after mercury pulse.

2.4 Pulse-expression of HrpZ

The distal growth pattern of the Hrp pilus strongly indicates that pilin monomers are travelling through a putative axial channel. Since the pilin monomer is a relatively small protein, we asked whether bigger proteins are able to use the same route. We cloned the HrpZ harpin, 36 kDa in size, under the mercury-inducible promoter, and introduced the plasmid in a *hrpZ*⁻ strain. HrpZ is dispensable for the secretion process hence the *hrpZ*⁻ strain could be used, and tagging of the protein was not necessary. The bacteria were again grown on EM grids in Hrp-inducing conditions in order to induce the pilus expression, after which mercury was added in order to induce HrpZ. HrpZ antibody was used to localize the site of HrpZ secretion. The pattern of labelling demonstrated clearly that harpin is secreted at the tip of the pilus (Fig. 3) and that the harpin most probably uses the same route as the pilin to reach the extracellular milieu.

3. DISCUSSION

Our results (Li *et al.*, 2002) and the work of Jin and He (2001) demonstrate that the Hrp pilus acts as a conduit through which the pilin subunits, harpins and Avr/Vir proteins travel. The finding explains how the effectors may cross the porous plant cell wall, and reach their target in an efficient concentration, without being lost to the intracellular space by diffusion. Notably, the type IV secretion used by *Agrobacterium* also requires a pilus for translocating T-DNA and virulence proteins into the plant cytoplasm. Although the two systems share no sequence homology, both have evolved to solve the same problem, long distance protein delivery through the plant cell wall, and it is tempting to suggest that the T-pilus would have a similar contribution to the translocation as the Hrp pilus.

The type III secretion of virulence proteins is an amazingly complicated process evolved among bacteria whose life style involves intimate relationship with eukaryotes. Our understanding of the secretion/translocation process at the molecular level is still shallow. The most important questions involve the events at both ends of the pilus, namely, the choice of the set of proteins that are entering the secretion channel, and the entry of Avr/Vir proteins to the plant cytoplasm through the plasma membrane. Also the mechanism of the movement of the proteins through the presumably crowded pilus lumen remains elusive. How is the energy that is released at the bacterial cytoplasmic membrane transferred to movement of proteins at the pilus lumen that can be up to 5 μm away from the bacterial surface? How can the narrow pilus channel accommodate the massive amount of proteins that are moving through it, and how does the system prevent the proteins from clotting and blocking the movement? Considering pilus dimensions, it is clear that the effectors must be at least partially unfolded during the translocation. Whether the proteins are actively pumped out or passively pushed out by new proteins entering the channel is not known. A tempting idea is that the pilus could acts as a nano-scale inside-out screw that would pump out the proteins by rotation. The homologous flagellar system functions by rotating, and the ability to rotate might be a common and ancient property of type III secretion systems.

REFERENCES

Blocker, A., Gounon, P., Larquet, E., Niebuhr, K., Cabiaux, V., Parsot, C., and Sansonetti, P., 1999, The tripartite type III secreton of Shigella flexneri inserts IpaB and IpaC into host membranes. *J. Cell Biol.* **147**: 683-693.

Brown, I.R., Mansfield, J.W., Taira, S., Roine, E., and Romantschuk, M., 2001, Immunocytochemical localisation of HrpA and HrpZ supports a role for a transfer of effector proteins from *Pseudomonas syringae* pv. *tomato* across the plant cell wall. *Mol. Plant-Microbe Interact.* **14**: 394-404.

Casper-Lindley, C., Dahlbeck, D., Clark, E., and Staskawicz, 2002, Direct biochemical evidence for type III secretion-dependent translocation of the AvrBs2 effector protein into plant cells. *Proc. Natl. Acad. Sci. USA*. **95**: 15849-15854.

Dale, C., Plague, G., Wang, B., Ochman, H., and Moran, N., 2002, Type III secretion systems and the evolution of mutualistic endosymbiosis. *Proc. Natl. Acad. Sci. USA* **99**: 12397-12402.

Emerson, S.U., Tokuyasu, K., and Simon, M., 1970, Bacterial flagella: Polarity of elongation. *Science* **169**: 190-192.

Hienonen, E., Roine, E., Romantschuk, M., and Taira, S., 2002, mRNA stability and the secretion signal of HrpA, a pilin secreted by the type III system in *Pseudomonas syringae*. *Mol. Gen. Genomics* **266**: 973-978.

Hoiczyk, E., and Blobel, G., 2001, Polymerisation of a single protein of the pathogen Yersinia enterocolitica into needles punctures eukaryotic cells. *Proc. Natl. Acad. Sci. USA* **98**: 4669-4674.

Jin, Q., and He, S.Y., 2001, Role of the Hrp pilus in type III protein secretion in *Pseudomonas syringae*. *Science* **294**: 2556-2558.

Jin, Q., Hu, W., Brown, I., McGhee, G., Hart, P., Jones, A.L. and He, S.Y., 2001, Visualisation of secreted Hrp and Avr proteins along the Hrp pilus during type III secretion in *Erwinia amylovora* and *Pseudomonas syringae*. *Mol. Microbiol.* **40**: 1129-1139.

Knutton, S., Rosenshine, I., Pallen, M.J., Nisan, I., Neves, B.C., Bain, C., Wolff, C., Dougan, G., and Frankel, G., 1998, A novel EspA-associated surface organelle of enteropathogenic *Escherichia coli* involved in protein translocation into epithelial cells. *EMBO J.* **17**: 2166-2176.

Kubori, T., Matsushima, Y., Nakamura, D., Uralil, J., Lara-Tejero, M., Sukhan, A., Galan, J.E., and Aizawa, S.I., 1998, Supramolecular structure of the Salmonella typhimurium type III protein secretion system. *Science* **280**: 602-605.

Leach, J.E., and White, F.F., 1996, Bacterial avirulence proteins. *Annu. Rev. Phytopathol.* **34**: 153-179.

Li, C.-M., Brown, I., Mansfield, J., Stevens, C., Boureau, T., Romantschuk, M., and Taira, S., 2002, Hrp pilus of *Pseudomonas syringae* elongates from its tip and acts as a conduit for translocation of the effector protein HrpZ. *EMBO J.* **21**: 1909-1915

Misra, T.K., Brown, N.L., Fritzinger, D., Pridmore, R., Barnes, W., and Silver, S., 1984, Mercuric ion-resistance operons of plasmid R100 and transposon Tn*501*: the beginning of the operon including the regulatory region and the first two structural genes. *Proc. Natl. Acad. Sci. U.S.A.* **81**: 5975-5979.

Roine, E., Wei, W., Yuan, J., Nurmiaho-Lassila, E.L., Kalkkinen, N., Romantschuk, M., and He, S.Y., 1997a, Hrp pilus: an hrp-dependent bacterial surface appendage produced by *Pseudomonas syringae* pv. *tomato* DC3000. *Proc. Natl. Acad. Sci. U S A* **94**: 3459-3464.

Roine, E., Saarinen, J., Kalkkinen, N., and Romantschuk, M., 1997b, Purified HrpA of Pseudomonas syringae pv. tomato DC3000 reassembles into pili. *FEBS Lett.* **417**: 168-172.

Taira, S., Tuimala, J., Roine, E., Nurmiaho-Lassila, E.L., Savilahti, H., and Romantschuk, M., 1999, Mutational analysis of the *Pseudomonas syringae* pv. *tomato hrpA* gene encoding Hrp pilus subunit. *Mol. Microbiol.* **34**: 737-744.

Wei, W., Plovanich-Jones, A., Deng, W.-L., Jin, Q.-L., Collmer, A., Huang, H.-C. and He, S.Y., 2000, The gene coding for the Hrp pilus structural protein is required for type III secretion of Hrp and Avr proteins in *Pseudomonas syringae* pv. *tomato*. *Proc. Natl. Acad. Sci.USA* **97**: 2247-2252.

Pseudomonas syringae Pathogenesis in *Arabidopsis*

S.Y. HE, S. BANDYOPADHYAY, E. BRAY, Q.-L. JIN, P. HAUCK, K. NOMURA, A. PLOVANICH-JONES, R. THILMONY, and J. ZWIESLER-VOLLICK
Department of Energy Plant Research Laboratory and Department of Plant Biology, Michigan State University, East Lansing, MI 48824, USA.

Abstract How microbial pathogens cause diseases in higher eukaryotes is a major unresolved question in biology. Our research is focused on the interaction between *Arabidopsis thaliana* and *Pseudomonas syringae* pv. *tomato* strain DC3000. In this model interaction, both the host and the pathogen are genetically and genomically amenable, making it an excellent system for dissecting the molecular details of bacterial virulence, host response, and host-pathogen co-evolution. Elucidation of how *P. s.* pv. *tomato* DC3000 manipulates the cellular functions of a susceptible *Arabidopsis* plant is the subject of our research. Here, we report our characterisation of the Hrp pilus, which is required for type III secretion of *P. s.* pv. *tomato* DC3000 effector proteins into plant cells, and the effects of type III effectors on *Arabidopsis* physiology.

Key words: Hrp pilus, disease resistance, type III protein secretion, microarray.

Author for correspondence: S.Y. He; e-mail: hes@msu.edu.

1. INTRODUCTION

Study of microbial modification of host cellular processes is an emerging field that promises to provide exciting opportunities for revealing fundamental cellular mechanisms in eukaryotic hosts. Despite impressive progress in the study of plant resistance response to avirulent pathogens (Dangl and Jones, 2001), our understanding of the molecular basis of plant susceptibility to virulent pathogens is in its infancy. A critical step during virulent infection of most Gram-negative bacterial pathogens is the injection

N.S. Iacobellis *et. al.* (eds.), *Pseudomonas syringae* and Related Pathogens, 293-300.

of a collection of "type III" effector proteins into plant or animal cells via the type III protein secretion system (He, 1997, 1998; Galan and Collmer, 1999; Staskawicz *et al.*, 2001). Type III effector proteins are believed to interact with key signalling or metabolic proteins in the host cell to suppress (or evade) host defence and, in the case of plant pathogenic bacteria, to cause host cells to release nutrients and water into the apoplast, where bacteria proliferate. Therefore, bacterial type III effector proteins, if their functions are clearly understood, are unique tools in the study of fundamental cellular mechanisms in the plant.

At present, how plant pathogenic bacteria inject type III effectors across the plant cell wall and the plasma membrane is not understood. Nor do we know the complete list of type III effector proteins in any given bacterial strain, or the cellular processes in the plant that are perturbed by the type III effector proteins. Martin Romantsuck's laboratory and our lab previously identified a novel bacterial pilus (the Hrp pilus) assembled during type III secretion in *Pseudomonas syringae* pv. *tomato* strain DC3000. We recently (i) obtained electron microscopic results showing extrusion of AvrPto from the tip of the Hrp pilus during type III secretion; (ii) analysed the secretion properties and biological function of a type III effector, VirPtoB, which is required for DC3000 to infect *Arabidopsis*; (iii) identified new type III effector proteins based on a functional genomic analysis of the DC3000 genome; and (iv) analysed the expression of 309 *Arabidopsis* genes that are affected by DC3000 infection.

2. THE TYPE III PROTEIN SECRETION SYSTEM IN *PSEUDOMONAS SYRINGAE* PV. *TOMATO* DC3000

The type III secretion system of *Pseudomonas syringae* pv. *tomato* DC3000 has been shown to secrete the many effector proteins (Yuan and He, 1996; Roine *et al.*, 1997; Charkowski *et al.*, 1998; van Dijk *et al.*, 1999; Wei *et al.*, 2000; Guttman *et al.*, 2002; Petnicki-Ocwieja *et al.*, 2002) and assembles a novel pilus named the Hrp pilus (Roine *et al.*, 1997). Our work on type III secretion in *P. syringae* has been focused mainly on the structure and function of the Hrp pilus and the HrpA protein. We showed that the HrpA protein is co-purified with the Hrp pilus in culture (Roine *et al.*, 1997), that mutation in the *hrpA* gene blocks the assembly of Hrp pili (Roine *et al.*, 1997), and that a HrpA antibody densely labels the entire Hrp pilus (Hu *et al.*, 2001). Based on these results, we concluded that HrpA is a structural protein of the Hrp pilus (Roine *et al.*, 1997; Hu *et al.*, 2001). The *hrpA* mutant was unable to cause disease in otherwise susceptible *Arabidopsis* and tomato plants or to elicit the hypersensitive resistance response (HR) in the resistant plants (Roine *et al*., 1997). Therefore, the *hrpA* gene is essential for DC3000 interaction with both susceptible and resistant plants. Genetic analysis suggested that the *hrpA* gene is required directly or indirectly for secretion of type III effector proteins and maximal expression of *hrp* and *avr* genes in *P. s.* pv. *tomato* DC3000 (Wei *et al*., 2000), providing a molecular explanation of the requirement of the *hrpA* gene for *P. s.* pv. *tomato* DC3000 interaction with plants.

Pili have also been shown to be required for type IV secretion of DNA and proteins in *Agrobacterium tumefaciens* (Fullner *et al*., 1996) and bacterial conjugation (Willetts and Skurray, 1980). However, it has never been shown directly that the secreted proteins and DNA are localised along or inside the pili, a key prediction of the "conduit/guiding filament" model, in which the pilus directly guides protein or DNA traffic. We performed in situ immunogold labeling of HrpW and AvrPto during active type III secretion and found that both HrpW and AvrPto were localised along the entire length of the Hrp pilus (Jin *et al*., 2001). The localisation of HrpW and AvrPto along the Hrp pilus is specific because these effectors are not found randomly in the intercellular space or along the flagellum. The Hrp pilus-specific localisation is also observed for two type III effectors, HrpN and DspA/E, of *Erwinia amylovora* (the causal agent of fire blight disease of apple and pear) (Jin *et al*., 2001). These results provide direct evidence that type III secretion occurs only at the site of Hrp pilus assembly, that secretion of effector proteins occurs during construction of the Hrp pilus, and that the Hrp pilus apparently guides the transfer of effector proteins outside the bacterial cell, favouring the 'conduit/guiding filament' model (Jin *et al*., 2001) The same conclusion was reached in the laboratories of John Mansfield and Martin Romantschuk (Brown *et al*., 2001)

The deposition of effector proteins along a continuously growing Hrp pilus occurs because in wild-type DC3000 cells expression of effector proteins and Hrp pilus assembly genes (*hrp*/*hrc* genes) is co-ordinately regulated. Deposition of effector proteins along the entire length of the Hrp pilus made it impossible to conclude whether effector proteins exit from the tip (the conduit model) or the base (the 'guiding filament' model) of a continuously growing pilus (Jin *et al*., 2001). To resolve the two type III secretion models, we made a critical modification of the previously described in situ immunogold labeling method (Brown *et al*., 2001; Jin *et al*., 2001) by incorporating a step to uncouple the synthesis of effector proteins from the Hrp pilus assembly before immunogold labeling. We reasoned that if the Hrp pilus serves as a conduit, the newly secreted effector proteins should accumulate near the tip of a sub-population of Hrp pili whose assembly begins before effector expression is induced. In contrast, if the Hrp pilus serves as a guiding filament, the effector proteins should accumulate at the base of the same sub-population of Hrp pili. Our immunogold labeling results with the AvrPto protein showed that the majority of newly made

AvrPto was deposited near the tip of the Hrp pilus, supporting the conduit model (Jin and He, 2002). Similar results were obtained in the laboratories of John Mansfield and Martin Romantschuk (Li *et al*., 2002).

3. ORF3$_{CEL}$: OVERCOMING SALICYLIC ACID(SA)-MEDIATED ARABIDOPSIS HOST DEFENSE?

In mammal-bacterium interactions, increasing evidence suggests that targets of type III effector proteins are the MAP kinase pathway, delivery of reactive oxygen-generating enzymes, ubiquitin-like molecules, and actin cytoskeleton (Galan and Collmer, 1999; Staskawicz *et al*., 2001). The plant "susceptibility" targets, however, are largely unknown. A successful plant pathogenic bacterium must produce defence suppressors and possibly water and nutrient releasers that aid its proliferation in the apoplast of a susceptible plant. One way to identify the defence-suppressor type of type III effectors is to examine the growth of *Pseudomonas syringae* pv. *tomato* DC3000 effector mutants in *Arabidopsis* plants that are compromised in defence, e.g., in *nahG* plants, which are deficient in the SA-mediated pathogen defence (Delaney *et al*., 1994). If a bacterial type III effector mutant is deficient in only defence suppressors, but retains other necessary positive virulence factors, it would be expected to grow and cause disease in *nahG* plants. In contrast, if a bacterium (e.g., a *hrp* mutant or any saprophytic bacterium) is deficient in both defence suppressors and water/nutrient releasers, it would not be expected to grow efficiently in *nahG* plants because of lack of factors to release water and nutrients from the host. Recently, a DC3000 type III effector mutant carrying a large deletion (Δ*avrE-avrF-orf3-orf4-hrpW-orf5*; called ΔCEL hereafter) in the conserved effector locus (CEL) was constructed (Alfano *et al*., 2000). This mutant has severely reduced virulence in wild-type tomato (Alfano *et al*., 2001) and *Arabidopsis* (Thilmony and He, unpublished results). The ΔCEL mutant has an interesting phenotype in *nahG* plants. Specifically, this mutant multiplied almost like *P. s.* pv. *tomato* DC3000 (a 400-fold increase compared with that in wild-type [ecotype Columbia] plants) and caused disease symptoms in *nahG* plants. In comparison, *P. s.* pv. *tomato* DC3000 and its *hrpH* mutant (Yuan and He, 1996) multiplied only about 5-fold more in *nahG* plants than in Columbia plants at this inoculum level (OD_{600} = 0.001). We also compared the growth of a non host bacterium, *P. s.* pv. *glycinea* race 4, in *nahG* and Columbia plants. This bacterium infects soybean, but not the wild-type Columbia plants. However, it multiplied about 10 times more in *nahG* plants compared with that in Columbia plants, suggesting that SA-mediated defence plays a role (albeit rather minor) in preventing it from multiplying in *Arabidopsis*. However, unlike the ΔCEL mutant, *P. s.* pv. *glycinea* must lack one or more critical positive virulence factors that are adapted to *Arabidopsis* to become fully virulent in *Arabidopsis* plants.

The restoration of virulence of the ΔCEL mutant in *nahG* plants suggests that the deleted effector genes (*avrE*, *avrF*, *orf3*, *orf4*, *hrpW*, and *orf5*) are dispensable for DC3000 infectivity in defence-compromised host plants. Through a series of genetic complementation analysis with *P. s.* pv. *tomato* DC3000 mutants and subclones of the CEL region, we have now determined that *orf3* and *orf4* are responsible for the CEL deletion phenotype. Further type III secretion assays showed that Orf3 is targeted into *Arabidopsis* cells and Orf4 is the cognate chaperone that is required for the translocation of Orf3 into *Arabidopsis* cells. Because of the clear demonstration of Or3 and Orf4 in DC3000 virulence in *Arabidopsis*, we have tentatively renamed Orf3 and Orf4 VirPtoB and VirPtoC, respectively. Our current effort is focused on identifying the protein targets and pathways in *Arabidopsis* that are modulated by VirPtoB.

4. IDENTIFICATION OF NEW HRP-REGULATED GENES IN *PSEUDOMONAS SYRINGAE* PV. *TOMATO* DC3000

The recent release of the DC3000 genomic sequence by The Institute of Genomic Research, Rockville, MD, has made it possible to estimate the total number of potential type III effector genes in this pathogen. In *Pseudomonas syringae*, all known *avr* genes are co-regulated with *hrp* genes owing to the presence of a conserved 'hrp box' in the promoter region (Innes *et al.*, 1993; Xiao *et al.*, 1994). We used the 'hrp box' motif as a search tool and identified 73 genes/operons. Among these genes/operons are all known *hrp*-regulated genes/operons in DC3000 and several *avr* genes known to be present in other *P. syringae* strains (*avrPto*, *avrE*, *virPphA*, *avrPpiB*, *avrPpiC*, *avrPphD*, *avrXv3*, and *hopPsyA/hrmA*).

To identify new type III effectors, we conducted microarray, RNA gel blotting, and AvrRpt2-based type III translocation assays (Mudgett *et al.*, 2000) of the 73 genes. From these analyses, we found eight new *hrp*-regulated genes, which we named *chp1-8* (Zwiesler-Vollick *et al.*, 2002). All eight Chp proteins are hydrophilic. The G+C contents of *chp6* (48%) and *chp8* (43%) are significantly lower than the G+C content of the *P. s.* pv. *tomato* DC3000 genome (about 60%; Alfano *et al.*, 2000), indicating the possibility of recent introduction of these genes by horizontal gene transfer. Chp1, Chp3, and Chp4 have no similarity to known proteins, whereas the C-terminus of Chp2 has significant similarity to the DnaJ family of proteins (BLAST score 72; E value 2.0E-12; 72% similar over 63 aa). Chp5 has a putative transglycosylase SLT domain (PF01464) and shares 25% identity with the N-terminal 259 amino acids of HrpW (424 amino acids total). Chp6 shows sequence similarity to Orf1 of the *avrPphF* locus in *P. s.* pv. *phaseolicola* (Tsiamis *et al.*, 2000). Chp7 shows similarity to proteins of the ApbE family (PF02424), some of which are involved in thiamine biosynthesis. Chp8 shows sequence similarity to several hypothetical proteins with the GGDEF domain. The most significant similarity is to a hypothetical 91.8-kD protein, AGR_L_1027p, of *Agrobacterium tumefaciens.* Several GGDEF domain (PF00990) proteins may possess diguanylate cyclase activity, modulating cyclic diguanylic acid levels in the cell (Zwiesler-Vollick *et al.*, 2002). Chp8 also contains the Pfam EAL domain (PF00563), which has no known function. No specific plant organelle-targeting sequences are present in these proteins.

5. *ARABIDOPSIS* GENE EXPRESSION SIGNATURE OF DC3000 INFECTION

A major roadblock to rapid progress in the study of the molecular basis of host susceptibility and functions of type III virulence proteins is lack of host cellular and molecular markers (in contrast to plant resistance). To identify these molecular markers, we examined the expression of about 11,000 *Arabidopsis* cDNA clones (representing about 8,000 unique sequences) from the *Arabidopsis* Functional Genomic Consortium (AFGC) during infection by *Pseudomonas syringae* pv. *tomato* DC3000 vs. by the *hrpS* mutant, which is defective in type III secretion (Yuan and He, 1996) and coronatine production (unpublished results). In four independent experiments, 309 genes reproducibly showed differential expression.

The most notable *P. s.* pv. *tomato* DC3000-induced genes in our current slide include, besides many (about 50%) novel genes, jasmonic acid (JA)-responsive genes; known defense genes; and genes encoding sugar transporters, sugar-sensing proteins, putative transcription factors, signal-transduction components, chlorophyll-degrading enzymes, and cell-wall-modifying enzymes. The induction of these genes is consistent with the fact that *P. s.* pv. *tomato* DC3000 (1) produces a JA-mimicking toxin, coronatine; (2) presumably releases sugars from the host cell; and (3) causes tissue chlorosis symptoms. The repressed genes include a family of the water channel genes, several defence genes, and genes encoding sugar-sensing proteins. The repression of these genes is likely related to *P. s.* pv. *tomato* DC3000-caused defence suppression, water trapping in the apoplast ('water soaking' symptom), and sugar production and release. Therefore, we believe that the observed changes in the expression of the 309 host genes represent a molecular signature of DC3000 infection in susceptible *Arabidopsis* plants.

ACKNOWLEDGEMENTS

Pseudomonas syringae pv. *tomato* DC3000 sequence data was obtained from The Institute for Genomic Research website at http://www.tigr.org/tdb/mdb/mdbinprogress.html. Sequencing of the *P. s.* pv. *tomato* DC3000 genome was accomplished with support from the National Science Foundation, as part of the tomato-*Pseudomonas syringae* functional genomics project led by Alan Collmer at Cornell University (co-PIs: Jim Alfano, Robin Buell, Samuel Cartinhour, Arun K. Chatterjee, Terrence P. Delaney, Sondra G. Lazarowitz, Gregory B. Martin, and Xiaoyan Tang). This work was supported by grants from the US Department of Agriculture, National Science Foundation, and Department of Energy. We thank Barbara Kunkel, Alan Collmer, John Mansfield, and Martin Romantschuk for their support and collaborations.

REFERENCES

Alfano, J.R., Charkowski, A.O., Deng, W.L., Badel, J.L., Petnicki-Ocwieja, T., van Dijk, K., and Collmer, A., 2000, The *Pseudomonas syringae* Hrp pathogenicity island has a tripartite mosaic structure composed of a cluster of type III secretion genes bounded by exchangeable effector and conserved effector loci that contribute to parasitic fitness and pathogenicity in plants. *Proc. Natl. Acad. Sci. USA* **97**: 4856-4861.

Brown, I.R., Mansfield, J.W., Taira, S., Roine, E., and Romantschuk, M., 2001, Immunocytochemical localisation of HrpA and HrpZ supports a role for the Hrp pilus in the transfer of effector proteins from *Pseudomonas syringae* pv. *tomato* across the host plant cell wall. *Mol. Plant-Microbe Interact.* **14**: 394-404.

Dangl, J.L., and Jones, J.D.G., 2001, Plant pathogens and integrated defence responses to infection. *Nature* **411**: 826-833.

Fullner, K.J., Lara, J.C., and Nester, E.W., 1996, Pilus assembly by *Agrobacterium* T-DNA transfer genes. *Science* **273**: 1107-1109.

Galan, J.E., and Collmer, A., 1999, Type III secretion machines: Bacterial devices for protein delivery into host cells. *Science* **284**: 1322-1328.

Guttman, D.S., Vinatzer, B.A., Sarkar, S.F., Ranall, M.V., Kettler, G., and Greenberg, J.T. 2002, A functional screen for the type III (Hrp) secretome of the plant pathogen *Pseudomonas syringae*. *Science* **295**: 1722-1726.

He, S.Y., 1997, Hrp-controlled interkingdom protein transport: learning from flagellar assembly? *Trends Microbiol.* **5**: 489-495.

He, S.Y., 1998, Type III secretion systems in animal and plant pathogenic bacteria. *Ann. Rev. Phytopathol.* **36**: 363-392.

Hu, W., Yuan, J., Jin, Q. L., Hart, P., and He, S.Y., 2001, Immunogold labeling of Hrp pili of *Pseudomonas syringae* pv. *tomato* assembled in minimal medium and in planta. *Mol. Plant-Microbe Interact.* **14**: 234-241.

Innes, R., Bent, A., Kunkel, B., Bisgrove, S., and Staskawicz, B., 1993, Molecular analysis of avirulence gene *avrRpt2* and identification of a putative regulatory sequence common to all known *Pseudomonas syringae* avirulence genes. *J. Bacteriol.* **175**: 4859-4869.

Jin, Q.–L., and He, S.Y., 2001, Role of the Hrp pilus in type III secretion in *Pseudomonas syringae*. *Science* **294**: 2556-2558.

Jin, Q.-L., Hu, W., Brown, I., McGhee, G., Hart, P., Jones, A., and He, S. Y., 2001, Visualization of effector Hrp and Avr proteins along the Hrp pilus during type III secretion in *Erwinia amylovora* and *Pseudomonas syringae*. *Mol. Microbiol.* **40**: 1129-1139.

Jin, Q.-L., Hu, W., Brown, I., McGhee, G., Hart, P., Jones, A., and He, S.Y., 2002, Secretion of Hrp and DspE proteins along the Hrp pilus during type III secretion in *Erwinia amylovora*. *Acta Hort.*, in press.

Mudgett, M.B., Chesnokova, O., Dahlbeck, D., Clark, E. T., Rossier, O., *et al.*, 2000, Molecular signals required for type III secretion and translocation of the *Xanthomonas campestris* AvrBs2 protein to pepper plants. *Proc. Natl. Acad. Sci. USA* **97**: 13324-13329.

Petnicki-Ocwieja, T., Schneider, D.J., Tam, V.C., Chancey, S.T., Shan, L., Jamir, Y., Schechter, L.M., Janes, M.D., Buell, C.R., Tang, X., Collmer, A., Alfano, J.R. 2002, Genomewide identification of proteins secreted by the Hrp type III protein secretion system of *Pseudomonas syringae* pv. *tomato* DC3000. *Proc. Natl. Acad. Sci. USA* **99**: 7652-7657.

Roine, E., Wei, W., Yuan, J., Nurmiaho-Lassila, E.-L., Kalkkinen, N., Romantschuk, M., and He, S.Y., 1997, Hrp pilus: An hrp-dependent bacterial surface appendage produced by *Pseudomonas syringae* pv. *tomato* DC3000. *Proc. Natl. Acad. Sci. USA* **9**: 3459-3464.

Staskawicz, B.J., Mudgett, M.B., Dangl, J.L., and Galan, J.E., 2001, Common and contrasting themes of plant and animal diseases. *Science* **292**: 2285-2289.

Tsiamis, G., Mansfield, J.W., Hockenhull, R., Jackson, R.W., Sesma, A., Athanassopoulos, E., Bennett, M.A., Stevens, C., Vivian, A., Taylor, J.D., and Murillo, J., 2000, Cultivar-specific avirulence and virulence functions assigned to *avrPphF* in *Pseudomonas syringae* pv. *phaseolicola*, the cause of bean halo-blight disease. *EMBO J.* **19**: 3204-3214.

van Dijk, K., Fouts, D.E., Rehm, A.H., Hill, A.R., Collmer, A., and Alfano, J.R., 1999, The Avr (effector) proteins HrmA (HopPsyA) and AvrPto are secreted in culture from *Pseudomonas syringae* pathovars via the Hrp (type III) protein secretion system in a temperature- and pH-sensitive manner. *J. Bacteriol.* **181**: 4790-4797.

Wei, W., Plovanich-Jones, A., Deng, W.-L., Jin, Q.-L., Collmer, A., Huang, H.-C., and He, S. Y., 2000, The gene coding for the structural protein of the Hrp pilus is required for type III secretion of Hrp and Avr proteins in *Pseudomonas syringae* pv. *tomato*. *Proc. Natl. Acad. Sci. USA* **97**: 2247-2252.

Willetts, N., and Skurray, R., 1980, The conjugation system of F-like plasmids. *Annu. Rev. Genet.* **14**: 41-76.

Xiao, Y., and Hutcheson, S., 1994, A single promoter sequence recognised by a newly identified alternate sigma factor directs expression of pathogenicity and host range determinants in *Pseudomonas syringae*. *J. Bacteriol.* **176**: 3089-3091.

Yuan, J., and He, S.Y., 1996, The *Pseudomonas syringae hrp* regulation and secretion system controls the production and secretion of multiple extracellular proteins. *J. Bacteriol.* **178**: 6399-6402.

Zwiesler-Vollick, J., Plovanich-Jones, A., Nomura, K., Bandyopadhyay, S., Joardar, V., Kunkel, B.N., and He, S.Y., 2002, Identification of novel *hrp*-regulated genes through functional genomic analysis of the *Pseudomonas syringae* pv. *syringae* DC3000 genome. *Mol. Microbiol.* **45**: 1207–1218.

Early Induced Resistance, a General, Symptomless Plant Response to Bacteria

Z. KLEMENT, Z. BOZSÓ, E. BESENYEI, A. CZELLENG, M.L. KECSKÉS and P.G. OTT

Plant Protection Institute of the Hungarian Academy of Sciences, Herman Ottó Street 15, H-1022 Budapest, Hungary.

Abstract: Many micro-organisms including pathogenic and saprophytic bacteria react with plant cells in the intercellular spaces inducing different defence responses. The local Early Induced Resistance (EIR) is a first line defence mechanism against bacteria. Here an overview will be given of this local, non-specific, symptomless defence mechanism as a separate entity from the incompatible-specific Hypersensitive Response (HR). The EIR operates 1-6 h after inoculation (hpi) for about one day depending on temperature and leaf age. The EIR can be inhibited by a short heat shock (50°C for 15 sec) of leaves or by a plant protein synthesis inhibitor, cycloheximide (5 μg ml^{-1}). In a compatible host-pathogen relationship (*Pseudomonas syringae* pv. *tabaci*/tobacco) the effect of EIR does not eventuate. However, the EIR develops simultaneously with the HR and sometimes is able to prevent it when the induction time of HR is longer than the time required for the development of the EIR (e.g. *P. s.* pv. *phaseolicola* does not induce HR in tobacco above 28°C). It seems that the EIR inhibits the metabolism of bacteria and the activity of *hrp* genes. Moreover, EIR activates the accumulation of H_2O_2 at the bacterial attachment site expressing new peroxidase isoenzymes in the initiated plant tissue. Further investigations, hopefully, will clarify the relationship of other complementary defence mechanisms like local late induced resistance (LIR) examined by Sequeira (1983) and Mazzucchi and co-workers (1979), Minardi (1995), Newman *et al.*, (2001).

Key words: local early induced resistance (EIR), bacterially induced plant responses, inhibition of hypersensitive response, interaction of host responses.

Author for correspondence: Z. Klement; e-mail: zklem@nki.hu.

N.S. Iacobellis *et. al.* (eds.), *Pseudomonas syringae* and Related Pathogens, 301-309.

1. INTRODUCTION

It is well known that plants can recognise pathogenic micro-organisms but it is less evident that plants perceive non-pathogenic bacteria as well. Many micro-organisms landing on the plant surface enter the intercellular spaces of mesophyll via open stoma, hydathodes or wounds. This is, where the plant cell surface reacts direct with the bacterial cells. Plant pathogenic bacteria either cause disease in susceptible hosts or they can induce a rapid hypersensitive necrosis (HR) in resistant cultivars or in non-host plants (Klement, 1982). However, not only pathogenic but also non-pathogenic bacteria, such as saprophytes, *hrp* or *avr* mutants and opportunistic pathogens, are recognised by plant cells, inducing local non-specific symptomless defence reactions like the local early induced resistance (EIR) (Klement *et al*., 1999; Fig. 1). The EIR is the first-line defence against bacteria apart from the incompatible pathogen-specific hypersensitive response (HR) (Klement *et al*., 1964). It is tempting to speculate that without

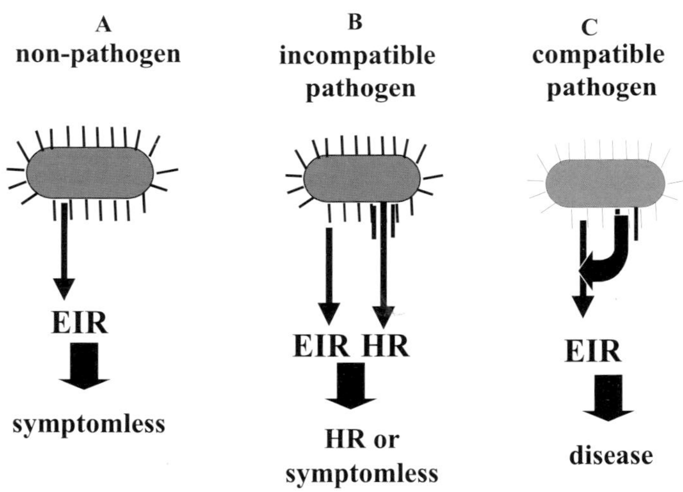

Figure 1. Defence reactions induced in different types of plant-pathogen interactions. A) If non-plant pathogenic bacteria enter the intercellular spaces of a plant tissue, only the EIR is activated by bacterial surface factors. Because the HR is not triggered there is no plant cell death and the plant tissue remains symptomless. B) In the cases of infection by incompatible bacteria both the EIR and HR may be induced. The EIR is induced by bacterial surface molecules as in the case of non-pathogenic bacteria and the HR is triggered by type III system secreted Avr protein(s). The results of this interactions may be HR or a symptomless response depending on the speed of the development of two defence reactions. C) In the case of a compatible situation the EIR is induced by bacterial surface molecules but the bacteria can inhibit the development of the EIR by injecting Vir proteins into plant cells by type III system.

this early common response plant tissues would provide nutrients for micro-organisms. Under natural conditions, presumably, a similar process starts in autumn in senescent leaves when EIR no longer works allowing the invasion of saprophytes. Similarly, at colder temperatures when EIR does not operate some opportunistic pathogens like *Pseudomonas syringae* pv. *syringae* may invade some plants causing serious diseases.

In the past decade we have shown that the earlier described local induced resistance (Lovrekovich and Farkas, 1963; Sequeira and Hill, 1974) is not a single and uniform response to bacteria, because an early (EIR) and a late form (LIR) can be distinguished. This paper is an overview of these investigations in connection with the EIR.

2. RESULTS AND DISCUSSION

2.1 Induction of EIR

The EIR can be activated in tobacco leaves by a wide range of bacteria, such as living and killed cells of pathogenic and saprophytic bacteria as well as *hrp*/*avr* mutants (Fig. 1A). This wide range of bacteria has some common features, e.g. surface components such as lipopolysaccharides (LPS) (Dow *et al.*, 2000; Kecskés *et al.*, unpublished), LPS-protein complex (Mazzucchi *et al.*, 1979) and flagellin (Gomez-Gomez and Boller, 2002), can probably elicit this common non-specific local induced resistance. At present it is not clear which component of bacterial surface plays the main role in the induction of EIR.

2.2 Detection of EIR

Since EIR is a symptomless defence mechanism, its detection is not as simple as that of the HR. Moreover, EIR presumably only inhibits the metabolism of bacteria but does not kill them. To overcome these difficulties we use an indirect method for the detection of EIR. The essence of this method is the fact that active metabolism of the pathogen is required for the induction of HR (Klement *et al.*, 1999) and that the EIR is able to inhibit the expression of *hrp* genes during the induction period of HR. (Bozsó *et al.*, 1997). Therefore, the lack of HR caused by the challenge pathogen indicates the effect of the EIR in the pre-inoculated plant tissue. Recently, we have followed the development of EIR by a peroxidase-dependent diamino-benzidine staining reaction (Bozsó *et al.*, 2001), which is also suitable for the detection of EIR.

2.3 Influence of Environmental Factors on the Development of EIR

The time required for the development of EIR greatly depends on the temperature namely 1÷2 h at 30°C and 6÷8 h at 15°C. At temperatures below 15°C the effectiveness of EIR rapidly decreases and at 5°C it does not develop until 48 h after inoculation (hpi) (Klement *et al.*, 2003). The decrease in the effectiveness of EIR during a quick cooling down period may be a determinative factor in the appearance of certain epidemics. EIR develops faster in younger tobacco leaves (2÷3 hpi) than in older ones (4÷6 hpi). In fully developed middle standing leaves EIR developed moderately. In yellowing senescent leaves EIR was no longer detectable (Klement *et al.*, 1999). It has already been established that the EIR does not require illumination but the LIR is strongly light-dependent (Burgyán and Klement, 1979).

2.4 Inhibition of EIR

The development of EIR can be blocked by a short heat-treatment of plant leaves (50°C for 15 sec) just before inoculation, or by injecting the eukaryotic protein synthesis inhibitor cycloheximide (5 $\mu g\ ml^{-1}$) together with the bacterial suspension (Bozsó *et al.*, 1999).

2.5 Inhibition of Bacterial Growth by EIR

Non-pathogenic bacteria or *hrp* and *avr* mutants usually do not increase their population in the long run in plant tissue, but when the EIR is inhibited, they persist, indicating that EIR is one of the defence mechanisms which are responsible for the inhibition of bacterial growth in the plant. We could not isolate any antibacterial compound from inoculated leaves that could be related to the EIR. It seems from our experiments that the EIR inhibits the metabolism of bacteria and the activity of *hrp* genes of pathogens (Bozsó *et al.*, 1997).

2.6 Plant Processes Associated with EIR

H_2O_2 is a possible direct and indirect antibacterial factor known to be produced by plant cells upon infection. Pre-treatment of leaves with the *hrpL* mutant of *Pseudomonas syringae* pv. *phaseolicola* – a mutant that only induces EIR but not HR - conditioned a very fast but moderate H_2O_2 response to challenge bacteria throughout the whole cell wall (Ott, 2002).

In the presence of H_2O_2, peroxidases react with the dye diamino-benzidine (DAB) to yield a brown polymer in tobacco leaves infiltrated with different bacteria (DAB staining). The intensity of staining correlates in time with the development of the EIR. When the EIR was inhibited either by short heat-treatment of tobacco leaves before bacterial inoculation or by living *P. s.* pv. *tabaci* (compatible relationship) then the DAB staining did not develop. The appearance of new peroxidase isoenzymes in the intercellular washing fluid could be detected after injection of EIR-inducing bacteria. Plant peroxidase genes were cloned from tobacco and *Arabidopsis*, which became activated during EIR. The transcription of tobacco peroxidase gene was induced both by incompatible and saprophytic bacteria and blocked by compatible *Pseudomonas* spp. but was slightly induced by water infiltration. The appearance of peroxidase proteins and its transcription activation may be a useful marker for detailed study of EIR. Accumulation of H_2O_2 and peroxidase may lead to fortification of plant cell wall at the attachment site of the bacteria and H_2O_2 may cause direct damage in bacterial cells (Bozsó *et al.*, 2001).

2.7 Immune Type Response Due to EIR

In certain cases plants inoculated with incompatible pathogens remain symptomless instead of showing the HR (Fig. 1B). For instance, when tobacco leaves are inoculated with *Pseudomonas syringae* pv. *phaseolicola* at temperatures higher than 26÷28°C the HR necrosis does not appear. The nature of this "immune" (symptomless) response was not clear earlier. According to our experiments, in certain host-pathogen relationships after a single inoculation of tobacco with *P. s.* pv. *phaseolicola* at above 26°C the EIR develops before the bacterial cells are able to induce the HR (Klement *et al.*, 1999). In these cases the induction of HR and the quick necrosis does not occur. Accordingly, when the HR induction-time of *P. s.* pv. *pisi* was extended by chloramphenicol treatment of the inoculum, it did not cause the HR at 30°C, as it does normally. However, when the EIR was inhibited the HR appeared in all experiments (Klement *et al.*, 1999). Generally, the immune response depends on two factors: 1) a quick development of EIR preceding the HR-induction and/or 2) an extended HR-induction time of the pathogen preceding the development of EIR. The first factor is influenced by temperature during inoculation and by the age of the inoculated leaf. The second factor depends on the HR-induction time of the bacterium species or strain. The EIR-mediated inhibition of bacterial cells during the HR induction period was also confirmed by the finding that harpin P_{SS} was able to cause tissue necrosis in the presence of the EIR (Klement *et al.*, 1999).

2.8 Inhibition of EIR in Compatible Relationships

In our earlier experiment we have shown that the *hrpK* mutant of *Pseudomonas syringae* pv. *syringae 61* causes HR when the EIR was previously inhibited (Bozsó *et al.*, 1999). A similar result was obtained when tobacco leaves were pre-treated with 5 x 10^5 *P. s.* pv. *tabaci* cells ml^{-1} (compatible relationship) and challenged with the 10^8 cells ml^{-1} *hrpK* mutant one day later. In this case in the *P. s.* pv. *tabaci* pre-treated leaves the *hrpK* mutant induced HR, similarly to the tissues when the EIR was blocked with protein synthesis inhibitors. However, water or heat-killed *P. s.* pv. *tabaci* did not cause this effect (Bozsó, 2000).

The population level of seven *hrp* mutants of *P. s.* pv. *syringae 61* was measured. The living cell number decreased continuously in non-treated or in water pre-treated tobacco leaf tissues. In contrast, when the tobacco leaf was pre-treated with the compatible pathogen one day earlier, the cell number of *hrp* mutants did not decrease, in some cases it even increased. Likewise the transcription level of most *hrp* genes also increased (Klement *et al.*, 2003).

These experiments make some alternative speculations possible. For instance, pre-treatment with the compatible pathogen may cause an activation of K^+H^+ exchange leading to increased extracellular pH, or outflow of nutrients from plant cells. We assume that the compatible bacterium may retard the development of EIR, (Fig. 1C) which normally blocks the HR induced by the *hrpK* mutant. This latter hypothesis seems to be supported by the fact that several EIR-related processes were also inhibited by compatible bacteria (Bozsó, 2000; Klement *et al.*, 2003).

3. CONCLUDING REMARKS

All living organisms have the ability to protect themselves against micro-organisms that try to invade them to use them as a nutrient source.

In mammals the immune system is based on highly specialised cells that are distributed by the circulatory system. In contrast, plants have main defence mechanisms such as EIR and HR that are localised in autonomous plant cells. Plants similarly to mammals have also developed their own defence systems during evolution.

Local early induced resistance (EIR) of plants is the first line of defence against bacteria (Fig. 2.) This non-specific defence mechanism may have developed in the early phase of evolution when plant cells first perceived bacterial surface elements and developed the EIR which protected them from bacteria (Klement *et al.*, 2003). However, certain cells of bacterial populations evolved a selective secretion and a transport system to inject virulence gene products into the invaded host cells to inhibit the effect of EIR and thus to be able to colonise the plant tissue (White *et al.*, 2000).

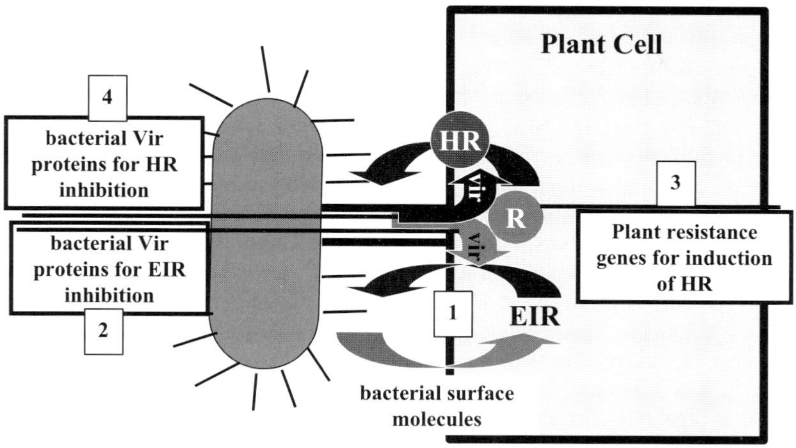

Figure 2. Possible development of defence reactions during evolution. See details in the text.

In response to this, plant cells developed a defence system (R genes) that could react with bacterial *avr* proteins causing the HR defence mechanism which was able to localise the pathogen attack (Klement, 1982).

In the following evolution step the pathogen developed a novel *vir* gene system to inhibit the HR defence. Finally only the virulent pathogens specialised selectively for susceptible cultivars able to cause disease (Jackson *et al.*, 1999).

We think that the first step in the evolution process was the development of the EIR which must be a determinative factor for the plant kingdom. Without this general defence, plants would be unable to exist because their plant tissues would be utilised as a nutrient source by many micro-organisms. Fortunately, plants have developed a common non-specific defence mechanism which lets them survive the continual attack.

ACKNOWLEDGEMENTS

The work of our laboratory was supported by grants from the OTKA TS 040835, OTKA T 038302 and OTKA F 037700.

REFERENCES

Bozsó, Z., 2000, PhD Thesis *Bacterial infection induced early defence reactions and their interactions.* Szent. István University, Gödöllő, Hungary

Bozsó, Z., Ott, P.G., and Klement, Z., 1997, HR-positive phenotype of the *Pseudomonas syringae* pv. *syringae hrpK* mutant and *hrp* gene superinduction in tobacco leaves treated with protein synthesis inhibitors. In Pseudomonas syringae *Pathovars and Related Pathogens* (K. Rudolph, T.J. Burr, J.W. Mansfield, D. Stead, A. Vivian and J. von Kietzell, eds.), Kluwer Academic Publishers, Dordrecht, The Netherlands, pp. 122-126.

Bozsó, Z., Ott, P.G., Kecskés, M.L., and Klement, Z., 1999, Effect of heat and cycloheximide treatment of tobacco on the ability of *Pseudomonas syringae* pv. *syringae* 61 *hrp/hrmA* mutants to cause HR. *Physiol. Mol. Plant Pathol.* **55**: 215-223.

Bozsó, Z., Ott, P.G., Kecskés, M.L., Czelleng, A., and Klement, Z., 2001, Non-specific, peroxidase and H_2O_2 associated reactions of tobacco leaves after infiltration with hrp/hrmA mutants of *P. syringae* pv. *syringae 61.* Proceedings of the 10th International Conference on Plant Pathogenic Bacteria, Charlottetown, Prince Edward Island, Canada, July 23-27, 2000. In *Plant Pathogenic Bacteria* (H. De Boer S.H. ed.), Kluwer Academic Publishers, Dordrecht, The Netherlands, pp.195-198.

Burgyán, J., and Klement, Z., 1979, Early induced selective inhibition of incompatible bacteria in tobacco plants. *Phytopathol. Medit.* **18**: 153-161.

Dow, M., Newman, M.A., and von Roepenack, E., 2000, The induction and modulation of plant defence responses by bacterial lipopolysaccharides. *Ann. Rev. Phytopathol.* **38**: 1-21.

Gomez-Gomez, L., and Boller, T., 2002, Flagellin perception: a paradigm for innate immunity. *Trends Plant Sci.* **7**: 251-256.

Jackson, R.W., Athanassopoulos, E., Tsiamis, G., Mansfield, J.W., Sesma, A., Arnold, D.L., Gibbon, M.J., Murillo, J., Taylor, J.D., Vivian, A., 1999, Identification of a pathogenicity island, which contains genes for virulence and avirulence, on a large native plasmid in the bean pathogen *Pseudomonas syringae* pv. *phaseolicola. Proc. Natl. Acad. Sci.* **96**: 10875-10880.

Klement, Z., 1982, Hypersensitivity. In *Phytopathogenic Prokaryotes II* (M.S. Mount and G.H. Lacy, eds.), Academic Press, New York, USA, pp.149-177.

Klement, Z., Bozsó, Z., Kecskés, M.L., Besenyei, E., Czelleng, A., and Ott, P.G., 2003, Local early induced resistance of plants as the first line of defence against bacteria. *Pest Manage. Sci.* (in press).

Klement, Z., Bozsó, Z., Ott, P.G., Kecskés, M.L., and Rudolph, K., 1999, Symptomless resistant response instead of the hypersensitive reaction in tobacco after infiltration of heterologous pathovars of *Pseudomonas syringae. J. Phytopathol.* **12**: 479-489.

Klement, Z., Farkas, G.l., and Lovrekovich, L., 1964, Hypersensitive reaction induced by phytopathogenic bacteria in the tobacco leaf. *Phytopathology* **54**: 474-477.

Lovrekovich, L., and Farkas, G.L., 1965, Induced protection against wildfire disease in tobacco leaves treated with heat-killed bacteria. *Nature* **205**: 823-824.

Mazzucchi, U., Pupillo, P., 1976, Prevention of confluent hypersensitive necrosis in tobacco leaves by a protein-lipopolysaccharide complex. *Physiol. Plant Pathol.* **9**: 101-102.

Minardi, P., 1995, Altered expression of *Erwinia amylovora hrp* genes in tobacco leaves pretreated with bacterial protein-lipopolysaccharide complexes. *J. Phytopathol.* **143**: 199-205.

Ott, P.G., 2002, PhD Thesis *Characterisation of plant processes of the early induced resistance (EIR) and the hypersensitive reaction (HR) and their interactions.* Szent István University, Gödöllő, Hungary.

Sequeira, L., and Hill, L.M., 1974, Induced resistance in tobacco leaves: The growth of *Pseudomonas solanacearum* in protected tissues. *Physiol. Plant Pathol.* **4**: 447-455.
Sequeira, Z., 1983, Mechanism of induced resistance in plants, fungi, viruses, bacteria. *Ann. Rev. Microbiol.* **37**: 51-79.
White, F.F., Yang, B., Johnson, L.B., 2000, Prospects for understanding avirulence gene function. *Curr. Opin. Plant. Biol.* **3**: 291-298.

Role of Flagella and Flagellin in Plant - *Pseudomonas syringae* Interactions

Y. ICHINOSE, R. SHIMIZU, F. TAGUCHI, K. TAKEUCHI[#], M. MARUTANI, T. MUKAIHARA[§], Y. INAGAKI, K. TOYODA and T. SHIRAISHI

Faculty of Agriculture, Okayama University, Tsushima 1-1-1, Okayama 700-8530, Japan; [#]National Institute of Agrobiological Sciences, Kannondai 2-1-2, Tsukuba 305-8602, Japan; [§]Research Institute for Biological Sciences (RIBS) Okayama, Yoshikawa 7459-1, Kayo-cho, Okayama 716-1241, Japan.

Abstract: The role of flagella and monomer flagellin of *Pseudomonas syringae* pv. *tabaci* in plant-bacteria interactions was investigated by using non-polar *fliC* or *fliD* mutants. These mutants deleted the open reading frames for *fliC* and *fliD*, respectively, and both mutants lost all flagella and motility. The Δ*fliC* mutant did not produce flagellin, whereas Δ*fliD* mutant, that lost HAP2 protein, secreted a large amount of monomer flagellin in the culture medium. Inoculation of tomato leaves with wild type and Δ*fliD* mutant of *P. s.* pv. *tabaci* induced HR, whereas the Δ*fliC* mutant caused symptom-like change and propagated as *P. s.* pv. *tomato*. In tomato suspension cultured cells, wild type *P. s.* pv. *tabaci* induced visible HR-like changes. The Δ*fliC* mutant did not induce the HR, but the response was activated by the Δ*fliD* mutant. The expression of typical defence genes such as *PAL* and *hsr203J* was rapidly and strongly induced by inoculation with the Δ*fliD* mutant compared to inoculation with wild type *P. s.* pv. *tabaci*. On the other hand, both *fliC* and *fliD* mutants were reduced in virulence when inoculated into host tobacco leaves. Furthermore, complementation of *fliC* gene in Δ*fliC* mutant restored motility and HR-inducing ability in tomato, and virulence in tobacco. These results suggest that the monomer flagellin of *P. s.* pv. *tabaci* is an essential factor in the elicitation of HR in non host tomato cells, and flagella are required for complete virulence in host tobacco cells.

Key words: Flagellin, *fliC*, *fliD*, HR Cell Death.

Author for correspondence: Y. Ichinose; e-mail: yuki@cc.okayama-u.ac.jp.

N.S. Iacobellis *et. al.* (eds.), *Pseudomonas syringae* and Related Pathogens, 311-318.

1. INTRODUCTION

Pseudomonas syringae species are Gram-negative, host-specific plant pathogens that cause typical plant reactions. The species are divided into pathogenic variants (pathovars) by their host plant species. For example, *P. s.* pv. *tabaci* is a causal agent for wild fire disease of tobacco, but it does not cause disease on non host tomato plants. Thus there are strict specificities between pathovars and plant species. It is known that pathogenicity and virulence determinants in *P. syringae* include *hrp* (*h*ypersensitive *r*esponse and *p*athogenicity) gene cluster, the production of phytotoxins and exopolysaccharides (Bender *et al.*, 1999; Penaloza-Vazquez *et al.*, 2000). *Hrp* genes control bacterial ability to elicit the HR in non host plants and cause disease (pathogenicity) in host plants. The *hrp* genes encode the components of the type III secretion pathway, regulatory factors for *hrp* gene expression and proteins to elicit the HR in non host plants, that are thought to be secreted through the type III pathway (Cornelis and Van Gijsegem, 2000). Harpin is a gene product of *hrpZ* in *P. syringae*, one of *hrp* gene clusters, and is able to elicit the HR. We have recently isolated *hrpZ* gene from several pathovars of *P. syringae*. Interestingly, *hrpZ* gene in *P. s.* pv. *tabaci* is defective by the internal deletion and frame shift (Taguchi *et al.*, 2001). However, the apparent HR-inducing ability in *P. s.* pv. *tabaci* in its non host tomato plant indicates the existence of novel HR-elicitors other than harpin. Thus we identified that a constituent of flagellar filament, flagellin, in *P. s.* pv. *tabaci* that strongly induces HR accompanying hypersensitive cell death in non host tomato cells (Taguchi *et al.*, 2003b).

Several studies on the characterisation of the elicitor activity of flagellin have been reported for *P. avenae* (*Acidovorax avenae*) (Che *et al.*, 2000) and *P. s.* pv. *tabaci* (Felix *et al.*, 1999). These flagellins are reported to induce plant defence responses in cultured cells of rice (Che *et al.*, 2000), tomato or *Arabidopsis thaliana* (Felix *et al.*, 1999). A 22 amino acid peptide (flg22) of a conserved domain near the N-terminus based on the sequence of *P. aeruginosa* flagellin was reported to induce plant defence responses, but not hypersensitive cell death (Felix *et al.*, 1999). Thus flg22 is known as general elicitor to induce innate immune response in plants (Gomez-Gomez and Boller, 2002). However, flagellins from *P. avenae* (Che *et al.*, 2000) and *P. s.* pv. *glycinea* and *P. s.* pv. *tomato* (Taguchi *et al.*, 2003b) are known to induce plant cell death in non host rice and tobacco plants, respectively.

In this paper, we introduce our recent findings, that is 1) flagellin is an active HR-elicitor as monomer molecules, 2) HR cell death inducing activity of flagellin is determined by the post-translational specific glycosylation. Thus, we propose that flagellin seems to be a major determinant for pathovars-plant species specificities, and *P. syringae* might be a potential pathogen in spite of the relationship between pathovars and plant species.

2. MATERIALS AND METHODS

2.1 Plant and Bacteria

Tobacco (*Nicotiana tabacum* L. cv. Xanthi NC) and tomato plants (*Lycopersicon esculentum* L. cv. Moneymaker) were grown at 25°C. Leaves of 4-weeks old plants were used for inoculation experiment. Suspension cultured cells of tobacco (BY-2 and Xanthi) and tomato (Syl-1) were routinely inoculated every week in MS, Gamborg B5 and No. 8 (LS) mediums, respectively. Each assay was performed using 3-days old suspension cultured cells.

Pseudomonas syringae pv. *tabaci* 6605, *P. s.* pv. *tomato* DC3000 (race 0) and *P. s.* pv. *glycinea* (race 4) were grown in King's B medium at 27°C. Bacteria were inoculated into plant leaves by infiltration method at 2×10^5 or 5×10^7 CFU ml^{-1} for the symptom assay or the HR assay, respectively. Plant cultured cells were also inoculated with bacteria at 5×10^7 CFU ml^{-1}.

2.2 Preparation of Flagellin Proteins

Flagella were released from the bacterial cells by vortexing, and flagellin proteins were further purified mainly by centrifugation. On the other hand, each *fliC* gene was inserted into a pET16b plasmid vector and expressed in *Escherichia coli* BL21. Briefly, 3 hr after induction of flagellin expression by the addition of isopropyl-1-thio-β-D-galactoside (IPTG), bacterial cells were harvested, dissolved in B-PER™ Reagent (Pierce, Rockford, IL, USA). Recombinant flagellin proteins were obtained from the inclusion bodies; they were further purified by dissolving and desalting procedures. Flagellins were used at 3.2 μM (100 μg ml^{-1}) in leaves and 0.32 μM (10 μg ml^{-1}) in BY-2 cells at final concentrations.

2.3 Generation of Flagella-Defective Mutants

Flagellin from *Pseudomonas syringae* pv. *tabaci* was identified as an elicitor of the HR in non host tomato leaves and cultured cells (Taguchi *et al.*, 2003b). To investigate the role of flagella and monomer flagellin of *P. s.* pv. *tabaci* in plant-bacteria interactions, we first isolated the flagella gene cluster containing the flagellin gene (*fliC,* Fig. 1). *FliC* constitutes an operon with *flaG* and *fliD*. Although the function of the *flaG* product was not clear, *fliD* (HAP2) was thought to polymerise monomer flagellins at the top of flagellar filament. To obtain two different flagella-defective mutants, we generated non-polar *fliC* or *fliD* mutants by PCR-mediated homologous recombination. DNA fragments for both upstream and downstream regions of *fliC* and *fliD* genes were amplified by PCR with two sets of primers. Each set of amplified DNA fragments was ligated at artificial *Bam*HI sites and inserted into the mobilizable cloning vector, pK18*mobsacB* (Schäfer *et al.*, 1994). The resultant plamids were transferred into *P. s.* pv. *tabaci via Escherichia coli* S17-1 by intergenetic conjugation. The Δ*fliC* and Δ*fliD* mutants lost all flagella and motility. Furthermore, the Δ*fliC* mutant did not produce flagellin, whereas Δ*fliD* secreted a large amount of monomer flagellins in the culture medium as expected (Shimizu *et al.*, 2003). Thus these mutants provide valuable tools to investigate the role of flagella and flagellin in plant-bacteria interactions.

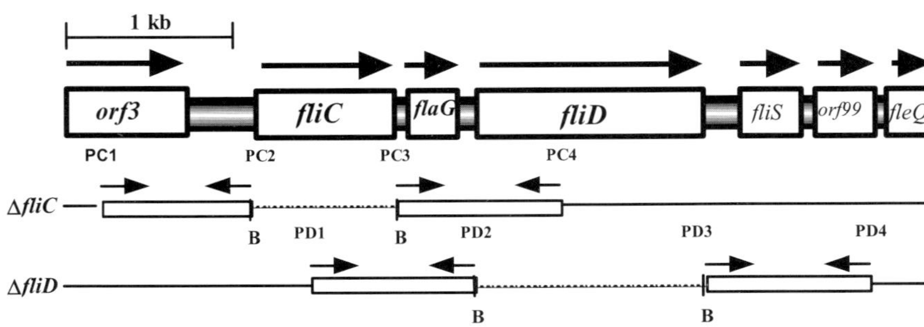

Figure 1. Schematic representation of part of flagella gene cluster and two deletion mutants.

2.4 Assessment of Plant Defence Responses

As active defence responses, we analysed the ratios of hypersensitive cell death, induction of defence-response related genes such as phenylalanine ammonia-lyase (*PAL*) and *hsr203J,* DNA fragmentation and the oxidative burst. The dead cells were detected by the staining with 0.05% (w v^{-1}) Evans blue (Sasabe *et al.*, 2000). Methods for chemiluminescence assay to measure hydrogen peroxide, TUNEL assay and Northern blot analysis were described by Sasabe *et al.* (2000).

2.5 Complementation Tests

For the complementation tests, we connected *fliC* promoter region and each open reading frame for *fliC* from different pathovars in a broad-host-range plasmid, pDSK519 (Keen *et al.*, 1988). The details were described elsewhere (Shimizu *et al.*, 2003; Taguchi *et al.*, 2003a).

Preliminary Investigations on the Role of Nitric Oxide in Systemic Acquired Resistance in the *Arabidopsis thaliana-Pseudomonas syringae* Pathosystem

R. BUONAURIO, C. MORETTI, C. CAGLIOTI, G. ARIENTI[#] and C.A. PALMERINI[#]

Dipartimento di Arboricoltura e Protezione delle Piante, Università degli Studi di Perugia, Borgo XX Giugno, 74-06121 Perugia (Italy); [#]Dipartimento di Scienze Biochimiche e Biotecnologie Molecolari, Via del Giochetto-06122 Perugia (Italy).

Abstract: There is evidence that nitric oxide (NO) plays a key role in the activation of disease resistance mechanisms in plants. It has recently been demonstrated that NO-releasing compounds induced resistance in tobacco plants against tobacco mosaic virus. We here report that NOC-18 [2,2'-(hydroxynitrosohydrazino)bis-ethanamine], a NO-releasing compound, systemically protected *Arabidopsis thaliana* (ecotype Columbia) plants against a virulent strain of *Pseudomonas syringae* pv. *maculicola*. The protection was detected as reduction in disease symptom severity and bacterial growth *in planta*. Using a specific and sensitive solid-state amperometric sensor assay for NO and its derivatives, we also observed significant increases in nitrite in phloem sap collected from plants in which resistance had been induced with an avirulent strain of *P. s.* pv. *maculicola* or NOC-18. The accumulation was detected in both inoculated and non-inoculated leaves from 3 to 12 hours post-inoculation. Further investigations are in progress to verify whether NOC-18 treatment induces transcript changes in SAR genes and to establish whether increases in phloematic NO-derivatives are associated with the induced resistance and whether NO derivatives are the systemic signals of SAR.

Key words: Nitric oxide, *Pseudomonas syringae* pv. *maculicola*, SAR, systemic signals.

Author for correspondence: R. Buonaurio; e-mail: buonaurio@unipg.it.

N.S. Iacobellis *et. al.* (eds.), *Pseudomonas syringae* and Related Pathogens, 319-325.

1. INTRODUCTION

In animals, nitric oxide (NO) is an important redox signalling molecule in both physiological and pathophysiological conditions, such as inflammatory, acute phase responses and programmed cell death (Stamler, 1994). Even though the functions of NO in plant systems are not still clear, NO in plants has recently received much attention. Plants accumulate and metabolise atmospheric NO and synthesise NO through enzymatic and non-enzymatic reactions (Leshem and Haramaty, 1996; Yamasaki and Sakihama, 2000). Mounting evidence suggests that NO is at the heart of several physiological functions, ranging from plant development to defence responses (Durner and Klessig, 1999; Beligni and Lamattina, 2000). Plants use NO as a signalling molecule through pathways remarkably similar to those of mammals, suggesting that there are many parallels between NO action in plants and animals. Indeed, there is evidence of the presence of a mammalian-type NO synthase (NOS) in plants (Wendehenne *et al.*, 2001) although NOS genes have not yet been identified. It is known that NO, as well as reactive oxygen species, stimulate promotion of the hypersensitive response (HR), provokes programmed cell death initiation and activates pathogenesis-related gene expression and phytoalexins production (Clarke *et al.*, 2000; Delledonne *et al.*, 1998; 2001; Durner *et al.*, 1998; Huang and Knopp, 1998; Molesini *et al.*, 2001; Noritake *et al.*, 1996). Song and Goodman (2001) have recently observed that NO-releasing compounds are able to induce systemic acquired resistance (SAR) to tobacco mosaic virus in tobacco plants. They also reported that NO-releasing compounds did not induce resistance in transgenic plants expressing the *nahG* gene and that NOS inhibitors or NO scavengers attenuated SAR induced by salicylic acid (SA). These findings are consistent with the hypothesis that NO plays an important role in SAR induction and that it is required for the complete function of SA as an SAR inducer. In addition, Durner *et al.* (1999) have hypothesised that NO, overproduced during the HR (Delledonne *et al.*, 1998), event often necessary to the establishment of SAR, binds to glutathione to form nitrosoglutathione (GSNO), which, in turn, acts as a long distance SAR signal. This hypothesis is based on the fact that GSNO is a powerful inducer of plant defence genes involved in the SAR (Durner *et al.*, 1998) and that glutathione is a major metabolite in the phloem, through which the SAR signal is transmitted. It is known that in mammals NO circulates in the blood as S-nitroso adducts, such as GSNO an that this molecule acts as an intra- and intercellular NO carrier.

The aims of this preliminary study were: i) to verify whether the NO-releasing compound [2,2'-(hydroxynitrosohydrazino)bis-ethanamine] (NOC-18) was able to systemically protect *Arabidopsis thaliana* plants from a virulent strain of *Pseudomonas syringae* pv. *maculicola*; ii) to determine the phloematic level of NO derivatives in *A. thaliana* plants in which resistance had been induced with an avirulent strain of *P. s.* pv. *maculicola.*

2. MATERIALS AND METHODS

2.1 Plant Material and Bacteria

Arabidopsis thaliana ecotype Columbia (Col-0) seeds were sown on flower potting compost. Seeds were vernalised for 2 days at 4°C after sowing. After 2 weeks of incubation in a growth chamber (8 h day at 24°C, 16 h night at 20°C, 90% RH, 140 µE m^{-2} s^{-1} illumination), seedlings were transplanted to pots containing potting compost and grown under the same conditions as given above. Fully expanded leaves of 5/6-week-old plants were used for all the experiments.

A virulent strain of *Pseudomonas syringae* pv. *maculicola* and an avirulent strain of the same pathovar carrying the avirulence gene *avrRpm 1*, kindly supplied by Prof. M. Delledonne, University of Verona (Italy), were used. The bacterial cultures were maintained as suspension in 15% glycerol at -80°C. To prepare the inocula, bacterial cultures were grown overnight at 27°C on King's medium B (KB) agar plates supplemented with 100 µg ml^{-1} rifampicin for the virulent strain and with 100 µg ml^{-1} rifampicin, kanamycin and streptomycin for the avirulent strain. Bacterial cells were suspended in 10 mM potassium phosphate buffer, pH 7.4 (PPB), centrifuged, washed twice, and resuspended in the same buffer at OD_{600}=0.1 corresponding to 10^8 CFU ml^{-1}.

2.2 Effect of NOC-18 on Pseudomonas syringae Infection

To verify the ability of the NO-releasing compound [2,2'-(hydroxynitrosohydrazino)bis-ethanamine] (NOC-18) in systemically protecting *Arabidopsis thaliana* plants against the virulent strain of *Pseudomonas syringae* pv. *maculicola*, three leaves of each plant (referred to as basal leaves) were infiltrated on the abaxial side (two spots per leaf) with 1 mM NOC-18 dissolved in PPB or for comparison with a bacterial suspension in PPB of the avirulent *P. s.* pv. *maculicola* strain at 10^8 CFU ml^{-1}, which it is known to induce SAR. Control plants were infiltrated with PPB. Three days after the treatment, 5 apical leaves of all plants were inoculated with the virulent strain of *P. s.* pv. *maculicola* (5 x 10^5 CFU ml^{-1}) suspended in 10 mM $MgCl_2$.

To estimate bacterial growth *in planta*, three leaf discs (0.75 cm in diameter) from three challenged leaves (one disc per leaf) were placed in ethanol for 2 min, rinsed in sterile water for 1 min and macerated in 150 μl sterile water. Leaf homogenates were diluted tenfold and 100 μl of each dilution were plated on KB supplemented with 100 μg ml^{-1} rifampicin for selection. After two days incubation at 27°C the number of colonies was counted.

2.3 NO-derivatives in Phloem Sap

Four basal leaves of *Arabidopsis thaliana* plants were inoculated with the avirulent *Pseudomonas syringae* pv. *maculicola* strain (10^8 CFU ml^{-1}) or treated with 1 mM NOC-18 or infiltrated with PPB. Phloem exudates were collected from basal and from 4 apical leaves 3 to 12 h after inoculation or treatment using the EDTA method (King and Zeevaart, 1974). Briefly, leaves were cut at the base of their petioles, recut under water, and subsequently rinsed to avoid contamination of cellular fluid. The petioles of four leaves (basal or apical) were immersed into 200 μl of 10 mM potassium phosphate buffer, 15 mM EDTA, pH 7.4 and incubated under fluorescent light (140 μE m^{-2} s^{-1}) for 4 hours at 24°C and 100% RH. Phloem sap samples were quickly frozen in liquid nitrogen and stored at -80°C until they were analysed. NO derivatives were determined in phloematic samples using a amperometric gas sensor according to the procedures described by Palmerini *et al.* (1998).

3. RESULTS AND DISCUSSION

3.1 Effect of NOC-18 on *Pseudomonas syringae* Infection

When apical leaves of *Arabidopsis thaliana* control plants were inoculated with the virulent strain of *Pseudomonas syringae* pv. *maculicola*, they showed the first symptoms of the disease 3÷4 days after the inoculation as yellowing of the leaf tissue in correspondence to the inoculated area. Sometimes the yellowed area was surrounded by a narrow water-soaked halo. By contrast, a very slight leaf yellowing or no symptoms were observed in apical leaves of plants that had been treated with NOC-18 or inoculated for comparison with the avirulent strain of *P. s.* pv. *maculicola*.
The protection exerted by NOC-18 was also observed as reduction in bacterial growth *in planta*. In fact, a 70-fold reduction in bacterial growth were detected in NOC-18 treated plants respect to the control plants, 7 days post-inoculation (Fig. 1). We have therefore demonstrated that, similarly to the results of Song and Goodmann (2001) for tobacco plants, NO is able to

systemically induce resistance in *A. thaliana*. We are investigating plant defence genes expression to establish which pathway leads to the systemic resistance development in NO treated plants.

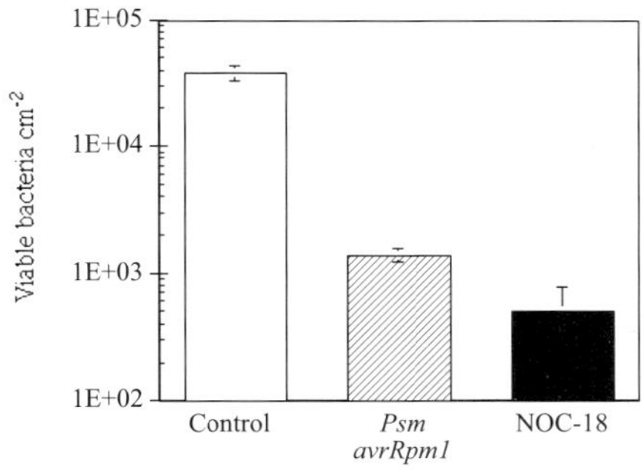

Figure 1. Effect of *Pseudomonas syringae* pv. *maculicola avrRpm1* (*Psm avrRpm1*) inoculation or NOC-18 treatment in basal *Arabidopsis thaliana* leaves on the bacterial growth of the virulent *P. s.* pv. *maculicola* strain in apical leaves, 7 days after the inoculation. Each value is the mean of 3 replicates±SE.

3.2 NO-derivatives in Phloem Sap

Increases in phloematic levels of nitrite (a NO derivative) were observed 3 to 12 hours post-inoculation both in basal leaves inoculated with the *Pseudomonas syringae* pv. *maculicola* avirulent strain and in apical leaves, where SAR to the virulent strain of *P. s.* pv. *maculicola* infection developed (Fig. 2). When resistance was induced with NOC-18, nitrite level increased in phloem sap only in apical leaves at 3, 9 and 12 hours after the treatment (Fig. 2A).

The very small amount of nitrosothiols (RSNO) recorded by the amperometric sensor in phloem sap extracted from four *Arabidopsis thaliana* leaves did not permit us to accurately detect changes in RSNO. To overcome this problem, we are planning further experiments in which higher amount of phloem sap will be used.

Although the results we reported here are preliminary, they seem to support the hypothesis that NO and/or its derivatives play a role in SAR as systemic signals. We hypothesised that NO generated during the HR induced by the avirulent strain of *P. s.* pv. *maculicola* or by NOC-18 treatment translocates in the phloem as NO and/or RSNO and then reach upper leaves, where it induces the expression of defence genes and resistance to *P. s.* pv. *maculicola* attacks.

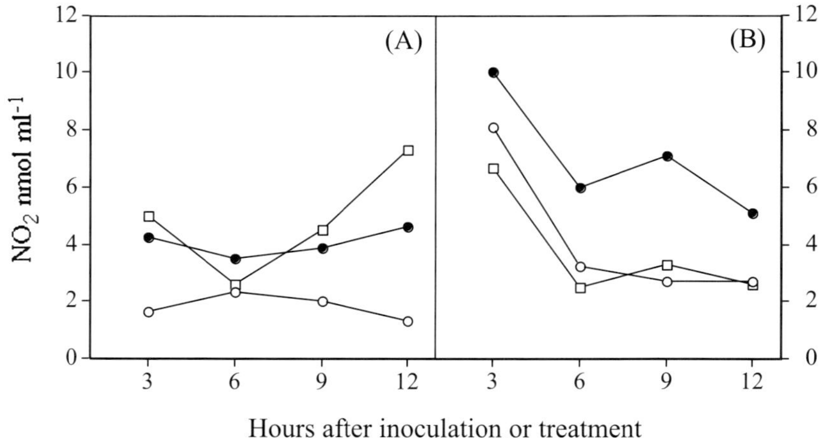

Figure 2. Nitrite level in phloem sap collected from apical (A) and basal (B) leaves of *Arabidopsis thaliana* plants whose basal leaves were inoculated with the avirulent *Pseudomonas syringae* pv. *maculicola* strain (black circle) or treated with 1 mM NOC-18 (white square). Basal leaves of control plants were treated with 10 mM potassium phosphate buffer, pH 7.4 (white circle). Each value is the mean of 3 replicates. Similar results were obtained in another experiment.

ACKNOWLEDGEMENTS

This work was supported by a grant from the 'Ministero dell'Università e della Ricerca Scientifica e Tecnologica (Cofin 2000)' to R.B., entitled: Role of nitric oxide in the hypersensitive reaction and in systemic acquired resistance in *Arabidopsis thaliana'*

Professor Roberto Palombari, University of Perugia, is thanked for providing the amperometric gas sensor.

REFERENCES

Beligni, M.V., Lamattina, L., 2000, Nitric oxide stimulates seed germination and de-etiolation, and inhibits hypocotyl elongation, three light-inducible responses in plants. *Planta* **210**: 215-221.

Clarke, A., Desikan, R., Hurst, R.D., Hancock, J.T., Neill, S.J., 2000, NO way back: nitric oxide and programmed cell death in *Arabidopsis thaliana* suspension cultures. *Plant J.* **24**: 667-677.

Delledonne, M., Xia, Y.J., Dixon, R.A., Lamb, C., 1998, Nitric oxide functions as a signal in plant disease resistance. *Nature* **394**: 585-588.

Delledonne, M., Zeier, J., Marocco, A., Lamb, C., 2001, Signal interactions between nitric oxide and reactive oxygen intermediates in the plant hypersensitive disease resistance response. *Proc. Natl. Acad. Sci. USA* **98**: 13454-13459.

Durner, J., Wendehenne, D., Klessig, D.F., 1998, Defence gene induction in tobacco by nitric oxide, cyclic GMP, and cyclic ADP-ribose *Proc. Natl. Acad. Sci. USA* **95**: 10328-10333.

Durner, J., Klessig, D.F., 1999, Nitric oxide as a signal in plants. *Curr. Opin. Plant Biol.* **2**: 369-374.

Huang, J.S., Knopp, J.A., 1998, "Involvement of nitric oxide in *Ralstonia solanacearum*-induced hypersensitive reaction in tobacco" In Bacterial wilt disease: molecular and ecological aspects, P. Prior, Elphinstone J. Allen C., ed. INRA and Springer Editions.

King, R.W., Zeevaart, J.A.D., 1974, Enhancement of phloem exudation from cut petioles by chelating agents. *Plant Physiol.* **53**: 96-103.

Leshem, Y.Y., Haramaty, E., 1996, The characterisation and contrasting effects of the nitric oxide free radical in vegetative stress and senescence of *Pisum sativum* Linn foliage. *J. Plant Physiol.* **148**: 258-263.

Molesini, B., Delledonne, M., Buonaurio, R., Pezzotti, M., Polverari, A., 2001, Studio del profilo trascrizionale modulato da ossido nitrico in *Arabidopsis thaliana*. Proceedings of the 'XI Meeting on Molecular and Physiological Aspects of the Plant-Pathogen Interactions; May 31 – July 1; Rome (Italy).

Noritake, T., Kawakita, K., Doke, N., 1996, Nitric oxide induces phytoalexin accumulation in potato tuber tissues. *Plant Cell Physiol.* **37**: 113-116.

Palmerini, C.A., Arienti, G., Mazzolla, R., Palombari, R., 1998, A new assay for the determination of low-molecular weight nitrosothiols (nitrosoglutathione), NO, and nitrites by using a specific and sensitive solid-state amperometric gas sensor. *Nitric Oxide* **2**: 375-380.

Song, F.M., Goodman, R.M., 2001, Activity of nitric oxide is dependent on, but is partially required for function of, salicylic acid in the signalling pathway in tobacco systemic acquired resistance. *Mol. Plant-Microbe In.* **14**: 1458-1462.

Stamler, J.S., 1994, Redox signalling: nitrosylation and related target interactions of nitric oxide. *Cell* **78**: 931-936.

Wendehenne, D., Pugin, A., Klessig, D.F., Durner, J., 2001, Nitric oxide: comparative synthesis and signalling in animal and plant cells. *Trends Plant Sci.* **6**: 177-183.

Yamasaki, H., Sakihama, Y., 2000, Simultaneous production of nitric oxide and peroxynitrite by plant nitrate reductase: *in vitro* evidence for the NR-dependent formation of active nitrogen species. *FEBS Lett.* **468**: 89-92.

Does the Harpin of *Pseudomonas syringae* Interact with a Host Protein?

C.M. LI, M. HAAPALAINEN, J. LEE[#], T. NÜRNBERGER[#], M. ROMANTSCHUK and S. TAIRA
Department of Biosciences, University of Helsinki, Finland; [#]Leibniz-Institut für Pflanzenbiochemie, Halle, Germany.

Abstract: Harpin HrpZ is one of the pathogenesis-associated proteins secreted through the inducible type III secretion system of *Pseudomonas syringae*. HrpZ elicits a hypersensitive response (HR) in some non-host plants, but the function of this protein in the compatible interaction with host plants is not known. HrpZ functions as a HR-elicitor only when it is applied outside of the plant cell, and its putative binding site is likely to be on the plasma membrane. To study the ability of HrpZ to interact with other proteins, we screened peptides that bind to HrpZ by using phage display technique. All the peptides that had a strong affinity to HrpZ contained a similar hydrophobic amino acid motif. We also mutagenised *hrpZ* with 15bp insertions created by a mini-Mu transposon and screened for mutants whose protein product did not bind the phage-displayed peptides. The mutations that prevented HrpZ from binding to those peptides concentrated to the central region of the protein. This region is separate from the region previously shown to possess the HR-inducing activity, which resides at the C-terminal part. Interestingly, this peptide-binding region includes the only region with sequence homology shared by HrpZ and HrpN harpin of *Erwinia* species. We suggest that the peptide-binding region with the conserved motif might play a role in a protein-protein interaction involved in the bacterial pathogenesis in host plants.

Key words: HrpZ, protein-protein interaction, phage display, mutagenesis.

Author for correspondence: M. Haapalainen; e-mail: mlhaapal@cc.helsinki.fi.

N.S. Iacobellis *et. al.* (eds.), *Pseudomonas syringae* and Related Pathogens, 327-333.

1. INTRODUCTION

Pseudomonas syringae can enter plant tissues both through wounds and natural openings, such as stomata. *P. syringae* does not invade living plant cells but multiplies in the intercellular spaces, and, depending on the pathovar-host combination, to varying extent on the plant surface. When bacteria get into close contact with plant cells, bacterial genes encoding the components of the type III secretion system (TTSS) are induced. Simultaneously induced are several genes encoding the effector proteins that are secreted through this system. The effector proteins are transported from the bacterial cell across the plant cell wall through a long pilus structure (Brown *et al.*, 2001; Jin and He, 2001; Li *et al.*, 2002). The presumed function of the proteins delivered to the host cell is to inhibit or suppress the host defence and thus promote bacterial survival and multiplication inside the host tissue. TTSS is essential for the ability of *P. syringae* to infect plants and to cause plant diseases (Roine *et al.*, 1997).

One of the pathogenesis-associated proteins delivered via the TTSS is harpin, encoded by *hrpZ* gene, a protein that elicits a hypersensitive response (HR) in some non-host plants (He *et al.*, 1993). During the HR, the plant is able to recognise some pathogen-derived component and to activate defence mechanisms that lead to a kind of programmed cell death. As a result, the cells attacked by the incompatible pathogen collapse and form a necrotic lesion arresting further growth of the pathogen. The function of HrpZ in the compatible interaction with host plants is not known. The picture is complicated by the fact that *P. syringae* produces several harpin-like proteins and a simple *hrpZ* mutant has no obvious phenotype.

Experiments with transgenic plants have previously shown that HrpZ elicits a HR only when localised into the apoplast, indicating an extracellular function (Tampakaki and Panopoulos, 2000). And, after application of partially purified harpin on tobacco cells in suspension culture, the harpin was localised in the outer part of the plant cell (Hoyos *et al.*, 1996). As HrpZ can bind to lipid bilayers and form ion-conducting pores (Lee *et al.*, 2001a; 2001b), its binding site is likely to be on the plasma membrane. Curiously, the association of harpins, both HrpZ and HrpN, with plant cell membrane seems to be a reversible event (Pike *et al.*, 1998; Lee *et al.*, 2001a). As a membrane-associated protein, HrpZ could be involved in the release of nutrients to the apoplast or it might assist in the delivery of other bacterial proteins that are targeted inside the host cell. We are looking for molecular components interacting with HrpZ in order to dissect the role of this effector protein in bacterial pathogenesis. For the present, no direct interaction with a plant protein receptor has been demonstrated for HrpZ, although some indirect results suggest that protein-protein interactions involving harpin may take place in the incompatible *P. syringae*-plant interactions (Lin *et al.*, 1997; Chen *et al.*, 1998).

2. MATERIALS AND METHODS

2.1 Phage Display

We screened peptides that bind to HrpZ by using a phage display technique (Koivunen *et al.*, 1999). Random peptides were inserted into the minor capsid protein PIII of a filamentous bacteriophage, which carries a tetracycline resistance gene, so that the peptides were displayed on the surface of the phage particles. Several different peptide libraries were screened for HrpZ-binding efficiency, followed by multiplication of the selected phage particles in *Escherichia coli* strain K91. After several rounds of panning and multiplication, the DNA inserts encoding the binding peptides were multiplied by PCR and sequenced. ELISA method, using monoclonal phage antibody for detection, was used to confirm that the selected phages truly bind to HrpZ. The ELISA colour reaction also allowed us to rate the peptides according to the strength of HrpZ affinity.

2.2 Insertion-mutagenesis of the Harpin

We mutagenised the *hrpZ* gene of *Pseudomonas syringae* pv. *phaseolicola* with 15 bp insertions created by the insertion and subsequent deletion of a mini-Mu transposon (MGS kit, Finnzymes). Mutated HrpZ proteins were expressed in *Escherichia coli* strain BL21 (DE3) and partially purified by a heat treatment (Wei *et al.*, 1992), with denaturation and precipitation of most other proteins. The mutant proteins were screened for both the peptide binding and for the ability to cause HR when infiltrated into tobacco leaves. The mutated genes whose protein product did not bind those phage-displayed peptides that strongly bound to the wild-type HrpZ were sequenced to localise the insertion.

2.3 Far-western Blotting

Binding of HrpZ to plant proteins was studied by far-western blotting. Proteins extracted from the leaf tissue of tomato, bean and *Arabidopsis thaliana* were separated by polyacrylamide gel-electrophoresis and blotted onto PVDF membrane, which was then blocked with 5% BSA. The blocked membranes were incubated overnight with HrpZ, and subsequently with anti-HrpZ antibody and the secondary antibody conjugated to alkaline phosphatase. The bound conjugate was detected by using BCIP/NBT colour-reaction.

3. RESULTS AND DISCUSSION

3.1 Peptide Ligands and Peptide-binding Region of the Harpin

By using the phage display method, we obtained several peptides with a strong affinity to HrpZ. The sequences of the HrpZ-binding peptides proved not to be unrelated but strikingly similar, all containing a hydrophobic amino acid motif with tryptophan and leucine (Fig. 1). Unfortunately, a protein with an exact match with any of these peptide sequences could not be found in the databases available. It is possible that the identified sequences are not present as a continuous stretch in the primary sequence of the putative HrpZ-binding protein.

The insertion mutations that prevented HrpZ from binding the peptide ligands all concentrated to the same, central region (Fig. 2). This supports a previous prediction that this region, being the most hydrophilic part of HrpZ, is probably surface-exposed (Lee *et al.*, 2001b). This region is separate from the region previously shown to possess the HR-inducing activity, which resides at the C-terminal part. Interestingly, this peptide-binding region includes the short region of similarity shared by HrpZ of all *Pseudomonas syringae* pathovars and HrpN harpin of *Erwinia* and *Pantoea* species (Table 1). The motif shared by HrpZ and HrpN was first described by He *et al.* (1993) when comparing the amino acid sequence of HrpZ of *P. s.* pv. *syringae* with HrpN of *E. amylovora*. Except for this 22 amino acid motif HrpZ and HrpN share no sequence homology. Yet both of these harpins are glycine-rich, contain no cysteine and are very heat-stable, suggesting that they may have some similarity in the protein folding. Unlike HrpZ proteins of different *P. syringae* pathovars, the different HrpN proteins have quite different N-terminal sequences, although the C-terminal part is conserved. The peptide-binding region with the conserved motif might play a role in a protein-protein interaction involved in the bacterial pathogenesis in host plants. The fact that this region is distinct from the HR-inducing domain suggests that these harpins would have several functional domains: the secretion signal, an elicitor/non-specific plant interaction domain and possibly a host-specific interaction domain.

PWLGL**W**LLGV
WAR**W**LLLTPA
PWLAR**W**LLGV
FVR**W**LLLGVP
DTWAR**W**ML
LGC**W**ILPRL
SWAL**W**MLG
WLGLMRW

Figure 1. Peptides with a strong affinity to HrpZ. The peptide sequences are arranged so that those at the top of the pile have the highest affinity and those at the bottom a weaker affinity to the harpin.

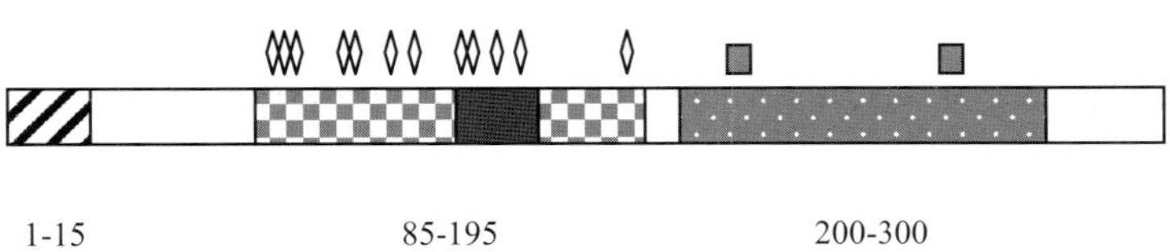

Figure 2. A schematic map of HrpZ protein of *Pseudomonas syringae* pv. *phaseolicola.* The insertion sites in mutants not binding the peptides shown in figure 1 are indicated by diamonds, and the "harpin motif" is marked as a solid block inside the peptide-binding region (checked). The HR-inducing region (amino acids 200-300, according to Lee *et al.*, 2001a) is dotted, and the locations of conserved repeats GGGLG(T/S)P found in all HrpZ proteins are indicated by squares. The serine-rich N-terminal region is marked by diagonal line pattern. The type III secretion signal supposedly resides at the N-terminus, either in the amphipathic amino acid sequence or in the messenger RNA.

Table 1. "Harpin motif" of HrpZ (*Pseudomonas*) and HrpN (*Erwinia* and *Pantoea*).

Bacterial species	Conserved sequence
Pseudomonas syringae pv. *tomato* DC3000	**D**MPTLEKVA**QFMD**DNKAQ**F**PTR
P. s. pv. *phaseolicola*	**D**MPMLKKIAE**FMD**DN**P**AQ**F**P**K**P
P. s. pv. *aptata*	**D**MPMLSKIA**QFMD**DN**P**AQ**F**P**K**P
P. syringae pv. *syringae*	**D**MPMLNKIA**QFMD**DN**P**AQ**F**P**K**P
P. s. pv. *pisi*	**D**MPMLNKIA**QFMD**DN**P**AQ**F**P**K**P
Erwinia amylovora	**D**RAMAKEIG**QFMD**QY**P**EV**F**G**K**P
E. carotovora	**D**RGMAKEIG**QFMD**QY**P**EV**F**G**K**A
E. chrysanthemi	**D**RGMAKEIG**QFMD**QY**P**EI**F**G**K**P
Pantoea agglomerans pv. *gypsophilae*	**D**RSLAKEVG**QFMD**QY**P**EI**F**G**K**P
P. stewartii subsp. *stewartii*	**D**RALAREVG**QFMD**QY**P**ET**F**G**K**P

3.2 Protein-protein Interaction

Results obtained by far-western method suggest that in addition to the peptide ligands, HrpZ has interaction with actual plant proteins. We found that HrpZ binds to several proteins extracted from tomato, bean and *Arabidopsis thaliana*, and, when the soluble and membrane-bound fractions were separated, the interacting *Arabidopsis* proteins were found in the membrane fraction. Rabbit antiserum raised against one of the HrpZ-binding peptides also recognised specifically one *Arabidopsis* membrane protein. The future goal of our project is to identify the plant proteins interacting with HrpZ and the peptide antibody. We aim to find epitopes associated with host-specificity by comparing the binding of peptide ligands by HrpZ proteins derived from different *Pseudomonas syringae* pathovars, and by comparing the set of HrpZ(pha)-binding proteins from several plant species; our preliminary results show that there are both similarities and differences. Yet the different interaction of the harpin with different plants does not determine the host range of the bacterium, as previously shown (Preston *et al.*, 1995). Each *P. syringae* pathovar produces a large set of effector proteins, HrpZ being only one component of the complex but tightly regulated system.

REFERENCES

Brown, I.R., Mansfield, J.W., Taira, S., Roine, E., and Romantschuk, M., 2001, Immunocytochemical localisation of HrpA and HrpZ supports a role for the Hrp pilus in the transfer of effector proteins from *Pseudomonas syringae* pv. *tomato* across the host plant cell wall. *Mol. Plant-Microbe Interact.* **14**: 394-404.

Chen, C.-H., Lin, H.-J., and Feng, T.-Y., 1998, An amphipathic protein from sweet pepper can dissociate $harpin_{Pss}$ multimeric forms and intensify the $harpin_{Pss}$-mediated hypersensitive response. *Physiol. Mol. Plant Pathol.* **52**: 139-149.

He, S.Y., Huang, H.-C., and Collmer, A., 1993, *Pseudomonas syringae* pv. *syringae* $harpin_{Pss}$: a protein that is secreted via the Hrp pathway and elicits the hypersensitive response in plants. *Cell* **73**: 1255-1266.

Hoyos, M.E., Stanley, C.M., He, S.Y., Pike, S., Pu, X.-A., and Novacky, A., 1996, The interaction of $harpin_{Pss}$ with plant cell walls. *Mol. Plant-Microbe Interact.* **9**: 608-616.

Jin, Q., and He, S.Y., 2001, Role of the Hrp pilus in type III protein secretion in *Pseudomonas syringae*. *Science* **294**: 2556-2558.

Koivunen, E., Restel, B.H., Rajotte, D., Lahdenranta, J., Hagedorn, M., Arap, W., and Pasqualini, R., 1999, Integrin-binding peptides derived from phage display libraries. *Meth. Mol. Biol.* **129**: 3-16.

Lee, J., Klessig, D.F., and Nürnberger, T., 2001a, A harpin binding site in tobacco plasma membranes mediates activation of the pathogenesis-related gene *HIN1* independent of extracellular calcium but dependent on mitogen-activated protein kinase activity. *Plant Cell* **13**: 1079-1093.

Lee, J., Klüsener, B., Tsiamis, G., Stevens, C., Neyt, C., Tampakaki, A.P., Panopoulos, N.J., Nöller, J., Weiler, E.W., Cornelis, G.R., Mansfield, J.W., and Nürnberger, T., 2001b, $HrpZ_{Psph}$ from the plant pathogen *Pseudomonas syringae* pv. *phaseolicola* binds to lipid bilayers and forms an ion-conducting pore *in vitro*. *Proc. Natl. Acad. Sci. USA* **98**: 289-294.

Li, C.-M., Brown, I., Mansfield, J., Stevens, C., Boureau, T., Romantschuk, M., and Taira, S., 2002, The Hrp pilus of *Pseudomonas syringae* elongates from its tip and acts as a conduit for translocation of the effector protein HrpZ. *EMBO Journal* **21**: 1909-1915.

Lin, H.-J., Cheng, H.-Y., Chen, C.-H. Huang, H.-C., and Feng, T.-Y., 1997, Plant amphipathic proteins delay the hypersensitive response caused by $harpin_{Pss}$ and *Pseudomonas syringae* pv. *syringae*. *Physiol. Mol. Plant Pathol.* **51**: 367-376.

Pike, S.M., Ádám, A.L., Pu, X.-A., Hoyos, M.E., Laby, R., Beer, S.V., and Novacky, A., 1998, Effects of *Erwinia amylovora* harpin on tobacco leaf cell membranes are related to leaf necrosis and electrolyte leakage and distinct from perturbations caused by inoculated *E. amylovora*. *Physiol. Mol. Plant Pathol.* **53**: 39-60.

Preston, G., Huang, H.-C., He, S.Y., and Collmer, A., 1995, The HrpZ proteins of *Pseudomonas syringae* pvs. *syringae*, *glycinea*, and *tomato* are encoded by an operon containing *Yersinia ysc* homologs and elicit the hypersensitive response in tomato but not soybean. *Mol. Plant-Microbe Interact.* **8**: 717-732.

Roine, E., Wei, W., Yuan, J., Nurmiaho-Lassila, E.-L., Kalkkinen, N., Romantschuk, M., and He, S.Y., 1997, Hrp pilus: an *hrp*-dependent bacterial surface appendage produced by *Pseudomonas syringae* pv. *tomato* DC3000. *Proc. Natl. Acad. Sci. USA* **94**: 3459-3464.

Tampakaki, A.P., and Panopoulos, N.J., 2000, Elicitation of hypersensitive cell death by extracellularly targeted $HrpZ_{Psph}$ produced in planta. *Mol. Plant-Microbe Interact.* **13**: 1366-1374.

Wei, Z.-M., Laby, R.J., Zumoff, C.H., Bauer, D.W., He, S.Y., Collmer, A. and Beer, S.V., 1992, Harpin, elicitor of the hypersensitive response produced by the plant pathogen *Erwinia amylovora*. *Science* **257**: 85-88.

Lanthanum Inhibits Programmed Cell Death but not Resistance in the Tobacco - *Pseudomonas savastanoi* pv. *phaseolicola* Incompatible Interaction

P.G. OTT, Z. KLEMENT, I. NAGY[#] and A.L. ÁDÁM[#]
Plant Protection Institute, Hungarian Academy of Sciences, P.O. Box 102, 1525 Budapest, Hungary; [#]Agricultural Biotechnology Center, P.O. Box 521, 2100, Gödöllő, Hungary.

Abstract: Avirulent pathogenic bacteria often cause a hypersensitive response (HR) in plants, an effective resistance phenomenon coupled to programmed cell death (PCD) and usually an increased production of reactive oxygen species (ROS) in the infected cells. HR induction is dependent on Ca^{2+} accumulation/influx within the plant cytoplasm. Here we show that Lanthanum (La^{3+}), a Ca^{2+} channel blocker, inhibited PCD and H_2O_2 accumulation in tobacco leaf tissue injected with *Pseudomonas savastanoi* pv. *phaseolicola*, an avirulent pathogen, even if it was added up to 4.5 h after the bacterium. This was 2 h after the HR induction phase and the beginning of an intense ROS accumulation in the control (HR-positive) leaf tissue. However, La^{3+} did not inhibit transcription of the pathogenesis-related protein PR-1a gene and only partially blocked the HR-associated resistance. We conclude that during HR: 1) Ca^{2+} influx is needed not only to trigger but also to maintain ROS production and processes leading to PCD; 2) There is a PCD- and ROS-independent resistance mechanism. The comparison of this "remaining" resistance with the early induced resistance (EIR) needs further analysis.

Key words: pathogenesis-related protein, hypersensitive response, resistance, calcium channel blocker.

Authors for correspondence: P.G. Ott and A.L. Ádám; e-mail: pott@nki.hu, aadam@netposta.net.

N.S. Iacobellis *et. al.* (eds.), *Pseudomonas syringae* and Related Pathogens, 335-344.

1. INTRODUCTION

The hypersensitive reaction (HR) represents one of the most effective resistance mechanisms of plants towards microbial pathogens in incompatible plant-pathogen interactions. HR is consistently associated with programmed cell death (PCD). Although no unique biochemical and genetic markers for HR or plant PCD were described until now, several lines of evidence suggest that hypersensitive cell death is a specific form of PCD (Heath, 2000). There is a strong association of hypersensitive response-related PCD with plant defence responses, and the interrelationship(s) of these two features of HR is still uncertain. The separation of these two features by mutational analysis suggests that hypersensitive cell death and defence gene activation are not tightly linked phenomena (Yu *et al.*, 1998). Even more striking problem is the role of cell death in resistance. For example, one of the putative HR marker genes (*HSR203J)* is also expressed during cell death caused by pathogenesis (Pontier *et al.*, 1999). Consequently, the exact function(s) of cell death during HR of different plant-pathogen interactions needs further studies.

One of the earliest steps in HR elicitation is the changes in ion fluxes, especially the entry of extracellular calcium into the cell (Pike *et al.*, 1998; Xu and Heath, 1998; Grant *et al.*, 2000). Subsequent extracellular Ca^{2+}- and/or Ca^{2+} entry-dependent events involve oxidative burst (production of reactive oxygen species, ROS) and activation of MAP (mitogen-activated protein) kinase pathways (Ligterink *et al.*, 1997; Romeis *et al.*, 1999). However, Ca^{2+}-independent activation of a MAPK pathway was reported during elicitation of HR by harpin (Ádám *et al.*, 1997; Lee *et al.*, 2001) and after hyperosmotic stress in tobacco (Hoyos and Zhang, 2000). Interestingly, activation of the same MAPK (SIPK) in tobacco by another elicitor of HR, beta-elicitin, was Ca^{2+}-dependent (Lebrun-Garcia *et al.*, 1998).

We applied La^{3+}, a known Ca^{2+} channel blocker to distinguish between Ca^{2+}-dependent and independent events in *Nicotiana tabacum* L. cv. Samsun - *Pseudomonas savastanoi* pv. *phaseolicola* interaction. Here we present further evidence for partial biochemical uncoupling of HR-mediated resistance from HR-associated cell death (PCD) and the oxidative burst. We have also dissected the induction of PR1a transcription during HR from the HR-associated oxidative burst and PCD. The basis for this PCD-independent "residual" resistance and its putative connection with other symptomless general resistance responses (early induced resistance, EIR) (Klement *et al.*, 1999) merits further investigation.

2. MATERIALS AND METHODS

Tobacco (*Nicotiana tabacum* L. cv. Samsun) plants were grown in a greenhouse. Experiments were conducted using growth chambers set to 20°C, 16/8 h light/dark period. Plants with 3÷5 fully developed leaves were transferred into the chambers one day before experiments. Leaf injections were performed with a hypodermic syringe and a 26-gauge needle. La^{3+} (as 1 mM $LaCl_3$) was injected concomitantly with or after bacterial inocula. Heat treatment was done by immersing leaves into a 51°C water bath for 15 sec, immediately before bacterial inoculation.

Pseudomonas savastanoi pv. *phaseolicola* race 6 (a gift of J.W. Mansfield), pathogenic on bean and causing HR on tobacco, was used. *hrp* gene activity was measured using a *P. s.* pv. *syringae hrpJ* mutant (strain 5112, a gift of S.W. Hutcheson). In this strain, the *hrpJ* gene contains an insertion with a promoterless GUS sequence, so the activity of the *hrpJ* promoter can be measured as GUS activity (Xiao *et al.*, 1992). Bacteria were injected at 10^8 cells ml^{-1} (*P. s.* pv. *phaseolicola*) or at 5 x 10^8 cells ml^{-1} (*P. s.* pv. *syringae hrpJ* mutant) in 10 mM $MgCl_2$.

Diaminobenzidine (5 mM DAB) staining of H_2O_2 in leaf panels was performed according to Thordal-Christensen *et al.* (1997).

RNA was extracted according to Stiekema *et al.* (1988). RNA prepared from 0.5 g frozen tissue was resuspended in 20 μl water. Routinely, 10 μg total RNA was applied to Northern blots and hybridised to alpha-^{32}P-dATP-labelled PR1a fragment (Sambrook *et al.*, 1989).

3. RESULTS AND DISCUSSION

3.1 La^{3+} Treatment Inhibits Hypersensitive Cell Death and ROS Production Even After the Initiation of the Oxidative Burst

Concomitant application of a known calcium channel blocker, La^{3+} inhibited the oxidative burst and subsequent PCD during the *Pseudomonas savastanoi* pv. *phaseolicola*-induced hypersensitive response (Figs. 1 and 2). Hydrogen peroxide accumulation started about 3 hrs after bacterial infiltration and a very strong DAB staining reaction was experienced 5 hrs after infiltration. La^{3+} treatment completely blocked H_2O_2 formation (Fig. 2). This effect was manifested even when La^{3+} was added separately, 3÷4 hours after bacterial infiltration (data not shown), when oxidative burst has already been initiated. KCN did not inhibit DAB staining (Fig. 2). This may indicate a peroxidase-independent H_2O_2 formation. Other inhibitors such as PMSF and a known

SOD inhibitor, diethyldithiocarbamate (DETC) produced less consistent results (data not shown). Our results suggest that Ca^{2+}-influx is required not only for initiating but also maintaining the oxidative burst.

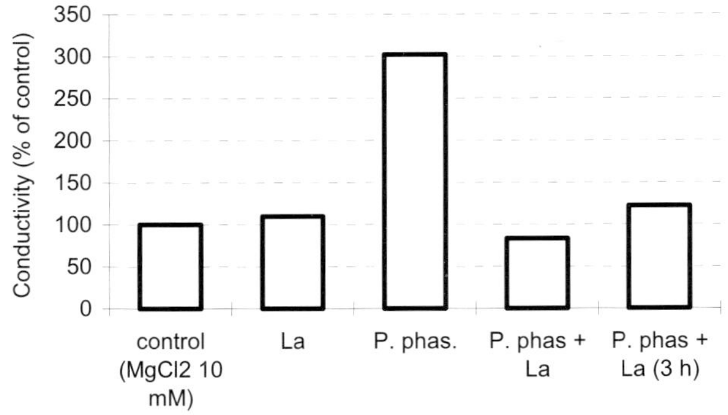

Figure 1. Effect of La^{3+} on programmed cell death (PCD) during *Pseudomonas savastanoi* pv. *phaseolicola* induced HR. Electrolyte leakage was measured in samples taken 6 h after bacterial infiltration. La = La^{3+} 1 mM; P.phas = *P. s.* pv. *phaseolicola*; + = mixed inoculum. In the rightmost column, La^{3+} was applied 3 h after bacterial infiltration.

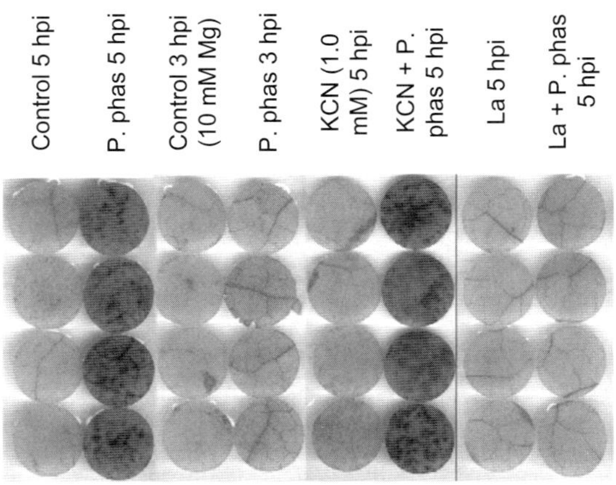

Figure 2. Diaminobenzidine (DAB) staining of H_2O_2 in leaf disks after different sampling times. Hpi = hours post inoculation; P. phas = *Pseudomonas savastanoi* pv. *phaseolicola*; La = La^{3+} 1 mM; + = mixed inoculum.

Similar to our present results, Blume *et al.* (2000) also concluded that sustained increase of cytoplasmic calcium concentrations rather than a transient increase may encode signal specificity towards activation of phytoalexin production, after elicitor treatment of parsley cells. Grant *et al.* (2000) also reported a sustained increase in cytoplasmic calcium concentrations of *Arabidopsis* leaves, specific for incompatible interaction.

An influx of Ca^{2+} also seems necessary for the PCD that occurs during the differentiation of tracheary elements in *Zinnia* (Groover and Jones, 1999). However, besides Ca^{2+} signature, additional signal(s) might be necessary for HR elicitation. Cytosolic calcium accumulation itself equivalent to those levels seen prior to the HR in cowpea did not result in cell death (Xu and Heath, 1998). Alternatively, Ca^{2+} signalling for HR needs a specific feature, other than its longevity.

3.2 La^{3+} Treatment Does not Interfere with Transcriptional Activation of a Pathogenesis Related Gene, PR1a During HR

We have also studied the expression pattern of the PR1a gene during HR in bacterially infiltrated tissue and in a 3÷5 mm wide border around the infiltrated area. As it is clearly seen in Fig. 3, treatment with 1 mM $LaCl_3$ applied concomitantly with bacteria did not inhibit the expression of PR1a gene. As compared to the 24 h salicylic acid treatment, however, the expression was relatively weaker. After a longer exposure time of Northern blots, La^{3+} treatment alone was found induce some PR1a expression but its level was comparable to absolute control and water-treated tissues (Fig. 3).

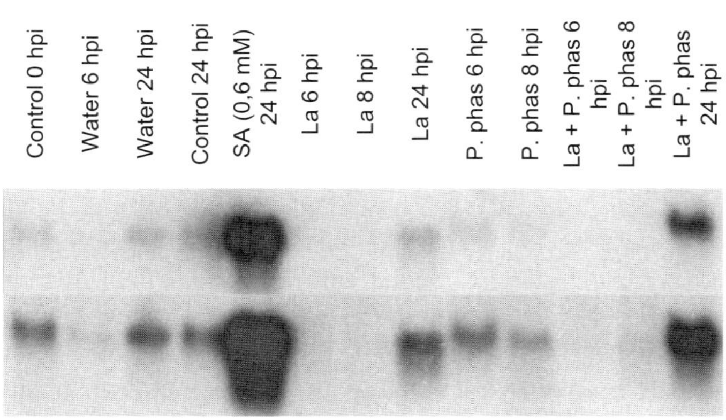

Figure 3. Effect of La^{3+} on transcription of the pathogenesis-related PR-1a gene in the infiltrated area. Upper and lower panels show Northern blots, 2 and 5 days exposure, respectively. Control = untreated tissue; hpi = hours post inoculation; SA = salicylic acid; La = La^{3+} 1 mM; P. phas = *Pseudomonas savastanoi* pv. *phaseolicola*; + = mixed inoculum.

Interestingly, in bacterially infiltrated tissues undergoing hypersensitive-related PCD, we could not detect PR1a expression. On the other hand, as depicted in Fig. 4, in the border zone surrounding the bacterially infiltrated tissue, after the manifestation of HR cell death PR1a was expressed to a high level. La^{3+} treatment also did not inhibit PR1a expression in this border tissue. Taken together these results indicate that PR1a expression is not regulated in an extracellular Ca^{2+}-dependent manner in this incompatible interaction.

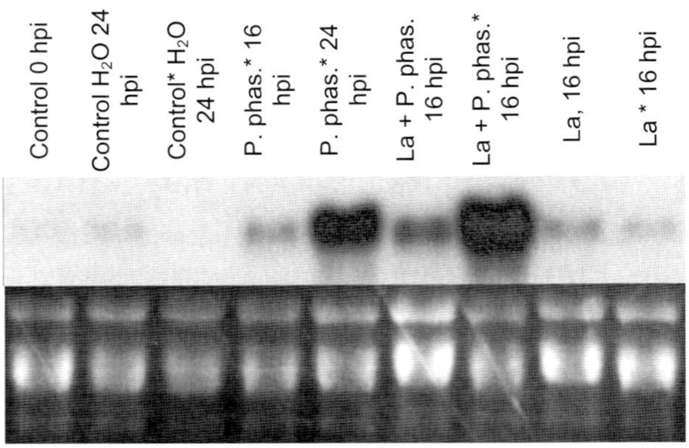

Figure 4. Effect of La^{3+} on transcription of the pathogenesis-related PR-1a gene. Asterisks indicate samples taken from tissues surrounding the inoculated area. Lower panel: 10 μg total RNA in each lane. Control = untreated tissue; hpi = hours post inoculation; La = La^{3+} 1 mM; P. phas = *Pseudomonas savastanoi* pv. *phaseolicola*; + = mixed inoculum.

As previously discussed, La^{3+}-treatment completely blocked the hypersensitive cell death and the oxidative burst (Fig. 1 and 2). The subsequent PR1a transcript accumulation was induced neither in a PCD- nor oxidative burst-dependent manner.

MAP and MAPK kinase activation was indicated as an upstream event in PR gene (*HIN1)* expression during harpin-induced HR in tobacco as well as in other systems (Lee *et al.*, 2001; Zhou *et al.*, 1998). Similar to our present results, PR gene expression seems independent of the oxidative burst during HR induced by *Phytophthora sojae* and *Cladosporium fulvum* elicitors (Ligterink *et al.*, 1997; Romeis *et al.*, 1999). However, the multiple signalling pathways that are responsible separately for defence gene activation and PCD, putatively converge to one point downstream of Ca^{2+}-influx. Another convergence point is obviously at the MAPKK level, e.g. *NPK1* in tobacco can interfere with the function of the disease resistance genes *N*, *Bs2* and *Rx* (Jin *et al.*, 2002). The diversity in convergence points can be one explanation for the differences in calcium-dependence of signalling pathways in different plant-pathogen interactions.

3.3 La^{3+} Treatment Inhibits Only in Part HR Mediated Resistance

Former studies focussed on the prevention of PCD during HR by La^{3+}-treatment (Levine *et al.*, 1996; Pike *et al.*, 1998; Grant *et al.*, 2000) but there are no studies until now determining whether or not resistance is maintained in the absence of hypersensitive related cell death. We have found that *Pseudomonas savastanoi* pv. *phaseolicola* was still inhibited in the absence of HR cell death (Fig. 5A). Because bacterial cells may also be influenced by La^{3+}, we made efforts to show that the effects of La^{3+} are imposed on the plant, rather than on bacterial cells (Fig. 5B, 6).

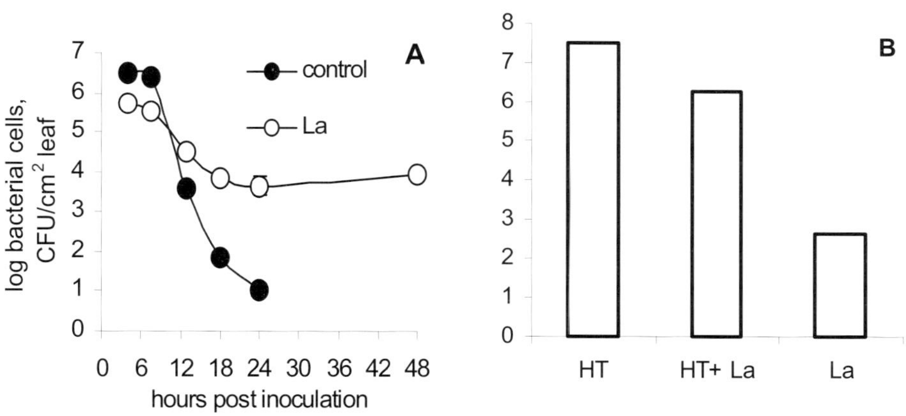

Figure 5. In planta numbers of *Pseudomonas savastanoi* pv. *phaseolicola* in La^{3+} -treated leaf tissue. A, time course. B, bacterial density at 19 h post inoculation. Control = bacterial suspension; La = bacterial suspension containing 1 mM La^{3+}; HT = heat treatment immediately before inoculation of bacteria; + = combined treatment.

During the HR, bacterial cells were killed more rapidly and efficiently than without plant cell death, i.e. with La^{3+}-treatment. Both *in planta* population curves followed similar inhibition kinetics, with a short lag at the beginning (Fig. 5A). In the La^{3+} -treated leaf tissue the inhibition of bacteria spanned about two orders of magnitude.

Heat treatment (HT) suppresses the HR temporally (for about 20 hours) and also early induced resistance, two early local defence responses known to affect bacterial cell numbers in tobacco (Klement *et al.*, 1999). As a result, incompatible bacteria can grow in HT tissue (Bozsó *et al.*, 1999). As compared to HT alone, the effect of La^{3+} on bacterial population in HT leaves was small (Fig. 5B). Thus, bacteria were not killed up to 19 h post inoculation, indicating that inhibition of bacteria is mostly due to an active plant response, rather than to a direct toxic effect of La^{3+}- on bacteria.

Apparently, La^{3+} and HT affect the HR in different ways: HT temporally inhibits both cell death and resistance, La^{3+} blocks only cell death completely, allowing a significant portion of resistance to operate.
In vitro, the La^{3+}-mediated effect on bacteria was very variable, depending on the particular cultivation medium: low on LB-medium, intermediate on minimal medium and strong in water (not shown). This *in vitro* effect had different kinetics as compared to the *in planta* effect (Fig. 5A). With increasing chemical complexity of the surrounding medium, bacterial cells seem to be more protected from the direct adverse effect of La^{3+}. Supporting the above data, La^{3+} did not inhibit *in planta* the induction of the pathogenicity gene, *hrpJ* after 19 hpi (Fig. 6.).

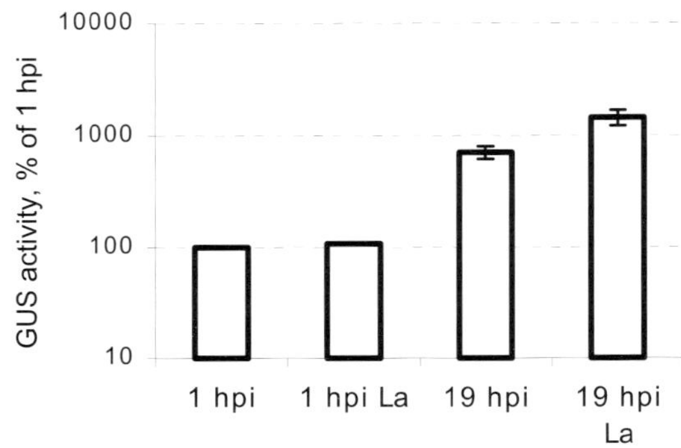

Figure 6. Effect of La^{3+} on *hrpJ* expression *in planta.* Before inoculation, bacteria were cultured on an amino acid-rich medium that suppresses the induction of pathogenicity genes. Leaves were heat-treated before inoculation to allow high level of *hrp* gene induction. 1 hpi = one hour post inoculation; 19 hpi = 19 hours post inoculation; La = bacterial suspension containing 1 mM La^{3+}.

In summary, La^{3+} inhibited only partly the HR-associated resistance toward bacteria. This is another example of uncoupling HR-resistance from HR-cell death. The "residual" antibacterial activity in La^{3+}-treated leaves resembles in its amplitude and timing the effect of EIR (Klement *et al.*, 2003). Further work is needed to determine by what means inhibition of bacteria in the absence of PCD. PR1a gene expression is probably not directly involved as only antifungal activity has been attributed to this PR protein (Rauscher *et al.*, 1999).

ACKNOWLEDGEMENT

This study was supported by grants OTKA T25363 and T038302 and AKP 2000-80 3,1.

REFERENCES

Ádám, A.L., Pike, S., Hoyos, M.E., Stone, J.M., Walker, J.C. and Novacky, A., 1997, Rapid and transient activation of a myelin basic protein kinase in tobacco leaves treated with harpin from *Erwinia amylovora*. *Plant Physiol.* **115**: 853-861.

Blume, B., Nürnberger, T., Nass, N. and Scheel, D., 2000, Receptor-mediated increase in cytoplasmic free calcium required for activation of pathogen defense in parsley. *Plant Cell* **12**: 1425-1440.

Bozsó, Z., Ott, P.G., Kecskés, M.L. and Klement, Z., 1999, Effect of heat and cycloheximide treatment of tobacco on the ability of *Pseudomonas syringae pv. syringae* 61 *hrp/hrmA* mutants to cause HR. *Physiol. Mol. Plant Pathol.* **55**: 215-223.

Grant, M., Brown, I., Adams, S., Knight, M., Ainslie, A. and Mansfield, J., 2000, The RPM1 plant disease resistance gene facilitates a rapid and sustained increase in cytosolic calcium that is necessary for the oxidative burst and hypersensitive cell death. *Plant J.* **23**: 441-450.

Groover, A. and Jones, A.M., 1999, Tracheary element differentiation uses a novel mechanism co-ordinating programmed cell death and secondary wall synthesis. *Plant Physiol.* **119**: 375-384.

Heath, M.C., 2000, Hypersensitive response-related death. *Plant Mol. Biol.* **44**: 321-334.

Hoyos, M.E., and Zhang, S., 2000, Calcium-independent activation of salicylic acid-induced protein kinase and a 40-kilodalton protein kinase by hyperosmotic shock. *Plant Physiol.* **122**: 1355-1363.

Jin, H., Axtell, M.J., Dahlbeck, D., Ekwenna, O., Zhang, S., Staskawicz, B. and Baker, B., 2002, NPK1, an MEKK1-like mitogen-activated protein kinase kinase, regulates innate immunity and development in plants. *Dev. Cell* **3**: 291-297.

Klement, Z., Bozsó, Z., Ott, P.G., Kecskés, M.L. and Rudolph, K., 1999, Symptomless resistant response instead of the hypersensitive reaction in tobacco after infiltration of heterologous pathovars of *Pseudomonas syringae*. *J. Phytopathol.* **12**: 479-489.

Klement, Z., Bozsó, Z, Kecskés, M.L., Besenyei, E., Czelleng, A. and Ott, P.G., 2003, Local early induced resistance of plants as the first line of defence against bacteria. *Pest Manag. Sci.* 59: 465-474.

Lebrun-Garcia, A., Ouaked, F., Chiltz, A. and Pugin, A., 1998, Activation of MAPK homologues by elicitors in tobacco cells. *Plant J.* **15**: 773-781.

Lee, J., Klessig, D.F., and Nürnberger, T., 2001, A harpin binding site in tobacco plasma membranes mediates activation of the pathogenesis-related gene HIN1 independent of extracellular calcium but dependent on mitogen-activated protein kinase activity. *Plant Cell* **13**: 1079-1093.

Ligterink, W., Kroj, T., zur Nieden, U., Hirt, H., and Scheel, D., 1997, Receptor-mediated activation of a MAP kinase in pathogen defence of plants. *Science* **276**: 2054-2057.

Pike, S.M., Ádám, A.L., Pu, X-A., Hoyos, M.E., Laby, R.J., Beer, S.V., and Novacky, A., 1998, Effects of *Erwinia amylovora* harpin on tobacco leaf cell membranes are related to leaf necrosis and electrolyte leakage and distinct from perturbations caused by inoculated *E. amylovora*. *Physiol. Mol. Plant Pathol.* **53**: 39-60.

Pontier, D., Gan, S., Amasino, R.M., Roby, D., and Lam., E., 1999, Markers for hypersensitive response and senescence show distinct patterns of expression. *Plant Mol. Biol.* **39**: 1243-1255.

Rauscher, M., Ádám, A.L., Wirtz, S., Guggenheim, R., Mendgen, K. and Deising, H., 1999, PR-1 protein inhibits the differentiation of rust infection hyphae in leaves of acquired resistant broad bean. *Plant J.* **19**: 625-633.

Romeis, T., Peidras, P., Zhang. S., Klessig, DF., Hirt, H. and Jones, J.D.G., 1999, Rapid Avr9- and Cf-9-dependent activation of MAP kinases in tobacco cell cultures and leaves:

convergence of resistance gene, elicitor, wound and salicylate responses. *Plant Cell* **11**: 273-287.

Sambrook, J., Fritsch, E.F., and Maniatis, T., 1989, *Molecular Cloning: a Laboratory Manual.* Vol. 2. Cold Spring Harbor Laboratory Press

Stiekema, W.J., Heidkamp, F., Dirkse, W.G., Van Beckum, J., De Haan, P., Bosh, T., and Lauwerse, J.D., 1988, Molecular cloning and analysis of four tuber specific mRNAs. *Plant Mol. Biol.* **11**: 255-269

Thordal-Christensen, H., Zhang, Z., Wei, Y., and Collinge, D.B., 1997, Subcellular localization of H_2O_2 accumulation in plants. H_20_2 accumulation in papillae and hypersensitive response during the barley-powdery mildew interaction. *Plant J.* **11**: 1187-1194.

Xiao, Y., Lu, Y., Heu, S., Hutcheson, S.W., 1992, Organization and environmental regulation of the *Pseudomonas syringae* pv. *syringae* 61 *hrp* cluster. *J. Bacteriol.* **174**: 1734-1741.

Xu, H-X., and Heath, M.C., 1998, Role of calcium in signal transduction during the hypersensitive response caused by basidiophore derived infection of the cowpea rust fungus. *Plant Cell* **10**: 585-597.

Yu, L.C., Parker, J., Bent, A.F., 1998, Gene-for-gene disease resistance without the hypersensitive response in *Arabidopsis dndl* mutant. *Proc. Natl. Acad. Sci. USA* **95**: 7819-7824.

Zhou, J., Tang, X., Frederick, R., and Martin, G., 1998, Pathogen recognition and signal transduction by the Pto kinase. *J. Plant Res.* **111**: 353-356.

Molecular Characterisation / Genomics

Pseudomonas in the Underworld:
The Secret Life of *Pseudomonas fluorescens* SBW25

G. PRESTON, A. SPIERS, X.-X. ZHANG, R. JACKSON, M. GAL, C. KNIGHT, S. GEHRIG, J. MALONE, C. MOON, S. GODFREY, Z. ROBINSON, N. BERTRAND*, D. FIELD* and P. RAINEY[#]
*Department of Plant Sciences, University of Oxford, South Parks Road, Oxford OX1 3RB, UK; *NERC Centre for Ecology and Hydrology, Mansfield Road, Oxford, OX1 3SR, UK; [#]School of Biological Sciences, University of Auckland, Auckland, PB92019, NZ.*

Abstract: *Pseudomonas fluorescens* SBW25 is a gram-negative, fluorescent pseudomonad that is able to colonise the roots and leaves of a wide range of plants. *Pf*SBW25 has been widely adopted as a model organism for investigations into the molecular basis of plant colonisation, plant growth promotion and biocontrol, and is also used as a model organism in studies of biofilm formation and experimental evolution. The 6.7 Mb genome of *Pf*SBW25 is currently being sequenced at the Sanger Institute. We have developed two *in vivo* expression technology (IVET) systems for *P. fluorescens* SBW25, and have used IVET as a functional genomics approach to identify and characterise more than a hundred plant-induced genes from *P. fluorescens* SBW25. The predicted functions of many of these genes can be assigned to broad functional categories such as nutrient acquisition, surface colonisation, stress tolerance, antibiotic biosynthesis, protein secretion and transcriptional regulation. We have begun the process of analysing the function and expression of some of these plant-induced genes in depth, focusing in particular on genes encoding a rhizosphere-induced type III protein secretion system (TTSS) that is similar to the Hrp TTSS of *P. syringae*. Functional genomic analyses of *P. fluorescens* SBW25 have provided insight into the biology and ecology of plant-colonising *Pseudomonas*, and into the utility and limitations of functional genomic techniques in understanding the biology of bacteria in complex environments.

Key words: functional genomics, IVET, type III secretion system, rhizosphere.

Author for correspondence: G. Preston; e-mail: gail.preston@plants.ox.ac.uk.

N.S. Iacobellis *et. al.* (eds.), *Pseudomonas syringae* and Related Pathogens, 347-353.

1. INTRODUCTION

Pseudomonas fluorescens SBW25 is a model plant-associated bacterium that is able to colonise the roots and foliage of a wide range of host plants. *P. fluorescens* SBW25 exhibits biocontrol activity against soil-borne pathogens such as *Rhizoctonia* and *Pythium*, and has been shown to stimulate induced systemic resistance (ISR) in *Arabidopsis thaliana* against *P. syringae* pv. *tomato* (Naseby *et al.* 2001; Hanson, Preston, Rainey and Pieterse, unpublished). The 6.7 Mb genome of *P. fluorescens* SBW25 is currently being sequenced in a BBSRC-funded project at the Sanger Institute and is predicted to be complete in 2003 (www.sanger.ac.uk/Projects/P_fluorescens). Genome sequence data are also available for the plant pathogen *P. s.* pv. *tomato* DC3000 and *P. s.* pv. *syringae* B728a, and for *P. aeruginosa* PAO1, *P. putida* KT2440, *P. fluorescens* Pf-5 and *P. fluorescens* PfO-1 (TIGR, www.tigr.org; Stover *et al.,* 2000; JGI/DOE, www.jgi.doe.gov). The data from these genome sequence projects provide a strong foundation for comparative analyses of the biology of *Pseudomonas*, the molecular basis of plant colonisation, and the evolution of host-microbe interactions.
We have undertaken the first steps in the functional genomic analysis of *P. fluorescens* SBW25 by developing and applying techniques known as IVET (*in vivo* expression technology) and RIVET (recombinase IVET) to study plant colonisation by *P. fluorescens* SBW25. We have used these techniques to identify phyllosphere and rhizosphere-induced genes, and to study the temporal and spatial expression of these genes (Rainey, 1999; Rainey and Preston, 2000; Gal, Preston, Massey and Rainey, in preparation; Zhang and Rainey, in preparation). We have also developed an online database of functional genomic data from *P. fluorescens* SBW25 and general information on *Pf*SBW25 (www.plants.ox.ac.uk/sbw25) (Spiers *et al.,* 2001).

2. MATERIALS AND METHODS

2.1 *In vivo* Expression Technology (IVET)

We have developed two complementary IVET systems based on *panB* and *dapB*, which are required for biosynthesis of pantothenate and diaminopimelate respectively (Rainey, 1999; Gal, Preston, Massey, Spiers and Rainey, submitted). We constructed genomic libraries of 3 kb fragments of *Pseudomonas fluorescens* SBW25 DNA in the integrative IVET vectors pIVETP and pIVETD, which contain a multiple cloning site (MCS) in front of a promoterless copy of *panB* or *dapB*, followed by a promoterless reporter cartridge encoding *lacZY* (β-galactosidase). IVET libraries were introduced into *panB* or *dapB* deletion mutants of *P. fluorescens* SBW25. Pools of 100÷300 recombinant strains were inoculated onto sugar beet seeds and selected on the basis of their ability to persist for two-three weeks in the phyllosphere or rhizosphere. Strains were subsequently screened *in vitro* for expression of β-galactosidase to exclude strains with constitutively expressed promoters and to identify a subset of fusions that are induced during plant colonisation. Plasmids were recovered into *Escherichia coli* by triparental mating and sequenced to identify putative plant-induced genes.

2.2 Functional Analysis of TTSS Expression

IVET, reporter fusions and HR elicitation were used to analyse the expression of type III secretion system (TTSS) genes in *Pseudomonas fluorescens* SBW25. Promoter sequences from each of the TTSS operons were cloned into pIVETD and the integrative reporter plasmid pGFPGUS, which carries a promoterless *gfp::uidA* (green fluorescent protein and β-glucuronidase) cartridge. TTSS reporter plasmids were used in conjunction with regulatory mutants, and plasmids constitutively expressing RspR and RspL to confirm that the TTSS system was expressed in the rhizosphere and regulated by RspR and RspL. Plant assays were performed as described in Preston *et al.* (2001).

3. RESULTS AND DISCUSSION

3.1 Functional Genomic Analysis of Plant-induced Genes

We have used *panB* and *dapB*-based IVET systems to identify plant-induced genes from *Pseudomonas fluorescens* SBW25. *panB* and *dapB* mutants of *P. fluorescens* SBW25 that are compromised in their ability to colonise the rhizosphere and phyllosphere but can be rescued *in vitro* by adding pantothenate or diaminopimelate and lysine to culture media. We have tailored our screens to specifically identify genes that are expressed during the first two-three weeks following germination of non-sterile sugar beet seedlings, but which are not expressed when bacteria are cultured on minimal medium (M9 agar). We have identified more than a hundred plant-induced gene fusions to date, and aim to complete a comprehensive screen of the *P. fluorescens* SBW25 genome in the near future.

The predicted functions of many of these plant-induced gene fusions can be assigned to eight broad functional categories: (1) nutrient acquisition; (2) surface colonisation; (3) stress tolerance; (4) non-ribosomal peptide synthesis (antibiotics, toxins, siderophores); (5) protein secretion (type II and type III); (6) transcriptional regulation; (7) motility and chemotaxis; (8) phytohormone biosynthesis. We have also identified a significant number of fusions with homology to hypothetical proteins of unknown function, and a small number of fusions with little homology to previously characterised genes. Analysis of plant-induced gene fusion sequences in conjunction with the unfinished genome sequence of *P. fluorescens* SBW25 has enabled us to reduce the number of fusions that fall into the latter category, and to more accurately assign predicted functions to other fusions.

Approximately 80% of the fusions identified to date are most similar to sequences from other *Pseudomonas*, but a significant number are most similar to sequences from other gram-negative and gram-positive bacteria, and in one instance, to a plant gene. *P. putida* KT2440 contains the greatest number of "top hits", but a number of fusions have homology to genes in *P. syringae* pv. *tomato* DC3000 and not in *P. putida*, or have highest homology to *P. syringae* and other plant-associated bacteria, which suggests that many plant-induced genes may have been acquired by horizontal gene transfer.

We have isolated fusions from both the rhizosphere and the phyllosphere, and we are systematically testing the ability of individual fusions to promote colonisation of both environments. While many genes are expressed in both environments, there appears to be a significant level of differential expression for many fusions, which suggests that some plant-induced gene fusions are responding to rhizosphere or phyllosphere-specific signals (Gal, Preston, Massey, Spiers and Rainey, manuscript submitted).

The collective results of the IVET screen present a detailed picture of life on plants in which *P. fluorescens* SBW25 survives as a saprophyte by degrading and scavenging a diverse range of nutrients, using chemotaxis and polymer synthesis to effectively spread through the rhizosphere and across plant surfaces. *P. fluorescens* SBW25 expresses genes that enable it to tolerate osmotic and oxidative stress and the damaging effects of ultra-violet light, and to defend itself against competition and against antibiotics. Intriguingly, the presence of genes with homology to genes from plant pathogens such as *P. syringae*, such as the TTSS genes described in Preston *et al.* (2001), suggests that there may be more intimate and direct communication between *P. fluorescens* SBW25 and host plants than has previously been suspected.

3.2 Type III Secretion in *Pseudomonas fluorescens* SBW25

Cell contact dependent interactions mediated by type III protein secretion systems (TTSSs) are one of the most important factors determining parasitism, host specificity and induction or suppression of host defences by plant pathogenic *Pseudomonas* such as *P. syringae* (Preston, 2000). The discovery of TTSS genes in *P. fluorescens* SBW25, a plant growth-promoting, ISR-inducing bacterium, suggests that TTSSs may play a significant role in interactions between some non-pathogenic rhizobacteria and plants (Rainey, 1999; Preston *et al.,* 2001).

The type III (Rsp) secretion system of *P. fluorescens* SBW25 is encoded by a 20 kb gene cluster that is similar to the *hrp* gene cluster of *P. syringae.* The cluster encodes components of the TTSS machinery, TTSS regulatory proteins and a putative TTSS-secreted protein, RopE, which is homologous to the virulence protein AvrE from *P. syringae* (Lorang *et al.,* 1995; Preston *et al.,* 2001). Regulatory proteins RspR, RspL and RspV are homologous to corresponding proteins from *P. syringae* (HrpR/S, HrpL, HrpV), and operate in a similar regulatory hierarchy, but are unresponsive to environmental signals that regulate TTSS expression in *P. syringae. rsp* genes are expressed at a low level in the rhizosphere and under a wide range of environmental conditions. Constitutive expression of RspL or RspR *in trans* elevates expression of *rsp* genes and *ropE* (Rainey, 1999; Preston *et al,* 2001; Preston, Jackson, Bertrand and Rainey, in preparation).

*Pf*SBW25 can deliver Avr proteins such as the *P. syringae* protein AvrB to plant cells and elicits the HR in *Arabidopsis thaliana* Col-0 leaves when AvrB and the positive regulators RspL or RspR are constitutively expressed *in trans*, albeit with greatly reduced efficiency compared to *P. syringae. Pf*SBW25 also elicits the HR in *Nicotiana clevelandii* when RspL or RspR is expressed *in trans*, indicating that *P. fluorescens* SBW25 carries a elicitor that may affect the interaction of the bacterium with the host plant (Preston *et al.,* 2001).

TTSS mutants of *P. fluorescens* SBW25 are able to reach high population levels in the rhizosphere and phyllosphere, which suggests that TTSS activity is not a critical factor in the overall ability of *P. fluorescens* SBW25 to colonise plants (Preston *et al.,* 2001). Factors such as competition for available nutrients and antibiotic biosynthesis or resistance may exert a stronger effect. The absence of TTSS-related effects on plant colonisation may reflect the fact that TTSS-mediated interactions with plant cells only have a significant effect on endophytic bacteria and bacteria that are in close contact with plant cells. It is also possible that the TTSSs of non-pathogenic

P. fluorescens, like those of *P. syringae*, may only have a marked effect on interactions with specific hosts.

3.3 Post-IVET Analyses of Gene Function

Functional genomic techniques such as IVET provide a powerful method of identifying interesting and potentially significant genes. However, the complexity of environments such as the rhizosphere, and the pleiotropic nature of traits such as plant colonisation present a substantial challenge to researchers seeking to further analyse the function and ecological significance of these genes. Genes such as *panB* and *dapB* play an important role in a wide range of environments, and mutants are strongly impaired in survival. However, a large number of genes will have variable expression and significance at the microenvironment level, or have ecological significance that varies as a result of host specificity or genetic background. It is inevitable that genomic and functional genomic analyses of bacteria will identify genes that are functionally redundant, or of minimal ecological significance, even though homologous genes may be highly significant in other genetic backgrounds, as has been observed for several Avr genes in *Pseudomonas syringae* (Vivian and Arnold, 2000). The identification and characterisation of TTSS genes in *P. fluorescens* SBW25 and other non-pathogenic *Pseudomonas* provides a good example of the opportunities and challenges facing researchers and of some of the techniques and approaches that can be used to address these challenges and investigate gene function.

We are already developing some of the technology needed to address the challenge posed by environmental heterogeneity. Recombinase IVET (RIVET) has been developed for *P. fluorescens* SBW25 by Zhang and Rainey (manuscript in preparation). This technique uses promoter fusions to a recombinase gene to identify bacteria in which promoters have been activated, and can be used in conjunction with fluorescent reporter proteins to map and measure gene expression at a spatial and temporal level with high resolution and low background (Lee and Camilli, 2000).

4. CONCLUSIONS

Functional genomic analyses have provided a description of plant-induced gene expression in *Pseudomonas fluorescens* SBW25, and a glimpse into the complexity of the environment inhabited by plant-associated *Pseudomonas.* The overlapping ecological niches of plant-pathogenic and plant growth-promoting bacteria provide these bacteria with the opportunity to trade genetic information and acquire traits that enhance their common ability to colonise plant hosts. Sequence analyses of IVET fusions support the suggestion that there is a high degree of horizontal gene transfer and recombination among plant-associated *Pseudomonas*. Comparative analyses of the complete genome sequences of several *Pseudomonas* in the coming months is likely to reveal many more intriguing similarities and differences between *Pseudomonas* with different lifestyles. There is a pressing need for the development of post-genomic technology that will meet the challenge of analysing and understanding the evolution, biology and ecology of this diverse and fascinating group of organisms.

ACKNOWLEDGEMENTS

We would like to thank Linda Hanson and Corné Pieterse for sharing data prior to publication. Research described here was supported by grants from the BBSRC, NERC and The Royal Society. The genome of *P. fluorescens* SBW25 is currently being sequenced as a BBSRC funded project at the Sanger Institute, Cambridge, UK (www.sanger.ac.uk/Projects/P_fluorescens).

REFERENCES

Lee, S.H., and Camilli, A., 2000, Novel approaches to monitor bacterial gene expression in infected tissue and host. *Curr. Opin. Microbiol.* **3**: 97-101.

Lorang, J.M., and Keen, N.T., 1995, Characterization of *avrE* from *Pseudomonas syringae* pv. *tomato*: a *hrp*-linked avirulence locus consisting of at least two transcriptional units. *Mol. Plant-Microbe Interact.* **8**: 49-57.

Naseby, D.C., Way, J.A., Bainton, N.J., and Lynch, J.M., 2001, Biocontrol of *Pythium* in the pea rhizosphere by antifungal metabolite producing and non-producing *Pseudomonas* strains. *J. Appl. Microbiol.* **90**: 421-429.

Preston, G.M., Bertrand, N., and Rainey, P.B., 2001, Type III secretion in plant growth-promoting *Pseudomonas fluorescens* SBW25. *Mol. Microbiol.* **41**: 999-1014.

Preston, G,M., 2000, *Pseudomonas syringae* pv. tomato: the right pathogen, of the right plant, at the right time. *Mol. Plant Pathol.* **1:** 263-275.

Rainey, P.B., 1999, Adaptation of *Pseudomonas fluorescens* to the plant rhizosphere. *Env. Microbiol.* **1**: 243-257.

Rainey, P.B., and Preston, G.M., 2000, *In vivo* expression technology strategies: valuable tools for biotechnology. *Curr. Opin. Biotechnol.* **11**: 440-444.

Spiers, A.J., Field, D., Bailey, M., and Rainey, P.B., 2001, Notes on designing a partial genomic database: The PfSBW25 Encylopaedia, a sequence database for *Pseudomonas fluorescens* SBW25. *Microbiol.* **147**: 247-249.

Stover *et al.,* 2000, Complete genome sequence of *Pseudomonas aeruginosa* PAO1, an opportunistic pathogen. *Nature* **147**: 947-948.

Vivian, A., and Arnold, D.L., 2000, Bacterial effector proteins and their role in host-pathogen interactions. *J. Plant Pathol.* **82**: 163-178.

Regulation of Coronatine Biosynthesis in *Pseudomonas syringae*

C.L. BENDER
Department of Entomology and Plant Pathology, Oklahoma State University, Stillwater, OK 740784, USA.

Abstract: Coronatine (COR) is a chlorosis-inducing phytotoxin produced by *Pseudomonas syringae* pv. *tomato* DC3000 and *P. s.* pv. *glycinea* PG4180. The *cfl* gene, which encodes *c*orona*f*acate *l*igase, is required for COR production in both strains. Sequence analysis indicated that the *cfl* gene is conserved in DC3000 and PG4180 strains, but the *cfl* upstream regions diverge in the two strains. The *cfl* promoter regions in DC3000 and PG4180 were fused to *uidA* (encoding glucuronidase), and transcriptional fusions were used to investigate the regulation of COR production. *In vitro* assays indicated that the DC3000 *cfl* promoter was activated within 6 to 24 h after incubation at 21°C in COR-inducing media, with expression declining thereafter. However, expression of *cfl* in strain PG4180 increased gradually beginning at 12 h, and continued to increase throughout the sampling period. Histochemical staining for GUS activity *in planta* demonstrated that *cfl* transcription is activated prior to symptom development in tomato and collard plants inoculated with strain DC3000. However, *cfl* expression in soybeans inoculated with PG4180 occurred more slowly and was most evident at the onset of chlorosis. Therefore, the temporal expression of *cfl* activity *in planta* is consistent with results obtained *in vitro*. The implication of these results with respect to the role of COR in disease development and symptom manifestation is discussed.

Key words: glucuronidase, histochemical staining, phytotoxin, virulence.

Author for correspondence: C. Bender; e-mail: cbender@okstate.edu.

1. INTRODUCTION

The genetic basis of pathogenicity and virulence in *Pseudomonas syringae* is complex and includes global regulators, a type III secretion system (the *hrp*

N.S. Iacobellis *et. al.* (eds.), *Pseudomonas syringae* and Related Pathogens, 355-362.

cluster) (Galán and Collmer, 1999), and virulence factors such as phytotoxins (Bender *et al.*, 1999). Coronatine (COR) is a chlorosis-inducing phytotoxin produced by several *P. syringae* pathovars, including *P. s.* pv. *glycinea* PG4180 (a pathogen of soybean) and *P. s.* pv. *tomato* DC3000 (a pathogen of *Arabidopsis thaliana*, *Brassica* spp., and tomato). COR consists of two parts: coronafacic acid (CFA), which is derived from the polyketide pathway, and coronamic acid (CMA), an ethylcyclopropyl amino acid derived from isoleucine (Bender *et al.*, 1999). The COR gene cluster in *P. s.* pv. *glycinea* PG4180 has been extensively characterised and is borne on a 90-kb plasmid (Bender *et al.*, 1999). In *P. s.* pv. *glycinea* PG4180, the biosynthetic genes for CMA and CFA are located on opposing ends of the COR gene cluster and are separated by a 3.4-kb regulatory region (Bender *et al.*, 1999). In contrast, the chromosomally-encoded COR genes in DC3000 have not been well characterised (Moore *et al.*, 1989); however, *P. s.* pv. *tomato* DC3000 has become a model strain for investigating plant-microbe interactions because of its genetic tractability, pathogenicity on *Arabidopsis*, and the effort to deduce its genomic sequence [http://www.tigr.org].
When compared *in vitro*, COR production by *P. s.* pv. *glycinea* PG4180 is ~25 to 40-fold higher than DC3000 (Peñaloza-Vázquez *et al.*, 2000); however, both DC3000 and PG4180 produce similar amounts of COR *in planta* (Zhao *et al.*, 2001). The latter finding is consistent with reports showing that COR synthesis is induced by plant signals in DC3000 (Li *et al.*, 1998). To further understand the differential regulation of COR production in strains PG4180 and DC3000, we undertook a comparative study of the *cfl* promoters from these two strains. Transcriptional fusions containing the PG4180 and DC3000 *cfl* promoter regions were utilised to evaluate how COR production is regulated *in vitro* and *in planta.*

2. MATERIALS AND METHODS

2.1 Bacterial Strains, Transcriptional Fusions, and Inoculation Procedures

The bacterial strains and plasmids used in this study are listed in Table 1. Oligonucleotide primers were synthesised by the OSU Recombinant DNA/Protein Resource Facility. The primers used to amplify the *cfl* upstream region from DC3000 have been described elsewhere (Wang *et al.*, 2002).

Table 1. Bacterial strains and Plasmids

Strain or plasmid	Characteristics	Reference
P. s. tomato DC3000	COR^+ Rif^r	Moore *et al.*, 1989
P. s. glycinea PG4180.N9	COR^+ Km^r	Ullrich *et al.*, 1994
Plasmids		
pBBR.Gus	Cm^r; 6.6-kb promoter probe vector containing the promoterless *uidA* gene	Peñaloza-Vázquez and Bender, 1998
pCFL.560a	Cm^r; *cfl-uidA* transcriptional fusion from DC3000 in pBBR.Gus	Wang *et al.*, 2002
pCFLP	Cm^r; contains the PG4180 *cfl* promoter	Peñaloza-Vázquez and Bender, 1998

Tomato, collard, and soybean seedlings were maintained in a growth chamber at 24°C, 40% RH, with a 12 h photoperiod. DC3000(pBBR.Gus) and DC3000(pCFLP) were grown for 36 h at 28°C, and cells were suspended (OD_{600}=0.3) in distilled water. Four-week-old tomato and collard plants were spray-inoculated with an airbrush (~8 psi) until leaf surfaces were uniformly wet. PG4180.N9(pBBR.Gus) and PG4180.N9(pCFLP) were suspended in sterile distilled water at OD_{600}=0.4 and used to inoculate soybean plants.

2.2 GUS assays

Prior to inoculation, bacterial strains were grown on agar for 48 h at 28°C. Cells were then removed and suspended to an OD_{600}=0.05 in Hoitink-Sinden minimal medium optimised for COR production (HSC) (Palmer and Bender, 1993). Aliquots were then inoculated to 10 ml HSC and incubated at 21°C, 250 rpm. At different time points (0, 6, 12, 24, and 48 h after inoculation), bacterial cells were removed from each tube for GUS assays (Peñaloza-Vázquez and Bender, 1998). The protein concentration of cell lysates was determined using the Bio-Rad (Richmond, Calif.) protein assay kit. Glucuronidase activity (GUS) was expressed in units (U) per mg of protein, with 1 U equivalent to 1 nmol of methylumbelliferone formed min^{-1}.

At each sampling time, three leaves were randomly removed from three separate plants. Leaves were either immediately weighed and macerated in 2 ml sterile water or kept in a sealed petri dish at 4°C prior to maceration. Leaves were vacuum-infiltrated with a substrate-surfactant solution (X-Gluc, 0.5 mg ml^{-1}; and L77 at 0.2 µl ml^{-1} in 0.5 mM sodium phosphate buffer, pH 7.0), incubated at 37°C overnight, and fixed and decolourised in 80% ethanol at 37°C (Hugouvieux *et al.*, 1998).

3. RESULTS

3.1 GUS Assays *in vitro*

The transcriptional activity of pCFLP (*cfl* promoter from PG4180) and pCFL.560a (*cfl* promoter from DC3000) was evaluated in both DC3000 and PG4180.N9 (Fig. 1). In DC3000, both fusions were activated within 12 h, with GUS activity decreasing rapidly thereafter (Fig. 1). In contrast, the expression of both fusions in PG4180 began to increase at 12 h and continued to rise throughout the sampling period (Fig. 1). It is important to note that the two *cfl* promoters had very different expression levels *in vitro*. Transcriptional activity of the *cfl* promoter from PG4180 (construct pCFLP) varied from a maximum of 2000 U in strain DC3000 to 9000 U in PG4180 (Fig. 1A). Although the maximal activity of the *cfl* promoter from DC3000 (construct pCFL.560a) was not significantly different in DC3000 or PG4180, overall expression of this construct was much lower than pCFLP (Fig. 1). These results indicate that both the bacterial host and the promoter construct function to impact the level of *cfl* expression. In all experiments, GUS activity in DC3000 and PG4180 containing the promoter probe vector, pBBR.Gus, was negligible.

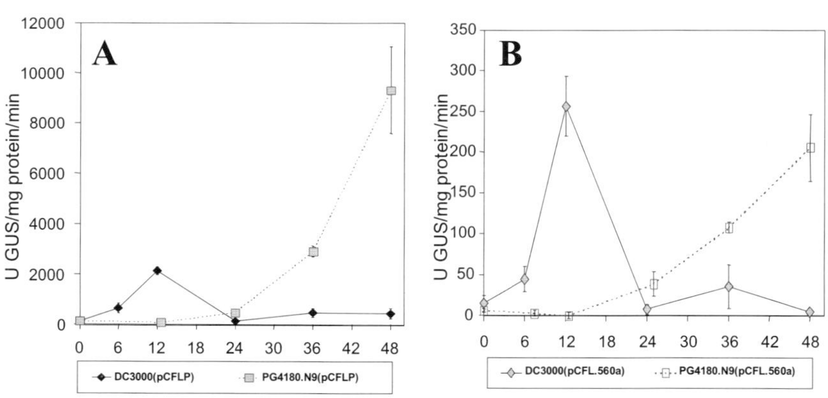

Figure 1. GUS activity in *Pseudomonas syringae* pv. *tomato* DC3000 and *P. s.* pv. *glycinea* PG4180.N9 containing pCFLP (**A**) and pCFL.560a (**B**). Constructs pCFLP and pCFL.560a contain *cfl-uidA* transcriptional fusions from PG4180 and DC3000, respectively. The values shown represent the means from one experiment of three replicates per strain, and vertical bars indicate the standard error.

3.2 Symptom Development and GUS Activity *in planta*

Histochemical staining was used to study expression of the *cfl* promoter *in planta*. When the *cfl* promoter is activated, β-glucuronidase will be produced, and leaves incubated with the chromogenic substrate X-Gluc will stain blue where the bacteria are located. When DC3000(pBBR.Gus) and DC3000(pCFLP) were inoculated to tomato seedlings, necrotic lesions were not readily visible until 96 h after inoculation (Fig. 2A). The chlorosis associated with COR production was readily visible at 168 h (Fig. 2A). On tomato leaves infected with DC3000(pCFLP), small blue spots were seen 6-12 h after inoculation (Fig. 2B), indicating that the *cfl* promoter was activated very quickly and before visible symptoms were evident. Maximal GUS staining was evident 48 h after inoculation (Fig. 2B), which coincided with the appearance of microscopic water-soaked lesions. In collard leaves inoculated with DC3000(pCFLP), a similar pattern was observed; the *cfl* promoter was activated within 24 h, which was before the onset of visible lesions (data not shown).

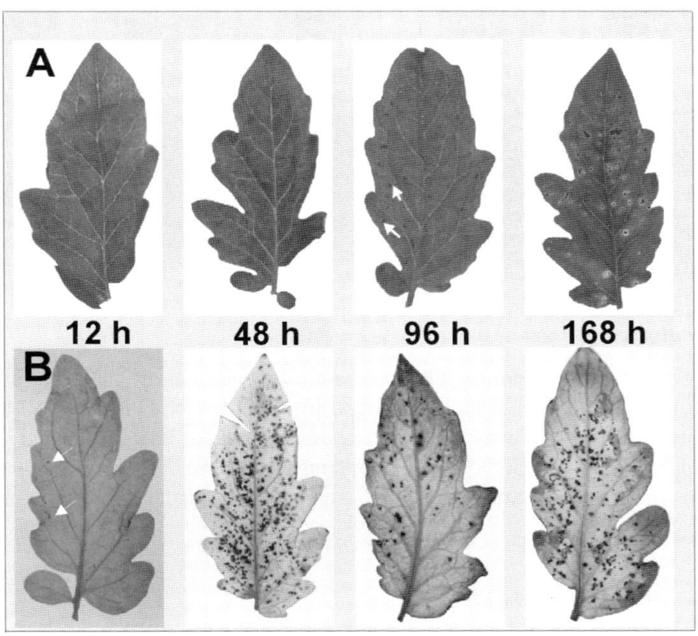

Figure 2. Symptom development and histochemical localisation of GUS activity in tomato leaves inoculated with *Pseudomonas syringae* pv. *tomato* DC3000(pCFLP). Inoculum was sprayed (OD_{600}=0.3) onto tomato leaves until surfaces were uniformly wet. After inoculation, plants were incubated in a growth chamber maintained at 24°C, 88% RH. Symptoms (A) and GUS activity (B) were monitored at 12, 48, 96, and 168 h after inoculation. Arrows at 96 h and 12 h indicate the location of necrotic lesions (panel A) and GUS activity (panel B), respectively.

When PG4180.N9(pBBR.Gus) and PG4180.N9(pCFLP) were inoculated to soybean seedlings, chlorotic haloes were evident at 96 h, and necrotic lesions were first observed at 168 h (Fig. 3A). In contrast to the results obtained with tomato and collard, GUS activity was not detected in soybean until 48 h post-inoculation and was only visible by light microscopy. The blue precipitate indicative of GUS activity was readily visible at 96 h (Fig. 3B) and was synchronised with the development of chlorotic lesions.
No GUS activity was evident when tomato and collard were inoculated with DC3000(pBBR.Gus) and soybeans were inoculated with PG4180(pBBR.Gus). Furthermore, the symptoms induced by DC3000 and PG4180 containing pBBR.Gus were indistinguishable from those observed with DC3000 and PG4180 containing pCFLP (data not shown).

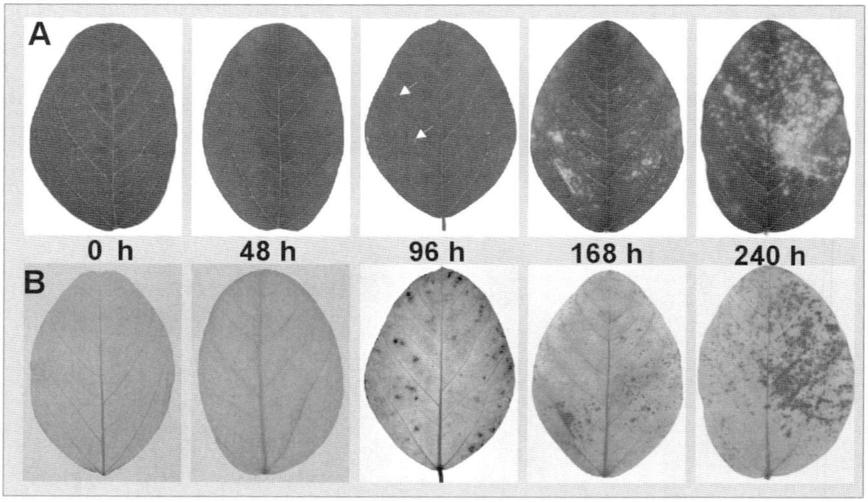

Figure 3. Symptom development and histochemical localisation of GUS activity in soybean leaves inoculated with *Pseudomonas syringae* pv. *glycinea* PG4180.N9(pCFLP). Inoculum was sprayed (OD_{600}=0.4) onto soybean leaves, and plants were incubated in a growth chamber maintained at 24°C, 88% RH. Symptoms (A) and GUS activity (B) were monitored at 0, 48, 96, 168, and 240 h after inoculation. Arrows show location of small chlorotic lesions.

4. DISCUSSION

Multiple factors, including osmolarity, temperature, carbon source, and host signal molecules, impact COR production and transcriptional activity (Palmer and Bender, 1993; Palmer *et al.*, 1997; Li *et al.*, 1998;). In the present study, the evaluation of *cfl* promoter activity *in vitro* indicated that *cfl* expression was dependent on the genetic background of the host strain. When monitored *in vitro*, *cfl* transcriptional activity was much lower in

DC3000 than PG4180 (Fig. 1A, B). This correlates well with the amount of COR synthesised *in vitro*, since DC3000 produces 25 to 40-fold less COR than PG4180 (Peñaloza-Vázquez *et al*., 2000). Furthermore, both pCFLP and pCFL.560a (the *cfl* promoters from PG4180 and DC3000, respectively) were activated very early in DC3000, whereas the expression of these two promoters was more gradual in PG4180 (Fig. 1). When transcriptional activity was monitored in DC3000(pCFLP) and PG4180(pCFLP) by histochemical staining, a similar pattern was observed. For example, GUS activity was evident very soon after inoculation when DC3000 was inoculated to tomato and collard (Fig. 2B); in contrast, GUS activity was delayed for PG4180 (Fig. 3B).

These results have greater implications when considered with respect to the importance of COR in different host-pathogen interactions. For example, COR^- mutants of DC3000 failed to multiply and cause disease in tomatoes when inoculated by dipping or spraying (Mittal and Davis, 1995; Peñaloza-Vázquez *et al*., 2000), indicating that COR may be required for invasion and symptom development in selected host-pathogen interactions. Furthermore, COR production by DC3000 was shown to decrease the expression of several genes involved in the plant defence response (Mittal and Davis, 1995). Many plant defence systems are activated very quickly after pathogen attack. Therefore, if COR is to be effective in suppressing such defence systems, it must be expressed very quickly after infection. In the present study, the *cfl* promoter in DC3000 was activated very quickly, which is consistent with a potential role for COR in suppressing host defence.

In contrast, COR^- mutants of PG4180 still cause some disease in soybean, although symptom severity is reduced in comparison with the wild-type PG4180 (Budde and Ullrich, 2000). Furthermore, *cfl* expression in PG4180 occurs later than in DC3000, possibly because PG4180 is able to establish a successful infection in the absence of COR. In the PG4180-soybean pathosystem, COR presumably plays a more important role in symptom development but is not required for successful invasion of the host plant, which provides a plausible explanation for why COR^- mutants still cause disease in soybean. In summary, the timing of *cfl* gene expression and COR synthesis may play a vital role in the outcome of selected host-pathogen interactions.

ACKNOWLEDGEMENTS

This research was funded in part by the Oklahoma Agricultural Experiment Station and National Science Foundation grants MCB-9807774 and IBN-0130693.

REFERENCES

Bender, C.L., Alarcón-Chaidez F., Gross, D.C., 1999, *Pseudomonas syringae* phytotoxins: mode of action, regulation and biosynthesis by peptide and polyketide synthetases. *Microbiol. Mol. Biol. Rev.* **63**: 266-292.

Budde, I.P., Ullrich, M.S., 2000, Interactions of *Pseudomonas syringae* pv. *glycinea* with host and nonhost plants in relation to temperature and phytotoxin synthesis. *Mol. Plant-Microbe Interact.* **13**: 951-961.

Galán, J.E., Collmer, A., 1999, Type III secretion machines: bacterial devices for protein delivery into host cells. *Science* **284**: 1322-1328.

Hugouvieux, V., Barber, C.E., Daniels, M.J., 1998, Entry of *Xanthomonas campestris* pv. *campestris* into hydathodes of *Arabidopsis thaliana* leaves: a system for studying early infection events in bacterial pathogenesis. *Mol. Plant-Microbe Interact.* **11**: 537-543.

Li, X-Z, Starratt, A.N., Cuppels, D.A., 1998, Identification of tomato leaf factors that activate toxin gene expression in *Pseudomonas syringae* pv. *tomato* DC3000. *Phytopathology* **88**: 1094-1100.

Mittal, S.M., Davis, K.R., 1995, Role of the phytotoxin coronatine in the infection of *Arabidopsis thaliana* by *Pseudomonas syringae* pv. *tomato*. *Mol. Plant-Microbe Interact.* **8**: 165-171.

Moore, R.A., Starratt, A.N., Ma, S.-W., Morris, V.L., Cuppels, D.A., 1989, Identification of a chromosomal region required for biosynthesis of the phytotoxin coronatine by *Pseudomonas syringae* pv. *tomato*. *Can. J. Microbiol.* **35**: 910-917.

Palmer, D.A., Bender, C.L., 1993, Effects of environmental and nutritional factors on production of the polyketide phytotoxin coronatine by *Pseudomonas syringae* pv. glycinea. *Appl. Environ. Microbiol.* **59**: 1619-1626.

Palmer, D.A., Bender, C.L., Sharma, S., 1997, Use of Tn*5-gusA5* to investigate environmental and nutritional effects on gene expression in the coronatine biosynthetic gene cluster of *Pseudomonas syringae* pv. glycinea. *Can. J. Microbiol.* **43**: 517-525.

Peñaloza-Vázquez, A., Bender, C.L., 1998, Characterisation of CorR, a transcriptional activator which is required for biosynthesis of the phytotoxin coronatine. *J. Bacteriol.* **180**: 6252-6259.

Peñaloza-Vázquez, A., Preston, G.M., Collmer, A., Bender, C.L., 2000, Regulatory interactions between the Hrp type III protein secretion system and coronatine biosynthesis in *Pseudomonas syringae* pv. tomato DC3000. *Microbiology* **146**: 2447-2456.

Ullrich, M., Guenzi, A.C., Mitchell, R.E., Bender, C.L., 1994, Cloning and expression of genes required for coronamic acid (2-ethyl-1-aminocyclopropane 1-carboxylic acid), an intermediate in the biosynthesis of the phytotoxin coronatine. *Appl. Environ. Microbiol.* **60**: 2890-2897.

Wang, X., Alarcón-Chaidez, F., Peñaloza-Vázquez, A., and Bender, C.L., 2002, Differential regulation of coronatine biosynthesis in *Pseudomonas syringae* pv. *tomato* DC3000 and *P. syringae* pv. *glycinea* PG4180. *Physiol. Mol. Plant Pathol.* **60**: 111-120.

Zhao, Y.F., Jones, W.T., Sutherland, P., Palmer, D.A., Mitchell, R.E., Reynolds, P.H.S., Damicone, J.P., and Bender, C.L., 2001, Detection of the phytotoxin coronatine by ELISA and immunolocalization in infected plant tissue. *Physiol. Mol. Plant Pathol.* **58**: 247-258.

Genomic Mining for Substrates of the Type III Secretion System of *Pseudomonas syringae* pv. *tomato* DC3000: New Insights into Mechanisms of Pathogenesis

J.R. ALFANO, C.R. BUELL[*], S.T. CHANCEY, A. COLLMER[#], A. ESPINOSA, M. GUO, M. HANKS, Y. JAMIR, M.D. JANES, D.J. SCHNEIDER[ϕ], L. SHAN[§], T. PETNICKI-OCWIEJA, L.M. SCHECHTER[#], V.C. TAM and X. TANG[§]

Plant Science Initiative and the Department of Plant Pathology, University of Nebraska, Lincoln, NE, USA; [*]Institute for Genomic Research, Rockville, MD, USA; [#]Department of Plant Pathology, Cornell University, Ithaca, NY, USA; [ϕ]USDA-Agricultural Research Service Center for Agricultural Bioinformatics, Cornell University, Ithaca, NY, USA; [§]D Department of Plant Pathology, Kansas State University, Manhattan, KS, USA.

Abstract: The ability of *Pseudomonas syringae* pv. *tomato* DC3000 to be pathogenic on plants is dependent upon the type III secretion system (TTSS) and the effector proteins it translocates into plant cells. A draft sequence of DC3000 is currently being analysed for *hop* (Hrp outer protein) genes that encode TTSS substrates. Several different approaches were used to mine *hop* genes from the DC3000 genome. These approaches included ones that were based on linkage to other *hop* genes, the presence of apparent Hrp promoters, indicators of horizontal transfer, and N-terminal export-associated patterns. Many candidate *hop* genes were confirmed to encode TTSS-secreted proteins based on secretion and/or translocation assays. These approaches have substantially enlarged the inventory of DC3000 TTSS substrates. Hop proteins include extracellular accessory proteins that aid in translocation and effectors that are injected into plant cells. Several novel Hops have a clear potential to modify plant signal transduction pathways, which is consistent with the predicted roles for TTSS effectors. Activities of several Hops and new insights into bacterial pathogenesis of plants will be discussed.

Key words: type III secretion, effectors, Avr proteins, Hops.

Author for correspondence: J. Alfano; e-mail: jalfano2@unl.edu.

N.S. Iacobellis *et. al.* (eds.), *Pseudomonas syringae* and Related Pathogens, 363-372.

1. INTRODUCTION

The determination of the genome sequence for the Gram-negative bacterial plant pathogen *Pseudomonas syringae* pv. *tomato* DC3000 has been carried out by The Institute for Genomic Research (TIGR) as part of a project funded by the U.S. National Science Foundation Plant Genome Research Program. The first uses of the DC3000 genome sequence have been to learn more about the type III secretion system (TTSS) and the effector proteins that it delivers into plant cells. Here we report some of the progress that we have made in our laboratories in identifying new TTSS effectors as well as other progress in understanding the TTSS of DC3000.

2. IDENTIFICATION OF TTSS EFFECTORS IN THE DC3000 HRP PATHOGENICITY ISLAND EEL REGION

The Hrp apparatus is encoded by the *hrp/hrc* genes, which make up a central region of a pathogenicity island (Pai) called the Hrp Pai (Alfano *et al.*, 2000). To the right of the *hrp/hrc* cluster in the Hrp Pai is a region that is conserved in *Pseudomonas syringae* pathovars called the conserved effector locus (CEL), which carries, among other likely effector genes, the well-characterized *avr* gene *avrE* and the *hrpW* harpin gene (Charkowski *et al.*, 1998; Lorang and Keen, 1995). Flanking the other side of the *hrp/hrc* cluster in *P. syringae* is a region that is dissimilar between even closely related *P. syringae* strains. This region appears to contain the products of different

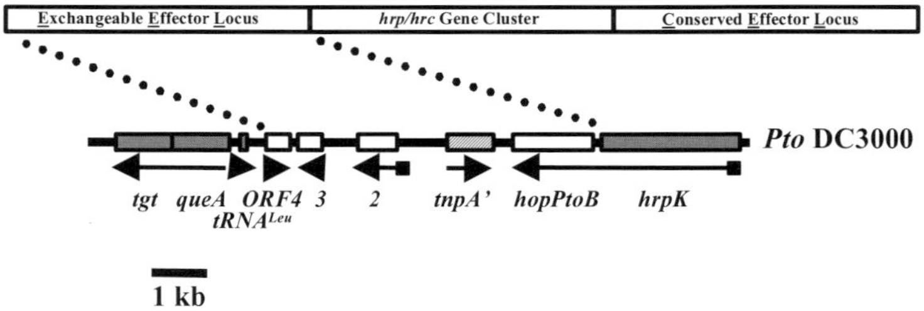

Figure 1. The exchangeable effector locus (EEL) of the DC3000 Hrp pathogenicity island. The EEL of DC3000 is bordered by the *hrpK* gene and a $tRNA^{leu}$ gene. Gray boxes denote border sequences, white boxes indicate novel genes, and striped boxes denote DNA associated with mobile genetic elements. The arrows indicate predicted directions of transcription and the box at the beginning of some of the operons indicate the presence of a TTSS promoter.

recombination events, remnants of mobile DNA, and is rich in effector genes. Because of these characteristics, the region was named the exchangeable effector locus (EEL) of the *P. syringae* Hrp Pai (Alfano *et al.*, 2000). The DC3000 EEL contains several ORFs that have Hrp boxes and two genes, *hrpK* and *hopB*, that we recently determined to encode proteins that are secreted via the TTSS (van Dijk, Petnicki-Ocwieja and Alfano, manuscript submitted) (Fig. 1). A DC3000 EEL deletion mutant was only slightly reduced in its ability to grow in tomato plants (Alfano *et al.*, 2000). This EEL deletion mutant lacked the entire EEL except for the *hrpK* gene, which is at the border between the EEL and the *hrp/hrc* regions. We constructed a nonpolar deletion mutation in *hrpK* and the resulting mutant, UNL111, was severely impaired in its ability to grow *in planta* and to produce disease symptoms. Thus, HrpK is likely performing an important role in bacteria-plant interactions.

3. GENOME-WIDE IDENTIFICATION OF GENES THAT ENCODE TYPE III-SECRETED PROTEINS

Because *Pseudomonas syringae* TTSS promoters contain a *cis* element, the Hrp box, it is possible to search for this sequence throughout the genome of DC3000. Three different reports identified Hrp promoters throughout the DC3000 genome (Fouts *et al.*, 2002; Guttman *et al.*, 2002; Zwiesler-Vollick *et al.*, 2002). An immediate goal of researchers working with the DC3000 genome is to determine which of the ORFs downstream of apparent Hrp promoters encode Hops. A recent report identified *P. syringae* effector genes with a transposon that made fusions to an Avr reporter (Guttman *et al.*, 2002). Only the fusions that can direct the secretion of the Avr reporter will elicit an HR on the appropriate test plants. We used a two-pronged genome-wide search for DC3000 Hops. In the first, we searched ORFs downstream of Hrp promoters for horizontal transfer indicators. Experimental testing confirmed that five of these encoded novel proteins secreted via the TTSS system in culture, and these were given the following Hop names: HopPtoE, HopPtoG, HopPtoH, HopPtoI, and HopPtoS1. Interestingly, HopPtoS1 shares similarity with ADP-ribosyltransferases, enzymes that modify signal transduction pathways. Five different proteins that showed homology to Avr proteins from different *P. syringae* pathovars proved also to be secreted or translocated. Since the avirulence activity of the DC3000 homologs was unknown these proteins were renamed HopPtoC (AvrPpiC2 homolog), HopPtoD1 and HopPtoD2 (AvrPphD homologs), HopPtoJ (AvrXv3), and HopPtoK (AvrRps4 homolog). When all of the known Hops of *P. syringae* were compared, shared characteristics within the amino termini were recognised (Petnicki-Ocwieja *et al.*, 2002). These Hop characteristics or "rules" allowed an algorithm to be written that was used to search the DC3000 draft genome sequence for ORFs that shared these characteristics. This allowed us to search for Hops independent of whether their genes possessed Hrp promoters, which resulted in a pool of 32 ORFs enriched in candidate *hop* genes. Two of the highly-ranked candidate *hop* genes within this pool encoded proteins that were secreted in culture and were given the names HopPtoS2 and HopPtoL (Petnicki-Ocwieja *et al.*, 2002). HopPtoL shares similarity with SrfC, a candidate type III effector from *Salmonella enterica.* HopPtoS2 possesses significant similarity to ADP-ribosyltransferases, as did two other ORFs identified in this genome-wide search. ADP-ribosyltransferases have been known for quite some time to be important in animal pathogens (Finlay and Falkow, 1997), but it now appears that these enzymes also play an important role in bacterial pathogenesis of plants. A summary of type III-secreted proteins is given in Figure 2.

Helper proteins

HopPmaG$^{i}_{Pto}$
HopPmaH$_{Pto}$ (Pectin lyase-related)
HopPtoP (HrpW-related)
HrpA$_{Pto}$
HrpK$_{Pto}$
HrpW$_{Pto}$
HrpZ$_{Pto}$

7 Helper proteins

Effectors

AvrE$_{Pto}$
AvrPphE$_{Pto}$
AvrPpiB1$_{Pto}$
AvrPpiB2$_{Pto}$
AvrPto$_{DC3000}$
AvrPtoB
HopPmaI$_{Pto}$
HopPmaJ$_{Pto}$
HopPsyA$_{Pto}$
HopPtoA1 (CEL ORF5)
HopPtoA2
HopPtoB1 (EEL ORF1)
HopPtoC (AvrPpiC2$_{Pto}$)
HopPtoD1 (AvrPphD1$_{Pto}$)
HopPtoD2 (AvrPphD2$_{Pto}$)
HopPtoE
HopPtoF (AvrPphFORF2$_{Pto}$)
HopPtoG
HopPtoH
HopPtoI
HopPtoJ (AvrXv3$_{Pto}$)
HopPtoK (AvrRps4$_{Pto}$)
HopPtoL
HopPtoM (CEL ORF3)
HopPtoN (CEL ORF7)
HopPtoS1 (HopPtoO)
HopPtoS2
HopPtoS3
HopPtoT1 (ORF16)

29 Effectors

36 Hop proteins and at least 15 other strong Hop candidates

Total Hop inventory may be over 50

Figure 2. Hop and Avr proteins identified in DC3000. The current inventory of type III secreted substrates from DC3000 is separated into two broad groups: Effectors that are "injected" into plant cells and helper proteins that assist in effector delivery.

4. *PSEUDOMONAS SYRINGAE* TYPE III CHAPERONES

The type III secretion of a subset of effectors from animal pathogens is assisted by type III chaperones. These proteins contain predicted α-helical structure in their C termini, are small, acidic, and are usually encoded by a gene adjacent to their cognate effector gene (Bennett and Hughes, 2000; Plano *et al.*, 2001; Wattiau *et al.*, 1996). Many chaperones have been shown to prevent aggregation and/or proteolysis of their cognate effectors within the bacterial cell. There is also evidence that chaperones may give their effectors a competitive edge over other effectors for entry into the type III pathway (Boyd *et al.*, 2000). However, the role of chaperones in type III secretion appears to be multidimensional because several chaperones have recently been shown to play a regulatory role as well (Darwin and Miller, 2001; Francis *et al.*, 2001; Tucker and Galán, 2000).

There were two early indications that there are type III chaperones in bacterial plant pathogens. An ORF (now named *shcA*) upstream of the *Pseudomonas syringae* pv. *syringae* 61 effector gene *hopPsyA* shared the general characteristics of type III chaperone genes (Alfano *et al.*, 1997), and the *Erwinia amylovora dspB* gene upstream of the *dspA* effector gene resembled chaperone genes and was required for the secretion of DspA (Gaudriault *et al.*, 1997). Recently, both ShcA and DspA were shown to act as chaperones (Gaudriault *et al.*, 2002; van Dijk *et al.*, 2002). ShcA binds HopPsyA in a region within the N-terminal 166 amino acids, consistent with the binding sites in animal type III effectors being in the N-terminal portion of the protein (Cornelis and van Gijsegem, 2000; van Dijk *et al.*, 2002).

Table 1. Putative chaperones identified in *Pseudomonas syringae* pv. *tomato* DC3000.

Chaperone or ORF name	Size (kDa)	pI	Effector or ORF name	Accession number	Reference
CEL ORF2	14.6	5.3	AvrE	AF232006	(Alfano *et al.*, 2000)
CEL ORF4	18.0	5.3	HopPtoM	AF232006	(Alfano *et al.*, 2000)
CEL ORF8	19.9	6.8	HopPtoN	AF232006	(Alfano *et al.*, 2000)
AvrPphF locus ORF1	15.6	6.1	HopPtoF	AF231452	(Petnicki-Ocwieja *et al.*, 2002)
ORF6	14.7	6.5	ORF17	None	(Petnicki-Ocwieja *et al.*, 2002)
ORF13	15.6	5.5	HopPtoS1	None	(Petnicki-Ocwieja *et al.*, 2002)
ORF14	17.0	5.9	ORF15	None	(Petnicki-Ocwieja *et al.*, 2002)
ORF25	17.2	5.3	ORF26	None	(Petnicki-Ocwieja *et al.*, 2002)

ShcA and HopPsyA are also present in DC3000, and we have confirmed that these proteins interact with each other as they do in *P. s.* pv. *syringae* 61.

Thus, DC3000 also possesses type III chaperones. Indeed, a search of the *avr* and *hop* genes represented in the draft genome sequence of DC3000 reveal that there are at least 8 candidate chaperone/effector pairs present (Table 1).

5. HOPPTOD2 IS AN ACTIVE PROTEIN TYROSINE PHOSPHATASE THAT SUPPRESSES THE HYPERSENSITIVE RESPONSE

We reported that HopPtoD2 was secreted in culture by the TTSS of *Pseudomonas syringae* pv. *tomato* strain DC3000 (Petnicki-Ocwieja *et al.*, 2002). HopPtoD2 is 468 amino acids in length, with a predicted molecular mass of 51.4 kDa. Based on comparisons with proteins in the databases, HopPtoD2 appears to have a modular organisation, with an N-terminal 200 amino acids that share high similarity with several AvrPphD homologs present in other *P. syringae* pathovars as well as other plant pathogens (Arnold *et al.*, 2001) and a C-terminal domain that showed similarity to protein tyrosine phosphatases (PTPs). HopPtoD2 has the invariant residues (HCxxGxxRS/T) present in the catalytic domains of functional PTPs (Denu *et al.*, 1996). An affinity-purified GST-HopPtoD2 hybrid was tested for its ability to dephosphorylate phosphotyrosine-containing peptides derived from either the insulin receptor (residues 1142-1024) or EGF receptor (residues 1014-1024). We also tested a GST-HopPtoD2 protein that contained a Ser in place of Cys_{378} (GST-$HopPtoD2_{C378S}$). Several studies have shown that this conserved Cys is required for PTP activity (Guan and Dixon, 1990; Zhang and Dixon, 1994). An important discovery was that HopPtoD2 was able to suppress the hypersensitive response (HR) elicited by *P. s.* pv. *phaseolicola* NPS3121 (*Pph*). The HR is a programmed cell death of plant tissue associated with a successful defence against pathogens. We found this by testing whether *Pph* containing pLN56 (which carries *hopPtoD2*) could alter the ability of this pathogen to elicit a defence-associated HR on *Nicotiana benthamiana.* We infiltrated *Pph*(pLN56) into leaves of *N. benthamiana* alongside *Pph* containing the control vector. *Pph* produced an HR on *N. benthamiana* within approximately 12 h. Interestingly, *Pph*(pLN56) expressing HopPtoD2 did not elicit an HR during the same time period. To determine if the HR suppression phenotype that occurred when *hopPtoD2* was provided *in trans* was due to PTP activity, we cloned an allele of *hopPtoD2* that encodes an inactive PTP domain ($HopPtoD2_{C378S}$) into the same broad-host-range vector. When *Pph(*pLN214) was infiltrated into *N. benthamiana* an HR occurred within 12 h similar to the response observed with the *Pph* control. This indicated that HopPtoD2 was capable of HR suppression and this suppression was dependent on its PTP activity.

6. OTHER TYPE III EFFECTORS THAT SUPPRESS THE HR

Our results suggested that it was possible to expand our screen for HR suppressors to include the recently identified DC3000 effectors (Collmer *et al.*, 2002). To do this, we turned to the pHIR11 system, which allows nonpathogens such as *Escherichia coli* and *Pseudomonas fluorescens* to elicit the HR and secrete effectors in culture via the TTSS. The HR elicited by pHIR11 is due solely to HopPsyA, the only Avr protein encoded by this cosmid. Therefore, this allowed us to test whether each candidate suppressor inhibited the HR elicited by HopPsyA. We cloned 19 confirmed effectors (i.e., known to be secreted or translocated via the TTSS) into a broad host-range plasmid. To do these experiments, we expressed each candidate suppressor gene in *P. fluorescens*(pHIR11) and infiltrated these strains into tobacco, a plant that normally responds to pHIR11-containing bacteria by producing an HR. Surprisingly, 7 of the 19 effectors tested were capable of suppressing the pHIR11-dependent HR in tobacco. We repeated these results with bacterial mixing experiments, where only half of the cells have the suppressor. These were still able to suppress the pHIR11-dependent HR. This indicates that the HR blocking activity is not occurring in the bacterial cell. Two of these suppressors were homologs of AvrPphF and VirPphA, two Avr proteins able to "block" the HR produced by *P. syringae* pv. *phaseolicola* (Jackson *et al.*, 1999; Tsiamis *et al.*, 2000). The VirPphA homolog, AvrPtoB, was recently reported to suppress the HR elicited by AvrPto (Abramovitch *et al.*, 2003). Unexpectedly, the HopPtoD2 PTP was not among the group of HR suppressors isolated in this screen, which suggests that this HR suppressor works by a different mechanism or that the pHIR11 assay was not sensitive enough to detect the suppressor activity of HopPtoD2.

7. CONCLUSIONS

It is an exciting time to be studying the TTSS of DC3000. We now have identified a large number of effectors that are delivered into plant cells. Many of these appear to be involved in suppressing host defences, but the mechanisms by which they accomplish this are not well understood. As the genome of DC3000 is further inspected, we will likely learn more about the DC3000 TTSS. For example, the genome sequence reveals that many of the DC3000 TTSS effectors appear to have cognate chaperones. The increased number of chaperones available for analysis should provide clues to how this group of proteins facilitates the secretion of effectors into plant cells. *Pseudomonas syringae* researchers have many research challenges ahead.

However, the prospect of discovering what TTSS effectors do to the plant to favour disease and what the associated plant targets are should encourage substantial ongoing research effort.

ACKNOWLEDGEMENTS

This work was supported in part by National Science Foundation Grants IBN-0096348 and DBI-0077622 and National Research Initiative Competitive Grants Program, US Department of Agriculture Grant 01-35319-10019.

REFERENCES

Abramovitch, R.B., Kim, Y.J., Chen, S., Dickman, M.B., and Martin, G.B., 2003, *Pseudomonas* type III effector AvrPtoB induces plant disease susceptibility by inhibition of host programmed cell death. *EMBO* **22**: 60-69.

Alfano, J.R., Kim, H.-S., Delaney, T.P., and Collmer, A., 1997, Evidence that the *Pseudomonas syringae* pv. *syringae hrp*-linked *hrmA* gene encodes an Avr-like protein that acts in an *hrp*-dependent manner within tobacco cells. *Mol. Plant-Microbe Interact.* **10**: 580-588.

Alfano, J.R., Charkowski, A.O., Deng, W., Badel, J.L., Petnicki-Ocwieja, T., van Dijk, K., and Collmer, A., 2000, The *Pseudomonas syringae* Hrp pathogenicity island has a tripartite mosaic structure composed of a cluster of type III secretions genes bounded by exchangeable effector and conserved effector loci that contribute to parasitic fitness and pathogenicity in plants. *Proc. Natl. Acad. Sci. USA* **97**: 4856-4861.

Arnold, D.L., Gibbon, M.J., Jackson, R.W., Wood, J.R., Brown, J., Mansfield, J.W., Taylor, J.D., and Vivian, A., 2001, Molecular characterisation of *avrPphD*, a widely-distributed gene from *Pseudomonas syringae* pv. *phaseolicola* involved in non-host recognition by pea (*Pisium sativum*). *Physiol. Mol. Plant Pathol.* **58**: 55-62.

Bennett, J.C., and Hughes, C., 2000, From flagellum assembly to virulence: The extended family of type III export chaperones. *Trends Microbiol.* **8**: 202-204.

Boyd, A.P., Lambermont, I., and Cornelis, G., 2000, Competition between the Yops of *Yersinia enterocolitica* for delivery in eukaryotic cells: Role of the SycE chaperone binding domain of YopE. *J. Bacteriol.* **182**: 4811-4821.

Charkowski, A.O., Alfano, J.R., Preston, G., Yuan, J., He, S.Y., and Collmer, A., 1998, The *Pseudomonas syringae* pv. *tomato* HrpW protein has domains similar to harpins and pectate lyases and can elicit the plant hypersensitive response and bind to pectate. *J. Bacteriol.* **180**: 5211-5217.

Collmer, A., Lindeberg, M., Petnicki-Ocwieja, T., Schneider, D., and Alfano, J.R., 2002, Genomic mining type III secretion system effectors in *Pseudomonas syringae* yields new picks for all TTSS prospectors. *Trends Microbiol.* **10**: 462-470.

Cornelis, G., and van Gijsegem, F., 2000, Assembly and function of type III secretion systems. *Annu. Rev. Microbiol.* **54**: 734-774.

Darwin, K.H., and Miller, V.L., 2001, Type III secretion chaperone-dependent regulation: activation of virulence genes by SicA and InvF in *Salmonella typhimurium*. *EMBO* **20**: 1850-1862.

Denu, J.M., Stuckey, J.A., Saper, M.A., and Dixon, J.E., 1996, Form and function in protein dephosphorylation. *Cell* **87**: 361-364.

Finlay, B.B., and Falkow, S., 1997, Common themes in microbial pathogenicity revisited. *Microbiol. Mol. Biol. Rev.* **61**: 136-169.

Fouts, D.E., Abramovitch, R.B., Alfano, J.R., Baldo, A.M., Buell, C.R., Cartinhour, S., Chatterjee, A.K., D'Ascenzo, M., Gwinn, M., Lazarowitz, S.G., Lin, N.-C., Martin, G.B., Rehm, A.H., Schneider, D.J., van Dijk, K., Tang, X., and Collmer, A., 2002, Genomewide identification of *Pseudomonas syringae* pv. *tomato* DC3000 promoters controlled by the HrpL alternative sigma factor. *Proc. Natl. Acad. Sci. USA* **99**: 2275-2280.

Francis, M.S., Lloyd, S.A., and Wolf-Watz, H., 2001, The type III secretion chaperone LcrH co-operates with YopD to establish a negative, regulatory loop for control of Yop synthesis in *Yersinia pseudotuberculosis*. *Mol. Microbiol.* **42**: 1075-1093.

Gaudriault, S., Malandrin, L., Paulin, J.-P., and Barny, M.-A., 1997, DspA, an essential pathogenicity factor of *Erwinia amylovora* showing homology with AvrE of *Pseudomonas syringae*, is secreted via the Hrp secretion pathway in a DspB-dependent way. *Mol. Microbiol.* **26**: 1057-1069.

Gaudriault, S., Paulin, J.-P., and Barny, M.-A., 2002, The DspB/F protein of *Erwinina amylovora* is a type III secretion chaperone ensuring efficient intrabacterial production of the Hrp-secreted DspA/E pathogenicity factor. *Mol. Plant Pathol.* **3**: 313-320.

Guan, K.L., and Dixon, J.E., 1990, Protein tyrosine phosphatase activity of an essential virulence determinant in *Yersinia*. *Science* **249**: 553-556.

Guttman, D.S., Vinatzer, B.A., Sarkar, S.F., Ranall, M.V., Kettler, G., and Greenberg, J.T., 2002, A functional screen for the type III (Hrp) secretome of the plant pathogen *Pseudomonas syringae*. *Science* **295**: 1722-1726.

Jackson, R.W., Athanassopoulos, E., Tsiamis, G., Mansfield, J.W., Sesma, A., Arnold, D.L., Gibbon, M.J., Murillo, J., Taylor, J.D., and Vivian, A., 1999, Identification of a pathogenicity island, which contains genes for virulence and avirulence, on a large native plasmid in the bean pathogen *Pseudomonas syringae* pathovar phaseolicola. *Proc. Natl. Acad. Sci. USA* **96**: 10875-10880.

Lorang, J.M., and Keen, N.T., 1995, Characterization of *avr*E from *Pseudomonas syringae* pv. tomato: A *hrp* -linked avirulence locus consisting of at least two transcriptional units. *Mol. Plant-Microbe Interact.* **8**: 49-57.

Petnicki-Ocwieja, T., Schneider, D.J., Tam, V.C., Chancey, S.T., Shan, L., Jamir, Y., Schechter, L.M., Janes, M.D., Buell, C.R., Tang, X., Collmer, A., and Alfano, J.R., 2002, Genomewide identification of proteins secreted by the Hrp type III protein secretion system of *Pseudomonas syringae* pv. *tomato* DC3000. *Proc. Natl. Acad. Sci. USA* **99**: 7652-7657.

Plano, G.V., Day, J.B., and Ferracci, F., 2001, Type III export: new users for an old pathway. *Mol. Microbiol.* **40**: 284-293.

Tsiamis, G., Mansfield, J.W., Hockenhull, R., Jackson, R.W., Sesma, A., Athanassopoulos, E., Bennett, M.A., Stevens, C., Vivian, A., Taylor, J.D., and Murillo, J., 2000, Cultivar-specific avirulence and virulence functions assigned to *avrPphF* in *Pseudomonas syringae* pv. *phaseolicola*, the cause of bean halo-blight disease. *EMBO* **19**: 3204-3214.

Tucker, S.C., and Galán, J.E., 2000, Complex function for SicA, a *Salmonella enterica* serovar *typhimurium* type III associated chaperone. *J. Bacteriol.* **182**: 2262-2268.

van Dijk, K., Tam, V.C., Records, A.R., Petnicki-Ocwieja, T., and Alfano, J.R., 2002, The ShcA protein is a molecular chaperone that assists in the secretion of the HopPsyA effector from the type III (Hrp) protein secretion system of *Pseudomonas syringae*. *Mol. Microbiol.* **44**: 1469-1481.

Wattiau, P., Woestyn, S., and Cornelis, G.R., 1996, Customized secretion chaperones in pathogenic bacteria. *Mol. Microbiol.* **20**: 255-262.

Zhang, Z.Y., and Dixon, J.E., 1994, Protein tyrosine phosphatases: mechanism of catalysis and substrate specificity. *Adv. Enzymol. Relat. Areas Mol. Biol.* **68**: 1-36.

Zwiesler-Vollick, J., Plovanich-Jones, A.E., Nomura, K., Bandyopadhyay, S., Joardar, V., Kunkel, B.N., and He, S.Y., 2002, Identification of novel *hrp*-regulated genes through functional genomic analysis of the *Pseudomonas syringae* pv. *tomato* DC3000 genome. *Mol. Microbiol.* **45**: 1207-1218.

Contribution of Virulence Determinants from *Pseudomonas* and Other Bacteria to *hrp*-dependent Gall Formation by *Erwinia herbicola* pv. *gypsophilae*

S. MANULIS and I. BARASH[#]
Department of Plant pathology, ARO, The Volcani Center, Bet-Dagan; [#]Department of Plant Sciences, Tel-Aviv University, Tel-Aviv, Israel.

Abstract: *Erwinia herbicola* is prevalent in nature on plant surfaces as well as in many diverse habitats. Strains of *E. herbicola* incite galls on several host plants including gypsophila and beet. The host range of *E. h.* pv. *gypsophilae* is restricted to gypsophila whereas *E. h.* pv. *betae* forms galls on beet and gypsophila. Pathogenicity of both pathovars is dependent on the presence of a similar plasmid designated as $pPATH_{Ehg}$ and $pPATH_{Ehb.}$ The $pPATH_{Ehg}$ (~150 kb) harbours a pathogenicity island in which the virulence genes can be classified into three groups: a) *hrp* gene cluster b) genes encoding Hop (Hrp outer protein) virulence effectors and c) genes encoding *hrp*-independent virulence effectors. Furthermore, both plasmids contain multiple and diverse IS elements from five different IS families that are specific to the pPATHs. In the present paper we describe the contribution of various characterized virulence effectors to gall formation by *E. h.* pv. *gypsophilae* as well as some preliminary results on putative effectors that are currently under investigation. Results obtained suggest that *Pseudomonas* and other bacteria, that presumably encountered *E. herbicola* in its various habitats, left their fingerprints on its genome and were instrumental in its conversion into a biotrophic gall-forming pathogen.

Key words: *Pantoea agglomerans*, *Gypsophila paniculata*, *Erwinia herbicola* pv. *betae*, Hop effectors.

Author for correspondence: S. Manulis and I. Barash; e-mail: shulam@volcani.agri.gov.il, isaaci@post.tau.ac.il.

N.S. Iacobellis *et. al.* (eds.), *Pseudomonas syringae* and Related Pathogens, 373-381.

1. INTRODUCTION

Erwinia herbicola (*Pantoea agglomerans*) is widespread in nature in many diverse habitats. In addition to its prevalence as an epiphyte and endophyte, it has been isolated from seeds, water, humans and animals (Manulis and Barash, 2003). The adaptation of *E. herbicola* to widely different microenvironments suggests that this species maintains high genome plasticity. This quality also allowed *E. herbicola* to acquire virulence determinants that were instrumental in its conversion into a plant pathogen. The transformation of *E. herbicola* into a gall forming bacterium appears to constitute a dominant evolutionary route. *E. herbicola* has so far been reported to cause crown and root galls on gypsophila (*Gypsophila paniculata* L.), beet (*Beta vulgaris* L.), Douglas-fir (*Pseudotsuga menziesii* L.), wisteria (*Wisteria sinensis* L.) and cranberry (*Vaccinium macrocarpon* Aiton) (Manulis and Barash, 2003).

Two pathovars of *E. herbicola* namely, *E. h.* pv. *gypsophilae* and *E. h.* pv. *betae* have been shown to induce *hrp*-dependent gall formation. The host range of *E. h.* pv. *gypsophilae* appears to be restricted to gypsophila (Volcani, 1985) whereas *E. h.* pv. *betae* infects both, beet and gypsophila (Burr *et al.*, 1991). *E. h.* pv. *gypsophilae* is a most destructive pathogen of gypsophila especially during its propagation from cuttings where it prevents root development (Cooksey, 1986; Volcani, 1985). Tumors on table beet results in economic loss because infected beet cannot be mechanically processed (Burr *et al.*, 1991). All the pathogenic strains of *E. h.* pv. *gypsophilae* and *E. h.* pv. *betae* carry a pathogenicity-associated plasmid that is absent in non-pathogenic strains (Manulis and Barash, 2003; Manulis *et al.*, 1991). This plasmid was designated as $pPATH_{Ehg}$ and $pPATH_{Ehb}$ for *E. h.* pv. *gypsophilae* and *E. h.* pv. *betae*, respectively. The pPATHs of *E. h.* pv. *gypsophilae* and *E. h.* pv. *betae* as well as of different pathogenic strains of the two pathovars may differ in size but share similar genes or DNA sequences (Manulis and Barash, 2003).

The $pPATH_{Ehg}$ from strain *Ehg*824-1 has been extensively investigated (Fig. 1). The plasmid size is approximately 150 kb and harbours a putative pathogenicity island of more than 60 kb. The various virulence genes that have so far been characterised on the $pPATH_{Ehg}$ can be divided into three groups according to their function in pathogenicity: a) Hrp gene cluster (Mor *et al.*, 2001; Nizan *et al.*, 1997), b) genes encoding Hop (Hrp-dependent outer protein) virulence effectors (Ezra *et al.*, 2000; Guo *et al.*, 2002; Mor *et al.*, 2001; Valinsky *et al.*, 1998) and c) genes encoding Hrp independent virulence effectors such as enzymes for biosynthesis of IAA and cytokinins (Clark *et al.*, 1993; Gafni *et al.*, 1997; Lichter *et al.*, 1995a; Lichter *et al.*, 1995b). Several copies of IS elements representing five different families have also been located on the pathogenicity island of $pPATH_{Ehg.}$ (Guo *et al.*,

2002; Lichter *et al.*, 1996). All these IS elements were found also on the $pPATH_{Ehb}$ and, with the exception of one IS element, they were specific to pPATH of the two pathovars (Guo *et al.*, 2002). The latter finding as well as the similarity in *hrp* and other virulence genes (Manulis and Barash, 2003) suggests a common ancestor for both plasmids.
The *hrp* gene cluster on the $pPATH_{Ehg}$ has been fully characterised (Mor *et al.*, 2001; Nizan *et al.*, 1997). It shows a remarkable similarity and co-linearity with the *hrp* genes of *Pantoea stewartii* subsp. *stewartii* and *Erwinia amylovora.* With few exceptions, the percentage identities of *E. h.* pv. *gypsophilae* nucleotides and of the deduced proteins of the various *hrp* operons, with those of *E. amylovora* and *P. s.* subsp. *stewartii* ranged from 60 to 90% (Mor *et al.*, 2001). The mode through which the *hrp* regulatory genes activate the *hrp* regulon has been recently demonstrated (Nizan-Koren *et al.*, 2002).
As indicated earlier, pathogenicity of *E. h.* pv. *gypsophilae* and *E. h.* pv. *betae* had most likely been evolved through genes acquisition by horizontal gene transfer. In the present paper we shall briefly review the characterised *hrp*-dependent and *hrp*-independent virulence effectors with emphasis on their homology to *Pseudomonas* and other bacteria. Preliminary results on some potential virulence effectors will also be presented.

2. MATERIALS AND METHODS

For isolation of virulence genes that reside on the $pPATH_{Ehg}$, "plasmid walking" was performed through which the order of overlapping cosmids that constitute the plasmid in a pLAFR3 library were identified. Each of these cosmids was subjected to DNA sequencing and analysis for the presence of ORFs. The ORFs were further analysed for the presence of homologous genes in GenBank, and suspected ORFs were used for generation of marker-exchange mutants and screening for pathogenicity.
If not otherwise mentioned, the bacterial strains of *Erwinia herbicola* pv. *gypsophilae* and *E. h.* pv. *betae* used throughout these studies were *Ehg*824-1 and *Ehb*4188. Genetic manipulations were carried out in *Escherichia coli* strains DH5α and HB101. DNA isolation, restriction enzyme digestions, agarose gel electrophoresis, subcloning, transformations, Southern blot analysis, and PCR were performed according to standard procedures (Ausubel *et al.*, 1995).
Pathogenicity tests were performed on cuttings of *Gypsophila paniculata* cv. *perfecta* essentially as described by Valinsky *et al.* (1998).

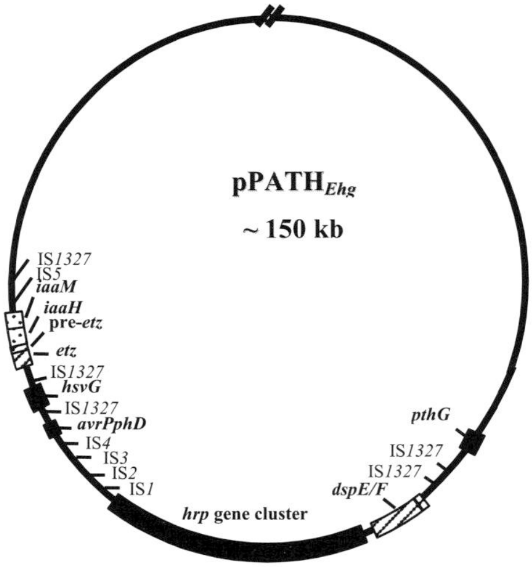

Figure 1. Schematic map of the pathogenicity plasmid of *Erwinia herbicola* pv. *gypsophilae*. The various genes and IS elements shown in the figure are described in the text.

3. RESULTS AND DISCUSSION

Several Hop virulence proteins have so far been characterised. *hsvG* (**h**ost-**s**pecific **v**irulence gene for **g**ypsophila) determines the ability of *Erwinia herbicola* pv. *gypsophilae* and *E. h.* pv. *betae* to incite galls on gypsophila (Valinsky *et al.*, 1998). This gene has been isolated and sequenced from $pPATH_{Ehg}$ and $pPATH_{Ehb}$. Mutations in *hsvG* eliminated the ability of *E. h.* pv. *gypsophilae* and *E. h.* pv. *betae* to cause galls on gypsophila, but mutants retained the full pathogenicity of *E. h.* pv. *betae* on beet and HR induction by both pathovars on tobacco. DNA sequence analysis of *hsvG* revealed an ORF of 2 kb with a "*hrp* box" upstream of the start codon. It encodes a protein of 72 kDa. A low homology (28% identity, 113/191 aa) has been observed with *psvA*, a plasmid-born virulence gene from *Pseudomonas syringae* pv. *eriobotryae* that is pathogenic on loquat trees (Table 1) (Kamiunten, 1999). The *hsvG* gene is Hrp regulated (Valinsky *et al.*, 2002a), and the protein could be translocated into human epithelial cells by the type III secretion system of enteropathogenic *Escherichia coli* (Valinsky *et al.*, 2002b). *pthG* (**path**ogenicity gene on **g**ypsophila) was isolated from the $pPATH_{Ehg}$. Only remnants of the gene sequences were identified on $pPATH_{Ehb}$ (Ezra *et al.*, 2000). *pthG* has a dual function, it acts as a *hop* virulence gene on gypsophila and an *avr* on beet. Marker-exchange mutation of this gene in *E. h.* pv. *gypsophilae* extended the host range of this pathovar on beet and thus acts as the major factor in the separation of *E. h.* pv. *gypsophilae* and *E. h.* pv. *betae* into two pathovars. PthG is plant-induced and *hrp*-regulated (Manulis and Barash, 2003). It shows low homology (54% identity, 19/33) with a hypothetical protein of *Ralstonia solanacearum* (Table 1). It has been recently demonstrated that PthG elicits HR on multiple beet species (Vintal *et al.*, unpublished) as opposed to the cultivar-specificity of many *avr* genes.

dspE constitutes an operon with *dspF* (Mor *et al.*, 2001). Mutation in *dspE* abolished pathogenicity on gypsophila but not HR on tobacco. The transcription of *dspE* is *hrp* regulated and its secretion is dependent on a functional type III system (Mor *et al.*, 2001). The identities of the amino acids of DspE with those of *P. s.* subsp. *stewartii* and *E. amylovora* were 66% (1217/1836) and 57% (1059/1080), respectively (Table 1). It also shows 26% identity (417/1602) with AvrE of *P. s.* pv. *tomato*.

$avrPphD_{Ehg}$ was isolated from the flanking region of the *hrp* gene cluster in $pPATH_{Ehg}$ (Guo *et al.*, 2002). It is an homologue of *avrPphD* that is widely distributed in various *P. syringae* pathovars (Arnold *et al.*, 2001). The highest identity of, 84% in its predicted protein, was found with *P. s.* pv. *savastanoi* (600/706) and *P. s.* pv. *glycinea* (597/706).

The *hrp*-independent virulence effectors on $pPATH_{Ehg}$ include the biosynthetic enzymes for IAA and cytokinins (Fig. 1) (Clark *et al.*, 1993; Gafni *et al.*, 1997; Lichter *et al.*, 1995a; Manulis *et al.*, 1998). Mutations in each of the two IAA biosynthetic genes that constitute the indoleacetamide pathway, namely, tryptophan-2-monooxygenase (*iaaM*) and indole-3-acetamide hydrolase (*iaaH*), substantially reduced gall formation but did not eliminate gall initiation (Manulis *et al.*, 1998). A similar effect on pathogenicity was obtained with a mutant of the cytokinin biosynthetic gene (*etz*) encoding isopentenyl transferase (Lichter *et al.*, 1995a), as well as simultaneous inactivation of all the IAA and cytokinin biosynthetic genes (Manulis *et al.*, 1998). As illustrated in Table 1 both IaaM (68%; 378/551) and IaaH (57%; 257/450) showed highest identity with *P. syringae* pathovars, although the similarity of IaaM was higher than IaaH. The Etz protein showed almost the same similarity to *R. solanacearum* (41%; 95/231) and *P. s.* pv. *savastanoi* (40%; 92/230).

Further sequencing of $pPATH_{Ehg}$ cosmids revealed several ORFs with and without homologies to sequences in the database. These ORFs are currently under investigation for determination of their role in pathogenicity. Some ORFs with homologies to putative virulence effectors of *Pseudomonas* are shown in Table 1. A truncated homologue of ORF4 that was originally isolated from the pathogenicity island of plasmid pAV511 of *P. s.* pv. *phaseolicola* (Jackson *et al.*, 1999), was also isolated from $pPATH_{Ehg}$. ORF4 was found to cause a clear delay in the onset of the HR normally observed in bean pods inoculated with *P. s.* pv. *phaseolicola* RW60. Two homologues of HopPmaB and HopPmaH of *P. s.* pv. *maculicola* (Guttman *et al.*, 2002) were also isolated from the $pPATH_{Ehg}$ (Table 1). HopPmaB is a putative ortholog of AvrPphE (AF458042). The latter possesses an eukaryotic NH_2-

myristoylation motif that may facilitate targeting to the host cell membrane. HopPmaH showed high similarity to pectate lyase (Guttman *et al.*, 2002).

Table 1: Virulence genes and ORFs on $pPATH_{Ehg}$ with homology to *Pseudomonas* and other bacterial pathogens.

Gene/ORF	Size (aa)	Function in *Ehg*	Similar sequence and accession #	% identity/ E value
hsvG	671	pathogenicity and host specificity	*P. syringae* pv. *eriobotryae* (AB018553)	28/ 2e-23
pthG	506	pathogenicity and host specificity	*R. solanacearum* (NC_003295)	57/ 0.01
dspE	1829	Pathogenicity	*Pantoea stewarti* (AF282857)	66/ 0.0
			E. amylovora (AAC04850)	57/ 0.0
			P. syringae pv. *tomato* (AF232004)	26/ e-134
$avrPphD_{Ehg}$	710	Pathogenicity	*P. s.* pv. *savastanoi* (AJ439730)	84/ 0.0
			P. s. pv. glycinea (AJ439731)	84/ 0.0
iaaM	562	Virulence	*P. s.* pv. *savastanoi* (M11035)	70/ 0.0
			P. s. pv. *syringae* (U04358)	67/ 0.0
iaaH	460	Virulence	*P. s.* pv. *savastanoi* (M11035)	57/ e-139
			P. s. pv. *syringae* (U04358)	57/ e-137
etz	237	Virulence	*R. solanacearum* (P14333)	41/1e-43 1e-43
			P. s. pv. *savastanoi* (P06619)	40/6e-39
ORF4 homologue	255	not determined	*P. s.* pv. *phaseolicola* (AF141883)	68/ 0.0
HopPmaB homologue	327	not determined	*P. s.* pv. *maculicola* (AF458041)	79/ e-174
HopPmaH homologue	529	not determined	*P. s.* pv. *maculicola* (AF458046)	62/ e-175
ORF_{Ehg}	1371	not determined	*X. campestris* pv. *campestris* (NC_003902)	56/ 0.0

Another ORF, designated as ORF_{Ehg} (Table 1), showed high identity to a hypothetical transducer protein of *Xanthomonas campestris* pv. *campestris* (Da silva *et al.*, 2002) that could be associated with chemotaxis .

It is quite plausible that traits that confer pathogenicity on gypsophila have been acquired by horizontal transfer of gene clusters rather than by slow adaptive evolution. It could be speculated that *P. syringae* pathovars acted as important contributors of Hop and none-Hop virulence effectors to *E.*

herbicola. This could arise from the prevalence of pseudomonads and *E. herbicola* in the same ecological niche and/or genetic relatedness that facilitates mobilisation of DNA from one organism to another. However, the origin of these genes is not at all clear and they could be acquired by *Pseudomonas* from other bacteria. The observation that these genes exhibit a wide range of GC content (Manulis and Barash, 2003) and some are significantly varied from the typical GC percentage of *E. herbicola* and *Pseudomonas* might support the later premise

ACKNOWLEDGEMENTS

This study was supported by research grant US-3265-01C from BARD, the US-Israel Binational Agricultural Research and Development Fund, and The Israel Academy of Sciences and Humanities (grant no. 578-01).

REFERENCES

Arnold, D.L., Gibbon, M.J., Jackson, R.W., Wood, J.R., Brown, J., Mansfield, J.W., Talor, J.D., and Vivian, A., 2001, Molecular characterisation of *avrPphD*, a widely-distributed gene from *Pseudomonas syringae* pv. *phaseolicola* involved in non-host recognition by pea (*Pisum sativum*). *Physiol. Mol. Plant Pathol.* **58**: 55-62.

Ausubel, F.M., Brent, R., Kingston, R.E., Moore, D.D., Seidman, J.G., Smith, J.A., Struhl, K., and eds. 1995, Current protocols in molecular biology. John Wiley & Sons, Inc., New York.

Burr, T.J., Katz, B.H., Abawi, G.S., and Crosier, D.C., 1991, Comparison of tumorigenic strains of *Erwinia herbicola* isolated from table beet with *E. h. gypsophilae*. *Plant Dis.* **75**: 855-858.

Clark, E., Manulis, S., Ophir, Y., Barash, I., and Gafni, Y., 1993, Cloning and characterisation of *iaaM* and *iaaH* from *Erwinia herbicola* pathovar *gypsophilae*. *Phytopathology* **83**: 234-240.

Cooksey, D.A., 1986, Galls of *Gypsophila paniculata* caused by *Erwinia herbicola. Plant Dis.* **70**: 464-468.

Da silva, A.C.R., Ferro, J.A., Reinach, F.C., Farah, C.S., and Furlan, L.R., e. a. 2002, Comparison of genomes of two *Xanthomonas* pathogens with differing host specificities. *Nature* **417**: 459-453.

Ezra, D., Barash, I., Valinsky, L., and Manulis, S., 2000, The dual function in virulence and host range restriction of a gene isolated from pPATH$_{Ehg}$ plasmid of *Erwinia herbicola* pv. *gypsophilae*. *Mol. Plant-Microbe Interact.* **13**: 693-698.

Gafni, Y., Manulis, S., Kunik, T., Lichter, A., Barash, I., and Ophir, Y., 1997, Characterisation of the auxin synthesis genes of *Erwinia herbicoola* pv. *gypsophilae*. *Israel J. of Plant Sciences* **45**: 279-284.

Guo, M., Manulis, S., Mor, H., and Barash, I., 2002, The presence of diverse IS elements and an *avrPphD* homologue that acts as a virulence factor on the pathogenicity plasmid of *Erwinia herbicola* pv. *gypsophilae*. *Mol. Plant-Microbe Interact.* **15**: 709-716.

Guttman, D.S., Vinatzer, B.A., Sarkar, S.F., Ranall, M.V., Kettler, G., and Greenberg, J.T. 2002, A functional screen for type III (Hrp) secretome of the plant pathogen *Pseudomonas syringae*. *Science* **295**: 1722-1726.

Jackson, J.F., Athanassopoulos, E., Tsiamis, G., Mansfield, J. W., Sesma, A., Arnold, D.L., Gibbon, M.J., Murillo, J., D., T. J., and Vivian, A., 1999, Identification of a pathogenicity island, which contains genes for virulence and avirulence, on a large native plasmid in the bean pathogen *Pseudomonas syringae* pathovar *phaseolicola*. *Proc. Natl. Acad. Sci. USA* **96**: 10875-10880.

Kamiunten, H., 1999, Isolation and characterisation of virulence gene *psvA* on a plasmid of *Pseudomonas syringae* pv. *eriobotryae*. *Ann. Phytopathol. Soc. Japan*. **65**: 501-509.

Lichter, A., Barash, I., Valinsky, L., and Manulis, S., 1995a, The genes involved in cytokinin biosynthesis in *Erwinia herbicola* pv. *gypsophilae*: characterisation and role in gall formation. *J. Bacteriol*. **177**: 4457-4465.

Lichter, A., Manulis, S., Sagee, O., Gafni, Y., Gray, J., Meilan, R., Morris, R. O., and Barash, I., 1995b, Production of cytokinins by *Erwinia herbicola* pv. *gypsophilae* and isolation of a locus conferring cytokinin biosynthesis. *Mol. Plant- Microbe Interact*. **8**: 114-121.

Lichter, A., Manulis, S., Valinsky, L., Karniol, B., and Barash, I., 1996, IS*1327*, a new insertion-like element in the pathogenicity-associated plasmid of *Erwinia herbicola* pv. *gypsophilae* . *Mol. Plant-Microbe Interact*. **9**: 98-104.

Manulis, M., and Barash, I., 2003, The Molecular Basis for Transformation of an Epiphyte into a Gall-Forming Pathogen as Exemplified by *Erwinia herbicola* pv. *gypsophilae*. Pages in: Plant-Microbe Interactions. Vol. 6. G. Stacey and N. Keen, eds. American Phytopathological Society, St. Paul, Minnesota.

Manulis, S., Gafni, Y., Clark, E., Zutra, D., Ophir, Y., and Barash, I., 1991, Identification of a plasmid DNA probe for detection of *Erwinia herbicola* pathogenic on *Gypsophila paniculata*. *Phytopathology* **81**: 54-57.

Manulis, S., Haviv-Chesner, A., Brandl, M.T., Lindow, S.E., and Barash, I., 1998, Differential involvement of indole-3-acetic acid biosynthetic pathways in pathogenicity and epiphytic fitness of *Erwinia herbicola* pv. *gypsophilae*. *Mol. Plant-Microbe Interact* **11**: 634-642.

Mor, H., Manulis, S., Zuc, M., Nizan, R., Coplin, D.L., and Barash, I., 2001, Genetic organization of the *hrp* gene cluster and dspAE/BF operon in *Erwinia herbicola* pv. *gypsophilae*. *Mol. Plant-Microbe Interact*. **14**: 431-436.

Nizan, R., Barash, I., Valinsky, L., Lichter, A., and Manulis, S., 1997, The presence of *hrp* genes on the pathogenicity- associated plasmid of the tumorigenic bacterium *Erwinia herbicola* pv. *gypsophilae*. *Mol. Plant-Microbe Interact*. **10**: 677-682.

Nizan-Koren, R., Manulis, S., Mor, H., Iraki, N.M., and Barash, I., 2002, The regulatory cascade that activates the Hrp regulon in *Erwinia herbicola* pv. *gypsophilae*. *Mol. Plant-Microbe Interact*. **15**:(submitted).

Valinsky, L., Barash, I., Chalupowicz, L., Ezra, D., and Manulis, S., 2002a, Regulation of *hsvG*, a host-specific virulence gene from *Erwinia herbicola* pv. gypsophilae. *Physiol. Mol. Plant Pathol*. **60**: 19-29.

Valinsky, L., Manulis, S., Nizan, R., Ezra, D., and Barash, I., 1998, A pathogenicity gene isolated from the pPATH of *Erwinia herbicola* pv. *gypsophila* determines host specificity. *Mol. Plant-Microbe Interact*. **11**: 753-762.

Valinsky, L., Nisan, I., Tu, X., Nisan, G., Rosenshine, I., Hanski, E., Brash, I., and Manulis, S., 2002b, A host-specific virulence protein of *Erwinia herbicola* pv. *gypsophilae* is translocated into human epithelial cells by the type III secretion system of enteropathogenic *Escherichia coli*. *Mol. Plant Pathol*. **3**: 97-101.

Volcani, Z., 1985, Bacterial diseases of plants in Israel. Agricultural Research Organization, The Volcani Center, Bet Dagan, Israel. 388 pp.

Strain-specific Sequence Alterations in the Gene Encoding the Histidine Protein Kinase CorS Might be Responsible for Temperature-dependent Production of the Phytotoxin Coronatine by *Pseudomonas syringae*

A.V. SMIRNOVA, S. AUFHAMMER, and M.S. ULLRICH
International University Bremen, School of Engineering and Sciences, Campus Ring 1, Research II, 28759 Bremen, Germany.

Abstract: A modified two-component regulatory system consisting of two response regulators, CorR and CorP, and the histidine protein kinase CorS, regulates the thermoresponsive production of the phytotoxin coronatine (COR) in *Pseudomonas syringae* PG4180, which is the causal agent of bacterial blight of soybean. COR is produced at 18°C but not at 28°C. It is assumed that CorS responds to a temperature change and subsequently transduces the signal to its cognate response regulator(s), which in turn facilitates transcriptional activation of COR biosynthetic genes. Based on results of experiments with translational fusions between CorS and either alkaline phosphatase (PhoA) or β-galactosidase (LacZ), CorS possesses six membrane-spanning domains and the location inside or outside of the membrane of certain regions of CorS is altered in dependence of temperature. The genome sequencing project of *P. syringae* DC3000 has revealed the presence of the CorRSP system in this strain as well. However, DC3000 produces significantly less COR than does PG4180, and the production is temperature-independent. We analysed COR production of various *P. syringae* strains at 18°C and 28°C and also deduced amino acid sequences of CorS from COR-producing strains. We suggest that PG4180-specific sequence alterations in the sequence of the temperature sensor kinase-encoding gene *corS* might be responsible for thermoresponsive production. We performed site-directed mutagenesis of CorS from PG4180 by introducing mutations identified in CorS of DC3000. Each of these mutations resulted in a change of the thermoresponsive phenotype for CorS of PG4180.

Key words: bacterial blight, infection, genome sequence, phytotoxin.

Author for correspondence: A.V. Smirnova; e-mail: a.smirnova@iu-bremen.de.

N.S. Iacobellis *et. al.* (eds.), *Pseudomonas syringae* and Related Pathogens, 383-391.

1. INTRODUCTION

Pseudomonas syringae pv. *glycinea* and the closely related pathovars *atropurpurea*, *morsprunorum*, *maculicola*, and *tomato* produce the polyketide phytotoxin coronatine (COR), which is not host-specific (Mitchell, 1982). The biological effects of COR include the induction of leaf chlorosis, hypertrophy of plant storage tissue, compression of thyllakoids, thickening of plant cell walls, and accumulation of plant-borne protease inhibitors (Sakai *et al.*, 1979; Mitchell, 1982; Palmer and Bender, 1995). COR can elicit symptoms on a wide variety of plant species and increase disease severity on a host plant. *P. s.* pv. *glycinea* PG4180, which causes bacterial blight of soybeans, produces significantly more COR than other pathovars. The production is temperature-dependent.

Enzymes involved in COR biosynthesis in *P. syringae* PG4180 are encoded by a plasmid-borne 32.8-kb DNA region consisting of two biosynthetic gene clusters required for synthesis of CFA and CMA, respectively, and a regulatory region (Ullrich *et al.*, 1995). Based on the nucleotide sequence analysis of CMA biosynthetic genes, it was suggested that CMA is synthesised via a mechanism similar to the biosynthesis of non-ribosomal peptides (Ullrich and Bender, 1994; Budde *et al.*, 1998). CFA synthesis resembles the biosynthesis of antibiotics with a polyketide structure in *Streptomyces* and *Bacillus* spp. (Rangaswamy *et al.*, 1998).

Transcription of both biosynthetic operons depends on a modified two-component regulatory system encoded within the COR biosynthetic gene cluster (Ullrich *et al.*, 1995). According to nucleotide sequence analysis, this two-component system consists of an environmental sensor, the histidine protein kinase CorS, and two response regulators, CorR and CorP. Insertion mutations in the *corR, corS,* and *corP* genes completely abolished transcriptional activation of the COR biosynthetic promoters. Moreover, a 3.4-kb DNA fragment containing these regulatory genes restored temperature-regulated transcriptional activation of the COR biosynthetic promoters and phytotoxin production in the insertion mutants (Ullrich *et al.*, 1995). These facts clearly highlight the importance of the regulatory genes *corS, corR,* and *corP* for thermoresponsive COR biosynthesis. The sensor kinase CorS is thought to respond to a temperature signal and transduce the signal to its cognate response regulators, CorR and CorP. We assume that membrane fluidity, which is affected by temperature change, might influence the function of the membrane-associated CorS. The hydrophobic N-terminus may function as the sensor domain of CorS in signal perception. Therefore, a precise determination of the structure of this region is crucial to an understanding of the signal perception. To elucidate the structure of the N-terminal region of CorS, we applied gene fusion techniques using the periplasmically localised alkaline phosphatase and the cytoplasmic β-galactosidase of *Escherichia coli* (Manoil and Becwith, 1986).

Based on the results with translational CorS-PhoA and CorS-LacZ fusions, we proposed a hypothetical model for a topological change of CorS. We also analysed whether the proposed model would fit to CorS proteins from various *P. syringae* strains which were known to produce COR. To find evidence supporting our assumption that a topology change of CorS results in crucial changes of CorS activity and the ability of CorS to be autophosphorylated, we performed site-directed mutagenesis of CorS.

2. TOPOLOGICAL ANALYSIS OF CORS

2.1 Membrane Topology Prediction

According to a hydrophilicity plot generated by the program PROTEAN, the N-terminal region of CorS (approximately 200 amino acid residues) contained multiple hydrophobic zones. This indicated that this portion of the protein may be localised in the membrane. According to the TopPred 2 prediction, CorS possesses 5, 6 or 7 transmembrane domains (TMDs) (Table 1). Interestingly, the 7th TMD was predicted to occur downstream of the conserved H-box, where the actual site of autophosphorylation, residue His_{254}, is located. This prediction seemed unlikely given that the His residue must be located in the cytoplasm in order to be phosphorylated. The 2nd, 6th, and 7th TMDs were predicted to be putative. The 6th TMD showed the lowest hydropathy index among all TMDs. Taking into account these prediction results, it is likely that CorS is a hydrophobic protein with six TMDs in its N-terminus.

Table 1. Membrane topology of CorS predicted by TopPred 2 program.

Domain	Amino acid residues	Hydropathy index	Certainty
1	23-43	1.405	certain
2	61-81	0.729	putative
2	104-124	1.230	certain
4	127-147	1.898	certain
5	155-175	1.123	certain
6	181-201	0.682	putative
7	257-277	0.786	putative

2.2 Analyses of CorS-PhoA and CorS-LacZ Translational Fusions

To investigate these predictions, a series of 7 translational fusions of hybrid proteins of CorS with either alkaline phosphatase (PhoA) or β-galactosidase

(LacZ) were constructed. The design of the hybrid proteins was such that at least one fusion was placed in each loop facing either the periplasm or the cytoplasm. Five additional CorS-PhoA fusions were placed upstream and downstream of the conserved H-box and the putative 7th TMD. Plasmids carrying hybrid constructs were mobilised into *Pseudomonas syringae* PG4180 from *Escherichia coli*. The activities of CorS-PhoA and CorS-LacZ hybrid fusion proteins were determined in qualitative and quantitative assays. Their expression levels in *P. syringae* were also analysed by immunoblotting.

High activities of PhoA fusions at amino acid residues A_{51}, L_{125} and V_{177} (Fig. 1 and Table 2) indicated their periplasmic location. LacZ fusions at these residues showed the expected lack of activity for Ala_{51} and Leu_{125}, whereas elevated LacZ activity was observed for the fusion at residue V_{177}. According to previous topological studies of various membrane proteins, LacZ is a less reliable reporter enzyme than is PhoA. It has been suggested that β-galactosidase fused to periplasmic domains sometimes exhibits high activity because of disruption of membrane integration of the fusion protein by this large-size reporter enzyme (Hennesssey and Broome-Smith, 1993).

Table 2. Specific alkaline phosphatase and β-galactosidase activities* for CorS-PhoA and CorS-LacZ fusion proteins in PG4180.

Fusion	PhoA activity [U mg^{-1} protein]		LacZ activity [U mg^{-1} protein]
	HSC medium	KB medium	
Control**	0	0	0
T_{27}	0	0.44±0.02	95.8±8.8
A_{51}	2.5±0.02	8.1±0.1	0
Y_{85}	0.4±0.3	0.16±0.1	34.1±6.6
L_{125}	2.2±0.7	9.6±0.4	0
L_{151}	0.5±0.3	0.21±0.06	157.8±16.9
V_{177}	2.2±0.1	12±1	34.8±7.6
R_{204}	0.13±0.06	0.5±0.2	48.5±4.6
Q_{214}	0.5±0.3	0.3±0.1	Not measured
D_{227}	0.6±0.3	0.51±0.01	Not measured
L_{249}	0	0	Not measured
Q_{281}	0	0	Not measured

* Cells for enzymatic measurements were grown at 28°C;

** As a control for enzymatic measurements, wild type cells lacking recombinant PhoA and LacZ were used.

In contrast, a PhoA fusion requires active translocation of the reporter enzyme moiety through the cytoplasmic membrane for activity. Since the PhoA fusion at V_{177} showed high specific activity in a quantitative assay, and in addition Western blot analysis clearly demonstrated its periplasmic location, we concluded that this site was indeed periplasmic. Absence of PhoA activities and presence of LacZ activities for fusions at amino acid residues T_{27}, Y_{85}, and L_{151} indicated a cytoplasmic location of these residues.

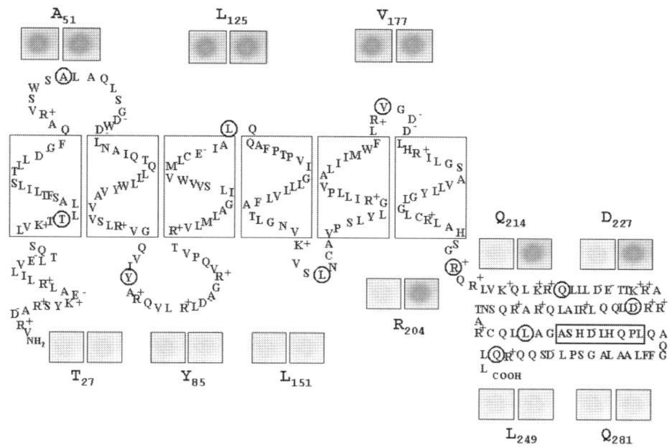

Figure 1. Schematic representation of the secondary structure of CorS derived from hydrophobicity, prediction analyses and experimental phenotypic characterisation of CorS-PhoA translational fusions on substrate-containing agar plates at 18°C (left rectangles) and 28°C (right rectangles). Circles indicate the residues of CorS that were fused to PhoA. The charged residues are indicated with "+" and "-". Boxed amino acids represent the conserved histidine kinase domain.

The N-and-C-termini of CorS were also shown to be located in the cytoplasm. Our experimental data confirmed the predicted topology of CorS, with six TMDs and cytoplasmic locations for the N- and C-terminus. However, since three PhoA fusions at amino acid residues R_{204}, Q_{214}, and D_{227} showed a temperature-dependent phenotype on MG-indicator plates (Fig. 1), we assume that the CorS topology might be thermoresponsive in this particular protein region. Colonies harboring PhoA fusions at R_{204}, Q_{214}, and D_{227} showed a $PhoA^{+}$ phenotype at 28°C, whereas a $PhoA^{-}$ phenotype was observed for these fusions at 18°C (Fig. 1). This indicates that the 6th TMD of CorS might be translocated into the periplasm at 28°C, while at the same temperature the putative 7th TMD downstream of the histidine kinase domain might be embedded into the membrane. The PhoA fusions at L_{249} and Q_{281}, located in close proximity upstream and downstream of the histidine kinase domain, both showed a $PhoA^{-}$ phenotype and were therefore cytoplasmic. As yet it has been impossible to confirm the temperature-sensitive phenotype of these fusions in a quantitative assay. To complicate things, the PhoA phenotypes were medium-dependent.

Minimal HSC medium optimised for COR production was used for bacterial growth in liquid culture because *P. syringae* does not grow well in MG liquid medium. In HSC medium, a quantitative difference between PhoA activities for fusions R_{204}, Q_{214}, and D_{227} was not observed. Both fusions showed a constitutively negative phenotype in this medium. The reason for this remains unknown. Complex KB medium was the best medium to differentiate periplasmic and cytoplasmic fusions in liquid culture as well as on agar plates (Table 2). However, there is no detectable COR production by *P. syringae* in KB medium. Thus, this medium cannot be used to study temperature-dependent phenomena mediated by CorS.

2.3 Model for the Topological Organisation of CorS

A thermoresponsive phenotype of CorS-PhoA fusions located in the linker region between the last TMD and the histidine kinase domain allows us to speculate that this region is involved in a temperature-dependent conformational change affecting autophosphorylation of the conserved histidine residue. According to our hypothesis, at 18°C CorS is fixed in the membrane and has six TMDs (Fig. 2). The linker region and the histidine kinase domain are located in the cytoplasm. In such a situation, the conserved His_{254} residue of CorS can be efficiently autophosphorylated and subsequently initiates the signal transduction pathway via phosphorylation of CorR. In contrast, at 28°C the 6th TMD and the linker region are translocated into the periplasm and the putative 7th TMD is embedded in the membrane, making the His_{254} residue inaccessible for autophosphorylation.

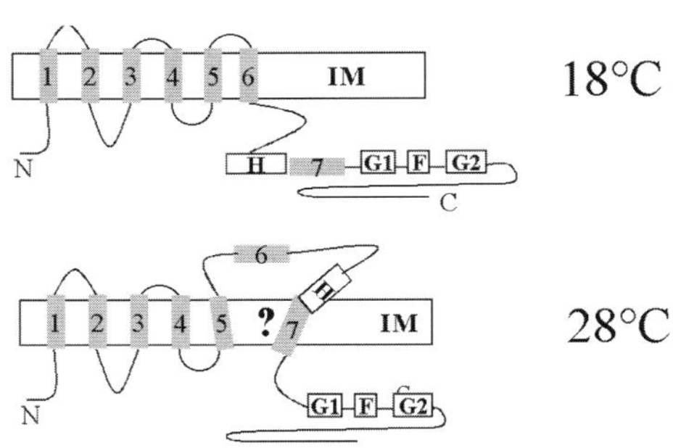

Figure 2. Hypothetical model for a topology change in CorS in response to temperature.

2.4 Strain-specific Sequence Alterations in the Region Encoding the 6th and 7th Transmembrane Domains

We derived our hypothetical model from topology prediction results and experimental data for CorS from *Pseudomonas syringae* PG4180. We then wanted to test whether the model fits to CorS proteins from other COR-producing strains.

We tested COR production at 18°C and 28°C in eleven COR-producing *P. syringae* strains of different geographic origins. Among them, 8 strains (including PG4180) produced COR in a temperature-dependent manner, whereas three of them (including DC3000) showed a temperature-independent COR production. *corS* genes from all tested strains were PCR amplified and subsequently sequenced. We analysed the deduced amino acid sequences of CorS from these *P. syringae* strains with respect to the hydrophobicity of TMDs. Interestingly, the three strains producing COR in a temperature-independent manner had specific amino acid substitutions in the sequences coding for the 6th and the putative 7th TMDs, such that there was an increase of the hydropathy index for the 6th TMD and a decrease for the putative 7th TMD (Table 3). The difference in the hydropathy index between the 6th and the putative 7th TMDs is negligible for temperature-independent COR producers as compared to the other strains such as PG4180 (Table 3).

Table3. Comparison of COR production and hydrophobicity of the 5th, 6th, and the putative 7th TMDs from *Pseudomonas syringae* strains of different geographical origins.

Strain	Geographic origin	COR production		Hydropathy index			Difference in hydropathy index (7th - 6th)
		18°C	28°C	5th TMD	6th TMD	7th TMD	
PG4180	New Zealand	39.7	0	1.123	0.682	0.786	0.104
Psg7a/90	Germany	28.5	0.3	1.123	0.682	0.786	0.104
Psg8aII/90	Germany	25.4	0	1.123	0.682	0.786	0.104
GSPB1545	Congo	17.1	0.2	1.123	0.682	0.786	0.104
GSPB1548	Rwanda	26.8	0	1.123	0.682	0.786	0.104
GSPB1993	Italy	21.0	0	1.123	0.682	0.786	0.104
GSPB2002	Hungary	32.7	0.3	1.123	0.682	0.786	0.104
GSPB2578	Germany	18.8	0	1.123	0.645	0.786	0.141
MAFF3013	Japan	6.4	7.1	1.123	0.839	0.709	-0.13
GSPB-mac	Germany	10.2	9.0	1.144	0.717	0.744	0.027
DC3000	Canada	7.9	7.3	1.144	0.717	0.744	0.027

3. SITE-DIRECTED MUTAGENESIS OF CORS

In order to find evidence to support our assumption that the 6th and the putative 7th TMDs are involved in a conformational change of CorS that results in a crucial change of CorS activity, we performed site-directed mutagenesis of CorS. A substitution of serine-187 (S_{187}) and alanine-188 (A_{188}) in the 6th TMD of CorS from PG4180, with isoleucine (I) and threonine (T) (the residues present in the respective TMD of CorS from DC3000) would result in an increased hydrophobicity of the 6th TMD. A substitution of serine-277 (S_{277}) with proline (P) in the putative 7th TMD of CorS would probably result in a destabilisation of α-helix for the 7th TMD.

Mutations ($S_{187}A_{188}$→IT and S_{277}→P) were introduced into CorS from PG4180 using the technique of overlap extension PCR. Mutants of CorS were tested in complementation analysis of *corS*$^-$ *corR*$^-$ mutant PG4180.D4 (pRGMU1). Plasmid pRGMU1 contains a promoterless β-glucuronidase reporter gene (*uidA*) fused to a 2.9 kb *Pst*I-fragment of the *cmaABT* promoter region from the COR gene cluster of PG4180. The PG4180 (pRGMU1) transconjugant allows the quantification of promoter activity *in trans* by measuring the β-glucuronidase (GUS) activity in a fluorescence assay. Transcriptional activation of the *cmaABT* promoter was previously shown to be temperature-dependent (Ullrich and Bender, 1994) and CorRSP-system-dependent (Ullrich *et al.*, 1995), since mutations in the regulatory genes resulted in a complete loss of promoter activity. Nevertheless, a 3.4-kb *Hin*dIII fragment containing *corRSP* genes could restore temperature-dependent promoter activity (Fig. 3) (Ullrich *et al.*, 1995). When a 3.4-kb *Hin*dIII fragment containing the three regulatory genes *corR*, *corP* and *corS* with mutation in the 6th TMD ($S_{187}A_{188}$→IT) of CorS, was introduced *in trans* into the *corS*$^-$ *corR*$^-$ mutant PG4180.D4 (pRGMU1), it resulted in a decrease of promoter activity at 18°C. Simultaneous mutations of the 6th and 7th TMDs ($S_{187}A_{188}$→IT and S_{277}→P) of CorS led to a complete loss of promoter activity at 28°C and to a further decrease of promoter activity at 18°C (Fig. 3). These results agree with an assumption that CorS of DC3000 is unable to change its conformation and to be efficiently autophosphorylated. However, we can not exclude the possibility that CorS dimerization ability was affected by introduced mutations.

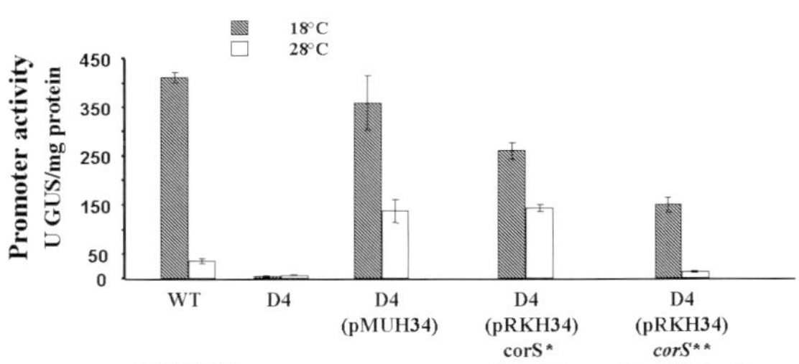

Figure 3. Complementation analysis of mutant PG4180.D4 (*corS*$^-$ *corR*$^-$) with plasmid pMUH34 and plasmids pRKH34 containing *corR*, *corP* and *corS* with site-directed mutations. *cmaABT::uidA* expression was determined in PG4180 (pRGMU1), PG4180.D4 (pRGMU1), and PG4180.D4 (pRGMU1) derivatives harboring plamids pMUH34, pRKH34:*corS* (SA→IT), and pRKH34:*corS* (SA→IT; S→P).

4. CONCLUSIONS

We investigated the membrane topology of the histidine protein kinase CorS from *Pseudomonas syringae* PG4180. Based on the results with translational PhoA and LacZ fusions, CorS is a membrane-associated protein with six TMDs. Three CorS-PhoA fusions showed a thermoresponsive phenotype suggesting that the membrane topology of CorS might be involved in signal perception. We could also show that mutations in the 6th TMD and simultaneous mutations in the 6th and 7th TMDs affected the transcriptional activation of COR biosynthetic genes.

ACKNOWLEDGEMENTS

This work was supported by grants from the Deutsche Forschungsgemeinschaft and the Max-Planck-Gesellschaft.

REFERENCES

Budde, I.P., Rohde, B.H., Bender, C., and Ullrich, M.S., 1998, Growth phase and temperature influence promoter activity, transcript abundance, and protein stability during biosynthesis of the *Pseudomonas syringae* phytotoxin coronatine, *J. Bacteriol.* **180**(6): 1360-1367.

Hennessey, E.S., and Broome-Smith, J.K., 1993, Gene-fusion techniques for determining membrane-protein topology, *Curr. Opin. Struct. Biol.* **3**: 524-531.

Manoil, C., and Beckwith, J., 1986, A genetic approach to analyzing membrane protein topology, *Science* **233**: 1403-1408.

Mitchell, R.E., 1982, Coronatine production by some phytopathogenic pseudomonads, *Physiol. Plant. Pathol.* **20**: 83-89.

Palmer, D.A., and Bender, C.L., 1995, Ultrastructure of tomato leaf tissue treated with the pseudomonad phytotoxin coronatine and comparison with methyl jasmonate, *Mol. Plant-Microbe Interact.* **8**: 683-692.

Rangaswamy, V., Jiralerspong, S., Parry, R., and Bender, C.L., 1998, Biosynthesis of the *Pseudomonas* polyketide coronafacic acid requires monofunctional and multifunctional polyketide synthase proteins, *Proc. Natl. Acad. Sci. USA* **95**: 15469-15474.

Sakai, R., Nishiyama, K., Ichihara, A., Shiraishi, K., and Sakamura, S., 1979, Studies on the mechanism of physiological activity of coronatine: Effect of coronatine on cell wall extensibility and expansion of potato tuber tissue, *Ann. Phytopathol. Soc. Japan* **45**: 645-653.

Ullrich, M.S., and Bender, C.L., 1994, The biosynthetic gene cluster for coromanic acid, an ethylcyclopropyl amino acid, contains genes homologous to amino acid-activating enzymes and thioesterases, *J. Bacteriol.* **176**(24): 7574-7586.

Ullrich, M.S., Peñaloza-Vázquez, A., Bailey, A.M., and Bender, C.L., 1995, A modified two-component regulatory system is involved in temperature-dependent biosynthesis of the *Pseudomonas syringae* phytotoxin coronatine, *J. Bacteriol.* **177**: 6160-6169.

Characterisation of Effector Genes of Pseudomonads Causing Disease on Hazelnut

S. LORETI, A. GALLELLI, R. JACKSON[#], D. BUTCHER[#], D. ARNOLD[#] and A. VIVIAN[#]

Istituto Sperimentale per la Patologia Vegetale, Via Bertero, 22, Roma Italy; [#]Centre for Research in Plant Science, University of West England, Bristol, UK.

Abstract: Effector genes, previously identified in the pathogenicity island (PAI) of *Pseudomonas savastanoi* pv. *phaseolicola* were characterised in two strains belonging to the two main groups of *Pseudomonas* pathogenic to hazelnut. ISPaVe-MCB-011, the type strain representing the typical *P. avellanae* group, causes severe cankers on trees in central Italy, while ISPaVe-MCB-596 belongs to the less aggressive *Pseudomonas* population, genetically distinct from typical *P. avellanae* strains. This study revealed striking differences between ISPaVe-MCB-011 and ISPaVe-MCB-596. Most genes of the *Pph*-PAI (*virPphA*, ORF2, ORF3, ORF4, *avrPphC*, *avrPphF*) were detected in ISPaVe-MCB-596, whereas only homologues of ORF3 and ORF2 were identified in *P. avellanae* ISPaVe-MCB-011. The ISPaVe-MCB-596 *virPphA* gene showed 99% DNA identity with *virPphA* and appears to be plasmid-borne. Although *virPphA* was absent from ISPaVe-MCB-011, a homologue of *avrPtoB* was found. The diversity between the two strains causing disease on hazelnut, the high identity showed between ISPaVe-MCB-596 with some *P. syringae* pathovars and the similarities between *P. avellanae* ISPaVe-MCB-011 and *P. syringae* pv. *tomato*, highlights the necessity to reconsider the taxonomy of the low aggressive pseudomonad group.

Key words: bacterial canker, *Pseudomonas avellanae*, *Corylus avellanae*, *avr* (avirulence genes).

Author for correspondence: S. Loreti; e-mail: mal.crit@ispave.it.

N.S. Iacobellis *et. al.* (eds.), *Pseudomonas syringae* and Related Pathogens, 393-398.

1. INTRODUCTION

Two main groups of pseudomonads are pathogenic to *Corylus avellana*. The first group is represented by the *Pseudomonas avellanae* the causal agent of bacterial canker of hazelnut, a dangerous disease present in central Italy and in Greece. A second group of pseudomonads, also previously identified as *P. avellanae,* has been reported in several Italian regions. This is characterised by a lower virulence, and seems to be genetically distinct from typical *P. avellanae* strains (Scortichini *et al*., 2000). One strain belonging to the virulent *P. avellanae* group (ISPaVe-MCB-011) and one to the less aggressive group (ISPaVe-MCB-596) were used for this study. A previous investigation reported that harpin encoding genes are more highly conserved in ISPaVe-MCB-596 than in *P. avellanae* strain ISPaVe-MCB-011, which only showed high similarity with *P. syringae* pv. *tomato* (Loreti *et al*., 2001). Moreover, Scortichini *et al*. (2002) recently proposed a distinct taxon for the less virulent *Pseudomonas* group.

In view of the recent observations that some effector genes affect in the fitness and pathogenicity of phytopathogenic bacteria, our investigation was extended to the identification and characterisation of effector genes found in the pathogenicity island (PAI) of *P. savastanoi* pv. *phaseolicola* (Jackson *et al*., 1999).

2. MATERIALS AND METHODS

2.1 DNA Extraction

Total genomic DNA was extracted from 1.5 ml broth cultures by the use of the Puregene DNA isolation kit (Gentra System-Flowgen, Leicestershire, UK). The DNA quantity was estimated comparing known standards in ethidium bromide stained 1% agarose gel.

2.2 PCR and DNA Sequencing

Polymerase chain reaction (PCR) reactions were conducted in a final volume of 50 μl. The final concentration of reagents: 1X PCR buffer (Life Technologies, Paisley, UK), 0.2 mM dNTPs, 0.04 μg μl^{-1} BSA, 1 μM each primer, 30÷50 ng genomic DNA. The thermal profile consisted of an initial hot start step, 70°C 3min and a initial denaturation step, 94°C 3min, followed by 30 cycles at 94°C (1 min), annealing at the proper temperature for each gene (1 min 30 sec) and extension at 72°C (from 1 min 30 sec to 2 min, depending on the gene length). All the amplification products were

analysed on 1% (w v^{-1}) agarose gel cast and run in TBE buffer (0.09 M Tris-borate, 0.002 M EDTA).
PCR primers for amplification of *virPphA*, ORF3, *avrF*, *avrC* ORF2, ORF4 and *avrPtoB* were designed from published sequences and are available upon request. All primers were synthesised by Life Technologies (Paisley, UK) and DNA sequencing was performed by M-Medical Genenco (Rome, Italy) and sequence were aligned using ClustalW (www.ebi.ac.uk/clustalw/).

3. RESULTS

Homologues of the *Pph*-PAI, *virPphA*, ORF2, ORF3, ORF4, *avrPphC*, *avrPphF*-ORF1, *avrPphF*-ORF2 were identified by PCR in strain ISPaVe-MCB-596; *Pseudomonas avellanae* ISPaVe-MCB-011 only showed the presence of ORF3 and ORF2 homologues. The sequencing of these genes revealed a high identity between the PAI homologues genes of ISPaVe-MCB-596 and those of *P. savastanoi* pv. *phaseolicola* PAI (Table 1).

Table 1. Identity (%) of effectors gene sequences of ISPaVe-MCB-596 and Pa ISPaVe-MCB-011 with *Pseudomonas savastanoi* pv. *phaseolicola* effector genes. The deduced amino acid homologies percentage are below in parenthesis.

	avrPphF	*avrPphC*	*ORF3*	*ORF2*	*virPphA*	*ORF4*	*avrPphD*
ISPaVe-MCB-596	99% (99%)	99% (95%)	96% (90%)	99% (100%)	99% (100%)	99% (99%)	100% (100%)
***Pa* ISPaVe-MCB-011**	-	-	95% (90%)	99% (99%)	-	-	91% (81%)

3.1 *virPphA*

Three genomic clones containing the *virPphA* homologues were detected in the ISPaVe-MCB-596 gene library. The ISPaVe-MCB-596 *virA* gene was 99% identical to *virPphA,* whereas the predicted amino acid sequence was 100% identical. Hybridisation experiments using a digoxigenin labelled *virPphA* probe showed that this gene appears to be plasmid borne as observed in the case of *Pseudomonas savastanoi* pv. *phaseolicola.*

3.2 ORF2 and ORF3

ORF2 of ISPaVe-MCB-596 was 100% identical to *virPphA. Pa* ISPaVe-MCB-011 showed 99% of identity with both *Pseudomonas savastanoi* pv. *phaseolicola* and ISPaVe-MCB-596. ORF3 of ISPaVe-MCB-596 contained a stretch of 10 base pairs longer than the homologous *P. s.* pv. *phaseolicola* gene. These two genes showed 96% of identity at the nucleotide level and

90% between the predicted amino acid sequences. Similarly, ISPaVe-MCB-011 ORF3 nucleotide sequence was 99% identical with ORF3 of ISPaVe-MCB-596 and 95% with that of *Pph*.

3.3 ORF4

ORF4 of ISPaVe-MCB-596 and *Pseudomonas savastanoi* pv. *phaseolicola* showed 99% identity at both nucleic acid and predicted amino acid sequences.

3.4 *avrPphC* and *avrPphF*

Homologues of these genes were identified in ISPaVe-MCB-596 only. *avrPphF* homologues retained 99% identity at both nucleic acid and predicted amino acid levels. Although the *avrPphC* homologue showed an identity of 99% with the corresponding gene, the predicted amino acid sequences showed a lower identity (95%).

3.5 *avrD*

Homologues of *avrD* were identified in both strains. The comparison of these gene sequences with those known in literature indicated an identity of 91% between *Pseudomonas avellanae* ISPaVe-MCB-011 and ISPaVe-MCB-596. ISPaVe-MCB-596 *avrD* sequences showed a high similarity with that of *P. savastanoi* pv. *phaseolicola* (100%) and *P. s.* pv. *glycinea* (99%). *P. avellanae* ISPaVe-MCB-011 showed high similarity with *P. syringae* pv. *tomato* (95%).

4. DISCUSSION

The presence of effector genes in the *Pph*-PAI has been previously identified in some pathovars of *Pseudomonas syringae* and *P. savastanoi.* This study revealed that not all these PAI genes were present in the genome of the surveyed bacterial strains (Jackson *et al*., 2002). When the presence of *Pph*-PAI genes where investigated in ISPaVe-MCB-596, it was found that it contained all the main effector genes of the PAI. Otherwise, only ORF2 and ORF3 genes where detected in *P. avellanae* ISPaVe-MCB-011. Amongst all the *Pph*-PAI genes, the *avrPphF* gene is not known to be widely distributed. It is only found in some *P. s.* pv. *phaseolicola* strains and in *Pto* (Jackson *et al*., 2002). The presence of this gene in ISPaVe-MCB-596 validate the homologies with *P. s.* pv. *phaseolicola*. Moreover, the characterisation of the

ISPaVe-MCB-596 effector gene sequences revealed a high percentage of identity with the homologous *P. s.* pv. *phaseolicola* genes.
The *avrD* gene sequence of ISPaVe-MCB-596 showed the highest homology with that of *P. s.* pv. *phaseolicola*. Conversely, *P. avellanae* ISPaVe-MCB-011 *avrD* revealed higher identity with *P. s.* pv. *tomato,* and, although *virPphA* was absent from this strain, a homologue of *avrPtoB* was found.
These data confirmed previous results obtained from the characterisation of harpin genes (Loreti *et al.*, 2001), that indicated a lower similarity between the two pathogens of hazelnut. These differences agree with the diversity detected in the molecular fingerprints, pathogenic and morphological characteristics (Scortichini *et al.*, 1998; Scortichini and Angelucci, 1999; Scortichini *et al.*, 2000).
Furthermore, these data highlight the high correlation between ISPaVe-MCB-596 and *P. s.* pv. *phaseolicola* and other *P. syringae* pathovars, in agreement with the previous study characterising harpin encoding genes (Loreti *et al.*, 2001). Recently, Scortichini *et al.* (2002) proposed that the less aggressive pseudomonad group, to which ISPaVe-MCB-596 belong, is a distinct taxon more closely related to *P. s.* pv. *syringae*. Conversely, ISPaVe-MCB-011 shares homology with *Pto*, as indicated previously from the characterisation of harpin genes (Loreti *et al.*, 2001).
The high homology between ISPaVe-MCB-596 and *P. syringae* pathovars, in particular *P. s.* pv. *phaseolicola*, highlights the necessity to reconsider the low aggressive pseudomonad group from a taxonomic point of view.
Further studies are under way on *P. avellanae* ISPaVe-MCB-011 to characterise the presumptive PAI region and the role of *avrPtoB* in establishment of disease in hazelnut by *P. avellanae*.

REFERENCES

Jackson, R.W., Athanassopoulos, E., Tsiamis, G., Manfield, J.W., Sesma, A., Arnold, D.L., Gibbon, MM.J., Murrillo, J., Taylor, J.D., and Vivian, A., 1999, Identification of a pathogenicity island, which contains genes for virulence and avirulence, on a large native plasmid in the bean pathogen *Pseudomonas syringae* pathovar *phaseolicola*. *Proc. Natl. Acad. Sci. USA* **96**: 10875-10880.

Jackson, R.W., Mansfield, J.W., Ammouneh, H, Dutton, L.C., Wharton, B, Ortz-Barredo, A, Arnold, D.L., Tsiamis, G., Sesma, A., Butcher, D., Boch, J., Kim, Y.J., Martin, G.B., Tegli, S., Murrillo, J. and Vivian, A., 2002, Location and activity of members of a family of *virPphA* homologues in pathovars of *Pseudomonas syringae* and *P. savastanoi*. *Mol. Plant Pathol.* **3**: 205-216.

Loreti, S., Sarrocco, S., and Gallelli, A., 2001, Identification of *hrp* gene, encoding harpin, in *Pseudomonas avellanae* (Psallidas) Janse *et al. J. Phytopathol.* **149**: 219-226.

Scortichini, M., Dettori, M.T., Marchesi, U., Palombi, M.A. and Rossi, M.P., 1998, Differentiation of *Pseudomonas avellanae* strains from Greece and Italy by rep-PCR genomic fingerprinting. *J. Phytopathol.* **146**: 417-420.

Scortichini, M., and Angelucci, U., 1999, Phenotypic characterisation of *Pseudomonas avellanae* (Psallidas) Janse et al. occurrence of colony variants. *J. Plant Pathol.* **81**: 55-61.

Scortichini, M., Marchesi, U., Angelucci, L., Rossi, M.P. and Dettori, M.T., 2000, Occurrence of *Pseudomonas avellanae* (Psallidas) Janse et al. and related pseudomonads on wild *Corylus avellana* trees and genetic relationship with strains isolated from cultivated hazelnuts. *J. Phytopathol.* **148**: 523-532.

Scortichini, M., Marchesi, U., Rossi, M.P., and Di Prospero, P., 2002, Bacteria associated with hazelnut (*Corylus avellana* L.) decline are two groups: *Pseudomonas avellanae* and strains resembling *P. syringae* pv. *syringae*. *App. Environ. Microbiol.* **68**: 476-484.

Generation of Microarrays for the Study of Gene Expression Patterns in *Ralstonia solanacearum*

S.A. SIMPKINS, A. FRISCINA[#], K. MADAGAN[§], J.G. ELPHINSTONE, and D.E. STEAD
Central Science Laboratory, Sand Hutton, York, UK; [#]Dipartimento di Scienze e Tecnologie Agroambientali (DISTA), Università degli Studi di Bologna, Italy; [§]Centre for Novel Agricultural Products, University of York, UK.

Abstract: This work represents the first steps in the development of functional genomic methods to identify variably expressed genes involved in the regulation of host specificity within *Ralstonia solanacearum*. Microarrays were designed that contained a subset of the genetic complement of *R. solanacearum*. These were spotted out onto glass slides using a commercial microarrayer, with sufficient replicates of each fragment included to allow for statistically significant expression level differences to be measured. Extraction of bacterial mRNA from *in vitro* cultures and infected plant material was performed using commercial total RNA extraction kits. Problems encountered included obtaining sufficient amounts of bacterial RNA in total plant extract, evaluating mRNA extraction efficiency, and reverse transcription of bacterial mRNA. This work has given us the techniques and experience in the use of microarrays to conduct functional genomic studies into the differential expression of a large number of functionally uncharacterised genes from the recently published genomic sequence of *R. solanacearum*.

Key words: *Ralstonia*, Microarrays, functional genomics.

Author for correspondence: D.E. Stead; e-mail: d.stead@csl.gov.uk.

1. INTRODUCTION

Bacterial wilt (causal agent *Ralstonia solanacearum*) has been described as the single most important plant pathogenic bacteria world wide, infecting more than 400 plant species including such economically important crop

N.S. Iacobellis *et. al.* (eds.), *Pseudomonas syringae* and Related Pathogens, 399-403.

plants as potato, tomato, banana, tobacco, eggplant, peppers, ginger and groundnut (Hayward, 1991).

Its genome has been sequenced in order to understand the mechanisms of virulence and its host range (Salanoubat *et al.*, 2002). Many virulence factors have been investigated/identified but much is still unknown (Genin and Boucher, 2002). For example, little is known about how *R. solanacearum* is able to infect latently and multiply to high levels within tolerant host plant cultivars, and especially the dependence of host range and latency on temperature.

This latent infectivity appears to break down with an increase in temperature and pathogenicity results (Hayward, 1991). Clearly this phenomenon is not wholly dependent on the bacterium. However, the study of this in the bacteria in relation to global gene regulation and expression will enable a better understanding of the mechanisms involved from which novel targets and strategies for control and breeding can be devised. Furthering knowledge of factors affecting host range is of particular importance in disease control, for example, where weed hosts have been shown to play an important role in epidemiology (Elphinstone, 2003).

In addition, this bacterium is able to cause disease over a wide temperature range in that it is a problem in temperate European climates (causing potato brown rot) as well as in sub-tropical climates, where it causes bacterial wilt of tomato, banana and tobacco (Elphinstone, 2003). The disease is therefore of importance in relation to basic, subsistence farming in many parts of the world.

Here we present the initial phase of a project to exploit microarray methods to identify variably expressed genes involved in the regulation of host specificity within *R. solanacearum*.

2. MATERIALS AND METHODS

In order to generate PCR products, the published sequences of individual *Ralstonia solanacearum* genes were interrogated and a subset of genes selected from a wide range of genes including those with housekeeping as well as virulence-related functions and both constitutively and inducibly expressed genes. PCR primers to amplify 200÷450 bp products at the 3' end of each of these genes were designed using Primer3 software (Rozen and Skaletsky, 2000). Fifty one products were amplified and then purified either directly from the PCR tube (Wizard PCR prep kit, Promega, Southampton, UK) or by gel excision and clean up (QIAquick gel extraction kit, Qiagen, Crawley, UK). In addition, a range of control PCR products were obtained including tomato plant ATPase and 25S rDNA gene fragments along with a number of human genes to serve as spiking controls.

The generation of microarrays was performed using commercial poly lysine or aminopropyltriethoxysilane-coated slides (Sigma, Poole, UK; Fisher Scientific, Loughborough, UK, (Corning CMT-GAPS II slides); VWR International, Lutterworth, UK (BDH Poly lysine slides)) and a Microgrid arrayer (Biorobotics, Cambridge, UK). Genepix pro software (Axon Instruments, Union City, USA) was used to design a grid layout such that spots were arrayed out in 4x4 squares each of 100. This gave 1600 per slide and therefore 25 replicates for each of 64 gene fragments.

Total bacterial RNA was extracted from cultures grown on TTC agar and from root inoculated tomato seedlings after 7÷10 days incubation at 28°C. Total RNA extraction was carried out using an RNAqueous Midi kit (Ambion, Huntingdon, UK). This has modifications for extracting from both plant material and from bacteria and typically gives $\geq$ 400 μg ml^{-1} of total RNA from an infected plant extract.

The RNA was then used to generate labelled cDNA. Briefly, this involved a first step in which amino allyl dUTP (AA-dUTP), an amine-modified nucleotide, was incorporated during reverse transcription, using an Amino Allyl cDNA Labelling Kit (Ambion, Huntingdon, UK). Subsequently, monofunctional forms of Cyanine 3 and Cyanine 5 dyes were reacted with the amine-modified cDNA. In order to prime the reverse transcription reaction a minimal set of primers were designed against repeated motifs lying at the 3' end of each gene, using the Genome directed primer (GDP) software, (Talaat *et al.,* 2000). These were used to reverse transcribe mRNA linearly into cDNA, in order to ensure that expression level differences between the two conditions were preserved.

Hybridisation was performed using commercial hybridisation solutions (Ambion, Huntingdon, UK) in small 25-30 μl volumes and proprietary hybridisation coverslips (Hybri-Slips, Sigma, Poole, UK) to retain the solution on the microarray surface. The microarrays were then incubated overnight in a humid chamber. A number of incubation temperatures were used ranging from 45÷55°C.

The following day washing of the slide was carried out. The Hybri-Slip was first removed being careful not to let the slide dry out until after the washing had been completed. A number of different washing protocols were used but the most appropriate method involved two washes at 55°C for 15 minutes in a low stringency solution (2X SSC, 0.5% SDS), followed by two washes under the same regime in a high stringency solution (0.5X SSC, 0.5% SDS).

The slides were then spun dry or dried with an air duster and transferred to a light-tight container until ready for scanning in an GenePix 4000B array scanner (Axon Instruments, Union City, USA) and analysed using GenePix pro software (Axon Instruments, Union City, USA).

3. RESULTS AND DISCUSSION

Initial hybridisation experiments revealed that the spiking controls worked well but we saw only weak signals for one of the *Ralstonia solanacearum* array spots; namely the 16S rRNA gene. The strength of this signal was improved by increasing the hybridisation stringency, primarily by way of an increase in the incubation temperature from 50 to 55°C. This resulted in less background and sharper spots. The most significant improvement in hybridisation signal from the target array spots was afforded by careful optimisation of starting RNA concentration prior to the reverse transcription and Cyanine dye labelling step. We have found that at least 3μg of total RNA is required in a reaction for detectable hybridisation signals to be obtained from the *R. solanacearum* arrayed genes. Initial experiments have used total RNA extracted from *in vitro* cultures of *R. solanacearum* in order for us to concentrate on optimising protocols for cDNA production and labelling, microarray hybridisation and post-hybridisation treatments. The accomplishment of this enabled us to detect the expression of a number of genes including the *pehB* gene (Huang and Allen, 1997), a recently sequenced porin gene implicated in pathogenicity (E. Stefani, personnal communication) as well as the highly expressed 16S rRNA gene. Therefore, we have been able to detect inducible as well as constitutively expressed *R. solanancearum* genes.

This is an initial functional genomic study that has served to equip us with a knowledge of microarray design and the requirements for arraying out PCR products, with protocols for bacterial mRNA extraction, reverse transcription and Cyanine labelling, and with the experience of microarray hybridisation methodologies. This will enable us to study the differential expression of a large number of functionally uncharacterised genes by expanding and redesigning our microarrays in order to make use of the complete genomic sequence of *R. solanacearum* that is now available.

ACKNOWLEDGEMENTS

The authors gratefully acknowledge the provision of Department of Environment, Food and Rural Affairs-Central Science Laboratory (DEFRA-CSL) seed corn research funds.

REFERENCES

Elphinstone, J.G., 2003, The current bacterial wilt situation: a global overview. In: *Bacterial wilt: The disease and the* Ralstonia solanacearum *species complex*. C. Allen, P. Prior and C.A. Hayward. (eds). APS Press, St. Paul, Minnesota, USA. In press.

Genin, S., and Boucher, C.A., 2002, *Ralstonia solanacearum*: secrets of a major pathogen unveiled by analysis of its genome. *Mol. Plant Pathol.* **3**: 111-118.

Hayward, A.C., 1991, Biology and epidemiology of bacterial wilt caused by *Pseudomonas solanacearum*. *Annual Review of Phytopathology* **29**: 65-87.

Huang, Q., and Allen, C, 1997, An exo-poly-α-D-galacturonosidase, PehB, is required for wild-type virulence of *Ralstonia solanacearum*. *J. Bacteriology* **179**: 7369-7378.

Rozen, S., and Skaletsky H.J., 2000, Primer3 on the WWW for general users and for biologist programmers. In: *Bioinformatics Methods and Protocols: Methods in Molecular Biology*. Krawetz S, Misener S. (eds). Humana Press, Totowa, NJ, pp 365-386.
(Code available at http://www-genome.wi.mit.edu/genome_software/other/primer3.html.)

Salanoubat, M., Genin, S., Artiguenave, F., Gouzy, J., Mangenot, S., Arlat, M., Billault, A., Brottier, P., Camus, J. C., Cattolico, L., Chandler, M., Choisne, N., Claudel-Renard, C., Cunna, S., Demange, N., Gaspin, C., Lavie, M., Moisan, A., Robert, C., Saurin, W., Schiex, T., Siguier, P., Thebault, P., Whalen, M., Wincker, P., Levy, M., Weissenbach, J. and Boucher, C. A., 2002, Genome sequence of the plant pathogen *Ralstonia solanacearum*. *Nature* **415**: 497-502. (The complete sequence can be accessed at http://sequence.toulouse.inra.fr/R.solanacearum.html).

Talaat, A. M., Hunter, P., and Johnston, S.A., 2000, Genome-directed primers for selective labelling of bacterial transcripts for DNA microarray analysis. *Nature Biotech.* **18**: 679-682.

Sequence Analysis of the *hrpC* Operon and the *hrpE* Gene of *Pseudomonas syringae* subsp. *savastanoi*

A. SISTO, M.G. CIPRIANI, and M. MOREA
Istituto di Scienze delle Produzioni Alimentari, C.N.R., Viale Einaudi 51, 70125 Bari, Italy.

Abstract: The virulence of *Pseudomonas syringae* subsp. *savastanoi,* which causes hyperplastic symptoms (knots) on olive plants, is dependent upon produced phytohormones (indole-3-acetic acid and cytokinins). Recent studies showed that a Tn*5*-induced mutant of the above-mentioned bacterium did not cause any disease symptom on olive plants although it was still able to produce phytohormones; in addition, this mutant also failed to elicit a hypersensitive response in a non host plant. Molecular characterisation revealed that DNA flanking Tn*5* insertion in the mutant encoded a protein highly similar to the HrcC protein of *P. s.* pv. *syringae*. Since the *hrcC* gene of *P. s.* pv. *syringae* is known to be part of an operon within the *hrp/hrc* gene cluster, in order to define the genetic organisation of the region and to identify other *hrp* genes in *P. s.* subsp. *savastanoi,* further sequence analysis of DNA regions flanking Tn*5* insertion in the mutant was carried out. The results revealed that the gene encoding the HrcC protein in *P. s.* subsp. *savastanoi* was part of an operon including five genes with the same arrangement as in other phytopathogenic bacteria. On the basis of their homologies to previously characterised genes of other *P. syringae* pathovars, these genes were named *hrpF*, *hrpG, hrcC, hrpT* and *hrpV.* Upstream of this operon an homolog of the *hrpE* gene was also found in *P. s.* subsp. *savastanoi*. The similarity between some of the above-mentioned genes of *P. s.* subsp. *savastanoi* and the corresponding genes of *P. s.* pv. *glycinea* or of *P. s.* pv. *phaseolicola* was particularly high. These results, together with the identification and characterisation of the *hrcC* mutant of *P. s.* subsp. *savastanoi,* suggest that at least some *hrp/hrc* genes play a key role in the pathogenicity of this plant pathogen which induces hyperplastic symptoms on its host plants.

Key words: *Pseudomonas syringae* subsp. *savastanoi,* olive knot disease, *hrpC* operon, *hrp/hrc* genes.

Author for correspondence: A. Sisto; e-mail: angelo.sisto@ispa.cnr.it.

N.S. Iacobellis *et. al.* (eds.), *Pseudomonas syringae* and Related Pathogens, 405-410.

1. INTRODUCTION

Pseudomonas syringae subsp. *savastanoi* (=*Pseudomonas savastanoi* pv. *savastanoi*) (Young *et al*., 1996) causes hyperplastic symptoms on olive (*Olea europaea* L.) oleander (*Nerium oleander* L.) and ash (*Fraxinus excelsior* L.) as well as on other minor host plants. Symptom development on olive and oleander plants depends on the production by the bacterium of indole-3-acetic acid and cytokinins that alter the physiological hormone balance in the infected tissues and cause the proliferation of plant cells surrounding the infection site (Surico *et al*., 1985; Surico and Iacobellis, 1992; Iacobellis *et al*., 1994). The ability of many plant pathogenic bacteria to cause disease in host plants and to induce a hypersensitive response (HR) in non host plants is also known to depend on *hrp* genes (Lindgren, 1997). The *hrp* genes are typically clustered; many of them encode components of a type III protein secretion system that is believed to be used by phytopathogenic bacteria for translocating virulence proteins into plant cells which are ultimately responsible for leakage of plant nutrients to the extracellular space of infected tissues (He, 1998; Collmer *et al*., 2000). The *hrp* genes have been identified and studied in plant pathogenic bacteria of the genera *Pseudomonas*, *Ralstonia, Xanthomonas* and *Erwinia* (Lindgren, 1997; Mor *et al*., 2001) and nine of them have been renamed *hrc* (for HR and conserved) to denote their conservation among the type III secretion systems of both plant and animal pathogens (Bogdanove *et al*., 1996). We have recently reported on the identification of mutant ITM317-1421, a new prototrophic Tn*5*-induced mutant of *P. s.* subsp. *savastanoi,* failing to elicit a hypersensitive response in non host plants, i.e. tobacco (Sisto *et al*., 2001). The mutant, which was still able to produce indole-3-acetic acid and cytokinins, did not cause any disease symptom (i.e. knot formation) on olive, its host plant, and was unable to multiply in host tissue. DNA regions flanking Tn*5* insertion in mutant ITM317-1421 were in part sequenced and database searches revealed that transposon insertion occurred within a DNA region encoding a predicted product that exhibited a very high similarity with the HrcC protein of *P. s.* pv. *syringae*. This protein is encoded by *hrcC* gene which in the latter plant pathogen is known to be part of an operon within the *hrp/hrc* gene cluster (Deng *et al*., 1998). Here we report on further sequence analysis of DNA regions flanking Tn*5* insertion in mutant ITM317-1421, carried out in order to define the genetic organisation of the region and to identify other *hrp* genes in *P. s.* subsp. *savastanoi*.

2. MATERIALS AND METHODS

2.1 DNA Sequencing and Analysis

Preparations of plasmid-DNA from *Escherichia coli* TG1 strain for sequence analysis were made using the Quantum Prep Plasmid Miniprep kit (Biorad, CA, USA).

Two subclones in pBluescript SK II plasmid vector containing the DNA regions flanking the Tn*5* element in mutant ITM317-1421 were sequenced by using, as a primer, a single oligonucleotide complementary to and extending outward from the ends of the inverted repeat of transposon (Rich and Willis, 1990) and by primer walking. PCR primers 5'-GACCGGCTTGGTCAGGAAT-3' (forward) and 5'-CGGCTTTTCCCGGATTTCT-3'(reverse) were used to amplify from the wild-type parental strain ITM317 a 593-base-pair (bp) DNA fragment that included the 9-bp sequence corresponding to Tn*5* insertion in the mutant. DNA sequencing was performed with BigDye™ Terminator Chemistry from Applied Biosystems on an ABI 310 sequencer (Applied Biosystems, Foster City, CA, USA). Sequenced DNA fragments were examined for similarity with deposited sequences using the BLASTx program (Altschul *et al.*, 1997), available from the National Centre for Biotechnology Information (http://www.ncbi.nlm.nih.gov/).

3. RESULTS AND DISCUSSION

Overall 4,050 base pairs were sequenced by primer walking from two subclones (Sisto *et al.*, 2001) which contained the DNA regions flanking Tn*5* insertion in mutant ITM317-1421 of *Pseudomonas syringae* subsp. *savastanoi*. In addition, by using two PCR primers located respectively 316 bp upstream and 268 bp downstream from the 9-bp sequence corresponding to Tn*5* insertion in the mutant, a DNA fragment of 593 bp was amplified from the wild-type parental strain ITM317. This fragment was sequenced and the results confirmed that it was identical to the DNA regions flanking the Tn*5* element in mutant ITM317-1421. Furthermore, sequence analysis and database searches confirmed that Tn*5* insertion occurred in mutant ITM317-1421 within an homolog of the *hrcC* gene and revealed that this gene of *P. s.* subsp. *savastanoi* was part of an operon that includes five genes. On the basis of their homologies to previously characterised *hrp*/*hrc* genes of other *P. syringae* pathovars, these *P. s.* subsp. *savastanoi* genes were named *hrpF*, *hrpG*, *hrcC*, *hrpT* and *hrpV* and formed the *hrpC* operon. Upstream of this operon, a homolog of the *hrpE* gene was also found in *P. s.*

subsp. *savastanoi*. All these genes in *P. s.* subsp. *savastanoi* were arranged co-linearly with their homologs of other *P. syringae* pathovars (Deng *et al.* 1998) and had the same orientation. A typical consensus promoter sequence (*hrp*-box), consisting of TGGAACC-16N-CCACNNA (Xiao and Hutcheson, 1994) as well as a putative ribosome-binding site (GGAG) were found upstream of the ATG translational start codon of the *P. s.* subsp. *savastanoi hrpF* gene. In addition, downstream of the *P. s.* subsp. *savastanoi hrpV* gene a typical rho-independent terminator characterised by an inverted repeat and a poly(T) was identified. The predicted products, HrpE, HrpF, HrpG, HrcC, HrpT and HrpV encoded by the above-mentioned *hrp/hrc* genes of *P. s.* subsp. *savastanoi* showed a high degree of homology with the deduced products encoded by the corresponding genes of *P. syringae* pathovars. The levels of identity were higher with the predicted proteins of the related *P. s.* pvs. *glycinea* and *phaseolicola*, which were included in the same genomospecies (Gardan *et al.*, 1999) than with those of *P. s.* pv. *syringae* and *P. s.* pv. *tomato.* In particular, HrpE, HrpF, HrpG, HrpT and HrpV predicted proteins of *P. s.* subsp. *savastanoi* showed levels of identity with the homologous proteins of *P. s.* pv. *glycinea* or *P. s.* pv. *syringae* ranging from 96 to 100% and from 82 to 92%, respectively. In the case of *P. s.* pv. *tomato*, HrpF seems to be an exception as it was divergent and only 36% identical to the HrpF of *P. s.* subsp. *savastanoi*; on the contrary, the percentage of identity between the other above-mentioned predicted proteins of *P. s.* subsp. *savastanoi* and those of *P. s.* pv. *tomato* ranged from 76 to 79%. Only the sequence of HrpE of *P. s.* pv. *phaseolicola* was available in database and it was 94% identical to the HrpE protein of *P. s.* subsp. *savastanoi.* HrcC protein of *P. s.* subsp. *savastanoi* was 92% identical to HrcC protein of *P. s.* pv. *syringae*, as previously reported (Sisto *et al.*, 2001), and 79 % identical to the HrcC protein of *P. s.* pv. *tomato.* The latter protein is a member of a superfamily of outer membrane proteins (PulD-pIV) involved in macromolecular traffic across the bacterial outer membrane, it is thought to multimerize and form a channel for the translocation of proteins and it has been shown to be required for both HrpZ secretion and the delivery of Avr signals (Deng *et al.*, 1998). The effects of non polar mutations in *hrpF*, *hrpG*, *hrcC*, *hrpT* and *hrpV* genes of *P. s.* pv. *syringae* have been studied (Deng *et al.*, 1998; Preston *et al.*, 1998). It is interesting to note that, exactly like the Tn*5*-induced *hrcC* mutant ITM317-1421 of *P. s.* subsp. *savastanoi*, a non polar *hrcC* mutant of *P. s.* pv. *syringae* did not induce any disease symptom or hypersensitive reaction and was markedly hampered in its ability to multiply within the tissues of the host plant. This suggests that the mutated phenotypes observed in mutant ITM317-1421 were not due to a polar effect on downstream *hrpT* and *hrpV* genes. The function of these genes has been proposed (Preston *et al.*, 1998): *hrpT* encodes a putative lipoprotein and it might function as a HrcC chaperone; *hrpV* encodes a negative regulator of *hrp* gene expression. It has been reported that, besides *hrcC*, also *hrpE* (Huang *et al*., 1995), *hrpG*, *hrpF*, and *hrpT* (Preston *et al*., 1998) of *P. s.* pv. *syringae*, play a role in HrpZ secretion, and these genes in *P. s.* subsp. *savastanoi*, on account of their high similarity with the corresponding genes in other *P. syringae* pathovars, might have the same function.

In conclusion, our results indicate that *hrpE, hrpF*, *hrpG*, *hrcC, hrpT* and *hrpV* genes are present in *P. s.* subsp. *savastanoi* and that they are organised as in other *P. syringae* pathovars; this, together with the identification and characterisation of the *hrcC* mutant ITM317-1421 of *P. s.* subsp. *savastanoi* (Sisto *et al*., 2001), suggests that at least some *hrc/hrp* genes are involved in the pathogenicity of *P. s.* subsp. *savastanoi* which causes hyperplastic symptoms on its host plants. In addition, since *hrc/hrp* genes are required for the secretion of proteins taking part in plant-microbe interaction, it is very likely that similar proteins might also be involved in the case of *P. s.* subsp. *savastanoi.*

ACKNOWLEDGEMENTS

Work supported by a grant from the Italian Ministry of Scientific and Technological Research (MURST), Plan for the Development of Research Networks, Law 488/92, Cluster C06 + 07, Project 1.1: Micro-organisms and Microbial Metabolites in Plant Protection.

REFERENCES

Altschul, S.F., Madden, T.L., Schäffer, A.A., Zhang, J., Zhang, Z., Miller, W., and Lipman, D.J., 1997, Gapped BLAST and PSI-BLAST: a new generation of protein database search programs. *Nucleic Acids Res.* **25**: 3389-3402.

Bogdanove, A.J., Beer, S.V., Bonas, U., Boucher, C.A., Collmer, A., Coplin, D.L., Cornelis, G.R., Huang, H.C., Hutcheson, S.W., Panopoulos, N.J., and van Gijsegem, F., 1996, Unified nomenclature for broadly conserved *hrp* genes of phytopathogenic bacteria. *Mol. Microbiol.* **20**: 681-683.

Collmer, A., Badel, J.L., Charkowski, A.O., Deng, W.L., Fouts, D.E., Ramos, A.R., Rehm, A.H., Anderson, D.M., Schneewind, O., van Dijk, K., and Alfano, J.R., 2000, *Pseudomonas syringae* Hrp type III secretion system and effector proteins. *Proc. Natl. Acad. Sci. USA* **97**: 8770-8777.

Deng, W.L., Preston, G., Collmer, A., Chang, C.J., and Huang, H.C., 1998, Characterization of the *hrpC* and *hrpRS* operons of *Pseudomonas syringae* pathovars *syringae*, *tomato*, and *glycinea* and analysis of the ability of *hrpF*, *hrpG*, *hrcC*, *hrpT*, and *hrpV* mutants to elicit the hypersensitive response and disease in plants. *J. Bacteriol.* **180**: 4523-4531.

Gardan, L., Shafik, H., Belouin, S., Broch, R., Grimont, F., and Grimont, P.A.D., 1999, DNA relatedness among the pathovars of *Pseudomonas syringae* and description of *Pseudomonas tremae* sp. nov. and *Pseudomonas cannabina* sp. nov. (ex Sutic and Dowson 1959). *Int. J. Syst. Bacteriol.* **49**: 469-478.

He, S.Y., 1998, Type III protein secretion systems in plant and animal pathogenic bacteria. *Annu. Rev. Phytopathol.* **36**: 363-392.

Huang, H.C., Lin, R.H., Chang, C.J., Collmer, A., and Deng, W.L., 1995, The complete *hrp* gene cluster of *Pseudomonas syringae* pv. *syringae* 61 includes two blocks of genes required for harpin$_{Pss}$ secretion that are arranged colinearly with *Yersinia ysc* homologs. *Mol. Plant-Microbe Interact.* **8**: 733-746.

Iacobellis, N.S., Sisto, A., Surico, G., Evidente, A., and DiMaio, E., 1994, Pathogenicity of *Pseudomonas syringae* subsp. *savastanoi* mutants defective in phytohormone production. *J. Phytopathology,* **140**: 238-248.

Lindgren, P.B., 1997, The role of *hrp* genes during plant-bacterial interactions. *Annu. Rev. Phytopathol.* **35**: 129-152.

Mor, H., Manulis, S., Zuck, M., Nizan, R., Coplin, D.L., and Barash, I., 2001, Genetic organization of the *hrp* gene cluster and *dspAE/BF* operon in *Erwinia herbicola* pv. *gypsophilae*. *Mol. Plant-Microbe Interact.* **14**: 431-436.

Preston, G., Deng, W.L., Huang, H.C., and Collmer, A., 1998, Negative regulation of *hrp* genes in *Pseudomonas syringae* by HrpV. *J. Bacteriol.* **180**: 4532-4537.

Rich, J.J., and Willis, D.K., 1990, A single oligonucleotide can be used to rapidly isolate DNA sequences flanking a transposon Tn*5* insertion by the polymerase chain reaction. *Nucleic Acids Res.* **18**: 6673-6676.

Sisto, A., Morea, M., and Cipriani, M.G., 2001, Identification of a functional *hrcC* gene in *Pseudomonas syringae* subsp. *savastanoi*. *Proc. 11th Cong. of the Mediterranean Phytopathological Union/3rd Congress of the Sociedade Portuguesa de Fitopatologia,* Evora (Portugal), 17-20 September 2001, 202-204.

Surico, G., and Iacobellis, N.S., 1992, Phytohormones and olive knot disease. In *Molecular signals in plant-microbe communications* (D.P.S. Verma, ed.), CRC Press, Boca Raton, pp. 209-229.

Surico, G., Iacobellis, N.S., and Sisto, A., 1985, Studies on the role of indole-3-acetic acid and cytokinins in the formation of knots on olive and oleander plants by *Pseudomonas syringae* pv. *savastanoi*. *Physiol. Plant Pathol.* **26**: 309-320.

Xiao, Y., and Hutcheson, S.W., 1994, A single promoter sequence recognized by a newly identified alternate sigma factor directs expression of pathogenicity and host range determinants in *Pseudomonas syringae*. *J. Bacteriol.* **176**: 3089-3091.

Young, J.M., Saddler, G.S., Takikawa, Y., De Boer, S.H., Vauterin, L., Gardan, L., Gvozdyak, R.I., and Stead, D.E., 1996, Names of plant pathogenic bacteria 1864-1995. *Rev. Plant Pathol.* **75**: 721-763.

Classification and Identification of Plant Pathogenic *Pseudomonas* species by REP-PCR Derived Genetic Fingerprints

D.E. STEAD, S.A. SIMPKINS, S.A. WELLER, J. HENNESSY, A. ASPIN, H. STANFORD, N.C. SMITH and J.G. ELPHINSTONE
Central Science Laboratory, Sand Hutton, York, UK.

Abstract: Repetitive sequence based-PCR fingerprints based on REP primers allow a classification of plant pathogenic *Pseudomonas* spp. that supports most of the recognised species. The classification also supports a pathovar-based system but indicates that some clusters of strains were isolated from unrelated hosts. Standardisation of methods is essential, and improved DNA extraction methods may be necessary before identification can be made based on comparisons with a database.

Key words: Phytopathogenic bacteria, repetitive extragenic palindromic PCR, *Pseudomonas* spp.

Author for correspondence: D.E. Stead, e-mail: d.stead@csl.gov.uk.

1. INTRODUCTION

It is widely accepted that genetic fingerprints based on repetitive sequence PCR methods (REP, BOX, ERIC) are useful in classification and identification of plant pathogenic bacteria (Louws *et al.*, 1994; Rademaker *et al.*, 2000; Weingart and Volksch, 1997). Our paper investigates the use of repetitive extragenic palindromic sequence primers for classification of most validly named plant pathogenic species and pathovars within *Pseudomonas*.

N.S. Iacobellis *et. al.* (eds.), *Pseudomonas syringae* and Related Pathogens, 411-420.

2. MATERIALS AND METHODS

Genetic fingerprints of approximately 500 strains representing almost all plant pathogenic taxa within *Pseudomonas* were prepared using standard methods for both growth on nutrient agar and for the generation of REP-PCR fingerprints. Most strains were from the National Collection of Plant Pathogenic Bacteria (NCPPB). Cell suspensions were boiled and the supernatant used directly for the PCR reaction. Some 350 fingerprints were compared in a dendrogram based on an UPGMA analysis using Bionumerics software (Applied Maths, Belgium) (Fig. 1).
Repeat analyses for some strains showed that reproducibility was often good but could vary with the taxon. Standardisation of methods was critical.

3. RESULTS AND DISCUSSION

3.1 Species Taxonomy

Species that are shown to be genetically heterogeneous: *Pseudomonas. cichorii*, *P. viridiflava, P. syringae, P. savastanoi, P. marginalis*[+] and *P. caricapapayae*.
The following species appear genetically homogeneous based on the range of strains tested: *P. avellanae, P. amygdali, P. tolaasii*[+], *P. fiscuserectae*, *P. corrugata*[+], *P. asplenii*, *P. agarici*, *P. flavescens*, and *P. fuscovaginae*.
P. fuscovaginae and *P. asplenii* may be synonymous.
P. cichorii and *P. viridiflava* have diverse subgroups that show some correlation with host and geographical location.
Pathovars from genomospecies 2, 3 and 4 (Gardan *et al*., 1999) tended to cluster together.

3.2 Pathovar and Host Taxonomy

The following pathovars appear genetically heterogeneous: *actinidiae*, *atrofaciens, berberidis, lachrymans, passiflorae, phaseolicola, pisi, ribicola, sesami, syringae* and *viburni*.
Pathovar *persicae* may have a broader host and geographic range than previously thought.
The following host plants have pathogens from different clusters within the dendrogram: apple, beans in the genus *Phaseolus*, cucumber, lilac,

[+] not all results are shown

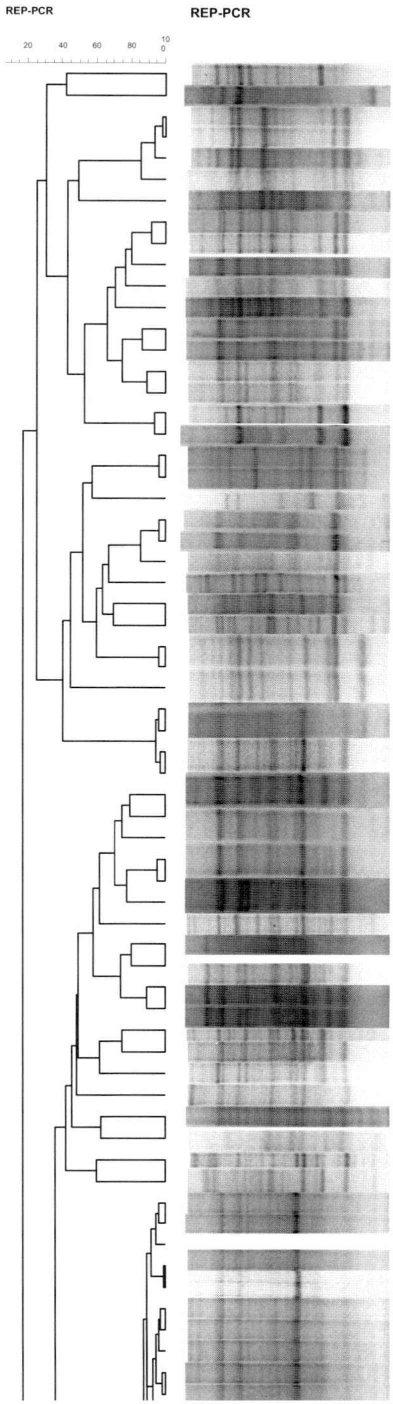

Pseudomonas	cichorii		NCPPB 3179	Pepper
Pseudomonas	cichorii		NCPPB 1512	Chrysanthemum
Pseudomonas	tolaasii		NCPPB 741	Mushroom
Pseudomonas	tolaasii		NCPPB 1116	Mushroom
Pseudomonas	tolaasii		NCPPB 2194	Mushroom
Pseudomonas	tolaasii		NCPPB 2192	Mushroom
Pseudomonas	syringae	dendropanacis	NCPPB 3464	
Pseudomonas	syringae	morsprunorum	NCPPB 2787	Cherry
Pseudomonas	syringae	morsprunorum	NCPPB 2787	Cherry
Pseudomonas	syringae	morsprunorum	NCPPB 2755	Sour Cherry
Pseudomonas	syringae	morsprunorum	NCPPB 560	
Pseudomonas	syringae	aesculi	NCPPB 3681	
Pseudomonas	savastanoi	fraxinii	NCPPB 1006	Ash
Pseudomonas	savastanoi	fraxinii	NCPPB 3471	Ash
Pseudomonas	savastanoi	fraxinii	NCPPB 2716	Ash
Pseudomonas	savastanoi	fraxinii	NCPPB 1464	Ash
Pseudomonas	cichorii		NCPPB 2474	Mexican Poppy
Pseudomonas	cichorii		NCPPB 3466	Chrysanthemum
Pseudomonas	syringae		NCPPB 3116	Coriander
Pseudomonas	syringae		NCPPB 3319	Coriander
Pseudomonas	cichorii		NCPPB 3153	Lettuce
Pseudomonas	amygdali		NCPPB 2608	Almond
Pseudomonas	amygdali		NCPPB 2607T	Almond
Pseudomonas	amygdali		NCPPB 2608	
Pseudomonas	cichorii		NCPPB 943	Endive
Pseudomonas	tremae		NCPPB 3465	
Pseudomonas	meliae		NCPPB 3033	
Pseudomonas	syringae	photiniae	NCPPB 3688	Photinia
Pseudomonas	syringae	photiniae	NCPPB 3689	Photinia
Pseudomonas	cichorii		NCPPB 1513	Celery
Pseudomonas	avellanae		NCPPB 3489	Hazel
Pseudomonas	avellanae		NCPPB 3488	Hazel
Pseudomonas	avellanae		NCPPB 3490	Hazel
Pseudomonas	avellanae		NCPPB 3487T	Hazel
Pseudomonas	syringae	rhaphiolepidis	NCPPB 3618T	Rhaphiolepis
Pseudomonas	savastanoi	savastanoi	NCPPB 639	Olive
Pseudomonas	syringae	eriobotryae	NCPPB 2801	Loquat
Pseudomonas	savastanoi	savastanoi	NCPPB 3471	Jasmine
Pseudomonas	savastanoi	nerii	NCPPB 3278	Oleander
Pseudomonas	savastanoi	nerii	NCPPB 640	Oleander
Pseudomonas	savastanoi	savastanoi	NCPPB 1506	Olive
Pseudomonas	savastanoi	nerii	NCPPB 1343	Oleander
Pseudomonas	savastanoi	nerii	NCPPB 1343	Oleander
Pseudomonas	savastanoi	nerii	NCPPB 3334	Oleander
Pseudomonas	savastanoi	nerii	NCPPB 3606	Oleander
Pseudomonas	savastanoi	savastanoi	NCPPB 1344	Olive
Pseudomonas	savastanoi	savastanoi	NCPPB 1342	Olive
Pseudomonas	savastanoi	savastanoi	NCPPB 2328	
Pseudomonas	savastanoi	fraxinii	NCPPB 3474	Ash
Pseudomonas	savastanoi	savastanoi	NCPPB 64	Olive
Pseudomonas	viridiflava		NCPPB 1252	Pear
Pseudomonas	caricapapa.		NCPPB 1873	Papaya
Pseudomonas	caricapapa.		NCPPB 1873	Papaya
Pseudomonas	syringae	lachrymans	NCPPB 1425	Cucumber
Pseudomonas	syringae	lachrymans	NCPPB 1428	Cucumber
Pseudomonas	syringae	lachrymans	NCPPB 1097	Cucumber
Pseudomonas	syringae	lachrymans	NCPPB 277	Cucumber
Pseudomonas	syringae	lachrymans	NCPPB 467	Cucumber
Pseudomonas	syringae	lachrymans	NCPPB 541	Cucumber
Pseudomonas	syringae	lachrymans	NCPPB 542	Cucumber
Pseudomonas	syringae	lachrymans	NCPPB 543	Cucumber
Pseudomonas	syringae	lachrymans	NCPPB 1436	Cucumber
Pseudomonas	syringae	lachrymans	NCPPB 2754	Cucumber

continue

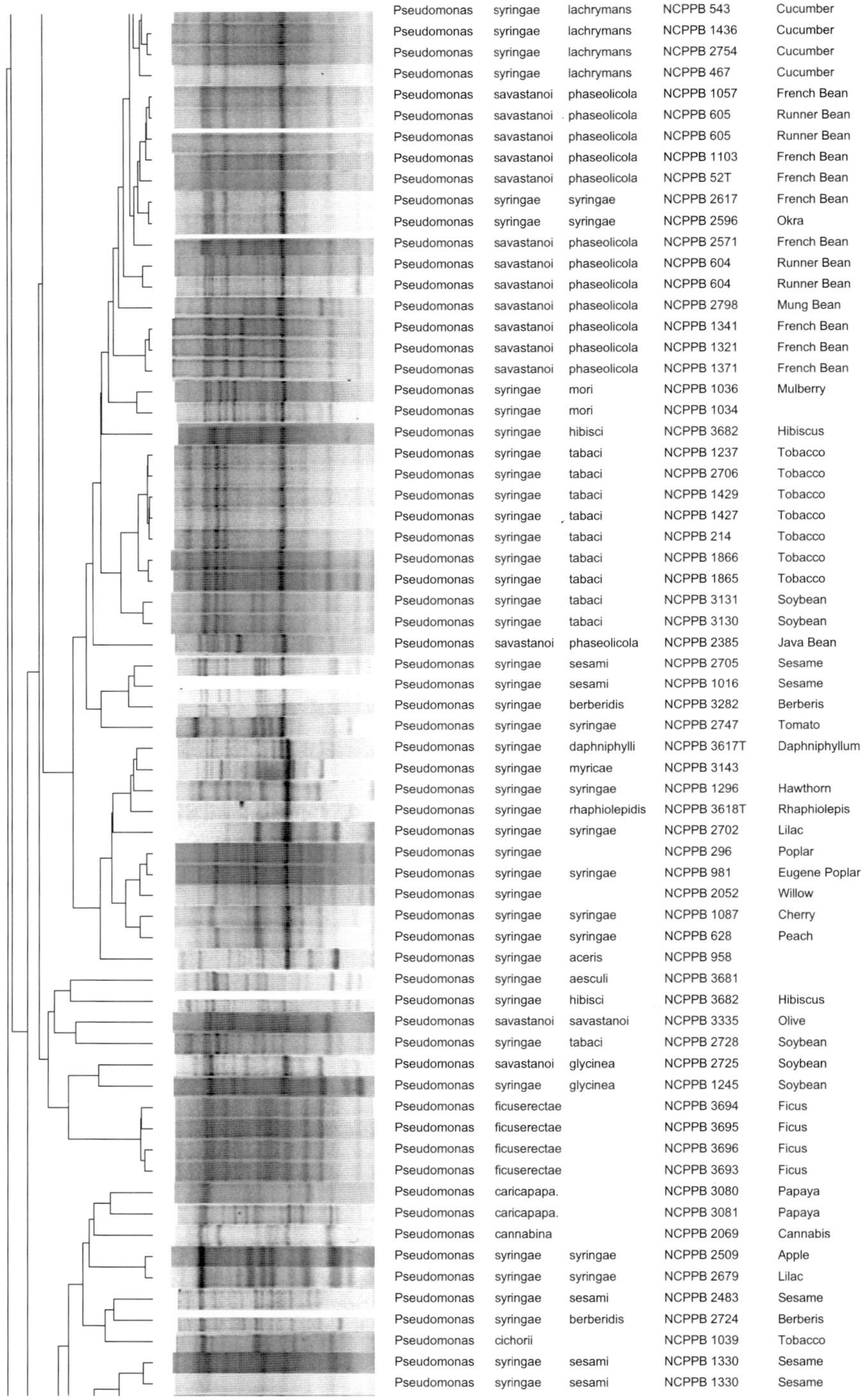
Pseudomonas syringae lachrymans NCPPB 543 Cucumber
Pseudomonas syringae lachrymans NCPPB 1436 Cucumber
Pseudomonas syringae lachrymans NCPPB 2754 Cucumber
Pseudomonas syringae lachrymans NCPPB 467 Cucumber
Pseudomonas savastanoi phaseolicola NCPPB 1057 French Bean
Pseudomonas savastanoi phaseolicola NCPPB 605 Runner Bean
Pseudomonas savastanoi phaseolicola NCPPB 605 Runner Bean
Pseudomonas savastanoi phaseolicola NCPPB 1103 French Bean
Pseudomonas savastanoi phaseolicola NCPPB 52T French Bean
Pseudomonas syringae syringae NCPPB 2617 French Bean
Pseudomonas syringae syringae NCPPB 2596 Okra
Pseudomonas savastanoi phaseolicola NCPPB 2571 French Bean
Pseudomonas savastanoi phaseolicola NCPPB 604 Runner Bean
Pseudomonas savastanoi phaseolicola NCPPB 604 Runner Bean
Pseudomonas savastanoi phaseolicola NCPPB 2798 Mung Bean
Pseudomonas savastanoi phaseolicola NCPPB 1341 French Bean
Pseudomonas savastanoi phaseolicola NCPPB 1321 French Bean
Pseudomonas savastanoi phaseolicola NCPPB 1371 French Bean
Pseudomonas syringae mori NCPPB 1036 Mulberry
Pseudomonas syringae mori NCPPB 1034
Pseudomonas syringae hibisci NCPPB 3682 Hibiscus
Pseudomonas syringae tabaci NCPPB 1237 Tobacco
Pseudomonas syringae tabaci NCPPB 2706 Tobacco
Pseudomonas syringae tabaci NCPPB 1429 Tobacco
Pseudomonas syringae tabaci NCPPB 1427 Tobacco
Pseudomonas syringae tabaci NCPPB 214 Tobacco
Pseudomonas syringae tabaci NCPPB 1866 Tobacco
Pseudomonas syringae tabaci NCPPB 1865 Tobacco
Pseudomonas syringae tabaci NCPPB 3131 Soybean
Pseudomonas syringae tabaci NCPPB 3130 Soybean
Pseudomonas savastanoi phaseolicola NCPPB 2385 Java Bean
Pseudomonas syringae sesami NCPPB 2705 Sesame
Pseudomonas syringae sesami NCPPB 1016 Sesame
Pseudomonas syringae berberidis NCPPB 3282 Berberis
Pseudomonas syringae syringae NCPPB 2747 Tomato
Pseudomonas syringae daphniphylli NCPPB 3617T Daphniphyllum
Pseudomonas syringae myricae NCPPB 3143
Pseudomonas syringae syringae NCPPB 1296 Hawthorn
Pseudomonas syringae rhaphiolepidis NCPPB 3618T Rhaphiolepis
Pseudomonas syringae syringae NCPPB 2702 Lilac
Pseudomonas syringae NCPPB 296 Poplar
Pseudomonas syringae syringae NCPPB 981 Eugene Poplar
Pseudomonas syringae NCPPB 2052 Willow
Pseudomonas syringae syringae NCPPB 1087 Cherry
Pseudomonas syringae syringae NCPPB 628 Peach
Pseudomonas syringae aceris NCPPB 958
Pseudomonas syringae aesculi NCPPB 3681
Pseudomonas syringae hibisci NCPPB 3682 Hibiscus
Pseudomonas savastanoi savastanoi NCPPB 3335 Olive
Pseudomonas syringae tabaci NCPPB 2728 Soybean
Pseudomonas savastanoi glycinea NCPPB 2725 Soybean
Pseudomonas syringae glycinea NCPPB 1245 Soybean
Pseudomonas ficuserectae NCPPB 3694 Ficus
Pseudomonas ficuserectae NCPPB 3695 Ficus
Pseudomonas ficuserectae NCPPB 3696 Ficus
Pseudomonas ficuserectae NCPPB 3693 Ficus
Pseudomonas caricapapa. NCPPB 3080 Papaya
Pseudomonas caricapapa. NCPPB 3081 Papaya
Pseudomonas cannabina NCPPB 2069 Cannabis
Pseudomonas syringae syringae NCPPB 2509 Apple
Pseudomonas syringae syringae NCPPB 2679 Lilac
Pseudomonas syringae sesami NCPPB 2483 Sesame
Pseudomonas syringae berberidis NCPPB 2724 Berberis
Pseudomonas cichorii NCPPB 1039 Tobacco
Pseudomonas syringae sesami NCPPB 1330 Sesame
Pseudomonas syringae sesami NCPPB 1330 Sesame

continue

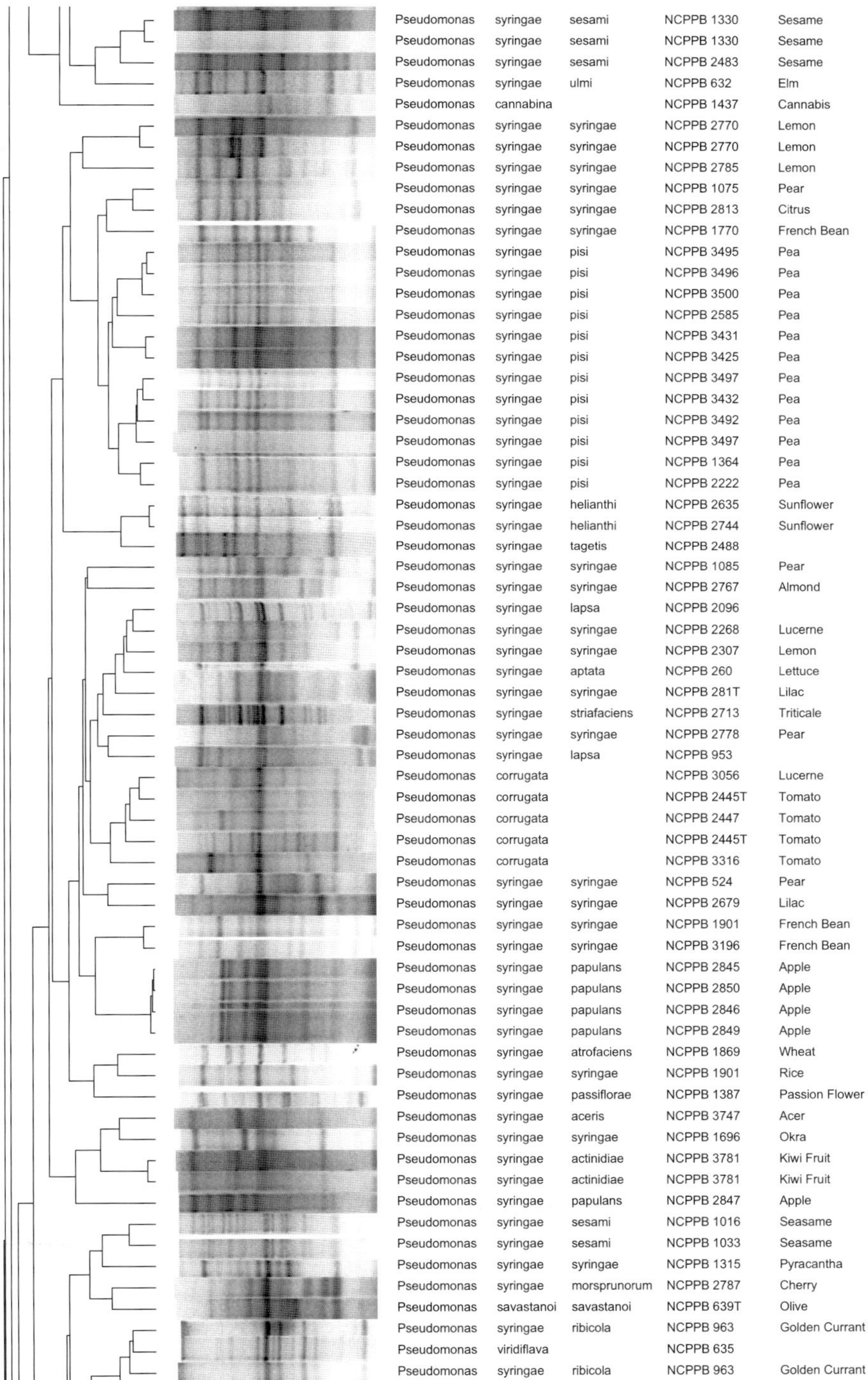
Pseudomonas syringae sesami NCPPB 1330 Sesame
Pseudomonas syringae sesami NCPPB 1330 Sesame
Pseudomonas syringae sesami NCPPB 2483 Sesame
Pseudomonas syringae ulmi NCPPB 632 Elm
Pseudomonas cannabina NCPPB 1437 Cannabis
Pseudomonas syringae syringae NCPPB 2770 Lemon
Pseudomonas syringae syringae NCPPB 2770 Lemon
Pseudomonas syringae syringae NCPPB 2785 Lemon
Pseudomonas syringae syringae NCPPB 1075 Pear
Pseudomonas syringae syringae NCPPB 2813 Citrus
Pseudomonas syringae syringae NCPPB 1770 French Bean
Pseudomonas syringae pisi NCPPB 3495 Pea
Pseudomonas syringae pisi NCPPB 3496 Pea
Pseudomonas syringae pisi NCPPB 3500 Pea
Pseudomonas syringae pisi NCPPB 2585 Pea
Pseudomonas syringae pisi NCPPB 3431 Pea
Pseudomonas syringae pisi NCPPB 3425 Pea
Pseudomonas syringae pisi NCPPB 3497 Pea
Pseudomonas syringae pisi NCPPB 3432 Pea
Pseudomonas syringae pisi NCPPB 3492 Pea
Pseudomonas syringae pisi NCPPB 3497 Pea
Pseudomonas syringae pisi NCPPB 1364 Pea
Pseudomonas syringae pisi NCPPB 2222 Pea
Pseudomonas syringae helianthi NCPPB 2635 Sunflower
Pseudomonas syringae helianthi NCPPB 2744 Sunflower
Pseudomonas syringae tagetis NCPPB 2488
Pseudomonas syringae syringae NCPPB 1085 Pear
Pseudomonas syringae syringae NCPPB 2767 Almond
Pseudomonas syringae lapsa NCPPB 2096
Pseudomonas syringae syringae NCPPB 2268 Lucerne
Pseudomonas syringae syringae NCPPB 2307 Lemon
Pseudomonas syringae aptata NCPPB 260 Lettuce
Pseudomonas syringae syringae NCPPB 281T Lilac
Pseudomonas syringae striafaciens NCPPB 2713 Triticale
Pseudomonas syringae syringae NCPPB 2778 Pear
Pseudomonas syringae lapsa NCPPB 953
Pseudomonas corrugata NCPPB 3056 Lucerne
Pseudomonas corrugata NCPPB 2445T Tomato
Pseudomonas corrugata NCPPB 2447 Tomato
Pseudomonas corrugata NCPPB 2445T Tomato
Pseudomonas corrugata NCPPB 3316 Tomato
Pseudomonas syringae syringae NCPPB 524 Pear
Pseudomonas syringae syringae NCPPB 2679 Lilac
Pseudomonas syringae syringae NCPPB 1901 French Bean
Pseudomonas syringae syringae NCPPB 3196 French Bean
Pseudomonas syringae papulans NCPPB 2845 Apple
Pseudomonas syringae papulans NCPPB 2850 Apple
Pseudomonas syringae papulans NCPPB 2846 Apple
Pseudomonas syringae papulans NCPPB 2849 Apple
Pseudomonas syringae atrofaciens NCPPB 1869 Wheat
Pseudomonas syringae syringae NCPPB 1901 Rice
Pseudomonas syringae passiflorae NCPPB 1387 Passion Flower
Pseudomonas syringae aceris NCPPB 3747 Acer
Pseudomonas syringae syringae NCPPB 1696 Okra
Pseudomonas syringae actinidiae NCPPB 3781 Kiwi Fruit
Pseudomonas syringae actinidiae NCPPB 3781 Kiwi Fruit
Pseudomonas syringae papulans NCPPB 2847 Apple
Pseudomonas syringae sesami NCPPB 1016 Seasame
Pseudomonas syringae sesami NCPPB 1033 Seasame
Pseudomonas syringae syringae NCPPB 1315 Pyracantha
Pseudomonas syringae morsprunorum NCPPB 2787 Cherry
Pseudomonas savastanoi savastanoi NCPPB 639T Olive
Pseudomonas syringae ribicola NCPPB 963 Golden Currant
Pseudomonas viridiflava NCPPB 635
Pseudomonas syringae ribicola NCPPB 963 Golden Currant
Pseudomonas viridiflava NCPPB 635 Bean

continue

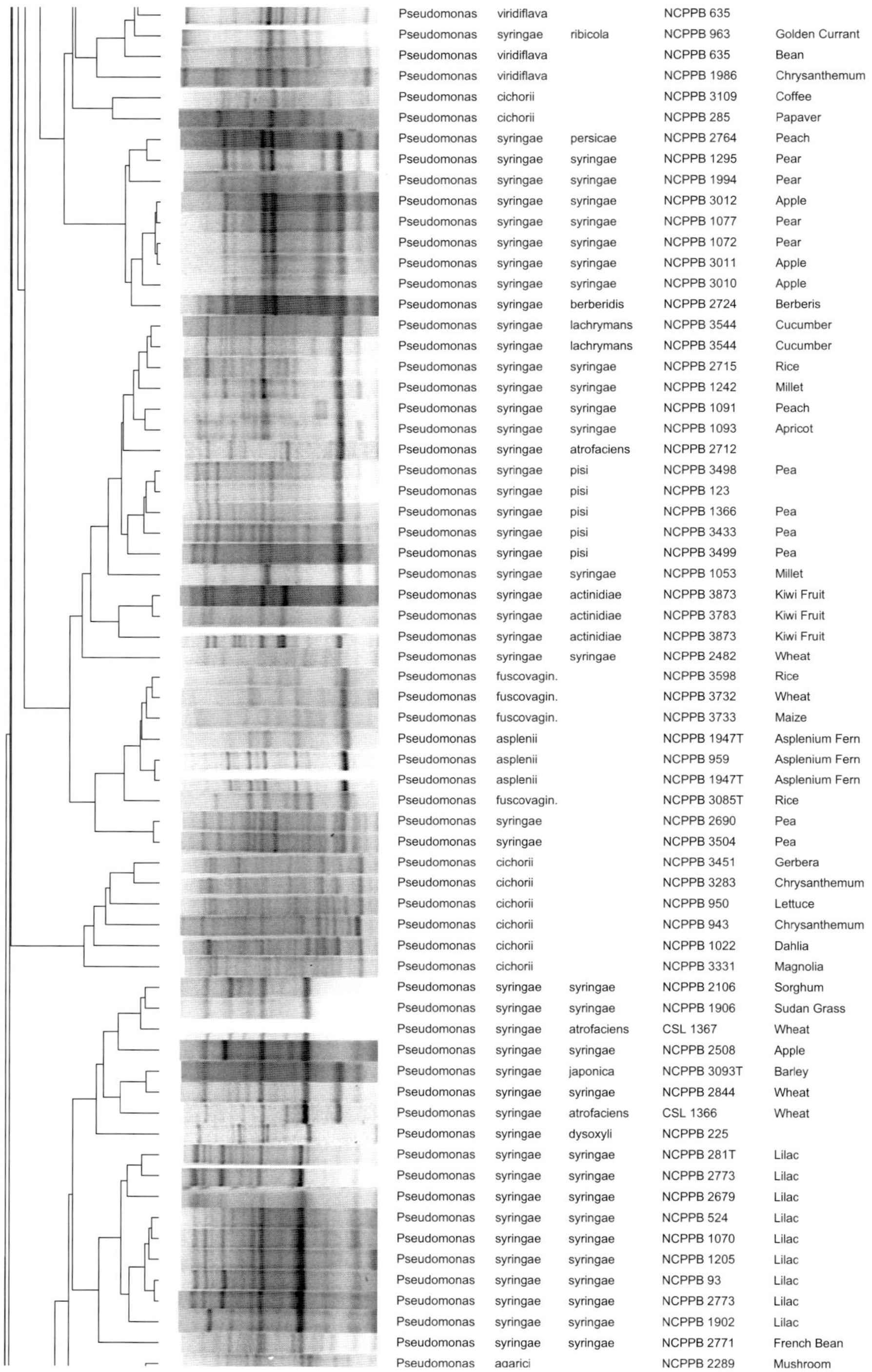
Pseudomonas viridiflava NCPPB 635
Pseudomonas syringae ribicola NCPPB 963 Golden Currant
Pseudomonas viridiflava NCPPB 635 Bean
Pseudomonas viridiflava NCPPB 1986 Chrysanthemum
Pseudomonas cichorii NCPPB 3109 Coffee
Pseudomonas cichorii NCPPB 285 Papaver
Pseudomonas syringae persicae NCPPB 2764 Peach
Pseudomonas syringae syringae NCPPB 1295 Pear
Pseudomonas syringae syringae NCPPB 1994 Pear
Pseudomonas syringae syringae NCPPB 3012 Apple
Pseudomonas syringae syringae NCPPB 1077 Pear
Pseudomonas syringae syringae NCPPB 1072 Pear
Pseudomonas syringae syringae NCPPB 3011 Apple
Pseudomonas syringae syringae NCPPB 3010 Apple
Pseudomonas syringae berberidis NCPPB 2724 Berberis
Pseudomonas syringae lachrymans NCPPB 3544 Cucumber
Pseudomonas syringae lachrymans NCPPB 3544 Cucumber
Pseudomonas syringae syringae NCPPB 2715 Rice
Pseudomonas syringae syringae NCPPB 1242 Millet
Pseudomonas syringae syringae NCPPB 1091 Peach
Pseudomonas syringae syringae NCPPB 1093 Apricot
Pseudomonas syringae atrofaciens NCPPB 2712
Pseudomonas syringae pisi NCPPB 3498 Pea
Pseudomonas syringae pisi NCPPB 123
Pseudomonas syringae pisi NCPPB 1366 Pea
Pseudomonas syringae pisi NCPPB 3433 Pea
Pseudomonas syringae pisi NCPPB 3499 Pea
Pseudomonas syringae syringae NCPPB 1053 Millet
Pseudomonas syringae actinidiae NCPPB 3873 Kiwi Fruit
Pseudomonas syringae actinidiae NCPPB 3783 Kiwi Fruit
Pseudomonas syringae actinidiae NCPPB 3873 Kiwi Fruit
Pseudomonas syringae syringae NCPPB 2482 Wheat
Pseudomonas fuscovagin. NCPPB 3598 Rice
Pseudomonas fuscovagin. NCPPB 3732 Wheat
Pseudomonas fuscovagin. NCPPB 3733 Maize
Pseudomonas asplenii NCPPB 1947T Asplenium Fern
Pseudomonas asplenii NCPPB 959 Asplenium Fern
Pseudomonas asplenii NCPPB 1947T Asplenium Fern
Pseudomonas fuscovagin. NCPPB 3085T Rice
Pseudomonas syringae NCPPB 2690 Pea
Pseudomonas syringae NCPPB 3504 Pea
Pseudomonas cichorii NCPPB 3451 Gerbera
Pseudomonas cichorii NCPPB 3283 Chrysanthemum
Pseudomonas cichorii NCPPB 950 Lettuce
Pseudomonas cichorii NCPPB 943 Chrysanthemum
Pseudomonas cichorii NCPPB 1022 Dahlia
Pseudomonas cichorii NCPPB 3331 Magnolia
Pseudomonas syringae syringae NCPPB 2106 Sorghum
Pseudomonas syringae syringae NCPPB 1906 Sudan Grass
Pseudomonas syringae atrofaciens CSL 1367 Wheat
Pseudomonas syringae syringae NCPPB 2508 Apple
Pseudomonas syringae japonica NCPPB 3093T Barley
Pseudomonas syringae syringae NCPPB 2844 Wheat
Pseudomonas syringae atrofaciens CSL 1366 Wheat
Pseudomonas syringae dysoxyli NCPPB 225
Pseudomonas syringae syringae NCPPB 281T Lilac
Pseudomonas syringae syringae NCPPB 2773 Lilac
Pseudomonas syringae syringae NCPPB 2679 Lilac
Pseudomonas syringae syringae NCPPB 524 Lilac
Pseudomonas syringae syringae NCPPB 1070 Lilac
Pseudomonas syringae syringae NCPPB 1205 Lilac
Pseudomonas syringae syringae NCPPB 93 Lilac
Pseudomonas syringae syringae NCPPB 2773 Lilac
Pseudomonas syringae syringae NCPPB 1902 Lilac
Pseudomonas syringae syringae NCPPB 2771 French Bean
Pseudomonas agarici NCPPB 2289 Mushroom

continue

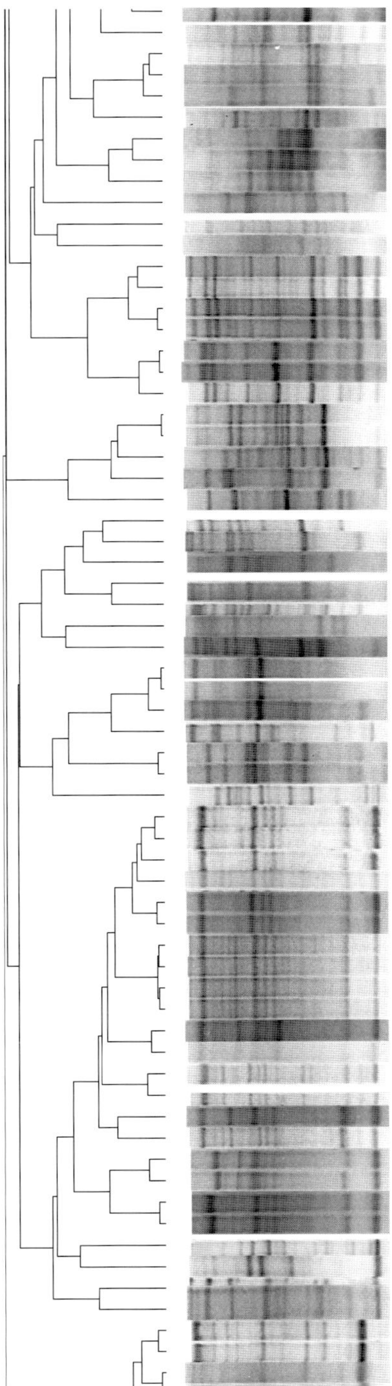

Pseudomonas	syringae	syringae	NCPPB 1902	Lilac
Pseudomonas	syringae	syringae	NCPPB 2771	French Bean
Pseudomonas	agarici		NCPPB 2289	Mushroom
Pseudomonas	agarici		NCPPB 2290	Mushroom
Pseudomonas	agarici		NCPPB 1999	Mushroom
Pseudomonas	cichorii		NCPPB 1511	Cabbage
Pseudomonas	syringae	syringae	CSL 1440	Apple
Pseudomonas	syringae	syringae	CSL 1405	Apple
Pseudomonas	syringae	syringae	CSL 1110	Apple
Pseudomonas	viridiflava		NCPPB 1282	Tomato
Pseudomonas	syringae	passiflorae	NCPPB 1390	Passion Flower
Pseudomonas	syringae	berberidis	NCPPB 3281	Berberis
Pseudomonas	syringae	philadelphi	NCPPB 3257T	Mock Orange
Pseudomonas	syringae	philadelphi	NCPPB 3257T	Mock Orange
Pseudomonas	syringae	philadelphi	NCPPB 3575	Mock Orange
Pseudomonas	syringae	philadelphi	NCPPB 3258	Mock Orange
Pseudomonas	syringae	"fabae"	NCPPB 3626	Broad Bean
Pseudomonas	syringae	"fabae"	NCPPB 3624	Broad Bean
Pseudomonas	syringae	"fabae"	NCPPB 3621	Broad Bean
Pseudomonas	syringae	primulae	NCPPB 2996	
Pseudomonas	syringae	primulae	NCPPB 2996	
Pseudomonas	viridiflava		NCPPB 2012	French Bean
Pseudomonas	viridiflava		NCPPB 2671	
Pseudomonas	marginalis		NCPPB 366	Sprouts
Pseudomonas	syringae	berberidis	NCPPB 1871	Barberry
Pseudomonas	syringae	viburni	NCPPB 1921T	
Pseudomonas	syringae	aesculi	NCPPB 3681	
Pseudomonas	syringae	passiflorae	NCPPB 224	Passion Flower
Pseudomonas	syringae	passiflorae	NCPPB 224	Passion Flower
Pseudomonas	caricapapa.		NCPPB 3439	Papaya
Pseudomonas	savastanoi	fraxinii	NCPPB 1465	Ash
Pseudomonas	marginalis		CSL 1646	Lettuce
Pseudomonas	marginalis		CSL 1647	Lettuce
Pseudomonas	marginalis		NCPPB 3183	Pea
Pseudomonas	syringae	mellea	NCPPB 2356	Tobacco
Pseudomonas	flavescens		NCPPB 3063	
Pseudomonas	flavescens		NCPPB 3064	
Pseudomonas	savastanoi	savastanoi	NCPPB 2327	Olive
Pseudomonas	syringae	coronafaciens	NCPPB 2395	Oat
Pseudomonas	syringae	coronafaciens	NCPPB 874	Oat
Pseudomonas	syringae	garcae	NCPPB 512	Coffee
Pseudomonas	syringae	passiflorae	NCPPB 1391	Passion Flower
Pseudomonas	syringae	garcae	NCPPB 3030	Coffee
Pseudomonas	syringae	garcae	NCPPB 588T	Coffee
Pseudomonas	syringae	porri	NCPPB 3545	Leek
Pseudomonas	syringae	porri	NCPPB 2692	Leek
Pseudomonas	syringae	porri	NCPPB 3334	Leek
Pseudomonas	syringae	porri	NCPPB 3336	Leek
Pseudomonas	syringae	striafaciens	NCPPB 2394	Oat
Pseudomonas	syringae	striafaciens	NCPPB 2394	Oat
Pseudomonas	syringae	striafaciens	NCPPB 2480	Oat
Pseudomonas	syringae	striafaciens	NCPPB 1898	
Pseudomonas	syringae	coronafaciens	NCPPB 2681	Oat
Pseudomonas	syringae	striafaciens	NCPPB 1898	
Pseudomonas	syringae	zizaniae	NCPPB 3690	Wild Rice
Pseudomonas	syringae	zizaniae	NCPPB 3691	Wild Rice
Pseudomonas	syringae	oryzae	NCPPB 3684	Rice
Pseudomonas	syringae	oryzae	NCPPB 3685	Rice
Pseudomonas	syringae	mellea	NCPPB 280	Tobacco
Pseudomonas	syringae	syringae	NCPPB 1302	White Bean
Pseudomonas	syringae	dysoxyli	NCPPB 1880	Dysoxylum
Pseudomonas	viridiflava		NCPPB 1080	Pear
Pseudomonas	syringae	actinidiae	NCPPB 3738	Kiwi Fruit
Pseudomonas	syringae	actinidiae	NCPPB 3740	Kiwi Fruit
Pseudomonas	syringae	actinidiae	NCPPB 3741	Kiwi Fruit
Pseudomonas	syringae	actinidiae	NCPPB 3739	Kiwi Fruit

continue

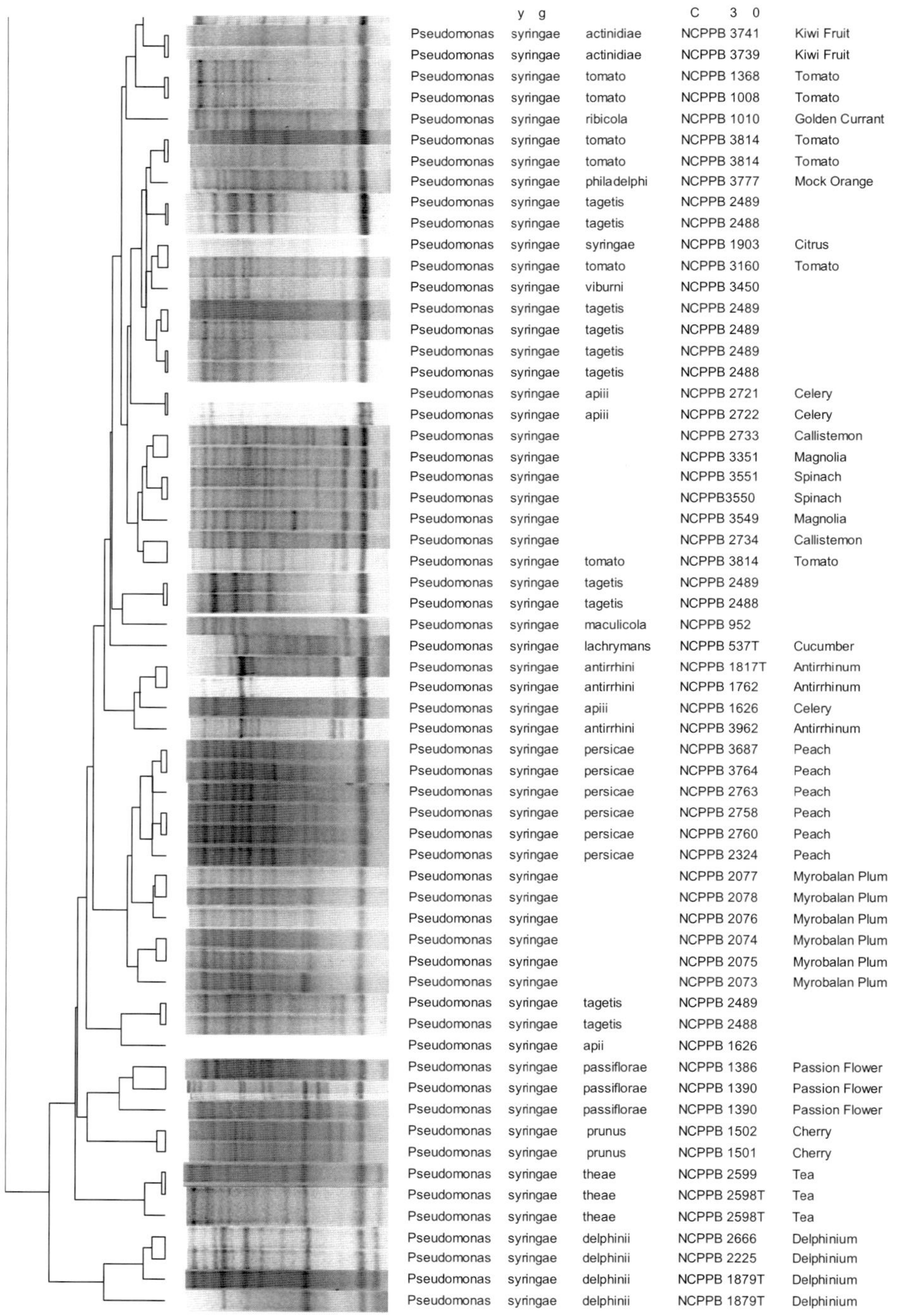

Figure 1. REP-PCR classification of plant pathogenic *Pseudomonas* spp.

stonefruits in the genus *Prunus*, pear, soybean, pea, currants, berberis and wheat.

The following groups of host plants have some pathogens that appear virtually indistinguishable:

asplenium, maize, rice and wheat;
*pear and apple;
*cucumber, rice, millet, peach, apricot, tobacco and soybean;
*poplar, willow, cherry and peach;
oat, coffee and passion flower;
tomato, viburnum, tagetes, magnolia, spinach and callistemon;
peach and myrobalan plum.

Pathovar *syringae* should be retained for lilac pathogens; pathovar *pisi* has 2 distinct genotypes; pathovar *japonica* is not part of pv. *syringae* but could contain strains from primarily cereal crops and may be synonymous with pv. *atrofaciens*.

4. CONCLUSIONS

REP-PCR is a useful tool for classification and identification. Problems of reproducibility need to be solved before identification can be made by comparison with a database. Identifications can be made by comparing fingerprints of a test bacterium with a panel of reference strains on a single gel. Pathovars within genomospecies 2, 3 and 4, which contain most of the currently recognised pathovars tended to cluster together, indicating the possibility of species-specific bands.

ACKNOWLEDGMENTS

This work was funded by CSL seed corn funding and by PHD, DEFRA project PH0178.

REFERENCES

Gardan, L., Shafik, H., Belouin, S., Broch, R., Grimont, F., and Grimont, P.A.D., 1999, DNA relatedness among the pathovars of *Pseudomonas syringae* and description of *Pseudomonas tremae* sp. nov. and *Pseudomonas cannabina* sp. nov.. *Int. J. Syst. Bacteriol.* **49**: 469-478.

* These fall into clusters without a pathotype strain.

Louws, F.J., Fulbright, D.W., Stephens, C.T., and de Bruijn, F.J., 1994, Specific genomic fingerprinting of phytopathogenic *Xanthomonas* and *Pseudomonas* pathovars and strains generated with repetitive sequences and PCR. *Appl. Environ. Microbiol.* **60**: 2286-2295.

Rademaker, J.L.W., Hoste, B., Louws, F.J., Kersters, K., Swings, J., Vauterin, L., Vauterin, P., and de Bruijn, F.J., 2000, Comparison of AFLP and rep-PCR genomic fingerprinting with DNA-DNA homology studies: *Xanthomonas* as a model system. *Int. J. Syst. Evol. Microbiol.* **50**: 665-677.

Weingart, H., and Volksch, B., 1997, Gentic fingerprinting of *Pseudomonas syringae* pathovars using ERIC-,REP-, and IS50-PCR. *J. Phytopathol.* **145**: 339-345.

Disease Management and Control

Control of *Pseudomonas syringae* Pathovars

L.E. CLAFLIN
Department of Plant Pathology, Kansas State University, Manhattan, KS, USA.

Abstract: Control of *Pseudomonas syringae* pathovars is similar to other phytopathogenic bacteria. Primary control protocols include quarantine, chemical control, host plant resistance, and seed certification. Recently, non protein amino acids such as β-aminobutyric acid and chemical activators including 2, 6-dichloroisonicotinic acid (INA) and benzo (1,2,3) thiadiazole-7-carbothoioic acid-S-methyl ester (BTH) add an additional facet for control. These chemicals have no or limited effect on the causal agent but invoke plant defence responses such as systemic acquired resistance (SAR).

Key words: Quarantine, BION, ACTIGARD 50 WG, β-aminobutyric acid.

Author for correspondence: L.E. Claflin; e-mail: claflin@plantpath.ksu.edu.

1. INTRODUCTION

Control of plant pathogenic bacteria has been a dilemma since the initial description by Erwin F. Smith nearly a century ago. Phytopathogenic bacteria have posed a constraint to crop yields, created inferior and blemished commodities, and resulted in serious political problems in the export trade. Various control protocols including eradication, quarantine, disinfection, cultural practices, chemical control, host plant resistance, seed certification and biological control have all been utilised. Control of these pathogens has been sporadic and each protocol offers advantages and disadvantages. This chapter cites examples of problems encountered when taxonomic names are changed and discusses the role of quarantine agencies, future control methodologies and the potential of *Pseudomonas syringae* pathovars as bioterrorist organisms.

N.S. Iacobellis *et. al.* (eds.), *Pseudomonas syringae* and Related Pathogens, 423-430.

2. EXCLUSION

Quarantines are most effective when movement of seed, propagative materials, soil and plant products or abiotic items are closely monitored. However, this avenue is increasingly difficult due to the volume of travellers and cargo in most countries. Introduction of various phytopathogenic bacteria have been lessened in the U.S. through the Animal and Plant Health Inspection Services (APHIS). APHIS has a staggering task with 650 plant quarantine inspectors. In the year 2000, 231.1 million travellers entered the U.S. Of these, 14.3 million were airline passengers accompanied with 40.2 million pieces of luggage and 308.4 thousand individual landings.

To assist in the control of phytopathogens, the World Trade Organisation in 1994 agreed on the Application of Sanitary and Phytosanitary Measures (SPS agreement). However, SPS is also an effective protectionist device and used as a trade barrier since producers are often protected by the importing country (Croome). International agencies such as the North American Plant Protection Organisation (Canada, Mexico, United States), International Plant Protection Convention (Developed in 1951 with FAO in Rome and updated as needed) (IPPC), and ICMP are charged with correcting misuse of phytosanitary issues and examination of the scientific justification of specific regulations. GATT rules [Article XX (b)] expressively state that regulations are necessary only when needed "to protect human, animal or plant and health".

Pathovars of *Pseudomonas syringae* are included in the list of quarantined organisms for most countries. For example, *Acidovorax avenae* pv. *avenae*, *Burkholderia andropogonis* and *Xanthomonas campestris* pv. *holcicola* are pathogens of sorghum (*Sorghum bicolor* L.) and used as trade barriers in the export trade of seed and feed grains. Changes in taxonomic nomenclature of these plant pathogenic bacteria have compounded problems for customs personnel. Several pathovars of *P. syringae* are very important including the pathovar *savastanoi*, which is of the ten most quarantined plant pathogenic organisms.

B. andropogonis, causal agent of bacterial leaf streak of sorghum, was formerly *P. andropogonis*. Following this taxonomic change in 1995 (Gillis, *et al.*), problems within the export trade occurred almost immediately and some countries have not changed their regulations to accommodate the name change. Also, the phytosanitary certificate may note that *B. andropogonis* was the target organism during field inspections whereas the importing country lists *P. andropogonis* as being the quarantined organism on certificates.

X. c. pv. *holcicola*, causal agent of bacterial leaf streak of sorghum, rarely is a constraint to yield or quality. Numerous countries list *X. c.* pv. *holcicola* as a plant pathogenic organism that has not been reported and consequently, place it on the quarantine list. A name change to *X. vasicola* pv. *holcicola* was proposed in 1995 (Vauterin *et al*., 1995). In a subsequent article, this change was refuted (Schaad *et al*., 2000) and that the epithet be retained as *X. c.* pv. *holcicola*. This specific example created havoc within the plant protection organisations of numerous countries including Argentina, Australia, South Africa, and Taiwan. Likewise, an important quarantine disease of corn, Stewart's wilt caused by *Pantoea stewartii*, resulted in confusion when the taxonomic name was changed from *Erwinia stewartii*. Taxonomic name changes are inevitable, but should be based on careful scientific merit, otherwise chaos may occur in regulatory agencies, export and import companies and the commercial seed companies who have invested a considerable amount of money and labour.

3. CHEMICAL

3.1 Antibiotics

Streptomycin and oxytetracycline are the principal antibiotics used in the U.S. for control of phytopathogenic bacteria and their use is mostly restricted for bacterial control in high income crops such as fruits and vegetables. Other countries may permit use of gentamycin, kasugamycin, novobiocin and terramycin. In general, a high number of applications are needed for control, resulting in an expensive input. Application of antibiotics at night-time is recommended due to their sensitivity to UV light. With prolonged applications, numerous reports of bacterial resistance to antibiotics have occurred (Roberts, 1996). There is also a serious cause for concern that antibiotics have become integrated within the food chain.

3.2 Sulphur and Copper Compounds

These chemicals have been utilised for many years and those commonly used include: wettable copper (29% basic copper sulphate); copper oxychloride [$3\ Cu(OH)_2 \cdot CuCl_2$]; tribasic copper sulphate [$3Cu(OH)_2 \cdot CuSO_4$)]; copper hydroxide [$Cu(OH)_2$]; and Bordeaux mixture [$CuSO_4 \cdot 3Cu(OH)_2 \cdot Ca(OH)_2$]. Sulphur and copper compounds have good efficacy particularly to foliar pathogens. Copper compounds are usually applied on a 7-day spray schedule. Efficacy depends on selection of chemical and timing of application. Some fungicides have been effective, e.g., benomyl for control of *Pseudomonas glumae* in rice. Disinfestation of contaminated surfaces are normally disinfected with a 5% solution of sodium hypochlorite by soaking 2-5 minutes.

A model was developed that utilised meteorological and biological variables to predict bacterial speck (*P. s.* pv. *tomato*) (Jardine *et al.* 1987). Use of the model resulted in three fewer sprays during the growing season in Michigan. A similar model was used to predict incidence and severity of brown spot (*P. s.* pv. *syringae*) in bean (Lindemann *et al.* 1984).
A wet seed treatment for control of *Acidovorax avenae* subsp. *citrulli*, causal agent of bacterial fruit blotch of watermelon, resulted in a reduction of disease from 61% to less than 1% (Hopkins, et al. 1996). The most effective treatment consisted of fermenting seeds in watermelon juice and debris for 24÷48 h followed by dipping in 1% HCl or 1% $CaOCl_2$ solutions for 15 min followed by washing and drying.

3.3 Biological Control

Pseudomonas fluorescens (marketed as Victus) was highly effective in controlling blotch disease (*P. tolaasii* or *P. "gingeri")* of mushrooms (Miller, 1995). This treatment resulted in significant increases in total yield and quality coupled with a decrease in second quality mushrooms in a large commercial operation. In China, a bactericide, Xi-JunQing 931, was 77÷93% more effective than streptomycin for controlling wildfire (*P. s.* pv. *tabaci*) on tobacco. The active ingredient was not reported (Y. MeiLi 1994). Plant growth-promoting rhizobacteria (PGRP) strains from *P. putida* and *Serratia marcescens* significantly decreased the number and size of angular leaf spot lesions (*P. syringae* pv. *lachrymans*) on cucumber (Liu, 1995). Populations of *P. s.* pv. *lachrymans* in inoculated plants were also lower with PGRP treatments than in controls. Efficacy of the PGRP treatment was attributed to induced systemic resistance (ISR) in cucumbers.
Basal kernel blight of barley, caused by *P. s.* pv. *syringae*, was reduced from 45 to 78% in field experiments and from 50 to 100% under greenhouse conditions with strains of *Pantoea agglomerans* (synanamorph *Erwinia herbicola*) (Braun-Kiewnick *et al.*, 2000). Efficacy of *P. agglomerans* was influenced by time and rate of application. A concentration of 10^7CFU ml^{-1} provided the best control. Storage at 4°C resulted in 90÷93% viability, whereas at 22°C, viability was 73÷79%. PGRP may promote growth of plants by producing plant growth regulators or enhancing nutrient uptake or by an indirect approach with their production of siderophores or antibiotics that provides protection to plants from various pathogens. Comprehensive long-term biocontrol experiments are very limited due to the extensive sampling required and personnel needed to perform such research. In addition, grant funds to support such research are rarely available.
Greenhouse systems offer the potential for biocontrol due to the more regimented cultural practices, confined environment and economical returns. The greenhouse market is over 300,000 ha world-wide with 50,000 ha in a

highly sophisticated production systems (Paulitz and Belanger, 2001). Biocontrol of insects is a standard control protocol.

3.4 Host Plant Resistance

Resistant germplasm is the only practical and economical control for low income crops such as cereal crops. Plant breeders have released excellent inbreeds, hybrids and cultivars; especially for small grain crops (Vassilev *et al.*, 1995). This protocol is economical and environmentally safe.

3.5 Seed Certification Programs

Effective control of those organisms that are primarily seedborne or in plant propagative materials have been greatly lessened through certified and foundation programs. Such a program assures buyers of seed or propagative materials that the germplasm is true to name of variety; materials are reasonably free of diseases and insect and contain a minimum of weed seeds and foreign debris. Halo blight of bean, caused by *Pseudomonas syringae* pv. *phaseolicola*, was endemic in the dry bean industry in the U.S. until such a certification program was developed. Bean seeds are produced in the more arid climates in the Western U.S. and are regularly inspected during the growing season. Some programs have a winter test in the greenhouse or areas that allow for bean production during winter.

4. PRESENT AND FUTURE CONTROL STRATEGIES

Plants react to plant pathogens by developing a defence response such as systemic acquired resistance (SAR). Recently, non protein amino acids such as β-aminobutyric acid (BABA) were shown to exhibit biological control of numerous causal agents such as bacteria, fungi, nematodes and virus(s) in plants (Cohen, 2002; Jakab *et al.*, 2001; Siegrist *et al.*, 2000). BABA was effective in control of foliar, stem and root diseases by either foliar applications, soil drenches and seed coating. BABA was effective in reducing disease 1÷3 days after application unlike other plant activators (SA, INA, BTH) (Cohen 2002). BABA also enhanced both vegetative and reproductive growth of pearl millet (Shailasree *et al.*, 2001). BABA occurs rarely in nature but is known as a partial agonist of the glycine receptor and imparts protection against pathogens in plants. Systemic acquired resistance (SAR) was first observed nearly 70 years (Chester, 1933) ago and the signal is now attributed to salicyclic acid (SA) (Siegrist *et al.*, 2000) in initiating the plant defence mechanisms followed by activation of genes encoding PR proteins. In contrast, ISR (rhizobacteria-mediated) appears to operate in an SA-independent manner.

Excellent control of fungal diseases through use of BABA have been chronicled (Cohen, 2002), but reports on control of plant pathogenic bacteria have been sparse. Zimmerli *et al.* (2000), showed that a 10-fold reduction in the titre of *Pseudomonas syringae* pv. *tomato,* and bacterial speck symptoms were greatly reduced in *Arabidopsis* plants after applying BABA. Treatment was only effective if applied prior to inoculation or preconditioning of the plants. Seemingly, BABA will control a multitude of plants pests under both field and greenhouse conditions (Cohen, 2002). Additional parameters such as toxicology and biodegradation data are vitally needed for registration of this compound. BABA is also a conditioner of plant defence mechanisms leading to a rapid response to pathogens and results in an incompatible interaction. e.g., HR, trailing necrosis and papilla formation (Zimmerli *et al.*, 2000).

4.1 Chemical Activators

Chemical activators used for disease resistance are 2,6-dichloroisonicotinic acid (INA) and benzo (1,2,3) thiadiazole-7-carbothoioic acid-S-methyl ester (BTH) which are synthetic compounds (Oostendorp *et al.*, 2001). BTH is a commercial product from Syngenta and Bion® (Europe) and Actigard® (US) are trademark names. These compounds provide plant resistance in many crops against fungi, bacteria and viruses through induced resistance based on induced proteins and dependence on SA. Resistance by BION is typically long-lasting in monocots but is less pronounced in dicots. Presently, Actigard 50 WG is labelled for control of downy mildew of leafy vegetables, bacterial leaf spots of tomatoes, and blue mold of tobacco. The mode of action mimics SAR, which is commonly found in most plant species. Bion and Actigard have no direct activity against target pathogens.

Recommendations for Actigard include:

- Should be applied to plants before disease is observed;
- Uniform spray coverage is essential;
- Maximum disease control is usually 4 days after application;
- May be tank mixed with other products if curative activity is desired;
- Is compatible with most registered fungicides and insecticides;
- Rate of application for bacterial speck is 1/3-3/4 oz/A.

ACKNOWLEDGEMENTS

Contribution No. 03-107-A from the Kansas Agricultural Experiment Station, Manhattan.

Funds for partial support of this manuscript were granted by the Sorghum/Millet Collaborative Research Program (INTSORMIL) AID/DAN-1254-g-00-1022-00, and the Kansas Agricultural Experiment Station, Manhattan.

Grateful appreciation is expressed to Drs. F. W. White and D. L. Stuteville for their critical review.

REFERENCES

Braun-Kiewnick, A., Jacobsen, B.J., and Sands, D.C., 2000, Biological control of *Pseudomonas syringae* pv. *syringae*, the causal agent of basal kernel blight of barley, by antagonistic *Pantoea agglomerans*. *Phytopathology* **90**: 368-375.

Chester, K.S., 1933, The problem of acquired physiological immunity in plants. *Q. Rev. Biol.* **8**: 275-324.

Cohen, Y.R., 2002, β-aminobutyric acid-induced resistance against plant pathogens. *Plant Dis.* **86**: 448-457.

Croome, J., 1999, Reshaping the world trading system: A history of the Uruguay round. 2nd ed. Kluwer Law International, The Hague, Netherlands.

Gillis, M., Van Van, T., Bardin, R., Goor, M., Hebbar, P., Willems, A., Segers, P., Kersters, K., Heulin, T., and Fernandez, M.P., 1995, Polyphasic taxonomy in the genus *Burkholderia vietnamiensis* sp. nov. for N_2 - fixing isolates from rice in Vietnam. *Int. J. Syst. Bacteriol.* **45**: 274-289.

Hopkins, D.L., Cucuzza, J.D., and Watterson, J.C., 1996, Wet seed treatments for the control of bacterial fruit blotch of watermelon. *Plant Dis.* **80**:529-532.

Jakab, G., Cottier, V., Toquin, V. Rigoli, G., Zimmerli, L., Metraux, J.-P., and Mauchi-Mani, B., 2001, β-Aminobutyric acid-induced resistance in plants. *Eur. J. Plant Pathol.* **107**: 29-37.

Jardine, D.J., and Stephens, C.T., 1987, A predictive system for timing chemical application to control *Pseudomonas syringae* pv. *tomato*, causal agent of bacterial speck. *Phytopathology* **77**: 823-827.

Lindemann, J., Arny, D.C., and Upper, C.D., 1984, Use of an apparent infection threshold of *Pseudomonas syringae* to predict incidence and severity of brown spot of bean. *Phytopathology* **74**: 1334-1339.

Liu, L., Kloepper, J.W., and Tuzun, S., 1995, Induction of systemic resistance in cucumber against bacterial angular leaf spot by plant growth-promoting rhizobacteria. *Phytopathology* **85**: 843-847.

Miller, F.C., and Spear, M., 1995, Very large scale commercial trial of a biological control for mushroom blotch disease. Science and cultivation of edible fungi, V. 2: In *Proceedings of the* 14th *International Congress*, Oxford. Edited by T.J. Elliott pp. 634-642.

Oostendorp, M., Kunz, W., Dietrich, B., and Staub, T., 2001, Induced disease resistance in plants by chemicals. *Eur. J. Plant Pathol.* **107**: 19-28.

Paulitz, T.C., and Belanger, R.R., 2001, Biological control in greenhouse systems. *Annu. Rev. Phytopathol.* **39**: 103-133.

Roberts, M.C., 1996, Tetracycline-resistant determinants: Mechanism of action, regulation of expression, genetic mobility, and distribution. *FEMS Microbiol. Review* **19**: 1-14.

Schaad, N.W., Vidaver, A.K., Lacy, G.H., Rudolph, K., and Jones, J.B., 2000, Evaluation of proposed amended names of several pseudomonads and xanthomonads and recommendations. *Phytopathology* **90**: 208-213.

Shailasree, S., Sarosh, B.R., Vasanthi, N.S., and Shetty, H.S., 2001, Seed treatment with β-aminobutyric acid protects *Pennisetum glaucum* systemically from *Sclerospora graminicola. Pest Manag. Sci.* **57**: 721-728.

Siegrist, J., Orober, M., and Buchenauer, H., 2000, β-Aminobutyric acid-mediated enhancement of resistance in tobacco to tobacco mosaic virus depends on the accumulation of salicylic acid. *Physiol. Mol. Plant Pathol.* **56**: 95-106.

Vauterin, L,, Hoste, B., Kersters, K., and Swings, J., 1995, Reclassification of *Xanthomonas*. *Int. J. Syst. Bacteriol.* **45**: 472-489.

Vassilev, V., Kolev, K., Zaharieva, M., and Sevov, V., 1995, Resistance of Aegilops, maize and wheat genotypes to *Pseudomonas syringae* pathovars *atrofaciens* and *syringae*. *Agronomie* **15**: 25-29.

Yuan, MeiLi, 1994, Controlling effect of bactericide on wildlife of tobacco. *Rev. Plant Pathology* **76**: 754.

Zimmerli, L. Jakab, G. Metraux, J., and Mauch-Mani, 2000, Potentiation of pathogen-specific defence mechanisms in *Arabidopsis* by β-aminobutyric acid. *Proc. Natl. Acad. Sci.* **97**: 12920-12925.

Enhancing the Efficacy of Bioherbicides

D.C. SANDS, A.L. PILGERAM, N.K. ZIDACK, B.J. JACOBSEN, and K.S. TIOUREBAEV
Department of Plant Science and Plant Pathology, Montana State University, Bozeman, MT, USA.

Abstract: Plant pathogenic pseudomonads are inhibited by certain amino acids due to feedback inhibition or repression of key biosynthetic enzymes in amino acid biosynthesis pathways. As it turns out, plants are similarly inhibited by certain amino acids. These inhibitions can play a large role in plant pathology. For instance, Frenching disease of tobacco is caused by rhizosphere bacteria that overproduce the amino acid isoleucine. We found that overproduction of an inhibitory amino acid would greatly enhance the virulence of a plant pathogen greatly increasing its efficacy as a bioherbicide for control of noxious weeds.

Key words: *Pseudomonas*, *Fusarium,* mycoherbicide, weeds.

Author for correspondence: D.C. Sands; e-mail: dsands@montana.edu.

1. INTRODUCTION: BIOLOGICAL CONTROL, THE MISSING COMPONENTS

If a scientist were funded today to develop a biological control agent against a specific weed, using the current technologies, he or she would probably fail to achieve efficacious and economically justifiable control of the weed. We in this field don't like to hear this, because it interferes with our vision and with our ability to compete for grants. We need new concepts, paradigms, and algorithms involving virulence enhancement, genetics of host range determinants, dissemination, and the biochemistry and evolution of the host/parasite interaction. We will attempt to cover these subjects in this paper. The more conventional incremental approach would be boring, fruitless and redundant.

N.S. Iacobellis *et. al.* (eds.), *Pseudomonas syringae* and Related Pathogens, 431-441.

2. AMINO ACID EXCRETION ENHANCES THE VIRULENCE OF PLANT PATHOGENS

The virulence and efficacy of bioherbicide can be greatly enhanced by selecting variants of the weed pathogen that overproduce and excrete amino acids that are inhibitory to the target plant (Tiourebaev, 1999). Our approach to enhancing the virulence of bioherbicides is modelled after "Frenching disease", a naturally occurring disease of tobacco (Steinberg, 1946). Steinberg *et al.* (1950) discovered that saprophytic bacteria (*Pseudomonas fluorescens* and *Bacillus subtilis*), growing on the roots of symptomatic plants overproduced a single amino acid, isoleucine. Isoleucine is synthesised in plants via the branched chain amino acid pathway. The end products of the pathway (valine, leucine, and isoleucine) allosterically regulate the activity of acetolactate synthase (ALS). The enzyme is differentially inhibited by these amino acids in different plant species. In "Frenching disease", overproduction of isoleucine by the saprophytic bacteria inhibited the activity of ALS in the tobacco, shutting down synthesis of valine and leucine, which disrupted essential protein metabolism. Free-living bacteria have complex systems of regulating their amino acid biosynthesis, permitting them to tolerate unusual imbalances in external amino acids. In contrast, plants are less well regulated, seemingly because they are self contained and independent of external amino acid sources. A plant's regulation may be considered "degenerate" in that one amino acid may be the regulatory factor for the whole group.

Interestingly, several modern chemical herbicides also target biosynthesis and inhibit single biosynthetic enzymes in plants, rendering treated plants incapable of producing a metabolite essential for plant growth. For example, ALS, the key enzyme in the branched chain amino acid biosynthesis, is inhibited by the sulfonylurea herbicides. The herbicide glyphosate inhibits 5'enolpyruvylshikimate 3-phosphate synthase (EPSPS), the key enzyme in the shikimic acid pathway (Amrhein, 1986)].

Given this knowledge, we screened our target weed for amino acid inhibition and determined that it was inhibited by mMolar amounts of the amino acid valine. We then selected variants of our pathogen, *Fusarium oxysporum* f. sp. *cannabis,* that were resistant to valine analogs. Some of these resistant variants excreted 10÷55 times more valine than their wild type parent and were more virulent to the target weed (Table 1) (Tiourebaev, 1999). The wild type strain controlled 25% of the target plant, while the valine excretion increased control to 70÷90%. In addition, the development of disease was more rapid in the plants infested with the valine overproducers. Limited studies on fourteen other plant species did not reveal a change in host range. Thus, overproduction of an essential amino acid provided a highly effective means of enhancing the virulence of a biocontrol agent (Sands *et al.* 1999).

This technology has also been used to enhance the virulence of *F. o.* f. sp. *papaveris*, *P. syringae* pv. *tagetis*, and *Xanthomonas campestris* pv. *poae* .

Table 1. Valine excretion and virulence of valine overproducing variants of *Fusarium oxysporum* f. sp. *cannabis*.

Strain	Description	Valine Excretion[a] ($mg\ l^{-1}$)	% Kill	Mortality Rate[b] (weeks)
C95	Wild-type	0-0.18	25	6-8
4nv	Norvaline[R]	2.84	70	2-3
6pa	Penicillamine[R]	2.48	90	2-3
8pa	Penicillamine[R]	9.93	90	2

a. Valine excretion. Valine excretion was determined by spectrophotometric analysis of growth of *Pediococcus cerevisiae* ATCC8042 in culture supernatant.
b. Mortality Rate. Mortality rate is the duration between inoculation and the first appearance of severe disease symptoms or death.

3. GENERAL METHODOLOGY

3.1 Determination of Amino Acids that are most Inhibitory to the Target Weed

Surface sterilised seed are placed on plates of water agar (1.5% agar, 1 mM Tris, pH 6.8) that have been supplemented with a single amino acid (2÷5 mM l-form). Amino acid(s) that decrease seed germination, inhibit shoot growth or cause necrosis are further evaluated (For example, tryptophan is inhibitory to growth in Fig. 1). Certain plants may be more inhibited by combinations of amino acids rather than single amino acids. Combinations of all the amino acids derived from a single pathway typically are not inhibitory.

3.2 Selection of Variants of the Bioherbicide Resistant to Analogs of the Selected Amino Acid

Amino acid overproducing strains of each fungus or bacterium are selected by exposure to specific amino acid analogs (Table 2). For example, if the target weed is inhibited by lysine, then pathogens for control of that weed will be exposed to lysine analogs to select mutants that overproduce lysine. Resistant colonies can be selected using a well zone-diffusion assay on CUTS minimal medium (Czapek-Dox Agar (Difco) (35 $g\ l^{-1}$) supplemented with ammonium sulphate (0.5 $g\ l^{-1}$), uracil (20 $mg\ l^{-1}$), thiamine (4 $mg\ l^{-1}$) and a vitamin mixture (100 mg of crushed Sesame Street Complete Vitamins). The zone diffusion plates are prepared by cutting a plug from the

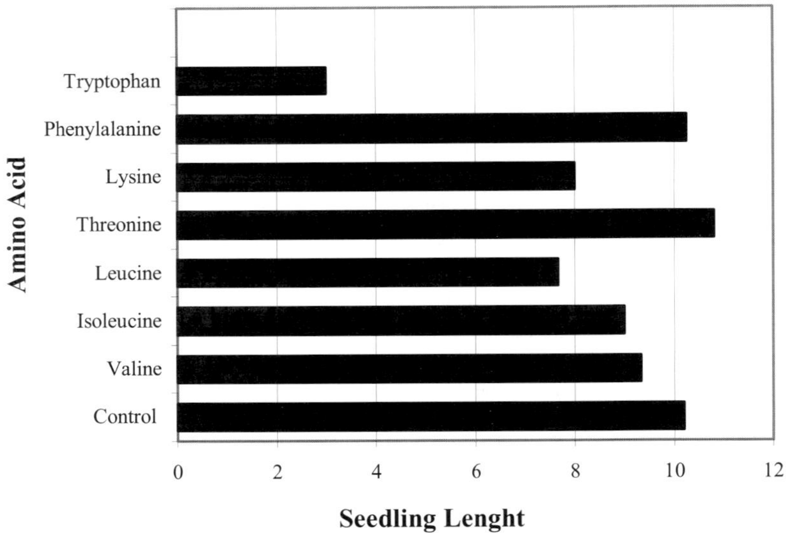

Figure 1. Inhibition of the growth of field bindweed seedlings by selected l-amino acids. Growth was measured 14 days after placing the seed on the amino acid plates.

Table 2. Amino acid analogs.

Amino Acid	Analog(s)
Valine	Valine hydroxamate, norvaline, penicillamine
Lysine	Aminoethyl-cysteine, hydroxylysine, aminocaproic acid
Methionine	selenoethionine, ethionine, selenomethionine,
Tryptophan	5-Methyltryptophane
Arginine	Canavanine
Proline	Hydroxy-proline, dehydro-proline
Leucine	Norleucine

centre of the plate. The plates are then inoculated with $10^6 \div 10^7$ fungal spores, a suspension of $10^3 \div 10^5$ mycelial fragments, or a suspension of $10^7 \div 10^8$ bacteria. A sterile solution of the amino acid analog (0.1 ml of a100 mM solution) is then added to the well. The plates are incubated in a laminar flow hood for 4 hours and an additional 0.1 ml of the analog solution is added to the well. The plates are then incubated at 28°C and monitored daily for the appearance of zones of inhibition and resistant colonies within the zone. Resistant colonies are isolated and analysed for amino acid excretion. This selection may be repeated several times using increasing concentrations of analog and/or different analogs.

3.3 Assay for Amino Acid Excretion

Bacterial auxotrophs are used to assay amino acid excretion (Sands and Hankin, 1974). The auxotroph is seeded into media lacking the amino acid required for growth. Subsequent growth of the auxotroph in the media is dependent upon and proportional to the quantity of added amino acid. For example, in order to assay valine, a valine auxotroph of *Escherichia coli* is seeded into CUTS medium. The auxotroph will not grow unless exogenous valine is added to the medium. Colonies of the plant pathogenic fungi or bacteria that are resistant to a valine analog are sub-cultured onto the seeded medium and incubated for 2÷3 days. If the resistant variants excrete valine, there will be a zone of auxotroph growth surrounding the colony. The size of the zone is proportional to the amount of valine excreted. A standard dose-response can be determined by placing discs containing various levels of amino acid onto the agar.

3.4 Testing Virulence and Host Range of the Amino Acid Overproducing Variants in Growth Chamber Studies

The virulence (rate of kill and % mortality) of amino acid producing variants of each pathogen are evaluated by inoculating target weed plants with each amino acid excreting variant and its respective wild type parent. Amino acid excreting variants that are more virulent than the parent should be further evaluated in host range experiments and scale-up experiments.

4. IMPROVING DISSEMINATION

The effectiveness of a soil-applied mycoherbicide depends upon the survival and movement of the pathogen from the formulation particle through the soil profile to the root and crown of the target plant (Fravel *et al.,*1996; Bailey *et al.*, 1997, Ciotola *et al.*, 2000). Spore suspensions and food-based formulations are common delivery systems for bioherbicide. However, this technique does not work well for *Fusarium oxysporum,* as spores become bound by soil particles in the top few millimetres of the soil (Connick *et al.*, 1998). *F. oxysporum* saprophytically colonises the roots of many non-host plants (Burgess, 1981). Thus, *F. oxysporum* mycoherbicide could be delivered to farmers' fields on non-host seed such as crops or grass, positioning the mycoherbicide directly in the rhizosphere of the target weed (Sands *et al,* 2002; Tiorebaev, 1999; Ciotola *et al.*, 2000). In addition, the multiplication of fungal biomass in the rhizosphere of the carrier seedling allows for application of low levels of the mycoherbicide, greatly reducing the cost of inoculum production. Many phytopathogenic bacteria such as *Pseudomonas syringae* and *Xanthomonas campestris* also have a saprophytic phase and could be delivered in a similar manner.

5. GENETICS: READY OR NOT

Genetic techniques have only recently been integrated into the field of biocontrol. When confronted with poor field trials, biocontrol researchers simply sought more efficacious formulations or looked for new pathogens. The following sections cover some of the efforts currently underway in the genetic improvement of mycoherbicides.

5.1 Genetic Marking

Researchers often want to track the fate of a released biocontrol strain. Three types of genetic markers have been used to label bioherbicide: auxotrophy (Miller *et al.*, 1989b; Chacko *et al.*, 1994; Yang and TeBeest, 1995), antibiotic resistance [Yang and TeBeest, 1995;Toyota *et al.*, 1992; Holmstrom-Ruddick and Mortensen, 1995; Brooker *et al.*, 1996) and visual markers such as *gus* and *gfp* (Cohen *et al,* 2002; Robinson and Sharon, 1999). Transformation with *lux* genes that confer bioluminescence is also a possibility for marking bioherbicidal bacteria (Kennedy, 1995; Kragelund *et al.*, 1997). Amino acid analog resistant mutants for virulence enhancement have an unanticipated value in that the strains are marked by their resistance and can be easily be followed using selective media. Care has to be taken to select for mutants with normal growth and optimal virulence characteristics.

5.2 Contained Systems

Biocontrol agents must be judged to be safe prior to release into a new environment, because once an organism is released, there may be no way to get it back. Host range evaluation, human toxicity and environmental risk analysis can be used to reach a probability-based decision on safety. Possible genetic-containment systems include temperature sensitivity (species or mutants unable to survive the heat of summer or the cold of winter), antibiotic requiring mutants that can only grow in the presence of antibiotic, auxotrophy, or mutants unable to produce sexual or survival structures (Gressell, 2001; Miller *et al.*, 1989a).

5.3 Virulence

Besides the already mentioned virulence enhancement by amino acid excretion, other ways to enhance virulence also include overproduction of known virulence genes or genes whose products disengage host responses. The genes encoding production of the natural herbicide bialaphos have been cloned from *Streptomyces hygroscopicus* (Murakami, 1986) and transferred into the plant pathogen *Xanthomonas campestris* pv. *campestris* (Charudattan *et al.*, 1996). This herbicide inhibits the enzyme glutamine synthetase and disrupts normal amino acid metabolism. The transformed bacterium retained pathogenicity to its natural hosts (broccoli and cabbage) but elicited a hypersensitive response on four non-host plants (Charudattan *et al.*, 1996). The number of genes and size of the gene cluster necessary for bialaphos production may limit transfer of this technology to mycoherbicidal strains of plant pathogenic fungi. Cohen *et al.* (2002) transformed two bioherbicidal strains of *Fusarium* with two genes from the indole-3-acetimide (IAA) pathway. Transgenic strains of the fungi were more effective in suppressing the plant parasitic weed, *Orobanche*, than the parental wild-types. Thus, stimulating an auxin imbalance also enhanced pathogen virulence.

5.4 Isolation of Promoters

DNA promoters can be isolated from plant pathogenic fungi (Weltring, 1995) or bacteria (Horwitz and Loeb, 1986) by screening libraries consisting of small DNA fragments that have been inserted into plasmids containing a promoter-less selectable marker. The plasmid library is transformed into the target pathogen and the transformants are screened for expression of the selectable marker. DNA fragments that drive expression of the marker are considered promoters. Inducible promoters can be isolated by differentially screening for expression under different conditions or in the presence of different inducers. Desirable promoters include host specific, tissue specific, temperature specific, or herbicide-inducible.

Plants produce a variety of secondary compounds. Some of these compounds are specific to certain plant species. For example, tomato produces an antifungal saponin called alpha-tomatine. *Fusarium oxysporum* f. sp. *lycopersici*, a pathogen of tomato, produces tomatinase, an enzyme that degrades the tomatine (Roldan *et al.*, 1999). The promoter regulating expression of the tomatinase is induced by the tomatine, and is therefore, tomato-specific. This tomato-specific promoter may not prove useful in biocontrol. However, spotted knapweed produces cnicin, a secondary metabolite that inhibits the growth of other plants. A cnicin-specific promoter would allow targeting of hypervirulence genes to spotted knapweed.

6. SPECULATIONS: BOOLEAN ALGORITHMS, EVOLUTION, AND BIOHERBICIDES

Many of the more than 200,000 known species of plants are attacked by host-specific fungi or bacteria. Given this situation, where the diversity of the plant kingdom is paralleled by near equivalent diversity in plant pathogens, there might be some underlying principles of evolution whereby the host-parasite relationship was established. If we assume that each host-parasite relationship is completely unique, then we might expect that genetic analysis of each pair would reveal unique sets of genes conferring unique biochemistry. Studying all of these differences would require a lot of biochemists. On the other hand, if the host parasite relationship could partially be explained by a defined series of biochemical "decisions or states" of the either/or type (i.e., Boolean) (Wolfram, 2001), then some general and predictable kinds of genetic adaptations could account for host specificity. We report evidence for some "either/or" losses of options in the host and concomitantly in the homologous pathogen. We hypothesise that, in the case of host-specific plant pathogen relationships, a handful of variations in the regulation at branch points of intermediary metabolism may account for a significant amount of the biochemical differences leading to host-specificity.

At first glance, the host-specific but non-obligate plant pathogenic microbes and strict saprophytic microbes are very similar in their metabolism in that both grow on minimal media with simple carbon and nitrogen sources. But in closer examination, there are significant differences. In strictly saprophytic microbes, amino acid biosynthesis is meticulously regulated because the external environment relative to nutrients such as amino acids can be quite variable. In the host-specific plant pathogenic pseudomonads much of this regulation does not exist (Sands and Zucker, 1976). Pathovars of *Pseudomonas syringae* are inhibited by certain single amino acids that shut down an entire pathway, even when not all end-products are available. This pattern appears to be pathovar dependent. Related saprophytes (*P. fluorescens* and *P. putida)* appear to be well regulated as they are not inhibited by single amino acids. Perhaps host-specific pathogens, limited to a specialised environment, do not have frequent occasion to regulate for certain amino acid imbalances. For instance, if a plant "milieu interior" is routinely high in the free amino acid lysine, then the lysine regulated aspartokinase in the pathogen would seldom if ever be active. One might presume that this unused enzyme, and the corresponding gene, might in time suffer losses in efficacy, leading eventually to deletion and permanent loss. This suggests that the key regulatory point of a biosynthetic pathway can serve as an "either/or" gate. In Boolean algebra, more commonly used in decision trees in computer programming, this branch point could be referred to as a "Boolean Gate". Unless all gates are set correctly within a pathogen, there is no chance for protein biosynthesis or growth of the pathogen to sufficiently high levels to cause disease. A handful of such gates operating in parallel would be sufficient to narrow the biosynthetic window of a pathogen to one or a few plants. Regardless of the relative merits of the structure of models, it is clear that both plants and their microbial pathogens are subjected to misregulation by an oversupply of some amino acids, usually at key branch points in their biosynthetic pathways. These points of misregulation may serve as determinants of host-pathogen specificity. For instance a plant always high in lysine might only be a host for pathogens without lysine regulation of their single aspartokinase. Possession of a single acetolactate synthase inhibitable by valine would render a pathogen only capable of attacking plants either low in free valine or high in Isoleucine and leucine, the other two amino acids in that biosynthetic family. Alternatively a microbe that invades the vascular system of a plant and/or resides in high numbers in its rhizosphere and excretes a single amino in sufficiently high amounts may be expected to inhibit plant growth.

7. CONCLUSION

Over the last thirty years, numerous pathogens have been investigated as potential bioherbicides. Despite this intensive research effort, few pathogens have been successful as biocontrol agents. The inherent constraints associated with biological species are largely responsible for this failure, yet our preconceived ideas about these agents are also at fault. The authors believe that a paradigm shift must occur if bioherbicides are to enjoy wider success as a weed control method. In the largely unsuccessful past, researchers have required lethality and host specificity as requirements for a successful agent. However, many pathogens that do not meet these criteria can be enhanced by simple selection as described in this paper, or through, perish the thought, recombinant genetics.

REFERENCES

Amrhein, N., 1986, Specific inhibitors as probes into the biosynthesis and metabolism of aromatic amino acids. *Re. Ad. Phytochem.* **20**: 83-117.

Bailey, B.A., Hebbar, K.P., Strem, M., Darlington, L.C. and Lumsden, R.D., 1997, An Alginate Prill Formulation of *Fusarium oxysporum* Schlechtend:Fr f. sp. *erythroxyli* for Biocontrol of *Erythroxylum coca* var. *coca*. *Biocontrol Sci. Technol.* **7**: 423-435.

Brooker, N.L., Mischke, C.F., Patterson, C.D., Mishke, S., Bruckart, W.L., and Lydon, J., 1996, Pathogenicity of bar-transformed *Colletotrichum gloeosporioides* f. sp. *aeschynomene*. *Biol. Control* **7**: 159-166.

Burgess, L.W., 1981, General Ecology of the Fusaria. In: *Fusarium: Diseases, Biology, and Taxonomy* (P.E. Nelson, T.A. Toussoun, and R.J. Cook, eds.), Pennsylvania State University Press, University Park, pp. 225-235.

Ciotola, M., DiTommaso, A., and Watson, A.K., 2000, Chlamydospore production, inoculation methods and pathogenicity of *Fusarium oxysporum* M12-4A, a biocontrol for *Striga hermonthica*. *Biocontrol Sci.Technol.* **10**: 129-145.

Chacko, R.J., Weidemann, G.J., TeBeest, D.O. and Correll, J.C., 1994, The use of vegetable compatibility and heterokaryosis to determine potential asexual gene exchange in *Colletotrichum gloeosporioides*, *Biol. Control* **4**: 382-389.

Charudattan, R., Prange, V.J., and DeValerio, J.T., 1996, Exploration of the use of the "Bialaphos Genes" for improving bioherbicide efficacy, *Weed Technol.* **2**: 625-636.

Cohen, B.A., Amesellem, Z., Maor, R., Sharon, A., and Gressel, J., 2002, Transgenically enhanced expression of indole-3-acetic acid confers hypervirulence to plant pathogens, *Phytopathology* **95**: 590-596.

Connick, W.J., Jr., Daigle, DJ., Pepperman, A.B., Hebbar, K.P., Lumsden, R.D., Anderson, T.W., and Sands, D.C., 1998, Preparation of stable granular formulations containing *Fusarium oxysporum* pathogenic to narcotic plants, *Biol. Control* **13**: 79-84.

Fravel, D.R., Stosz, S.K. and Larkin, R.P., 1996, Effect of Temperature, Soil Type, and Matrix Potential on Proliferation and Survival of *Fusarium oxysporum* f. sp. *erythroxyli* from *Erythroxylum coca*, *Phytopathology* **86**: 236-240.

Gressel, J., 2001, Potential failsafes against spread and introgression of transgenic hypervirulent biocontrol fungi, *Trends Biotechnol.* **19**: 149-154.

Holmstrom-Ruddick, B. and Mortensen, K., 1995, Factors affecting pathogenicity of a benomyl-resistant strain of *Colletotrichum gloeosporioides* f. sp. *malvae, Mycol. Re.* **99**: 1108-1112.

Horwitz, M. S. Z. and Loeb, L.A., 1986, Promoters selected from random sequences, *Proc. Natl. Acad. Sci. USA* **83**: 7405-7409.

Kennedy, A.C., 1995, Molecular Biology of Bacteria and Fungi for Biological Control of Weeds. In *Molecular Biology of the Biological Control of Pests and Diseases of Plants* (M. Gunasekaran and D.J. Weber, eds.), CRC Press, Boca Raton, pp. 155-172.

Kragelund, L., Hosbond, C., and Nybroe, O., 1997, Distribution of metabolic activity and phosphate starvation response of Lux-tagged *Pseudomonas fluorescens* reporter bacteria in the barley rhizosphere, *Appl. Environ. Microbiol.* **135**: 4920-4928.

Miller, R. V., E.J. Ford, and Sands, D.C., 1989a, A nonsclerotial pathogenic mutant of *Sclerotinia sclerotiorum*. *Can. J. Microbiol.* **35**: 517-520.

Miller, R.V., E.J. Ford, N.J. Zidack, and Sands, D.C., 1989b, A pyrimidine auxotroph of *Sclerotinia sclerotiorum* for use in biological weed control. *J. Gen. Microbiol.* **135**: 2085-2091.

Murakami, T., 1986, The bialaphos biosynthetic genes of *Streptomyces hygroscopicus*: molecular cloning and characterization of the gene cluster, *Mol. Genet.* **205**: 42-50.

Robinson, M. and Sharon, A., 1999, Transformation of the bioherbicide *Colletotrichum gloeosporioides* f. sp. *aeschynomene* by electroporation of germinated conidia, *Curr. Genet.* **36**: 98-104.

Roldan, A., Perez-Espinosa, A., and Ruiz-Rubio, M., 1999, Tomatinase from *Fusarium oxysporum* f. sp. *lycopersici* defines a new class of saponinases, *Mol. Plant-Microbe Interact.* **12**: 852-86.

Sands, D.C. and Hankin, L., 1974, Selecting lysine-excreting mutants of Lactobacilli for use in food and feed enrichment, *Appl. Microbiol.* **28**: 523-524.

Sands, D.C. and Zucker, M., 1976, Amino acid inhibition of pseudomonads and its reversal by biosynthetically related amino acids. *Physiol. Plant Pathol.* **9**: 127-133.

Sands, D.C., Tiourebaev, K.S., Pilgeram, A.L., and Anderson, T.W., 2002, Carrier methodology for aerial dispersal and soil penetration of bioactive agents. US Patent 6,403,530.

Sands, D.C., Tiourebaev, K.S., Pilgeram, A.L., and Anderson, T.W., 1999, Virulence Enhancement of Bioherbicides, U.S. Provisional Patent No. 60/172,771.

Steinberg, R.A., 1946, A "Frenching" Response of Tobacco Seedlings to Isoleucine, *Science* **103:** 329-330.

Steinberg, R.A., Bowling, J.D., and McMurtrey, J.E., 1950, Accumulation of free amino acids as a chemical basis for morphological symptoms in tobacco manifesting frenching and mineral deficiency symptoms, *Plant Physiol.* **25**: 279-288.

Tiourebaev, K.S., 1999, PhD Thesis *Virulence and Dissemination Enhancement of a Mycoherbicide.* Montana State University.

Toyota, K., Tsuge, T. and Kimura, M., 1992, Potential application of genetic transformants of *Fusarium oxysporum* f. sp. *raphani* for assessing fungal autecology, *Soil Biol. Biochem.* **24**: 489-494.

Weltring, K.M., 1995, A method for easy isolation of promoter fragments from promoter-probe libraries of filamentous fungi, *Curr. Genet.* **28**: 190-196.

Wolfram, S., 2001, A New Kind of Science, Wolfrom Media, Inc. Champaign.

Yang, X.B. and TeBeest, D.O., 1995, Competiveness of mutant and wild-type isolates of *Colletotrichum gloeosporioides* f. sp. *aeschynomene*, *Phytopathology* **85**: 705-710.

Integrated Management of Bacterial Streak and Bulb Rot of Onion

R.D. GITAITIS, F.H. SANDERS, J.C. DIAZ-PEREZ and R.R. WALCOTT
Department of Plant Pathology, University of Georgia, Coastal Plain Experiment Station, Tifton, GA 31793-0748, USA.

Abstract: Significantly higher levels of bacterial streak and bulb rot of onion (BSBR), caused by *Pseudomonas viridiflava*, occurred in onions receiving excessive applications of fertiliser and were correlated with higher nitrogen content in plant tissues. Based on previous work that weeds, particularly cutleaf evening primrose (*Oenothera laciniata*), were the primary source of initial inoculum, oxyfluorfen was evaluated as a post-emergence herbicide after transplanting. The herbicide not only reduced weed levels as expected, but also reduced the incidence of BSBR. Lower levels of BSBR also occurred in plots treated with weekly applications of cupric hydroxide and maneb when compared to non sprayed plots, or plots treated only with the fungicide chlorothalonil. Harvesting onions with cutting shears contaminated with *P. viridiflava* increased the incidence of postharvest rots in onions clipped immediately after uprooting but not in onions allowed to cure in the field for 48 h. Finally, lower levels of postharvest disease occurred when onion bulbs were stored under controlled-atmospheric (CA) conditions (3% oxygen, 5% carbon dioxide, and 92% nitrogen) at 1°C. The various control procedures were combined into an integrated management strategy for the control of BSBR in Georgia.

Key words: *Pseudomonas viridiflava, Allium cepa,* postharvest rot of onion, controlled-atmospheric conditions.

Author for correspondence: R.D. Gitaitis; e-mail: gitaitis@tifton.uga.edu.

1. INTRODUCTION

Bacterial streak and bulb rot (BSBR) of onion (*Allium cepa* L.) is caused by the bacterium, *Pseudomonas viridiflava* (Burkholder) Dowson. BSBR was first observed on short-day, sweet onions in Georgia in 1990 (Gitaitis *et al.*,

N.S. Iacobellis *et. al.* (eds.), *Pseudomonas syringae* and Related Pathogens, 443-449.

1991) and has been responsible for intermittent and sporadic losses since its first occurrence. Conditions most favourable for epidemic development occur between 15 January and 30 March. Generally, new lesions are small (≤1.0 cm), oval, olive-green to tan, and surrounded by an area of chlorosis and water-soaking. As lesions develop, they become dark-green to black, extensively water-soaked, and develop into streaks that encompass the entire leaf. Sometimes a single leaf is affected, however under favourable conditions entire plants become blighted. Eventually interveinal areas collapse, causing the veins to appear prominent. Often, soft rot symptoms develop at the base of the leaf and in the neck of the bulb. Bulbs of severely infected plants are difficult to harvest as infected leaves become detached from necks and the bulbs remain in the ground. We previously demonstrated that the primary source of inoculum for BSBR epidemics was most likely from epiphytic populations of *P. viridiflava* on several winter weeds (Gitaitis *et al.*, 1998), including cutleaf evening primrose (*Oenothera laciniata* Hill), which is the most significant weed problem in onion fields in Georgia. Our objective was to develop an integrated management strategy based on these and additional results regarding the ecology of *P. viridiflava* and the epidemiology of BSBR.

2. MATERIALS AND METHODS

To confirm the presence of BSBR, samples of diseased tissues were taken periodically from the various experiments and isolations were made on King's Medium B (King *et al.*, 1954) or T5 (Gitaitis *et al.*, 1997). Suspect colonies were characterised by biochemical and physiological reactions (Jones *et al.*, 1986), fatty acid methyl ester profiles (Gitaitis *et al.*, 1991), and by the polymerase chain reaction (PCR) (Gitaitis *et al.*, 1998). Primers used for the PCR tests were developed from the pectate lyase gene [RDG1 (5'-GTATTGCTGGTG TTACCC-3') and RDG2 (5'-GTATCCAGAAACGAAC-3')] and from the internally transcribed spacer (ITS) region [Fwd pv1 (5'-AGTGAAAGT GGTCAACAGTC-3') and Rev pv1 (5'-ATAGCTCAGCTGGGAG-3')] between the 16S and 23S rRNA genes.

Effects of fertility on development of BSBR were evaluated. Onions, cv. Granex 33, were planted at three different locations at a population of about 370,650 seeds ha^{-1} in double line rows with four rows per 1.83 m bed using a standard and excessive fertility program. The standard fertiliser program consisted of 1120.82 kg ha^{-1} of 5-10-15 (NPK) and 224.17 kg ha^{-1} of 10-34-0 applied prior to planting. Side-dressings of 224.17 kg ha^{-1} of 15-0-14 were applied at 8, 12, and 16 weeks after planting, and 224.17 kg ha^{-1} of calcium nitrate were applied at 20 weeks after planting. The excessive fertiliser program was the same, except that rates of 15-0-14 at 8, 12, and 16 weeks after planting were increased to 1120.82 kg ha^{-1} to simulate some grower's practices. Plants were evaluated for BSBR incidence over time. Tissue analysis was done by lyophilising onion leaves, grinding them into a fine powder and determining nitrogen content by the total combustion method (Campbell, 1991) and calcium, phosphorus and potassium content by the dry ash procedure (Dahlquist and Knoll, 1978).

Effects of weed control on BSBR incidence were evaluated at four different locations (sites A, B, C, and D). Onions, cv. Granex 33, were transplanted at a population of about 370,650 plants ha^{-1}. Plants were planted in double line rows with four rows per 1.83 m bed using a standard fertility program. At sites A and B, treatment one (DCPA [dimethyl tetrachloroterephthalate 75 WP]) at 10.09 kg ha^{-1} applied as a pre-plant herbicide) was compared with treatment two (DCPA) at 10.09 kg ha^{-1} and oxyfluorfen applied as a postemergence herbicide at 2.34 l ha^{-1}). At site C, treatment one (DCPA) at 10.09 kg/ha) was compared with treatment three (oxyfluorfen at 2.34 l ha^{-1}). At site D, treatment three (oxyfluorfen at 2.34 l ha^{-1}) was compared with a control of no herbicide. Final BSBR incidence was recorded.

Use of fixed-copper sprays were evaluated for control of BSBR. Onions, cv. Granex 33, were direct seeded using standard fertility and weed management procedures as mentioned above. Treatments were arranged in a complete block design with four replications. Each treatment consisted of a 4-row bed 15.24 m in length. Treatments were separated with a single, non-sprayed bed of onions. Alleys between blocks were 3.05 m in length. Treatments were: 1) 1.75 l ha^{-1} chlorothalonil (54%) applied weekly; 2) 1.75 l ha^{-1} chlorothalonil alternated with 1.68 kg ha^{-1} Mankocide (300 g kg^{-1} copper present as cupric hydroxide and 150 g kg^{-1} mancozeb), (Griffin LLC., Valdosta, GA) on a weekly basis; 3) 1.75 l ha^{-1} chlorothalonil alternated with 2.98 kg ha^{-1} Mankocide on a weekly basis; 4) 3.36 kg ha^{-1} Mankocide applied weekly; 5) non-sprayed control. A total of 16 sprays were applied between 13 January and 28 April. Eight disease ratings were conducted between 15 January and 30 April. In addition, percent loss of stand, leaf diameter and yield data were recorded.

Onion maturity and harvest practices were evaluated for control of BSBR. Granex 33 onions were direct seeded and managed as described above. Onions were uprooted when bulbs were immature (necks hard and tops break 8.0 cm above the bulb), mature (necks soft with ~20% tops down), and over-mature (all tops down and beginning to dry). A subset of onions was clipped immediately using shears dipped in a suspension of *P. viridiflava* (1 x 10^8 CFU ml^{-1}) or sterile water. Another subset of onions was clipped in a similar manner after field drying for 48 h. Onions were segregated by treatment, cured with forced hot air (~38°C) for 72 h and stored at ~ 26°C. After 1 wk, onion bulbs were cut and evaluated for BSBR severity.

The effect of onion bulb storage conditions on postharvest levels of BSBR were evaluated. Granex 33 onions were uprooted at optimum maturity. After field drying for 48 h, foliage and roots were removed by clipping with shears. Bulbs were cured with forced hot air (~38°C) for 72 h. Treatments applied after curing consisted of 1) stab-inoculation to a depth of 1.0 cm with a dissecting needle contaminated with *P. viridiflava* and storage at 4 °C, 2) stab-inoculation to a depth of 1.0 cm with a sterile dissecting needle and storage at 4°C, 3) stab-inoculation to a depth of 1.0 cm with a dissecting needle contaminated with *P. viridiflava* and storage in a controlled atmosphere (CA) cell, and 4) stab-inoculation to a depth of 1.0 cm with a sterile dissecting needle and storage in a CA cell. Conditions in the cold room consisted of a normal atmosphere at 4°C. The temperature and relative humidity in CA cells were established at 1°C and 75%, respectively. The atmosphere in CA cells consisted of 3% oxygen, 5% carbon dioxide, and 92% nitrogen. Temperature, relative humidity, carbon dioxide and oxygen concentrations were determined hourly and corrected as necessary. After a 10-wk storage period, bulbs were removed, split, and evaluated for disease severity.

3. RESULTS

Suspect colonies from onion leaves with BSBR symptoms were fluorescent on King's medium B and T5; negative for oxidase, arginine dihydrolase, and levan production; positive for potato rot and tobacco hypersensitivity; and had a fatty acid methyl ester profile similar to known strains of *P. viridiflava*. Furthermore, DNA from all strains were amplified by primers RDG1 and RDG2, which are specific for the pectate lyase gene. DNA from test strains also was amplified by primers Fwd pv1 and Rev pv1, which are specific for the ITS region between the 16S and 23S rRNA genes in *P. viridiflava*.

Increased levels of BSBR were observed in onions receiving an excessive rate of fertiliser compared to onions grown under a standard fertility program (Table 1). The mean level of BSBR for the excessive fertility program for the three locations was 37.2% in contrast to only 14.2% BSBR for onions grown under the standard fertility program. Tissue analysis indicated that nitrogen was significantly higher in leaf tissues of plants grown using the excessive fertility program and correlated with increased levels of disease.

Applications of oxyfluorfen as a post-emergence herbicide alone or in combination with DCPA significantly reduced BSBR levels (Table 2). BSBR levels ranged from as high as 24.0 % and 20.3 % in plots not receiving oxyfluorfen, whereas plots at the same sites receiving oxyfluorfen had only 11.0 and 7.6 % incidence of BSBR, respectively.

Table 1. Tissue analysis and bacterial streak and bulb rot levels of onions grown using two fertility programs.

Site	Fertility Program	%N	%P	%K	%Ca	% Disease
A	Excessive	3.2 a	0.4 a	2.3 a	0.7 a	42.0 a
A	Standard	2.3 b	0.3 a	2.3 a	0.6 a	10.5 b
B	Excessive	2.7 a	0.4 a	2.1 a	0.6 a	38.5 a
B	Standard	2.1 b	0.5 b	2.1 a	0.5 a	16.0 b
C	Excessive	2.6 a	0.3 a	2.1 a	0.7 a	31.0 a
C	Standard	2.4 b	0.3 a	2.0 a	0.7 a	16.0 b

Paired values within the same column with different letters are significantly different at P = 0.01.

Table 2. Incidence (%) bacterial streak and bulb rot of onion at four sites with different weed control treatments.

Treatment	Site A	Site B	Site C	Site D
DCPA	11.6 a	2.0 a	24.0 a	ND
DCPA + Oxyfluorfen	4.6 b	0.6 b	ND	ND
Oxyfluorfen	ND	ND	11.0 b	7.6 b
No Herbicide	ND	ND	ND	20.3 a

Paired values within the same column with different letters are significantly different at P = 0.05; ND = No data, as treatment was not at that site.

Mankocide applied at 3.36 kg ha^{-1} on a weekly basis provided the best disease control of the various compounds and spray schedules tested (Table 3).

Table 3. Effect of fixed-copper sprays on bacterial streak and bulb rot, onion leaf diameter, stand loss and yield.

Treatment	Rate/ha	% Disease	Leaf Diam.(mm)	% Stand Loss	Yield (kg ha^{-1})
1. chlorothalonil	1.75 l	50.0 c	23.8 c	24.8 c	9,968 d
2. chlorothalonil	1.75 l	40.5 b	24.5 bc	10.8 b	13,588 c
Mankocide *	1.68 kg				
3. chlorothalonil	1.75 l	44.0 b	25.9 ab	10.1 b	16,111 b
Mankocide *	2.98 kg				
4. Mankocide	3.36 kg	29.5 a	26.6 a	6.1 a	18,268 a
5. No treatment	-	54.0 c	21.1 d	24.2 c	13,054 c

Values within the same column with different letters are significantly different at P = 0.05;
* Treatments applied on alternate weeks.

Plants treated with the higher rate of Mankocide on a weekly schedule were more robust, had a significantly larger leaf diameter, and had a higher survival rate (stand count). In addition, plots receiving the higher rate of Mankocide had yields increased by 5,214 kg ha^{-1} and 8,300 kg ha^{-1} when compared to non-sprayed and chlorothalonil treatments, respectively.

Harvesting onions at the proper stage of maturity and drying them for a minimum of 48 h, prior to removal of leaves and roots with shears contaminated with *P. viridiflava*, provided a significant level of disease control (Table 4). Onions harvested at an immature stage, when necks were hard, had significantly higher levels of postharvest disease both in onions that were clipped immediately as well as those that were cured in the field for 48 h.

Table 4. Effects of onion maturity and hours of field-curing on % postharvest bacterial streak and bulb rot (BSBR).

Onion Maturity	Field-Curing (hours)	% BSBR
Immature	0	79 c
Immature	48	68 c
Mature	0	41 b
Mature	48	15 a
Over-Mature	0	41 b
Over-Mature	48	18 a

Values with different letters are significantly different at $P = 0.05$.

In onions harvested at an immature level, BSBR levels were 79% and 68% when there were 0 and 48 h of field-curing, respectively. However, when onions were harvested at an optimum level of maturity or when over-mature, field-curing for 48 h significantly reduced BSBR levels to 15% and 18%, respectively.

CA storage was effective in retarding the development of BSBR (Table 5). Mean BSBR levels in water-inoculated, control onions were 10.5% and 4.5% after 10 wk of storage in a standard cold room or CA cell, respectively. When onion bulbs were inoculated with *P. viridiflava* prior to storage, the level of BSBR in onions stored in a standard cold room increased to 55% but remained at a relatively low level of 7% in onions stored in a CA cell.

Table 5. Levels of rot in inoculated and control onions stored conventionally or in ca cells.

Type of Storage	Inoculated	% Bulb Rot
Cold Room	Water Control	10.5 a
CA	Water Control	4.5 a
Cold Room	*P. viridiflava*	55.0 b
CA	*P. viridiflava*	7.0 a

Paired values within the same column with different letters are significantly different at $P = 0.05$.

4. DISCUSSION

Cultural practices can have a significant impact on plant disease incidence and development. In this study we demonstrated that weed management, fertility management, proper decision-making regarding the stage of maturity of onion plants selected for harvest, utilising field-curing and storing onions in CA conditions all reduced BSBR levels. When combined with chemical control in an integrated pest management strategy, they provide excellent levels of disease control for BSBR of onion.

ACKNOWLEDGEMENTS

Funding for this research was provided by the Vidalia Onion Committee and USDA/CSREES Grant No. 99-34389-7437, entitled "Integrated Disease Management to Improve the Quality of Sweet Onions in Georgia". We also would like to acknowledge Reid Torrance, county agent for Tattnall county for his contributions to the fertility and weed management studies.

REFERENCES

Campbell., C.R., 1991, Determination of total nitrogen in plant tissue by combustion. p. 21-23. In C.O. Plank (ed.) Plant Analysis Reference Procedures for the Southern Region of the United States. Southern Cooperative Series Bulletin 368. The Georgia Agricultural Experiment Station, The University of Georgia. Athens, GA.

Dahlquist, R.L., and Knoll, J.W., 1978, Inductively coupled plasma atomic emission spectrometer: analysis of biological materials and major, trace and ultra-trace elements. *Appl. Spectros.* **32** (1): 1-29.

Gitaitis, R.D., Baird, R.E., Beaver, R.W., Sumner, D.R., Gay, J.D., and Smittle, D.A., 1991, Bacterial blight of sweet onion caused by *Pseudomonas viridiflava* in Vidalia, Georgia. *Plant Dis.* **75**: 1180-1182.

Gitaitis, R., MacDonald, G., Torrance, R., Hartley, R., Sumner, D.R., Gay, J.D., and Johnson, W.C., III. 1998, Bacterial streak and bulb rot of sweet onion II: Epiphytic survival of *Pseudomonas viridiflava* in association with multiple weed hosts. *Plant Dis.* **82**: 935-938.

Gitaitis, R.D., Sanders, F.H., and Walcott, R.R., 1998, Polymerase chain reaction and ELISA for the identification of *Burkholderia cepacia* and *Pseudomonas viridiflava*, bacterial pathogens of onion. pp 3.3.33 in: Offered Papers Abstracts vol. 3 (Themes 3,4,5 and Miscellaneous) 7th International Congress of Plant Pathology, Edinburgh, Scotland 9-16, August 1998.

Gitaitis, R., Sumner, D., Gay, D., Smittle, D., MacDonald, G., Maw, B., Johnson, W.C., III, Tollner, B., and Hung, Y., 1997, Bacterial streak and bulb rot of onion I. A diagnostic medium for the semiselective isolation and enumeration of *Pseudomonas viridiflava*. *Plant Dis.* **81**: 897-900.

Jones, J.B., Gitaitis, R.D., and McCarter, S.M., 1986, Fluorescence on single-carbon sources for separation of *Pseudomonas syringae* pv. *syringae*, *P. syringae* pv. *tomato* and *P. viridiflava* on tomato transplants. *Plant Dis.* **70**: 151-153.

King, E.O., Ward, M.K., and Raney, D.E., 1954, Two simple media for the demonstration of pyocyanin and fluorescin. *J. Lab. Clin. Med.* **44**: 301-307.

Control of Olive Knot Disease with a Bacteriocin

P. LAVERMICOCCA, F. VALERIO, S.L. LONIGRO, S. LAZZARONI[#], A. EVIDENTE[#] and A. VISCONTI

Istituto di Scienze delle Produzioni Alimentari (ISPA), C.N.R. Via L. Einaudi 51, 70125 Bari, Italy; [#]Dipartimento di Scienze del Suolo, della Pianta e dell'Ambiente, Università di Napoli 'Federico II', Portici (Napoli), Italy.

Abstract: A bacteriocin produced by *Pseudomonas syringae* pv. *ciccaronei*, used at three purification levels and several concentrations, inhibited the multiplication in olive tissues of *P. s.* subsp. *savastanoi*, the causal agent of olive knot disease and affected its epiphytic survival on leaves and twigs of olive plants. We determined that 6,000 Active Units ml^{-1} of crude bacteriocin, obtained by a simple purification step, is a concentration suitable to obtain an effective reduction of olive knot formation. Moreover, treatments of olive phylloplane with 6,000 AU ml^{-1} of crude bacteriocin were able to prevent bacterial multiplication by controlling the epiphytic survival of the pathogen. The inhibitory effect lasted until the end of observation (30 days). *P. s.* pv. *ciccaronei* bacteriocin, purified by ammonium sulphate precipitation, membrane ultrafiltration, gel filtration and gel electrophoresis when analysed by SDS-PAGE revealed three protein bands with molecular weights of 76, 63 and 45 kDa, respectively.

Key words: Biological control, olive, bacteriocin, *Pseudomonas syringae* pv. *ciccaronei*.

Author for correspondence: P. Lavermicocca; e-mail: paola.lavermicocca@ispa.cnr.it.

1. INTRODUCTION

Bacteriocins are antibacterial proteins produced by bacteria that kill or inhibit the growth of other bacteria. Many phytopathogenic bacteria belonging to the genera *Pseudomonas*, *Xanthomonas*, *Erwinia*, *Agrobacterium* produce bacteriocins, but only a few were characterised (Chuang *et al.* 1999; Roberts *et al.* 1977). Among plant pathogenic

N.S. Iacobellis *et. al.* (eds.), *Pseudomonas syringae* and Related Pathogens, 451-457.

pseudomonads, about 10 species or subspecies are reported to produce bacteriocins and syringacins 4-a and W-1 were identified and purified from two strains of *Pseudomonas syringae* pv. *syringae* isolated from corn and bean, respectively (Haag and Vidaver, 1974; Smidt and Vidaver, 1986). Applicative studies on disease control using bacteriocin-producing bacteria have been reported by Sakthivel and Mew (1991). They observed that treatment with nonpathogenic bacteriocin-producing strains of *Xanthomonas campestris* pv. *oryzae* reduces the incidence and severity of bacterial leaf streak on rice plants.

Olive knot disease, caused by *P. s.* subsp. *savastanoi*, is characterised by overgrowths (knots) and/or hyperplastic cankers on several hosts (olive, oleander, privet, jasmine, ash). The disease can cause severe damage in most olive-growing regions and *P. s.* subsp. *savastanoi* is included in the list of transmissible agents of olive diseases thus its absence in propagating material is advisable for the certification of olive mother plants in Italy (Martelli *et al.*, 1995). The use of copper compounds could be one of the conventional practices employed to reduce symptoms and to control epiphytic survival of the pathogen. The diffuse resistance to copper bactericides among pathovars of *P. syringae* prompts the development of alternative methods for the control of olive knot disease. Recently, Lavermicocca *et al.* (1999; 2002) found that *P. s.* pv. *ciccaronei* and a bacteriocin produced in culture strongly and selectively inhibits the *P. s.* subsp. *savastanoi* strains isolated from several host plants. The objectives of this research were: (i) to estimate the effectiveness of *P. s.* pv. *ciccaronei* bacteriocin preparations and concentrations for reducing symptom expression in olive plants inoculated with *P. s.* subsp. *savastanoi* strains, (ii) to determine the efficacy of bacteriocin for reducing epiphytic populations of the pathogen, and (iii) to define purification steps that allow to obtain relevant amounts of bacteriocin preparations.

2. MATERIALS AND METHODS

2.1 Bacteriocin Purification

Bacteriocin production and purification were obtained as reported in Lavermicocca *et al.* (1999). Bacteriocin preparations used in this study were: culture supernatant (cs) of strain NCPPB 2355 of *Pseudomonas syringae* pv. *ciccaronei* grown for 4 days in liquid medium; crude bacteriocin (cb) (dialysed ammonium sulphate precipitate) and fraction IV (fIV) (the most active fraction group recovered from Sephadex G150 column). Fraction IV was analysed by SDS PAGE (Agilent 2100 Bio-analyser, Agilent

Technologies GmbH, Waldbronn, Germany), fractionated by preparative native gel and active fraction recovered and analysed by SDS PAGE. Growth inhibition of the pathogenic strain was expressed in active units (AU ml^{-1}). An active unit is the amount of bacteriocin in an end-point dilution which completely inhibits the growth of the test strain in the area of application of a 10 μl droplet preparation.

2.2 Wounds Inoculations

Leaf scars and V-shaped slits made in the bark of two-year old olive stems cv. Nocellara del Belice were inoculated with 10 μl of bacterial suspension containing 10^5 or 10^8 CFU ml^{-1} of *Pseudomonas syringae* subsp. *savastanoi* strains PVBa229 or PVBa304 (= ITM 304). Three days after inoculation, 10 μl of sterile distilled water (SDW), bacteriocin preparations at different concentrations (2,000, 6,000 and 12,000 AU ml^{-1}) or Kocide solution (0.3 %) were added to the wounds. Experiments were performed three times in duplicate on two different plants. Plants were observed for symptom development and after 30 days the overgrowths were excised from stems and weighted.

2.3 Effect of Epiphitic Populations

A bacterial suspension containing 10^8 CFU ml^{-1} of strain PVBa229 or PVBa304 of *Pseudomonas syringae* subsp. *savastanoi* and 0.03% of a wetting agent (Adiplant N, Bayer) was sprayed on twigs and leaves of olive plants. One day after, inoculated phylloplane was treated with SDW, 6,000 AU ml^{-1} crude bacteriocin or 0.3% Kocide solution. Fifteen leaves and 20 cm of twigs were randomly collected at each sampling (after 0, 2, 9, 15, 30 days) and added to 0.85% NaCl solution; the resulting suspensions obtained by shaking samples (200 rpm) were plated on KB agar. The experiment was carried out in duplicate on two different plants and was performed three times.

3. RESULTS AND DISCUSSION

3.1 Bacteriocin Purification

Bacteriocin was purified by ammonium sulphate precipitation of the culture supernatant, membrane ultrafiltration, gel filtration and gel electrophoresis. The above purification procedure gave a recovery of antimicrobial activity

of at least 62% at each step. The molecular mass of the bacteriocin was estimated approximately 100 kDa (Lavermicocca *et al.*, 1999).
An electropherogram in denaturing conditions of fraction IV, obtained by gel filtration, is shown in Fig. 1. Analysis indicates the presence of 16 main peaks relevant to at least 16 proteins with sizes ranging from 10.3 to 86.8 kDa. When fraction IV was analysed by PAGE in non denaturing conditions, eight protein bands were obtained. Only a single band was related to the antimicrobial activity. The active band was recovered from the gel and revealed three bands on SDS-PAGE with molecular masses of 76, 63 and 45 kDa, respectively (Lavermicocca *et al.*, 1999). Since in denaturing conditions no bands retained antibacterial activity, it remains to be ascertained if the three bands are sub-units of a single active protein.

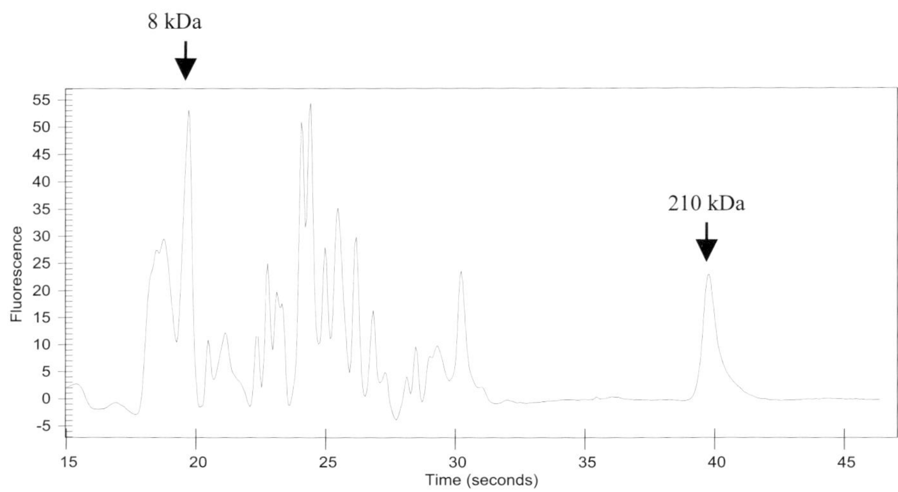

Figure 1. Electropherogram of the separation of fraction IV on the Protein 200 LabChip. Arrows indicate internal standards.

3.2 Wound Inoculations

Bacteriocin preparations reduced symptom expression in olive stems inoculated with the pathogen. Thirty days after inoculation of strains PVBa229 and PVBa304 of *Pseudomonas syringae* subsp. *savastanoi,* typical knots appeared at the inoculation sites. The knot weights ranged from about 40 to 80 mg at V-shaped slits inoculated wounds and from 20 to 50 mg at leaf scars ones. When bacteriocin preparations were added onto the inoculated wounds, the formation of knots with reduced size by 81 and 61% for strain PVBa229 and by 65 and 51% for strain PVBa304 was observed (Fig. 2). In Table 1 knot reduction obtained at inoculated sites treated with bacteriocin preparations with respect to the untreated wounds is reported.

For both strains, 6,000 AU ml^{-1} of all bacteriocin preparations determined a weight reduction of more than 50% with the exception of knots caused by both strains on leaf scars treated with culture supernatant (43%). Treatment with Kocide101 (0.3%) always resulted in almost complete inhibition of knot formation (Table 1).

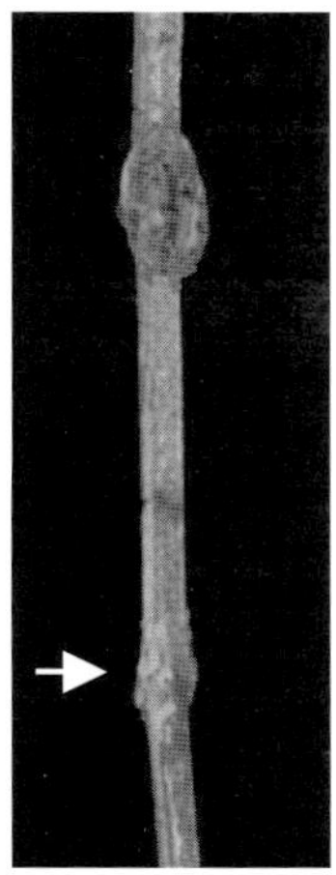

Figure 2. Knot formation at leaf scar 30 days after inoculation with *Pseudomonas syringae* subsp. *savastanoi* strain PVBa304. Arrow indicates the inoculation site treated with crude bacteriocin solution containing 6,000 AU ml^{-1}.

Table 1. Effects of *Pseudomonas syringae* pv. *ciccaronei* NCPPB2355 bacteriocin preparations (culture supernatant, cs; crude bacteriocin, cb and fraction IV, fIV), Kocide 101, k101 and sterile distilled water, sdw on olive plants 30 days after inoculation with *P. s.* subsp. *savastanoi* strains PVBa229 and 304.

Treatments		*Percentage of knot reduction*			
		V-shaped slits		*Leaf scars*	
		PVBa229	*PVBa304*	*PVBa229*	*PVBa304*
sdw		0	0	0	0
cs	2,000 AU ml^{-1}	9.3 ± 0.6	10 ± 1.1	29.8 ± 4.6	20 ± 4.5
	6,000 AU ml^{-1}	63.7 ± 2.9	46 ± 3,4	43.3 ± 1.2	43.4 ± 6.8
	12,000 AU ml^{-1}	79 ± 0.5	54 ± 6.5	N.D.	N.D.
cb	2,000 AU ml^{-1}	10 ± 2.1	10.6 ± 4.7	52,4 ± 8.7	31.7 ± 1.6
	6,000 AU ml^{-1}	80.8 ± 4.9	50.7 ± 6.1	60.8 ± 3.8	64.6 ±3.4
	12,000 AU ml^{-1}	75.2±3	72.6 ± 3.3	N.D.	N.D.
f IV	2,000 AU ml^{-1}	15.6 ± 2.3	29.3 ± 5.8	59.6 ± 3.9	54 ± 1.6
	6,000 AU ml^{-1}	64.3 ± 1.6	52.5 ± 5.3	79.9 ± 3.8	66.7 ± 2.7
	12,000 AU ml^{-1}	84.9 ± 2.1	56 ± 3.9	N.D.	N.D.
k 101		98.9 ± 0	96.9 ± 0	100 ± 0	100 ± 0

Standard errors were calculated at a confidence level of 95% (n=18).
N.D. = not determined.

3.3 Epiphitic Survival

Both *Pseudomonas syringae* subsp. *savastanoi* strains sprayed on olive phylloplane and treated with SDW survived and multiplied until the end of observations on twigs and on leaves, reaching a population density of about 5 x 10^4 CFU cm^{-2} on twigs and 6 x 10^3 CFU cm^{-2} on leaves (PVBa229) and 5 x 10^3 CFU cm^{-2} on twigs and 4.5 x 10^4 CFU cm^{-2} on leaves for the strain PVBa304 (not shown). Treatment with 6,000 AU ml^{-1} of crude bacteriocin or Kocide 101 reduced or eliminated the presence of bacteria (Fig. 3). For both strains, already at inoculation time, a high rate of bacterial growth inhibition was observed either on twigs or leaves. After 15 and 30 days the growth inhibition was almost complete. Treatment with Kocide 101 solution caused the death of epiphytic bacteria.

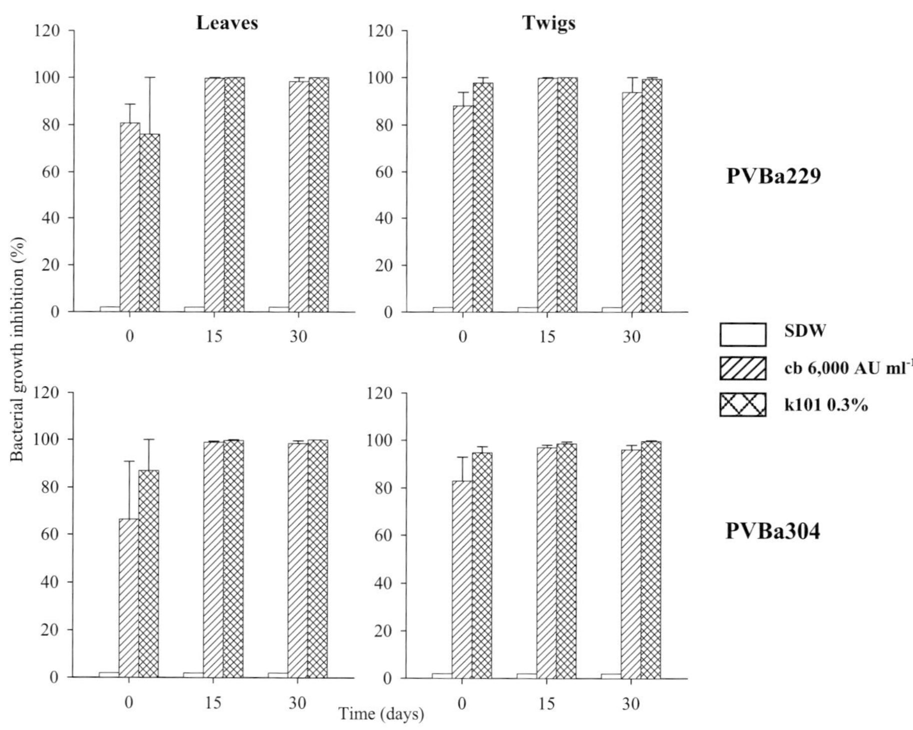

Figure 3. Epiphytic survival of *Pseudomonas syringae* subsp. *savastanoi* strains PVBa229 and PVBa304 on leaves and twigs of olive plants treated with *P. s.* pv. *ciccaronei* NCPPB2355 crude bacteriocin (cb), Kocide 101 (k101) and sterile distilled water (SDW). Bars indicate standard errors.

4. CONCLUSION

The results obtained in this study led us to conclude that crude bacteriocin preparations containing 6,000 AU ml^{-1} are suitable to obtain an effective inhibition of olive knot formation. Furthermore, the above bacteriocin preparation was shown to be efficient in controlling the epiphytic populations of the pathogen. Moreover, purification procedure allowed to obtain by a simple step (ammonium sulphate precipitation and dialysis) 7,000,000 AU of crude bacteriocin per litre of *Pseudomonas syringae* pv. *ciccaronei* culture. Since this bacteriocin is highly specific and can be easily produced it appears a valid alternative control system for olive knot disease.

ACKNOWLEDGEMENTS

This work was supported by the Italian Ministry of University and Scientific and Technological Research (MURST), Development of Research Networks n°488/92, Cluster C06+07, Project 1.1 "Micro-organisms and Microbial Metabolites for Plant Protection".

REFERENCES

Chuang, D.Y., Kyeremeh, A.G., Gunji, Y., Takahara, Y., Ehara, Y., and Kikumoto, T., 1999, Identification and cloning of an *Erwinia carotovora* subsp. carotovora bacteriocin regulator gene by insertional mutagenesis. *J. Bacteriol.* **181**: 1953–1957.

Haag, W.L., and Vidaver, A.K., 1974, Purification and characterisation of syringacin 4-A, a bacteriocin from *Pseudomonas syringae* 4-A. *Antimicrob. Agents and Chemother.* **6**: 76-83.

Lavermicocca, P., Lonigro, S.L., Evidente, A., and Andolfi, A., 1999, Bacteriocin production by *Pseudomonas syringae* pv. *ciccaronei* NCPPB2355. Isolation and partial characterization of the antimicrobial compound. *J. Appl. Microbiol.* **86**: 257-265.

Lavermicocca, P.,. Lonigro, S.L, Valerio, F., Evidente, A. and Visconti, A., 2002, Reduction of olive knot disease by a bacteriocin from *Pseudomonas syringae* pv. *ciccaronei*. *Appl. Environ. Microbiol.* **68**: 1403-1407.

Martelli, G.P., Savino, V., Di Terlizzi, B., Catalano, L., Sabanadzovic, S., and Greco, N., 1995, Viruses and certification of olive in Apulia (Southern Italy). *Acta Hortic.* **386**: 569-573.

Roberts, W.P., Tate, M.E., and Kerr, A., 1977, Agrocin 84 is a 6-*N*-phosphoramidate of an adenine nucleotide analogue. *Nature* **265**: 379-381.

Sakthivel, N., and Mew, T.W., 1991, Efficacy of bacteriocinogenic strains of *Xanthomonas oryzae* pv. *oryzae* on the incidence of bacterial blight disease of rice (*Oryza sativa* L.). *Can. J. Microbiol.* **37**: 764-768.

Smidt, M.L., and Vidaver, A.K., 1986, Isolation and characterisation of syringacin W-1, a bacteriocin produced by *Pseudomonas syringae* pv. *syringae*. *Can. J. Microbiol.* **32**: 231-236.

Use of Oxos, a Complex of Hydrogen Peroxide, Acetic Acid and Silver Ion, to Control Bacterial Speck of Tomato (*Pseudomonas syringae* pv. *tomato*) and Angular Leaf Spot of Melon (*P. s.* pv. *lachrymans*)

M. FATMI
Institut Agronomique et Vétérinaire Hassan II, Complexe Horticole d'Agadir, BP. 18/S, Agadir; Morocco.

Abstract: This paper reports the efficiency of a complex of hydrogen peroxide, acetic acid, and silver ion, called hereafter Oxos, to control bacterial speck of tomato and angular leaf spot of melon. In the *in vitro* trials, Oxos was added at different concentrations to bacterial suspensions containing about 10^8 to 10^9 CFU ml^{-1} of either *Pseudomonas syringae* pv. *tomato* or *P. s.* pv. *lachrymans.* The bacterial population density, monitored over time, showed a significant decrease for all the concentrations tested. The death of both pathogens, on the contrary of the control which remained almost unchanged after two hours, occurred after 15 min in solutions containing 0.5% Oxos. Similar results were obtained when tomato and melon plants were sprayed with 10^8 CFU ml^{-1} suspension of each pathogen followed, thirty minutes later, by treatments with Oxos solutions. Oxos at 0.25% induced the total death of *P. s.* pv. *tomato* and *P. s.* pv. *lachrymans* after 1 and 24 hours, respectively. No significant decrease of the two bacterial populations was detected after 24 hours in the control. In addition, trials under greenhouse conditions on inoculated and diseased tomato and melon plants showed that Oxos has both preventive and curative effects. Moreover, no phytotoxicity was observed even at higher concentrations of Oxos (1%). This work showed that Oxos has a good potential to control bacterial speck of tomato and angular leaf spot of melon and could be an important alternative to standard compounds currently used.

Key words: chemical control, curative agents, preventive agents, phytopathogenic bacteria.

Author for correspondence: M. Fatmi; e-mail: fatmi@iavcha.ac.ma.

N.S. Iacobellis *et. al.* (eds.), *Pseudomonas syringae* and Related Pathogens, 459-466.

1. INTRODUCTION

Bacterial angular leaf spot is a serious disease on several species of the *Cucurbitaceous* family including cucumber, melon, watermelon, zucchini, squash, pumpkin, cantaloupe (Van Gundy and Walker, 1957; Hopkins and Schenk, 1972; Scortichini and Tropiano, 1991; El-Sadek *et al.* 1992). The causal agent, *Pseudomonas syringae* pv. *lachrymans* has a world-wide distribution and is responsible for several outbreaks resulting in severe losses of the crop production (Bradbury, 1986). The bacterium induces necrotic spots on leaves, stems, and fruits. The pathogen is transmitted through both contaminated and infected seeds (Leben, 1981) and persists in soil within the plant debris (Kritzman and Zutra, 1983).

Bacterial speck of tomato, caused by *P. s.* pv. *tomato*, is one of the most serious bacterial disease problems encountered in the tomato growing regions where favourable conditions are met (Bonn, 1980; Goode and Sasser, 1980; Cuppels and Elmhirst, 1999). In addition to its effect on yield decrease through foliar necrosis, the pathogen affects fruit quality. In addition to being a seedborne pathogen, *P. s.* pv. *tomato* can survive as an epiphyte on tomato plants for extended periods of time (Smitley and McCarter, 1982; McCarter *et al.*, 1983).

The control of these diseases is mainly based on the use of pathogen-free seed and cupric treatments in the field. However, seed sanitation depends on the technique used as well as on the location of the bacterium on or in the seed. Cupric based compounds are bacteriostatic and should be used regularly which will increase the production costs. In addition, appearance of copper resistant bacterial strains resulted in limitation of the use of cupric treatments (Alexander and Graves, 2001). Thus, other alternatives are needed to control these diseases. This report describes for the first time, the use of a complex of hydrogen peroxide, acetic acid, and silver ion, named Oxos, to control foliar bacterial diseases and particularly bacterial speck of tomato and angular leaf spot of melon.

2. MATERIALS AND METHODS

2.1 *In Vitro* Evaluation of the Efficiency of Oxos on the Survival of *Pseudomonas syringae* pv. *lachrymans* and *P. s.* pv. *tomato*

Bacterial strains of *Pseudomonas syringae* pv. *tomato* and *P. s.* pv. *lachrymans* were obtained respectively from naturally infected tomato and melon plants in the Souss-Massa valley in Agadir, Morocco. The obtained strains were identified using morphological, chemical, and pathogenicity tests (Schaad *et al.*, 2001). For inoculum preparation, log phase culture was used and bacterial suspensions were made in sterile phosphate buffer (0.05 M pH 7.2) (SPB) and adjusted using a spectrophotometer to an optical density of 0.1 at 640 nm, corresponding to about 10^9 CFU ml^{-1}. Then 100 ml from each bacterial suspension was added separately to 15 separate 250 ml sterile bottles. The bottles were then divided randomly into five sets of 3 bottles each.

Oxos was used at four different concentrations (v v^{-1}) (0.1%, 0.25%, 0.5% and 1.0%) at one concentration per set and per bacterial species. Bottles without Oxos were used as control. The bottles, a total of 15 per bacterial species, were then shaken at 100 rpm at 22 °C for 2 hours.

To monitor the survival of the bacteria in each of the solutions, samples of 5 ml from each bottle were first taken just prior to adding Oxos and 15, 30, 60, and 120 min after adding Oxos. Ten-fold serial dilutions were made in SPB and triplicate 0.1 ml quantities from undiluted, 10^{-2}, and 10^{-4} dilutions were plated onto three Petri plates containing King B medium (Schaad *et al.*, 2001). The plates were then incubated at 22 °C and checked daily and the number of typical colonies per plate was determined.

2.2 Evaluation of the Effects of Oxos on the Survival of *Pseudomonas syringae* pv. *tomato* and *P. s.* pv. *lachrymans* on Tomato and Melon Leaves

Tomato (cv. Daniela) and melon (cv. Alma) plants at the second true leaf stage were first sprayed with a 10^8 CFU ml^{-1} bacterial suspension of either *Pseudomonas syringae* pv. *tomato* or *P. s.* pv. *lachrymans*. Thirty minutes later, the initial bacterial populations on both tomato and melon leaves were determined. Then the plants were divided into three sets of 4 plants each. Two sets were treated separately by spraying the leaves with two different concentrations of Oxos (0.25 and 0.50 %). The last group of plants was treated with sterile distilled water (SDW) as a control. The plants were kept under greenhouse conditions. The evolution of the bacterial population was determined 1 and 24 hours after application of Oxos. To monitor the bacterial population on the leaves, three discs of about 1 cm^2 were taken randomly from leaves of each plant and ground in 3 ml of SPB using sterile mortars and pestles. Ten-fold serial dilutions were made in SPB and triplicate 0.1 ml quantities from undiluted, 10^{-2}, and 10^{-4} dilutions were plated onto three Petri plates containing King B medium (Schaad *et al.*, 2001). The plates were then incubated at 22 °C and checked daily and the number of typical colonies per plate was determined.

2.3 Preventive and Curative Effects of Oxos on Development of Bacterial Speck of Tomato and Angular Leaf Spot of Melon

2.3.1 Preventive Effect

Tomato and melon plants were grown under greenhouse conditions. At the second true leaf stage, the plants were sprayed with a 10^8 CFU ml^{-1}bacterial suspension of either *Pseudomonas syringae* pv. *tomato* or *P. s.* pv. *lachrymans*. One hour after the inoculation, the plants were divided randomly into three lots of 4 plants each. Then two lots were treated by spraying separately the leaves with a solution containing 0.25 or 0.50% of Oxos. The remaining lot was sprayed with SDW as control. The treated plants were kept under greenhouse conditions with high humidity and a temperature of about 22°C and observed regularly for symptoms.

2.3.2 Curative Effect

In this case, the plants were first inoculated as above with either *Pseudomonas syringae* pv. *tomato* or *P. s.* pv. *lachrymans*. Then the inoculated plants were kept under greenhouse conditions and observed regularly. Once the typical symptoms of bacterial speck of tomato and angular leaf spots of melon started to appear, the plants were divided randomly into three lots of four plants each. Two lots were treated separately with a solution containing 0.25 and 0.50% of Oxos. The remaining lot was treated with SDW as a control. Then the treated plants were kept under greenhouse conditions and observed regularly for symptoms.

3. RESULTS AND DISCUSSION

3.1 *In vitro* Evaluation of the Effects of Oxos on the survival of *Pseudomonas syringae* pv. *tomato* and *P. s.* pv. *lachrymans*

The results of the effect, *in vitro*, of Oxos on the survival of the bacterial populations of *Pseudomonas syringae* pv. *tomato* and *P. s.* pv. *lachrymans* are presented in Fig.1. The data showed that the bacterial populations decreased significantly over time as compared to the control. This decrease was more important with higher concentrations of Oxos. In the case of *P. s.*

pv. *tomato*, the bacterial populations recorded after 60 min were 7.3 x 10^4 and 3.5 x 10^2 respectively for concentrations of Oxos of 0.1 and 0.25%. Concentrations of 0.5 and 1.0% induced total death of this bacterium respectively after 60 and 15 min, respectively (Fig. 1A). Similar results were obtained with *P. s.* pv. *lachrymans.* The bacterial populations recorded after 60 min were 1.4 x 10^4 and 2.0 x 10 respectively for Oxos concentrations of 0.1 and 0.25%. The concentrations 0.5 and 1.0% induced total death of this bacterium after 15 min. In contrast, in the case of the control, the bacterial density remained almost unchanged even after 120 min (Fig. 1B).

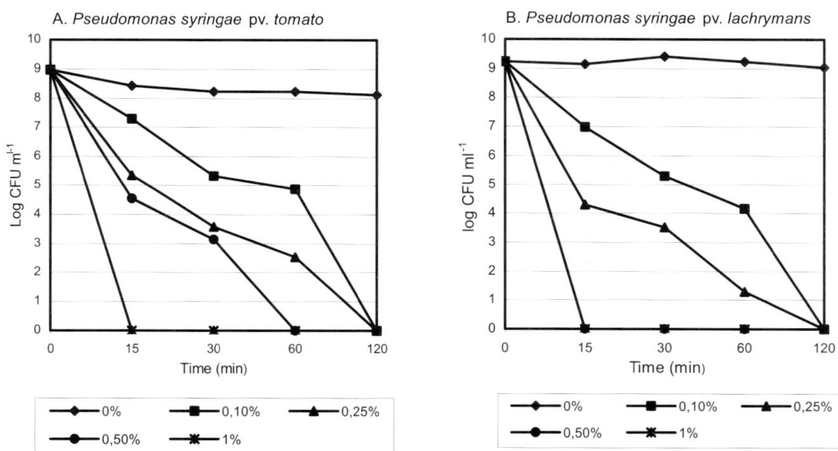

Figure 1. In vitro effect of different concentrations of Oxos on the survival of *Pseudomonas syringae* pv. *tomato* (A) and *P. s.* pv. *lachrymans* (B).

3.2 *In Vivo* Effects of Oxos on the Survival of *Pseudomonas syringae* pv. *tomato* and *P. s.* pv. *lachrymans* on Tomato and Melon Leaves

The results of the effects of Oxos on the survival of *Pseudomonas syringae* pv. *tomato* and *P. s.* pv. *lachrymans* on tomato and melon leaves, respectively, are presented in Table 1. Oxos at 0.25% induced total death of *P. s.* pv. *tomato* and *P. s.* pv. *lachrymans* after 1 and 24 hours, respectively. No significant decrease of the bacterial densities were detected after 24 hours in the case of the control (Table 1).

Table 1. Effect of different concentrations of Oxos on the survival of *Pseudomonas syringae* pv. *tomato* and *P. s.* pv. *lachrymans*, respectively, on tomato and melon leaves.

Bacterial species	Time (hours)	Average bacterial population (in CFU cm^{-2}) on tomato or melon leaves		
		0%	0.25%	0.50%
Pseudomonas syringae pv. *tomato*	0*	6.9×10^7	6.2×10^7	6.0×10^7
	1	5.2×10^7	0	0
	24	3.6×10^7	0	0
P. s. pv. *lachrymans*	0*	3.6×10^7	2.0×10^7	1.7×10^7
	1	1.1×10^6	4.7×10^5	1.7×10^3
	24	1.1×10^4	0	0

*: Time zero is 30 min after spraying the bacterial suspension and just before spraying Oxos.

3.3 Preventive and Curative Effects of Oxos on Development of Bacterial Speck of Tomato and Angular Leaf Spot of Melon

The results of trials under greenhouse conditions on development and evolution of bacterial speck of tomato and angular leaf spot of melon showed that Oxos has both preventive and curative effects. The application of Oxos at 0.25% and 0.50% on diseased tomato and melon plants stopped disease evolution. While in the control, the diseases continued their progression. In addition the application of Oxos, at the same concentrations as above, on 1 hour-inoculated tomato and melon plants resulted in no disease and the plants remained healthy. On the contrary, in the case of the control, more than 80% of the leaves showed typical symptoms.
Moreover, no phytotoxicity was observed even at higher concentrations of Oxos (1%).

4. CONCLUSION

Bacterial speck of tomato and angular leaf spot of melon constitute a threat to tomato and melon production during cool and wet seasons. Several treatments using cupric compounds, antibiotics and other products have been used, but with unsatisfactory results (Alexander and Graves, 2001). The results obtained in this study showed that Oxos (a complex of hydrogen peroxide, acetic acid and silver ion) has a significant efficacy under both *in vitro* and *in vivo* conditions. Under laboratory conditions Oxos at 0.5% induced total death of both *Pseudomonas syringae* pv. *tomato* and *P. s.* pv. *lachrymans* after an exposure of 15 min. The application of Oxos at 0.25%

on inoculated leaves resulted in the cleaning of the pathogens from the leaves. In addition, the treatment of inoculated and diseased plants with Oxos at 0.25% stopped disease development and evolution. Furthermore, no phytotoxicity was observed even at higher concentrations of Oxos (1%). These findings show that Oxos has a bactericide effect and can provide significant disease control compared to the standard compounds currently used to control bacterial speck of tomato and angular leaf spot of melon. It may also have an important potential to control other bacterial blights of other crops.

ACKNOWLEDGEMENTS

I thank CELTIS France for supplying the compound Oxos and for financial support. I also thank Bougsiba M. for his valuable technical assistance and Prof. M.C. Harrouni for reviewing this manuscript.

REFERENCES

Alexander, S.A., and Graves, A.S., 2001, Controlling bacterial speck with acibenzolar-S-methyl in staked tomatoes. *Phytopathology* **91**: S2. Publication no. P-2001-0010-AMA.

Bonn, W.G., 1980, Incidence and severity of bacterial speck of tomato in southwestern Ontario in 1979. *Plant Dis.* **64**: 586-587.

Bradbury, J.F., 1986, Guide to plant pathogenic bacteria. C.A.B. International Mycological Institute, Ferry Lane, Kew, England.

Cuppels, D.A., and Elmhirst, J., 1999, Disease development and changes in the natural *Pseudomonas syringae* pv. *tomato* populations on field tomato plants. *Plant Dis.* **83**: 759-764.

El-Sadek S.A.M., Abdel-Latif, M.R., Abdel-Gawad, T.I., and Hussein, N.A., 1992, Occurrence of angular leaf spot disease in greenhouse cucumbers in Egypt. *Egypt. J. Microbiol.* **27** (2): 157-175.

Goode, M.J., and Sasser, M., 1980, Prevention - the key to controlling bacterial spot and bacterial speck of tomato. *Plant Dis.* **64**: 831-834.

Hopkins, D.L., and Schenk, N.C., 1972, Bacterial leaf spot of watermelon caused by *Pseudomonas syringae* pv *lachrymans*. *Phytopathology* **62**: 542-545.

Kritzman, G., and Zutra, D., 1983, Survival of *Pseudomonas syringae* pv. *lachrymans* in soil, plant debris, and the rhizosphere of the non-host plants. *Phytoparasitica* **11**: 99-108.

Leben, C., 1981, Survival of *Pseudomonas syringae* pv. *lachrymans* with cucumber seed. *Can. J. Plant Pathol.* **3**: 247-249.

McCarter, S.M., Jones, J.B., Gitaitis, R.D., and Smitley, D.R., 1983, Survival of *Pseudomonas syringae* pv. *tomato* in association with tomato seed, soil, host tissue, and epiphytic weed hosts in Georgia. *Phytopathology* **73**: 1393-1398.

Schaad, N.W., Jones, J.B., and Chun, W., 2001. Laboratory Guide for Identification of Plant Pathogenic Bacteria, 3rd Edition. APS press.

Scortichini, M., and Tropiano, F.G., 1991, Occurrence of *Pseudomonas syringae* pv *lachrymans* in black zucchini squash. *Petria* **1**: 177-182.

Smitley, D.R., and Mc Carter, S.M., 1982, Spread of *Pseudomonas syringae* pv. *tomato* and role of epiphytic populations and environmental condition in disease development. *Plant Dis.* **66**: 713-717.

Van Gundy, S.D., and Walker, J.C., 1957, Seed transmission, overwintering, and hostrange of the cucurbit angular leaf spot pathogen. *Plant Dis. Re.* **41**: 137-140.

Screening Wild Cherry Micropropagated Plantlets for Resistance to Bacterial Canker

J.G. VICENTE and S.J. ROBERTS
Horticulture Research International, Wellesbourne, Warwick CV35 9EF, UK.

Abstract: Bacterial canker of cherry (*Prunus avium* L.) caused by *Pseudomonas syringae* pv. *syringae* and *P. s.* pv. *morsprunorum* is a serious disease in wild cherry plantations and in sweet cherry orchards. A screening method using micropropagated plantlets of cherry was developed. Wounding the plantlets before inoculation and freezing after inoculation were not necessary for symptom development. Leaf spots were only observed in plantlets inoculated by dipping in suspensions of *P. s.* pv. *morsprunorum* isolates; this method was selected for further screening experiments. Seventeen clones of wild cherry and two clones of sweet cherry were screened for resistance with seven *P. s.* pv. *morsprunorum* and four *P. s.* pv. *syringae* strains. Some clones showed partial levels of resistance to *P. s.* pv. *morsprunorum* and to a *P. s.* pv. *syringae* strains, but all clones were highly susceptible to three *P. s.* pv. *syringae* strains.

Key words: *Pseudomonas syringae*, *Prunus avium*, micropropagation, pathogenicity.

Author for correspondence: J.G. Vicente; e-mail: joana.vicente@hri.ac.uk.

1. INTRODUCTION

Bacterial canker of cherry (*Prunus avium* L.) caused by *Pseudomonas syringae* is a major limitation in the use of cherry for timber production in farm woodlands in the UK (Nicoll, 1993). This disease is also one of the major diseases associated with sweet cherry production and causes significant losses in the Hardy Nursery Stock industry. The search for disease resistant genotypes has been limited by lack of understanding of the pathogens and by lack of a reliable and consistent inoculation method.

N.S. Iacobellis *et. al.* (eds.), *Pseudomonas syringae* and Related Pathogens, 467-474.

Two pathovars of *P. syringae* can cause bacterial canker in cherry: *P. s.* pv. *syringae* van Hall and *P. s.* pv. *morsprunorum* (Wormald) Young, Dye and Wilkie. In the UK, previous studies have been mainly concerned with sweet cherry and bacterial canker was traditionally considered to be mainly caused by *P. s.* pv. *morsprunorum* (Wormald, 1937; Crosse, 1955; Garrett *et al.*, 1966; Burkowicz and Rudolph, 1994). Isolates obtained from recent surveys of wild cherry woodland plantations and nurseries were predominantly *P. s.* pv. *syringae*, but strains of *P. s.* pv. *morsprunorum* races 1 and 2, and intermediate strains were also found (Vicente and Roberts, unpublished). Resistance to bacterial canker has been assessed in field conditions, over several years, following natural infection and leaf scar or branch wound inoculations (Crosse and Garrett, 1966; Garrett, 1986); resistance has also been assessed in controlled conditions using a detached-twig assay (Garrett and Fletcher, 1989; Krzesinska and Azarenko, 1992). These methods are dependent on the number of branches/trees available, are time-consuming and the results obtained are often variable and inconsistent. Micropropagated plantlets of pear (*Pyrus communis* L.) and lilac (*Syringa vulgaris* L.) have been used successfully to evaluate the pathogenicity of *P. s.* pv. *syringae* strains (Yessad *et al.*, 1992; Scheck *et al.*, 1997). The objectives of this work were to develop a rapid screening method using micropropagated plantlets of cherry and to use this method to screen a number of wild cherry clones for resistance.

2. MATERIALS AND METHODS

2.1 Tissue Culture

Clones of 17 wild cherry accessions were obtained from the commercial micropropagation unit at Horticulture Research International (HRI)-Efford. Cultures were maintained at 24.6±2°C with a 16 h photoperiod. For long-term maintenance, the cultures were kept in an incubator at 4°C with 16 h photoperiod. In addition, micropropagated cultures of the sweet cherry cultivars Napoleon (universal susceptible) and Roundel (reported to be resistant to race 1 of *Pseudomonas syringae* pv. *morsprunorum*) were established. Plant cultures were maintained with periodic sub-culturing in 71 mm diameter transparent plastic pots (Insulpak C71-141). Cherry cultures were maintained in a shoot culture medium based on that of Hammatt and Grant (1993) consisting of Murashige and Skoog (1962) salts supplemented with 87.6 mM sucrose, 2.2 μM benzyladenine, 0.49 μM indole-3-butyric acid, 1.28 mM phloroglucinol, 6 g of agar, pH 5.65. Lilac cv. Sensation cultures were maintained in a lilac tissue culture medium based on that of

Scheck *et al.* (1997) consisting of Murashige and Skoog (1962) salts supplemented with 87.6 mM sucrose, 8.9 μM benzyladenine, 6 g of agar, pH 5.2.

2.2 Development of a Routine Screening Test

Two *Pseudomonas syringae* pv. *morsprunorum* race 1 strains (HRI 5239 and HRI 5270), a *P. s.* pv. *morsprunorum* race 2 strain (HRI 5255) and a *P. s.* pv. *syringae* strain (HRI 5357) were used in this experiment (Table 1). Strains were grown on King's medium B (King *et al.*, 1954) for 24 h. Bacterial growth was scraped from the plates and suspended in approximately 15 ml of sterile tap water to produce a suspension containing 10^8 to 10^9 CFU ml^{-1}. Two to three week-old cultures of five wild cherry clones (FD1-57-4/17, 1900, 1912, F12/1 and Charger) were inoculated using three different methods: 1) dipping in bacterial suspensions without wounding; 2) dip in suspensions after wounding the plantlets by cutting some leaves; 3) stabbing the stem with an entomological pin charged with bacterial growth from a culture plate. These methods were used with and without freezing 24 h after inoculation (pots kept at -5°C for 10 min). Symptoms were recorded weekly until the fourth week. The severity of symptoms was assessed on a six point scale of 0 to 9: 0, no symptoms; 1, small necrotic or chlorotic areas in one or two leaves; 3, less than half of the plantlet with necrotic or chlorotic lesions; 5, half of the plantlet with necrotic or chlorotic lesions; 7, more than half of the plantlet with necrotic stem and leaves; 9, dead plantlet. The presence of leaf spots was also recorded. Plants with a score of 0 to 1 were considered resistant; plantlets with a score of 3 to 5 were considered intermediate and plants with 7 or 9 were considered susceptible. Data in the form of the frequency of plantlets in each response category (0, 1, 3, 5, 7 and 9) were analysed by fitting a multinomial or proportional-hazards (McCullagh and Nelder, 1983) model using the generalised linear modelling facilities of GenStat (Payne *et al.*, 1993).

2.3 Resistance Screening

Micropropagated plantlets of 17 wild cherry clones, two sweet cherry clones and lilac cv. Sensation were tested with a selection of 11 key isolates of *Pseudomonas syringae* pv. *syringae* and *P. s.* pv. *morsprunorum* (Table 1). Two to three week-old cultures of the cherry clones and lilac were inoculated by dipping the plantlets in bacterial suspensions: plantlets were submerged, gently swirled in the suspension and returned to their pots. Plantlets of each clone were dipped in sterile tap water as controls. Each isolate was inoculated into one or two pots containing three plants of each clone. Results were recorded and analysed as described above.

Table 1. *Pseudomonas syringae* pv *syringae* (*Pss*) and *P.s.* pv. *morsprunorum* (*Psm*) isolates used in this study.

Isolate number	Origin (date)	Host	Pathovar, race
5239	UK (1957)	Sweet cherry	*Psm*, race 1
5244	UK (1960)	Sweet cherry	*Psm*, race 1
5270	UK (1990)	Wild cherry	*Psm*, race 1
5833	UK (1994)	Wild cherry	*Psm*, race 1
5255	UK	Sweet cherry cv. Napoleon	*Psm*, race 2
5260	UK	Sweet cherry cv. Roundel	*Psm*, race 2
5271	UK (1990)	Wild cherry	*Psm*, race 2
5264	UK (1990)	Wild cherry	*Pss*
5275	UK (1990)	Wild cherry	*Pss*
5355	France (1978)	Sweet cherry	*Pss*
5357	UK (1979)	Sweet cherry cv. Van	*Pss*

3. RESULTS AND DISCUSSION

3.1 Development of a Screening Method

Statistical analysis for the overall response showed that the bacterial strain was the most important treatment factor; clone, freezing, method of inoculation and the interactions between these factors were not important. Thus freezing was not necessary for symptom development and the clones reacted the same way regardless of the inoculation method. Table 2 presents the predicted probabilities for different response categories (0 to 9) four weeks after inoculation. The wild cherry clones were very susceptible to the *Pseudomonas syringae* pv. *syringae* strain and showed some resistance or partial levels of resistance to the *P. s.* pv. *morsprunorum* strains.

Table 2. Probabilities for response categories 0 to 9 following inoculations of five wild cherry clones with three *Pseudomonas syringae* pv. *morsprunorum* (*Psm*) strains, a *P. s.* pv. *syringae* (*Pss*) strains and sterile tap water (stw) as control using three different methods of inoculation.

Inoculum	Probability for each response category[a]					
	0	1	3	5	7	9
stw control	0.26	<u>0.44</u>	0.26	0.03	0.01	0.01
5239 (*Psm* race 1)	0.10	0.31	<u>0.46</u>	0.08	0.03	0.02
5270 (*Psm* race 1)	0.03	0.13	<u>0.49</u>	0.19	0.10	0.06
5255 (*Psm* race 2)	0.20	<u>0.42</u>	0.32	0.04	0.01	0.01
5357 (*Pss*)	0.00	0.00	0.01	0.02	0.04	<u>0.93</u>

[a] Highest probability is underlined; *Psm* = *Pseudomonas syringae* pv. *morsprunorum*; *Pss* = *P. s.* pv. *syringae;* stw=sterile tap water.

Leaf spots were only observed in cherry plantlets inoculated with *P. s.* pv. *morsprunorum* strains and therefore the data were analysed separately. The analysis of deviance for the three *P. s.* pv. *morsprunorum* strains showed that the method of inoculation was the most important factor: the probability of the presence of leaf spots was very low (less than 10%) in plantlets inoculated by stabbing with a pin compared with plantlets that were dipped in a suspension with or without previous wounding (higher than 65%). Therefore the fastest and simplest method, dipping plantlets in a suspension (without wounding or freezing), was selected for the screening studies.

3.2 Resistance Screening

Statistical analysis showed that the pathovar/race (*Pseudomonas syringae* pv. *morsprunorum* race 1, *P. s.* pv. *morsprunorum* race 2, typical *P. s.* pv. *syringae* and atypical *P. s.* pv. *syringae*) and the control (sterile tap water) were the most important factors followed by clone; the bacterial strain (within a pathovar/race) and the interactions between factors were relatively unimportant. Table 3 presents the predicted probabilities for different response categories (0 to 9) four weeks after inoculation. The results obtained with lilac cv. Sensation clearly differentiated *P. s.* pv. *morsprunorum* (not pathogenic on lilac) and *P. s.* pv. *syringae* (pathogenic on lilac) strains. The *P. s.* pv. *syringae* strain HRI 5264 was less aggressive on lilac than the other three *P. s.* pv. *syringae* strains used. These results are in agreement with those obtained from inoculations of rooted lilac plants (Vicente and Roberts, unpublished results). Scheck *et al.* (1997) also showed that the pathogenicity of *P. s.* pv. *syringae* strains could be evaluated through a rapid bioassay using lilac cv. Sensation plantlets.

The cherry clones showed partial levels of resistance to the *P. s.* pv. *morsprunorum* strains and, in general, were more susceptible to *P. s.* pv. *morsprunorum* race 1 than to *P. s.* pv. *morsprunorum* race 2. The wild cherry clones FD1-57-4/166 and Cobtree showed the highest level of resistance to *P. s.* pv. *morsprunorum* race 1 and race 2 strains respectively. There was no clear difference between the reactions of sweet cherry Napoleon and Roundel when inoculated with strains of the two *P. s.* pv. *morsprunorum* races. These results contrast with previous work that suggested that Roundel was resistant to race 1 and susceptible to race 2 (Freigoun and Crosse, 1975).

All cherry clones were very susceptible to the three typical *P. s.* pv. *syringae* strains, but some clones (e.g. Arger Fen E, Treswell, FD1-57-4/122, FD1-57-4/166), were resistant or partially resistant to the *P. s.* pv. *syringae* strain HRI 5264. These results suggest that HRI 5264 is a different type/race of *P. s.* pv. *syringae*.

Table 3. Probabilities for response categories 0 to 9 following inoculations of seventeen wild cherry clones, two sweet cherry clones (Napoleon and Roundel) and lilac cv. Sensation with *Pseudomonas syringae* pv. *morsprunorum* (*Psm*) races 1 and 2, typical *P. s.* pv. *syringae* (*Pss*), one atypical *P. s.* pv. *syringae* isolate (HRI 5264) and sterile tap water (stw) as control.

Clone	Probability for each response category[a]														
	stw control			*Psm* race 1 (4 isolates)			*Psm* race 2 (3 isolates)			*Pss* (3 isolates)			*Pss* isolate HRI 5264		
	0-1	3-5	7-9	0-1	3-5	7-9	0-1	3-5	7-9	0-1	3-5	7-9	0-1	3-5	7-9
Lilac Sensation	0.90	0.09	0.00	0.92	0.08	0.00	0.97	0.02	0.00	0.00	0.00	1.00	0.02	0.28	0.70
Napoleon	0.74	0.24	0.01	0.08	0.61	0.31	0.69	0.30	0.02	0.00	0.00	1.00	0.03	0.45	0.51
Roundel	0.74	0.25	0.01	0.12	0.67	0.22	0.56	0.41	0.03	0.00	0.00	1.00	0.12	0.66	0.22
Arger Fen E	1.00	0.00	0.00	0.24	0.66	0.11	0.69	0.30	0.02	0.01	0.14	0.85	0.83	0.16	0.01
Boxted Hall	0.40	0.55	0.05	0.14	0.68	0.18	0.40	0.55	0.05	0.00	0.00	1.00	0.03	0.40	0.57
Cobtree	0.74	0.24	0.01	0.42	0.53	0.05	0.82	0.17	0.01	0.00	0.07	0.92	0.18	0.68	0.14
Groton[b]	0.01	0.23	0.76	0.06	0.58	0.35	0.04	0.50	0.46	0.00	0.00	1.00	0.09	0.64	0.27
Howley Hood	0.53	0.44	0.03	0.02	0.34	0.64	0.22	0.66	0.11	0.00	0.03	0.96	0.02	0.34	0.64
Studley Royal	0.38	0.56	0.06	0.07	0.61	0.32	0.04	0.51	0.45	0.00	0.03	0.96	0.03	0.44	0.53
Treswell	0.85	0.14	0.01	0.24	0.65	0.10	0.56	0.41	0.03	0.01	0.23	0.75	0.85	0.14	0.01
FD1-57-4/17	0.94	0.06	0.00	0.21	0.67	0.12	0.19	0.67	0.14	0.00	0.00	1.00	0.89	0.10	0.00
FD1-57-4/122	0.94	0.06	0.00	0.16	0.68	0.16	0.18	0.68	0.14	0.00	0.00	1.00	0.35	0.59	0.07
FD1-57-4/166	0.85	0.14	0.01	0.55	0.42	0.03	0.37	0.57	0.06	0.02	0.28	0.70	0.90	0.09	0.00
1900	0.53	0.44	0.03	0.10	0.65	0.25	0.23	0.66	0.11	0.00	0.04	0.96	0.00	0.00	1.00
1904	0.90	0.09	0.00	0.21	0.67	0.12	0.70	0.28	0.02	0.00	0.00	1.00	0.50	0.47	0.04
1909	0.25	0.65	0.10	0.08	0.62	0.30	0.20	0.67	0.13	0.00	0.00	1.00	0.09	0.64	0.28
1912	0.68	0.30	0.02	0.16	0.68	0.16	0.11	0.66	0.23	0.00	0.00	1.00	0.14	0.67	0.19
1919	0.53	0.44	0.03	0.20	0.67	0.13	0.57	0.40	0.03	0.00	0.03	0.96	0.06	0.59	0.35
F12/1	0.97	0.02	0.00	0.35	0.58	0.06	0.38	0.56	0.06	0.00	0.11	0.88	0.02	0.31	0.67
Charger	0.87	0.13	0.01	0.31	0.62	0.08	0.51	0.46	0.03	0.00	0.02	0.98	0.46	0.50	0.04

[a] Highest probability is underlined. [b] Sterile tap water pot with fungal contamination; .

Leaf spots were only observed in cherry plantlets inoculated with *P. s.* pv. *morsprunorum* strains and therefore the data were analysed separately. The analysis of the results showed that pathovar was the most important factor followed by clone. Leaf spots were more frequent in plantlets inoculated with *P. s.* pv. *morsprunorum* race 1 (probability of 54%) than with *P. s.* pv. *morsprunorum* race 2 strains (probability of 34%).
Although some resistance to *P. s.* pv. *morsprunorum* was found in the wild cherry clones tested, no resistance was found to *P. s.* pv. *syringae* strains. The method used here has many advantages over inoculations in the field or inoculations of twigs: plantlets are easily multiplied and can be available all year round and this method is quicker, more consistent and clearly differentiates *P. s.* pv. *syringae* and *P. s.* pv. *morsprunorum* strains. Inoculations on rooted plants of five wild cherry clones also showed that they are more susceptible to *P. s.* pv. *syringae* than *P. s.* pv. *morsprunorum* strains (data not shown). Future screenings of cherries for resistance should always include *P. s.* pv. *morsprunorum* and *P. s.* pv. *syringae* strains.

ACKNOWLEDGEMENTS

This work was funded by the Department for Environment, Food and Rural Affairs (DEFRA) of the United Kingdom. We thank L.W. Moore for providing micropropagated plantlets of lilac cv. Sensation, L. Andrews for supplying plantlets of cherry and lilac, C.M.E. Garrett and J.P. Luz for providing *P. syringae* isolates, N.J. Grant and D.O. Sage for technical support, and J. P. Alves and B. Everett for laboratory assistance.

REFERENCES

Burkowicz, A., and Rudolph, K., 1994, Evaluation of pathogenicity and of cultural and biochemical tests for identification of *Pseudomonas-syringae* pathovars *syringae*, *morsprunorum* and *persicae* from fruit-trees. *J. Phytopathol.* **141**: 59-76.

Crosse, J.E., 1955, Bacterial canker of stone-fruits. I. Field observations on the avenues of autumnal infection of cherry. *J. Hortic. Sci.* **30**: 131-142.

Crosse, J.E., and Garrett, C.M.E., 1966, Bacterial canker of stone-fruits. VII. Infection experiments with *Pseudomonas mors-prunorum* and *P. syringae*. *Ann. Appl. Biol.* **58**: 31-41.

Freigoun, S.O., and Crosse, J.E., 1975, Host relations and distribution of a physiological and pathological variant of *Pseudomonas morsprunorum*. *Ann. Appl. Biol.* **81**: 317-330.

Garrett, C.M.E., 1986, Influence of rootstock on the susceptibility of sweet cherry scions to bacterial canker, caused by *Pseudomonas syringae* pvs *morsprunorum* and *syringae*. *Plant Pathol.* **35**: 114-119.

Garrett, C.M.E., and Fletcher, D.A., 1989, *In vitro* screening for bacterial canker resistance in cherry. *Proc. 7th Int. Conf. Plant Path. Bact.*: 991-993.

Garrett, C.M.E., Panagopoulos, C.G., and Crosse, J.E., 1966, Comparison of plant pathogenic Pseudomonads from fruit trees. *J. Appl. Bacteriol.* **29**: 342-356.

Hammatt, N., and Grant, N.J., 1993, Apparent rejuvenation of mature wild cherry (*Prunus avium* L.) during micropropagation. *J. Plant Physiol.* **141**: 341-346.

King, E.O., Ward, M.K., and Raney, D.R., 1954, Two simple media for the demonstration of pyrocanin and fluorescin. *J. Lab. Clin. Med.* **44**: 301-307.

Krzesinska, E.Z., and Azarenko, A.N.M., 1992, Excised twig assay to evaluate cherry rootstocks for tolerance to *Pseudomonas syringae* pv. *syringae*. *Hort. Sci.* **27**: 153-155.

McCullagh, P., and Nelder, J.A., 1983, *Generalized linear models*. Chapman and Hall, London.

Murashige, T., and Skoog, F., 1962, A revised medium for rapid growth and bioassays with tobacco tissue cultures. *Physiol. Plant.* **15**: 473-497.

Nicoll, F.J., 1993, Genetic improvement of cherry for farm woodlands. *Q. J. For.* **87**: 187-194.

Payne, R.W., Lane, P.W., Digby, P.G.N., Harding, S.A., Leech, P.K., Morgan, G.W., Todd, A.D., Thompson, R., Wilson, G.T., Welham, S.J., and White, R.P., 1993, *Genstat 5 Release 3 Reference manual*. Clarendon Press, Oxford.

Scheck, H.J., Canfield, M.L., Pscheidt, J.W., and Moore, L.W., 1997, Rapid evaluation of pathogenicity in *Pseudomonas syringae* pv. *syringae* with a lilac tissue culture bioassay and syringomycin DNA probes. *Plant Dis.* **81**: 905-910.

Wormald, H., 1937, Bacteriosis of stone fruit trees in Britain. VI. Field observations on bacteriosis of sweet cherry trees. *J. Pomol. Hortic. Sci.* **15**: 35-48.

Yessad, S., Manceau, C., and Luisetti, J., 1992, A detached leaf assay to evaluate virulence and pathogenicity of strains of *Pseudomonas syringae* pv. *syringae* on pear. *Plant Dis.* **76**: 370-373.

Investigations on the Systemic Acquired Resistance Induced by Acibenzolar-S-Methyl in Tomato Plants Against *Pseudomonas syringae* pv. *tomato*

R. BUONAURIO, C. MORETTI, A. BERTONA[#], L. SCARPONI[§]

Dipartimento di Arboricoltura e Protezione delle Piante, Università degli Studi di Perugia, Borgo XX Giugno, 74-06121 Perugia (Italy); [#]Syngenta Protezione Piante S.p.A., Via S. Sofia, 21-20122 Milano (Italy); [§]Centro di Studio sulla Chimica e Biochimica dei Fitofarmaci del CNR, Borgo XX Giugno, 72-06121 Perugia (Italy).

Abstract: Growth chamber experiments showed that acibenzolar-S-methyl (ASM) (50÷250 μM) treatments systemically protect tomato plants against *Pseudomonas syringae* pv. *tomato*. Evidence for this was the reduction in bacterial spot diameter and bacterial growth *in planta*. The efficacy of ASM was also found in open field conditions, where both leaves and fruit were protected from the disease. The best efficacy (about 80%) was obtained by spraying the plants 6÷8 times with a mixture of ASM and copper hydroxide (2.5 + 80 g hl^{-1} active ingredient). Open field experiments revealed that neither quality nor quantity of fruit yield were negatively affected by ASM treatments. By using the HPLC method of Scarponi *et al.* (2001), we showed that ASM rapidly translocates from treated to untreated tomato leaves, with the maximum translocation occurring 8 h after the treatment. ASM residues decreased as the time elapsed in both treated and untreated tomato leaves, reaching negligible values 72 h after treatment. We also demonstrated that the acidic derivative of ASM (CGA 210007), which systemically protected tomato plants against *P. s.* pv. *tomato*, was formed in tomato plants as early as 2 h after ASM treatment, albeit only in the treated leaves. CGA 210007 residues decreased in treated tomato leaves with a trend similar to that observed for ASM. Since neither ASM nor CGA 210007 inhibited bacterial growth *in vitro* and the protection against bacterial speck of tomato has been observed after the two compounds were completely degraded, we can conclude that the protection must be due to the activation of the plant's defence mechanisms.

Key words: Bacterial speck, bacterial diseases control, SAR, persistence.

Author for correspondence: R. Buonaurio; e-mail: buonaurio@unipg.it.

N.S. Iacobellis *et. al.* (eds.), *Pseudomonas syringae* and Related Pathogens, 475-482.

1. INTRODUCTION

Bacterial speck of tomato caused by *Pseudomonas syringae* pv. *tomato* (Okabe) Young *et al.* is an important disease in many tomato-growing areas of the world (Goode and Sasser, 1980), including Italy (Sanguineti and Bazzi, 1993).

As for other bacterial diseases, the control of bacterial speck in the field is difficult. Chemical control, essentially carried out using copper compounds, is not completely satisfactory (Conlin and McCarter, 1983), because copper resistant strains of the bacterium frequently appear (Pernezny *et al.*, 1995). Although a number of genes for resistance to the disease have been identified, few resistant tomato cultivars are available. In addition, the cultivation of resistant tomato cultivars could trigger the development of new bacterial races capable of overcoming resistance (Buonaurio *et al.*, 1996).

The use of systemic acquired resistance (SAR) is a promising environment-friendly strategy for controlling plant diseases, including those caused by bacteria. Besides being biologically induced, SAR can be triggered by some chemicals including salicylic acid and its synthetic analogues, such as a benzothiadiazole derivative acibenzolar-S-methyl [ASM: benzo(1,2,3)thiadiazole-7-carbothioic acid S-methyl ester], which was developed by Syngenta (Kessmann *et al.*, 1996). Generally, ASM has no antimicrobial activity and has been reported to protect dicotyledonous and monocotyledonous plant species against a number of plant diseases, including those caused by bacteria (Friedrich *et al.*, 1996; Lawton *et al.*, 1996; Siegrist *et al.*, 1997; Cole, 1999; Brisset *et al.*, 2000; Romero *et al.*, 2001; Buonaurio *et al.*, 2002; Werner *et al.*, 2002). ASM has been registered in Italy (2001) as Bion® to control a number of bacterial diseases spread on tomato, hazelnut and pear.

In this paper we report the results of some investigations carried out in tomato plants: i) to verify the efficacy of ASM in protecting plants from *P. s.* pv. *tomato* attacks, both in growth chambers and open field conditions; ii) to study the persistence and translocation of ASM in the plant; iii) to evaluate the effect of ASM treatments on fruit yield.

Part of the data here presented have been reported elsewhere (Scarponi *et al.*, 2000, 2001).

2. MATERIALS AND METHODS

To verify the ability of ASM in protecting plants from *Pseudomonas syringae* pv. *tomato* attack and to study the persistence and translocation of ASM in growth chamber conditions, tomato (*Lycopersicon esculentum* Mill.) plants, cv. Bonny Best, at the $5^{th}\div6^{th}$ true leaf stage, grown in a growth

chamber programmed for a 12 h day at 25±0.5°C with 65÷70% R.H. and 12 h night at 18±0.5°C with 85÷90% R.H. were used. Cool white fluorescent lamps provided 240 μE m^{-2} s^{-1} illumination.
Plant treatments were performed by spraying aqueous suspensions of ASM (50% water dispersible granules) onto cotyledons, the 1st and the 2nd leaves (referred to as basal leaves) of tomato plants. The concentrations used were: 250 μM to determine persistence and translocation, and from 50 to 250 μM of ASM to evaluate the effect on bacterial speck disease.
Five days after treatment, the 3rd÷5th leaves (referred to as apical leaves) of treated and control plants were airbrush sprayed with a suspension (10^8 CFU ml^{-1}) of the *P. s.* pv. *tomato* race 0 (strain DAPP-PG 214). After inoculation, plants were covered with plastic bags for the first 2 days.
The severity of the bacterial infection was evaluated by determining bacterial spot diameter and the bacterial growth *in planta*, as previously reported by Scarponi *et al.* (2001).
ASM and its acidic derivative (CGA 210007) were identified by GC-MS and determined in basal ASM treated- and apical-non inoculated leaves according to the HPLC method described by Scarponi *et al.* (2001).
Field trials were carried out in different locations and years to evaluate the ability of ASM to protect tomato plants and to evaluate the effect of ASM on fruit yield both at quantitative and qualitative levels. The trials were arranged in a completely randomised block design with four replicates. In all trials, the treatment comparison included ASM and copper hydroxide alone or in combination. The doses used are reported in the Tables 1 and 2. Four to 8 sprays were applied every 8-12 days and disease symptoms on leaves and fruits were periodically assessed.

3. RESULTS AND DISCUSSION

Experiments performed in growth chamber conditions showed that treatment of basal tomato leaves with ASM at concentrations ranging from 50 to 250 μM systemically protected apical leaves against *Pseudomonas syringae* pv. *tomato* attacks. In fact, ASM treatments led to significant reductions in lesion diameter and bacterial growth *in planta* (Fig. 1).

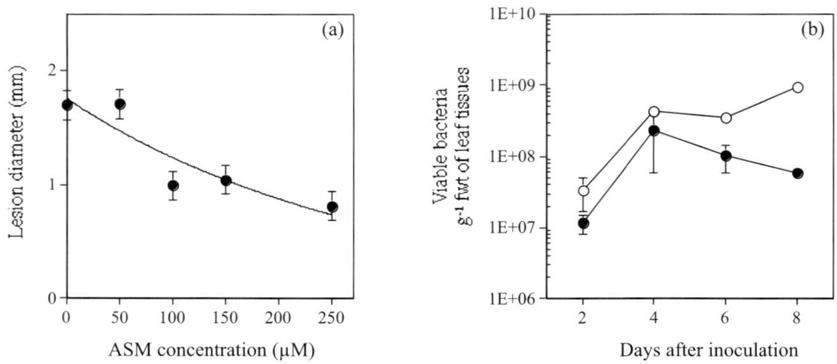

Figure 1. (a) Lesion diameter of *Pseudomonas syringae* pv. *tomato* in tomato plants treated at different concentrations of ASM, 10 days after the inoculation. Data are the means of six replicates±SEM, and have been fitted as follows: $y= 1.756e^{-0.00348x}$. (b) Effect of ASM treatment (250 μM) on *P. s.* pv. *tomato* growth in treated (black circle) and untreated (white circle) tomato plants. Data are the means of two independent experiments ± SE.

Our data are in agreement with those previously reported in other plant-bacterium combinations (Friedrich *et al.*, 1996; Lawton *et al.*, 1996; Siegrist *et al.*, 1997; Cole, 1999; Brisset *et al.*, 2000; Romero *et al.*, 2001; Buonaurio *et al.*, 2002; Werner *et al.*, 2002) and, according to the above studies, reduced disease severity is correlated with a significant decrease in bacterial growth *in planta*.

ASM residues decreased in basal (treated) leaves as time elapsed showing an abrupt diminution between 2 and 4 h after treatment (Fig. 2). ASM rapidly translocated from basal to apical leaves and its level was always greater in the basal than in the apical leaves. Translocation was rapid since the level of ASM in the apical leaves was 30.5% of the inducer by 2 h after treatment. However, maximum relative translocation occurred 8 h after treatment, reaching 40.0% of the total quantity found. The acidic derivative of ASM (CGA 210007) was formed in the tomato plants, however it was detected only in the basal ASM-treated leaves and not in the apical leaves. ASM and its acidic derivative decreased as time elapsed to undetectable levels 72 h after treatment (Fig. 2). Although CGA 210007 was not detected in the apical leaves, we cannot exclude that small amounts of this metabolite might translocate to these leaves and be rapidly degraded. Like ASM, the acidic derivative of ASM was able to systemically protect tomato plants from bacterial speck disease (data not shown). This metabolite has been reported to protect cucumber plants from a number of fungal and bacterial pathogens (Kunz *et al.*, 1997) and pepper plants from *Xanthomonas campestris* pv. *vesicatoria* (Buonaurio *et al.*, 2002).

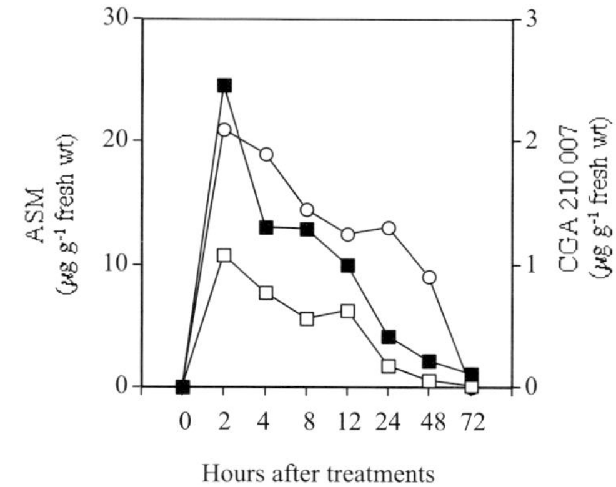

Figure 2. Residues of ASM in basal (white square) and apical (black square) tomato leaves; residues of the acidic derivative of ASM (CGA 210 007) in basal tomato leaves (white circle). Data are the means of 3 determinations.

Since ASM and its acidic derivative showed very poor inhibition of *P. s.* pv. *tomato* growth *in vitro* (EC_{50} were higher than 100 μg ml^{-1}) and the protection was observed when they were completely degraded in the plant, the protection must be due to the activation of the plant's defence mechanisms.

It may be hypothesised that in the leaf intercellular space of ASM-treated tomato plants, where *P. s.* pv. *tomato* grows, an unfavourable environment for bacterial growth occurs (e.g.: low amount of nutrient, formation of antimicrobial compounds) as a consequence of the treatment and may be responsible for the reduction in disease severity. The hypothesis that antimicrobial compounds could be produced in the leaf intercellular space of ASM-treated tomato plants seems to be supported by the electron microscopy study of Guiderdone *et al.* (2001). In fact, they found ultrastructural changes of *P. s.* pv. *tomato* cells probably associated with their reduced viability, such as detachment and undulation of the cell envelope and increase in cytoplasm granulation.

Similarly to the results obtained by Louws *et al.* (2001), the protective effect of ASM was also confirmed in open field conditions. From preliminary field trials carried out by Syngenta, it was already clear that one application was not sufficient to induce disease resistance throughout the growing season. In a trial performed in the 1998, ASM alone or in combination with copper hydroxide, sprayed 6 times, significantly protected tomato plants against leaf and fruit infections (68÷85% efficacy) (Table 1). The protection was significantly higher than that obtained with copper hydroxide alone. The efficacy of ASM in protecting tomato plants from bacterial speck disease was confirmed in other field trials (data not shown).

Table 1. Control efficacy of ASM alone or in combination with copper hydroxide on tomato bacterial speck disease in open field trials.

Treatment	Doses (g a.i. hl^{-1})	Leaf infection LSI (%)	Leaf infection Efficacy[(1)] (%)	Fruit infection Infection index	Fruit infection Efficacy[(1)] (%)
Control	-	16.25 a	0	52.50 a	0
Copper hydroxide (CH)	80	10.00 b	39	45.00 b	14
ASM	1.25	5.25 cd	68	16.75 c	68
ASM	2.5	2.75 ef	83	11.00 de	79
ASM+CH	1.25+40	3.75 de	77	10.50 de	80
ASM+CH	2.5+80	2.50 f	85	9.50 e	82

LSI= Leaf surface infected. Each value is the mean of 4 replicates. Data followed by the same letters are not significantly different at $P<0.05$ (Duncan's multiple range test).
[(1)]Disease severity on untreated leaves-Disease severity on treated leaves/Disease severity on untreated leaves x 100.

Since it is known that induced resistance is a transient phenomenon, especially in open field conditions, more applications of ASM are needed in order to increase efficacy and duration of the crop protection. In our conditions, we observed that 4÷8 treatments provided an acceptable protection of tomato. The number of recommended applications per season depends on several factors such as the level of infection pressure, plant susceptibility and environmental conditions.

In addition to the good protection from bacterial speck disease, ASM alone or in combination with copper hydroxide slightly increased marketable yield (Table 2).

Table 2. Effect of ASM alone or in combination with copper hydroxide on marketable tomato yield in open field condition

Treatment	Doses (g a.i. hl^{-1})	% of leaf surface infected	Yield (Kg/plot)
Control	-	51.25 a	21.60 a
Copper hydroxide	80	35.00 b	26.98 b
ASM	2.50	26.25 cd	27.40 b
ASM+copper hydroxide	0.62+20	28.75 c	27.15 b
ASM+copper hydroxide	1.25+40	26.25 cd	25.65 ab
ASM+copper hydroxide	2.50+80	23.75 cd	28.23 b
ASM+copper hydroxide	3.75+120	22.50 d	27.35 b

Each value is the mean of 4 replicates. Data followed by the same letters are not significantly different at $P<0.05$ (Duncan's multiple range test).

Furthermore, some parameters of fruit quality, such as Brix values, pH and colour a/b (Gardner), were not affected in ASM treated plants (data not shown).

In conclusion, the use of ASM to control bacterial speck disease of tomato may represent a new environment friendly strategy within an integrated crop protection system. It is worth noting that ASM may be able to control the attacks of other important tomato pathogens, such as cucumber mosaic virus (Anfoka, 2000), *X. c.* pv. *vesicatoria* (Louws *et al.*, 2001), *Clavibacter michiganensis* subsp. *michiganensis* (Werner *et al.*, 2002) and *Fusarium oxysporum* f. sp. *radicis-lycopersici* (Benhamou and Belanger, 1998).

REFERENCES

Anfoka, G.H., 2000, Benzo-(1,2,3)-thiadiazole-7-carbothioic acid S-methyl ester induces systemic resistance in tomato (*Lycopersicon esculentum*. Mill cv. Vollendung) to cucumber mosaic virus. *Crop Prot.* **19**: 401-405.

Benhamou, N., Belanger, R.R., 1998, Benzothiadiazole-mediated induced resistance to *Fusarium oxysporum* f. sp. *radicis-lycopersici* in tomato. *Plant Physiol.* **118**: 1203-1212.

Brisset, M.N., Cesbron, S., Thomson, S.V., Paulin, J.P., 2000, Acibenzolar-S-methyl induces the accumulation of defense- related enzymes in apple and protects from fire blight. *Eur. J. Plant Pathol.* **106**: 529-536.

Buonaurio, R., Stravato, V.M., Capelli, C., 1996, Occurrence of *Pseudomonas syringae* pv. *tomato* race 1 in Italy on *Pto* gene-bearing tomato plants. *J. Phytopathol.* **144**: 437-440.

Buonaurio, R., Scarponi, L., Ferrara, M., Sidoti, P., Bertona, A., 2002, Induction of systemic acquired resistance in pepper plants by acibenzolar-S-methyl against bacterial spot disease. *Eur. J. Plant Pathol.* **108**: 41-49.

Cole, D.L., 1999, The efficacy of acibenzolar-S-methyl, an inducer of systemic acquired resistance, against bacterial and fungal diseases of tobacco. *Crop Prot.* **18**: 267-273.

Conlin, K.C., McCarter, S.M., 1983, Effectiveness of selected chemicals in inhibiting *Pseudomonas syringae* pv. *tomato* and controlling bacterial speck. *Plant Dis.* **67**: 639-644.

Friedrich, L., Lawton, K., Ruess, W., Masner, P., Specker, N., Rella, M.G., Meier, B., Dincher, S., Staub, T., Uknes, S., Metraux, J.P., Kessmann, H., Ryals, J.A., 1996, Benzothiadiazole derivative induces systemic acquired resistance in tobacco. *Plant J.* **10**: 61-70.

Goode, M.J., Sasser, M., 1980, Prevention-the key to controlling bacterial spot and bacterial speck of tomato. *Plant Dis.* **64**: 831-834.

Guiderdone, S., Quaglia, M., Marte, M., 2001, Indagini ultrastrutturali sulla resistenza a *Pseudomonas syringae* pv. *tomato* indotta in pomodoro da BTH. Proceedings of the 'XI Meeting on Molecular and Physiological Aspects of the Plant-Pathogen Interactions; May 31 – July 1; Rome (Italy), p. 6

Kessmann, H., Oostendorp, M., Staub, T., Goerlach, J., Friedrich, L., Lawton, K., Ryals, J., 1996, CGA 2455704: Mode of action of a new plant activator. Proceedings of the Brighton Crop Protection Conference; p. 12.

Kunz, W., Schurter, R., Maetzke, T., 1997, The chemistry of benzothiadiazole plant activators. *Pestic. Sci.* **50**: 275-282.

Lawton, K.A., Friedrich, L., Hunt, M., Weymann, K., Delaney, T., Kessmann, H., Staub, T., Ryals, J., 1996, Benzothiadiazole induces disease resistance in *Arabidopsis* by activation of the systemic acquired resistance signal transduction pathway. *Plant J.* **10**: 71-82.

Louws, F.J., Wilson, M., Campbell, H.L., Cuppels, D.A., Jones, J.B., Shoemaker, P.B., Sahin, F., Miller, S.A. 2001, Field control of bacterial spot and bacterial speck of tomato using a plant activator. *Plant Dis.* **85**: 481-488.

Pernezny, K., Kudela, V., Kokoskova, B., Hladka, I., 1995, Bacterial diseases of tomato in the Czech and Slovak Republics and lack of streptomycin resistance among copper-tolerant bacterial strains. *Crop Prot.* **14**: 267-270.

Romero, A.M., Kousik, C.S., Ritchie, D.F., 2001, Resistance to bacterial spot in bell pepper induced by acibenzolar-S-methyl. *Plant Dis.* **85**: 189-194.

Sanguineti, M.C., Bazzi, C., 1993, "*Pseudomonas syringae* pv. *tomato* pomodoro" In Miglioramento genetico delle piante per resistenza a patogeni e parassiti, P., Crinò, Sonnino, A., Saccardo, F., Buiatti, M., Porta-Puglia, A., Surico G., ed. Edagricole, pp. 433-450.

Scarponi, L., Buonaurio, R., Bertona, A., 2000, Persistence of acibenzolar-S-methyl and SAR induction in tomato and pepper plants. Proceedings of the 5th Congress of the European Foundation for Plant Pathology. Biodiversity in Plant Pathology; 18-22 September, Taormina (Italy), p. 67.

Scarponi, L., Buonaurio, R., Martinetti, L., 2001, Persistence and translocation of a benzothiadiazole derivative in tomato plants in relation to systemic acquired resistance against *Pseudomonas syringae* pv. *tomato*. *Pest Manag. Sci.* **57**: 262-268.

Siegrist, J., Glenewinkel, D., Kolle, C., Schmidtke, M., 1997, Chemically induced resistance in green bean against bacterial and fungal pathogens. *J. Plant Dis. Protect.* **104**: 599-610.

Werner, N.A., Fulbright, D.W., Podolsky, R., Bell, J., Hausbeck, M.K., 2002, Limiting populations and spread of *Clavibacter michiganensis* subsp. *michiganensis* on seedling tomatoes in the greenhouse. *Plant Dis.* **86**: 535-542.

Integrated Management of Bacterial Decline of Hazelnut, by Using Bion as an Activator of Systemic Acquired Resistance (SAR)

M. SCORTICHINI and R. LIGUORI[#]
Istituto Sperimentale per la Frutticoltura, Via di Fioranello, 52, I-00040 Ciampino Aeroporto (Roma), Italy; [#]Syngenta Crop Protection S.p.A., Via Gallarate, 139 I-20151 Milano, Italy.

Abstract: The field trials performed in Italy to test the effectiveness of acibenzolar-S-methyl, registered as Bion, to control bacterial canker and decline of hazelnut are summarised. Five sprays from late April to September allowed to significantly reduce the number of dead branches and trees. Bion was more effective in orchards where diseased trees were regularly removed.

Key words: acibenzolar-S-methyl, bacterial canker, *Pseudomonas avellanae,* bacterial disease control.

Author for correspondence: M. Scortichini; e-mail: mscortichini@hotmail.com.

1. INTRODUCTION

Bacterial canker and decline caused by *Pseudomonas avellanae* (Psallidas) Janse *et al.* is a serious threat for hazelnut (*Corylus avellana* L.). In northern Greece, young plantings were almost completely destroyed by this bacterium. In central Italy (Latium, province of Viterbo), where the disease is referred as "moria", since it was first discovered it has resulted in the dead of more than 40,000 trees, and, currently, it is damaging approximately 1,000 hectares. When the disease was first discovered during 80's, it was not aggressively controlled and as a consequence was allowed to spread, also because hazelnut was retained an hard species and, traditionally, sprayings to control pests and diseases were seldom carried out. In Italy, there is a

N.S. Iacobellis *et. al.* (eds.), *Pseudomonas syringae* and Related Pathogens, 483-487.

national law to partially compensate the farmers whose trees have been seriously damaged by the disease.
After the discovery of the causative agent of "moria" (Scortichini and Tropiano, 1994), copper compounds started to be applied either during the vegetative season or in autumn during the leaf fall. However, studies aiming to elucidate the disease cycle (Scortichini and Lazzari, 1996), pointed out that *P. avellanae* enters the tree in autumn through the leaf scars and, subsequently, it systemically colonises the plant. Consequently, we realised that from spring to early autumn, the copper compounds resulted ineffective for controlling the disease.

2. MATERIALS AND METHODS

To find out other solutions, on 1997, we started preliminary investigations by using acibenzolar-S-methyl (CGA 245704), a compound inducing systemic acquired resistance (SAR) to herbaceous plants (Reuss *et al.*, 1996) aiming to point out the minimum dose having a possible satisfactory control of the disease without inducing any phytotoxic effect. In an hazelnut orchard severely damaged by *Pseudomonas avellanae* (more than 1,600 trees out of 3,600 were already dead), on October 1996 we chosen trees, judged as dying, to be utilised for the trial in the following year. Doses of 2.5, 5.0 and 20.0 g hl^{-1} of CGA 245704 at 50% of active ingredient (a.i.), were applied three times, once a month, from late April to June. The compound was applied on three trees per each thesis. Five trees did not receive any treatment and served as control. On June 1997, per each thesis, the total number of leaves on three branches randomly sampled on the canopy of the tree, was counted.

3. RESULTS AND DISCUSSION

The dose inducing a partial control of the disease without causing any phytotoxic effect to the foliage was 5.0 g hl^{-1} (1,238 leaves). At 2.5 g.hl^{-1} (247 leaves) and 20.0 g hl^{-1} (474 leaves), a significant reduction of the number of leaves was observed, with the latter dose inciting a slight chlorosis. However, all trees sprayed with acibenzolar-S-methyl survived over the whole season and the following years, whereas the five control-trees died in June and July 1997.
After this encouraging result, during the following years, we investigated the field effectiveness of acibenzolar-S-methyl on a larger number of trees and by choosing two different kinds of orchard management: a) farms where the removal and destruction of infected branches and trees was carried out

regularly year by year (type A); b) farms where the trees were almost abandoned (type B). In five hazelnut orchards (i.e. three farms of type A and two farms of type B) 5.0 g hl^{-1} (i.e. 25 g ha^{-1} of a.i.) were applied once a month, for three times in 1998, and five times in 1999 and 2000. The initial spraying was in late April. Copper-oxychloride (50%)-treated trees, receiving the treatments at the same time than acibenzolar-S-methyl, and, in addition, during the leaf fall, served as standard reference, whereas untreated trees as control.
The number of trees and branches that resulted dead over the season were counted. In addition, in the orchard where we performed the preliminary trial, all visibly infected trees were removed and CGA 245704 was applied five times as the sole compound to control the decline (type C). The results are shown in Table 1.

Table 1. Effectiveness of acibenzolar-S-methyl (Bion) compared with copper oxychloride at 50% in controlling the bacterial decline of hazelnut caused by *Pseudomonas avellanae*. In 1998 the compounds were applied three times, whereas in 1999 and 2000 five times. Due to the heavy spread of the pathogen in 1998 after spring frosts, in 1999 and 2000 other nearby orchards were chosen.

Year	Farm	Number of trees	Dead trees			Dead branches		
			Ac	Ox	C	Ac	Ox	C
1998	A	150	11	20	39	70a	98b	138c
	B	100	10	15	44	65a	90b	139c
	C	410	1	NT	39	4	NT	67
1999	A	150	4	11	29	42a	64b	88c
	B	100	9	12	25	48a	60ab	82c
	C	410	4	NT	34	16	NT	56
2000	A	150	7	9	12	9a	29b	31b
	B	100	8	14	18	14a	27b	34b
	C	410	5	NT	18	11	NT	51

Ac = Acibenzolar-S-methyl (Bion); Ox = copper oxychloride at 50%; C = control.
A, B, C = different kinds of farm (see the text). NT = Not tested.
Data showing different letters are significantly different at P=0,05 using the Student's *t*-test.

On 1998, a spring frost allowed the considerable spread of the pathogen by causing wounds along the branches and trunk of the trees, as assessed by the local agricultural services. In all orchards (types A and B), acibenzolar-S-methyl reduced either the number of dead trees or dead branches. The effectiveness was higher in the orchards of type A, where among the trees treated with acibenzolar-S-methyl, 11 trees (7.3% of the total) resulted dead compared to 20 (13.3%) of copper oxychloride and 39 (26.0%) of control trees. A similar and statistically different result was obtained by counting the number of dead branches. In the orchards of type B, the number of dead trees and branches was higher (see Table 1), and also in this case, acibenzolar-S-methyl reduced the number of dead branches and trees. We performed also isolations from symptomatic trees, and we found that in many cases *Pseudomonas avellanae* had colonised the root system.

On 1999, by increasing the number of treatments from three to five, the effectiveness of acibenzolar-S-methyl was higher. In this year, however spring frosts did not occur. In the orchards of type A, 4 (2.6% of the total) and 11 trees (7.3%) out of 150 and 100 trees treated, respectively, with acibenzolar-S-methyl or copper oxychloride resulted dead. Control plots showed 29 (19.0%) of dead trees. Also the number of dead branches was significantly reduced. In the orchards of type B, the effectiveness was lower. However, also in these cases acibenzolar-S-methyl reduced both the number of dead trees and branches. In farm C, we counted only 4 dead trees and 16 wilted branches, respectively. In the untreated plot, 34 and 56 dead trees and branches, respectively, were recorded.

In 2000, we obtained similar results in both types of hazelnut farms and the efficacy of acibenzolar-S-methyl was confirmed. In farms of type A and B, dead trees in the acibenzolar-treated plots were 7 (4.8%) and 8 (8.7%), respectively. In copper oxychloride-treated plots, we counted 9 (6.4%) dead trees in farms of type A, and 14 (15.9%) in the farms of type B. Twelve (9.9%) (type A) and 18 (24.0%) (type B) were the percentages scored for the control trees. In farm C, only 5 trees (1.2% of the total) resulted dead.

From these trials, it is evident that acibenzolar-S-methyl reduced both the number of dead trees and branches. It is also clear that the removal and destruction of infected plant tissues, reducing the inoculum of the pathogen, enhances the effectiveness of the disease management. In addition, if the pathogen reaches the roots, it would seem that such a compound cannot counteract the bacterium. It is also evident that predisposing factors, such as spring frost, play a significant role in the spread of the pathogen and, consequently in the effectiveness of acibenzolar-S-methyl. In fact, in the farm C, where spring frosts never occur, the percentage of dead trees after the treatment of acibenzolar was always very low. Late April (i.e. leaf diameter of 2÷3 cm) is the period to start the sprayings, and a total of five treatments would seem indicated. In addition, we also verified in other field trials (unpublished data), that a treatment after the harvest in late September would enhance the efficacy of the compound during the following year.

Acibenzolar-S-methyl has been registered by Syngenta Crop Protection with the trade name of Bion in spring 2002, and it has been introduced in the market as a wet table powder containing 50% of the active ingredient. It can be used for the control of bacterial decline of hazelnut, bacterial speck of tomato, caused by *P. syringae* pv. *tomato* and fire blight of pome fruit trees, caused by *Erwinia amylovora*.

The decline of hazelnut, caused by *P. avellanae*, can be now more effectively controlled by combining agronomic practises aiming to reduce the amount of bacterial inoculum and spray treatments at key times (Scortichini *et al.*, 2000). The latter include copper sprays during the leaf fall to prevent the colonisation of the leaf scars by the pathogen and the application of Bion during the vegetative season. The treatments with copper compounds are retained very important also after the spring frost occurrence to, possibly, avoid the colonisation of the wounds by the wind-driven cells of the pathogen. Surveys in the orchards to check the initial wilting of the trees, are retained fundamental to avoid the spread of the bacterium within the planting. Wilted twigs, branches and trees have to be removed and burnt. The uprooting of the heavily infected trees must be included also the root system. In fact, *P. avellanae* can latently infect the suckers let in place to re-build the tree; in these cases, infection reappear after one-three years.

REFERENCES

Reuss, W., Mueller, K., Knauf-Beiter, G., Kunz, W. and Staubt, T, 1996, Plant activator CGA 245 704, an innovative approach for disease control in cereals and tobacco. In: Proceedings. Brighton Crop Protection Conference-Pests and Diseases, BCPC; Farnham, Surrey, UK, pp 53-60.

Scortichini, M. and Tropiano, F.G., 1994, Severe outbreak of *Pseudomonas syringae* pv. *avellanae* on hazelnut in Italy. *J. Phytopath.* **140:** 65-70.

Scortichini. M., and Lazzari, M., 1996, Systemic migration of *Pseudomonas syringae* pv. *avellanae* in twigs and young trees of hazelnut and symptom development. *J. Phytopath.* **144**: 215-219.

Scortichini, M., Liguori, R., Nobile, M., Moretti, D., Valentini, B., Egitto, M. and Leonelli, A., 2000, Moria del nocciolo: un triennio di prove di campo con acibenzolar-S-methyl, induttore di resistenza sistemica acquisita. Atti Giornate Fitopatologiche, 2000, Vol. 2, 379-384.

Response of Some Olive Cultivars, Hybrid and Open Pollinated Seedlings to *Pseudomonas savastanoi* pv. *savastanoi*

D. HASSANI, R. BUONAURIO[#] and A. TOMBESI[#]

Horticulture Department, Seed and Plant Improvement Institute, 31585 Karaj, Iran; [#]Dipartimento di Arboricoltura e Protezione delle Piante, Università degli Studi di Perugia, Borgo XX Giugno, 74-06121 Perugia, Italy.

Abstract: Olive knot disease caused by *Pseudomonas savastanoi* pv. *savastanoi* is spread in all olive cultivation area including Italy where it can cause significant yield losses. The aim of the present study was to screen in greenhouse some olive genotypes for resistance/tolerance to olive knot disease. In a first experiment, the olive cultivars Frantoio, Leccino, Moraiolo and Nostrale di Rigali were inoculated with 5 bacterial strains collected in Umbria (Central Italy) from cv. Frantoio. In a second experiment, 25 olive genotypes, in particular: three half-sib families [open pollinated seedlings of Ascolana (6 genotypes), Leccino (6 genotypes) and Moraiolo (3 genotypes)], three full-sib families [hybrid seedlings of Frantoio x Maurino (3 genotypes), Leccino x Maurino (3 genotypes) and Maurino x Frantoio (1 genotype)] and three olive cultivars (Frantoio, Leccino and Moraiolo) were inoculated with a bacterial strain which in the first experiment showed the highest virulence. Olive shoots were inoculated (5 inoculation sites per shoot) with a suspension containing 5 x 10^8 CFU ml^{-1}. Symptom severity was evaluated about three months after inoculation, determining knot size and weight. The first experiment showed that, independently from the bacterial strains used, the cv. Frantoio was the most susceptible, followed by Nostrale di Rigali, Moraiolo and Leccino, which showed a knot weight which was 86% lower respect to the cv. Frantoio. The result of second experiment confirmed the cultivars Frantoio and Leccino as the most and the less susceptible, respectively. Among the hybrid and open pollinated seedlings, it seems interesting the Mo.6.105 genotype, which was more tolerant to the disease than cv. Leccino.

Key words: olive knot disease, olive genotypes, resistance.

Author for correspondence: D. Hassani; e-mail: hassanida@yahoo.it.

N.S. Iacobellis *et. al.* (eds.), *Pseudomonas syringae* and Related Pathogens, 489-494.

1. INTRODUCTION

Olive knot disease caused by *Pseudomonas savastanoi* pv. *savastanoi* is widespread in all olive cultivation area in the world. Like many bacterial diseases, olive knot is difficult to control. In fact, cultural practices and chemical sprays, essentially carried out with copper compounds, are not completely effective. The use of resistant olive cultivars could be the best way to control the disease. However, up to now only tolerant olive genotypes have been found and few and sometimes contradictory information are available on susceptibility of olive cultivars to knot disease (Varvaro and Surico, 1981; Benjama *et al.*, 1992; Marcelo *et al.*, 1999; Sisto *et al.*, 2001; Iannotta *et al.*, 2002).

Due to the relevance of the olive knot disease in Italy, we planned to evaluate the response of several olive hybrid and open pollinated seedlings to olive knot disease, as an important starting point for future breeding programs.

2. MATERIALS AND METHODS

Olive genotype responses to *Pseudomonas savastanoi* pv. *savastanoi* were evaluated in two separate experiments. In the first experiment, cutting propagated plants of the olive cultivars: Frantoio, Leccino, Moraiolo and Nostrale di Rigali (about 70 cm in height) were inoculated with 5 *P. s.* pv. *savastanoi* strains (I.2, I.6, I.8, I.10, I.11) collected in Umbria (Central Italy) from infected trees of Frantoio cultivar. Strain identity was confirmed by biochemical, physiological and pathogenicity tests (Schaad *et al.*, 2001), and by rep-PCR, using the primer BOX (Rademaker and De Bruijn, 1997). To prepare inoculum, bacterial strains were grown on Nutrient Agar for 24 h and suspended in sterile deionised water (SDW). The suspensions were spectrophotometrically adjusted to 5×10^8 CFU ml^{-1}. For the inoculation, olive shoots were wounded (5 wounds per shoot) with a sterile scalpel and 0.1 ml of bacterial suspensions were put into wounds, which were covered with Parafilm for 10 days. For control plants it was used SDW instead of bacterial suspensions. In the second experiment, cutting propagated plants of 25 olive genotypes (about 30 cm in height), were inoculated with a bacterial strain as in the first experiment. Genotypes were from nine groups: three half-sib families [open pollinated seedlings of Ascolana (6 genotypes), Leccino (6 genotypes) and Moraiolo (3 genotypes)], three full-sib families [hybrid seedlings of Frantoio x Maurino (3 genotypes), Leccino x Maurino (3 genotypes) and Maurino x Frantoio (1 genotype)] and three olive cultivars (Frantoio, Leccino and Moraiolo).

In both experiments, control and inoculated plants were kept under a plastic tunnel in a greenhouse at 20÷25°C and 80% R.H. for the duration of the experiment. Symptom severity (referred to as disease index) was evaluated three months after inoculation, determining knot size, by a arbitrary visual scale from 0 to 5 respectively with 0, 20% (1), 40% (2), 60% (3), 80% (4) and 100% (5) of maximum knot development, and fresh and dry weight of knots.

3. RESULTS AND DISCUSSION

Data of the first experiment, reported in Table 1, showed that independently from the bacterial strains used, the cv. Frantoio is the most susceptible to olive knot disease, followed by Nostrale di Rigali, Moraiolo and Leccino, which showed a reduction in knot weight more than 80% respect to the cv. Frantoio.

Table 1. Responses of 4 olive cultivars (A) to inoculation with 5 strains of *Pseudomonas savastanoi* pv. *savastanoi* (B) .

Cultivars/strains		Disease index		Knot fresh wt (mg)		Knot dry wt (mg)	
A. Cultivars	Frantoio	2.10	a	912	a	324	a
	Leccino	1.26	b	132	b	58	b
	Moraiolo	1.82	ab	280	b	117	b
	N. Rigali	1.98	a	489	b	182	b
B. Strains	I.2	1.40	a	326	a	123	a
	I.6	2.10	a	616	a	206	a
	I.8	1.63	a	321	a	123	a
	I.10	1.88	a	398	a	157	a
	I.11	1.95	a	607	a	243	a

Values are the mean of 5 and 4 replications for cultivars and strains, respectively. Data within a column for cultivars and strains followed by the same letters are not statistically different at P = 0.05 (Duncan's multiple range test)

In general, Nostrale di Rigali, Moraiolo and Leccino showed values of disease index, fresh and dry weight of knots statistically lower respect to those recorded for cv. Frantoio. It was also showed that independently from the cultivars tested no significant differences in virulence were observed between the five *Pseudomonas savastanoi* pv. *savastanoi* strains.

Results obtained from the second experiment are showed in Tables 2 and 4. Among the three olive cultivars, Frantoio, with 373 mg knot fresh weight, was the most susceptible, while Leccino, with 67 mg knot weight (82 percent less than Frantoio), the least susceptible.

The differences in both disease index and knot weight among the different crosses in full-sib (FS) families were not statistically significant. The same results were also observed for the differences among three half-sib (HS) groups (Table 2).

Table 2. Responses of olive cultivars, full-sib (FS) and half-sib (HS) families to inoculation with one strain of *Pseudomonas savastanoi* pv. *savastanoi*.

	Cultivars and Family	Disease index		Knot fresh wt (mg)		Knot dry wt (mg)	
Cultivar	Frantoio	2.80	a	373	a	189	a
	Leccino	0.93	b	67	b	47	b
	Moraiolo	1.87	ab	243	ab	153	ab
	Mean of cultivars	1.87		228		137	
FS	Frantoio x Maurino	1.85	a	273	a	140	a
	Leccino x Maurino	1.97	a	302	a	157	a
	Maurino x Frantoio	2.60	a	380	a	190	a
	Mean of FS seedlings	1.98		298		154	
HS	Ascolana	1.81	a	308	a	153	a
	Leccino	1.71	a	275	a	145	a
	Moraiolo	1.75	a	308	a	156	a
	Mean of HS seedlings	1.76		294		150	

Values are the mean of at least 3 replications. Data within a column for cultivars, FS and HS seedlings followed by the same letters are not statistically different at P=0.05 (Least Square Means).

There was a significant difference in disease index and knot fresh and dry weight of different genotypes. The Mo.2.108, with knot fresh weight of 850 mg, was the most susceptible genotype and Mo.6.105, with knot fresh weight of 30 mg (i.e. 96 percent knot fresh weight less than Mo.2.108), was the most tolerant one (Table 4).

Table 3. Estimate of standard error of mean for replication number from 3 to 10 and 5 and 10 inoculation sites.

5 inoculation/shoot		10 inoculation/shoot	
Replication number	SEM*	Replication number	SEM*
3	0.321	3	0.297
4	0.277	4	0.257
5	0.249	5	0.230
6	0.228	6	0.210
7	0.210	7	0.195
8	0.197	8	0.182
9	0.184	9	0.170
10	0.176	10	0.161

*Standard Error of Mean

The correlation coefficients between estimated disease index and knot fresh and dry weight are 0.91 and 0.94 respectively, showing that knot weight could be substituted by disease index without loosing significant information.

Table 4. Responses of olive cultivars, full-sib and half-sib genotypes to inoculation with *Pseudomonas savastanoi* pv. *savastanoi*.

Genotype	Disease index		Knot fresh wt (mg)		Knot dry wt (mg)	
As.1.101	1.6	cdefg	220	defg	120	efghi
As.4.119	1.4	efg	100	efg	61	fghi
As.4.122	1.53	defg	227	defg	115	efghi
As.6.105	1.4	efg	230	defg	116	efghi
As.6.121	1.6	cdefg	210	defg	113	efghi
As.6.123	2.65	abcd	640	abc	299	abc
Frantoio	2.8	abc	373	cdef	189	bcdefgh
Fr x Ma.1.107	2.7	abcd	470	bcd	245	bcde
Fr x Ma.4.104	0.6	g	60	fg	26	hi
Fr x Ma.6.109	1.4	efg	90	efg	45	ghi
Le.1.107	1.4	efg	140	efg	98	efghi
Le.1.133	1.8	cdefg	248	defg	130	defghi
Le.4.120	1	fg	63	fg	43	ghi
Le.6.108	2.1	bcdef	400	cde	220	bcdef
Le.6.120	2	bcdef	370	cdef	203	bcdefg
Le.6.122	2.4	abcde	710	ab	290	abcd
Leccino	0.93	fg	67	fg	47	ghi
Le x ma.6.302	1.85	bcdef	218	defg	129	defghi
Le x Ma.4.301	1.4	efg	190	defg	102	efghi
Le x Ma.4.303	3	ab	750	ab	324	ab
Ma x Fr.3.114	2.6	abcde	380	cdef	190	bcdefgh
Mo.2.108	3.4	a	850	a	412	a
Mo.6.105	0.6	g	30	g	16	i
Mo.6.118	1.5	defg	175	defg	99	efghi
Moraiolo	1.87	bcdef	243	defg	153	cdefghi

Values are the mean of at least 3 replications. Data within a column followed by the same letters are not statistically different at P=0.05 (Least Square Means).

To find the optimal number of replication and inoculation sites, sample size estimate was performed using experimental error and sampling error variance obtained from the analysis of variance of disease index. The standard error of mean of 5 or 10 inoculation points per shoot, for increasing replication numbers from 3 to 10 is shown in Table 3. The reduction in inoculation points from 10 to 5 could be compensated by increasing the replication numbers from 5 to 6 or 6 to 7. Therefore the experimental labours could be reduced quite to half, with the same statistical precisions.

In conclusion, all the olive genotypes tested showed typical olive knot symptoms when inoculated with *P. s.* pv. *savastanoi* and, similarly to other studies, resistant genotypes were not found. However in spite of small number of genotypes assayed, some interesting results were obtained that are promising for accessing to desired genotypes.

REFERENCES

Benjama, A., Walali, L., Janati, L., and Moukhali, A., 1992, Field reaction of different varieties of olive (*Olea europaea* L.) to olive knot disease caused by *Pseudomonas syringae* pv. *savastanoi* (Smith) Stevens. *Al-Awamia* **75**: 41-52.

Iannotta, N., Lombardo, N., Monardo, D., and Perri, L., 2002, Prime osservazioni sulla suscettibilità di cultivar di olivo alla "Rogna" (*Pseudomonas savastanoi*). Atti del Convegno Internazionale di Olivicoltura. 22-23 Aprile 2002.

Marcelo, A., Fernandes, M., Potes, M.F., and Serrano, J.F., 1999, Reaction of some cultivars of *Olea europaea* L. to experimental inoculation with *Pseudomonas syringae* pv. *savastanoi*. *Acta Hort.* **474**: 581-584.

Rademaker, J.L.W., and de Bruijn, F.J., 1997, Characterisation and classification of microbes by rep-PCR genomic fingerprinting and computer-assisted pattern analysis, p.151-171. In: Protocols, Application and Overviews (G. Caetano-Anolles, P. Gressfoff, ed.), J. Wiley and Sons, NY, USA.

Schaad, N.W., Jones, J.B., and Chun, W., 2001, Plant pathogenic bacteria. Third Edition The American Phytopathological Society, St. Paul, Minnesota, USA.

Sisto, A., Lo Cantore, P., and Iacobellis, N.S., 2001, Risultati preliminari sulla risposta di varietà di olivo alla inoculazione artificiale con *Pseudomonas savastanoi* pv. *savastanoi*. VII Convegno Annuale della SIPaV- Potenza, 3-5 Ottobre 2001.

Varvaro, L., and Surico, G., 1981, Titolazione dell'infettività di *Pseudomonas syringae* pv. *savastanoi* (Smith) su alcune cultivar di olivo (*Olea europaea* L.). *Phytopathol. Medit.*, **20**(1):72-78.

Preliminary Results on the Antibacterial Activity of Essential Oils on Some Pathovars of *Pseudomonas syringae*

P. LO CANTORE, N.S. IACOBELLIS, F. SENATORE[#] and F. CAPASSO[§]

Dipartimento di Biologia, Difesa e Biotecnologie Agro Forestali, Università degli Studi della Basilicata, C.da Macchia Romana, 85100 Potenza, Italy; [#]Dipartimento di Chimica delle Sostanze Naturali, Università di Napoli, Italy; [§]Dipartimento di Farmacologia Sperimentale, Università di Napoli, Italy.

Abstract: Essential oils extracted from fruits of cumin, caraway, fennel and coriander, were assayed *in vitro* for antibacterial activity toward strains of ten *Pseudomonas syringae* pathovars. Significant antibacterial activity was shown by essential oils of cumin, caraway and coriander. In general a lower activity was shown by essential oil from fennel. These preliminary results indicated the potential use of the above essential oils in the control of diseases caused by *P. syringae* pathovars on important crops.

Key words: Bactericides, Bacterial disease control, Phytopathogenic bacteria, *Pseudomonas syringae* pathovars.

Author for correspondence: N.S. Iacobellis; e-mail: iacobellis@unibas.it.

1. INTRODUCTION

The limited availability of bactericides useful for the control of bacterial diseases of plants is a real problem in agriculture since bacterial diseases of cultivated plants under certain environmental conditions can be highly destructive. Furthermore, a high number of phytopathogenic bacteria can be diffused also at long distances by contaminated and/or infected seeds and these traits may have very high importance in bacterial diseases management and control. Beside antibiotics and copper compounds, no other active compounds are available. However, either antibiotics or copper compounds,

N.S. Iacobellis *et. al.* (eds.), *Pseudomonas syringae* and Related Pathogens, 495-499.

though for different aspects, present undesirable attributes for human and animal health, and for the environment (Ambrige and Haines, 1987). As a consequence, antibiotics have been banned in agricultural practices in many countries (Barnard *et al.*, 1997) and the use of copper compounds is on the way to be restricted and controlled in the European Union. Furthermore, quite recent studies have shown that some phytopathogenic bacterial species have developed resistance toward copper and antibiotics commonly used in agriculture (May, 1985; Urech *et al.*, 1997; Williams and Heymann, 1998). This situation prompts the need for the development of alternative active compounds and/or methods for the control of plant bacterial diseases to be used in integrated crop management (Bolkan and Reinert, 1994; Rice *et al.*, 1998) as well as in bio-organic agriculture. Plant extracts and/or essential oils and/or their components - already known, also from alternative and traditional medicine, to have antimicrobial activities (Cox *et al.*, 2001; Cimanga *et al.*, 2002) - capable to inhibit the growth of phytopathogenic bacteria may represent new active compounds with potentially reduced or no environmental impact and phytotoxicity, and safe for consumers (Satish *et al.*, 1999). In this paper, preliminary results on the antibacterial activity shown *in vitro* by essential oils obtained by hydrodistillation from fruits of cumin, caraway, fennel and coriander is reported.

2. MATERIALS AND METHODS

2.1 Isolation of Essential Oils

Aliquots of 25 g of dried fruits of cumin (*Cuminum cyminum* L.), caraway (*Carum carvi* L.), fennel (*Foeniculum vulgare* Mill.) and coriander (*Coriandrum sativum* L.) were grounded and then subjected to hydrodistillation for 3 h following the procedure reported in the *Farmacopea Europea* (1997). Before their use the essential oils were stored at 4°C.

2.2 Antimicrobial Assays

Aliquots (1, 2, 4 or 8 μl) of each essential oil and 10 μl serial dilutions of a stock solution of 1.6 mg ml^{-1} of rifampicin were added to 6 mm diameter sterile blank disks (Oxoid). Thereafter these were deposited onto the surface of Petri plates containing 10 ml King medium B (King *et al.*, 1954) covered with 3 ml soft agar (0.7%) seeded with 500 μl of suspensions of the target bacteria to reach a final population of about 10^7 CFU ml^{-1}. *Pseudomonas syringae* pathovar strains used as target micro-organisms were routinely grown for 72 h at 25°C on glycerol nutrient agar. After 2, 4 and 7 days incubation at 25°C the minimal inhibitory quantity for each compound, expressed in µl, was recorded. The minimal inhibitory quantity is the oil aliquot which does not cause an apparent inhibition zone out of the 6 mm diameter disk. The assays were performed twice with three disks.

3. RESULTS AND DISCUSSION

In this paper, preliminary results on the antibacterial activity of essential oils of cumin, caraway, fennel and coriander toward 22 strains of 10 *Pseudomonas syringae* pathovars, namely *P. s.* pv. *syringae*, *P. s.* pv. *phaseolicola*, *P. s.* pv. *pisi*, *P. s.* pv. *aptata*, *P. s.* pv. *apii*, *P. s.* pv. *atrofaciens*, *P. s.* pv. *lachrymans*, *P. s.* pv. *maculicola*, *P. s.* pv. *tomato* and *P. s.* pv. *glycinea* are reported. Essential oils of cumin, caraway and coriander showed in general a higher significant antibacterial activity than essential oil of fennel (Table 1).

Table. 1. Antibacterial activity of essential oils of cumin, caraway, fennel and coriander toward strains of *Pseudomonas syringae* pathovars.

Bacteria	N° strains	Minimal inhibitory quantity of essential oil (µl)			
		Cumin	Caraway	Fennel	Coriander
Pseudomonas syringae pv. *phaseolicola*	4	2	2	8**	2
P. s. pv. *pisi*	2	4	4	8**	2
P. s. pv. *syringae*	3	2	2	8**	4
P. s. pv. *aptata*	2	8	2	8**	4
P. s. pv. *apii*	1	2	2	8**	ND
P. s. pv. *atrofaciens*	2	1*	1*	4	8
P. s. pv. *lachrymans*	2	1*	1*	8**	8**
P. s. pv. *maculicola*	2	2	2	8**	1*
P. s. pv. *tomato*	2	8	4	8**	4
P. s. pv. *glycinea*	2	1*	1*	1	1*

*The deposition of 1 µl of essential oils led to a measurable inhibition zone around the 6 mm diameter disks. **The deposition of 8 µl of essential oils did not lead to an inhibition zone around the 6 mm diameter disks. ND = not determined.

In particular, interest is the fact that 1 µl of essential oils of cumin, caraway and coriander are still active in inhibiting the growth of strains of *P. s.* pv. *atrofaciens, P. s.* pv. *lachrymans, P. s.* pv. *maculicola* and *P. s.* pv. *glycinea.* Data on quantitative composition of essential oils are not yet available and consequently no conclusion on their specific activity may be drown. However, of interest is the fact that the minimal inhibitory quantity of rifampicin against the same bacteria ranged from 1 to 4 μg. Results of this study, though preliminary, confirmed the antimicrobial activity of essential oils already reported mainly for food-borne and animal pathogens (Kim *et al.*, 1995; Cox *et al.*, 2001; Cimanga *et al.*, 2002) and for some phytopathogenic bacteria and fungi (Scortichini and Rossi, 1989; 1993; Caccioni *et al.*, 1995; Peana e Moretti, 1996; Arras e Arru, 1998; Satish *et al.*, 1999; Deena and Thoppil, 2000).

P. syringae pathovars are responsible for hundreds of diseases of important crops whose control and management may be difficult due to the lack of effective bactericides that at the same time present low or no environmental impact. The results reported here appear promising for a possible use of the above essential oils or their components as bactericides for the control and management of plant diseases caused by *P. syringae* pathovars. Of particular interest is the possibility to use these substances for seeds treatment. In this regard, it is necessary to remember that all pathovars of *P. syringae* considered in this study as well as a high number of phytopathogenic bacteria are disseminated at long distances by the trade of infected and/or contaminated seeds (Bazzi, 1991; Claflin, 2003). Hence there is a need for active substances for seed treatment and/or coating. So far acid compounds (i.e. HCl, acetic acid, etc.), copper compounds or chlorine derivatives as well as fermentation of fruits from which seeds are extracted are used with a certain efficacy in seed treatments for the sanitation and/or reduction of some phytopathogenic bacteria (Claflin, 2003). The availability of new active principles such as essential oils or their components is of great interest. Of course, other studies are necessary to evaluate the toxicity of the substances of interest toward seeds and/or plants and to set the appropriate formulations useful for the purpose.

ACKNOWLEDGEMENTS

This research has been supported in part by "Agenzia Lucana di Sviluppo e di Innovazione in Agricoltura", Italy.

REFERENCES

Ambridge, E.A., and Haines, T.H., 1987, Some aspects of pesticide use and human safety in South East Asia. In *Proceedings of 11th International Congress of Plant Protection*, October 5-9 ed. Magallona, E.D. pp. 219-224. Manila, Philippines.

Arras, G., e Arru, S., 1998, Attività fungitossica di oli essenziali nei confronti di patogeni degli agrumi in postraccolta. Saggi in ambiente ipobarico sull'essenza di *Thymus capitatus*. Atti VI Convegno Annuale SIPaV, Campobasso, 17-18 settembre 1998, p. 38.

Bazzi, C., 1991, I batteri fitopatogeni trasmissibili per seme. *Petria* **1**: 19-30.

Bolkan, H.A., and Reinert, W.R., 1994, Developing and implementing IPM strategies to assist farmers: an industry approach. *Plant Disease* **78**: 545-550.

Caccioni, D.R.L., Deans, S.G., and Ruberto, G., 1995, Inhibitory effect of citrus fruit essential oil components on *Penicillium italicum* and *P. digitatum*. *Petria* **5** (2): 177-182.

Cimanga, K., Kambu, K., Tona, L., Apers, S., De Bruyne, T., Hermans, N., Totté, J., Pieters, L., and Vlietinck, A.J., 2002, Correlation between chemical composition and antibacterial activity in the Democratic Republic of Congo. *J. Ethnopharmacol.* **79**: 213-220.

Cox, S.D., Mann, C.M., and Markham, J.L., 2001, Interaction between components of the essential oil of *Melaleuca alternifolia*. *J. Appl. Microbiol.* **91**: 492-497.

Claflin, L., 2003, Control of *Pseudomonas syringae* Pathovars, 423-430, (this volume).

Deena, M.J., and Thoppil, J.E., 2000, Antimicrobial activity of the essential oil of *Lantana camara*. *Fitoterapia* **71**: 453-455.

European Pharmacopoeia, 1997, 3th Ed., Council of Europe, Strasbourg, p. 121.

Kim, J., Marshall, M.R., and Cheng-i Wei, 1995, Antibacterial Activity of Some Essential Oil components against the Foodborne Pathogens. *J. Agric. Food Chem.* **43**: 2839-2845.

King, E.O., Ward, M.K., and Raney, D.E., 1954, Two simple media for the demonstration of pyocyanin and fluorescein. *J. Lab. and Cl. Med.* **44**: 301-307.

May, R.M., 1985, Evolution of pesticide resistance. *Nature* **15**: 12-13.

Peana, A.T., e Moretti, M.D.L., 1996, Attività "in vitro" degli oli essenziali di *Salvia desoleana* Atzei and Picci e *Salvia sclarea* L. nei confronti di fitopatogeni. *Acta Phytother.* **1**: 25-29.

Rice, M.J., Legg, M., and Powell, K.A., 1998, Natural products in agriculture - a view from the industry. *Pesticides Sci.* **52**: 184-188.

Satish, S., Raveesha, K.A., and Janardhana, G.R., 1999, Antibacterial activity of plant extracts on phytopathogenic *Xanthomonas campestris* pathovars. *Lett. Appl. Microbiol.* **28**: 145-147.

Scortichini, M., and Rossi, M.P., 1989, *In vitro* activity of some essential oils toward *Erwinia amylovora* (Burril) Winslow. *Acta Phytopatholog. Entomol.* **24**: 423.

Scortichini, M., and Rossi, M.P., 1993, *In vitro* behaviour of *Erwinia amylovora* towards some natural products showing bactericidal activity. *Acta Hortic.* **338**: 191.

Urech, P.A., Staub, T., and Voss, G., 1997, Resistance as a concomitants of modern crop protection. *Pesticide Sci.* **51**: 227-234.

Williams, R.J., and Heymann, D.L., 1998, Containment of antibiotic resistance. *Science* **279**: 1153-1154.

Molecular Techniques for Identification and Detection

Pathovars of *Pseudomonas syringae* are Structured in Genetic Populations Allowing the Selection of Specific Markers for their Detection in Plant Samples

C. MANCEAU and C. BRIN
U.M.R. Pathologie Végétale INRA-INH-Université d'Angers, 42 rue Georges Morel, 49071 Beaucouzé, France.

Abstract: DNA relatedness was assessed between strains representative of all genomospecies of oxydase-negative fluorescent pseudomonads by *f*-AFLP analysis (fluorescent labelled-Amplified Fragments Length Polymorphism analysis). Our results indicated that AFLP taxa fit well with genomospecies previously described by DNA pairing analysis and delineated sub-groups which matched with pathovars. We used several approaches to clones pathovar-specific DNA fragments which could be used to design pathovar-specific primers. We identified PCR primers for the detection of *Pseudomonas syringae* pathovar *pisi*, *papulans*, *tomato*, *lachrymans* in plant samples.

Key words: AFLP, DNA/DNA hybridisation, *hrpL*, PCR.

Author for correspondence: C. Manceau; e-mail: manceau@angers.inra.fr.

1. INTRODUCTION

Pseudomonas syringae pathovars cause consequent losses on numerous and often high added value crops. In front of the lack of efficient chemicals and resistant varieties in many cases, plant certification and crop management are basic ways to control bacterial disease. The development of reliable and specific detection procedures that are the indispensable tools for plant certification and epidemiological studies, is based on pathovar specific markers. Molecular techniques have been available for three decades.

N.S. Iacobellis *et. al.* (eds.), *Pseudomonas syringae* and Related Pathogens, 503-512.

Several different approaches have been used to identify molecular markers including gene encoding pathovar specific traits, ribosomal operon sequence analysis and random selection of markers (Manceau and Horvais, 1998). Determination of intraspecies relatedness has been facilitated by rapid DNA typing methods. Among those which target whole genomes, AFLP (Amplified Fragment Length Polymorphism) seems to fit well to assess the relatedness between pathovars of *P. syringae* (Clerc *et al.,* 1998). DNA relatedness was assessed between strains representative of all genomospecies by *f*-AFLP analysis (fluorescent labelled fragments analysis). Our results indicated that AFLP taxa fit well with genomospecies previously described by DNA pairing analysis and delineated sub-groups which matched with pathovars. The success in the selection of pathovar-specific molecular markers within *P. syringae* depends on the genetics of host and non-host specificity and the possibility of genetic transfer of virulence traits between plant associated bacteria. Current genomic sequence similarity deduced from AFLP data was compared with nucleotide sequence similarity obtained from *hrpL* and *hrpS* sequences previously described by Sawada *et al.* (1999). The data were found to be highly correlated supporting the hypothesis that *hrp* gene may have been introduced into an ancestor of *P. syringae* and followed the same evolution process.

These results suggest that it might be possible to clone pathovar-specific DNA fragments either by global approaches such as cloning of RAPD fragments (Trebaol *et al.,* 2001), subtractive hybridisation (Manceau *et al,* 2000) or by sequence comparison of housekeeping genes (Smid *et al.,* 1995) and pathogenicity related genes (Leite *et al.*, 1994). We used such approaches to clones pathovar-specific DNA fragments which could be used to design pathovar-specific primers. We identified PCR primers for the detection of *P. syringae* pv. *pisi*, *P. s.* pv. *papulans*, *P. s.* pv. *tomato*, *P. s.* pv. *lachrymans* in plant samples (Manceau *et al*., unpublished data).

2. HOW IS *PSEUDOMONAS SYRINGAE* GENETICALLY STRUCTURED?

The first question that is raised when we attempt to select genomic markers of pathovars is: How are pathovars genetically structured? Two possibilities are either pathovar determinants might be clustered on special sites to form kinds of pathogenicity islands or they are scattered throughout the genome as a consequence of the evolution into clonal populations.

The consequence of an organisation in a single cluster would be the potential occurrence of horizontal gene transfer and an high importance of mutations in pathogenicity variations. In contrast, the evolution of pathovars into clonal populations would result in a vertical transfer of pathogenicity determinants, this could result in the occurrence of phylogenetic markers correlating with pathovar grouping.

So, to understand how pathovars might be delineated, we have undertaken a phylogenetic analysis of *Pseudomonas syringae.* We use the AFLP (Amplified Fragment Length Polymorphism) technique (Vos *et al.*, 1995) to investigate the genetic structure of *P. syringae* (Clerc *et al.*, 1998).The studies were based on the pathovar nomenclature of Gardan *et al*, (1999), consisting in 9 closely related genomic species within the cytochrome-oxydase-negative fluorescent pseudomonads.

2.1 AFLP Analysis

The different steps of our AFLP protocol are the following:

- Total bacterial DNA samples were extracted by using the Dneasy tissue kit of Qiagen S.A. (Courtaboeuf, France).
- The DNA samples were digested with either *EcoR*I or *Pst*I endonuclease and *Msp*I.
- Complementary double stranded DNA adaptators were ligated to both ends of restriction fragments
- A selective amplification was performed using two primers complementary to the adaptators. One of them contained 2 selective nucleotides at the 3' end.

DNA fragments were separated by an ABI prism 310 genetic analyser following methods described by the manufacturer. Sample data were then analysed with the GenScan Analysis software (Applied Biosystems, Foster City, CA.) (Fig. 1) and electrophoregrams analysed for each sample.AFLP patterns were transformed into tabular binary matrices and the program Dist-AFLP was used to calculate similarities with the DICE index (http://pbil.univ-lyon1.fr/ADE-4/microb/). Phylogenic analyses were performed with the phylogenic inference package PHYLIP (Feselstein, 1993). Bootstrapped dendrograms, constructed by using the Neighbour Joining method, showed that strains grouped with high bootstrap values (<70%). These groups were highly correlated to genomic groups delineated by DNA/DNA hybridisation technique. The strains of *Pseudomonas syringae* pv. *actinidiae* were grouped with *P. s.* pv. *theae.* This grouping is in accordance with the results published by Scortichini *et al.,*(2002) and reinforces the proposition that has suggested to place *P. s. pv. actinidiae* in the genomic group 8 defined by Gardan *et al.,* (1999).

In addition, the AFLP analysis displayed pathovars as genetic structured populations at the intra specific level. A strong structuration with significative bootstraps (100%) was observed at the pathovar level within the genomic species 2 (Fig. 2). The two strains of *P. s.* pv. *morsprunorum* and the eleven strains of *P. s.* pv. *lachrymans* as well are strongly grouped.

Groups of strains were delineated within *P. s.* pv. *lachrymans* but no correlation was observed with any geographic origin or phenotypic trait.

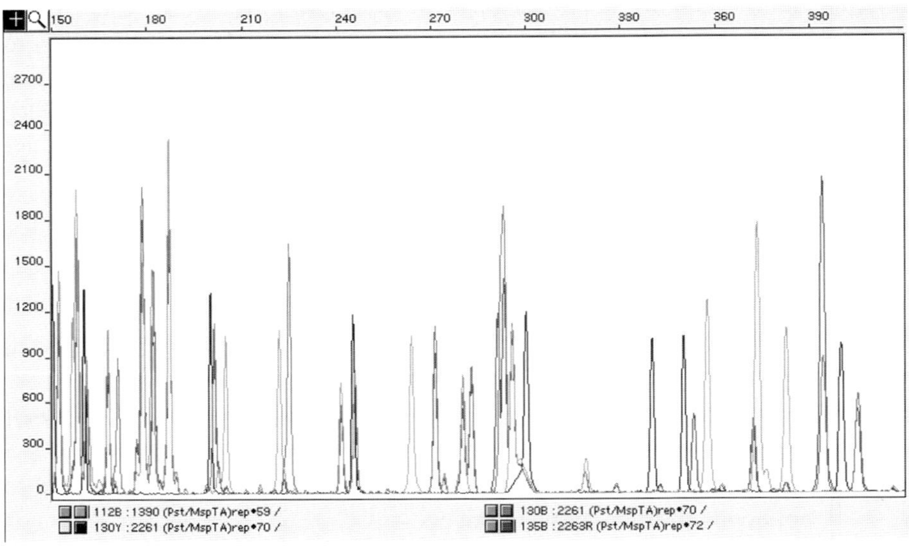

Figure 1. Parts of electrophoregrams (150-400 bp) of three AFLP samples (one strain of *Pseudomonas syringae* pv. *phaseolicola* and two strains of *P. s.* pv. *lachrymans*). The *P. s.* pv. *phaseolicola* strain displayed 40 picks and the *P. s.* pv. *lachrymans* strains displayed 35 and 36 picks, respectively. They displayed 20 picks in common and 35 polymorphic picks. Only five picks were polymorphic between the strains of *P. s.* pv. *lachrymans*.

2.2 AFLP Versus DNA/DNA Hybridisation Data

Our AFLP data were compared to the DNA/DNA reassociation values obtained previously by Gardan *et al.* (1999). An high correlation score and significant linear regressions were obtained by considering the whole data set and by considering only pairs of bacteria belonging to the same genomic group. However, there was no significant correlation among distances for more distantly related strains (Fig 3).

2.3 Diversity of the *hrpL* Gene

The *hrpL* is a gene involved in the regulation of numerous genes that are themselves involved in plant/bacteria interactions. It is located in the hrp cluster (Lindgren *et al.*, 1986) that is common to all phytopathogenic fluorescent pseudomonads. Sawada *et al.* (1999) reported that these *hrp* genes have been stable and evolved along with the *Pseudomonas syringae*

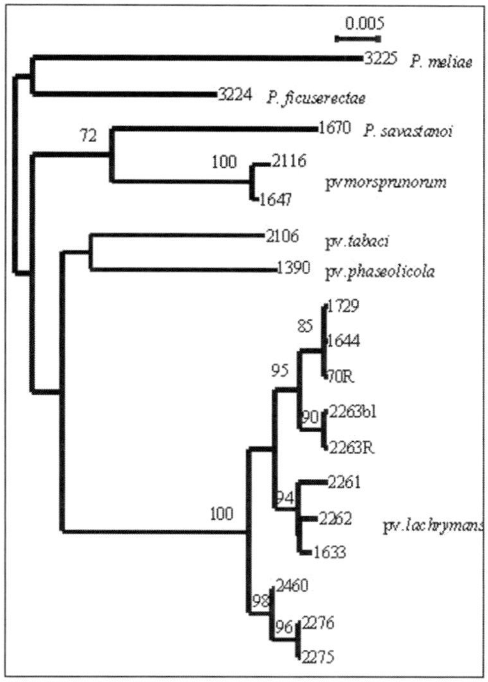

Figure 2. Dendrogram obtained by comparison of AFLP data from *Pseudomonas syringae* strains belonging to the genomic group 2 defined by Gardan *et al.*, (1999). *P. meliae* and *P. ficuserectae* formed two separated lineages. The two strains of *P. s.* pv. *morsprunorum* and the eleven strains of *P. s.* pv. *lachrymans* formed lineages with bootstrap scores of 100.

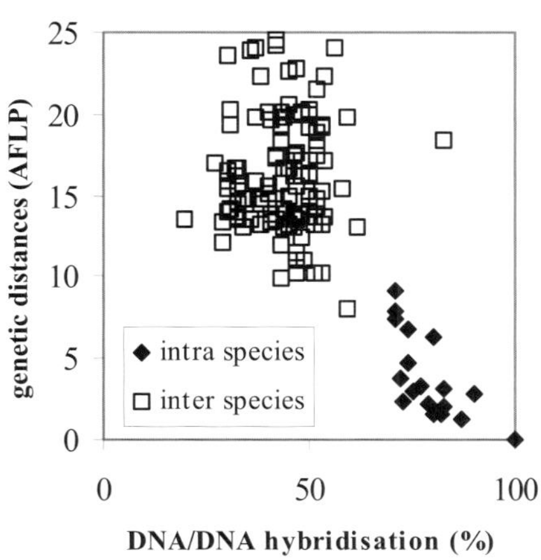

Figure 3. Relation between genetic distances calculated with AFLP data and genetic distances calculated with sequence data of *hrpL* gene of *Pseudomonas syringae* strains belonging to the same genomic group (intra specific) and to different genomic groups (inter specific).

genome. They used 2 genes: *gyrB* and *rpoD* as index genes to determine the course of bacterial genomes evolution. Again, when we compared the data obtained by AFLP with data obtained by sequence comparison of the *hrpL* gene, we observed linear regression between the two sets of data (Fig. 4).
However, the data obtained from different techniques result in different conclusions. Data from *hrpL* analysis showed a wide gap between the intra specific and the inter specific distances. So this type of data is probably useful to identify bacterial species. In contrast, AFLP analysis shows data from strains within the same genomic species and strains belonging to different genomic groups ranged more regularly. So, although AFLP is a better technique to analyse the diversity within a genomic species than *hrpL* sequence analysis, both techniques can be used to select genomic markers of either species or pathovars.

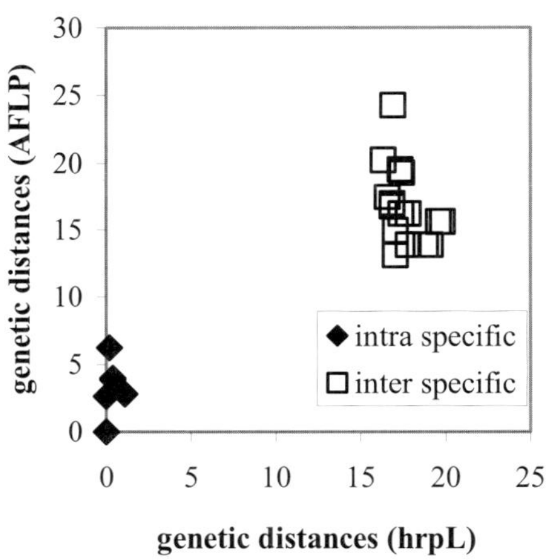

Figure 4. Relations between genetic distances calculated with AFLP data and genetic distances calculated with sequence data of *hrpL* gene of *Pseudomonas syringae* strains belonging to the same genomic group (intra specific) and to different genomic groups (inter specific).

3. HOW TO SELECT PATHOVAR MARKERS

Four useful strategies to select genomic markers of *Pseudomonas syringae* pathovars include the following:

- The use of genes which encode specific pathogenesis traits, for example phaseolotoxin genes to detect *P. s.* pv. *phaseolicola* (Prosen *et al.,* 1993; Marques *et al.*, 2000).

- Designing of primers after comparison of widely conserved DNA region such as *rrn* operon (Horvais, 1995) .
- Cloning of selected DNA fragments from differential banding patterns (AFLP, RAPD,) or at random by subtractive hybridisation.
- Designing of primers after comparison of *hrp* gene sequences.

The two last strategies make sense only because the pathovars of *P. syringae* are structured as clonal populations. Hereafter, we are going to illustrate these strategies by reporting two examples of selection of pathovar specific primer sets.

3.1 Cloning of Selected DNA Fragments

There are two strategies to select DNA fragments specific to pathovars either by cloning DNA fragment that have been detected by electrophoresis of DNA fragments or by using subtractive hybridisation technique to clone specific DNA fragments at random. We used a subtractive hybridisation technique adapted from the procedure described by Seal *et al.* (1992) and Manceau *et al.* (2000) to isolate DNA markers specific to *Pseudomonas syringae* pv. *pisi*. Three probes specific to strains of *P. s.* pv. *pisi* were selected and three primer sets were designed and tested for specificity. One set of primers (A/B) detected all *P. s.* pv. *pisi* strains whatever the race. The other sets detected only certain strains of *P. s.* pv. *pisi*. The patterns of detection of these two primer sets were different from those described by Arnold *et al.* (1996). These results suggest that the *P. s.* pv. *pisi* is probably quite polymorph.

3.2 Comparison of DNA Sequences in the *hrpL* Gene

The identification and detection of *Pseudomonas syringae* pv. *papulans,* the causal agent of blister spot on apple, in apple leaf and fruit by PCR amplification with specific DNA fragment in *hrpL* sequence has been reported (Kerkoud *et al.,* 2002). Consensus primers hrpL1 and hrpL2 were designed based on the alignment of the pseudomonad *hrpL* gene sequences available (GenBank Accession No. U03864 and L3653). This primer set produced a 631-bp amplicon with 33 out of the 53 pseudomonad strains tested. All the strains belonged to genomospecies 1 and 2 as described by Gardan *et al.* (1999). The amplicon obtained with 30 of these 33 strains was digested with eight restriction enzymes. Three different patterns, A1, A2 and B, were produced. Patterns A1 and A2 were observed with strains belonging to genomospecies 1 only and pattern B characterised strains belonging to genomospecies 2 only. Patterns A1 and A2 differed at only two sites, a *Bsp 143*I site located at nucleotide 360 and a *Mse*I site located at nucleotides 22-24. Group A2 consisted solely of *P. s.* pv. *papulans* strains. The *hrpL* gene in

P. s. pv. *papulans* strain CFBP3323 was sequenced. Two primer sets, Pap1/Pap2 and Pap1/Pap3, were designed and tested for specificity to *P. s.* pv. *papulans*. These primers amplified fragments of 242 and 303 bp, respectively. Pap1/Pap2 amplified only with *P. s.* pv. *papulans* DNA (Fig. 5) while Pap1/Pap3 amplified all tested strains belonging to genomospecies 1. A diagnostic procedure based on the Pap1/Pap2 primer set was evaluated for the detection of *P. s.* pv. *papulans* in diseased fruits and inoculated leaves.

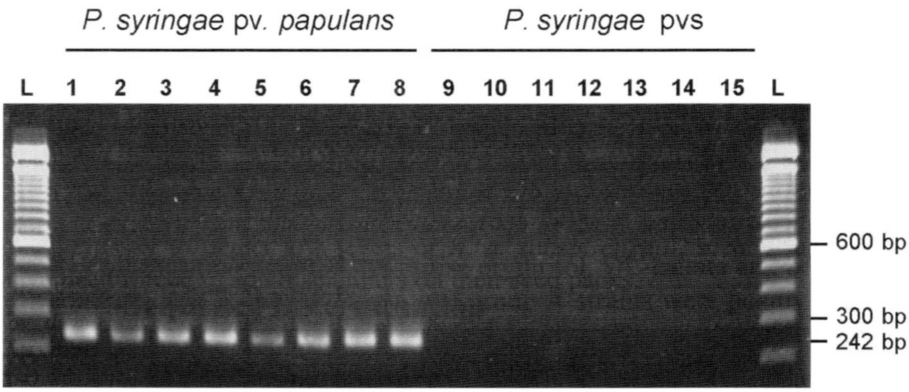

Figure 5. PCR amplification of *Pseudomonas syringae* strains with primer set Pap1/Pap2. Lanes 1 to 8, *P. s.* pv. *papulans*: CFBP1754, CFBP3324, CFBP3323, CFBP5043, CFBP5134, CFBP5135, CFBP5132, CFBP5133. Lane 9, negative control; lane 10, *P. s.* pv. *syringae* CFBP1392; lane 11, *P. savastanoi* CFBP1670; lane 12, *P. s.* pv. *tomato* CFBP2212; lane 13, *P. s.* pv. *garcae* CFBP1634; lane 14, *P. viridiflava* CFBP2107; lane 15, *P. s.* pv. *tagetis* CFBP1694; lane L, DNA size markers (100-bp ladder, Life Technologies).

4. CONCLUSION

The AFLP technique is a useful technique for determining genomic differences within strains of *Pseudomonas syringae*. AFLP analysis could be used along with complete genome sequence analysis to estimate genome divergence (Mougel *et al.*, 2002). Whole genomes probably diverge more rapidly than is reported for genes. Evolutionary genome divergence was used for phylogenetic inferences. Members of the same genomic species and the same pathovar clustered consistently. The AFLP analysis fits very well with intra-specific level analysis and displayed that pathovars are structured as clonal populations.

Thus, simple genomic markers can be selected to design reliable molecular reagents useful in the rapid identification and in the detection procedures of *P. syringae* pathovars.

ACKNOWLEDGEMENTS

The work from our laboratory was supported by grants from the Concil of "La Région des Pays de la Loire", France. We would also like to thank Dr. L. Sutra for helpful discussions on taxonomic aspects.

REFERENCES

Arnold, D.L., Atheh-Pollord, A., Gibbon, M.J., Taylor, J.D., and Vivian, A., 1996, Specific oligonucleotide primers for the identification of *Pseudomonas syringae* pv. *pisi* yield one of two possible DNA fragments by PCR amplification: evidence for phylogenetic divergence. *Physiol. Mol. Plant Pathol.* **49**: 233-245.

Clerc, A., Manceau, C., and Nesme, X., 1998, Comparison of randomly amplified polymorphic DNA with amplified fragment length poymorphism to assess genetic diversity and genetic relatedness within genospecies III of *Pseudomonas syringae. Appl. Environ. Microbiol.* **64**: 1180-1187

Felsenstein, J., 1993, PHYLIP: Phylogenetic Inference Package version 3.5c Seattle: University of Washington.

Gardan, L., Shafik, H., Belouin, S., Broch, R., Grimont, F., and Grimont, P.A.D., 1999, DNA relatedness among the pathovars of *Pseudomonas syringae* and description of *Pseudomonas tremae* sp. nov. and *Pseudomonas cannabina* sp. nov. (*ex* Sutic and Dowson 1959). *Int. J. Syst. Evol. Microbiol.* **49**: 469-478.

Horvais, A., 1995, Fragment de l'AND génomique de *Clavibacter michiganensis*, sonde d'hybridation, amorces d'amplification, réactifs et procédé de détection de *Clavibacter michiganensis*. French patent n° 95.05416.

Kerkoud, M., Manceau, C., and Paulin, J.P., 2002, Rapid diagnosis of *Pseudomonas syringae* pv. *papulans*, the causal agent of blister spot of apple, by polymerase chain reaction using specifically designed *hrpL* gene primers. *Phytopathology* **92**: 1077-1083.

Leite, R.P.Jr., Minsavage, G.V., Bonas, U., and Stall, R.E., 1994, Detection and identification of phytopathogenic *Xanthomonas* strains by amplification of DNA sequences related to the *hrp* genes of *Xanthomonas campestris* pv. *vesicatoria. Appl. Environ. Microbiol.* **60**: 1068-1077.

Lindgren, P.B., Peet, R.C., and Panopoulos, N.J., 1986, Gene cluster of *Pseudomonas syringae* pv. *phaseolicola* controls pathogenicity on bean plants and hypersensitive reaction on non-host plants. *J. Bacteriol.* **168**: 515-522.

Manceau, C., and Horvais, A., 1998, Strategies for the identification of oligonucleotidic sequences specific to plant pathogenic bacteria useful to design primers for PCR. *In* COST 823 - New technologies to improve phytodiagnosis. Advances in the detection of plant pathogens by polymerase chain reaction. 154 pp. C. Manceau and J. Spak ed., Office for official publications of the European Communities, Luxembourg.

Manceau, C., Coutaud, D., and Guyon, R., 2000, Assessment of substractive hybridization to select species and subspecies specific DNA fragments for the identification of *Xylophilus ampelinus* by polymerase chain reaction (PCR). *Eur. J. Plant Pathol.* **106**: 243-253.

Marques, A.S.A., Corbière, R., Gardan, L., Tourte, C., Manceau, C., Taylor, J.D., and Samson R., 2000, Multiphasic approach for the identification of the different classification levels of *Pseudomonas savastanoi* pv. *phaseolicola. Eur. J. plant Pathol.* **106**: 715-734.

Mougel, C., Thioulouse, J., Perrière, G. and Nesme, X., 2002, A mathematical method for determining genome divergence and species delineation using AFLP. *Int. J. Syst. Evol. Microbiol.* **52**: 573-586.

Prosen, D., Hatziloukas, E., Schaad, N.W. and Panopoulos, N.J., 1993, Specific detection of *Pseudomonas syringae* pv. *phaseolicola* DNA in bean seed by polymerase chain reaction-based amplification of a phaseolotoxin gene region. *Phytopathology* **83**: 965-970.

Sawada, H., Suzuki, F., Matsuda, I., and Saitou, N., 1999, Phylogenetic analysis of *Pseudomonas syringae* pathovars suggests the horizontal gene transfer of *argK* and the evolutionary stability of *hrp* gene cluster. *J. Mol. Evol.* **49**: 627-644.

Scortichini, M., Marchesi, U., and Di Prospero, P., 2002, Genetic relatedness among *Pseudomonas avellanae*, *P. syringae* pv. *theae* and *P. s.* pv. *actinidia*e, and their identification. *Eur. J. Plant Pathol.* **108**: 269-278.

Seal, S.E., Jackson, L.A., and Daniels, M.J., 1992, Isolation of a *Pseudomonas solanacearum*, specific DNA probe by substractive hybridisation and construction of species-specific oligonucleotide primers for sensitive detection by the polymerase chain reaction. *Appl. Environ. Microbiol.* **58**: 3751-3758.

Smid, E.J., Jansen, A.H.J., and Gorris L.G.M., 1995, Detection of *Erwinia carotovora* subsp. *atroseptica* and *Erwinia chrysanthemi*in potato tubers using polymerase chain reaction. *Plant Pathol.* **44**: 1058-1069.

Trebaol, G., Manceau, C. Tirilly, Y., and Boury S., 2001, Assessment of the genetic diversity among strains of *Xanthomonas cynarae* by randomly amplified polymrorphic DNA analysis and development of specific characterised amplified regions for the rapid identification of *X. Cynarae. Appl. Environ. Microbiol.* **67**: 3379-3384.

Vos, P.R., Hogers, R., Bleeker, M., Reijans, M., Van der Lee, T., Hornes, M., Fridjers, A., Pot, J., Peleman, J., Kuiper, M., and Zabeau M., 1995, AFLP: a new technique for DNA fingerprinting. *Nucleic Acids Res.* **23**: 4407-4414.

Real - Time PCR for *Ralstonia solanacearum*

D.E. STEAD, S.A. SIMPKINS, J.E. ELPHINSTONE, N.C. SMITH, J. HENNESSY and S.A. WELLER
Central Science Laboratory, Sand Hutton, York, YO41 1LZ, UK.

Abstract: Two TaqMan PCR assays have been developed to test for *Ralstonia solanacearum*. One assay (RS) was genus specific, reacting with all strains of *R. solanacearum* but also the Banana blood disease bacterium and *R. syzygii*. Another assay (B2) was specific for Biovar 2 (Race 3) strains which cause brown rot of potatoes. The selective media SMSA and Willbrinks medium have been modified into broth form to detect the pathogen in complex substrates, such as sewage and river water, where population numbers are often low. Using these media in a pre-enrichment real-time bio PCR assay facilitates the robust, highly sensitive detection of the potato brown rot pathogen in natural samples including tubers and waste water.

Key words: Bio PCR Potato Brown Rot, selective media, *Ralstonia* spp..

Author for correspondence: D.E. Stead; e-mail: d.stead@csl.gov.uk.

1. INTRODUCTION

Potato brown rot caused by *Ralstonia solanacearum* was first found in the UK in 1992 (Walker and Stead, 1993). We now have a good understanding of the sources of infection and the epidemiology of the pathogen. The role of aquatic *Solanum dulcamara* plants in the pathogen life cycle is well documented (Stead *et al*., 1996). Control of the pathogen, including its eradication from rivers, is dependant on cost effective monitoring of the pathogen in a number of complex substrates including river water, soil, sewage effluents and potato wastes. The methods approved for use as part of the European Commission (EC) Control Directive (Anonymous., 1998) are not always appropriate for some of these substrates. A recent EC funded

N.S. Iacobellis *et. al.* (eds.), *Pseudomonas syringae* and Related Pathogens, 513-521.

project to validate suitable methods (Elphinstone *et al.,* 2000) clearly showed that few of the available methods included in the validation were sensitive, specific or robust enough for detection of small populations in some of these substrates. A single cost-effective method capable of high sample throughput was an additional criterion for an ideal method. A fluorescent-based real-time PCR assay using the TaqMan system (Applied Biosystems, Foster City, CA, USA) has been developed for detecting the potato ring rot pathogen *Clavibacter michiganensis* subsp. *sepedonicus* in potato (Schaad *et al*., 1999) Although development work on real-time TaqMan PCR was being done at the time of this validation exercise, the methods were not ready for inclusion. This paper reviews the progress made since then and shows that real-time TaqMan PCR offers a potentially very useful single diagnostic method especially for specific, sensitive and robust detection in complex substrates, appropriate for large scale surveys.

2. MATERIALS AND METHODS

2.1 TaqMan Probe Design

Two TaqMan assays to detect *Ralstonia solanacearum* have been designed by staff at the Central Science Laboratory (CSL) (Weller *et al*., 2000a). Sequences of primers and TaqMan probes used are given in Table 1. The broad host range *R. solanacearum* probe (RS-P) is partially homologous to 16S rRNA gene primer OLI1 (Seal *et al*., 1993), with primers (RS-I and RS-II) flanking this region. Primers and TaqMan probe for the biovar 2A (Race 3) specific assay were elucidated from a biovar 2A-specific DNA fragment described by Fegan *et al*. (1998) retaining the original forward primer 631 (B2-II), and selecting probe (B2-P) and reverse primer (B2-I) from the upstream region. RS-P was covalently labelled at the 5′ terminal nucleotide with the (FAM) reporter dye and at the 3′ terminal nucleotide with the (TAMRA) quencher dye. B2-P was labelled with the (VIC) reporter dye at the 5′ terminal nucleotide and again with the (TAMRA) quencher dye at the 3′ terminal nucleotide. Both probes were made by Applied Biosystems.

2.2 5’ Nuclease PCR Assay

PCR was performed in 25 μl volumes using MicroAmp Optical 96-well reaction plates and MicroAmp Optical Caps (PE Biosystems, Foster City,

Table 1. Characteristics of primers and TaqMan probes used to detect *Ralstonia solanacearum.*

Primer/Probe[a]	Sequence (5′-3′)	Length (nt)	Dye
RS-I - F	GCA TGC CTT ACA CAT GCA AGT C	22	
RS-II - R	GGC ACG TTC CGA TGT ATT ACT CA	23	
RS-P	AGC TTG CTA CCT GCC GGC GAG TG	23	FAM
B2-I - F	TGG CGC ACT GCA CTC AAC	18	
B2-II - R	AAT CAC ATG CAA TTC GCC TAC G	22	
B2-P	TTC AAG CCG AAC ACC TGC TGC AAG	24	VIC

[a] F = Forward, R = Reverse, P = Probe

CA, USA) for each well. All reagents were obtained from the TaqMan Core PCR Reagent Kit (PE Biosystems, Foster City, CA, USA). The PCR mixture for the RS and B2 multiplex was; 2.0 μl of cell lysate; 2.5 μl 10X TaqMan Buffer A, 200 nM (each) dATP, dCTP, dGTP; 400 nM dUTP; 0.025 U μl^{-1} AmpliTaq Gold DNA polymerase; 0.01U μl^{-1} uracil-*N*-glycosylase (AmpErase UNG); 300 nM (each) primers RS-I, RS-II, B2-I, B2-II; 25 nM TaqMan probe RS-P; 50 nM TaqMan probe B2-P. An ABI Prism 7700 Sequence Detection System, (PE Biosystems; Foster City, CA, USA) was used for amplification and fluorescence measurement. All cycles began with 2 min at 50°C then 10 min at 95°C, followed by 40 two-step cycles of 10 s at 95°C then 1 min at 60°C.

2.3 Pre-enrichment in Selective SMSA Broth

Modified Selective medium South Africa (SMSA) broth (Elphinstone *et al.*, 1996) has been used in a pre-enrichment, Bio-PCR protocol (Schaad *et al.*, 1995) prior to the TaqMan assay to increase the sensitivity of the reaction and facilitate ease of DNA extraction (Weller *et al.*, 2000b). In addition Willbrinks medium can also be modified into broth form (Caruso *et al.*, 2002) and used to enrich samples prior to PCR. Briefly one hundred μl of any sample substrate (i.e. potato waste, river water, sewage) is added to 10 ml of selective broth and incubated at 28°C. After 24, 48, 72 and 96 hours, 100 μl of culture is removed and heated to 96°C for 4 minutes. This lysate is then used as the template in a TaqMan PCR reaction. A range of closely related bacterial strains, including strains that had cross-reacted with *Ralstonia solanacearum* polyclonal antisera in IF or ELISA were also evaluated (Table 2).

The threshold of detection of *R. solanacearum* bv 2 in sewage with a pre-enrichment was determined by comparing cycle threshold values for a dilution series of the pathogen in de ionised water and a sample of sewage effluent.

2.4 Source of Cultures

Ten strains of *Ralstonia solanacearum* bv 2, race 3 and single strains of bv 1, 2T, 3, 4 and 5; five *R. pickettii*, three *Ralstonia* spp., one Banana blood disease bacterium, three *R. syzygii* and eight other bacteria were used for specificity studies (Table 2).

Table 2. Ralstonia solanacearum strains, strains related to *R. solanacearum*, or having previously cross reacted with polyclonal antisera to *R. solanacearum,* including *Ralstonia* spp., *Burkholderia* spp., *Pseudomonas* spp. and *Bacillus polymyxa* in IF or ELISA, evaluated during PCR using RS-I and RS-II primers with RS-P fluorogenic probe and B2-I and B2-II primers with B2-P fluorogenic probe.

Organism	Strain	Bv.	Origin	Source	RS PCR	B2 PCR
R. solanacearum	NCPPB1824	2A	Egypt	Potato	+	+
R. solanacearum	CSL465	2A	Egypt	Potato	+	+
R. solanacearum	CSL1328	2A	Egypt	Potato	+	+
R. solanacearum	CSL3467	2A	England	River Water	+	+
R. solanacearum	CSL3468	2A	England	Tomato	+	+
R. solanacearum	NCPPB1584	2A	Cyprus	Potato	+	+
R. solanacearum	NCPPB2505	2A	Sweden	Potato	+	+
R. solanacearum	RVP502	2A	Belgium	Potato	+	+
R. solanacearum	NCPPB4069	2A	Slovenia	Potato	+	+
R. solanacearum	CIP310	2A	Colombia	Potato	+	+
R. solanacearum	NCPPB325	1	USA	Tomato	+	-
R. solanacearum	CIP226	2T	Brazil	Potato	+	-
R. solanacearum	CIP49	3	Australia	Potato	+	-
R. solanacearum	R470	4	Philippines	Ginger	+	-
R. solanacearum	R288	5	China	Mulberry	+	-
R. picketti	NCPPB3839		Unknown	Unknown	-	-
R. picketti	NCPPB4076		Unknown	Unknown	-	-
R. picketti	CSL3466		Unknown	Unknown	-	-
R. picketti	CFBP2459		Unknown	Unknown	-	-
Ralstonia sp.	NCPPB928		Jamaica	Sugarcane	+	-
Ralstonia sp.	NCPPB927		USA	Sugarcane	-	-
Ralstonia sp.	NCPPB1695		Tanzania	Sugarcane	-	-
Ralstonia sp.	PD2778		Unknown	Unknown	-	-
BBD Bacterium	NCPPB3726		S.Sulawesi	Banana	+	-
R. syzygii	NCPPB3790		N. Sumatra	Clove	+	-
R. syzygii	NCPPB3791		West Java	Clove	+	-
R. syzygii	NCPPB3792		Central Java	Clove	+	-
B. caryophylli	NCPPB353		USA	Unknown	-	-
B. cepacia	NCPPB945		Unknown	Unknown	-	-
B. cepacia	CFBP2227		Unknown	Unknown	-	
B. glumae	NCPPB3708		Japan	Rice	-	-
B. plantarii	NCPPB3590		Japan	Rice	-	-
Bacillus polymyxa	CFBP1954		Unknown	Unknown	-	-
P. m. marginalis	CFBP1538		Unknown	Unknown	-	-
P. chloraphis	NCPPB4082		Unknown	Unknown	-	-

3. RESULTS

3.1 Specificity of Probes

All *Ralstonia solanacearum* strains were positive using RS-P (Table 2). The B2-P 5′ nuclease assay resulted in positive results for all biovar 2A strains tested. With biovar 2T and biovar 1, 3, 4, and 5 strains no fluorescence was detected during the B2-P assay indicating that amplification had not occurred (Table 2). Several of the other bacterial strains produced fluorescence during the RS-P assay, namely the banana blood disease bacterium, *R. syzygii* strains, and one (from three tested) *Ralstonia* sp.(syn. *Pseudomonas* sp. (Taxon B) strain). No fluorescence was detected from the B2-P assay on any of the other strains. No positive results were obtained for any of the other reference bacteria from either assay.

3.2 Pre-enrichment of Sample Types

Both modified liquid SMSA (Elphinstone *et al.*, 1996) and modified liquid Willbrinks medium (Caruso *et al.*, 2002) have successfully been used as a pre-enrichment prior to TaqMan PCR. Pre-enrichment in selective broth facilitates detection of *Ralstonia solanacearum* from complex substrates (Fig. 1), which might otherwise inhibit the PCR or possess too few numbers of the pathogen, to be detected. In addition by observing decreases in cycle threshold (C_T) values between samples before and after enrichment, the viability of the pathogen in samples can also be demonstrated (Fig. 2).

3.3 Detection in Sewage Effluent

Pre-enrichment allowed very sensitive detection of *Ralstonia solanacearum* in sewage effluent. Even with a total of 3 cells per reaction, the CT value was 38. With a population of 30 cells per reaction the CT value was 31. The results in Fig. 2 clearly show that the CT values for the same population in water and sewage were the same.

4. DISCUSSION

As expected, the RS TaqMan primer/probe set detected all bacteria tested with known homology to the specific region of the 16S rRNA gene used for probe and primer design (Seal *et al.*, 1993). These included isolates representing all biovars and other infra-subspecific strains of *Ralstonia solanacearum* as well as the closely related banana blood disease bacterium

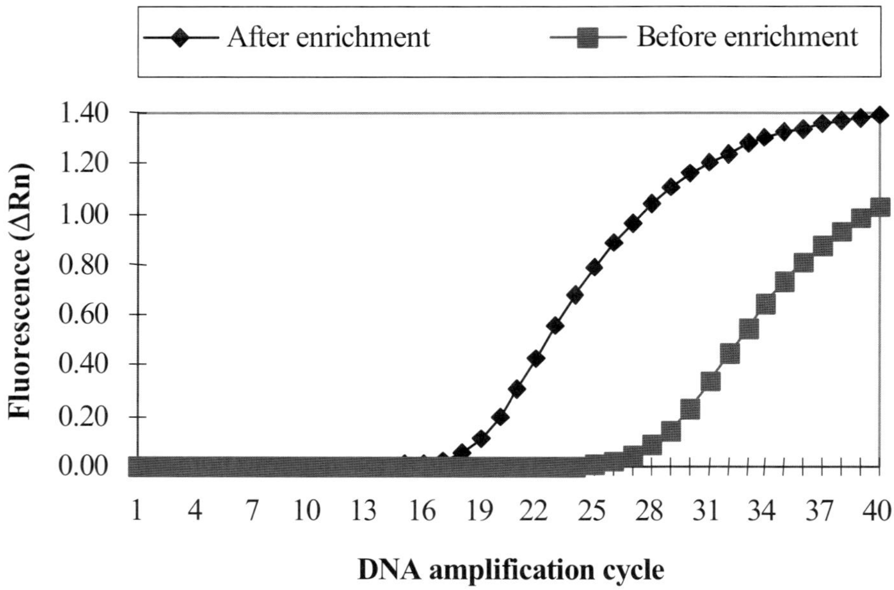

Figure.1. Detection of *Ralstonia solanacearum* in potato waste using enrichment in SMSA broth.

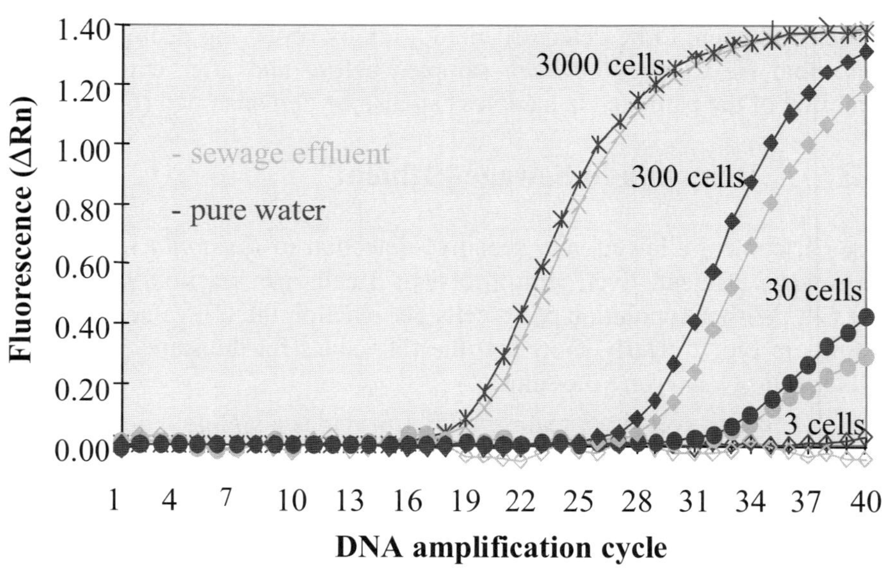

Figure 2. Comparison of detection of known concentrations of *Ralstonia solanacearum* in deionised water and sewage effluent.

and *R. syzygii* (the distribution of which has not been recorded outside of Indonesia). In addition, one isolate (from three tested) of *Ralstonia* sp. previously listed as *Pseudomonas* sp. Taxon B in the National Collection Of Plant Pathogenic Bacteria (NCPPB), commonly associated with sugar cane (Hayward, 1962), also produced a positive result during the RS assay. This bacterium is known to have cellular fatty acid profiles closely resembling those of *R. solanacearum.* Other known species of *Ralstonia* not sharing homology with this specific region of the 16S rRNA gene, were not detected.

In contrast, the B2 TaqMan primer/probe set was *R. solanacearum* bv2 specific, detecting only strains of the biovar 2A phenotype, including strains of both RFLP groups 26 and 27 as described by Cook and Sequeira (1991). As predicted, the DNA sequence used for probe and primer design in this assay is apparently unique to the biovar 2A genotype (Fegan *et al.,* 1998). However, the sensitivity of detection of *R. solanacearum* biovar 2A was significantly higher when the TaqMan rather than the original PCR protocol was used (N.C. Smith; unpublished data). The B2 assay therefore ensures sensitivity and specificity of detection of the race 3 phenotype. Biovar 2T, (which is believed to have originated in South America) has a significantly different genetic composition from that of other biovar 2A strains (Cook and Sequeira, 1991) accounting for the failure of the TaqMan probe B2-P to detect this strain.

The addition of an enrichment step or Bio-PCR to detect *R. solanacearum* cells using TaqMan PCR increases the sensitivity and robustness of testing procedures for the brown rot pathogen direct from potato tissue, as has also been shown for *Clavibacter michiganensis* subsp. *sepedonicus* (Schaad *et al.*, 1999) and for post enrichment ELISA to detect *R. solanacearum* (Caruso *et al.*, 2002). Although detection of as few as 10^2 cells ml^{-1} of *R. solanacearum* in potato extract using direct TaqMan™ PCR has been observed (Weller *et al.*, 2000a), experience has shown that repeat assays sometimes give negative results at this (and higher) concentrations, presumably due to the presence of PCR inhibitors present in potato tuber tissue extracts. In our laboratory we have detected the pathogen in potato tissues at a concentration of 10^1 cells per ml after 96 hours incubation at 28 °C (Weller *et al.*, 2000b). In addition by measuring the amount of DNA (and therefore cells) over time, real-time PCR provides a simple indication of the viability of the pathogen isolated from samples.

This and other post enrichment TaqMan PCR or TaqMan Bio-PCR assays demonstrates the potential for robust detection of plant pathogenic bacteria, typically present in low population numbers-especially where PCR inhibitors are known to be a problem e.g. soil (Ito *et al.*, 1998), seeds (Schaad *et al.*, 1995). potato waste and sewage. From some of these sample types, other detection techniques such as immunofluorescence are not appropriate methods. Pre-enrichment has been shown to be equally effective in detecting a known concentration of cells in sewage effluent as it is in detecting the same cell concentrations in deionised water, demonstrating the effectiveness of the method. This method is now the method of choice for the detection of the brown rot pathogen in our laboratory.

ACKNOWLEDGEMENTS

This work was funded by Plant Health Division, DEFRA, Project PH0154.

REFERENCES

Anonymous, 1998, Council Directive 98/57/EC. *Official Journal of the European Communitiies* **L-235**: 1-39.

Caruso, P., Gorris, M.T., Cambra, M., Palomo, J.L., Collar, J., and Lopez, M.M., 2002, Enrichment double-antibody sandwich indirect enzyme-linked immunosorbent assay that uses a specific monoclonal antibody for sensitive detection of *Ralstonia solanacearum* in asymptomatic potato tubers. *Appl. Environ. Microbiol.* **68**: 3634-3638.

Cook, D., and Sequeria, L., 1991, The use of subtractive hybridisation to obtain a DNA probe specific to *Pseudomonas solanacearum*. *Mol. General Gen.* **227**: 401-410.

Elphinstone, J.G., Hennessy, J., Wilson, J.K., and Stead, D.E., 1996, Sensitivity of different methods for the detection of *Ralstonia solanacearum*. *OEPP/EPPO Bulletin* **26**: 674-678.

Elphinstone, J.G., Stead, D.E., Caffier, D., Janse, J.D., Lopez, M.M., Mazzucchi, U., Muller, P., Persson, P., Rauscher, E., Schiessendoppler, S., Sousa Santos, M., Stefani, E., and van Vaerenbergh, J., 2000, Standardisation of methods for detection of *Ralstonia solanacearum* in potato. *OEPP/EPPO Bulletin* **30**: 391-395.

Hayward, A.C., 1962, Differentiation, taxonomy and nomenclature of the bacteria causing red stripe and mottled stripe diseases. *Mauritius Sugar Industry Research Institute Occasional Paper* **13**: 13-27.

Fegan, M., Holoway, G., Hayward, A.C., Timmis, J., 1998, Development of a diagnostic test based on the polymerase chain reaction (PCR) to identify strains of *R. solanacearum* exhibiting the biovar 2 genotype, p. 34-43. *In* Prior, P.H., Allen, C., Elphinstone, J.G. (eds.), Bacterial Wilt Disease: molecular and ecological aspects. Springer-Verlag, Berlin Heidelberg, Germany.

Ito, S., Ushijima, Y., Fujii, T., Tanaka, S., Kameya-Iwaki, M., Yoshiwara, S., and Kishi, F., 1998, Detection of viable cells of *Ralstonia solanacearum* in soil using a semi-selective medium and a PCR technique. *J. Phytopathol.* **146**: 379-384. 1998.

Schaad, N.W., Cheong, S.S., Tamaki, S., Hatziloukas, E., and Panopoulos, N.J., 1995, A combined biological and enzymatic amplification (BIO-PCR) technique to detect *Pseudomonas syringae* pv. *phaseolicola* in bean seed extracts. *Phytopathology* **85**: 243-248.

Schaad, N.W., Berthier-Schaad, Y., Sechler, A., and Knorr, D., 1999, Detection of *Clavibacter michigsanensis* subsp. *sepedonicus* in potato tubers by BIO-PCR and an automated Real time fluorescence detection system. *Plant Dis.* **83**: 1095-1100.

Seal, S.E., Jackson, L.A., Young, J.P.W., and Daniels, M.J., 1993, Differentiation of *Pseudomonas solanacearum, Pseudomonas syzygii, Pseudomonas pickettii* and the blood disease bacterium by partial 16S rRNA sequencing: construction of oligonucleotide primers for sensitive detection by polymerase chain reaction. *J. Gen.Microbiol.* **139**: 1587-1594.

Stead, D.E., Elphinstone, J.G., and Pemberton, A.W., 1996, Potato Brown Rot in Europe. *Brighton Crop Protection Conference Proceedings* 9B-3, **3**: 1145-1152.

Walker, D.R.I., and Stead, D.E., 1993, Potato brown rot: a new threat to potato production in the EC. *In* Plant Health and the European Single Market (D. Ebbels, ed.) BCPC Monograph **54**: 345-348. British Crop Protection Council, Farnham, UK

Weller, S.A., Elphinstone, J.G., Smith, N.C., Boonham, N., and Stead, D.E., 2000a, Detection of *Ralstonia solanacearum* strains using a quantitative, multiplex, real-time, fluorogenic PCR (TaqMan) assay. *Appl.Environ. Microbiol.* **66**: 2853-2858.

Weller, S.A., Elphinstone, J.G., Smith, N.C., Stead, D.E., 2000b, Detection of *Ralstonia solanacearum* from potato tissue by post-enrichment TaqMan PCR. *OEPP/EPPO Bulletin* **30**: 381-393.

Optimising PCR Detection of *Ralstonia solanacearum* and *Pseudomonas savastanoi* pv. *savastanoi*: Two Models, Two Approaches

E. BERTOLINI, P. CARUSO, R. PENYALVER, A. OLMOS, J.M. QUESADA, M. CAMBRA and M.M. LÓPEZ

Instituto Valenciano de Investigaciones Agrarias (IVIA). Apartado oficial. 46113 Moncada, Valencia, Spain.

Abstract: Available protocols and primers for PCR detection of *Ralstonia solanacearum* on water samples and *Pseudomonas savastanoi* pv. *savastanoi* on asymptomatic olive trees have relatively low reliability because of sensitivity and specificity problems.

In order to improve PCR detection of *R. solanacearum* in water samples a recently described Co-operational amplification technique (Co-PCR) was applied. The simultaneous use of OLI-1, OLI-2 and JE2 primers improved the detection, giving an amplification product of 408 bp and a sensitivity of < 1 CFU ml^{-1} of water. This method was used for analysis of water samples from several origins and it was the most efficient for *R. solanacearum* detection when comparing to isolation and to previously described PCR protocols.

The PCR detection of *P. s.* pv. *savastanoi* by using IAALF and IAALR primers was improved with an efficient DNA extraction protocol (after pre-enrichment of the target bacterial population in PVF-1 liquid medium), designing IAALN1 and IAALN2 internal primers for a nested PCR in a single closed tube and by the colorimetric detection of the amplicon of 338 bp. Moreover, the four primers were included in a multiplex nested RT-PCR protocol that allows the simultaneous detection of the bacterial target and four RNA viruses in olive plants. Sensitivity in spiked samples was up to 1 CFU ml^{-1} of plant extract by either method and both were successfully applied to the detection of *P. s.* pv. *savastanoi* on asymptomatic shoots from olive plants grown in orchards and nurseries.

Key words: Co-PCR, nested PCR, multiplex nested-PCR.

Author for correspondence: M.M. López; e-mail: mlopez@ivia.es.

N.S. Iacobellis *et. al.* (eds.), *Pseudomonas syringae* and Related Pathogens, 523-529.

1. INTRODUCTION

Ralstonia solanacearum and *Pseudomonas savastanoi* pv. *savastanoi* are phytopathogenic bacteria responsible of bacterial wilt and brown rot of solanaceous plants and olive knot disease, respectively. *R. solanacearum* is a quarantine bacterium in the EU and *P. s.* pv. *savastanoi* is included in the certification programs of olive trees in several EU countries. Consequently, efforts are required to improve the sensitivity and specificity of the available PCR protocols of both targets. We describe here two different approaches used to optimise their PCR detection improving sensitivity and specificity, reducing the effect of inhibitors and facilitating routine applications.

2. OPTIMISING PCR DETECTION OF *RALSTONIA SOLANACEARUM* IN WATER SAMPLES

To date, there is no a validated PCR protocol recommended for the detection of *Ralstonia solanacearum* in water (Anonymous, 1998). Most of the detection protocols for this quarantine organism in environmental water samples are time consuming and too laborious for its routine use with a large number of samples. Besides, the results are frequently negative, specially in cooler months when the bacterial population is very low (Janse *et al.*, 1998; Biosca *et al.*, 2002).

A new and highly sensitive method for the amplification of viral RNA targets from infected plants called Co-operational amplification (Co-PCR) has been recently developed. It is based on the simultaneous action of four primers in a single reaction (Olmos *et al.*, 2002). In order to increase the sensitivity in routine detection of *R. solanacearum* from water samples, a three primers Co-PCR approach has been applied (Caruso et al., 2003) and compared with the standard PCR protocols described by Seal *et al.* (1993) and Boudazin *et al.* (1999).

R. solanacearum strain IVIA 2528.55 (biovar 2) was used to prepare spiked water samples in sterile PBS or autoclaved river water to test the sensitivity of the different PCR protocols. For sensitivity determination, 5 µl of each heat treated water sample was analysed by the two standard PCR protocols and by Co-PCR as described below.

Fifty-six water samples were taken from September 2001 to May 2002 from various sources in Spain. They were tested for the presence of *R. solanacearum* by plating onto modified SMSA (Elphinstone *et al.*, 1996), enrichment-DASI-ELISA with specific monoclonal antibodies (Caruso *et al.*, 2002), standard PCRs and Co-PCR.

The Co-PCR approach was performed with primers: OLI1, OLI2 (Seal *et al.*, 1993; Boudazin *et al.*, 1999) and JE2 (Elphinstone *et al.*, 1996). The reaction was done in a final volume of 25 µl with the following reagents: 1 x *Taq* buffer (GIBCO-BRL), 4 mM $MgCl_2$, 0.2 mM of each deoxynucleoside triphosphate (Pharmacia LKB), 5% dimethyl sulfoxide, 0.15 µM of external primers, 0.075 µM of the internal primer, 1 U of *Taq* DNA polymerase (GIBCO-BRL), 3% formamide and 5 µl of each heat treated water sample. The amplification was done at 95°C for 3 min followed by 55 cycles at 94°C for 30 s, 65°C for 30 s and 72°C for 45 s, and then one cycle at 72°C for 10 min. The PCR products were visualised after electrophoresis on agarose gel and by colorimetric detection, according to Bertolini *et al.* (2003) and Caruso *et al.* (2003).

The sensitivity threshold of pure *R. solanacearum* suspensions in spiked samples of autoclaved river water and in natural water samples, using all assayed amplification protocols, is shown in Table 1.

Table 1. Sensitivity of the detection of *Ralstonia solanacearum* (Rs) in spiked samples and detection of the target in 56 environmental water samples by several PCR protocols.

PCR primers	Sensitivity in spiked samples (CFU ml^{-1})	Number of Rs positive water samples/number of analysed samples
OLI1/Y2	10^2	5/56
OLI1/Z	10	7/56
OLI1/OLI2/JE2 (Co-PCR)*	10^{-2}	21**/56

*Co-PCR with colorimetric detection.
**= 12 positive samples confirmed by isolation or enrichment DASI-ELISA.

The sensitivity level reached by Co-PCR amplification was higher than those obtained using any of the other methods (Seal *et al.*, 1993; Elphinstone *et al.*, 1996; Boudazin *et al.*, 1999). Co-PCR was able to detect *R. solanacearum* up to the 10^{-8} dilution from a suspension of 10^6 CFU ml^{-1}. The detection limit was estimated in 1 cell ml^{-1} after a total cells count, corresponding to cultivable, non cultivable and dead cells.

The usefulness of the protocol designed for the detection of *R. solanacearum* was confirmed using 56 environmental water samples (Table 1): as show in this table, 21 out of the analysed samples were positive by Co-PCR and only 5 or 7 by standard PCRs. All positive samples were collected from river water at points where *R. solanacearum* had been previously isolated (Caruso *et al.*, 2003), while the remaining 35 negative samples were taken from localities where *R. solanacearum* was not reported. The attempt to detect *R. solanacearum* by isolation or enrichment-ELISA using a modified protocol from Caruso *et al.* (2002) was successful only in 12 out of 56 analysed samples. It is not surprising that many samples positive by Co-PCR were undetected by the other PCR methods since they proved to be less sensitive in the comparative assays.

The application of Co-PCR to detect low populations of *R. solanacearum* in watercourses showed greater sensitivity and specificity than conventional

PCR protocols. We observed that the reaction of the three primers (OLI1/OL12/JE2) allows concomitant work in a single amplification reaction.

3. OPTIMISING PCR DETECTION OF *PSEUDOMONAS SAVASTANOI* PV. *SAVASTANOI* IN OLIVE SAMPLES

Conventional detection of *Pseudomonas savastanoi* pv. *savastanoi* is based on bacterial isolation followed by pathogenicity tests and biochemical or serological techniques, that are time consuming and expensive. For these reasons specific primers to *iaaL* gene were designed for PCR detection of *P. savastanoi*. A PCR detection protocol was developed with a previous enrichment of the bacteria in the semi selective medium PVF-1 in liquid preparation, reaching a sensitivity of 10 to 10^2 CFU ml^{-1} in plant extract (Penyalver *et al*., 2000). We report here on the development of a nested-PCR in a single closed tube coupled with dot-blot hybridisation to improve the sensitivity of the detection of *P. s.* pv. *savastanoi* in asymptomatic plant material.

The IAALF/IAALR primer pair (Penyalver *et al*., 2000) was used as external primers and two new internal primers (IAALN1 5'-3' CTC CCT CTC CAA CGT CTT C, and IAALN2 5'-3' GCC TGA TGA TTT TCT TCT G) were designed (Bertolini *et al*., 2003). As these primers give amplification products at 50ºC but not at 62ºC (annealing temperature of external primers), it was thus possible to separate internal and external primer pair activity. PCR reactions were performed in a final volume of 25 µl by combining: 1X *Taq* buffer (GIBCO-BRL), 2.0 mM $MgCl_2$, 3% formamide, 0.2 mM of each deoxynucleoside triphosphate (Pharmacia LKB), 0.005 µM of each external primer, 2 µM of each internal primer, 10% glycerol, 2U of *Taq* DNA polymerase (GIBCO-BRL). The template was 5 µl. Amplification conditions were 94°C for 5 min, followed by 25 cycles at 94°C for 30 s, 62°C for 30 s and 72°C for 30 s, immediately followed by a second round of 40 cycles at 94°C for 30 s, 50°C for 30 s 72°C for 30 s and an additional step of 72°C for 10 min. The specificity of the amplification was confirmed by Bertolini *et al.* (2003) using more than two hundred strains of different species and origins.

To evaluate the nested-PCR sensitivity level in plant samples, spiked olive plant extracts were prepared with known amounts of *P. s.* pv. *savastanoi* strain IVIA 1628 from ca. 1 to 10^7 CFU ml^{-1} of the plant extract. In addition, to determine the efficiency of this nested-PCR in detecting the target in asymptomatic olive tissues, 213 samples were collected from commercial orchards in all seasons taking shoots from branches without knots from

naturally infected olive trees. In addition, 30 samples were taken from symptomless plants grown in nurseries.
The newly developed nested-PCR was able to detect from 1 to 10 CFU ml^{-1} in the same amended extracts with or without a pre-enrichment. Consequently, a 10-fold improvement in sensitivity was observed over the previously developed PCR assay. The results of the detection of the target bacterium in asymptomatic olive samples by several techniques and PCR protocols are shown in Table 2.

Table 2. Detection of *Pseudomonas savastanoi* pv. *savastanoi* in asymptomatic olive tissues from naturally infected or healthy olive trees.

Technique	% positive samples
Isolation*	15.5
PCR	10.3
Enrichment-PCR	21.5
Nested-PCR	22.2
Enrichment nested-PCR (gel)	30.5
Enrichment nested-PCR (colorimetric detection)	35.2

* King's medium B and PVF-1 according to (Bertolini *et al.*, 2003).

This new nested-PCR coupling with a colorimetric detection was significantly different from the bacterial isolation and from the conventional PCR in detecting *P. s.* pv. *savastanoi* (Bertolini *et al.,* 2003). A pre-enrichment in PVF-1 medium was necessary for optimising the detection of the bacterium with all PCR techniques.
Furthermore, the nested-PCR with a pre-enrichment step detected *P. s.* pv. *savastanoi* in 9 out of 30 asymptomatic nursery plants confirming the presence of such bacterium as epiphytic or endophytic in asymptomatic olive plants.
Finally, a multiplex nested RT-PCR in a single closed tube was developed for the simultaneous detection of four RNA viruses (*Cucumber mosaic virus, Cherry leaf roll virus*, strawberry latent ringspot virus and *Arabis mosaic virus*) and the bacterium *P. s.* pv. *savastanoi* (Bertolini *et al.*, 2003). The method enabled for the first time the sensitive and simultaneous detection of RNA and DNA targets from plant viruses and a bacterium, saving time, decreasing risks of self-contaminations and reducing costs, compared with conventional monospecific nested amplifications. It was successfully coupled with colorimetric detection of amplicons using specific oligo-probes to simplify routine detection. Two hundred and forty five olive trees from 15 different varieties were analysed by nested-PCR and multiplex nested RT-PCR with colorimetric detection. The bacterium was detected in 28.7% of the analysed trees by this new multiplex method and in the same samples by the nested-PCR previously developed.
Sanitary selection and certification programs for olive plants that include *P. s.* pv. *savastanoi* in the schemes, require sensitive and specific methods for

the detection of this bacterium and have recently been started in Spain, Italy, Portugal, Israel and Argentina. In this context, the nested and multiplex nested PCR protocols have been developed for routine applications.

4. CONCLUSION

The performance of the designed PCR protocols for detection of *Ralstonia solanacearum* in water and *Pseudomonas savastanoi* pv. *savastanoi* in asymtomatic olive samples is better than other previous methods in sensitivity and specificity. Both are simple to perform and constitute useful tools that could also be used for epidemiological studies.

ACKNOWLEDGMENTS

The research of E. Bertolini and P. Caruso was supported by PhD grants from the Agencia Española de Cooperación Internacional and Instituto Valenciano de Investigaciones Agrarias, respectively. This work was supported by project FAIR 5-CT97-3632 of the European Union, and 1 FD 1997-2279, OLI96-2179 and CAOOO-007 from MEC, CICYT and MAPA, Spain.

REFERENCES

Anonymous, 1998, Council Directive 98/57/EC of 20 July 1998 on the control of *Ralstonia solanacearum* (Smith) Yabuuchi *et al.* Official Journal of the European Communities, No. L235/1, 21.08.98.

Bertolini, E., Olmos, A., López, M.M., and Cambra, M., 2003, Multiplex nested RT-PCR in a single tube for sensitive and simultaneous detection of four RNA viruses and *Pseudomonas savastanoi* pv. *savastanoi* in olive trees. *Phytopathology* **93**: 286-292.

Bertolini, E., Penyalver, R., García, A., Olmos, A., Quesada, J.M., Cambra, M., and López, M.M., 2003, Highly sensitive detection of *Pseudomonas savastanoi* pv. *savastanoi* in asymptomatic olive plants by nested-PCR in a single closed tube. *J. Microbiol. Methods* **52**: 261-266.

Biosca, E.G., Caruso, P., Bertolini, E., Álvarez, B., Palomo, J.L., Gorris, M.T., and López, M.M., 2002, Improved detection of *Ralstonia solanacearum* in culturable and VBNC state from water samples at low temperature. 3rd International Bacterial Wilt Symposium (in press).

Boudazin, G., Le Roux, A.C., Josi, K., Labarre, P., and Jouan, B., 1999, Design of division specific primers of *Ralstonia solanacearum* and application to the identification of European isolates. *Eur. J. Plant Pathol.* **105**: 373-380.

Caruso, P., Gorris, M.T., Cambra, M., Palomo, J.L., Collar, J., and López, M.M., 2002, Enrichment Double-Antibody Sandwich Indirect Enzyme-Linked Immunosorbent Assay that uses a specific monoclonal antibody for sensitive detection of *Ralstonia solanacearum* in asymptomatic potato tubers. *Appl. Environ. Microbiol.* **68**: 3634-3638.

Caruso, P., Bertolini, E., Cambra, M., and López, M.M., 2003, A new and sensitive Co-operational polymerse chain reaction (Co-PCR) for a rapid detection of *Ralstonia solacearum* in water. *J. Microbiol. Methods* (in press).

Elphinstone, J.G., Hennessy, J., Wilson, J.K., and Stead, D.E., 1996, Sensitivity of different methods for the detection of *Ralstonia solanacearum* in potato tubers extracts. *Bull. OEPP* **26**: 663-678.

Janse, J.D., Araluppan, F.A.X., Schans, J., Wenneker, M., and Westerhuis, W., 1998, Experiences with bacterial brown rot *Ralstonia solanacearum* biovar 2, race 3 in the Netherlands. In Prior, P., Allen, C., Elphinstone, J. (Eds.), Bacterial wilt disease. Molecular and ecological aspects, Heidelberg, Germany. Springer Verlag editions, pp. 146-152.

Penyalver, R., García, A., Ferrer, A., Bertolini, E., and López, M.M., 2000, Detection of *Pseudomonas savastanoi* pv. *savastanoi* in olive plants by enrichment and PCR. *Appl. Environ. Microbiol.* **66**: 2673-2677.

Olmos, A., Bertolini, E., and Cambra, M., 2002, Simultaneous and Co-operational amplification (Co-PCR): a new concept for detection of plant viruses. *J. Virol. Methods* **106**: 15-23.

Seal, S.E., Jakson, L.A., Young, J.P.W., and Daniels, M.J., 1993, Differentiation of *Pseudomonas solanacearum*, *Pseudomonas syzygii, Pseudomonas pickettii*, and Blood Disease Bacterium by partial 16S sequencing: construction of oligonucleotide primers for sensitive detection by polymerase chain reaction. *J. Gen. Microbiol.* **139**: 1587-1594.

Use of PCR for Rapid Identification of *Acidovorax avenae* and *A. avenae* subsp. *citrulli*

W.Y. SONG, A.J. SECHLER[#], E. HATZILOUKAS[§], H.M. KIM, and N.W. SCHAAD[#]

Dept. Agricultural Biology, Chonbuk National University, Chonju, Korea; [#]Foreign Disease-Weed Science Research Unit, ARS/USDA, 1301 Ditto Avenue, Ft. Detrick MD, 21702 USA; [§]University Ioannia, Ionnia, Greece.

Abstract: The species *Acidovorax avenae* contains bacteria pathogenic to several crops: *A. a.* subsp. *avenae* infects oats, corn, rice, millet, and sugarcane; subsp. *citrulli* infects watermelon and melon; and subsp. *cattleyae* infects orchid. Because these bacteria are closely related their proper identification requires expensive, time-consuming pathogenicity tests. To develop PCR assays for rapid identification of all species and subsp. *citrulli*, we sequenced the intergenic spacer region of the 16S-23S rDNA and designed *avenae*-specific and subsp. *citrulli*-specific classical primers and real-time primers and probes. Classical *avenae*-specific primers Oaf1/Oar1 amplified a DNA product from all 46 strains of subsp. *avenae* originating from foxtail, oats, corn, rice, sugarcane, and millet; the type strains of subsp. *cattleyae* from orchid; and 11 strains of subsp. *citrulli* from watermelon. Real-time *avenae*-specific primers Oaf1/Oar2 and probe AaP1 reacted with all 40 strains of subsp. *avenae*, the type strain of subsp. *cattleyae*, and 8 strains of subsp. *citrulli*. Classical *citrulli*-specific primers Aacf2/Aacr2 and real-time primers Aacf3/Aacr2 and probe AaP2 amplified a PCR product from all 11 strains of subsp. *citrulli*, but not from any other strain of subsp. *avenae*. None of over 50 other bacteria tested generated a PCR product with either classical primers or real-time primers and probe. The species-specific classical primers Oaf1/Oar1 and real-time primers Oaf1/Oar2 and probe AaP1 should prove useful for rapid identification of *A. avenae* strains from all hosts, whereas the *citrulli*-specific classical primers Aacf2/Aacr2 and real-time primers Aacf3/Aacr2 and probe AaP2 should be useful for specific identification of subsp. *citrulli*.

Key words: Real-time PCR, bacteria, rice, watermelon.

Author for correspondence: N.W. Schaad; e-mail: schaad@ncifcrf.gov.

N.S. Iacobellis *et. al.* (eds.), *Pseudomonas syringae* and Related Pathogens, 531-544.

1. INTRODUCTION

Acidovorax avenae (*Pseudomonas avenae*), is the causal agent of disease of several important crops including bacterial stripe of rice (Goto, 1964), leaf blight of oats (Manns, 1909), red stripe of sugarcane (Martin and Wismer, 1989) and millet (Goto and Okabe, 1952), and bacterial leaf spot and bud rot of orchid (Ark and Thomas, 1946). The organism was originally described by Manns as the causal agent of blade blight of oats in Ohio in 1909 (Manns, 1909). *Pseudomonas alboprecipitans* was later described as causing brown stripe of *Setaria italica* (foxtail) (Rosen, 1922), and as the causal agent of stalk rot of corn (Rosen, 1926), but the organism was shown later to be synonymous with *P. avenae* (Schaad *et al*., 1975). On rice the organism has been previously referred to as *P. setariae* or *P. panici* (Goto, 1964), and on sugarcane as *P. rubrilineans* and *P. rubrisubalbicans* (Martin and Wismer, 1989). These non-fluorescent pseudomonads have been shown to be closely related and reclassified as *P. avenae* based on numerical and genomic analysis (Hu *et al*., 1991). Another proposal places these organisms in the new 'acidovorans' DNA-rRNA homology group as *Acidovorax* (Willems *et al*., 1992). This group contains many pathogens previously classified in the non-fluorescent pseudomonad group, including *P. avenae* (Manns, 1909), *P. rubrilineans* (Martin and Wismer, 1989), *P. cattleyae* (Savulescu, 1947), and *P. pseudoalcaligenes* subsp. *citrulli* (Schaad *et al*., 1978). All these bacteria have been reclassified as a subspecies of *A. avenae* on the basis of results of DNA-DNA hybridisation, DNA-rRNA hybridisations, polyacrylamide gel electrophoresis of whole-cell proteins, and a numerical analysis of carbon assimilation tests (Willems *et al*., 1992). Those strains infecting rice, sugarcane, corn, millet, wheat, and oats are classified as *A. a.* subsp. *avenae*. Those infecting orchid are classified as *A. a.* subsp. *cattleyae* and those strains infecting watermelon and melon are classified as *A. a.* subsp. *citrulli*.

A. a. subsp. *avenae* is perhaps most serious on rice (Goto, 1964; Shakya, 1987). Isolation of the organism is difficult, because it is frequently overgrown by other bacteria such as *Pantoea* (*Erwinia*) *agglomerans* and *Pseudomonas fluorescens*. These pathogens are recognised as producing distinct disease symptoms on rice but field diagnosis is very difficult. In such cases, detection and identification is based on pathogen isolation, through the use of available semi-selective media, and pathogenicity tests to discriminate between the various pathogens. However, the presence of coexisting epiphytes such as *P. putida* and *P. fluorescens* makes isolation difficult. Furthermore, pathogenicity tests are very time consuming and expensive because rice seedlings are difficult to grow and symptoms on seedlings are often non-discriminating. Phenotypically, strains of *A. a.* subsp. *avenae* cannot be differentiated from strains of *A. a.* subsp. *citrulli*. A PCR method has been described for identification of subsp. *citrulli* (Minsavage *et al*., 1995), but the primers are not highly specific and the method has not gained acceptance. Recently, variations in length and sequences of 16S-23S rDNA intergenic spacer (ITS) regions of several bacteria have been used to discriminate strains at the species (Barry *et al.*, 1991; van der Giessen *et al.*, 1994; Jensen *et al.*, 1993; Kim and Song, 1996; Shakya, 1987) and subspecies level (Dolzani *et al.*, 1994; Gurtler and Barrie, 1995; Kostman *et al.*, 1992; Leblond-Bourget *et al.*, 1996; Matar *et al.*, 1993; Navarro *et al.*, 1992; Regnery *et al.*, 1991; Song *et al.*, 2002). These spacers are short stretches of DNA located between the 16S and 23S genes in prokaryotic rRNA loci (Nakagawa *et al.*, 1994) and usually contain one or more tRNA genes (Loughney *et al.*, 1992). Non-coding elements important to transcription of DNA that apparently are not functional should exhibit a considerable degree of sequence variation. Such variable regions make sequences of ITS good markers to measure short-term phylogeny. The characterisation of the ITS region of *A. avenae* and other bacterial pathogens of rice suggested that ITS might contain unique sequences (Kim and Song, 1996).

For identification of *A. avenae* strains from oats, rice, corn, millet, wheat, sugarcane, orchid, or watermelon, we describe a set of species-specific classical primers and real-time PCR primers and probe derived from ITS sequences of *A. a.* subsp. *avenae.* We also describe a second set of classical primers and real-time primers and probe that are subspecies *citrulli*-specific for identification of the watermelon fruit blotch pathogen, *A. a.* subsp. *citrulli.* Details on selection of classical (Schaad *et al.*, 2000) and real-time (Song *et al.*, 2002) PCR primers and probes to *A. a.* subsp. *avenae* and subsp. *citrulli* have been published.

2. MATERIALS AND METHODS

2.1 Bacterial Strains and Confirmation of Identity

The strains and sources of *Acidovorax avenae* and other bacteria used in this study are shown in Table 1. Cultures were maintained on yeast extract-dextrose-calcium carbonate (YDC) or King *et al.* medium B (KB) agars and stored at -85°C (Schaad, 2001). Colony morphology of *A. avenae* strains was checked on YDC and accumulation of poly-β-hydroxybytyrate was determined as described (Jensen *et al.*, 1993). Final identity of all strains of *A. a.* subsp. *avenae* and *A. a.* subsp. *citrulli* was confirmed by inoculating corn (cv Ida Sweet) seedlings and watermelon (cv. Charleston Gray) seedlings, as described (Schaad *et al.*, 1975; 1978), respectively. Pathogenicity of strains of *A. a.* subsp. *cattleyae* was not determined.

Table 1. Species and strains used in this study and results of classical and real-time PCR with classical primers Oafl/Oarl and real-time primers Oafl/Oar2 and probe AaP1 of *Acidovorax avenae* subsp. *avenae* and classical primers Aacf2/Aacr2 and real-time primers Aacf3/Aacr2 and probe AaP2 of *A. a.* subsp. *citrulli.*

Bacterial Strains	No	Host	Source[a]	Classical[b,c]		Real-time[d]	
				Of/Or	Af/Ar	AaP1	AaP2
Acidovorax avenae subsp. *avenae* (*Pseudomonas avenae*)							
COA1-3	3	Rice	1	+	-	+	-
19860	1	Corn	2	+	-	+	-
PA135	4	Corn	3	+	-	nt	nt
PA136-138	3	Teosinte	3	+	-	+[e]	-[e]
3307-8PAv (C-A, C-2)	2	Corn	4	+	-	nt	nt
3403PAv (Vs-1)	1	Vaseygrass	4	+	-	+	-
3406PAv (76-1)	1	Corn	4	+	-	+	-
Nepal	1	Rice	5	+	-	+	-
39123a	1	Rice	5	+	-	+	-
39459 - 39463	5	Rice	5	+	-	+	-
30150-30161	10	Rice	6	+	-	+[f]	-[f]
311056-311058	3	Rice	6	+	-	+	-
(*Pseudomonas rubrilineans*)							
19307	1	Sugarcane	2	+	-	+	-
3431PR	1	Millet	4	+	-	+	-
931	1	Sugarcane	7	+	-	nt	nt
3107	1	Corn	7	+	-	nt	nt
3111	1	*Canna panicula*	7	+	-	nt	nt
XR119	1	Sugarcane	3	+	-	+	-
XR3	1	Sugarcane	1	+	-	+	-
(*P. setariae*)							
19882	1	Rice	2	+	-	+	-
PS177	1	Rice	3	+	-	+	-
(*P. rubrisubalbicans*)							
XR105, 106	2	Sugarcane	3	+	-	+	-
A. a. subsp. *citrulli* (*P. pseudoalcaligenes* subsp. *citrulli*)							
HIB, H17, B164	3	Watermelon	8	+	+	+[g]	+[g]
1214, wwG, Mie3	3	Watermelon	8	+	+	+	+
29625 (FC-247)	1	Watermelon	2	+	+	+	+
9217, 8408	2	Watermelon	9	+	+	nt	+
2576, 2578	2	Watermelon	8	+	+	nt	+
A. a. subsp. *cattleyae* (*P. avenae*)							
33619 (Pc 21)	1	Orchid	2	+	-	+	-
PC107, 112, 145	3	Orchid	3	-	-	-	-
A. konjaci (*P. pseudoalcaligenes* subsp. *konjaci*)							
33996	1	Konjac	2	-	-	+	-
A. facilis (*P. facilis*)							
11228	1	Soil	2	-	-	-	-
Burkholderia caryophylli (*P. caryophylli*)							
PC131	1	Carnation	3	-	-	nt	-

continue

continue

B. cepacia (*P. cepacia*)							
PC22, PC142	2	Onion	3	-	-	-[h]	-[h]
B. gladioli pv. *allicola* (*P. allicola*)							
PA7	2	Onion	3	-	-	-	-
PA16	1	Gladioli	3	-	-	-	-
B. g. pv. *gladioli*							
25417	1	Gladioli	2	-	-	nt	-
B. glumae (*P. glumae*)							
COG1-4	4	Rice	1	-	-	-	-
33617	1	Rice	2	-	-	nt	-
P. fuscovaginae							
Fed1259-3 IR VM	1	Rice	10	-	-	nt	-
P. oryzicola							
PO101	1	Rice	3	-	-	nt	-
P. syringae pv. *syringae*							
Chili3 1-3	1	Rice	1	-	-	nt	-
B728a	1	Beans	11	-	-	-	-
Xanthomonas oryzae pv. *oryzae*							
CXO105, 211, 315	3	Rice	1	-	-	-[i]	-[i]
35932	1	Rice	2	-	-	-	-
X. o. pv. *oryzicola*							
P501	1	Rice	12	-	-	-	-
40972	1	Rice	2	-	-	-	-
Pantoea agglomerans							
CEh1	1	Rice	1	-	-	-	-
23375	1	Milletia japonica	2	-	-	nt	-
P. s. pv. *phaseolicola*							
C199	1	Beans	13	-	-	-	-
P. fluorescens							
13525	1		2	-	-	-	-
Other bacteria							
UPL1-6, UXL 1-7	13	Rice seeds	14	-	-	-	-
UWPL1-8, UWB 1-8	14	Watermelon Seeds	14	-	-	-	-

[a]1, W. Y. Song, Dept. Agric. Biology, Chonbuk National University, Chonju, Korea; 2, ATCC, American Type Culture Collection, Manassas, VA, USA; 3, ICPPB, International Collection of Plant Pathogenic Bacteria, USDA, Frederick, MD, USA; 4, L. E. Claflin, Dept. Plant Pathology, KSU, Manhatten, KS, USA; 5, C. Mortensen, Danish Govt. Inst. of Seed Pathology for Developing Countries, Copenhagen, DK; 6, MAFF, Ministry of Agriculture, Forestry and Fisheries of Japan, Tsukuba, Japan; 7, NCPPB, National Collection of Plant Pathogenic Bacteria, Jork, UK; 8, P. S. Randhawa, California Seed & Plant Lab., Elverta, California USA; 9, R. Latin, Dept. Plant Path., Purdue, Univ. West Lafayette, IN, USA; 10, F. Correa, CIAT, Cali, Colombia; 11, S. Hutchenson, Plant Biology Dept, Univ. Maryland, College Park, MD, USA; 12, T. Mew, IRRI, Manila, Philippines; 13, N. W. Schaad, ARS/USDA, Frederick, MD, USA; 14, this study.

[b]Positive result (+) for classical primers OAF1/Oar1 of *Acidovorax avenae* means presence of 550 bp band and negative means no visible band in ethidium bromide stained agarose gel.

[c]Positive result (+) for classical primers Aacf2/Aacr2 of *A. avenae* means presence of 450 bp band and negative means no visible band in ethidium bromide stained agarose gel.

[d]Positive result (+) and cycle threshold value for real-time primers OAF2/OAR2 and probe AaP1 of *A. avenae* and Aacf3/Aacr2 and probe AaP2 of *A. citrulli*. A negative (-) result means no product detected after 40 cycles. Cycle threshold is the cycle fluorescence rose above baseline. NT = not tested by real-time PCR.

[e] Strain CA-137 not tested.

[f] Strains 30150, 55, 59, 60 and 61 not tested.

[g] Strain B-164 not tested.

[h] Strain PC-22 not tested.

[i] Strain CXO-315 not tested.

2.2 Growth of Bacteria, Extraction of Genomic DNA, Preparation of PCR Templates, and Source of Chemicals

Cells were grown in Luria-Bertain (LB) liquid medium overnight on a rotary shaker at 30°C. Cell suspensions were centrifuged, suspended in 10 mM Tris-HC1 solution (pH 8.0), and adjusted to a concentration of 0.1 O.D. at A_{600}. Either suspensions of cells or single colonies growing on YDC agar or KB agar were used for PCR amplification without prior DNA extraction or other sample processing. Genomic DNA was extracted from washed cells as described (Prosen *et al.*, 1993). Chemicals and antibiotics were purchased from Sigma Chemical Co. (St. Louis, MO, USA), restriction and DNA modifying enzymes were purchased from Life Technologies (Bethesda, MD, USA), and all supplies for PCR were purchased from Perkin Elmer (Norwalk, CT, USA).

2.3 Amplification and Cloning of the 16S-23S ITS Region of *Acidovorax avenae* subsp. *avenae* and subsp. *citrulli*

Amplification of the ITS region was performed using primers R16-1 and R23-2R (Nakagawa *et al.*, 1994) and genomic DNA of *Acidovorax avenae* subsp. *avenae* strain COA1 and *A. a.* subsp. *citrulli* strain 29625. Amplification conditions were as described (Kim and Song, 1996). The amplified product was detected by electrophoresis in a 2% Metaphore gel (FMA BioProducts, Rockland, MD, USA). The resulting target DNA band was excised from the gel and the DNA fragment extracted using QIAquick gel extraction kit, following the manufacturers recommendations (Qiagen, Valencia, CA, USA). The gel-purified 950-bp products were ligated into pCR2.1 with TA cloning kit (Invitrogen, Carlsbad, CA, USA). These resulting clones, pOA3 and pCIT7, which contain a 950-bp insert of full length ITS, were then sequenced.

2.4 Sequencing and Primer Selection

The sequencing of the cloned 16S-23S ITS fragment was performed by the dideoxy-chain termination method (Sanger *et al.*, 1977) using ABI PRISM 310 Genetic Analyser and Dye Terminator Cycle Sequencing Core Kit (Perkin Elmer, Foster City, CA, USA) following the manufacturer's protocol. From the ITS sequences, candidate primers were selected using the program OLIGO (National Biosciences, Hamel, MN, USA) and synthesised commercially by GIBCO BRL (Frederick, MD, USA). Real-time primers and fluorescent probes were selected using the Primer Design software program (Primer Express I.O., Foster City, CA, USA) from ITS sequences, as described (Schaad *et al.*, 1978). Several combinations of forward and reverse PCR primers and fluorescent probes were tested using *Acidovorax avenae* subsp. *avenae* PA-138 and *A. a.* subsp. *citrulli* FB-247, as described below.

2.5 Testing Specificity of Primers

Classical primers were screened for specificity in 25 ml reaction volumes containing 2÷5 μl of the bacterial cell suspensions, 1.5 mM $MgC1_2$, 200 μM each of dNTPs, 4 pmole of each primer, 1x PCR reaction buffer II and 1.0 U of AmpliTaq Gold polymerase (Perkin Elmer, Norwalk, CT, USA). All classical PCR amplifications were conducted in a Perkin Elmer 9600 thermocycler (Perkin Elmer). For *Acidovorax avenae*-specific primers, the amplifications were conducted with an initial DNA denaturation at 94°C for 10 min followed by 25 cycles of 94°C for 30 S; 55°C for 30 S; 72°C for 1 min and a final extension step of 7 min at 72°C. For *A. a.* subsp. *citrulli*-specific primers, the amplifications were conducted with an initial DNA denaturation at 94°C for 10 min followed by 5 cycles of 94°C for 30 S; 56°C for 30 S; 72°C for 1 min, then 25 cycles of 94°C for 45 S; 55°C for 45 S; 72°C for 1 min and a final extension step as above. To detect a product, 5 to 10 μl of the reaction mixture were subjected to electrophoresis on 1.0% agarose gels and stained with ethidium bromide as described (Maniatis *et al.*, 1989). Cells of 43 strains of *A. a.* subsp. *avenae*, 11 strains of *A. a.* subsp. *citrulli,* four strains of *A. a.* subsp. *cattleyae*, single strain of *A. konjaci* and *A. facilis* (type strain of the genus), 26 other known bacteria, and 27 unknown bacteria isolated from rice or watermelon seeds, were used to screen primers of *A. avenae* and *A. a.* subsp. *citrulli* (Table 1). For real-time PCR, one μl of cell suspension or DNA was added to 24 μl of master mix (Cepheid, Sunnyvale, CA), as described (Schaad *et al.*, 2002). Starting with an annealing temperature of 55°C and the parameters as described (Schaad *et al.*, 2002), annealing temperatures of 56, 57, 58, 59, and 60°C were tested simultaneously using strain PA-138 for *A. a.* subsp. *avenae* and FB-247 for subsp. *citrulli.* The most stringent temperature that gave the highest endpoint fluorescence for both organisms was chosen (60°C) although little variation in end point fluorescence occurred. Also, little or no variation in cycle threshold (Ct) occurred with the different temperatures. The Ct value is defined as the PCR cycle number at which time the signal (fluorescence) of the probe rises above background. Real-time primers and probe were screened against most of the same organisms (Table 1) as used for classical PCR using a Smart Cycler (Cepheid, Sunnyvale, CA, USA). The primers and probe were added just prior to loading the reaction tube. PCR reactions were run as follows: 95°C hot start for 5 min. followed by 40 cycles at 95°C for 5 s, and 60°C for 45 s.

3. RESULTS

3.1 Confirmation of Identity of *Acidovorax avenae* Strains

Colonies of all strains received as *Acidovorax avenae* subsp. *avenae* (including *Pseudomonas rubrilineans*, *P. setariae*, and *P. rubrisubalbicans*) were non-fluorescent on KB medium and round, convex, non-mucoid, and light tan in colour on YDC medium. Colonies of *A. a.* subsp. *citrulli*, *A. a.* subsp. *cattleyae* (ATCC 33619) and *A. a.* subsp. *avenae* were indistinguishable on both media. All strains were positive for poly-β-hydroxybutyrate accumulation. The colony morphologies of subsp. *cattleyae* strains PC-107, 112, and 145 were somewhat different and poly-β-hydroxybutyrate did not accumulate. These results suggested that these cultures might not have been subsp. *cattleyae.* All strains of *A. a.* subsp. *avenae* and subsp. *citrulli* tested for pathogenicity were positive on corn and watermelon, respectively. The symptoms started out as water-soaked lesions which eventually extended for 3÷5 cm after 7÷10 days; some seedlings collapsed.

3.2 Sequencing and Primer Selection

The size of the ITS fragment of *Acidovorax avenae* subsp. *avenae* we sequenced was 631 bp. It had a GC content of 48.87%, and a single $tRNA^{Ala}$ gene of 73 bp. The size of the sequenced ITS fragment of *A. a.* subsp. *citrulli* was 623 bp. It had a GC content of 48.94%, and the same single $tRNA^{Ala}$ gene. The sequences were first aligned by using the DNASIS program (Hitachi SEA, Ltd.) and the alignments were corrected manually. By comparing the gene sequence of *A. a.* subsp. *avenae* to *Leuconostoc oenos* (Zavaleta *et al.*, 1996), *Stapthylcoccus aureus* (Gurtler and Barrie, 1995), *Bacillus subtilis* (Loughney *et al.*, 1982), and *Escherichia coli* (Harvey *et al.*, 1988), classical primers, Oafl (5'-GTCGGTGCTAACGACATGG-3') and Oarl (5'-AGACATCTCCGCTTTCTTTCAA-3') for *A. a.* subsp. *avenae* and Aacf2 (5'-GGAAGAATTCGGTGCTACCC-3') and Aacr2 (5'-TCGTCATTACTGAATTTCAACA-3') for *A. a.* subsp. *citrulli*, were selected (Song *et al.*, 2002). Minor errors in PCR and sequencing were eliminated by sequencing both strands from two independently cloned fragments and running two separate PCR experiments. No sequence variability between the two 16S-23S ITS copies was found. For real-time PCR, classical primer Oafl (see above) and primer Oar2 (5'-TCCTCGCATCTTATGTTCGGAA-3') were used as the forward and reverse primers, respectively, for subsp. *avena*e and AaP1 (6FAM-TCAGCTGGTTAGAGCACCGTCTAGA-TAMRA) was used as the probe. For *A. a.* subsp. *citrulli* forward primer Aacf3 (5'CCTCCACCAACCAATACGCT-3') and reverse primer Aacr2 (5'-TCGTCATTACTGAATTTCAACA-3') and probe AaP2 (6FAM-CGGTAGGGCGAAGAAACCAACACC-TAMRA) were used (Song *et al.*, 2002.

3.3 Testing Specificity of Primers

For classical primers Oafl and Oar1, a PCR product of approximately 550 bp (Fig. 1) was obtained for all strains of *Acidovorax avenae* subsp. *avenae* and *A. a.* subsp. *citrulli* and the ATCC strain of *A. a.* subsp. *cattleyae* (Table 1). The other three strains of subsp. *cattleyae* were negative. None of the other bacteria tested, including 12 strains of *Burkholderia* (non-fluorescent pseudomonad), produced a PCR product. Classical primers Aacf2 and Aacr2 produced a 450-bp product (Fig. 1) with all strains of *A. a.* subsp. *citrulli* but no other bacteria (Table 1). The real-time primers and probe were highly specific (Fig. 2; Table 1). The three morphologically different strains of subsp. *cattleyae* were negative by both classical and real-time PCR.

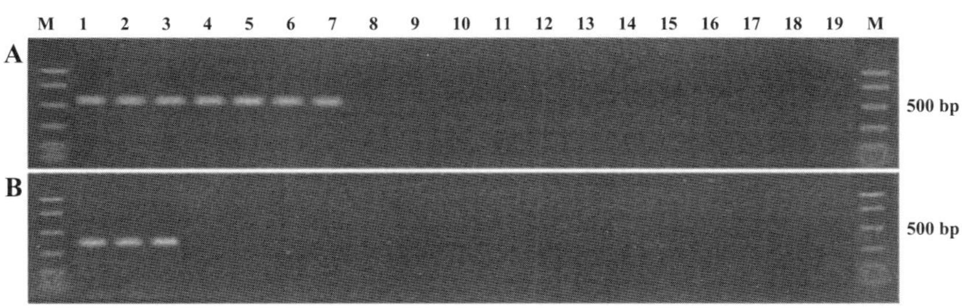

Figure 1. Ethidium bromide stained gel showing 550 bp PCR product formed by primers OAfl/OAr1 with cells of *Acidovorax avenae* subsp. *avenae*, *A. a.* subsp. *citrulli*, and *A. a.* subsp. *cattleyae* (A) and 450 bp PCR product formed by primers Aacf2/Aacr2 with cells of *A. a.* subsp. *citrulli* strains (B). Strains (lane) were as follows: *A. a.* subsp. *citrulli* 29625 (1), 1214 (2) and 2578 (3); *A. a.* subsp. *avenae* COA1 (4), 19860 (5) and 19882 (6); *A. a.* subsp. *cattleyae* 33619 (7); *A. konjaci* 33996 (8); *A. facilis* 11228 (9); *Burkholderia caryophylli* PC131 (10); *B. cepacia* PC142 (11); *B. gladioli* pv. *alliicola* PA16 (12); *B. g.* pv. *gladioli* 25417 (13); *B. glumae* 33617 (14); *Pseudomonas fuscovaginae* Fed1259-3 (15); *Xanthomonas oryzae* pv. *oryzae* 35932 (16); *X. o.* pv. *oryzicola* 40972 (17); *Pantoea agglomerans* 23375 (18); *P. fluorescens* 13525 (19). M = means molecular weight marker (Promega).

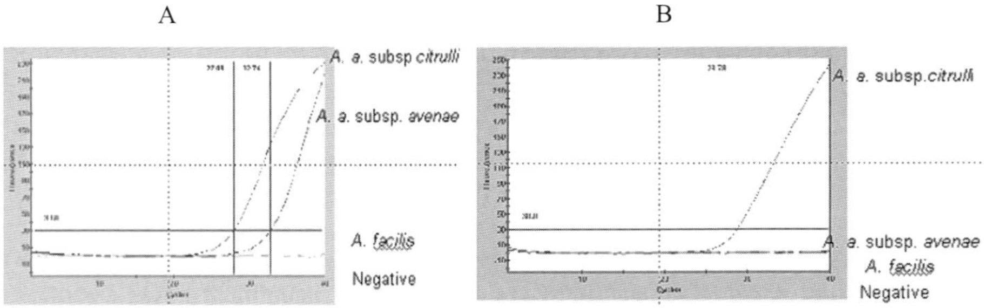

Figure 2. A, B. Real-time PCR amplification of cells of *Acidovorax avenae* subsp. *avenae* ATTC 19860, *A. a.* subsp. *citrulli* ATTC 29625, *A. facilis* ATTC 17695, and water control using A) avenae-specific primers Oafl/Oar2,and probe AaP1 and B) citrulli-specific primers Aacf3/Aacr2 and probe AaP2 and a portable Smart Cycler (Cepheid, Sunnyvale, CA, USA) for 40 cycles. The Y axis is the change in the fluorescence that measures probe cleavage and the X axis is the cycle number. The cycle threshold (Ct) value (number on vertical line) is the cycle when the fluorescence rises above background (baseline). Water control and *A. facilis* are below the baseline.

4. DISCUSSION

Identification of plant pathogenic bacteria at the subspecies level is sometimes not possible using classical phenotypic tests alone (Schaad, 2001). This is certainly true for all the subspecies of *Acidovorax avenae*, including *avenae*, *cattleyae*, and *citrulli*. *A. a.* subsp. *avenae* and subsp. *citrulli* can only be identified through a combination of phenotypic and time-consuming pathogenicity tests (Jones *et al.*, 2001). Moreover, the pathogenicity tests for the rice strains of subsp. *avenae* can be difficult to interpret because similar symptoms often occur on rice as a result of other bacterial pathogens such as *Pseudomonas fuscovaginae*, *P. glumae*, and *P. syringae* pv. *syringae* (H.M. Kim, personal observations). Semi-selective media are useful for presumptive identification of *A. avenae* but more sensitive and specific methods are needed to confirm the identification, especially in seed health situations. Serology assays are available for identification of *A. avenae*. However the assays are not specific for *A. avenae* (Jones *et al.*, 2001; Shakya, 1987). PCR primers are available for a large number of plant pathogenic bacteria and PCR is fast becoming a standard tool for identification of plant pathogenic bacteria (Schaad, 2001). Classical PCR primers have been described for detection of *A. a.* subsp. *citrulli* in watermelon seeds (Minsavage *et al.*, 1995; Walcott and Gitaitis, 2000). However, neither primers are specific for *A. a.* subsp. *citrulli*.

In this study, we utilise *A. avenae*-specific classical PCR primers (Oafl/Oar1) described previously for detection of strains of *A. avenae* in rice seeds (Schaad *et al.,* 2000). We show that these primers are species-specific and can be used to identify all strains of *A. avenae*, regardless of host of origin. The value of these primers for identification was shown by the results with *A. a.* subsp. *cattleyae* strains PC-107, 112, and 147. PCR confirmed that these strains were not *A. a.* subsp. *cattleyae* as suggested by the colony morphology and negative poly-β-hydroxybutyrate results. There was 100 % between the results for PCR and accumulation of poly-β-hydroxybutyrate. To be able to utilise the more robust, user friendly real-time PCR, we designed a new reverse primer (OAR2) and a probe (AaP1) specific for *A. a.* subsp. *avenae*. For identification of *A. a.* subsp. *citrulli*, we designed classical primers (Aacf2/Aacr2) and real-time PCR primers (Aacf3/Aacr2) and probe (AaP2) that were specific to subsp. *citrulli*. We screened fewer other bacteria against our real-time primers and probe then against our classical primers. However, the results for the real-time PCR were more conclusive because the results were also based on hybridization whereas the classical PCR results were not. No Southern blots were performed for any of the classical PCR results. Although subtractive hybridization techniques have been needed to successfully design PCR primers to other closely related plant pathogenic bacteria (Cook and Sequeira, 1991; Mills *et al.*, 1997), we found that such a labour intensive technique was not needed. The ITS region worked very well for designing primers and probe specific to subsp. *avenae* as well as subsp. *citrulli*, the causal agent of watermelon fruit blotch (WFB).

Real-time PCR has been shown to be highly reliable and cost effective and is becoming the method of choice for detection and identification of many bacteria. Real-time protocols have been developed for several plant pathogenic bacteria, including *Clavibacter michiganensis* subsp. *sepedonicus* (Schaad *et al.*, 1999), *Ralstonia solanacearum* (Weller and Stead, 2002; Weller *et al.*, 2000a; 2000b), *Agrobacterium tumefaciens* (Walcott and Gitaitis, 2000) and the Pierce's disease bacterium *Xylella fastidiosa* (Schaad *et al.*, 2002). Using our *A. a.* subsp. *avenae*-specific real-time primers and probe and the portable Smart Cycler, we have shown that WFB can be diagnosed in watermelon seedlings in less then one hour (Schaad *et al.*, 2001). By using the newly described *A. a.* subsp. *citrulli*-specific real-time primers and probe, one can now be certain that a positive WFB diagnosis is due to the presence of *A. a.* subsp. *citrulli* and not to other *Acidovorax* strains. The potential risk of the dissemination of *A. a.* subsp. *avenae* in international exchange of germplasm of rice, corn, and *A. a.* subsp. *citrulli* in watermelon is very high (Walcott and Gitaitis, 2000). The availability of PCR-based methods to identify these two closely related bacteria represents a valuable tool for routine diagnosis, monitoring natural disease spread, and rapid identification of these bacteria, especially in seed lots.

NOTES

The use of trade, firm, or corporation names in this publication (or page) is for the information and convenience of the reader. Such use does not constitute an official endorsement or approval by the United States Department of Agriculture or the Agricultural Research Service of any product or service to the exclusion of others that may be suitable.

ACKNOWLEDGEMENTS

We thank L. E. Claflin, Kansas State University, Carmen Mortensen, Royal Veterinary and Agricultural University, Copenhagen, P. Randhawa, California Plant and Seed Laboratory, Elverta, CA, R. Latin, Purdue University, F. Correa, CIAT, Cali, Columbia, S. Hutchenson, Plant Biology, U. Maryland, and T. Mew, IRRI, Manilia, Philippines for providing cultures. Also, we thank B. Schneider and M. Bonde for reviewing the manuscript.

REFERENCES

Ark, P.A., and Thomas, H.E., 1946, Bacterial leaf spot and bud rot of orchids caused by *Phytomonas cattleyae*. *Phytopathology* **36**: 695-698.

Barry, T., Colleran, G., Glennon, M.M., Dunican, L.K., and Gannon, F., 1991, The 16S/23S ribosomal spacer region as a target for DNA probes to identify eubacteria. *PCR Methods Appl.* **1**: 51-56.

Cook, D., and Sequeira, L., 1991, The use of subtractive hybridization to obtain a DNA probe specific for Pseudomonas solanacearum race 3. *Mol. Gen. Genet.* **227**: 401-410

Dolzani, L., Tonin, E., Lagatolla, C., and Monti-Bragadin, C., 1994, Typing of *Staphylococcus aureus* by amplification of the 16S-23S rRNA intergenic spacer sequences. *FEMS Microbiol. Lett.* **119**: 167-174.

van der Giessen, J.W.B., Haring, R.M., and Van der Zeijst, B.A.M., 1994, Comparison of the 23S ribosomal RNA genes and the spacer region between the 16S and 23S rRNA genes of the closely related *Mycobacterium avium* and *Mycobacterium paratuberculosis* and the fast-growing *Mycobacterium phlei*. *Microbiology*. **140**: 1103-1108.

Goto, M., 1964, Nomenclature of the bacteria causing bacterial leaf streak and bacterial stripe of rice. Bull. of Faculty of Agriculture, Shizuoka University, **14**: 3-10.

Goto, M., and Okabe, N., 1952, Studies on the causal organisms of bacterial stripe disease of millet and the brown stripe disease of Italian millet. Bull. Fac. Agric. Shizuoka Univ. **2**: 15-24

Gurtler, V., and Barrie, H.D., 1995, Typing of *Staphylococcus aureus* strains by PCR-amplification of variable-length 16S-23S rDNA spacer regions: characterisation of spacer sequences. *Microbiology* **141**: 1255-1265.

Harvey, S., Hill, C.W., Squires, C., and Squires, C.L., 1988, Loss of the spacer loop sequence from the *rrn*B operons in the *Escherichia coli* K-12 subline that bears the *rel*A1 mutation. *J. Bacteriol.* **170**: 1235-1238.

Hu, F.P., Young, J.M., and Trigg, C.M., 1991, Numerical analysis and determination tests for non-fluorescent plant-pathogenic *Pseudomonas* spp. and genomic analysis and reclassification of species related to *Pseudomonas avenae* Manns 1909. *Int. J. Syst. Bacteriol.* **41**: 516-523.

Jensen, M.A., Webster, J.H., and Straus, N., 1993, Rapid identification of bacteria on the basis of polymerase chain reaction-amplified ribosomal DNA spacer polymorphism. *Appl. Environ. Micro.* **59**: 945-952.

Jones, J.B., Gitaitis, R.D., and Schaad, N.W., 2001, *Acidovorax* and *Xylophilus*. *In*: N.W. Schaad, J.B. Jones, and W. Chun (eds). Laboratory Guide for Identification of Plant Pathogenic Bacteria. APS Press. St. Paul, MN.

Kim, H.M., and Song, W.Y., 1996, Characterisation of ribosomal RNA intergenic spacer region of several seedborne bacterial pathogens of rice. *Seed Sci. & Technol.* **24**: 571-580.

Kostman, J.R., Edlind, T.D., LiPuma, J.J., and Stull, T.L., 1992, Molecular epidemiology of *Pseudomonas cepacia* determined by polymerase chain reaction ribotyping. *J. Clin. Microbiol.* **30**: 2048-2087.

Leblond-Bourget, N., Philippe, H., Mangin, I., and Decaris, B., 1996, 16S rRNA and 16S to 23S internal transcribed spacer sequence analyses reveal inter- and intraspecific B*ifidobacterium* phylogeny. *Int. J. Syst. Bacteriol.* **46**: 102-111.

Loughney, K., Lund, E., and Dahlberg, J.E., 1982, tRNA genes are found between 16S and 23S rRNA genes in *Bacillus subtilis*. *Nucleic Acids Res.* **10**: 1607-1624.

Maniatis, T., Fritsch, E.F., and Sambrook, J., 1989, Molecular Cloning, a Laboratory Manual. Cold Spring Harbor, N.Y.

Manns, T.F., 1909, The blade blight of oats-a bacterial disease. *Ohio Agric. Exp. Stn. Res. Bull.* **210**: 91-167.

Martin, J.P., and Wismer, C.A., 1989, Red Stripe. *In*: Diseases of Sugarcane. pp. 80-91. (Eds) C. Ricaud, B.T. Egan, A.G. Gillespie, Jr., and C.G. Hughes. Elsevier, The New York, N.Y.

Matar, G.M., Swaminathan, B., Hunter, S.B., Slater, L.N., and Welch, D.F., 1993, Polymerase chain reaction-based restriction fragment length polymorphism analysis of a fragment of the ribosomal operon from *Rochalimaea* species for subtyping. *J. Clin. Microbiol.* **31**: 1730-1734.

Mills, D., Russell, B.W., and Hanus, J.W., 1997, Specific detection of *Clavibacter michiganensis* subsp. *sepedonicus* by amplification of three unique sequences isolated by subtractive hybridisation. *Phytopathology* **87**: 852-861.

Minsavage, G.V., Hoover, R.J., Kucharek, T.A., and Stall, R.E., 1995, Detection of the watermelon fruit blotch pathogen on seeds with the polymerase chain reaction. *Ann. Meeting Amer. Phytopath. Soc. Abstract*, no.379.

Nakagawa, T., Shimada, M., Mukai, H., Asada, K., Kado, I., Fujino, K., and Sato, T., 1994, Detection of alcohol-tolerant Hiochi bacteria by PCR. *Appl. Environ. Microbiol.* **60**: 637-640.

Navarro, E., Simonet, P., Normand, P., and Bardin, R., 1992, Characterisation of natural populations *Nitrobacter* spp. using PCR/RFLP analysis of the ribosomal intergenic spacer. *Arch. Microbiol.* **157**: 107-115.

Prosen, D., Hatziloukas, E., Schaad, N.W., and Panopoulous, N.J., 1993, Specific detection of *Pseudomonas syringae* pv. *phaseolicola* DNA in bean seed by polymerase chain reaction-based amplification of a phaseolotoxin gene region. *Phytopathology*. **83**: 965-970.

Regnery, R.L., Spruill, C.L., and Plakaytis, B.D., 1991, Genotypic identification of rickettsiae and estimation of intraspecies sequence divergence for portions of two rickettsial genes. *J. Bacteriol.* **173**: 1576-1589.

Rosen, J.R., 1922, A bacterial disease of foxtail (*Chaetochloa lutescens*). *Ann. Mo. Bot. Gard.* **9**: 333-402.

Rosen, J.R., 1926, Bacterial stalk rot of corn. *Phytopathology*. **16**: 241-267.

Sanger, F., Nicklen, S., and Coulsen, A.R., 1977, DNA sequencing with chain-terminating inhibitors. *Proc. Natl. Acad. Sci. U.S.A.* **74**: 5463-5467.

Savulescu, T., 1947, Contribution a la classification des bacteriacees phytopathogenes. *An. Acad. Romane Ser. III* **22**: 135-160.

Schaad, N.W., 2001, 3rdEd. Initial Identification of Common Genera. *In*: N.W. Schaad, J.B. Jones, and W. Chun (eds), Laboratory Guide for Identification of Plant Pathogenic Bacteria. APS Press. St. Paul, MN.

Schaad. N.W., Berthier-Schaad, Y., Sechler, A., and Knorr, D., 1999, Detection of *Clavibacter michiganensis* subsp. *sepedonicus* in potato tubers by BIO-PCR and an automated real-time fluorescence detection system. *Pl. Dis.* **83**: 1095-1100.

Schaad, N.W., Gaush, P., Postnikova, E., and Frederick, R., 2001, On-site one hour PCR diagnosis of bacterial diseases. *Phytopathology*, **91**: S79 (Abstr).

Schaad, N.W., Kado, C.I., and Summer, D.R., 1975, Synonymy of *Pseudomonas avenae* Manns 1905 and *Pseudomonas alboprecipitans* Rosen 1922. *Int. J. Syst. Bacteriol.* **25**: 133-137.

Schaad, N.W., Opgenorth, D., and Gaush, P., 2002, Real-time PCR for one-hour on-site diagnosis of Pierce's disease of grape in early season asymptomatic vines. *Phytopathology*, **92**: 721-718.

Schaad, N.W., Song, W.Y., and Hatziloukas, E., 2000, PCR primers for detection of plant pathogenic species and species of *Acidovorax*. *United States patent number* **6**: 146,834.

Schaad, N.W., Sowell Jr., G., Goth, R.W., Colwell, R.R., and Webb, R.E., 1978, *Pseudomonas pseudoalcaligenes subsp. citrulli* subsp. nov. *Int. J. Syst. Bacteriol.* **28**: 117-125.

Shakya, D.D., 1987, Rapid diagnosis of *Pseudomonas avenae* by pathogenicity and serology. Korean. *J. Pl. Path.* **3**: 300 (Abst).

Song, W.Y., Kim, H.M., and Schaad, N.W., 2002, PCR primers for detection and identification of plant pathogenic species, subspecies, and strains of *Acidovorax*. *United States patent number* **6**: 423-499.

Walcott, R.R., and Gitaitis, R.D., 2000, Detection of *Acidovorax avenae* subsp. *citrulli* in watermelon seed using immunomagnetic separation and the polymerase chain reaction. *Plant Dis.* **84**: 470-474.

Weller, S.A., and Stead, D.E., 2002, Detection of root mat associated *Agrobacterium* strains from plant material and other samples types by post-enrichment TaqMan PCR. *J. Appl. Microbiol.* **92**: 118-126.

Weller, S.A., Elphinstone, J.G., Smith, N.C., Boonham, N., and Stead, D.E., 2000a, Detection of *Ralstonia solanacearum* strains with a quantitative, multiplex, real-time fluorogenic PCR (TaqMan) assay. *Appl. Environ. Microbiol.* **66**: 2853-2858.

Weller, S.A., Elphinstone, J.G., Smith, N.C., and Stead, D.E. 2002b, Detection of *Ralstonia solanacearum* from potato tissue by post-enrichment TaqMan PCR. OEPP/EPPO Bull. 30: 381-383.

Willems, A., Goor, M., Thielemans, S., Gillis, M., Kersters, K., and De Ley, J. 1992, Transfer of several phytopathogenic *Pseudomonas* species to *Acidovorax* as *Acidovorax avenae* subsp. nov., comb. nov., *Acidovorax avenae* subsp. *citrulli*, *Acidovorax avenae* subsp. *cattleyae*, and *Acidovorax konjaci*. *Int. J. Syst. Bacteriol.* **42**: 107-119.

Zavaleta, A.I., Martinez-Murcia, A.J., and Rodriguez-Valera, F., 1996, 16S-23S rDNA intergenic sequences indicate that *Leuconostoc oenos* is phylogenetically homogeneous. *Microbiology* **142**: 2105-2114.

Limitations of the Current Methods for the Detection of Spanish Strains of *Pseudomonas syringae* pv. *phaseolicola*

A. RICO, R. LOPEZ[#], M.T. AIZPUN, C. ASENSIO[#] and J. MURILLO
Departamento de Producción Agraria, Universidad Pública de Navarra, Pamplona, Spain;
[#]Servicio de Investigación y Tecnología Agraria, Junta de Castilla y León, Valladolid, Spain.

Abstract: Bean plants showing symptoms of halo blight and/or brown spot in fields of the county of Castilla y León, Spain, were examined for the presence of *Pseudomonas syringae* pv. *phaseolicola* and *P. s.* pv. *syringae*. A 91 % of the strains were identified as *P. s.* pv. *phaseolicola* using standard biochemical tests, while the remaining were identified as *P. s.* pv. *syringae*. Around two thirds of the *P. s.* pv. *phaseolicola* strains failed to produce phaseolotoxin and were not detected with phaseolotoxin-specific PCR primers. These strains also reacted weakly to a commercial antibody in DAS-ELISA experiments. Our results indicate that phaseolotoxin-minus strains of *P. s.* pv. *phaseolicola* can be of epidemiological significance in field conditions, and that current methods for the certification of pathogen-free seeds based on phaseolotoxin-specific primers can be of limited use in Spain and in any other countries where such strains might occur frequently.

Key words: BIO-PCR, brown spot, halo blight, phaseolotoxin.

Author for correspondence: J. Murillo; e-mail: jesus@unavarra.es.

1. INTRODUCTION

In Spain, most of the dry bean crop is produced in the County of Castilla y León, in Central Northeastern Spain. Among others, high-quality local dry bean varieties are extensively cultivated in this county, and their high susceptibility to bacterial diseases is an important limitation for bean production. Halo blight, caused by *Pseudomonas syringae* pv. *phaseolicola*, is one of the most important bean diseases world-wide and is present in all places where beans are grown (Saettler, 1991; Taylor *et al.*, 1996). The

N.S. Iacobellis *et. al.* (eds.), *Pseudomonas syringae* and Related Pathogens, 545-552.

control of this pathogen is difficult and management is based on the use of healthy seed and resistant cultivars. The certification of seed as free of the pathogen is routinely done by serology with mono- or polyclonal antibodies, or by biochemical and molecular methods based on the production of the phytotoxin phaseolotoxin or the possession of genes for its biosynthesis (Rudolph, 1995; Schaad *et al.*, 2001). Since some production areas, such as Idaho, USA, have zero tolerance for the halo blight pathogen (Webster *et al.*, 1983), methods have to be highly sensitive and specific (Schaad *et al.*, 1995). On the other hand, the utilisation of resistant cultivars is based on the identification of nine races of the pathogen, whose distribution and abundance is variable among countries (Taylor *et al.*, 1996; Fourie, 1998). Brown spot, another disease of beans caused by the closely related but unregulated *P. s.* pv. *syringae*, is particularly important in certain locations of Europe (Smith *et al.*, 1988), including Asturias, Northern Spain (González *et al.*, 1998).

A main goal of the breeding programs at the Servicio de Investigación y Tecnología Agraria, Spain, is to improve the management of halo blight and brown spot by introducing durable resistance into the local bean cultivars. In order to support the breeding programme and to develop appropriate molecular analytical tools, in this work we describe a population of *P. s.* pv. *phaseolicola* strains that are unable to synthesise phaseolotoxin (Tox-minus) and lack at least part of the phaseolotoxin biosynthesis cluster.

2. MATERIAL AND METHODS

Fluorescent bacteria were isolated on King's medium B (KMB) (King *et al.*, 1954) between 1993 and 2001 from diseased bean plants in different locations of the county of Castilla y León, Spain. Strains were examined following the LOPAT scheme (Lelliott and Stead, 1987) and further characterised by different biochemical and genetic tests. The capability to use various compounds as the sole carbon source was tested using the minimal medium of Ayers, Rupp and Johnson (Lelliott and Stead, 1987) supplemented with 0.1% of each compound. Growth on the semiselective medium MT, and the hydrolysis of casein on the same medium were tested as described (Goszczynska and Serfontein, 1998). Other tests were carried out as described (Lelliott and Stead, 1987; Schaad *et al.*, 2001).

PCR amplification was performed using primers specific for *virPphA* (Jackson *et al.*, 1999); the *efe* gene (Sato *et al.*, 1997), for ethylene production, and for gene *cfl*, which is essential for the biosynthesis of coronatine (Bereswill *et al.*, 1994). The amplification of DNA specific for the biosynthesis of phaseolotoxin was tested using the primer pairs P 5.1–P 3.1, which amplify an internal fragment of ORF6 from locus *phtE* (Prosen *et al.*, 1993; Zhang and Patil, 1997), and PHA19–PHA95, which amplify an internal fragment of the amidinotransferase gene *amtA* (Marques *et al.*, 2000; Hernández-Guzmán and Alvarez-Morales, 2001). In all cases, conditions for amplification were as described for each particular primer pair.

Pathogenicity tests were carried out by stab inoculation of detached, immature green bean (*Phaseolus vulgaris* L.) pods of cv. Canadian Wonder (Harper *et al.*, 1987); inoculated pods were kept under high humidity at room temperature (20÷23°C), and reactions were scored daily for up to nine days.

The *in vitro* detection of antimicrobial toxins and phaseolotoxin was carried out essentially as described (Gasson, 1980; Staskawicz and Panopoulos 1979) using *Escherichia coli* CECT831 (Colección Española de Cultivos Tipo, Burjasot, Spain) as an indicator strain. Serological detection was carried out using a commercial DAS-ELISA kit (Loewe Phytodiagnostica, Sauerlach, Germany) according to the manufacturer's recommendations.

3. RESULTS AND DISCUSSION

3.1 The most Abundant Fluorescent Pseudomonad Pathogenic on Beans is *Pseudomonas syringae* pv. *phaseolicola*

All bacterial strains obtained from either leaves or pods of field-grown beans showing typical symptoms of bacterial diseases, produced green fluorescent pigments on KMB. All strains belonged to LOPAT group Ia (+–––+) (Lelliott and Stead, 1987), which suggested they were *Pseudomonas syringae*.

Results of pathogenicity tests showed that 9% of the strains produced sunken brown lesions similar to those produced by *P. s.* pv. *syringae*, while 91% produced water-soaked lesions indistinguishable from those produced by control *P. s.* pv. *phaseolicola* strains (Fig. 1A). These latter strains were unable to utilise inositol, D-sorbitol, erythritol, L+tartrate, D+tartrate and L-lactate as a sole carbon source, although 70% of them utilised D-mannitol; none of the strains hydrolysed esculin, and none produced syringomycins. The biochemical and pathogenicity data suggested they were *P. s.* pv. *phaseolicola*, which has been confirmed by analysis of nucleotide sequences from the chromosomal genes *gyrB* and *rpoD* (A. Rico, and J. Murillo, unpublished results). Additionally, all the *P. s.* pv. *syringae*-type strains hydrolysed casein whereas the *P. s.* pv. *phaseolicola* strains did not (Fig. 1B). All the *P. s.* pv. *phaseolicola* strains produced specific amplification bands by PCR using *virPphA* and *avrPphE*-specific primers. No amplification bands were obtained, however, using *cfl*-specific primers, common to strains of *P. s.* pv. *glycinea*, or *efe*-specific primers, characteristic of kudzu strains of *P. s.* pv. *phaseolicola*.

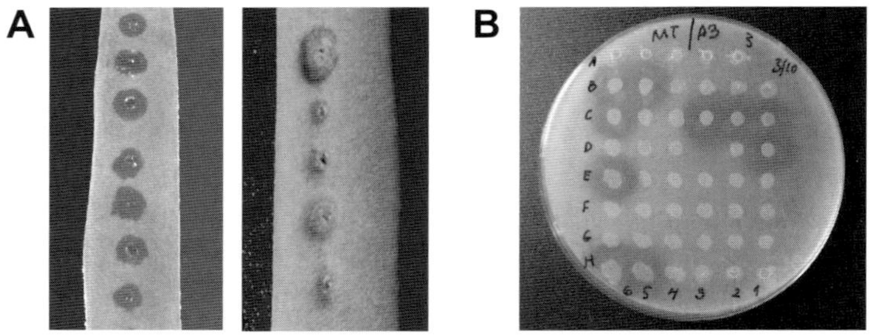

Figure 1. **A**: Pathogenicity test on bean pods of cv. Canadian Wonder. The left bean shows typical water soaking lesions produced by different *Pseudomonas syringae* pv. *phaseolicola* field strains in comparison with the control strain, 1449B, in the bottom position. The pod at right shows the typical sunken brown lesions produced by *P. s.* pv. *syringae* strains; B728a was used as control and was inoculated in the fourth position from the top. The pictures were taken 72 h after inoculation. **B**: Differentiation of *P. s.* pv. *phaseolicola* and *P. s.* pv. *syringae* on MT medium (Goszczynska and Serfontein, 1998). The clear haloes are produced only by the *P. s.* pv. *syringae* strains, due to the degradation of casein.

3.2 Most Field *Pseudomonas syringae* pv. *phaseolicola* Strains Lack the Phaseolotoxin Gene Cluster

The use of pathogen-free seeds is one of the most successful ways of controlling halo blight (Smith *et al.*, 1988; Webster *et al.*, 1983). Because of the rapid spread of the pathogen, the presence of even a 0.1% of diseased seeds can result in severe epidemics (Grogan and Kimble 1967) and, therefore, detection methods must be highly sensitive and specific. Currently, the most sensitive method is a PCR-enrichment technique using phaseolotoxin genes (*tox*)-specific primers (Schaad *et al.*, 1995). Additionally, the pathogen can be detected and diagnosed by cultural techniques and by serology, using different commercially available antibodies.

Unexpectedly, over 60% of the field strains of *Pseudomonas syringae* pv. *phaseolicola* did not produce phaseolotoxin *in vitro* and did not yield the expected PCR amplification band using *tox*-primers P 5.1 and P 3.1 (Fig. 2), which amplify a 0.5 kb DNA fragment specific to the phaseolotoxin gene cluster (Schaad *et al.*, 1995). These strains were therefore considered Tox-minus. When the DNA amplicon obtained with primers P 5.1 and P 3.1 from a Tox-plus reference strain (1449B, Taylor *et al.*, 1996) was used in hybridisation experiments with genomic DNA, only the Tox-plus strains showed significant hybridisation signals (Fig. 2 B and C). Taken together, these results suggest that the Spanish Tox-minus strains do not posses the cluster for the biosynthesis of phaseolotoxin. This finding implies that the certification of pathogen-free seeds can not be carried out in Spain following current PCR procedures that rely on the amplification of the phaseolotoxin gene cluster.

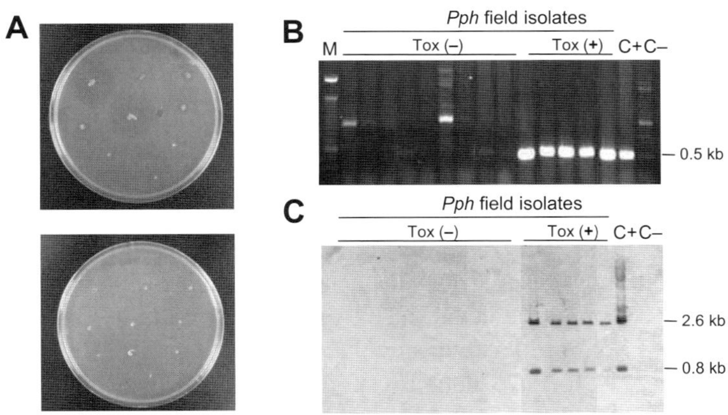

Figure 2. The Tox-minus strains of *Pseudomonas syringae* pv. *phaseolicola* (*Pph*) do not contain sequences for the biosynthesis of phaseolotoxin. **A.** Bioassay for phaseolotoxin production. Inhibition of *Escherichia coli* by Tox-plus strains is evident in minimal medium (supplemented with ornithine here), but is reverted by the addition of arginine (bottom). Tox-minus strains of *P. s.* pv. *phaseolicola* do not inhibit *E. coli.* **B.** PCR amplification products obtained with different Tox-minus and Tox-plus *P. s.* pv. *phaseolicola* field strains using primers P 5.1 and P 3.1, which are specific for the *tox* cluster. C+, reference strain *P. s.* pv. *phaseolicola* 1449B, that yields the expected 0.5 kb band; C–, negative control, strain PG4180 of *P. s.* pv. *glycinea*; M, 1-kb DNA marker. **C.** Southern blot hybridisation of genomic DNA from the same strains analysed in panel B. The probe is a mixture of the DNA fragments amplified with the *tox*-specific primer pairs P 5.1-P 3.1 and PHA19-PH95. Other symbols are as in B.

3.3 Can the Tox-minus Strains be Detected by Serology Using Current Antibodies for *Pseudomonas syringae* pv. *phaseolicola*?

Due to the limitations of the current PCR techniques for the detection of the Spanish Tox-minus strains, we determined if such strains could be detected by DAS-ELISA using a commercial polyclonal antibody (Loewe Phytodiagnostica, Sauerlach, Germany). The absorbance values at 405 nm were consistently higher that 2.5 for the ten Tox-plus strains analysed, but lower than 0.8 for ten Tox-minus strains and two strains each of *Pseudomonas syringae* pv. *glycinea* and *P. s.* pv. *syringae*. Although it would be necessary to test other antibodies, as well as the efficacy of other serological techniques, our results suggest that the presence of Tox-minus strains in seed lots can remain undetected using certain serological techniques. Therefore, the suitability of the detection techniques for the halo blight pathogen should be validated in each country depending on the local structure of the *P. s.* pv. *phaseolicola* population.

4. CONCLUSION

We have found a large proportion of Tox-minus strains of *Pseudomonas syringae* pv. *phaseolicola* among those bacteria that produce halo blight symptoms in field conditions in Spain. These strains can not be detected using available PCR primers, and our preliminary results suggest that serological techniques are also of limited value. The implementation of efficient control methods for this disease in Spain requires, therefore, assays that can detect the presence of both Tox-plus and Tox-minus strains of *P. s.* pv. *phaseolicola*. We are currently evaluating if Tox-minus strains of *P. s.* pv. *phaseolicola* also occur frequently in other countries, and it would be of prime interest to determine if they are also causing epidemics of economic importance in places other than Spain.

Others have reported the existence of Tox-minus strains, either containing or not the *tox* cluster (Mitchell, 1978; Schaad *et al.*, 1995; Völksch and Weingart, 1998); however, naturally occurring toxin-negative strains of *P. s.* pv. *phaseolicola* have been described only rarely (Rudolph, 1995). Phaseolotoxin was shown to be necessary for *in planta* systemic movement of *P. s.* pv. *phaseolicola* (Patil *et al.*, 1974), and it is generally believed that Tox-minus strains are of little or no epidemiological importance in halo blight of beans (Smith *et al.*, 1988; Saettler, 1991; Schaad *et al.*, 1995). The high frequency of Tox-minus strains that we have found argues against a relevant role for phaseolotoxin in pathogenesis, although it is possible that the prevalence of Tox-minus strains in the county of Castilla y León is favoured by the bean cultivars that are routinely used in this area, or because of the particular weather conditions. Nevertheless, our understanding of the role of toxins in virulence is rather limited because of the plethora of biochemical reactions that they can alter, and because of the different forms and intermediates of the toxin that can result in biological activity. For instance, the analysis of different plasmid-cured mutants has only recently shown (Sesma *et al.*, 2001) that the proposed role of coronatine in virulence (Bender *et al.*, 1987) has been traditionally overestimated due to the participation of coronafacic acid in virulence. Therefore, it is possible that phaseolotoxin itself plays only a limited role in virulence, or that its importance is more related to the ecology of the bacterium, as, for instance, in the exclusion of other epiphytes from the plant leaf surface.

ACKNOWLEDGEMENTS

We thank Alan Vivian and Antonio de Vicente for strains and suggestions, and Theresa Osinga for critical reading of the manuscript.
A. R. was funded by a fellowship from the Spanish Instituto de Investigaciones Agrarias (INIA). This work was supported with grant RTA01-005-C2, from the INIA.

REFERENCES

Bender, C.L., Stone, H.E., Sims, J.J., and Cooksey, D.A., 1987, Reduced pathogen fitness of *Pseudomonas syringae* pv. *tomato* Tn*5* mutants defective in coronatine production. *Physiol. Mol. Plant Pathol.* **30**: 273-283.

Bereswill, S., Bugert, P., Völksch, B., Ullrich, M., Bender, C.L., and Geider, K., 1994, Identification and relatedness of coronatine-producing *Pseudomonas syringae* pathovars by PCR analysis and sequence determination of the amplification products. *Appl. Environ. Microbiol.* **60**: 2924-2930.

Fourie, D., 1998, Characterisation of halo blight races on dry beans in South Africa. *Plant Dis.* **82**: 307-310.

Gasson, M.J., 1980, Indicator technique for antimetabolic toxin production by phytopathogenic species of *Pseudomonas*. *Appl. Environ. Microbiol.* **39**: 25-29.

González, A.J., Asensio, C., Ferreira, J.J., and Fueyo, M.A., 1998, Bacteriosis de mayor incidencia en el cultivo de la judía (*Phaseolus vulgaris* L.) en la zona norte de la península ibérica. *IX Congreso de la Sociedad Española de Fitopatología*, Salamanca, p. 81.

Goszczynska, T., and Serfontein, J.J., 1998, Milk-tween agar, a semiselective medium for isolation and differentiation of *Pseudomonas syringae* pv. *syringae*, *Pseudomonas syringae* pv. *phaseolicola* and *Xanthomonas axonopodis* pv. *phaseoli*. *J. Microbiol. Methods* **32**: 65-72.

Grogan, R.G., and K.A. Kimble, 1967, The role of seed contamination in the transmission of *Pseudomonas phaseolicola* in *Phaseolus*. *Phytopathology* **57**: 28-31.

Harper, S., Zewdie, N., Brown, I.R., and Mansfield, J.W., 1987, Histological, physiological and genetical studies of the responses of leaves and pods of *Phaseolus vulgaris* to three races of *Pseudomonas syringae* pv. *phaseolicola* and to *Pseudomonas syringae* pv. *coronafaciens*. *Physiol. Mol. Plant Pathol.* **31**: 153-172.

Hernández-Guzmán, G., and Alvarez-Morales, A., 2001, Isolation and characterisation of the gene coding for the amidinotransferase involved in the biosynthesis of phaseolotoxin in *Pseudomonas syringae* pv. *phaseolicola*. *Mol Plant Microbe Interact.* **14**: 545-554.

Jackson, R.W., Athanassopoulos, E., Tsiamis, G., Mansfield, J.W., Sesma, A., Arnold, D.L., Gibbon, M.J., Murillo, J., Taylor, J.D., and Vivian, A., 1999, Identification of a pathogenicity island, which contains genes for virulence and avirulence, on a large native plasmid in the bean pathogen *Pseudomonas syringae* pathovar phaseolicola. *Proc. Natl. Acad. Sci.* **96**: 10875-10880.

King, E.O., Ward, N.K., and Raney, D.E., 1954, Two simple media for the demonstration of pyocyanin and fluorescein. *J. Lab. Clin. Med.* **44**: 301-307.

Lelliott, R.A., and Stead, D.E., 1987, *Methods for the diagnosis of bacterial diseases of plants*. Blackwell Scientific Publications, Oxford.

Marques, A.S., Corbiere, R., Gardan, L., Tourte, C., Manceau, C., Taylor, J.D., and Samson, R., 2000, Multiphasic approach for the identification of the different classification levels of *Pseudomonas savastanoi* pv. *phaseolicola*. *Eur. J. Plant Pathol.* **106**: 715-734.

Mitchell, R.E., 1978, Halo blight of beans: Toxin production by several *Pseudomonas phaseolicola* isolates. *Physiol. Plant Pathol.* **13**: 37-49.

Patil, S.S., Hayward, A.C., and Emmons, R., 1974, An ultraviolet-induced non-toxigenic mutant of *Pseudomonas phaseolicola* of altered pathogenicity. *Phytopathology* **64**: 590-595.

Prosen, D., Hatziloukas, E., Schaad, N.W., and Panopoulos, N.J., 1993, Specific detection of *Pseudomonas syringae* pv. *phaseolicola* DNA in bean seed by polymerase chain reaction-based amplification of a phaseolotoxin gene region. *Phytopathology* **83**: 965-970.

Rudolph, K.W.E., 1995, *Pseudomonas syringae* pathovars. In *Pathogenesis and host specificity in plant diseases* (U.S. Singh, R.P. Singh and K. Kohmoto, eds.), Elsevier Science Ltd., Oxford, U.K., Vol. 1, pp. 47-138.

Saettler, A.W., 1991, Diseases caused by bacteria. In *Compendium of bean diseases* (R. Hall, ed.), APS Press, St. Paul.

Sato, M., Watanabe, K., Yazawa, M., Takikawa, Y., and Nishiyama, K., 1997, Detection of new ethylene-producing bacteria, *Pseudomonas syringae* pvs. *cannabina* and *sesami*, by PCR amplification of genes for the ethylene-forming enzyme. *Phytopathology* **87**: 1192-1196.

Schaad, N.W., Cheong, S.S., Tamaki, S., Hatziloukas, E., and Panopoulos, N.J., 1995, A combined biological and enzymatic amplification (BIO-PCR) technique to detect *Pseudomonas syringae* pv. *phaseolicola* in bean seed extracts. *Phytopathology* **85**: 243-248.

Schaad, N.W., Jones, J.B., and Chun, W. (eds.), 2001, *Laboratory guide for identification of plant pathogenic bacteria*. APS Press, St. Paul, Minnesota, USA.

Sesma, A., Aizpún, M.T., Ortiz, A., Arnold, D., Vivian, A., and Murillo, J., 2001, Virulence determinants other than coronatine in *Pseudomonas syringae* pv. *tomato* PT23 are plasmid-encoded. *Physiol. Mol. Plant Pathol.* **58**: 83-93.

Smith, I.M., Dunez, J., Lelliot, R.A., Phillips, D.H., and Archer, S.A., 1988, *European handbook of plant diseases*. Blackwell Scientific Publications, London.

Staskawicz, B., and N. J. Panopoulos, 1979, A rapid and sensitive microbiological assay for phaseolotoxin. *Phytopathology* **69**: 663-666.

Taylor, J.D., Teverson, D.M., Allen, D.J., and Pastor-Corrales, M.A., 1996, Identification and origin of races of *Pseudomonas syringae* pv. *phaseolicola* from Africa and other bean growing areas. *Plant Pathol.* **45**: 469-478.

Völksch, B., and Weingart, H., 1998, Toxin production by pathovars of *Pseudomonas syringae* and their antagonistic activities against epiphytic microorganisms. *J. Basic Microbiol.* **38**: 135-145.

Webster, D.M., Atkin, J.D., and Cross, J.E., 1983, Bacterial blights of snap beans and their control. *Plant Dis.* **67:** 935-940.

Zhang, Y.X., and Patil, S.S., 1997, The *phtE* locus in the phaseolotoxin gene cluster has ORFs with homologies to genes encoding amino acid transferases, the AraC family of transcriptional factors, and fatty acid desaturases. *Mol. Plant-Microbe Interact.* **10**: 947-960.

Detection of *Pseudomonas syringae* pv. *tomato* by PCR

M. ZACCARDELLI, A. SPASIANO, M. MERIGHI[*] and C. BAZZI[#]
Istituto Sperimentale per le Colture Industriali, Mi.P.A.F., Battipaglia (SA), Italy; [*]*Department of Plant Pathology/Plant Molecular Biology and Biotechnology Program, Ohio State University, Columbus, USA;* [#]*Dipartimento di Scienze e Tecnologie Agroambientali (DiSTA) – Patologia Vegetale. Università di Bologna, Italy.*

Abstract: *Pseudomonas syringae* pv. *tomato* is the causal agent of bacterial speck of tomato. This bacterium may cause spots on leaves, with or without a chlorotic halo, and on fruits, with consequent economic damage. The identification of the *P. s.* pv. *tomato* is mainly based on isolation of bacteria on semiselective media (e.g. VBTar) coupled to DNA hybridisation using coronatine synthesis genes and on serological techniques such as indirect fluorescent antibody staining (IFAS) and enzyme-linked immunosorbent assay (ELISA). A method based on PCR amplification of coronatine gene sequences was reported as well. Unluckily not only are *cor* genes conserved among several *P. syringae* pathovars, but also Cor$^-$ *P. s.* pv. *tomato* strains have been described, leading to potential false negative or positive misdiagnosis. In this study we present a molecular detection method based on PCR amplification of stable chromosomal sequences that are located in the Hrp_{PST} pathogenicity island. Pathovar-specific oligonucleotide primers were designed on nonconserved regions of the *hrp*Z gene. A specific amplification product of 532 bp was obtained from all of 35 isolates of *P. s.* pv. *tomato* but not from 98 pathogenic and symbiotic bacterial isolates belonging to the genera *Pseudomonas*, *Xanthomonas*, *Erwinia*, and *Rhizobium*. Moreover, no amplicons were obtained from a plethora of unidentified bacterial saprophytes isolated from tomato plants in the field. On experimentally and naturally infected tomato plant materials, the expected amplicon was always obtained using crud extracts from leaves and fruit spots.

Key words: Harpin, Hrp Type III secretion system, pathovar-specific, molecular diagnosis.

Author for correspondence: M. Zaccardelli; e-mail: iscibattipaglia@tiscalinet.it

N.S. Iacobellis *et. al.* (eds.), *Pseudomonas syringae* and Related Pathogens, 553-558.

1. INTRODUCTION

Pseudomonas syringae pv. *tomato* is the causal agent of bacterial speck of tomato. This bacterium causes necrotic spots on leaves and fruits; on leaves, they are angular, with or without a chlorotic halo (Fig. 1). The chlorotic halo is due to the phytotoxin coronatine, which plays an important role in pathogen fitness (Bender *et al.*, 1987). The coronatine gene is located on a specific plasmid of *P. s.* pv. *tomato* (Bender *et al.*, 1989). Economic losses arise from specks on fruits destined for fresh markets and processing.

Figure 1. Bacterial speck lesions on naturally infected tomato fruits.

P. s. pv. *tomato* grows as an epiphyte on tomato leaves and seedlings and survives soil-borne in infected plant debris (Cuppels and Elmhirst, 1999). Seeds are another important source of inoculum (Bazzi, 1991).

The identification of *P. s.* pv. *tomato* is mainly based on: (1) isolation of bacteria on semiselective media (e.g. VBTar) coupled to DNA hybridisation using coronatine synthesis genes as probes (Cuppels and Elmhirst, 1999); (2) serological techniques such as indirect fluorescent antibody staining (IFAS) and enzyme-linked immunosorbent assay (ELISA). A method based on PCR amplification of coronatine gene sequences was reported as well (Bereswill *et al.*, 1994). However, not only are coronatine genes (Cor) conserved among several others *P. syringae* pathovars, but also Cor negative (Cor^-) *P. s.* pv. *tomato* strains have been described, leading to potential false positive or negative results and a misdiagnosis.

In this study, we present preliminary results on a molecular detection method for identification and detection of *P. s.* pv. *tomato in planta*. The protocol is based on PCR amplification of a virulence-associated chromosomal sequence that is located in the hrp_{PST} pathogenicity island of *P. s.* pv. *tomato*.

2. MATERIALS AND METHODS

2.1 Bacterial Isolates

We used 35 *Pseudomonas syringae* pv. *tomato* strains from different geographic origins, ca. one hundred pathogenic and symbiotic bacterial strains from the genera *Pseudomonas*, *Xanthomonas*, *Erwinia* and *Rhizobium*, including 17 other pathovars with the *P. syringae* complex and more than fifty unidentified bacterial isolates obtained from tomato plants.

2.2 Experimentally Inoculated Plant Material

For production of bacterial lesions, two commercially available tomato varieties were inoculated with virulent *Pseudomonas syringae* pv. *tomato* strains OMP-BO 407/91 and LNPV 1741 (four treatments per strain per variety). For each treatment, plants were inoculated as described. Plants sprayed with sterile water were used as negative control. Upon symptom development, two spot samples (ten leaf spots/sample) per treatment or asymptomatic tissue from the controls, were collected for analysis.

2.3 Naturally Infected Plant Material

Plant material presenting naturally occurring symptoms consisted of ten samples of ten leaf spots each and ten samples of ten fruit spots each. Such samples were collected from different tomato varieties grown in two locations of the Salerno province (Southern Italy), with both fields showing heavy bacterial speck infections during the spring of 2002.

2.4 Plant Crude Extract Preparation

Each leaf or fruit spot sample was ground in a mortar with a pestle after adding 1.5 ml of antioxidant extraction buffer (Gorris *et al.*, 1996). Homogenates were filtered through paper and filtrates used for direct isolations and PCR analyses.

2.5 Reisolation of *Pseudomonas syringae* pv. *tomato* from Crude Extracts

Direct isolations of *Pseudomonas syringae* pv. *tomato* on KB-agar were performed by plating crude extracts from spots of naturally infected leaves and fruits as described above. Putative *P. s.* pv. *tomato* colonies were purified from a single colony and PCR analyses were performed on cell suspensions.

2.6 Primers Design and PCR Analysis

Two 20-mer pathovar-specific oligonucleotide primers were designed on non-conserved regions of the *hrp*Z gene of *Pseudomonas syringae* pv. *tomato* (Fig. 2). The reaction mixture was prepared as described in Ausubel *et al.,* 1987, and details on reaction cycles and primer sequences will be presented elsewhere. As a template, we used aqueous bacterial suspensions or crude extracts. The amplicons were visualised by standard agarose gel electrophoresis and ethidium bromide staining.

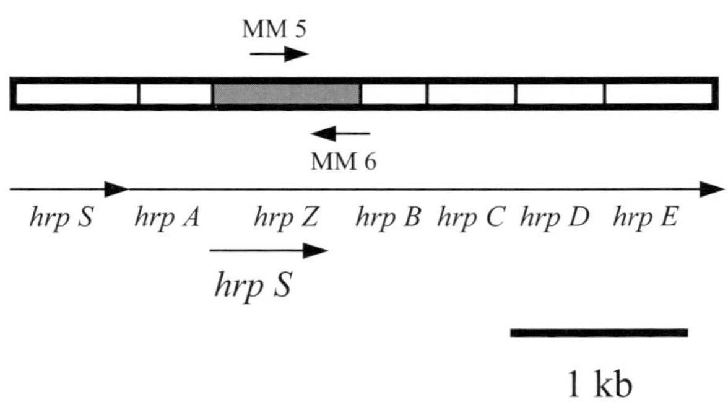

Figure 2. Physical map of the *hrpZ* operon with approximate positions of the MM5 and MM6 primers on the *hrpZ* ORF (Map adapted from Preston *et al.*, 1995).

3. RESULTS

The multiple alignment of the nucleotide sequences of several *hrpZ* genes available from Genbank, showed that this gene is highly conserved among *Pseudomonas syringae* pathovars. Nevertheless, we were able to find short pathovar-specific sequences, potentially suitable for primer design (primers MM5 and MM6; Fig. 2).

The expected pathovar-specific 532 bp amplification product was produced from all the *P. s.* pv. *tomato* isolates but not from the other pathogenic and symbiotic bacterial isolates belonging to the genera *Pseudomonas*, *Xanthomonas*, *Erwinia*, and *Rhizobium*. Moreover, no amplicons were obtained from all the unidentified bacterial isolates from tomato plants. Finally, we always amplified the expected PCR product using crude extracts of leaves and fruit spots collected from experimentally contaminated and naturally infected tomato plants (Fig. 3). PCR analyses performed on pure cultures obtained by direct isolation confirmed in all cases the presence of *P. s.* pv. *tomato* in the spot samples.

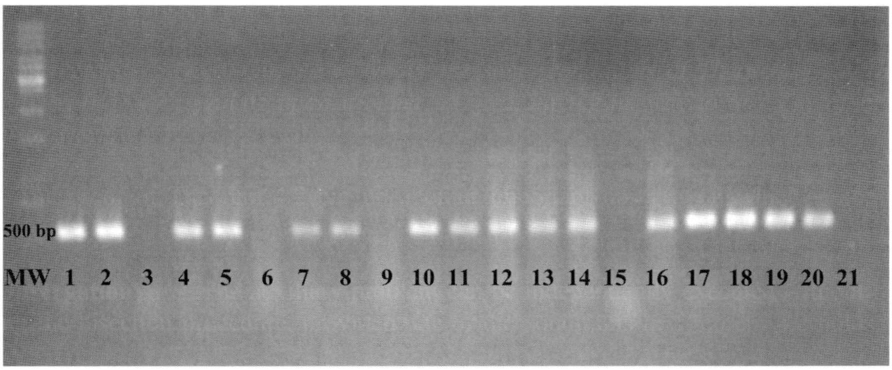

Figure 3. Gel electrophoresis analysis of PCR products amplified with primers MM5 and MM6. MW = molecular weight marker; 1, 2 = *Pseudomonas syringae* pv. *tomato* cells; 3 = water control; 4, 5, 7, 8 = experimentally inoculated plant material; 6, 9 = healthy leaves; 10, 11,12, 13, 14 = naturally infected leaves; 15 = healthy leaves; 16, 17, 18, 19, 20 = naturally infected fruits; 21 = healthy fruit.

4. CONCLUSION

The results show that our PCR protocol is suitable for specific detection of *Pseudomonas syringae* pv. *tomato* both in pure culture and in naturally infected tomato plant material, avoiding time consuming direct isolation and classical identification assays.

ACKNOWLEDGEMENTS

These investigations were supported by the Ministero delle Politiche Agricole e Forestali (MIPAF, ome, Italy) in the frame of the "Piano Nazionale Biotecnologie Vegetali", Programme no. 454, Intervention Area 4.2.10 "I Diagnostici".

REFERENCES

Ausubel, F.M., Brent, R., Kingston, R.E., Moore, D.D., Seidman, J.G., Smith, J.A., and Struhl, K. (eds). 1987-2002, Current Protocols in Molecular Biology, Vol. 1-4. New York: John Wiley & Sons.

Bazzi, C., 1991, I batteri fitopatogeni trasmissibili per seme. *Petria* **1**: 19-30.

Bender, C.L., Malvick, D.K., and Mitchell, R.E., 1989, Plasmid mediated production of the phytotoxin coronatine in *Pseudomonas syringae*. *J. Bacteriol.* **171** (2): 807-812.

Bender, C.L., Stone, H.E., Sims, J.J., and Cooksey, D.A., 1987, Reduced pathogen fitness of *Pseudomonas syringae* pathovar *tomato* Tn*5* mutants defective in coronatine production. *Physiol. Mol. Plant Pathol.* **30** (2): 273-283.

Bereswill, S., Bugert, P., Volksch, B., Ullrich, M., Bender, C.L., and Geider, K., 1994, Identification and relatedness of coronatine-producing *Pseudomonas syringae* pathovars by PCR analysis and sequence determination of the amplification products. *Appl. Environ. Microbiol.* **60** (8): 2924-2930.

Cuppels, D.A., and Elmhirst, J., 1999, Disease development and changes in the natural *Pseudomonas syringae* pv. *tomato* populations on field tomato plants. *Plant Dis.* **83** (8): 759-764.

Gorris, M.T., Cambra, M., Llop, P., López, M.M., Lecomte, P., Chartier, R., and Paulin, J.P., 1996, A sensitive and specific detection of *Erwinia amylovora* based on ELISA-DASI enrichment method with monoclonal antibodies. *Acta Hort.* **411**: 41-45.

Preston, G., Huang, H., He, S.Y., and Collmer, A., 1995, The HrpZ proteins of *Pseudomonas syringae* pv. *syringae*, *glycinea* and *tomato*, are encoded by an operon conteining yersinia ysc homologs and helicit the hypersensitive response in tomato but not in soybean. *Molec. Plant-Microbe Interact.* **8** (5): 717-732.

New Emerging Pathogens

Current Status of Some New and Some Old Plant Pathogenic Pseudomonads

D.E. STEAD, H. STANFORD, A. ASPIN and J. HEENEY
Central Science Laboratory, Sand Hutton, York, YO41 1LZ, UK.

Abstract: The current status of many economically important pseudomonads is discussed. Most are in *Pseudomonas* but *Acidovorax*, *Burkholderia* and *Ralstonia* spp. are also included. Some relevant taxonomic issues are included. Emphasis is given to possible emerging threats.

Key words: *Pseudomonas, Acidovorax, Burkholderia, Ralstonia.*

Author for correspondence: D.E. Stead; e-mail: d.stead@csl.gov.uk.

1. INTRODUCTION

This is a review of the relative importance of plant pathogenic pseudomonads with emphasis on those that are novel. It inevitably concentrates on the situation in Europe and covers those pathogens which are causing commercial crop loss but does not attempt to quantify these losses. Spotting new pathogens or new disease trends depends largely on the accuracy of nomenclature and classification and this in turn depends on the diagnostic methods available. Some of the taxonomic implications are discussed.

Modern methods for characterising bacteria such as genetic fingerprinting are improving the quality of classification and identification of many bacteria. We are now also able to monitor individual populations of some strains and to spot earlier emergence / spread of novel strains. There are many reasons why new strains are emerging including development of new cultivars of crop plants, changes in agronomic practices and changes in climate.

N.S. Iacobellis *et. al.* (eds.), *Pseudomonas syringae* and Related Pathogens, 561-572.

The increasing movement of plants and plant products around the world also contributes to the introduction of new pathogens into new regions.
This paper reviews some of these issues. It treats pseudomonads as genera of plant pathogenic bacteria that at one time were considered to belong to *Pseudomonas* including *Acidovorax, Burkholderia, Pseudomonas, Ralstonia* and other as yet unclassified bacteria.

2. ACIDOVORAX

Acidovorax is a relatively recently described genus for a group of Beta sub class *Proteobacteria*, to which several of the non-fluorescent plant pathogenic *Pseudomonas* species were later transferred (Willems *et al.*, 1992). *A. avenae* subsp. *avenae* comprises strains that were included in *Pseudomonas avenae, P. setariae,* and *P. rubrilineans. A. a.* subsp. *citrulli*, previously classified as *P. pseudoalcaligenes* subsp. *citrulli* is an increasingly important pathogen of melon and water melon which was first described in 1976 in the USA (Schaad *et al.*, 1978) and is now spreading to other parts of the world. It is discussed in detail elsewhere (Schaad *et al.*, 2003). Initially described as a seedling disease of watermelon, there is increasing spread to fruit and melons. An important question is what are the risks to other cucurbit crops such as cucumber? *A. konjaci*, previously described as *P. p.* subsp. *konjaci* (Goto, 1983) causes a leaf blight of the root vegetable, konjac (*Amorphophallus konjac*) but is currently known only from Japan, where it is considered widespread.
A. anthurii is a new species from the West Indies on *Anthurium*, which has the potential to cause economic damage (Gardan *et al.,* 2000).
Until recently, all plant pathogenic *Acidovorax* spp. were found in the warmer parts of the world. Increasingly however, new species and new hosts are being reported from cooler areas. A novel pathogen of the salad vegetable Lamb's lettuce or corn salad (*Valerianella locusta*) has been found in France and the name *A. valerianellae* has been submitted (Gardan *et al.*, 2002).
Several strains causing a leaf spot in *Primula* spp. in Italy have been shown to have fatty acid profiles typical of *Acidovorax*, in which the main and often the only hydroxy fatty acid is 3 hydroxydecanoic acid (10:0 3OH).
BOX-PCR derived genetic fingerprints of all the *Acidovorax* strains in the National Collection of Plant Pathogenic Bacteria (NCPPB) indicate that most strains from new hosts have unique profiles and are thus likely to be new taxa. These studies cannot in themselves determine whether these taxa warrant species status. However they have shown that the main subspecies *A. a.* subsp. *avenae* is heterogeneous, perhaps with strains that are specific for their hosts e.g. rice, pelargonium, canna. Based on DNA homology

studies, some of these subgroups perhaps represent new species (Stead, unpublished results; Postnikova and Schaad, 2002). Jones *et al*. (2003) also indicate a new taxon from geranium (*Pelargonium* spp.).

3. BURKHOLDERIA AND RALSTONIA

These are considered together because they are closely related genera. The most important pathogen is *Ralstonia solanacearum*. It was the subject of a recent international conference and thus is not discussed widely here, despite its world-wide importance. Current key issues are the recent findings of Race 3 / biovar 2A in North America for the first time (Kim *et al*., 2002), perhaps surprisingly not on potato or other solanaceous crop but on pelargoniums imported from Central America. This has become a recent major new commercial host for this pathogen and there have also been recent findings in Europe and Africa, where many of the pelargoniums sold in Europe are grown. In Europe, the last decade has seen an increasing realisation that bv. 2A is much more widespread than had previously been thought. In most countries where it has been found there has been an association with the roots of the weed species *Solanum dulcamara* plants within rivers (Stead *et al*., 1996). Race 1 strains continue to pose threats to the USA tobacco industry. Race 2 strains become ever more severe in banana through both Bugtok and Moko diseases. The Asian biovars 3, 4, and 5 are perhaps less well known. Ginger continues to be an important host with severe losses recorded (Elphinstone, 2002). In sum, *R. solanacearum* continues to have serious economic effect world-wide and its geographic and economic host range continue to increase.
Until recently, *Burkholderia*, like *Ralstonia*, was a genus of plant and mammalian pathogens. This is not now the case, since many new species have recently been described which are not associated with plant pathogenicity. The taxonomic situation has been confused by the inability to differentiate species within the *B. cepacia* complex and the fact that many of the taxonomic clusters within this complex and indeed within some other *Burkholderia* spp. contain strains that are associated with clinical conditions such as cystic fibrosis. There are no recently described plant pathogenic taxa. *B. gladioli* strains continue to cause occasional problems for the mushroom industry. *B. g.* pv. *agaricicola* is more widely distributed than previously recognised. *B. g.* pv. *alliicola* is now well established as a common and often damaging storage pathogen of onions and is probably very widely distributed. *B. cepacia* is now in process of being divided into species based largely on DNA:DNA homology studies, usually supported with some phenotypic evidence. Even onion pathogenic strains perhaps fall into up to three of these species (Richardson *et al.,* 2002). One of the more recently

described plant pathogenic species, *B. glumae* is now recognised as a damaging pathogen of paddy rice especially in Japan (Hikichi *et al.*, 1994).

4. PSEUDOMONAS

The major species, *Pseudomonas syringae* is increasingly recognised as a complex of species, although there is currently little phenotypic evidence to support some of the proposed genomospecies of Gardan *et al.* (1999). The nomenclature used here reflects that which has been validly published. Identification of species tends to rely on identification at the pathovar level, which then allows allocation to the species. Nomenclature in the literature is confused and there is still a trend to place any pathogen for which there isn't a pathovar into *P. s.* pv. *syringae*. Genetic fingerprinting e.g. by REP and BOX-PCR tend to support the pathovar system since many clusters are correlated to host. However, clusters of strains from unrelated hosts occur. The largest of these is the cluster which contains most of the strains of pv. *tomato* and pv. *maculicola* in genomospecies 3 (Stead, unpublished results). It is possible that these multi-host clusters are explained by groups of avirulence genes which increase host range. These taxonomic considerations are important when considering the status of pathogens the context of new disease trends. There are very few reports of problems caused by *Pseudomonas* spp. in the European Plant Protection Organisation Reporting Service since the time of the last *P. syringae* meeting in 1995.

4.1 Pseudomonas Species

Pseudomonas cichorii and *P. viridiflava* are both genetically diverse species. In REP-PCR derived dendrograms each comprises several distinct profile types. Within the limited number of strains tested so far, clusters may be correlated with host and with geographic location. *P. avellanae* currently has a limited distribution in hazelnut orchards in Greece and Italy, but seems to be spreading. It also poses a threat to native hazels throughout Europe (Scortichini, 2002). Other *P. syringae*-like strains also cause disease on hazelnut in Italy (see below) (Scortichini *et al.*, 2003). *P. fuscovaginae* is becoming more important in rice and other cereals and is considered widespread. *P. tremae* and *P. cannabina* are not widely distributed. *P. cannabina* can cause economic damage on cannabis crops and its greatest economic impact may be as a potential biocontrol agent

4.2 Pathovars of Pseudomonas Species

Several new pathovars have been described recently. These include *Pseudomonas savastanoi* pv. *retacarpa* from *Retama sphaerocarpa* (Garcia de los Rios, 1999), *P. syringae* pv. *broussonetiae* from paper mulberry in Japan (Takahashi *et al.*,1996), *P. s.* pv. *spinaciae* from spinach in Japan (Ozaki *et al.*, 1998) and *P. s.* pv. *coriandricola*, a well known pathogen of coriander widely distributed in Europe but not previously validly published (Toben and Rudolph, 1996). The newly proposed *P. s.* pv. *avii* is a pathogen of wild cherry in France (Sutra *et al.*, 2003). Some are considered to be causing economic damage or are pathogens of economically important crops, perhaps especially *P. s.* pv. *spinaciae*, now known from Japan and possibly Italy (Bazzi *et al.,* 1987) and USA (Koike *et al.,* 2002). One pathogen that probably deserves pathovar and possibly species rank is a pathogen of broad bean (*Vicia faba*). Once known as *P. fabae* but not accepted either on the approved lists or as a pathovar because of inadequate description and lack of reference strains in international collections, this pathogen is known in several provinces of China. A pathogen of broad bean is also reported from Russia and it may be that these are the same organism. In our host tests it is destructive causing a rapidly developing blight and black rot of leaves and stems, which often results in death of plants. It has unique fatty acid and REP-PCR profiles and does not cause symptoms on *Phaseolus* beans (Stead, unpublished). This is one pathogen that may deserve consideration as a quarantine pathogen in those regions where it does not currently occur.

The last 10 years has seen much research in Europe on several pathovars which have recently come to the fore. These include *P. s.* pv. *pisi* (Roberts, 1992) causing bacterial blight of pea, *P. s.* pv. *porri* causing leaf spots on leeks (Samson *et al.*, 1998; Koike *et al.*, 1999) and *P. s.* pv. *atrofaciens* causing glume blotch of wheat (von Kietzell and Rudolph, 1997).

P. s. pv. *pisi* has remained an occasionally serious pea pathogen and is increasingly widespread in Europe. It has become very widespread in the dry harvested protein pea cultivars. It is now well documented that it has an epiphytic stage in its life cycle (Grondeau and Samson, 1997). However, it causes economic damage in some years, most probably linked to the extent of seed infection and the soil moisture levels (Roberts *et al.*, 1996). *P. s.* pv. *porri* is most associated with French leek production and most of the reports of outbreaks are from France, where it appears to be common but only occasionally of economic importance.

P. s. pv. *atrofaciens* is well known from eastern Europe in recent years and often appears to be damaging. It is noted as a damaging disease of wheat in Russia (Matveeva *et al.,* 2002).It is recorded from western Europe but more often as an epiphyte with few occurrences of economic damage. It appears to be widespread in Germany (von Kietzell and Rudolph, 1999) but there are no reports of commercial damage.

P. s. pv *aptata* is perhaps more widespread especially in Europe than thought. It is well studied in Belgium and some other European countries (Maraite and Weyns, 1997), but with relatively little information on its economic importance on sugar beet yields.

Of the pathogens that have been problematic for longer periods, *P. s.* pv. *syringae* is the most complex because of its large host range and inadequate classification, having been used for many pathogens for which a pathovar is not known. This situation perhaps masks the risks some new emerging pathogens present. One example is that of a serious disease of pepper, often more prevalent in young plants in eastern Europe from Macedonia (Mitrev *et al.,* 2000) to Serbia and Croatia (Arsenijevic and Obradovic, 1997). This pathogen has been attributed to *P. s.* pv. *syringae* even though it is not pathogenic to lilac.

P. s. pv *morsprunorum* continues to be a major problem particularly on cherry and plum. Problems are often associated with new cultivars, grafting and young trees. A new pathovar (*P. s.* pv. *avii*) causing bacterial canker of wild cherry (*Prunus avium*) in France is also proposed (Sutra *et al.,* these proceedings) *P. s.* pv. *savastanoi* continues to be a commercial problem on olive in many parts of the Mediterranean region. *P. s.* pv. *phaseolicola* continues to be a widespread, common and often serious pathogen of *Phaseolus* beans world-wide. One interesting finding has been the existence of at least 9 races based on host resistance genes. Of these, races 1,2 5, 6, and 7 are of world-wide distribution with race 6 the most widespread. Races 3, 4, 8 and 9 are known only from east and southern Africa (Taylor *et al.* 1996). These may pose risks to countries importing beans from these regions. Exports from these countries are an expanding trade. *P. s.* pv. *glycinea* has a more limited distribution in countries which grow soybeans and is often under some form of legislative control. There are occasional reports of its spread into new areas often with economic damage.

P. syringae is a serious pathogen on hardy ornamental nursery stock throughout Europe and North America. Although several of the commercially important plant species are known to have their own pathovars e.g. *P. s.* pv. *philadelphi* on *Philadelphus* spp., most are attributed to *P. s.* pv. *syringae* itself. REP-PCR studies on isolates from different plant genera with symptoms on a nursery show a classification based on host specific clusters. This implies that under natural conditions, there is no movement from one host to another, but that all host types are carrying their own host specific populations, all of which tend to cause serious disease under the same conditions, often through overhead irrigation especially during periods of frost. The case of *P. s.* pv. *syringae* itself is an example. In REP-PCR dendrograms containing virtually all described pathovars (Stead *et al,* 2003), the cluster containing *P. s.* pv. *syringae* NCPPB 281, the type strain, contains almost all other strains from *Syringa* spp. (lilacs) and strains from no other hosts with the exception of *Forsythia suspensa*, which is closely related to lilacs in the *Oleaceae*. It appears therefore that under natural conditions *P. s.* pv. *syringae* is a pathogen only of *Syringa* spp. and perhaps a few other genera in the family *Oleaceae*. These results also question the value of artificial host tests. In REP-PCR studies, we see many so-called *P. s.* pv. *syringae* strains from other hosts clustering separately or with other pathovars usually in genomospecies 3, implying that they may indeed be separate taxa. REP-PCR studies cannot accurately identify at species level or be used to prove that a group of strains represent a novel species. However, a classification based on all pathovars, with a wide range of strains for many, showed fairly good correlation with the genomospecies defined by Gardan *et al.* (1999). The purpose of this taxonomic discussion is to illustrate the problems with risk assessment for particular crops and with assessment of economic impact of a particular pathogen. Returning to the nursery situation above, a heavily infected lot of one host is probably only a risk for spreading onto closely related plant species, unless of course the pathogen belongs to one of the clusters which does have strains from multiple hosts. Again the genetic fingerprint studies cannot predict the virulence or host range but can be used to select appropriate hosts for host studies. It goes without saying that these should be done under as natural conditions as possible.

The biggest single cluster comprising many pathovars comprises those pathovars from genomospecies 3. These include *P. s.* pv. *tomato*, *P. s.* pv. *maculicola* , *P. s.* pv. *apii* and *P. s.* pv. *tagetes*. There are also many strains from other hosts. All magnolia strains fall in this group, as do one *Viburnum* strain from the UK (Stead, unpublished). Recent *Viburnum* strains from the UK are different however and these have identical REP-PCR fingerprints to the pathotype strain of *P. s.* pv. *viburni*, previously known from North America with only spurious records from Europe. The Italian strains from *Spinacia* (spinach) (Bazzi *et al.*, 1988), although not pathogenic to *Brassica* spp also cluster with several *P. s.* pv. *maculicola* and *P. s.* pv. *tomato* strains. Novel hosts from which *P. syringae*-like strains have been recently deposited in the NCPPB and the CSL research collection include species from the following genera:

Amelanchier, Anemone, Bambusa, Betula, Camellia, Caryopteris, Choisya, Cordyline, Cotoneaster, Eruca, Euonymus, Hebe, Holodiscus, Lactuca, Lavandula, Photinia, Physocarpus, Purshia, Pyracantha, Rosa, Sagaretia, Spirea, Trigonella, Ulmus. Most of these are woody ornamentals. Almost all are grown on nurseries. REP-PCR studies showed that the *Camellia* strains were not *P. s.* pv. *syringae* or *P. s.* pv. *theae* (Stead, unpublished).

There are several recent reports of unusual outbreaks of *P. syringae* induced diseases occurring on crops including apricots and hazel nuts in Italy. In both cases the disease has been noticeably different from usual outbreaks, either because of new cultivars or novel symptoms. In both cases (Stefani and

Stead, 2002; Scortichini *et al.*, 2003), genetic fingerprinting has shown that several pathogens cause disease of each tree and that the outbreak is associated with what seems to be a novel strain. There are now several distinct pathovars causing diseases of *Prunus* spp. including pv. *syringae*, pv. *morsprunorum,* pv. *persicae*, together with some as yet undescribed taxa. A similar situation has occurred with *P. syringae* strains on crucifers in California (Cintas *et al.,* 2001). Here, three new and sometimes damaging bacterial leaf blights have recently been recorded on rappini, broccoli and broccolini. Whereas the broccolini pathogen is very similar to *P. s.* pv. *maculicola*, the isolates from rappini and broccoli appear to be very similar to each other but possibly representing a new pathovar (Cintas *et al.*, 2002). Another example is the *P. syringae* strains causing dieback of Ballerina apple cultivars in the UK some 7÷8 years ago. Ballerina apples are derived from a sport of cv. Mackintosh which has enhanced apical dominance, which makes them suitable for small gardens. These strains were typical LOPAT group1 strains (Lelliott *et al.,* 1966) and were occasionally pathogenic to Ballerina apple cultivars in host tests but not to other apple cultivars. They also had genetic fingerprints most similar to epiphytic strains from apples belonging to a wide range of cultivars (Stead, unpublished).

It remains to be seen whether these actually represent new pathovars but these studies do show the potential value of genetic fingerprints and the need for well characterised taxa even at these infra-specific levels if effective control is to be developed especially through such methods as breeding for resistance or through certification by seed testing for those pathogens that are seed borne. The presence of 2 pathogens on a single host, perhaps each with a different potential host range does not facilitate risk assessment and control.

5. UNCLASSIFIED PATHOGENS

One potentially interesting pathogen is a pathogen causing sectored wilt of *Pelargonium zonale* hybrids. Known from the UK in the 1950's and tentatively named *Pseudomonas fastidiosa* (Stead, 1988), this undescribed pathogen has a fatty acid profile similar to that of *Xylophilus ampelinus.* Further work is under way to further determine its taxonomic position.

6. CONCLUSIONS

Improved techniques for classification and identification of bacteria now allow improved assessment of risks and monitoring of emerging pathogens. It is becoming increasingly clear that some crops have several genetically distinct but closely related pathogens within the *P. syringae* complex. Such crops / hosts include hazelnut, apricot, cherry, peach, beans, brassicas, pea and viburnums. The emergence of some of these new pathogens may be due to selection through the development of new cultivars with different patterns of resistance. An improved understanding of the mechanisms of selection, for example, whether they represent a large, diverse group of epiphytes is required. Their presence has great significance for control strategies such as breeding for resistance and control of seed pathogens by seed certification, since control of one pathogen may simply create a niche for another.

ACKNOWLEDGMENTS

This work was supported by the Department for Environment, Food and Rural Affairs for provision of advice and diagnostic services for plant pathogenic bacteria.

REFERENCES

Arsenijevic, M., and Obradovic, A., 1997, A pathovar of *Pseudomonas syringae* causal agent of bacterial leaf spot and blight of pepper transplants. In: Pseudomonas syringae *pathovars and related pathogens* (Rudolph, K., Burr, T.J., Mansfield, J.W., Stead, D., Vivian, A., and von Kietzell, J., eds.). Kluwer Academic Publishers, Dordrecht. pp. 61-66.

Bazzi, C., Gozzi, R., Stead, D.E., and Sellwood, J., 1988, A bacterial leaf spot of spinach (*Spinacia oleracea* L) caused by a non-fluorescent *Pseudomonas syringae* Van Hall (pathovar). *Phytopath. Medit.* **27**: 103-107.

Bazzi, C., Stead, D.E., Alexandrova, M., and Stefani, E., 1997, Identification and classification of fluorescent *Pseudomonas* species from cereals in Italy. In: Pseudomonas syringae *pathovars and related pathogens* (Rudolph, K., Burr, T.J., Mansfield, J.W., Stead, D.E., Vivian, A., von Kietzell, J., eds.). Kluwer Academic Publishers, Dordrecht, pp. 509-514.

Cintas, N.A., Koike, S.T., and Bull, C.T., 2001, Emerging bacterial pathogens of crucifers in the Salinas Valley of California. In: *Plant Pathogenic Bacteria* (De Boer, S., ed.). Kluwer Academic Publishers, Dordrecht, pp. 272-274.

Cintas, N.A., Koike, S.T., and Bull, C.T., 2002, A new pathovar, *Pseudomonas syringae* pv. *alisalensis* pv. nov. proposed for the causal agent of bacterial blight of broccoli and broccoli raab. *Plant Dis.* **86**: 992-998.

Davies, J.M.L., and Stead, D.E., 1993, Cauliflower cultivar susceptibility and the effect of copper sprays on bacterial leaf spot. In: *Brighton Crop Protection Conference Pests and Diseases - 1992*. Farnham: British Crop Protection Council, 1992. Volume 3, pp. 1163-1168.

Elphinstone, J.G., 2003, The current bacterial wilt situation: a global overview. In: *Bacterial wilt: the disease and the* Ralstonia solanacearum *complex.* (Allen, C., Prior, P., and Hayward, C.A., eds). APS Press, St. Paul, USA., In Press.

Garcia de los Rios, J.E., 1999, *Retama sphaerocarpa* (L.) Boiss., new host of *Pseudomonas savastanoi*. *Phytopath. Medit.*, **38**: 54-60.

Gardan, L., Shafik, H., Belouin, S., Broch, R., Grimont, F., and Grimont, P.A., 1999, DNA relatedness among the pathovars of *Pseudomonas syringae* and description of *Pseudomonas tremae* sp. nov. and *Pseudomonas cannabina* sp. nov. *Int. J. Syst. Bacteriol.* **49**: 469-478.

Gardan, L., Dauga, C., Prior, P., Gillis, M., and Saddler, G.S., 2000, *Acidovora anthurii* sp. nov., a new phytopathogenic bacterium which causes bacterial leaf spot of anthurium. *Int J. Syst. Evol. Microbiol.* **50**: 235-246.

Gardan, L., Stead D.E., Dauga, C., and Gillis, M., 2002, *Acidovorax valerianellae* sp. nov., a novel pathogen of lamb's lettuce [*Valerianella locusta* (L.) Laterr.] *Int. J. Syst. Evol. Microbiol.* **53**: 795-800.

Goto, M., 1983, *Pseudomonas pseudoalcaligenes* subsp. *konjaci* subsp. nov., the causal agent of bacterial leaf blight of konjac (*Amorphophallus konjac* Koch.). *Int. J. Syst. Bacteriol.* **33**: 539-545.

Grondeau, C., and Samson, R., 1997, Effect of host genotype and non-host plants on epiphytic life of *Pseudomonas syringae* pv. *pisi*. In: Pseudomonas syringae *pathovars and related pathogens* (Rudolph, K., Burr, T.J., Mansfield, J.W., Stead, D.E., Vivian, A., and von Kietzell, J., eds.). Kluwer Academic Publishers, Dordrecht., pp. 22-25.

Hikichi, Y., Tsujiguchi, K., Maeda, Y., and Okuno, T., 1994, Isolation of the oxolinic acid resistance related-gene from genomic DNA of *Burkholderia glumae*. In: *Plant Pathogenic Bacteria* (De Boer, S., ed.). Kluwer Academic Publishers, Dordrecht, pp. 220-223.

Janse, J.D., Elphnstone, J.G., and Van Vaerenbergh, J., 2003, Introduction of *Ralstonia solanacearum* biovar 2 / race 3 in *Pelargonium* cuttngs from Kenya. In: *Bacterial wilt: the disease and the* Ralstonia solanacearum *complex.* (Allen, C., Prior, P., and Hayward, C.A., eds). APS Press, St. Paul, USA., In Press.

Jones, J.B., Stall, R.E., Minsavage, G.V., Dickstein, E.R., Bouzar, H., Roberts, P.D., Strickler, P.M., and Engelhard, A.W., 2003, Characterisation of an *Acidovorax* sp. Associated with Geranium and Petunia, 667-673, (this volume).

Kim, S.H., Olson, T.N., and Schaad, N.W., 2002, *Ralstonia solanacearum* Biovar 2 / Race 3 in geraniums imported from Guatemala to Pennsylvania in 1999. *Phytopathology*, **92** (6): S42.

Koike, S.T., Barak, J.D., Henderson, D.M., and Gilbertson, R.L., 1999, Bacterial blight of leek: a new disease in California caused by *Pseudomonas syringae*. *Plant Dis.* **83**: 165-170.

Lelliott, R.A., Billing, E., and Hayward, A.C., 1966, A determinative scheme for the fluorescent plant pathogenic pseudomonads. *J. Appl. Bacteriol.* **29**: 470-489

Maraite, H., and Weyns, J., 1997, *Pseudomonas syringae* pv. *aptata* and pv. *atrofaciens*, specific pathovars or members of pv. *syringae*? In: Pseudomonas syringae *pathovars and related pathogens* (Rudolph, K., Burr, T.J., Mansfield, J.W., Stead, D.E., Vivian, A., and von Kietzell, J., eds.). Kluwer Academic Publishers, Dordrecht., pp. 515-520.

Mitrev, S., Gardan, L., and Samson, R., 2000, Characterisation of bacterial strains of *Pseudomonas syringae* pv. *syringae* isolated from pepper leaf spot in Macedonia. *J. Plant Pathol.* **82**: 227-231.

Matveeva, E.V., Polytiko, V.A., Ignatov, A.N., Nikolaeva, E.V., Pekhtereva, E., Sh., and Schaad, N.W., 2002, Bacterial pathogens on cereals in the Russian Federation. *Phytopathology* **92** (6): S52

Ozaki, K., Kimura, T., and Matsumoto, K., 1998, *Pseudomonas syringae* pv. *spinaceae* pv. nov., the causal agent of bacterial leaf spot of spinach in Japan. *Annals of the Phytopathological Society*, Japan **64**: 264-269.

Postnikova, E., and Schaad, N.W., 2002, Phylogenetic studies of corn and rice strains of *Acidovorax avenae* subsp. *avenae* by DNA/DNA hybridisation. *Phytopathology* **92** (6): S66.

Richardson, J., Stead, D.E., Elphinstone, J.G., and Coutts, R.H.A., 2002, Diversity of *Burkholderia* isolates from woodland rhizosphere environments *J. Appl. Microbiol.* **93**: 616-630.

Roberts, S.J., 1992, Effect of soil moisture on the transmission of pea bacterial blight (*Pseudomonas syringae* pv. *pisi*) from seed to seedling. *Plant Pathol.* **41**: 136-140.

Roberts, S.J., Ridout, M.S., Peach, L., and Brough, J., 1996, Transmission of pea bacterial blight (*Pseudomonas syringae* pv. *pisi*) from seed to seedling: effects of inoculum dose, inoculation method, temperature and soil moisture. *J. Appl. Bacteriol.* **81**: 65-72.

Samson, R., Shafik, H., Benjama, H., and Gardan, L., 1998, Description of the bacterium causing blight of leek as *Pseudomonas syringae* pv. *porri* (pv.nov.). *Phytopathology* **88**: 844-850.

Schaad, N.W., Poostnikova, E., and Randhawa, P., 2003, Emergence of *Acidovorax avenae* subsp. *citrulli* as a Crop Threatening Disease of Watermelon and Melon, 573-581, (this volume).

Schaad, N.W., Sowell, G., Goth, R.W., Colwell, R.R., and Webb, R.E., 1978, *Pseudomonas pseudoalcaligenes* subsp. *citrulli* subsp. nov. *Int. J. Syst. Bacteriol.* **28**: 117-125.

Scortichini, M., 2002, Bacterial canker and decline of European Hazelnut. *Plant Dis.* **86**: 704-709.

Scortichini, M., Marchesi, U., Rossi, M.P., Janse, J.D., and Stead, D.E., 2003, The Pseudomonads Associated with Bacterial Canker and Decline of Hazelnut (*Corylus avellana* L.), 583-593, (this volume).

Stead, D.E., 1988, *Catalogue of the National Collection of Plant Pathogenic Bacteria*, Central Science Laboratory, Harpenden, UK.

Stead, D.E., Elphinstone, J.G., and Pemberton, A.W., 1996, Potato Brown Rot in Europe. *Brighton Crop Protection Proceedings* (1966) **933**: 1145-1152.

Stead, D.E., Simpkins, S.A., Weller, S.A., Hennessy, J., Aspin, A., Stanford, H., Smith, N.C., and Elphinstone, J.G., Classification and Identification of Plant Pathogenic *Pseudomonas* Species by REP-PCR Derived Genetic Fingerprints, 411-420, (this volume).

Stefani, E., and Stead, D.E., 2002, Isolation and characterisation of *Pseudomonas syringae* strains related to recent outbreaks of a severe decay affecting apricot trees. In: *Book of abstracts of 6th International Conference on* Pseudomonas syringae *pathovars and related pathogens*, 43.

Sutra, L., Menard, M., Luisetti, J., Prunier, J.P., and Gardan, L., 2003, Bacterial Canker of Wild Cherry Tree in France Caused by a New Pathovar of *Pseudomonas syringae* pv. *avii* (pv. nov.), 675-679, (this volume).

Takahashi, K., Nishiyama, K., and Sato, M., 1996, *Pseudomonas syringae* pv. *broussonetiae* pv. nov. the causal agent of bacterial blight of paper mulberry (*Broussonetia kazinoki* X *B.papyrifera*). *Annals of the Phytopathological Society of Japan* **62**: 17-22.

Taylor, J.D., Teverson, D.M., Allen, D.J., and Pastor-Corrales, M.A., 1996, Identification and origin of races of *Pseudomonas syringae* pv. *phaseolicola* from Africa and other bean growing areas. *Plant Pathol.* **45**: 469-478.

Toben, H., and Rudolph, K., 1996, *Pseudomonas syringae* pv. *coriandricola*, incitant of bacterial umbel blight and decay of coriander (*Coriandrum sativum* L.) in Germany. *J. Phytopathol.*, **144**: 169-178.

von Kietzell, J., and Rudolph, K., 1997, Epiphytic occurrence of *Pseudomonas syringae* pv. *atrofaciens*? In: Pseudomonas syringae *pathovars and related pathogens* (Rudolph, K.,

Burr, T.J., Mansfield, J.W., Stead, D.E., Vivian, A., and von Kietzell, J., eds.). Kluwer Academic Publishers, Dordrecht., pp. 29-34.

Willems, A., Goor, M., Thielmans, S., Gillis, M., Kersters, K., and De Ley, J., 1992, Transfer of several phytopathogenic *Pseudomonas* species to *Acidovorax* as *Acidovorax avenae* subsp. *avenae* comb. nov., *Acidovorax avenae* subsp. *citrulli*, *Acidovorax avenae* subsp. *cattleyae* and *Acidovorax konjaci*. *Int. J. Syst. Bacteriol.* **42**: 107-119.

Emergence of *Acidovorax avenae* subsp. *citrulli* as a Crop Threatening Disease of Watermelon and Melon

N.W. SCHAAD, E. POSTNIKOVA and P. RANDHAWA[#]
Foreign Disease-Weed Science Research Unit, ARS/USDA, 1301 Ditto Avenue, Ft. Detrick, MD, USA; [#]California Plant and Seed Labs, Elverta, CA.

Abstract: *Acidovorax avenae* subsp. *citrulli* (*Pseudomonas pseudoalcaligenes* subsp. *citrulli*), the causal agent of a watermelon seedling blight and fruit blotch (WFB), has emerged as a serious seedborne pathogen of watermelon and melons world-wide. The disease was first reported in 1965 in seedlings of several USDA Plant Introduction (PI) lines in Georgia. The pathogen was described 13 years later in diseased seedlings from two of the same PI seed sources in which Web and Goth had originally identified the disease. The disease remained known only in seedlings at the introduction station until Wall and Santos isolated a non-fluorescent pseudomonad from rotting fruit in Guam in 1987 and concluded it was the same organism. WFB was reported in Florida in 1989 and a year later emerged as a major economic disease of watermelon in the Southeast and Midwest USA. Since then, WFB has been reported in Turkey, Brazil, Australia, Japan, Korea, Thailand, and China infecting cantaloupe, melon, honeydew, pumpkin, and citron under field conditions. Because the original seedling strains were reported as giving a negative tobacco hypersensitive reaction and failed to rot fruit like the field disease strains, some have suggested that the originally described organism was not the same organism as that causing WFB. DNA fingerprinting of a collection of 121 strains using pulse field gel electrophoresis revealed considerable genetic diversity. All strains typed into two groups; over 80% were in Group I and the remaining, including the original type strain, in Group II. Control of the disease is based primarily on use of pathogen-free seed. Although attempts have been made to develop a routine seed assay to detect the organism in seeds, no reliable method is available. Contaminated seeds remain a problem for the seed industry.

Key words: fruit blotch, seedling blight, seedborne pathogen, seeds.

Author for correspondence: N.W. Schaad; e-mail: schaad@ncifcrf.gov.

N.S. Iacobellis *et. al.* (eds.), *Pseudomonas syringae* and Related Pathogens, 573-581.

1. INTRODUCTION

Acidovorax avenae subsp. *citrulli,* the causal agent of a seedling blight (Schaad *et al*., 1978; Webb and Goth, 1965) (Fig. 1A) and fruit rot (Wall and Santos, 1988a; 1988b) (Fig. 1B, C), has emerged as a serious seedborne pathogen of watermelon world-wide. Watermelon is one of the most important fresh market vegetables in the US with 250,000 acres and an estimated wholesale crop value of over $450,000,000 and a potential hybrid seed value of $75,000,000 (A. Abbott, personal communication). Losses associated with possible infested seeds have threatened the existence of the watermelon seed and transplant industries. Continued crop losses have resulted in litigation against several seed companies during the past 10 years. Pending litigation over crop losses resulting from suspected seed, contamination in one state alone totals more than $240,000,000 (A. Abbott, personal communication). Because of legal liabilities and the lack of a reliable assay to detect low levels of *A. a.* subsp. *citrulli*, seed companies have been forced to add disclaimers on their seed packs.

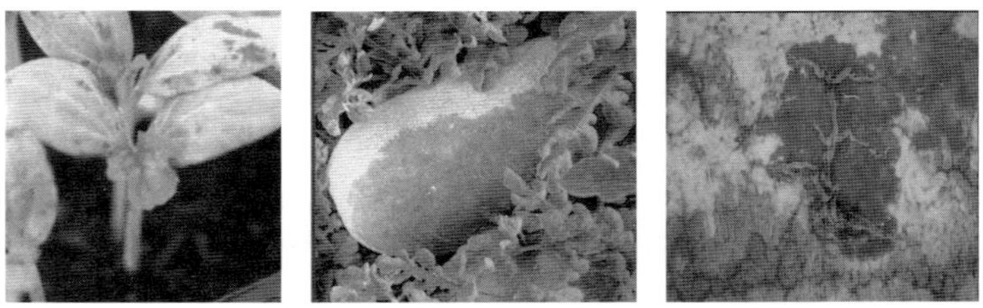

A. Water-soaking on cotyledons | B. Water-soaking on fruit | C. Cracking and water-soaking on fruit

Figure 1. Watermelon fruit blotch symptoms: A, seedlings under greenhouse conditions; B and C, natural infection of fruit under field conditions.

The disease was first recognised in watermelon seedlings at the USDA Regional Plant Introduction Station in Griffin, GA in early 1965 (Webb and Goth, 1965). Seedlings of several lines growing in the greenhouse for routine seed increase were observed with extensive water-soaking lesions that later turned grey to black. Webb and Goth (1965) consistently isolated a bacterium which produced white to tan coloured colonies on Difco Nutrient agar. They completed Koch's postulates in seedlings but did not attempt to identify the organism. The seedling disease may have been observed as early as 1963 at the University of Florida Experiment Station in Leesburg when Muller and Schneck reported bacterial lesions on watermelon leaves in their plots (Mullen and Schnek, 1963). A second report of a bacterial disease of watermelon in Florida occurred in 1972 when Hopkins and Schneck observed water-soaked lesions in the Leesburg plots and identified the causal agent as *Pseudomonas syringae* pv. *lachrymans* (Hopkins and Schneck, 1972). *P. s.* pv. *lachrymans* has never been reported to cause a disease on watermelon since that report, therefore, it is possible that the organism was misidentified and that the real causal organism was the same as that described by Webb and Goth in Georgia (Webb and Goth, 1965). The fruit symptoms of the disease may have been observed for the first time in the same University of Florida field plots in Leesburg. Crall and Schneck, (1969) reported a bacterial fruit rot of watermelon in their experimental plots. The symptoms presented in their paper are very typical of fruit blotch. The authors described a bacterium that was fluorescent when streaked onto King et al's medium B (King *et al.*, 1954) but they were unable to produce fruit symptoms similar to those observed in the field. In addition to the fruit symptoms, the authors reported that bacterial leaf spots similar to those described by Mullin and Schneck (1963) had been observed in their watermelon plots sporadically since 1963. The true causal organism of the seedling and leaf blight disease was finally described in watermelon seedlings of several plant introduction (PI) lines in 1978 in Griffin, Georgia (Schaad *et al.*, 1978). A non-fluorescent pseudomonad was isolated from infected seedlings of several PI lines including two of the same lines, 174103 and 174104, from which Webb and Goth described isolation of an unknown bacterium 13 years earlier (Webb and Goth, 1965). As in 1965, Grover Sowell, the ARS Plant Pathologist at the Introduction Station, was increasing various PI lines and observed several seedlings with water-soaked lesions. Seeds of those lots originated in Turkey and were first increased at the Introduction Station in 1957. Whether the pathogen originally came from the seeds imported from Turkey or from the Experiment Station in Florida where similar symptoms had been observed earlier is not known. The problem remained solely a disease of seedlings at the Plant Introduction Station until Wall and colleagues (Wall and Santos, 1988a; 1988b; Wall *et al.*, 1990) isolated a non-fluorescent pseudomonad from rotting fruit in Guam and made the connection between the fruit disease and the seedling disease described earlier.

Wall observed a large number of rotting watermelon fruit in Guam and Tinian in 1987. According to Wall and Santos (1988a; 1988b), "Symptoms similar to those observed are described from Australia (Volk and McCarthy, 1982) and the disease is referred to as fruit blotch". The name "fruit blotch" for the fruit phase of the disease was used for the first time in the Australian publication. Apparently many growers in Guam had planted the watermelon cultivar 'Glory' during the off-season in 1987 in an attempt to obtain higher prices. However, the fruit matured during a prolonged period of heavy rains and wide epidemics of rotting fruit occurred (Wall and Santos, 1988a). Wall and Santos completed Koch's postulates with both fruit and seedlings. They observed the same symptoms on seedlings as those described in the original descriptions of the seedling disease (Schaad *et al.*, 1978; Webb and Goth, 1965) and found that the phenotypic characteristics of the bacterium causing fruit blotch in Guam agreed with the original description of the organism described earlier in Georgia. They further showed that the causal organism was seed transmitted, infected cotyledons, leaves, and fruit, and survived in soil for at least one week.

Less then a year after Wall and Santos presented a paper at the 1988 American Phytopathology meetings in San Diego, CA (Wall *et al.*, 1990) describing the fruit disease, fruit blotch was diagnosed in Florida (Hopkins, 1989). Wall coined the name watermelon fruit blotch (WFB) in 1989 (Walcott *et al.*, 2000). Soon WFB was recognised as a major economic problem in the major U.S. watermelon production areas of the Southeast (Somodi *et al.*, 1991), Atlantic coast (Somodi *et al.*, 1991), and Midwest (Rane and Latin, 1992). In Texas, WFB has been reported in honeydew melons and citron melon, a common weed in watermelon production areas (Isakeit, 1997; 1998). The disease continues to be a problem in Texas when watermelon and melons are grown under centre pivot sprinkler irrigation (T. Isakeit, personal communication). In contrast, the disease has never been observed in California where the climate is very dry (P. Randhawa, unpublished). Under high, prolonged periods of rainfall in Central Florida and south-eastern Texas infections can make the fruit unmarketable. The disease has been found in several countries, including, Guam, Turkey, Australia, Japan, Thailand, and China (Schaad, unpublished).

2. CLASSIFICATION OF THE ORGANISM

The unknown bacterium isolated from infected seedlings in Georgia in 1977 was quite different from any existing plant pathogenic bacterium. Based on phenotypic characters and cell morphology, the bacterium was most related to the non-fluorescent pseudomonads. The new organism was Gram-negative, non-pigmented, polar-flagellated, aerobic, oxidase positive, and non-fluorescent on KB medium (Schaad *et al.*, 1978). Based on pathogenicity and several biochemical tests, the new bacterium could be differentiated from all known non-fluorescent pseudomonads. Important key characters that differentiated the watermelon pathogen from other pseudomonads included growth at 41°C, hydrolysis of gelatine, and a positive oxidase reaction (Schaad *et al.*, 1978). The new bacterium utilised β-alanine, n-propanol, and ethanol but not sucrose or glucose (Schaad *et al.*, 1978). These results were very different from any known plant pathogenic bacterium. The most closely related plant pathogenic pseudomonad was *Pseudomonas avenae*. Both bacteria were non-fluorescent, grew at 41°C, were positive for nitrate reductase, negative for hypersensitive reaction on tobacco, failed to peptomize litmus milk, hydrolysed lipid, and utilised β-alanine but not sucrose (Schaad *et al.*, 1978). Because of these and other characters, the watermelon bacterium was most similar to the non-fluorescent pseudomonad, *P. pseudoalcaligenes* and was therefore named as a subspecies, *P. pseudoalcaligenes* subsp. *citrulli* (Schaad *et al.*, 1978). Eight strains, C-38, C-41, C-42 (Type strain), C-43, C-45, C-98, C-130, and C-178 were deposited in the International Collection of Micro-organisms from Plants (ICMP) in Auckland, New Zealand. Strain C-42 was deposited in the American Type Culture Collection as 29625.

The organism was reclassified as *Acidovorax avenae* subsp. *citrulli* in 1992 (Willems *et al.*, 1992) based primarily on DNA:rRNA hybridisation data. Since the original strains isolated from seedlings of PI lines in Georgia were reported as negative for the hypersensitive test in tobacco, generally less virulent on seedlings then the fruit rotting strains, and failed to rot fruit like the field disease strains, some have suggested that the two organisms are different pathogens. Somodi *et al.* (1991) reported that the original type strain failed to infect fruit and that fatty acid profiles of the WFB strains matched more closely with *Chromobacterium violaceum* than *P. pseudoalcaligenes*. They suggested that the originally described strains from watermelon seedlings were not the same organism as the organism responsible for WFB. However, Latin and Hopkins, (1995) used three of the eight original strains included for describing *A. a.* subsp. *citrulli* and deposited in the ICMP to inoculate fruit under Florida conditions. In contrast to the negative fruit infection described above, Hopkins observed typical fruit rotting symptoms and concluded that the original negative results of inoculating fruit described by Schaad *et al.* (1978) was, most likely, due to the difference in weather. That is, most likely, the reason because the weather during the late summer of 1977 was hot and dry in Georgia when the fruit were inoculated (Schaad, personal observations). DNA fingerprinting of a collection of 121 strains using pulse field gel electrophoresis revealed considerable genetic diversity among the strains. However, they all could be typed into two groups (Walcott *et al.*, 2000). Over 80% of the strains typed into Group I and the remaining, including the original type strain from seedlings and those strains from pumpkin and the cantaloupe strain from Australia, typed into Group II. Walcott *et al.* (2000) concluded that although the original type strain was not typical of WFB strains, it should be retained as the representative strain of *A. a.* subsp. *citrulli*.

3. THE DISEASE CONTROL PROGRAM, SEED HEALTH TESTING

Many US seed companies produce hybrid seed overseas in countries such as China and Thailand due to lower labour costs. The seed then is sold in the US as well as other countries. Seed produced overseas has been found to contain *Acidovorax avenae* subsp. *citrulli* and have, most likely, resulted in epidemics in the US. Recently, due to a disease outbreak of WFB in cantaloupes in the US from a seed lot produced in China, one country, Costa Rica, has refused to purchase seed from the US that originated in China (Randhawa, personal information).

It is well known that WFB is seed transmitted (Rane and Latin, 1992; Schaad *et al.*, 1978; Wall and Santos, 1988a) and symptomatic fruit generally contain infected seeds. However, infections of flowers at pollination stage can lead to seed infections with no obvious fruit symptoms (Walcott *et al.*, 2002). The pathogen has been recovered from pulp of such infected but asymptomatic fruits. Transmission of the pathogen from infected seeds to seedlings is very high (Rane and Latin, 1992; Schaad *et al.*, 1978; Wall and Santos, 1988a). As few as 1 or 2 infected seedlings can lead to wide-spread disease under high temperatures and overhead watering such as occurs in greenhouse transplant operations (Kucharek *et al.*, 1993). Similar spread can occur in the field during high rainfall and high temperatures. It is generally recommended that only disease-free seeds be planted (Hopkins *et al.*, 1997). The seed industry has adopted a zero tolerance of infected seed. Although seed treatments have been attempted, none will eradicate the pathogen (Hopkins, 1996; Sowell and Schaad, 1979; Wall, 1998). Several seeds assay protocols are available. However, all have limitations (Gitaitis, 1993; Minsavage *et al.*, 1995; Randhawa *et al.*, 2001; Walcott and Gitaitis, 2000). A simple grow-out test in the greenhouse is very sensitive in summer but can result in false negatives in winter due to a variable environment (Randhawa, personal observations). By using the BIO-PCR (Schaad *et al.*, 1995) enrichment technique and assay with *A. a.* subsp. *citrulli*-specific real-time PCR primers and probe (Schaad *et al.*, 2000; Song *et al.*, 2002), the pathogen can be detected from washings of agar plates. Success as a seed assay depends on efficiency of extraction of pathogen and a minimal population of *A. a.* subsp. *citrulli* of 1000-10,000 cells/ml in the seed extract. Similarly, classical PCR (Minsavage *et al.*, 1995) and an IMS-PCR protocol (Walcott and Gitaitis 2000) require minimal populations of the pathogen. A sweat box technique, where seedlings are grown in closed containers under growth chamber conditions for symptom development is subject to interference by other pathogens and saprophytic organisms (Randhawa, personal observations).

Due to limitations of seed-testing methods, seed companies have adopted an additional measure of field inspections to increase their chances of producing disease free seed. By using a portable real-time PCR instrument, such as the Cepheid Smart Cycler (Cepheid, Sunnyvale, CA), WFB can be diagnosed on-site in one hour or less (Schaad *et al.*, 2001). An excellent review of WFB by Latin and Hopkins (1995) has been published.

NOTES

The use of trade, firm, or corporation names in this publication (or page) is for the information and convenience of the reader. Such use does not constitute an official endorsement or approval by the United States Department of Agriculture or the Agricultural Research Service of any product or service to the exclusion of others that may be suitable.

ACKNOWLEDGEMENTS

We thank Bob Goth for sharing in the recollections of the early work on describing the seedling disease.

REFERENCES

Crall, J.M., and Schneck, N.C., 1969, Bacterial fruit rot of watermelon in Florida. *Plant Dis. Rep.* **53**: 74-75.

Gitaitis, R.D., 1993, Development of a seed assay for watermelon fruit blotch. pp 9-18 *In*: J.W. Sheppard (Ed.) Proc. 1st ISTA Pl. Dis. Comm. Symp. Seed Health Test., Ottawa, Canada, Aug. 9-11, 1993.

Hopkins, D., 1989, Bacterial fruit blotch of watermelon: a new disease in the eastern USA. *Proc. Cucurbitaceae* **89**: 74-75.

Hopkins, D.L., 1996, Wet seed treatments for the control of bacterial fruit blotch of watermelon. *Plant Dis.* **80**: 529-532.

Hopkins, D., and Schneck, N.C., 1972, Bacterial leaf spot of watermelon caused by *Pseudomonas lachrymans*. *Phytopathology* **62**: 542-545.

Hopkins, D., Stall, R.E., Latin, R., Rushing, J., Cook, W., and Keinanth, A.P., 1997, Bacterial fruit blotch of watermelon: research update. Citrus and Vegetable Magazine. February, 1997.

Isakeit, T., 1997, First report of infection of honeydew with *Acidovorax avenae* subsp. *citrulli*. *Pl. Dis.* **78**: 831.

Isakeit, T., 1998, Natural infection of citron melon with *Acidovorax avenae* subsp. *citrulli*. *Pl. Dis.* **82**: 351.

King, E.O., Ward, M.K., and Raney, D.E., 1954, Two simple media for the demonstration of pyocyanin and fluorescein. *J. Lab. Clin. Med.* **44**: 301-307.

Kucharek, T., Perez, Y., and Hodge, C., 1993, Transmission of watermelon fruit blotch bacterium from infested seed to seedlings. (Abstr.) *Phytopathology* **83**: 466.

Langston, D.B., Jr., Walcott, R.D., Gitiatis, R.D., and Sanders, F.H., Jr., 1999, First report of a fruit rot of pumpkin caused by *Acidovorax avenae* subsp. *citrulli* in Georgia. *Pl. Dis.* **84**: 199.

Latin, R., and Hopkins, D.L., 1995, Bacterial fruit blotch of watermelon. *Pl. Dis.* **79**: 761-765.

Minsavage, G.V., Hoover, R.J., Kucharek, T.A., and Stall, R.E., 1995, Detection of watermelon fruit blotch pathogen on seeds with the polymerase chain reaction. (Abstr.) *Phytopathology* **85**: 1162.

Mullin, R.S., and Schneck, N.C., 1963, Bacterial leaf spot on watermelon. *Plant. Dis. Rep.* **47**: 848.

Randhawa, P.S., Pannu, S.S., and Schaad, N.W., 2001, Improved Bio-PCR test for detection of *Acidovorax avenae* subsp. *citrulli* in watermelon and cantaloupe seeds. (Abstr.) *Phytopathology* **91**: S74.

Rane, K.K., and Latin, R.X., 1992, Bacterial fruit blotch of watermelon: Association of the pathogen with seed. *Plant Dis.* **76**: 509-512.

Schaad, N.W., Cheong, S.S., Tamaki, S., Hatziloukas, E., and Panopoulos, N.J., 1995, A combined biological and enzymatic amplification (BIO-PCR) technique to detect *Pseudomonas syringae* pv. *phaseolicola* in bean seed extracts. *Phytopathology* **85**: 243-248.

Schaad, N.W., Gaush, P., Postnikova, E., and Frederick, R., 2001, On-site one hour PCR diagnosis of bacterial diseases. (Abstr). *Phytopathology* **91**: S79.

Schaad, N.W., Song, W-Y., and Hatziloukas, E., 2000, PCR primers for detection of plant pathogenic species and subspecies of *Acidovorax*. Patent number 6,146,834, United States Patent Office, Washington D.C.

Schaad, N.W., Sowell, G.Jr., Goth, R.W., Colwell, R.R., and Webb, R.E., 1978, *Pseudomonas pseudoalcaligenes* subsp. *citrulli* subsp. nov. *Int. J. Syst. Bacteriol.* **28**: 117-125.

Somodi, G.C., Jones, J.B., Hopkins, D.L., Stall, R.E., Kucharek, T.A., Hodge, N.C., and Watterson, J.C., 1991, Occurrence of bacterial watermelon fruit blotch in Florida. *Plant Dis.***75**: 1053-1056.

Song, W-Y., Sechler, A., Hatziloukas, E., Kim, H.M., and Schaad, N.W., 2003, Use of PCR for Rapid Identification of *Acidovorax avenae* subsp. *avenae* and subsp. *citrulli*, (this volume).

Sowell, G., Jr., and Schaad, N.W., 1979, *Pseudomonas pseudoalcaligenes* subsp. *citrulli* on watermelon: Seed transmission and resistance of plant introductions. *Plant Dis. Rep.* **63**: 437-441.

Vock, N.T., and McCarthy, G.J.P. (eds.) 1982, A handbook of plant diseases in colour. Vol. 1 Fruit and vegetables 2nd ed. Information Publication Q182011, Government Printer, Queensland, Australia.

Walcott, R.R., and Gitaitis, R.D., 2000, Detection of *Acidovorax avenae* subsp *citrulli* in watermelon seed using immunomagnetic separation and the polymerase chain reaction. *Plant Dis.* **84**: 470-474.

Walcott, R.R., Castro, A.C., and Gitaitis, R.D., 2003, Role of blossoms in watermelon seed infestation by *Acidovorax avenae* subsp. *citrulli*. *Phytopathology* **93**: 528-534.

Walcott, R.R., Langston, Jr., D.B., Sanders, Jr., and Gitaitis, R.D., 2000, Investigating intraspecific variation of *Acidovorax avenae* subsp. *citrulli* using DNA fingerprinting and whole fatty acid analysis. *Phytopathology* **90**: 191-196.

Wall, G.C., 1989, Control of watermelon fruit blotch by seed heat-treatments. (Abstr.) *Phytopathology* **79**: 1191.

Wall, G.C., and Santos, V.M., 1988a, A new bacterial disease of watermelon in the Mariana Islands. Proc. CAS Res. Conf. University of Guam.

Wall, G.C., and Santos, V.M., 1988b, A new bacterial disease on watermelon in the Mariana islands. (Abstr.) *Phytopathology* **78**: 1605.

Wall, G.C., Santos, V.M., Cruz, F.J., and Nelson, D.A., 1990, Outbreak of watermelon fruit blotch in the Mariana Islands. *Plant Dis*. **74**: 80.

Webb, R.E., and Goth, R.W., 1965, A seed-borne bacterium isolated from watermelon. *Plant Dis. Rep*. **49**: 818-821.

Willems, A., Goor, M., Thielemans, S., Gillis, M., Kersters, K., and De Ley, J., 1992, Transfer of several phytopathogenic *Pseudomonas* species to *Acidovorax* as *Acidovorax avenae* subsp. nov., comb. Nov., *Acidovorax avenae* subsp. *citrulli*, *Acidovorax avenae* subsp. *cattleyae* and *Acidovorax konjaci*. *Int. J. Syst. Bacteriol*. **42**: 107-119.

The Pseudomonads Associated with Bacterial Canker and Decline of Hazelnut (*Corylus avellana* L.)

M. SCORTICHINI, U. MARCHESI, M.P. ROSSI, J.D. JANSE[#], and D.E. STEAD[§]

Istituto Sperimentale per la Frutticoltura. Via di Fioranello 52, I-00040 Ciampino aeroporto (Roma), Italy; [#]Plant Protection Service, Department of Bacteriology, P.O. Box 9102, 6700 HC Wageningen, The Netherlands; [§]Central Science Laboratory, DEFRA, Sand Hutton, YO 41 1LZ York, United Kingdom.

Abstract: Bacterial canker and decline, caused by *Pseudomonas avellanae*, is seriously damaging the hazelnut cultivation in central Italy and in northern Greece. Surveys carried out in the other main areas of cultivation in Italy ascertained the presence of strains similar to *P. syringae* pv. *syringae* van Hall. In these areas severe epidemics of bacterial decline were not observed. The taxonomic relationships among the pseudomonads associated with hazelnut decline and the other plant pathogenic *Pseudomonas* spp. has been established by using a collection of more than 200 strains by means of ARDRA, repetitive PCR fingerprints using ERIC, REP and BOX primer sets with UPGMA analysis, 16S rRNA gene sequencing as well as by analysis of fatty acid methyl ester and whole-cell protein profiles, and by pathogenicity tests. *P. avellanae* belongs to genomospecies 8, *sensu* Gardan *et al.*, which also includes *P. s.* pv. *theae* and *P. s.* pv. *actinidiae*. *P. avellanae* shows a pathogenicity restricted to *Corylus avellana* and is very virulent to hazelnut germplasm. The *P. s.* pv. *syringae*-like strains isolated from hazelnut fit well with the main features of the pathovar, including the presence of the *syrB* gene. A population found in Piedmont appears different from the *P. s.* pv. *syringae*-like strains when analysed by fatty acid profiling. In addition, the strains do not have the *syrB* gene and show a very restricted pathogenicity. This population might be a new pathovar within the *P. syringae* complex.

Key words: polyphasic taxonomy, bacterial populations, genomospecies, decline.

Author for correspondence: M. Scortichini; e-mail: mscortichini@hotmail.com.

N.S. Iacobellis *et. al.* (eds.), *Pseudomonas syringae* and Related Pathogens, 583-593.

1. INTRODUCTION

Hazelnut (*Corylus avellana* L.), also known as filbert or European hazelnut, is native of eastern Asia and southern Europe. It is mainly cultivated in Turkey, Italy, U.S.A., Spain and Iran. In Italy, where the crop spans over more than 100.000 hectares, it is commercially grown only in particular areas, such as Campania, Latium, Piedmont, Sicily and Sardinia, Each region cultivates its own typical germplasm introduced since ancient times. Concerning bacterial diseases, hazelnut can suffer from *Xanthomonas arboricola* pv. *corylina,* a world-wide spread pathogen particularly dangerous only in the juvenile stage of the tree. However, during last decades, severe epidemics of hazelnut canker and decline has been reported firstly in northern Greece (Psallidas and Panagopoulos, 1979) and, some years later, in central Italy (Latium) (Scortichini and Tropiano, 1994). In both cases, the causative agent was identified as a pathovar of *Pseudomonas syringae*, namely, *P. s.* pv. *avellanae*. Subsequently, further studies allowed to elevate this bacterium at species level, as *P. avellanae* (Psallidas) Janse *et al.* (Janse *et al.*, 1996). During surveys carried out in Italy in all regions where hazelnut is commercially grown, we isolated fluorescent pseudomonads associated with hazelnut canker and decline. Then, comparisons were made among more than 200 isolates, by using molecular (16S rRNA gene sequencing, ARDRA, repetitive PCR and UPGMA analysis), phenotypical (fatty acids methyl esters and whole-cell protein analysis, biochemical tests) and pathogenicity tests. This paper summarises the results obtained until now and points out that, currently, there are at least three different pseudomonads associated with canker and decline of hazelnut in Europe.

2. MATERIAL AND METHODS

2.1 Isolation, Strains and Bacterial Culture Medium

All isolations were made from hazelnut specimens (i.e. twigs, branches, roots) showing symptoms of bacterial canker and decline, by following the techniques described elsewhere (Psallidas and Panagopoulos, 1979; Scortichini and Tropiano, 1994). The strains were collected from the regions and cultivars reported in Table 1. Some representative strains used for comparison purposes are shown in Figs. 1-5. Strains from Greece were kindly supplied by Dr. P.G. Psallidas (Benaki Phytopathological Institute, Kiphissia-Athens, Greece). Strains were routinely cultured on nutrient sucrose agar, at 25÷27°C.

Table 1 Origin of strains used in the assessment of relationships among pseudomonads associated with hazelnut decline. Further details can be found in the related papers.

Pathogen	Country	Region	Years of isolation
Pseudomonas avellanae	Greece	Drama	1976, 1987
	Greece	Katerini	1977
	Greece	Kavala	1987, 1990
	Greece	Kilkis	1976, 1987, 1988
	Italy	Latium-Viterbo	1992-1999
	Italy	Latium-Rome	1991-1998
P. syringae pv *syringae*	Italy	Latium-Rome	1997-1999
	Italy	Piedmont-Cuneo	1995-1999
	Italy	Campania-Avellino	1997
	Italy	Sicily-Palermo	1999
	Italy	Sardinia-Nuoro	1999
P. s. pv. *corylina*	Italy	Piedmont-Cuneo	1995

2.2 Characterisation of the Strains

The performing of ARDRA (Amplified Ribosomal DNA Restriction Analysis), repetitive PCR with ERIC, BOX and REP primer sets and UPGMA analysis, presence of *syrB* gene, FAME (Fatty Acid Methyl Esters) and SDS-PAGE of whole-cell protein extracts analysis were described elsewhere (Janse *et al.*, 1996; Scortichini *et al.*, 2002a,b).

2.3 Pathogenicity Tests

With strains representative of different regions, cultivars and year of isolation, pathogenicity tests to lemon fruits, bean pods and corn leaves were performed (Scortichini and Lazzari, 1996; Scortichini *et al.*, 2002b).

3. RESULTS

3.1 Comparison of *Pseudomonas avellanae* with *P. syringae* Pathovars

Repetitive PCR performed with many representative strains of phytopathogenic pseudomonads, indicated that *Pseudomonas avellanae*, *P. syringae* pv. *theae* and *P. s.* pv. *actinidiae* are closely related (Scortichini *et al.*, 2002a). The subsequent ARDRA analysis revealed that among the 12 endonucleases used, only *Tru* 9I differentiated *P. avellanae* from *P. s.* pv. *theae* and *P. s.* pv. *actinidiae* (Fig. 1). Genetic fingerprintings obtained with repetitive PCR and the subsequent UPGMA analysis are shown in Figs. 2

and 3. In addition, *P. avellanae*, *P. s.* pv. *theae* and *P. s.* pv. *actinidiae* do not have the *syrB* gene and thus they might be included in the genomospecies 8, *sensu* Gardan *et al.* (1999). The repetitive PCR profile of *P. s.* pv. *tomato* (genomospecies 3) was the most similar match to that of the *P. avellanae* strains (Fig. 2).

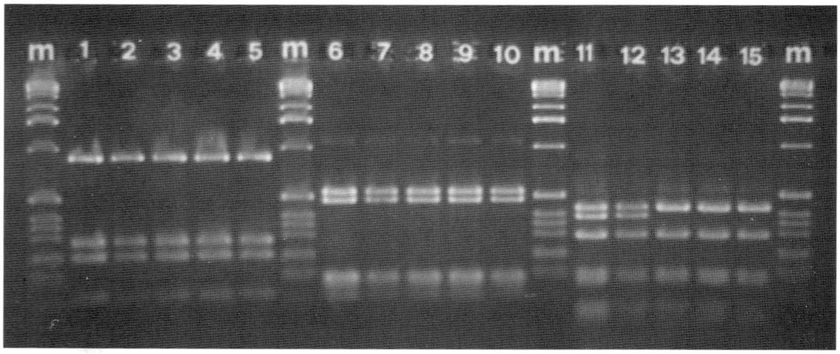

Figure 1. ARDRA banding patterns of amplified 16S rDNA using *Mbo* I, *Msp* I and *Tru* 9I restriction endonucleases of *Pseudomonas avellanae* (lanes 1, 6, 11: NCPPB 3487^{T}; lanes 2, 7, 12: NCPPB 3875); *P. syringae* pv. *theae* (lanes 3, 8, 13: NCPPB 2598^{T}) and *P. s.* pv. *actinidiae* (lanes 4, 9, 14: NCPPB 3739^{T}; lanes 5, 10, 15: NCPPB 3871*). P. avellanae* can be differentiated from *P. s.* pv. *theae* and *P. s.* pv. *actinidiae* by using *Tru* 9I. ^{T}: type-strain. m: molecular size marker (1-kb ladder; Gibco BRL) in base pairs (From: Scortichini *et al.*, 2002a).

3.2 Comparison of Pseudomonads Associated with Bacterial Decline of Hazelnut

By using nine different endonucleases in ARDRA analysis, only *Hin* fI pointed out two groups of pseudomonads associated with hazelnut decline (Fig. 4). Group A includes *Pseudomonas avellanae*; group B includes the strains isolated in Italy from all regions where hazelnut is cultivated. Repetitive PCR and UPGMA analysis revealed two populations of *P. avellanae*: one from northern Greece and the other from central Italy (Fig. 5). When ERIC primer sets are used, two PCR products of approximately 300 and 800 bp, present only in the strains of northern Greece, enable the differentiation of the two populations. The strains of group B resemble *P. syringae* pv. *syringae.* They belong to group Ia, *sensu* Lelliott *et al.* (1966), have the *syrB* gene, incite pathogenic symptoms to orange fruits, bean pods and corn leaves other than on hazelnut. *P. avellanae* resulted very aggressive to hazelnut germplasm also at doses of 100 CFU leaf^{-1} scar, whereas *P. s.* pv.

syringae strains incite only twig die-back with the highest doses (i.e. 100.000 CFU $leaf^{-1}$ scar).

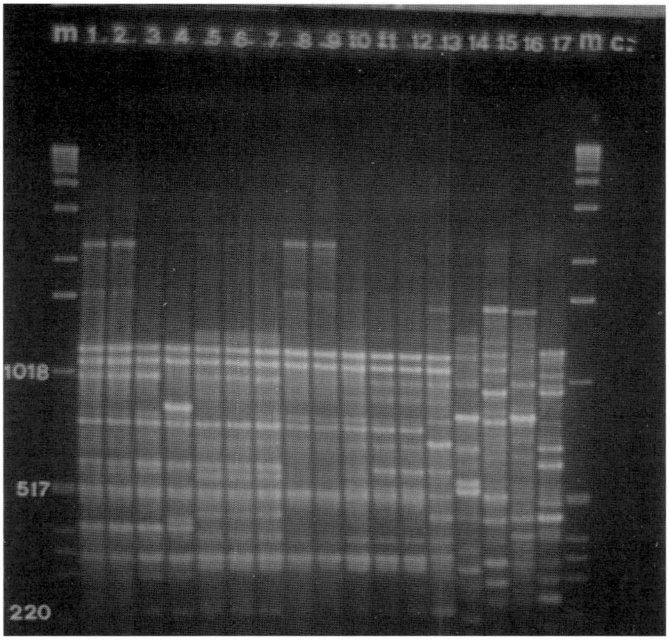

Figure 2. Repetitive PCR fingerprinting patterns from genomic DNA of *Pseudomonas avellanae*, *P. syringae* pv. *theae, P. s.* pv. *actinidiae* and other phytopathogenic pseudomonads strains obtained by using BOX primer set. m: molecular size marker in base pairs (1-kb ladder; Gibco BRL). Lane 1: *P. avellanae* ISF C1; lane 2: *P. avellanae* NCPPB 3875; lane 3: *P. avellanae* ISPaVe 056; lane 4: *P. avellanae* NCPPB 3487^{T}; lane 5: *P. s.* pv. *theae* NCPPB 2598^{T}; lane 6: CFBP 4096; lane 7: *P. s.* pv. *theae* 4097; lane 8: *P. s.* pv. *actinidiae* NCPPB 3739^{T}; lane 9: *P. s.* pv. *actinidiae* NCPPB 3740; lane 10: *P. s.* pv. *actinidiae* KW1; lane 11: *P. s.* pv. *actinidiae* KACC 10754; lane 12: *P.s.* pv. *actinidiae* KACC 10592; lane 13: *P. s.* pv. *tomato* NCPPB 1106^{T}; lane 14: *P. s.* pv. *ulmi* NCPPB 632; lane 15: *P. s.* pv. *syringae* NCPPB 3869; lane 16: *P. s.* pv. *helianthi* NCPPB 2460T; lane 17: *P. viridiflava* BPIC 394; C: control (From: Scortichini *et al.*, 2002a).

3.3 Comparison of *Pseudomonas syringae* Strains Associated with Bacterial Decline of Hazelnut

One population found in Piedmont on cultivar Tonda Gentile delle Langhe clustered apart from the other *Pseudomonas syringae* pv. *syringae* strains collected in the same region as well as in the other regions upon FAME analysis (Fig. 6) and repetitive PCR. Such strains do not have the *syrB* gene and can cause necrotic lesions to orange fruits, bean pods and corn leaves. This population, recently found also in Sardinia on Tonda Gentile delle

Langhe introduced from Piedmont during 70's (Fiori *et al.*, 2002, present book). The sequencing of 16S rRNA gene of this pathogen and other pseudomonads is currently under way to confirm such an hypothesis.

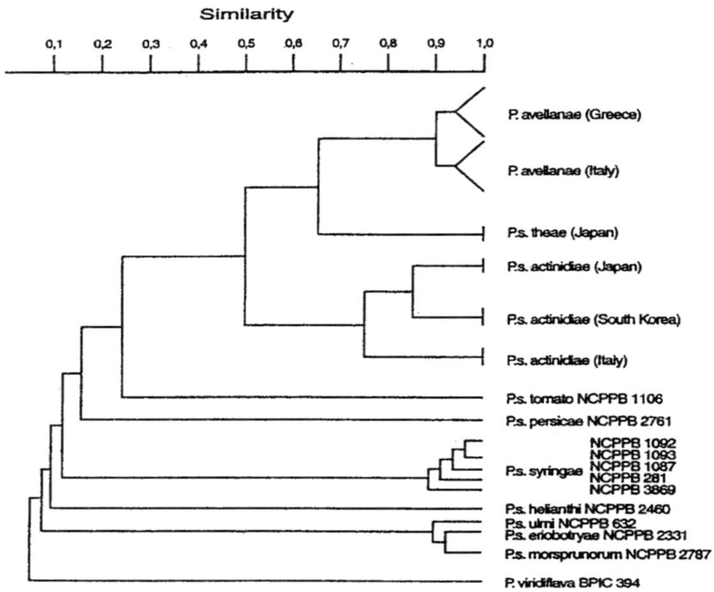

Figure 3. Dendrogram of relationships among *Pseudomonas avellanae*, *P. syringae* pv. *theae*, *P. s.* pv. *actinidiae* and other fluorescent phytopathogenic pseudomonads. A similarity matrix was produced from analysis of repetitive PCR data using ERIC, BOX and REP primer sets. Cluster analysis was performed by UPGMA on matrix calculated with the Dice's coefficients. (From: Scortichini *et al.,* 2002a).

4. DISCUSSION

Three different fluorescent pseudomonads are associated with canker and decline of hazelnut in Europe. *Pseudomonas avellanae* is the most aggressive species and, until now, it appears restricted to northern Greece and central Italy (Latium province). It typically infects only *Corylus avellana*. This species comprises two different populations which can be easily differentiated by means of repetitive PCR using ERIC primer sets. The two populations can be also discriminated into two serotypes by using monoclonal antibodies raised up against the cell wall polysaccharides (Ovod *et al.*, 1999). Both populations, that have the *hrp* genes encoding for the harpin proteins in phytopathogenic bacteria (Loreti *et al.*, 2001), are very

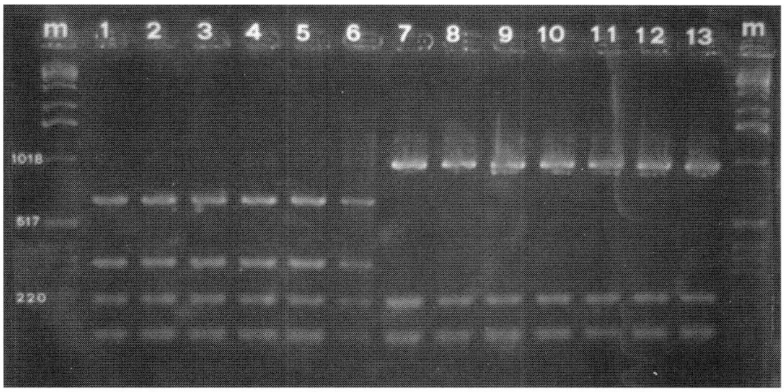

Figure 4. ARDRA band patterns from amplified 16S rDNA obtained by using *Hin* fI restriction endonuclease for *Pseudomonas avellanae* (lanes 1 to 6) and *P. syringae* pv. *syringae* strains (lanes 7 to 13). m: molecular size marker in base pairs. (From: Scortichini *et al.*, 2002b).

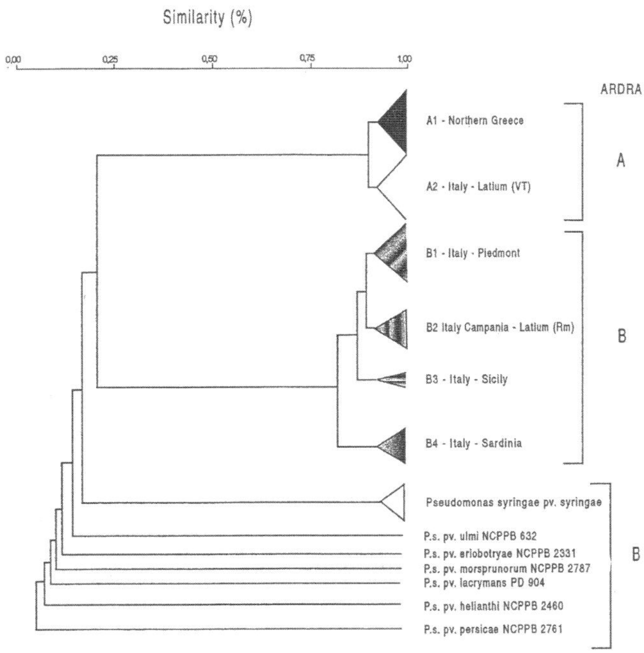

Figure 5. Simplified dendrogram showing relationships among *Pseudomonas avellanae, P. syringae* pv. *syringae* strains associated with hazelnut decline, and other *P. syringae* pathovars. Cluster analysis was performed using the results obtained with repetitive PCR and ERIC, BOX and REP primer sets, by UPGMA on matrix calculated with Dice's coefficients. (From: Scortichini *et al.*, 2002b).

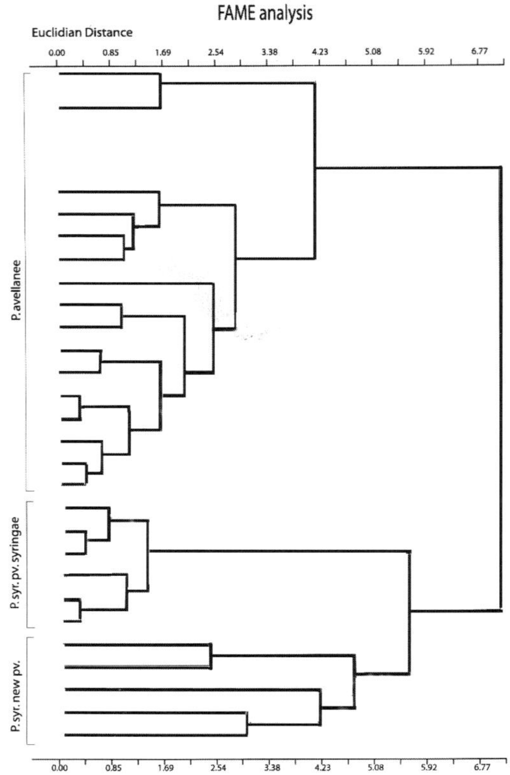

Figure 6. Dendrogram of relationships inferred by FAME (fatty acids methyl esters analysis) among pseudomonads associated with hazelnut canker and decline.

virulent to hazelnut germplasm being able to kill adult trees within few months of inoculation, in early autumn, of 25 leaf scars at a dose of approximately 1.000 CFU leaf^{-1} scar (Scortichini *et al.*, 2002b). Primers targeted to *hrpW* gene sequences, enables the discriminations between *P. avellanae* and other pseudomonads associated with hazelnut decline (Loreti *et al.*, 2002). When compared with other *P. syringae* pathovars by using SDS-PAGE of protein extracts and repetitive PCR (Scortichini *et al.*, 2002a), *P. avellanae* can be allocated in the genomospecies 8, *sensu* Gardan *et al.* (1999), together with *P. syringae* pv. *theae* and *P. s.* pv. *actinidiae*. The closest *P. syringae* pathovar is *P. s.* pv. *tomato* showing less than 30% of genetic similarity with *P. avellanae* by using both techniques. These findings, obtained by comparing other *P. syringae* pathovars, confirmed previous studies using FAME and SDS-PAGE proteins analyses that pointed out a relevant diversity between *P. avellanae* and the following *P. syringae* pathovars, *maculicola*, *pisi* and *porri* (Janse *et al.*, 1996). By using PCR and the primers PAV1 and PAV22, targeted to the 16S rRNA gene of *P. avellanae* and developed for the detection of this pathogen in apparently healthy and visibly infected hazelnut specimens (Scortichini and Marchesi, 2001), all strains of genomospecies 8 yielded a PCR product of 752 bp. These primers might be used for the detection and identification of this genomospecies. *P. avellanae*, *P. s.* pv. *theae* and *P. s.* pv. *actinidiae* do not have the *syrB* gene that encodes the production of cyclic lipodepsinonapeptides that are among the virulence factors of *P. s.* pv. *syringae*. The production of specific toxic compound(s) by *P. avellanae* is currently under study. In fact, preliminary investigations clearly indicate that filtrates of liquid cultures incited necrosis to lemon fruits, and string bean pods as well as the hypersensitivity reaction to tobacco leaves and an inhibitory effect to *Rhodotorula pilimanae* (Scortichini and Marchesi, unpublished data).

Strains resembling *P. s.* pv. *syringae* are present in the other areas of hazelnut cultivation in Italy. These strains can be differentiated from *P. avellanae* by means of ARDRA and *Hin* fI endonuclease as well as with repetitive PCR and FAME analysis. Although the strains can be grouped according to their geographic origin upon repetitive PCR and UPGMA analysis, all of them show a relevant similarity. When artificially inoculated on hazelnut, they resulted less aggressive than *P. avellanae* and, with the exception of one population found in Piedmont, they have the *syrB* gene. In hazelnut orchards, they cause twig and branches die-backs and the death of the tree is rarely observed.

The third group of pseudomonads is composed by strains isolated in Piedmont and Sardinia (Fiori *et al.*, present book), and obtained from the cultivar Tonda Gentile delle Langhe. These strains appear to be related but different to the other *P. s.* pv. *syringa*e strains isolated from hazelnut in Italy. FAME analysis clearly indicated that they can be grouped separately. It is interesting to note that in Piedmont, in the same area and on the same cultivar, two different populations of phytopathogenic pseudomonads do exist. When artificially inoculated, they cause symptoms to hazelnut and, to a lesser extent, to apricot, tomato and pepper plants as well as to lemon fruits. They did not incite symptom to lilac (Fiori *et al.*, 2003). In hazelnut orchards they incite mild symptoms in Piedmont. By contrast, in Sardinia, where Tonda Gentile delle Langhe was introduced from Piedmont during 70', the symptoms are more severe and, sometimes, the death of the tree is observed. Apparently, the pathogen was latently introduced in Sardinia with the hazelnut propagative material. This population might be a new pathovar within the *P. syringae* complex and the sequencing and comparison of the 16S rRNA gene is currently under way to confirm such an hypothesis. One possible name could be *P. s.* pv. *corylina*. However, *virPphA,* a major determinant in the pathogenicity of *P. savastanoi* pv. *phaseolicola*, is also present in one representative strain of this population (Jackson *et al.*, 2002).

This fact could implicate an horizontal gene transfer or a common ancestry. Consequently, the choice of the epithet should have to take into account also the possibility to indicate a potential wider host range for the new pathovar. This possibility would seem confirmed by the pathogenicity tests performed in the present studies. To circumvent the possibility of choosing an epithet that only partially indicate the potential host range of the pathogen, a non-descriptive name might be also chosen. This name could also indicate that work is continuously in progress to better characterise the new pathogen. Exchange of opinions are under way to select the most appropriate name.

REFERENCES

Fiori, M., Cicconi, L., and Scortichini, M., 2003, Bacterial Canker of Hazelnut (*Corylus avellana* L.) in Sardinia: Occurrence of *Pseudomonas syringae* Strains, 617-625, (this volume).

Gardan, L., Shafik, H., Belouin, S., Brosch, R., Grimont, F., and Grimont, P.A.D., 1999, DNA relatedenees among the pathovars of *Pseudomonas syringae* and description of *Pseudomonas tremae* sp. nov. and *Pseudomonas cannabina* sp. nov. (ex Sutic and Dowson). *Int. J. Syst. Bacteriol.* **49**: 469-478.

Jackson, R.W., Mansfield, J.W., Ammouneh, H., Dutton, L.C., Wharton, B., Ortiz-Barredo, A., Arnold, D.L., Tsiamos, G., Sesma, A., Butcher, D., Boch, J., Kim, Y-J, Martin, G.B., Tegli, S., Murillo, J., and Vivian, A., 2002, Location and activity of member of a family of *virPphA* homologues in pathovars of *Pseudomonas syringae* and *P. savastanoi*. *Mol. Pl. Pathol.* **3**(4): 205-216.

Janse, J.D., Rossi, M.P., Angelucci, L., Scortichini, M., Derks, J.H.J., Akkermans, A.D.L., De Vrijer, R., and Psallidas, P.G., 1996, Reclassification of *Pseudomonas syringae* pv. *avellanae* as *P. avellanae* (spec. nov), the bacterium causing canker of hazelnut (*Corylus avellana* L.). *Syst. Appl. Microbiol.* **19**: 589-595.

Lelliott, R.A., Billing, E., and Hayward, A.C., 1966, A determinative scheme for the fluorescent plant pathogenic *Pseudomonas*. *J. Appl. Bacteriol.* **29**: 470-489.

Loreti, S., Sarrocco, S., and Gallelli, A., 2001, Identification of *hrp* gene, encoding harpin, in *Pseudomonas avellanae* (Psallidas) Janse *et al. J. Phytopath.* **149**: 219-226.

Loreti, S., and Gallelli, A., 2002, Rapid and specific detection of virulent *Pseudomonas avellanae* strains by PCR amplification. *Eur. J. Pl. Pathol.* **108**: 237-244.

Ovod, V.V., Knirel, Y.A., Samson, R., and Krohn, K.J., 1999, Immunochemical characterization and taxonomic evaluation of the O polysaccharide of the lipopolysaccharides of *Pseudomonas syringae* serogroup O1 strains. *J. Bacteriol.* **181**: 6937-6947.

Psallidas, P.G., and Panagopolous C.G., 1979, A bacterial canker of hazelnut in Greece caused by *Pseudomonas syringae* pv. *avellanae*. *Phytopath. Z.* **94**: 103-111

Scortichini, M., and Tropiano, F.G., 1994, Severe outbreak of *Pseudomonas syringae* pv. *avellanae* on hazelnut in Italy. *J. Phytopath.* **140**: 65-70.

Scortichini, M., and Lazzari, M., 1996, Systemic migration of *Pseudomonas syringae* pv. *avellanae* in twigs and young trees of hazelnut and symptom development. *J. Phytopath.* **144**: 215-219.

Scortichini. M., and Marchesi, U., 2001, Sensitive and specific detection of *Pseudomonas avellanae* strains using primers based on 16S rRNA gene sequences. *J. Phytopath*. **149**: 527-532.

Scortichini, M., Marchesi, U., and Di Prospero, P., 2002a, Genetic relatedeness among *Pseudomonas avellanae*, *P. syringae* pv. *theae* and *P. s*. pv. *actinidiae*, and their identification. *Eur. J. Pl. Pathol*. **108**: 269-278.

Scortichini, M., Marchesi, U., Rossi, M.P. and Di Prospero, P., 2002b, Bacteria associated with hazelnut (*Corylus avellana* L.) decline are of two groups: *Pseudomonas avellanae* and strains resembling *P. syringae* pv. syringae *Appl. Env. Microbiol*. **68**: 476-484.

Pseudomonas "reactans" a New Pathogen of Cultivated Mushrooms

N.S. IACOBELLIS and P. LO CANTORE
Dipartimento di Biologia, Difesa e Biotecnologie Agro Forestali, Università degli Studi della Basilicata, C.da Macchia Romana, 85100 Potenza, Italy.

Abstract: Recent investigations have shown that brown blotch of *Agaricus bisporus* and the yellowing of *Pleurotus ostreatus,* caused by *Pseudomonas tolaasii,* are actually complex diseases. In fact, besides *P. tolaasii,* which may be considered the main causal agent of the above diseases, also *P. "reactans"* and not yet characterised fluorescent pseudomonads participated to the expression of the diseases symptoms. Furthermore, *P. "reactans"* has been demonstrated to be the causal agent of *P. eryngii* yellowing. The demonstration that *P. "reactans"*, a bacterial entity considered a saprotroph associated to cultivated mushrooms and useful for the specific identification of *P. tolaasii* in the "white line" assay, is a pathogen of *A. bisporus* and *Pleurotus* spp. cultivated mushrooms is a novelty. However, a strong variability in the pathogenicity, antagonistic activity and response in "white line" assay was observed among the strains of *P. "reactans"*. This behaviour is apparently due to the formation in the cultures of morphological variants which were avirulent, did not produce the White Line Inducing Principle (WLIP) and were negative in the "white line" assay. The presence of an high *ratio* of the above variants in the cultures may explain the unravelled pathogenicity of this bacterium.

Key words: *Pseudomonas tolaasii*, *Agaricus bisporus*, *Pleurotus* spp., lipodepsipeptides.

Author for correspondence: N.S. Iacobellis; e-mail: iacobellis@unibas.it.

1. INTRODUCTION

Cultivated mushrooms, due to the particular environment where cultivation are performed, are attacked by a series of parasites which cause significant crop losses. Among them bacteria, mainly fluorescent pseudomonads (Gill,

N.S. Iacobellis *et. al.* (eds.), *Pseudomonas syringae* and Related Pathogens, 595-605.

1995), cause a series of diseases considered highly detrimental for the mushrooms cultivation (Fermor, 1986; Gill, 1995).

Brown blotch of *Agaricus bisporus* and the yellowing of *Pleurotus ostreatus*, both caused by *Pseudomonas tolaasii,* are considered the main bacterial diseases of the two cultivated mushrooms species (Gill, 1995).

Brown blotch disease, observed the first time in the United States of America (Tolaas, 1915), has been reported in Italy in a mushroom house in Apulia by Ercolani (1970). The disease, on the contrary of brown discoloration of *A. bisporus* caused by *P. agarici,* which is characterised by superficial brown discoloured areas on the caps (Geels, 1994), causes irregular depressed brown lesions either on caps or stipes. In favourable environment conditions lesions, either on young or developed sporophores, may coalesce and large portions or the sporophores may turn brown. The symptoms of the disease may appear after harvest also when the mushrooms are maintained at low temperature (Wells *et al.*, 1996).

The yellowing of *P. ostreatus,* a disease more recently described (Ferri, 1985; Fermor, 1986), is characterised by a variety of symptoms either on primordia or developed sporophores. The infected primordia, which appear yellow-orange discoloured, stop to grow and soon they wither. The disease may interest all the primordia of a mushrooms cluster or only a few of them. On the cap or stipes of developed mushrooms irregular depressed yellow-orange lesions surrounded by yellow-orange discoloured areas occur. In addition, yellow-orange discoloured superficial lesions which may interest the whole cap may also be observed. At high temperature and humidity conditions the infected mushrooms may rot with a consequent unpleasant smell.

A similar disease occurs on *P. eryngii* a mushroom species recently introduced for the cultivation in some regions in Southern Italy. The disease, called the yellowing of *P. eryngii,* was thought to be caused by *P. tolaasii* (Ferri, 1985), but subsequent studies ascertained that fluorescent pseudomonads different from *P. tolaasii* were associated to the diseased mushrooms (Iacobellis and Lavermicocca, 1990). The disease is characterised by light brown discoloration that soon turn to brown-reddish on either developed sporophores and/or primordia. These stop to grow and soon they wither. The above lesions appear to interest the superficial layers of the mushrooms tissues and, on the contrary of the yellowing of *P. ostreatus*, depressed lesions have never been observed. High temperature and relative humidity conditions appear to lead to the disease establishment and, furthermore, in the above condition the infected mushrooms may rot with the consequent development of unpleasant smell which render the crop unmarketable.

The frequent occurrence of a series of alterations, possibly caused by bacteria, on several *A. bisporus* and/or *Pleurotus* spp. mushrooms cultivation in Apulia and Basilicata prompt us to start studies with the aim to establish the causes.

Here following some the results obtained in the last period are reported (Iacobellis and Lo Cantore, 1997; 1998a; 1998b; Lo Cantore and Iacobellis, 2000a; 2000b; Lo Cantore, 2001).

2. BROWN BLOTCH DISEASE OF *AGARICUS BISPORUS*

The isolation of bacteria from brown blotches on sporophores of *Agaricus bisporus* gave rise to bacteria that on the medium B of King *et al.* (1954) produced fluorescent pigments and formed colonies with different morphologies. These colonies were selected, purified and the resulting pure cultures were further characterised. The most representative isolates were identified as strains of *Pseudomonas tolaasii.* In fact, in the "white line" assay (Wong and Preece, 1979) they formed a white precipitate in agar when grown side by side with strain NCPPB1311 of *P. "reactans*". This is the typical feature of *P. tolaasii* strains in the assay which is considered useful for the specific identification of *P. tolaasii* (Wong and Preece, 1979). However, recent investigations raised doubts on the above mentioned specificity since other bacteria associated/pathogen to cultivated mushrooms, but different from *P. tolaasii,* showed the capability to form white precipitates on agar when grown near strains of *P. "reactans*" (Godfrey *et al*., 2001). Nevertheless, in our experience all the isolates positive in the above assay showed the LOPAT profile (Lelliott and Stead, 1987), the nutritional traits (Goor *et al*., 1986; Anonymous, 1999), and the pathogenicity feature of *P. tolaasii.* The above strains, assayed for pathogenicity on tissue blocks of *A. bisporus* (Gandy, 1968; Ercolani, 1970), caused sunken brown lesions after 24÷48 h incubation, depending on the strains, and then after the complete rot of the mushrooms tissue blocks was in most of the cases observed. A similar behaviour was shown by authentic virulent strains of *P. tolaasii* used for comparison. Also other fluorescent pseudomonads, distinguishable from strains of *P. tolaasii* for the colony morphology, caused brown lesions on tissue blocks of *A. bisporus* though the intensity of discoloration was variable among strains and lower when compared to the ones caused by strains of *P. tolaasii.* The above virulent isolates showed the LOPAT characters of bacteria belonging to the group V of fluorescent pseudomonads (Lelliott and Stead, 1987). Part of them, showed the nutritional characters of *P. "reactans*" strains (Goor *et al.*, 1986) and, as expected, were positive in the "white line" assay (Wong and Preece, 1979) when grown side by side with the type strain NCPPB2192 of *P. tolaasii.* Strains of *P. "reactans*" assayed for pathogenicity on tissue blocks of *A. bisporus* (Gandy, 1968; Ercolani, 1970) confirmed to be less virulent than *P. tolaasii* strains. Strains of *P. "reactans"* caused brown lesions after 48÷72 h and, furthermore, on the contrary of *P. tolaasii* strains, they did not cause neither sunken lesions nor the rot of the block tissues. The contemporary presence of both *P. tolaasii* and *P. "reactans"* in the isolation plates was shown by the formation of white precipitates between colonies which after their purification were demonstrated to belong to the above bacterial entities.

3. YELLOWING OF *PLEUROTUS OSTREATUS*

The isolation of bacteria from the different symptoms of the yellowing of *Pleurotus ostreatus* gave rise to apparently different results. In particular, from irregular depressed yellow-orange lesions surrounded by yellow-orange discoloured areas on either caps or stipes of developed mushrooms, bacteria with the pathogenetic features on *Agaricus bisporus* tissue blocks (Gandy, 1968; Ercolani, 1970), the biochemical (Lelliott and Stead, 1987) and nutritional characters (Goor *et al.*, 1986), as well as the behaviour in the white line assay (Wong and Preece, 1979) of both *Pseudomonas tolaasii* and *P. "reactans"* were ever obtained. Also in this case in the isolation plates white precipitates between colonies of the above bacteria were regularly observed. Furthermore, in the same isolation plates other bacterial isolates, ascertained to belong to the V group of fluorescent pseudomonads (Lelliott and Stead, 1987), were also obtained. These bacterial isolates were negative in the "white line" assay and showed a pathogenicity feature on *A. bisporus* tissue blocks which was comparable to that of *P. "reactans"* strains. Some of the latter bacteria, when characterised for the nutritional feature (Goor *et al.*, 1986), resulted to share more characters with authentic strains of *P. tolaasii* than of *P. "reactans"*, *P. agarici, P. "gingeri"* and *Pseudomonas* spp. used for comparison (Iacobellis e Lo Cantore, 1998b; Lo Cantore, 2001). The latter strains were the causal agent of mummy disease. Furthermore, a few of them showed the same nutritional characters of *P. tolaasii* (Goor *et al.*, 1986) and appeared to be morphological hypovirulent variants of *P. tolaasii* (Rainey *et al.*, 1992; Grewall *et al.*, 1994; 1995; Iacobellis e Lo Cantore, 1998b; Lo Cantore, 2001).

Strains of *P. tolaasii*, inoculated on whole sporophores of *P. ostreatus* reproduced the typical disease symptoms but they differed in relation to the inoculation procedure used in the pathogenicity assays. In particular, the injection with a syringe of bacterial suspensions into the cap tissue of sporophores induced the formation of depressed lesions. On the contrary, the deposition of the above suspensions on the cap surface caused the yellow discoloration of the interested area which was only slightly depressed. When suspensions of *P. tolaasii* were sprayed on developing young mushrooms cluster they turned yellow and then they withered (Iacobellis e Lo Cantore, 1998b; Lo Cantore, 2001).

In the latter pathogenicity assays on *P. ostreatus* sporophores, *P. "reactans"* strains, on the contrary of *P. tolaasii,* induced the yellow discoloration but never caused depressed lesions (Iacobellis e Lo Cantore, 1998b; Lo Cantore, 2001).

From superficial yellow-orange discoloured lesions on developed sporophores as well as from yellow discoloured primordia of *P. ostreatus* bacteria with the features of either *P. "reactans"* and/or fluorescent pseudomonads negative in the white line assay were mainly isolated. Only in a few cases also bacteria identified as strains of *P. tolaasii* have been obtained (Iacobellis e Lo Cantore, 1998b; Lo Cantore, 2001).

4. YELLOWING OF *PLEUROTUS ERYNGII*

From sporophores of *Pleurotus eryngii* showing the symptoms of yellowing, different bacteria belonging to the group V of fluorescent pseudomonads (Lelliott and Stead, 1987) and capable to cause the brown discoloration of *Agaricus bisporus* blocks (Gandy, 1968; Ercolani, 1970) have been obtained. Among them bacteria with the nutritional feature (Goor *et al.,* 1986) and the behaviour in the white line assay (Wong and Preece, 1979) of *P. "reactans"* were constantly present. The pathogenicity of strains of *P. "reactans"* on *P. eryngii* cultivation was not an easy task. In pathogenicity assays the disease symptoms were observed, as expected, after the inoculation with *P. "reactans"* suspension but and in some cases the typical symptom of disease were also observed in the control after treatment with sterile distilled water. From the latter diseased mushrooms bacteria with the biochemical, nutritional and pathogenetic features of *P. "reactans"* were ever isolated. It is possible that the bacterium is already present in the compost bags colonised by *P. eryngii* or in the casing soil used to cover the production surface (Iacobellis e Lo Cantore, 1998a; Lo Cantore, 2001). Although the above limitations in Koch postulate demonstration should taken in account, the results reported strongly indicate that *P. "reactans"* is the causal agent of yellowing of *P. eryngii.* The fact that from yellow discoloured areas on the cap of a wild sporophore of *P. eryngii* have been isolated fluorescent pseudomonads with the characters of *P. "reactans"* strongly support the above results. However, it is not excluded that also the other fluorescent pseudomonads may be responsible and/or contribute, as reported for the yellowing of *P. ostreatus* (Iacobellis e Lo Cantore, 1998b), to the symptom development. Though previous investigations claimed that *P. tolaasii* was the causal agent of the yellowing of *P. eryngii* (Ferri, 1985) no bacteria with the features of the latter pathogen (Goor *et al.,* 1986; Lelliott and Stead, 1987; Wong and Preece, 1979) have been isolated from lesions on *P. eryngii* sporophores. Furthermore, strains of *P. "reactans"*, assayed for the pathogenicity on whole sporophores of *P. eryngii,* on the contrary of strains of *P. tolaasii,* never caused depressed lesions. The latter behaviour in the alteration induction is likely the one observed in the natural infections.

5. CHARACTERISATION OF *PSEUDOMONAS "REACTANS"*

The finding that *Pseudomonas "reactans"* is the cause of the yellowing of *Pleurotus eryngii* and participate to the symptom development in the brown blotch disease of *Agaricus bisporus* and in the yellowing of *P. ostreatus* (Iacobellis and Lo Cantore, 1997; Iacobellis and Lo Cantore, 1998a; 1998b; Lo Cantore, 2001) prompt the need for its characterisation also in comparison with *P. tolaasii.*

The pathogenicity of *P. "reactans"* may be considered a novelty though indications in this regard have been previously reported in independent studies (Kim *et al.*, 1995; Wells *et al.*, 1996). Remain to be established the reason why the pathogenicity of this bacterium, considered a saprotroph associated with cultivated mushrooms and known by long time in literature for its use in the "white line" assay (Wong and Preece, 1979), was neglected for so long time.

The above considerations prompt us to characterise representative strains of *P. "reactans"* for some phenotypic traits also in comparison to strains of *P. tolaasii*. For the purpose 20 strains of both *P. tolaasii* and *P. "reactans"*, obtained from diseased *Pleurotus* spp. and *A. bisporus* in different geographic areas, were evaluated for the biochemical and nutritional features (Goor *et al.,* 1986; Lelliott e Stead, 1987; Anonymous, 1999), for the pathogenicity on whole sporophores and tissue blocks of *A. bisporus* (Gandy, 1968; Ercolani, 1970), for the behaviour in the "white line" assay (Wong and Preece, 1979) and for the antagonistic activity against bacteria and fungi (Lavermicocca *et al.*, 1997).

5.1 Pathogenicity Assays

The assays on *Agaricus bisporus* tissue blocks confirmed the pathogenicity of strains of *Pseudomonas "reactans"* though the intensity of the alteration was lower than the ones caused in average by strains of *P. tolaasii.* Furthermore, a strong variability among the virulence of the strains was also observed.

Further assays showed that the density of the bacterial suspensions causing visible alterations on the tissue blocks was at least 10^6 CFU ml^{-1} in the case of *P. tolaasii* strains and between $10^7 \div 10^8$ CFU ml^{-1} for strains of *P. "reactans"*. The virulence of strains of *P. "reactans"* was also confirmed when the assays were performed, also in comparison with strains of *P. tolaasii,* on whole sporophores of *A. bisporus.* The application by spray of 10^8 CFU ml^{-1} suspensions caused, after 7 days incubation at 4°C, brown depressed lesions on the whole caps. In general also in this assay the intensity of symptoms caused by strains of *P. "reactans"* was lower when compared to the one induced by strains of *P. tolaasii* strains confirming the higher virulence of the latter pathogen.

5.2 Biochemical and Nutritional Features

Strains of *Pseudomonas tolaasii* and *P. "reactans"*, as expected, showed the characters of fluorescent pseudomonads belonging to the LOPAT group V (Lelliott and Stead, 1987) and the nutritional profiles reported in the identification scheme of bacteria pathogen/associated to cultivated mushrooms by Goor *et al.* (1986). In particular, strains of *P. tolaasii* grew on minimal media supplemented of erythritol, sorbitol, L-arabitol, 2-ketogluconate and *n*-valerate but not on those containing L-arabinose, L-rhamnose, D-tartrate and histamine. Strains of *P. "reactans"* utilised erythritol, sorbitol, L-arabinose, L-rhamnose, L-arabitol, 2-ketogluconate e *n*-valerate but not D-tartrate and histamine. The above results confirmed the reproducibility and validity of the above identification scheme.
Strains of *P. tolaasii*, grown on 96 different carbon sources on GN2 micro-plates and evaluated for the results with the computer assisted Biolog system (Biolog Inc., Hayward, Ca, USA), were identified, as expected, as strains of *P. tolaasii* though a certain variability among the strains was observed. In the same experimental conditions, strains of *P. "reactans"*, a bacterial entity not yet classified and hence not included in the data base, have been identified in a different manner. Nine out twenty strains were identified as strains of *P. fluorescens*, three as strains of *P. marginalis*, and two as strains of *P. tolaasii*, *P. synxantha* and *Pseudomonas* spp., respectively. Other two strains were not identified possibly because of the esopolysaccharides interference in the assay (Franco-Buff *et al.*, 1998). The comparative evaluation of the nutritional features data base of *P. "reactans"* with the one of the Biolog Inc., showed that strains of *P. "reactans"* were included in a cluster which was distinct from the one containing strains of *P. tolaasii, P. fluorescens* and *P. marginalis,* respectively.

5.3 "White Line" Assay

All the strains of *Pseudomonas tolaasii* and *P. "reactans"* showed the capability to form white precipitates in agar when grown side by side with strains NCPPB2192 and NCPPB1311 of *P. tolaasii* and *P. "reactans"*, respectively. However, differences in the intensity as well as on the number (1 or 2) of white precipitation bands among the strains were observed.

5.4 Antagonistic Assays

Antagonistic assays performed either on the medium B of King (KB) (King *et al.*, 1954) or on a minimal medium (MMA) (Lavermicocca *et. al.*, 1997), also in presence of 10 μg ml^{-1} of $FeCl_3$, showed that strains of *P. tolaasii* inhibited the growth of strains ITM100, ATCC26423 and K12 ITM103 of *Bacillus megaterium*, *Rhodotorula pilimanae* and *Escherichia coli*, respectively. On the contrary strains of *P. "reactans"* inhibited the growth of *B. megaterium* and *E. coli* but not that of *R. pilimanae*.

6. MORPHOLOGICAL VARIANTS

During the characterisation of strains of *Pseudomonas "reactans"* either for the pathogenicity, the features in the "white line" assay or the antagonistic activities a strong variability in the above features among strains was observed. Furthermore, for a given strain the attenuation of the above characters were observed when the assays were repeated with the time. A similar behaviour, tough less pronounced, was also observed with strains of *P. tolaasii.* The latter behaviour was apparently correlated to the formation in the pure cultures of both *P. "reactans"* and *P. tolaasii* of morphological variants. These, in fact, were avirulent, did not form white precipitates in the "white line" assay and did not inhibit the growth of the target micro-organisms. In *P. tolaasii* the above phenomenon is already known (Rainey *et al.*, 1992; Grewal *et al.,* 1994; 1995) and the loss of the above phenotypes by the morphological variants is due to the lack of the production of tolaasin (Nutkins *et al.*, 1991), a lipodepsipeptide involved in the virulence of the pathogen (Rainey *et al.*, 1992). Also strains of *P. "reactans"* produce a lipodepsipeptides, called the White Line Inducing Principle (WLIP) (Mortishire-Smith *et al.*, 1991), which is known to interact with tolaasin in the formation of precipitate in agar in the "white line" assay (Wong and Preece, 1979; Rainey *et al.*, 1992) and recently demonstrated to be toxic against red blood cells, bacteria and fungi (Lo Cantore *et al.*, 2003b). Though further studies are necessary, it appear that the toxicity of WLIP - as in the case of other toxic lipodepsipeptides, including tolaasin, produced by phytopathogenic *Pseudomonas* spp. (Bender *et al.*, 1999; Menestrina *et al.*, 2003) is apparently due to its capability to form transmembrane pores (Lo Cantore *et al.*, 2003a). In conclusion it appear that the loss of WLIP production by variants of *P. "reactans"* is apparently responsible for the above phenotypes. The fact that in previous studies (Kim *et al.*, 1995) virulent and avirulent strains of *P. "reactans"* have been reported support the above possibility. The mechanism of the loss of WLIP production in *P. "reactans"* is not known but it is not excluded that the regulation of the lipodepsipeptide production, as observed for other phytopathogenic *Pseudomonas* spp. (Hrabak and Willis, 1992; Grewal *et al.*, 1995), may be involved.

7. CONCLUSIONS

The results of investigations here reported clearly demonstrated that brown blotch of *Agaricus bisporus* and the yellowing of *Pleurotus ostreatus,* caused by *Pseudomonas tolaasii,* are actually complex diseases. In fact, besides *P. tolaasii,* which may be considered the main causal agent of the above diseases, also *P. "reactans"* and not yet characterised fluorescent pseudomonads participated to the expression of the diseases symptoms. Furthermore, *P. "reactans"* has been demonstrated to be the causal agent of *P. eryngii* yellowing. The demonstration that *P. "reactans"*, a bacterial entity considered a saprotroph associated to cultivated mushrooms and useful for the specific identification of *P. tolaasii* in the "white line" assay, is a pathogen of *A. bisporus* and *Pleurotus* spp. cultivated mushrooms is a novelty. The formation of avirulent variants in the cultures of *P. "reactans"* strains, possibly with a high rate after several subcultures, appear to be responsible for the attenuation/loss of the bacterial virulence and may explain why the pathogenicity of *P. "reactans"* has not been correctly evaluated in the past. Furthermore, the fact that avirulent variants did not produce WLIP as well as the recent ascertained antifungal activity of this substance (Lo Cantore *et al.*, 2003a; Lo Cantore *et al.*, 2003b) strongly support that this lipodepsipeptide is involved, as in the case of tolaasin, in the symptom induction on the cultivated mushrooms.

ACKNOWLEDGEMENTS

The financial supports of the project "Giovani ricercatori" *"Biological characterisation of the tolaasin and WLIP, lipodepsipeptides of Pseudomonas tolaasii and P. "reactans"*, MURST ex 60% and Regione Basilicata are acknowledged.

REFERENCE

Anonymous, 1999, Biolog Microlog™ System, Release 4.01A, User Guide, Biolog. Inc.

Bender, C.L., Alarcón-Chaidez, F., and Gross, D.C., 1991, *Pseudomonas syringae* phytotoxins: mode of action, regulation, and biosynthesis by peptide and polyketide synthetases. *Microbiol. Mol. Biol. Rev.* **63**: 266-292.

Ercolani, G.L., 1970, Primi risultati di osservazioni sulla maculatura batterica dei funghi coltivati [*Agaricus bisporus* (Lange) Imbach] in Italia: identificazione di *Pseudomonas tolaasii* Paine. *Phytopath. Medit.* **9**: 59-61.

Fermor, T.R., 1986, Bacterial diseases of edible mushrooms and their control. *Proceedings of International Symposium Scientific and Technical Aspects of Cultivation Edible Fungi. Pennsylvania State University*, University Park, PA, USA, July, 1986, 361.

Ferri, F., 1985, *I funghi. Micologia, isolamento coltivazione*. Edagricole, Bologna, 398 pp.

Franco-Buff, A., Domenico, P., and Boohner, B.R., 1998, Inhibition of Capsule Production in Bacteria by Thioglycolate. *98th Annual Meeting of the American Society for Microbiology*, May 1998.

Gandy, D.G., 1968, A technique for screening bacteria causing brown blotch of cultivated mushrooms. In: *Report of the Glasshouse Crops Research Institute*, 150-154.

Geels, F.P., Hesen, L.P.W., and Van Griensven, L.J.D., 1994, Brown discoloration of mushrooms caused by *Pseudomonas agarici*. *J. Phytopathol.* **140**: 249-259

Gill, W.M., 1995, Bacterial Disease of Agaricus Mushrooms. *Rep. Tottori Mycol. Inst.* **33**: 34-55.

Godfrey, S.A.C., Marshall, J.W., and Klena, J.D., 2001. Genetic characterisation of *Pseudomonas* "NZI7" a novel pathogen that results in a brown blotch disease of *Agaricus bisporus*. *J. Appl. Microbiol.*, **91**: 412-420.

Goor, M., Vantomme, R., Swings, J., Gillis, M., Kersters, K., and De Ley, J., 1986, Phenotypic and genotypic diversity of *Pseudomonas tolaasii* and white line reacting organisms isolated from cultivated mushrooms. *J. Gen. Microbiol.* **132**: 2249-2264.

Grewal, S.I.S., Han, B., and Johnstone, K., 1994, Regulation of toxin synthesis and phenotypic variation in *Pseudomonas tolaasii*, cause of brown blotch disease of mushroom. Advances in *Molecular Genetics of Plant-Microbe Interactions* **3**: 25-32.

Grewal, S.I.S., Han, B., and Johnstone, K., 1995, Identification and characterisation of a locus which regulates multiple functions in *Pseudomonas tolaasii*, the cause of brown blotch disease of *Agaricus bisporus*. *J. Bacteriol.* **177**: 4658-4668.

Hrabak, E.M., and Willis, D.K., 1992, The *lem*A gene required for pathogenicity of *Pseudomonas syringae* pv. *syringae* on bean is a member of a family of two-component regulators. *J. Bacteriol.* **174**: 3011-3020.

Iacobellis, N.S., and Lavermicocca, P., 1990, Batteriosi del cardoncello: aspetti eziologici e prospettive di lotta. *Professione Agricoltore* **2**: 32-33.

Iacobellis, N.S., and Lo Cantore, P., 1997, Bacterial diseases of cultivated mushrooms in Southern Italy. In: *Proceedings of X Congress of Mediterranean Phytopathological Union*, Montpellier, 2-6 June, 33-37.

Iacobellis, N.S., and Lo Cantore, P., 1998a, Recenti acquisizioni sul determinismo della batteriosi del cardoncello (*Pleurotus eryngii*), *Agricoltura Ricerca* **176**: 51-54.

Iacobellis, N.S., and Lo Cantore, P., 1998b, Studi sull'eziologia dell'ingiallimento dell'ostricone (*Pleurotus ostreatus*), *Agricoltura Ricerca* **176**: 55-60.

Kim, J.W., Kwon, S.I., and Kang, H.J., 1995, Studies on the pathogenic *Pseudomonas* causing bacterial diseases of cultivated mushrooms in Korea. 2. Bacteriological characteristics of *P. tolaasii* causing mushroom brown blotch and white line reacting organisms. *Korean J. Plant Pathol.* **11**: 353-360.

King, E.O., Ward, M.K., and Raney, D.E., 1954, Two simple media for the demonstration of pyocyanin and fluorescein. *J. Lab. and Cl. Med.* **44**: 301-307.

Lavermicocca, P., Iacobellis, N.S., Simmaco, M., and Graniti, A., 1997, Biological properties and spectrum of activity of *Pseudomonas syringae* pv. *syringae* toxins. *Physiol. and Mol. Plant Pathol.* **50**: 129-140.

Lelliott, R.A., and Stead, D.E., 1987, Methods for the diagnosis of bacterial diseases of plants. Methods in Plant Pathology, vol 2, T.F. Preece ed. Blackwell Scientific Publications, Oxford, UK, 216 pp.

Lo Cantore, P., 2001, Dottorato Thesis, Aspetti patogenetici, fisiologici e molecolari di *Pseudomonas tolaasii* e *P. "reactans"*. Università degli Studi della Basilicata, Potenza, Italia.

Lo Cantore, P., and Iacobellis, N.S., 2000a, Phenotypic variability in *Pseudomonas "reactans"* and *P. tolaasii*. *5thCongress of the European Foundation for Plant Pathology*, Taormina (Ct) Italy, 18-22 September 2000, 46.

Lo Cantore, P., and Iacobellis, N.S., 2000b, Caratterizzazione di *Pseudomonas "reactans"* agente dell'"Ingiallimento" del cardoncello (*Pleurotus eryngii*). *Agricoltura Ricerca*, **188**: 59 - 64.

Lo Cantore, P., Coraiola, M., Dalla Serra, M., Menestrina, G., Lazzaroni, S. Evidente A., and Iacobellis, N.S., 2003a, Interaction of Tolaasin I and WLIP, Lipodepsipeptides of *Pseudomonas tolaasii* and *P. "reactans"*, with Biological and Model Membranes, 263-273, (this volume).

Lo Cantore, P., Lazzaroni, S., Evidente, A., Cafarchia, C., and Iacobellis, N.S., 2003b, Production of Tolaasin I and WLIP by *Pseudomonas tolaasii* and *P. "reactans"*, Their Antimicrobial Activity and Possible Role in the Virulence of the Pathogens, 255-262, (this volume).

Menestrina, G., Coraiola, M., Fogliano, V., Fiore, A., Grgurina, I., Carpaneto, A., Gambale, F., and Dalla Serra, M., 2003, Antimicrobial Lipodepsipeptides from *Pseudomonas* spp: a Comparison of Their Activity on Model Membranes, (this volume).

Mortishire-Smith, R.J., Nutkins, J.C., Packman, L.C., Brodey, C.L., Rainey, P.B., Johnstone, K., and Williams, D.H., 1991, Determination of the structure of an extracellular peptide produced by the mushroom saprotroph *Pseudomonas "reactans"*. *Tetrahedron*, **47**: 3645-3654.

Nutkins, J.C., Mortishire-Smith, R.J., Packman, L.C., Brodey, C.L., Rainey, P.B., Johnstone, K., and Williams, D.H., 1991, Structure determination of tolaasin, an extracellular lipodepsipeptide produced by the mushroom pathogen *Pseudomonas tolaasii* Paine. *J. Amer. Chem. Soc.* **113**: 2621-2627.

Rainey, P.B., Brodey, C.L., and Johnstone, K., 1992, Biology of *Pseudomonas tolaasii*, cause of brown blotch disease of the cultivated mushroom. *Plant Pathol.* **8**: 95-117.

Tolaas, A.G., 1915, A bacterial disease of cultivated mushrooms. *Phytopathology* **5**: 51.

Wells, J.M., Sapers, G.M., Fett, W.F., Butterfield, J.E., Jones, J.B., Bouzar, H., and Miller, F.C., 1996, Postharvest discoloration of the cultivated mushrooms *Agaricus bisporus* caused by *Pseudomonas tolaasii*, *P. "reactans"*, and *P. "gingeri"*. *Phytopathology* **86**: 1098-1104.

Wong, W.C., and Preece, T.F., 1979, Identification of *Pseudomonas tolaasii*: the white line in agar and mushroom tissue block rapid pitting tests. *J. Appl. Bacteriol.*, **47**: 401-407.

Identification of *Pseudomonas* Species from a Variety of Hosts in the Salinas Valley of California

C.T. BULL, P.H. GOLDMAN, N.A. CINTAS, and S.T. KOIKE[#]
US Department of Agriculture, Agricultural Research Service, 1636 E. Alisal St., Salinas, CA, 93905; [#]University of California Cooperative Extention, 1432Abbot St., Salinas, CA, 93905.

Abstract: Bacterial diseases have occurred on crucifers in the Salinas Valley since 1999, particularly on arugula (*Eruca vesicaria* ssp. *sativa)*, kale (*Brassica oleracea* var. *acephala*) and Brussels sprouts (*Brassica oleracea* var. *gemmifera*). Also, bacterial diseases have occurred on members of the *Chenopodiaceae* namely Swiss chard and table beet (both *Beta vulgaris* ssp. *vulgaris*). Isolates from arugula, Brussels sprouts, kale, table beet, and Swiss chard were Gram negative, and positive for levan production and the hypersensitive response on tobacco. Furthermore, they were negative for oxidase production, potato rot, and arginine dehydrolyase activity. The arugula, Brussels sprouts, and kale isolates were fluorescent on KMB, but the table beet, and Swiss chard isolates were not fluorescent. The pathotype strain of *Pseudomonas syringae* pv. *aptata* was not fluorescent in our laboratory. Isolates from all hosts were consistently identified as strains of *P. syringae* by fatty acid analysis. Strains of *P. syringae* from all hosts have been evaluated by BOX-PCR and banding patterns were compared to several pathotypes of *P. syringae*. BOX-PCR analysis indicated that the pathogen from Swiss chard and table beet were identical to *P. s.* pv. *aptata*. The isolates from arugula were similar to *P. s.* pv. *alisalensis* and the strains from Brussels sprouts and kale were very similar to *P. s.* pv. *maculicola*. Complete phenotypic and host range testing needs to be conducted to conclusively identify these strains.

Key words: *Pseudomonas syringae* pv. *alisalensis, P. s.* pv. *aptata, P. s.* pv. *maculicola,* crucifers.

Author for correspondence: C.T. Bull; e-mail: bactsalinas@aol.com.

N.S. Iacobellis *et. al.* (eds.), *Pseudomonas syringae* and Related Pathogens, 607-615.

1. INTRODUCTION

Salinas, California lies at the entrance to a rich agricultural valley (the Salinas Valley) that supplies the US with green vegetables. For example, 95% of US broccoli (*Brassica oleracea* L. var *italica/botrytis*) production occurs in California, with Monterey County (in which the entire Salinas Valley is located) producing 50% of the state acreage. This makes Monterey County by far the leading broccoli producing county in the US (Anonymous, 2002). Other crucifers represent 16% of the 2.3 billion dollar agricultural industry in the county (Lauritzen, 2000).

Bacterial leaf spot of cauliflower, caused by *Pseudomonas syringae* pv. *maculicola*, was initially documented in 1911 (McCulloch, 1911). This pathogen is now known to infect many crucifers (Bradbury, 1986; Zhao *et al.*, 2000), and in general bacterial diseases of crucifers were thought to be caused by this pathogen. In the Salinas Valley, this pathogen causes bacterial leaf blight on broccolini (*Brassica oleracea* L. var. *botrytis* x *Brassica alboglabra*) (Cintas *et al.*, 2002). However, a novel pathogen, *P. s.* pv. *alisalensis* (named for the Alisal region of the Salinas Valley) has been shown to cause bacterial leaf blights of broccoli raab (*Brassica rapa* subsp. *rapa*) and broccoli. These diseases can result in significant losses (Cintas *et al.*, 2002). The symptoms from these three diseases resemble each other on their respective hosts. Initial symptoms consist of small water soaked flecks on the lower foliage. These flecks expand and become surrounded by bright yellow borders, eventually which may eventually coalesce and result in large necrotic areas (Cintas *et al.*, 2002).

In addition to these diseases, several new disease have appeared on vegetables in the Salinas Valley. Since 1999, diseases have been found on crucifers, including arugula (*Eruca vesicaria* ssp. *sativa)*, kale (*Brassica oleracea* var. *acephala*), and Brussels sprouts (*Brassica oleracea* var. *gemmifera*) and members of the *Chenopodiaceae*, including Swiss chard and table beet (*Beta vulgaris* ssp. *vulgaris*). We are interested in understanding the relationships among these bacterial pathogens and how their ability to infect multiple hosts influences disease development in the current vegetable production systems. Thus, we are characterising the bacteria that cause these newly emerging diseases.

2. METHODS AND MATERIALS

2.1 Isolation of the Causal Agents

Symptomatic leaves were surface-sterilised in sodium hypochlorite (0.5%) for 1 min followed by rinsing in sterile distilled water 3 times. Small (3 x 3 mm) sections of tissue were aseptically excised from leaf spot margins and macerated in 40 μl each of sterile distilled water. The resulting suspensions or a dilution were streaked onto Sucrose Peptone Agar (SPA) (Canteros de Echenique *et al.*, 1985) and incubated at 24°C. After 3 to 5 days, representative bacterial isolates were purified by streaking single colonies onto SPA and successfully purified cultures were stored. To test for possible fungal pathogens, small leaf sections were aseptically removed and placed onto acidified potato dextrose agar (2 ml l^{-1} of lactic acid). Plates were incubated under lights at 24°C and examined after 3 to 7 days for fungal growth.

To test pathogenicity, strains were prepared by growing nutrient broth (Difco, 8 g l^{-1}) shake cultures for 48 h. Tween 20 (0.05%) was added to the resulting bacterial suspensions (containing 1 x 10^7 CFU ml^{-1} as determined by dilution plating) and each suspension was sprayed onto test plants until runoff using a hand mister. Control plants were treated with sterile nutrient broth plus Tween 20. Additional controls consisting of known pathogens were included. Plants were placed in a humid chamber for 48 h and then maintained in a greenhouse at 20 to 24°C and evaluated for symptom severity after 7, 14, and 28 days. Plant response was evaluated for symptoms and strains causing symptoms similar to those expressed in the original host were reisolated. Koch's postulates were completed as described (Cintas *et al.*, 2002).

2.2 Characterisation of Causal Agents

The pathotype strains of several authentic pathogens that have been shown to cause bacterial diseases on the hosts of interest in other locations were used as positive controls. The positive controls were *Pseudomonas syringae* pv. *aptata* strain CFBP 2280, *P. s.* pv. *syringae* CFBP 1392, *P. s. pv. tomato* CFBP 2212, *P. s.* pv. *maculicola* strain CFBP 1657, and *P. s.* pv. *alisalensis* BS91. *Escherichia coli* strain DH5α was used as a negative control. The KOH test to determine Gram character and levan production, oxidase reaction, potato rot, arginine dihydrolase production, and tobacco hypersensitivity (LOPAT tests) were conducted using usual methods (Schaad, 1988). The bacteria were evaluated for fluorescence on King's medium B (KMB) (King *et al.*, 1954).

For fatty acid analyses, bacteria were cultured on trypticase soy broth agar at 28°C for 24 h, then extracted for fatty acid methyl esters using a standard method (Paisley, 1995). Fatty acids were analysed with the Sherlock Microbial Identification System Version 2.11 (MIDI Inc., Newark, DE) that used an automated GC 6890 Hewlett-Packard gas chromatograph fitted with a 25 x 0.2 mm phenyl methyl silicone-fused silica capillary column, an HP 7673 automatic sampler and HP Chem Station Software. The analyses included assessing the degree of similarity of fatty acid composition to known standards.

Ice nucleating ability was tested as described (Fahy and Hayward, 1983). Cultures grown for 4 days at 21°C on Glucose-Yeast-Peptone agar (glucose, 5 g; yeast extract, 5 g; peptone, 5 g, agar 5 g, for 1 l) were tested for their ability to nucleate ice at -4 °C. Cells were removed from the plates and were suspended in 5 ml of water at 4 °C. Suspensions (> 1×10^9 CFU ml^{-1}) were held on ice until tested. After adjusting an ice-ethanol water bath to -4°C, the bacterial suspensions were immersed in the bath. Samples that froze within five minutes were considered ice nucleation positive.

The new pathogens were tested for lysis by PBS1, a bacteriophage which lyses *P. s.* pv. *alisalensis* (Cintas *et al.*, 2002). Standard procedures were used to determine if the other pathogens were sensitive to the phage (Klement *et al.*, 1990).

2.3 Genotypic Characterisation Using rep-PCR

The polymerase chain reaction of repetitive bacterial sequences, rep-PCR, was used to determine the relationships among pathotypes of *Pseudomonas syringae* and the newly isolated causal agents. PCR amplification of repetitive sequences was accomplished with the BOXA1R primer (5'-CTA-CGG-CAA-GGC-GAC-GCT-GAC-G-3'), synthesised by Operon Technologies, Inc. (Alameda, CA), that is designed to prime DNA synthesis from the BOXA subunit of the BOX element (Rademaker *et al.*, 1998; Versalovic *et al.*, 1994; Weingart and Völksch, 1997). The PCR was carried out according to published methods for BOX-PCR using purified genomic DNA or single colonies (Rademaker *et al.*, 1998). DNA samples were prepared by large-scale chromosomal DNA isolations (Ausubel *et al.*, 1987). The PCR was carried out in an MJ Research DNA-Engine thermo-cycler with a heated lid in the 'block' mode (MJ Research, Watertown, MA) with the use of Taq DNA polymerase (Perkin Elmer, Foster City, CA). Amplified DNA fragments were examined by agarose gel electrophoresis in 1.5% agarose gels in 0.5x Tris acetate EDTA buffer, TAE (Sambrook *et al.*, 1989). Gels were stained with ethidium bromide and DNA was photographed on a UV transilluminator with Polaroid type 55 film. Fingerprints generated from different strains were compared visually.

3. RESULTS

3.1 Isolation of the Causal Agent

No fungi were isolated from lesions on arugula, kale, Brussels sprouts, table beet, or Swiss chard. Large numbers of Gram negative bacteria were consistently isolated from lesions on these plants (Table 1). Isolated suspect bacteria were used to inoculate potential host plants. Disease resulted, bacteria were reisolated, and Koch's postulates were completed for all diseases. Tests to assess of the host range for each pathogen are currently in progress.

3.2 Phenotypic Characterisation of the Pathogens

All the bacteria isolated were Gram negative, and the majority of them were fluorescent on KMB (Table 1). Isolates from arugula, kale, and Brussels sprouts were fluorescent on KMB but, on the contrary, those obtained from table beet, Swiss chard were not fluorescent. In our laboratory the control strain of *Pseudomonas syringae* pv. *aptata* was not fluorescent. However, results from LOPAT reactions for all the isolates were consistent with results for *P. syringae*. The isolates were positive for levan production, produced a hypersensitive response on tobacco, were oxidase negative, arginine dihydrolase negative and did not rot potato slices (Table 1).

Table 1. Characterisation of pathogens isolated from crucifers.

Host or control strain[b]	Gram[c]	Fluorescent[d]	LOPAT testing[a]				
			L	O	P	A	T
Pseudomonas syringae pv. *syringae*	-	+	+	-	-	-	+
P. s. pv. *tomato*	-	+	+	-	-	-	+
P. s. pv. *maculicola*	-	+	+	-	-	-	+
P. s. pv. *aptata*	-	-	+	-	-	-	+
P. s. pv. *alisalensis*	-	+	+	-	-	-	+
Arugula	-	+	+	-	-	-	+
Kale	-	+	+	-	-	-	+
Brussels sprouts	-	+	+	-	-	-	+
Table beet	-	-	+	-	-	-	+
Swiss chard	-	-	+	-	-	-	+

[a]LOPAT evaluation scheme includes testing for levan production (L), oxidase reaction (O), potato rot (P), arginine dihydrolase production (A), and tobacco hypersensitivity (T); [b]Host refers to isolates obtained from particular hosts; [c]Gram stain was evaluated using the KOH testing method; [d]Fluorescence was evaluated on KMB agar.

The *P. s.* pv. *syringae, P. s.* pv. *alisalensis*, and *P. s.* pv. *aptata* pathotype strains and isolates from table beet, and Swiss chard nucleated ice (Table 2).

Isolates from arugula, kale, and Brussels sprouts, as well as strains of *P. s.* pv. *tomato* and *P. s.* pv. *maculicola* were negative for the above character. Bacteriophage PBS1 was originally isolated because *P. s.* pv. *alisalensis* is sensitive to it (Cintas *et al.*, 2002). Of all the strains tested in this study only isolates from arugula and *P. s.* pv. *alisalensis* were sensitive to this bacteriophage.

When isolates were compared by their fatty acid composition they were identified as strains of *P. syringae* with a similarity index of 0.75 or higher.

3.3 Genotypic Characterisation Using rep-PCR

Primers corresponding to BOX1A subunits of BOX elements annealed to genomic DNA and distinctive profiles were generated for all of the previously identified pathovars of *Pseudomonas syringae* (Cintas *et al.*, 2002). Fragments ranging in size from 0.4 to 2.5 kilobase pairs were amplified with the BOXA1R primer. Some of the unknown isolates had profiles that were identical to the profiles of known pathovars of *P. syringae* (Table 2). For example, the profiles of the Swiss chard and table beet isolates were identical to *P. s.* pv. *aptata*. Profiles from kale and Brussels sprouts isolates were nearly identical to *P. s.* pv. *maculicola* while isolates from arugula looked similar to *P. s.* pv. *alisalensis*.

Table 2. Additional characterisation of pathogens.

	MIDI identification[a]		BOX-PCR[b]			
Host[c]	ID	Confidence	ID	Comments	Phage[d]	Ice[e]
Arugula	*Pseudomonas syringae*	0.758-0.804	*P. s.* pv. *alisalensis*	S	+	-
Kale	*P. syringae*	0.907-0.948	*P. s.* pv. *maculicola*	S	-	-
Brussels sprouts	*P. syringae*	0.846-0.946	*P. s.* pv. *maculicola*	S	-	-
Table beet	*P. syringae*	0.927-0.964	*P. s.* pv. *aptata*	I	-	+
Swiss chard	*P. syringae*	0.921-0.977	*P. s.* pv. *aptata*	I	-	+

[a]MIDI identification refers to analysis of fatty acids and comparison to the Sherlock Microbial Identification System; [b]ID refers to similarities in BOX-PCR profile to pathotypes given; S=similar and I=identical profiles; [c]Host refers to strains isolated from particular hosts; [d]Lysis by bacteriophage PBS1 which lyses *Pseudomonas syringae* pv. *alisalensis*; [e]Ice nucleation at -4 °C.

4. DISCUSSION AND CONCLUSIONS

A number of new bacterial diseases are emerging in the Salinas Valley. These diseases may be appearing due to increased world-wide movement and marketing of seed and associated pathogens, or perhaps in response to changing cultural practices (including less fumigation), novel crops, and/or other factors. In addition to bacterial diseases on broccoli, broccoli raab and broccolini (Cintas *et al.*, 2002), new diseases on arugula, kale, Brussels sprouts, table beet, and Swiss chard have been brought to the attention of our county diagnostic laboratory. We have isolated the causal agents for these bacterial diseases and report initial characterisation of the pathogens here. The organisms isolated were Gram negative and, in all but two cases, fluorescent on KMB. Fatty acid characterisation indicated that all the bacterial isolates were strains of *Pseudomonas syringae*. Reactions to LOPAT tests by all the bacteria were consistent with reactions by authentic *P. syringae* strains.

P. s. pv. *alisalensis* is a recently described pathovar (Cintas *et al.*, 2002). To date it has been detected in plantings of broccoli and broccoli raab in commercial settings. This pathogen is similar to *P. s.* pv. *maculicola* in that it infects many crucifers, including cauliflower. It is unique among pathovars of *P. syringae* because of its host range (*P. s.* pv. *maculicola* did not infect broccoli raab), BOX-PCR banding pattern, ice nucleation activity, and reaction to phage PBS1. The isolates obtained from arugula have been found to have of the characteristics as this pathogen although they differ with respect to colony morphology. They have BOX-PCR banding patterns very similar to those of *P. s.* pv. *alisalensis*, are lysed by phage PBS1 thought to be specific for *P. s.* pv. *alisalensis* and did not nucleate ice at -4°C. The isolates from arugula may belong to *P. s.* pv. *alisalensis* or a related pathogen, but complete phenotypic characterisation and host range studies (specifically testing this strain on broccoli raab) are needed before an accurate and definitive identification can be made.

Because kale and Brussels sprouts are crucifers it is likely that either *P. s.* pv. *alisalensis* or *P. s.* pv. *maculicola* are the causal agents for these diseases. The BOX-PCR profiles for the isolates obtained from Brussels sprouts and kale resemble the profile for *P. s.* pv. *maculicola,* but do not appear to be identical. Other authors have found two distinct banding patterns among strains of *P. s.* pv. *maculicola* (Zhao *et al.*, 2000). Perhaps the slight differences in banding between these strains and *P. s.* pv. *maculicola* is representative of variation within this taxonomic grouping. Further phenotype characterisation and a complete host range study with these isolates will help to determine if they are strains of *P. s.* pv. *maculicola* as our working hypothesis suggests.

Genetic and phenotypic data suggest that the isolates from Swiss chard and table beet are strains of *P. s.* pv. *aptata*. For example, neither pathogen from table beet or Swiss chard nor *P. s.* pv. *aptata* produced a fluorescent water soluble pigment on KMB in our laboratory although *P. s.* pv. *aptata* is reported to produce fluorescent pigments (Bradbury, 1986). The Swiss chard and table beet isolates were identical with regard to the profile generated by

BOX-PCR to each other and to *P. s.* pv. *aptata*. Also in this case further phenotypic tests and completing host range studies are necessary.

Our results are important to vegetable growers in the Salinas Valley because of the cropping system used here. High land values and long production seasons encourage the production of 2 crops per year on a given piece of land. The second crop can be planted as soon as four days after incorporation of the unharvested portion of the previous crop. Our laboratory has demonstrated that pathogens incorporated into the soil along with the first planting can cause disease on the second crop (Cintas *et al.*, 2000). Our hope is to be able to help growers to choose non hosts for the second planting when they have had a bacterial disease outbreak in the first. To do this we need to know the host range of all potential pathogens and rapid methods for distinguishing among pathogens with overlapping host ranges. Studying the aetiology of the bacterial diseases of vegetables is the first step toward this goal.

ACKNOWLEDGEMENTS

Our thanks to N. Morris, J. Ford, and E.J. Rowland for their help conducting these experiments and B. Lewellen, K. Zydner and J. Briscoe for providing plant samples.

REFERENCES

Anonymous, 2002, Vegetables and Melons Outlook. U.S. Department of Agriculture, Economic Research Service, Bulletin VGS-289.

Ausubel, F.M., Brent, R., Kingston, R.E., Moore, D.D., Seidman, J.G., Smith, J.A., Struhl, K., 1987, *Current Protocols in Molecular Biology*. Wiley-Interscience, New York.

Bradbury, J.F., 1986, *Guide to Plant Pathogenic Bacteria*. CAB International Slough, UK.

Canteros de Echenique, B.I., Zagory, D., Stall, R.E., 1985, A medium for cultivation of the B-strain of *Xanthomonas campestris* pv. *citri*, cause of cancrosis B in Argentina and Uruguay. *Plant Dis*. **69**: 122-123.

Cintas, N.A., Koike, S.T., and Bull, C.T., 2000, Rappini bacterial blight declines with delayed replanting in the Salinas Valley of California. *Phytopathology* **90**: S15.

Cintas, N.A., Koike, S.T., and Bull, C.T., 2002, A new pathovar, *Pseudomonas syringae* pv. *alisalensis* pv. nov., proposed for the causal agent of bacterial blight of broccoli and broccoli raab. *Plant Dis* **86**: 992-998.

Fahy, P.C., and Hayward, A.C., 1983, Media and methods for isolation and diagnostic tests, p.337-378. *In*: Fahy, P. C. and Persley, G. J. (ed.), Plant Bacterial Diseases.

King, E.O., Ward, M.K., and Raney, D.E., 1954, Two simple media for the demonstration of pyocyanin and fluorescin. *J. Lab. Clin. Med.* **44**: 301-307.

Klement, Z., Rudolph, K., Sands, D.C., 1990, *Methods in Phytobacteriology*. Akadémiai Kiadó, Budapest.

Lauritzen, E., 2000, Monterey County Agricultural Commissioner 2000 Crop Report, CA.

Lelliott, R.A., Billing, E., and Hayward, A.C., 1966, A determinative scheme for the fluorescent plant pathogenic Pseudomonads. *J. Appl. Bacteriol.* **29**: 470-489.

McCulloch, L., 1911, A spot disease of cauliflower, *U. S. Dep. Agric. Bur. Plant Ind. Bull.* **225**: 1-15.

Paisley, R., 1995, Page M-5. In *MIS Whole Cell Fatty Acid Analysis by Gas Chromatography*. MIDI, Inc., Newark, DE.

Rademaker, J.L.W., Louws, F.J., and De Bruijn, F.J., 1998, Characterisation of the diversity of ecologically important microbes by rep-PCR genomic fingerprinting. *Mol. Microb. Ecol. Man.* **3.4.3**:1-27.

Sambrook, J., Fritsch, E.F., and Maniatis, T., 1989, *Molecular cloning: a laboratory manual*, 2nd ed. Cold Spring Harbor Laboratory Press, Cold Spring Harbor, N.Y.

Schaad, N.W., (ed.). 1988, *Laboratory Guide for Identification of Plant Pathogenic Bacteria*. American Phytopathological Society, St. Paul, MN.

Versalovic, J., Schneider, M., De Bruijn, F.J., and Lupski, J.R. 1994, Genomic fingerprinting of bacteria using repetitive sequence-based polymerase chain reaction. *Meth. Mol. Cell. Biol.* **5**: 25-40.

Weingart, H., and Völksch, B., 1997, Genetic fingerprinting of *Pseudomonas syringae* pathovars using ERIC-, REP-, and IS50-PCR. *J. Phytopathol.* **145**: 339-345.

Zhao, Y.F., Damicone, J.P., Demezas, D.H., Rangaswamy, V., Bender, C.L., 2000, Bacterial leaf spot of leafy crucifers in Oklahoma caused by *Pseudomonas syringae* pv. *maculicola*. *Plant Dis.* **84**: 1015-1020.

Bacterial Canker of Hazelnut (*Corylus avellana* L.) in Sardinia (Italy): Occurrence of *Pseudomonas syringae* Strains

M. FIORI, L. CICCONI[#] and M. SCORTICHINI[#]

Dipartimento di Protezione delle Piante, Università degli Studi di Sassari, Via E. De Nicola, 07100 Sassari (Italy); [#]Istituto Sperimentale per la Frutticoltura, Via Fioranello, 52, I-00040 Ciampino aeroporto, Roma (Italy).

Abstract: Surveys carried out since the end of 1998 in Sardinia, allowed to ascertain severe die-backs in hazelnut (*Corylus avellana* L.) orchards. Longitudinal cankers were observed along the twigs and the main branches. Sometimes, the death of the whole tree was also observed. The hazelnut cultivars recently introduced from the Italian peninsula (Piedmont) such as Tonda Gentile delle Langhe was attacked more than the local cultivars which showed only cankers. Isolations were performed in spring and autumn from symptomatic tissues. Fluorescent colonies on King's *et al.* medium B revealed their pathogenicity to hazelnut, pepper, tomato and pear seedlings, and, to a lesser extent, to apricot, peach and lemon fruits. The isolates did not incite any disease in lilac and apple. Biochemical and physiological tests allocated the isolates to the *Pseudomonas syringae* group Ia *sensu* Lelliott *et al.*. Slide agglutination and ELISA tests carried out using an antiserum toward *P. s.* pv. *syringae* gave positive results. Strains were also compared with other pseudomonads associated with hazelnut decline by means of repetitive PCR using BOX primers. The comparison revealed that the isolates obtained in Sardinia were different from *P. avellanae*, the causative agent of hazelnut decline in northern Greece and central Italy, and from the *P. s.* pv. *syringae* strains previously isolated in Sardinia from local cultivars of *C. avellana*. The isolates herein studied resemble the *P. syringae* population recently found in Piedmont and isolated from Tonda Gentile delle Langhe. Other studies are under way to confirm these findings.

Key words: *Pseudomonas avellanae* Hazelnut decline, pathogenicity, rep-PCR.

Author for correspondence: M. Fiori; e-mail: fiorim@uniss.it.

N.S. Iacobellis *et. al.* (eds.), *Pseudomonas syringae* and Related Pathogens, 617-625.

1. INTRODUCTION

Since winter 1998, surveys carried out in hazelnut (*Corylus avellana* L.) orchards in the province of Nuoro (central Sardinia) allowed to ascertain the presence of a hazelnut disease similar to the bacterial canker and decline caused by *Pseudomonas* spp., already observed in other areas of hazelnut cultivation in Italy (Scortichini, 2002). The hazelnut cultivar Tonda Gentile delle Langhe, introduced from Piedmont (north-west of Italy) during the 70' was particularly attacked. The main symptoms were longitudinal cankers on twigs and branches. On some occasions twig and branch die-back as well as the death of the whole plants were also observed. In spring the female buds did not swell and shoots, after an initial elongation, often withered. Exudates were frequently observed on cankers.
This paper aims to identify the causal agent of the above hazelnut disease in Sardinia.

2. MATERIALS AND METHODS

2.1 Isolation

The technique used by Psallidas and Panagopoulos (1979) was followed. Symptomatic hazelnut specimens of Tonda Gentile delle Langhe were collected in spring and autumn. Isolations were carried out from twigs, branches and shoots. The specimens were surface-sterilised with 0.5% sodium hypochloride, rinsed with sterile distilled water and small pieces of tissue adjacent to the necrosis were excited, placed into sterile mortars containing phosphate buffer (pH 7.2) and crushed. Ten-fold serial dilutions were prepared and 0.1 ml aliquots were spread onto medium B of King *et al.* (KB) (1954), plates incubated at 25°C for 48÷72 h. Single colonies were further purified on nutrient agar (NA).

2.2 Pathogenicity Tests

Three isolates which produced fluorescent pigments on KB were used in the pathogenicity tests. The isolates were grown on NA for 48 h at 25°C and then, suspensions in phosphate buffer, turbidimetrically adjusted at $1 \div 3 \times 10^6$ CFU ml^{-1}, were prepared. The following plant species were inoculated: hazelnut, apple, apricot, peach, pear, lilac, pepper, tomato and lemon fruits. Hazelnut plants were inoculated either in autumn or spring. For each isolate three two-year-old pot-cultivated plants of Tonda Gentile delle Langhe were used. For each plant, four lateral one-year-old twig were chosen. In autumn (early October), inoculations were performed by placing 10 μl of the suspension on leaf scar immediately after the leaf removal. In spring (early May), shoots and stems of the test-plant were gently wounded with a sterile scalpel and drops of the suspension were placed on the wounds. Control plants were treated with sterile phosphate buffer. After the inoculation plants were covered with plastic bags for 48 h and kept in open air for further observation. Re-isolations were performed following the techniques described above.

For each isolate three plants of apple, apricot, peach, pear, and lilac, and five of pepper and tomato were inoculated. They were wounded as above and stems were inoculated by spraying with the bacterial suspension. Two plants of each of the above mentioned species treated with sterile phosphate buffer were used as control.

Lemon fruits were surface-sterilised with 0.5% sodium hypochloride, rinsed with sterile distilled water and inoculated by wounding the epidermis with a syringe and by placing 10-20 μl of the suspension on the wounds. After inoculation, fruits were placed in a humid chamber for 48 h at room temperature. Fruits treated with sterile distilled water were used as control.

2.3 Bacteria Characterisation

The biochemical and nutritional characterisation of isolates was performed by following the methods described by Lelliott and Stead (1987) and by Schaad *et al.* (2001).

Suspension of the isolates of the interest were observed at transmission electron microscope (Zeiss TEM 109).

Ice nucleation activity was checked at -5 and at -10° C with a refrigerated circulating bath (Endocal RTE 210, Neslab, Newington, USA) following the technique described by Lindow (1990). Utilisation of carbon compounds was assessed on the medium of Ayers *et al.* (1919) as described by Lelliott and Stead (1987).

Slide agglutination and ELISA tests were performed by using an anti-*Pseudomonas syringae* pv. *syringae* van Hall NCPPB 281 (type-strain) polyclonal serum (titre 1:2560). For ELISA test, the methods described by Clark and Adams (1977) and Clark (1981) were followed. Preliminary titration tests, indicated the following optimal dilutions: coating: 1:200; ratio enzyme/serum: 1:1; conjugate dilution: 1:200. A reaction was scored as positive when the value of absorbance, measured with a photometer (GLV, Immunella S, Rome, Italy), was 2.5 times higher than the blank. Strains NCPPB 281 (isolated from *Syringa vulgaris* L.), CFBP 700 (isolated from *Juglans regia* L.) and CFBP 3072 (isolated from *Pyrus communis* L.) of *P. s.* pv. *syringae* as well strains BPIC 1077 (isolated in northern Greece) and

ISPaVe 2059 (isolated in central Italy) of *P. avellanae* were used for comparison.

2.4 Repetitive PCR

Isolates were evaluated for rep-PCR fingerprints following the procedure previously described (Scortichini *et al.*, 2000; Scortichini *et al.*, 2002). Strains of *Pseudomonas avellanae* (Psallidas) Janse *et al.* and *P. syringae* pv. *syringae* isolated from hazelnut either in Sardinia or in other areas of hazelnut cultivation (Scortichini *et al.*, 2002) as well as with pseudomonads strains previously isolated in Piedmont from Tonda Gentile delle Langhe (Scortichini *et al.*, 2000), were used for comparison.

3. RESULTS AND DISCUSSION

3.1 Isolation

Bacterial isolates producing fluorescent pigments on KB and inducing the hypersensitivity reaction on tobacco leaves, were mainly obtained during spring from longitudinal cankers on twigs and branches. In autumn, a saprophytic microflora was obtained. Three representative isolates, namely DPP48, DPP51 and DPP52e, were chosen for the further characterisation.

3.2 Pathogenicity Tests

The three isolates resulted pathogenic to hazelnut. On plants inoculated in autumn through the leaf scars, the formation of small cankers was observed in the following spring. On the shoots inoculated in spring, a twig die-back was frequently observed ten days after inoculation. Sometimes, the infection reached the branch. Pepper and tomato plants showed, four days after the inoculation, necrotic areas surrounding the inoculation site. Sometimes, wilted leaves were observed. Pear shoots, and to a lesser extent apricot and peach showed a light brownish area surrounding the inoculation site. Lemon fruits showed a light brownish lesion surrounding the inoculation site. Lilac and apple shoots as well as the control plants did not show sign of infection. Re-isolations allowed to obtain the same isolates.

3.3 Bacteria Characterisation

At the electronic microscope observation, the bacterial cells appeared as elongated rods with 1-3 polar flagella. They had the following dimensions: length: from 1.19 to 2.09 μm; width: from 0.43 to 1.13 μm.

The results of biochemical and nutritional tests are shown in Table 1. According to Lelliott *et al.* (1966), the isolates belong to *Pseudomonas syringae* group Ia. After the slide agglutination test, the isolates reacted positively with the serum anti-*P. s.* pv. *syringae* NCPPB 281. The ELISA tests gave the following Optical Density values: blank 0.03÷0.04; strains NCPPB 281, CFBP 700 and CFBP 3072 ranged from 0.8 to 1.8 OD; whereas strains BPIC 1077, ISPaVe 2059, DPP48, DPP51 and DPP52e gave OD values between 0.3 and 0.4.

Table 1. Biochemical and physiological characteristics of hazelnut isolates (DPP48, DPP51, DPP52e) and strains of *Pseudomonas avellanae* (BPIC 1077, ISPaVe 2059) and *P. syringae* pv. *syringae* (NCPPB 281).

Characteristics	Bacterial isolates					
	DPP48	DPP51	DPP52e	BPIC 1077	ISPaVe 2059	NCPPB 281
Levan	+	+	+	+	+	+
Oxidase	-	-	-	-	-	-
Potato rot	-	-	-	-	-	-
Arginine	-	-	-	-	-	-
Tobacco HR	+	+	+	+	+	+
Fluorescent pigment	+	+	+	+	+	+
Poly-β-hydroxybutyrate	-	-	-	-	-	-
Catalase	+	+	+	+	+	+
Nitrate reduction	-	-	-	-	-	-
Pectate liquifaction	-	-	-	-	-	-
2-ketogluconate	+	+	+	+	+	-
Lypolisis of Tween 80	-	-	-	-	-	-
Hydrolysis of: starch	-	-	-	-	-	-
Arbutin	+	+	+	-	+	+
Aesculin	+	+	+	+	+	+
Gelatin	+	+	+	+	+	+
Tyrosine	+	+	+	-	+	+
Urease	+	+	+	-	-	+
Acids from sucrose	+	+	+	+	+	+
Glucose: oxidation	+	+	+	+	+	+
Fermentation	-	-	-	-	-	-
NaCl tolerance: 4%	+	+	+	+	+	nt

continue

continue

Tetrazolium tolerance: 0.02%	+	+	+	+	+	nt
0.1%	+	+	+	+	+	nt
Indole production	-	-	-	-	-	-
Max. temperature growth (°C)	35	35	35	35	35	35
Ice nucleation	+	+	+	-	-	+
Utilisation of: L-arabinose	+	+	+	+	+	+
Cellobiose	+	+	+	+	+	-
Erythrose	+	+	+	+	+	-
Fructose	+	+	+	+	+	+
D-glucose	+	+	+	+	+	+
Inulin	+	+	+	+	+	nt
Lactose	+	+	+	+	+	-
Maltose	+	+	+	+	+	+
Mannose	+	+	+	nt	nt	nt
Sucrose	+	+	+	+	+	+
Salicin	+	+	+	+	+	-
Amygdalin	+	+	+	+	+	nt
Histidine	+	+	+	+	+	nt
Adonitol	+	+	+	-	+	nt
Dulcitol	+	+	+	+	+	-
Erythritol	+	+	+	-	-	+
Glycerol	+	+	+	+	+	+
Inositol	+	+	+	+	+	+
Mannitol	-	-	-	nt	nt	+
Adipate	-	-	-	nt	nt	+
Caprate	+	+	+	nt	nt	+
Citrate	+	+	+	-	-	+
Fumarate	+	+	+	+	+	nt
Gluconate	+	+	+	nt	nt	+
DL-lactate	+	+	+	+	+	+
Malate	+	+	+	-	-	+
L-tartrate	-	-	-	nt	nt	-
DL-tartrate	-	-	-	+	+	-

+ = positive reaction; - = negative reaction; nt = not tested.

3.4 Repetitive PCR

Repetitive PCR using BOX primer allowed a comparison between the isolates obtained in Sardinia and the other pseudomonads previously isolated from declining hazelnut in Greece and Italy.

Representative banding patterns are shown in Figs. 1, 2 and 3. DNA polymorphism was evident in the region between 200 and 2300 bp. Isolates DPP48, DPP51 and DPP52e are clearly different from *Pseudomonas avellanae* BPIC 631 (type-strain) and ISPaVe 2056 (Fig. 1). Qualitative difference are also evident with *P. syringae* pv. *syringae* strain (SarB3) previously isolated in Sardinia from local hazelnut cultivars (Fig. 1 and 2) as well as with other similar strains (Lab. 7 and Sic. 3) obtained from Latium and Sicily (Fig. 3) (Scortichini *et al.*, 2002).

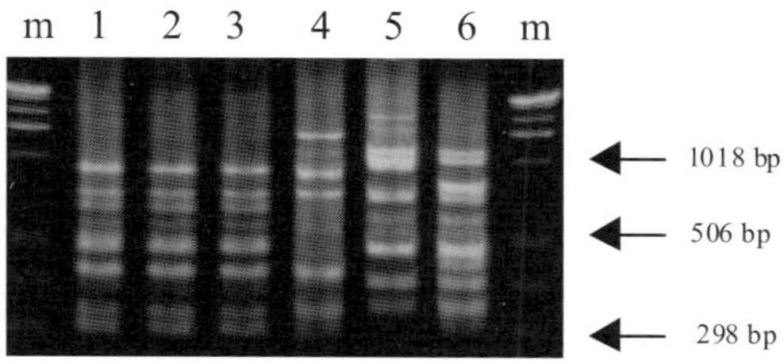

Figure 1. Rep-PCR fingerprinting patterns by using BOX primers obtained from genomic DNA of Sardinian isolates PPD48, PPD51 and PPD52e (Lines 1, 2 and 3), *Pseudomonas syringae* pv. *syringae* SarB3 (line 4) and *P. avellanae* BPIC631 and ISPaVe2056 (lines 5 and 6). m = molecular size marker 1 Kb ladder (Gibco BRL, Carlsbad, USA).

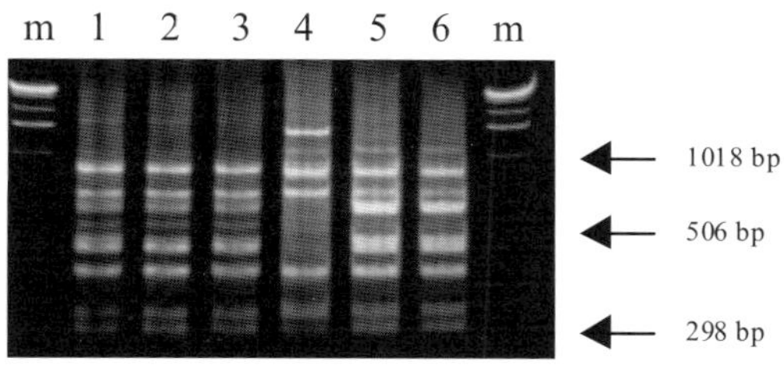

Figure 2. Rep-PCR fingerprinting patterns by using BOX primers obtained from genomic DNA of sardinian isolates PPD48, PPD51 and PPD52e (Lines 1, 2 and 3), *Pseudomonas syringae* pv. *syringae* SarB3 (line 4) and two pseudomonads isolated in Piedmont 595 and 598 (lines 5 and 6). m = molecular size marker 1 Kb ladder (Gibco BRL, Carlsbad, USA).

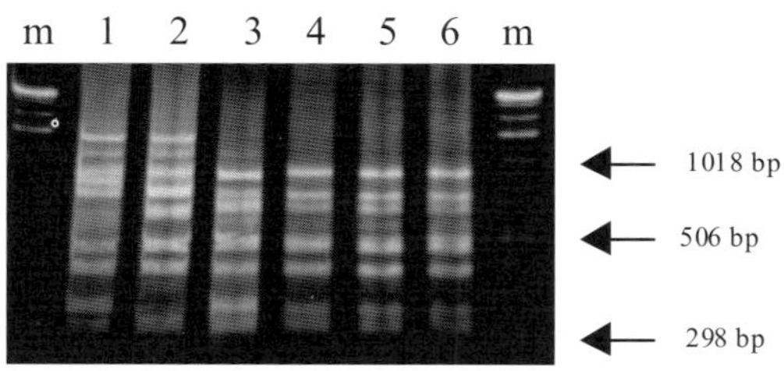

Figure 3. Rep-PCR fingerprinting patterns by using BOX primers obtained from genomic DNA of sardinian isolates PPD48, PPD51 and PPD52e (Lines 4, 5 and 6), a pseudomonad 595 isolated in Piedmont (line 3) and *Pseudomonas syringae* Lab. 7 (line 2) and Sic. 3 (line 1). m = molecular size marker 1 Kb ladder (Gibco BRL, Carlsbad, USA).

A very high similarity has been revealed with pseudomonads (595 and 598) previously isolated in Piedmont from Tonda Gentile delle Langhe (Fig. 2 and 3).

On the basis of the results obtained from pathogenicity, biochemical, nutritional and serological tests as well as from the comparison of rep-PCR fingerprinting, we conclude that strains causing canker to hazelnut trees cultivar Tonda Gentile delle Langhe in Sardinia, belong to *P. syringae, sensu* Lelliott *et al.* (1966). They showed some dissimilarities with the typical *P. s.* pv. *syringae* strains mainly in being not able to cause infection to lilac and inciting weak reaction on lemon fruits. The results of the rep-PCR indicate that strains isolated in Sardinia strongly resemble those previously obtained in Piedmont from the same hazelnut cultivar Tonda Gentile delle Langhe (Scortichini *et al.*, 2002) (see Fig 2 and 3).

The pathogen is clearly different from *P. avellanae* as shown by rep-PCR, and belong to *P. syringae sensu* Lelliott *et al.* (1966). It is very similar to the strains previously isolated in Piedmont from the same hazelnut cultivar (Scortichini *et al.*, 2000). This group of strains might represent a new pathovar, and further studies are under way to verify this.

The bacterial disease found in Sardinia is considered to be a serious threat for the cultivation of the high-prized hazelnut cultivar Tonda Gentile delle Langhe. Since its introduction in the early 70', the farmers observed the symptoms described in the present paper. The decline of the trees was initially attributed to the acidic soils characterising the Sardinian sites. This cultivar, indeed, is traditionally cultivated in the calcareous soils of the Langhe district in Piedmont. Subsequently, especially when the agronomical practises have not been regularly carried out, the severity of the disease was higher. It is not to be excluded that the pathogen has been introduced with the asyntomatic propagative material. Then, the bacterium has found a favourable environment to display its pathogenicity.

REFERENCES

Ayers, S.H., Rupp, P., and Johnson, W.T., 1919, A study of the alcali-forming bacteria in milk. *U. S. Dept. Agric. Bull.* 782.

Clark, M.F., and Adams, A.N., 1977, Characteristics of the microplate method of enzyme-linked immunosorbent assay for the detection of plant viruses. *J. Gen. Virol.* **34**: 475-483.

Clark, M.F., 1981, Immunosorbent assay in plant pathology. *Ann. Rev. Phytopath.* **19**: 83-106.

King, E.O., Ward, M.K., and Raney, D.E., 1954, Two simple media for the demonstration of pyocyanin and fluorescin. *J. Lab. Clin. Med.* **44**: 301-307.

Lelliott, R.A., Billing, E., and Hayward, A.C., 1966, A determinative scheme for the fluorescent plant pathogenic Pseudomonads. *J. Appl. Bacteriol.* **29**: 470-489.

Lelliott, R.A., and Stead, D.E., 1987, *Methods for the diagnosis of bacterial diseases of plants*. Oxford. Blackwell Scientific Publications.

Lindow, S.E., 1990, Bacterial ice-nucleation activity. 428-434 in: Klement, Z., Rudolph K., and Sands, D.C., (Eds.) *Methods in Phytobacteriology*. Akademiai Kiado, Budapest.

Psallidas, P.G., and Panagopoulos, C.G., 1979, A bacterial canker of *Corylus avellana* in Greece. *Phytopath. Zeitscrift* **94**: 103-111.

Schaad, N.W., Jones, J.B., and Chun, W., (Eds.) 2001, *Laboratory Guide for Identification of Plant Pathogenic Bacteria* (3rd ed.). St. Paul, Minnesota.

Scortichini, M., 2002, Bacterial canker and decline of European hazelnut. *Plant Dis.* **86**: 704-709.

Scortichini, M., Marchesi, U., Dettori, M.T., Angelucci, L., Rossi, M.P., and Morone, C., 2000, Genetic and pathogenic diversity of *Pseudomonas avellanae* strains isolated from *Corylus avellana* trees in north-west of Italy, and comparison with strains from other regions. *Eur. J. Plant Path.* **106**: 147-154.

Scortichini, M., Marchesi, U. Rossi, M.P. and Di Prospero, P., 2002, Bacteria Associated with Hazelnut (*Corylus avellana* L.) decline are of Two groups: *Pseudomonas avellanae* and strains resembling *P. syringae* pv. *syringae*. *Appl, Environ. Microbiol.* **68**: 476-484.

Pseudomonas huttiensis Associated with Leaf Necrosis and Blighting of Tomato Seedlings in the Greenhouse

A. OBRADOVIC, J.B. JONES, G.V. MINSAVAGE, E.R. DICKSTEIN, and T.M. MOMOL[#]

Plant Pathology Department, University of Florida, PO Box 110680, Gainesville 32611, Florida, USA; [#]NFREC, Quincy, Florida, USA.

Abstract: In October 2001, tomato foliage with blighting and a leaf spot was received from a transplant producer in Florida. The seedlings manifested apical or marginal leaf necrosis or discrete lesions along the leaf veins. Non-fluorescent bacterial strains forming viscous, creamy white colonies on King's medium B and causing a hypersensitive reaction in tomato and pepper leaves was consistently isolated from lesions. Pathogenicity was checked on three-week old tomato plants. The plants were incubated in high humidity for 24 h before and after inoculation. Similar symptoms were observed on the inoculated seedlings. We characterised 12 strains using bacteriological tests. According to the fatty acid profile, the strains displayed highest similarity with the bacterium *Pseudomonas huttiensis* Leifson 1962. Sequence analysis of the 16S rRNA indicated that this bacterium shows 98.7÷98.8% homology with two *Herbaspirillum* species. Thus, our strains were compared with one *P. huttiensis* and five *H. rubrisubalbicans* strains. The strains from tomato were Gram-negative, non-fluorescent and oxidase positive, but levan, pectate hydrolysis and arginine dihydrolase negative, and grew at 40°C. Based on these results, they had a high degree of similarity with both *P. huttiensis* and *H. rubrisubalbicans*.

Key words: *Lycopersicon esculentum*, *Herbaspirullum rubrisubalbicans*, *Xanthomonas campestris* pv. *vesicatoria*.

Author for correspondence: J.B. Jones; e-mail: JBJones@ifas.ufl.edu.

N.S. Iacobellis *et. al.* (eds.), *Pseudomonas syringae* and Related Pathogens, 627-630.

1. INTRODUCTION

In October 2001, a leaf spot and blight were observed on 3÷4 week old tomato seedlings grown in a greenhouse nursery in southern Florida. The seedlings manifested mixed symptoms. Some plants showed discrete lesions along the leaf veins and isolated leaf spots. The spots were irregular and varied in size from 2 to 4 mm. Affected tissue was discoloured and necrotic. Most of the seedlings had apical or marginal leaf necrosis. The diseased tissue collapsed, turned brown and narcotised. Blighting of younger plants caused unrecoverable loss of foliage and those had to be discarded. However, the disease remained localised on individual plants or on plants within one tray. Symptom development stopped within a week of the initial appearance. The symptoms did not occur on older plants. Less than 5% of the plants grown in the greenhouse were affected. Due to concern of possible spread of pathogens on tomato transplanting material and unusual symptomatology of the disease we decided to investigate the causal agent of the seedling leaf spot and marginal leaf necrosis.

2. MATERIALS AND METHODS

2.1 Isolation and Identification

Diseased transplants collected from one greenhouse were used for isolation. Lesions from leaves were excised and triturated in drops of sterile tap water. Loopfuls of the suspension were streaked onto nutrient agar (NA) plates and incubated at 28°C for 48 h. Representative colonies were transferred to new NA plates and used for further testing. Cell morphology was studied from a nutrient broth culture of a representative strain (SE1), grown on a shaker at 28°C for 24 h, and observed by transmission electron microscopy (TEM). To identify the pathogen we studied some bacteriological characteristics (Schaad *et al.*, 2001), carbon substrate utilisation (Biolog, Inc., Hayward, CA, USA) and membrane fatty acid composition (Microbial Identification System, MIDI, Newark, DE, USA) of 12 isolated obtained from diseased seedlings.

2.2 Pathogenicity Test

Hypersensitivity of the strains was checked by infiltration of bacterial suspensions (10^8 CFU ml^{-1}) into tomato, pepper and tobacco leaves. In order to reproduce the symptoms of natural infection, three-week old tomato seedlings cv. Bonny Best were inoculated either by spraying (10^8 CFU ml^{-1})

or by dipping [10^6 CFU ml^{-1} + the organosilicone surfactant, Silwet L77 (Loveland industries, Inc.) 0.025%] with suspensions of the investigated strains. The plants were kept in high-humidity conditions at 28°C for 24 h before and after inoculation in a growth room, before being transferred to the greenhouse for symptom development.

3. RESULTS AND DISCUSSION

Health management of transplanting material is a major concern of tomato transplant producers. Due to high plant density and conditions favourable for disease development, the presence of plants infected by bacteria could easily compromise production. Therefore, we studied the aetiology of the leaf spot and marginal leaf necrosis found on tomato transplants grown in the greenhouse in southern Florida.

3.1 Characterisation of the Pathogen

The consistent isolation of non-fluorescent bacterial isolates forming creamy white colonies indicated that the bacterium was different than known pathogens causing symptoms on tomato foliage (*Xanthomonas campestris* pv. *vesicatoria*, *Pseudomonas syringae* pv. *tomato*, *P. s.* pv. *syringae*, and *P. viridiflava*) (Jones *et al.*, 1991; Schaad *et al.*, 2001). When observed by TEM, cells of the representative strain (SE1) were slightly curved, with one or two flagella at one or both poles. The strains were Gram negative, non-fluorescent and oxidase positive, but levan, pectate hydrolysis and arginine dihydrolase negative, and grew at 40°C. According to the fatty acid profile, the strains displayed highest similarity with the bacterium *Pseudomonas huttiensis* (Leifson, 1962), isolated from distilled water. Since this organism was suggested to be transferred into the genus *Herbaspirillum* (Anzai *et al.*, 2000), we included in the analysis one *P. huttiensis* and four *H. rubrisubalbicans* strains obtained from the Bacterial Collection, University of Gent, Belgium. When the data generated by fatty acid analysis were processed by MIDI software, the strains isolated from tomato and the control strains clustered into one group (ED≤6) indicating similarity in fatty acid composition. However, within the group, the type strain and two other strains of *H. rubrisubalbicans* (LMG2286, LMG1278, and LMG6418, respectively) formed a separate subgroup.

Based on differential oxidation of the 95 carbon substrates available in the Biolog GN MicroPlate, three out of six tomato strains tested were identified as *P. huttiensis*, and the remaining three as *Burkholderia*-like species.

Identification was confirmed for both type strains of *P. huttiensis* (LMG2199) and *H. rubrisubalbicans* (LMG2286).

3.2 Pathogenicity Assay

With the exception of the control *Pseudomonas huttiensis* strain, both investigated and control strains caused a hypersensitive reaction in tomato, pepper and tobacco leaves. However, in inoculation tests only the isolates from tomato seedlings were able induce symptoms similar to those observed in natural infection. Apical tissue of leaves of inoculated plants collapsed and became necrotic three days after inoculation. Necrotic lesions occurred along the leaf margins. Small pinpoint spots were predominant along the veins five days after inoculation. The symptoms were more severe when a 24 h high-humidity incubation period preceded the inoculation. Younger plants were more susceptible. No secondary spread of symptoms was noticed after the plants were transferred to the greenhouse.
According to the bacteriological characteristics, metabolism of carbon compounds and fatty acid composition, a bacterium similar to *P. huttiensis* was responsible for the sudden outbreak of disease on tomato seedlings grown in the greenhouse. Based on pathogenicity tests it was demonstrated that this weak pathogen is able to endanger tomato transplant production at early plant stages under high-humidity conditions.

REFERENCES

Anzai, Y., Kim, H., Park, J., Wakabayashi, H., and Oyaizu, H., 2000, Phylogenetic affiliation of the pseudomonads based on 16S rRNA sequence. *Int. J. Syst. Evol. Microbiol.*, **50**: 1563-1589.

Jones, J.B., Jones, J.P., Stall, R.E., and Zitter, T.A., 1991, *Compendium of Tomato Diseases*. The American Phytopathological Society, St. Paul, MN.

Leifson, E., 1962, The bacterial flora of distilled stored water. III. New species of the genera *Corynebacterium*, *Flavobacterium*, *Spirullum*, and *Pseudomonas*. *Int. Bull. Bacteriol. Nomencl. Taxon.* **12**: 161-170.

Schaad, N.W., Jones, J.B., and Chun, W., 2001, *Laboratory Guide for Identification of Plant Pathogenic Bacteria*. The American Phytopathological Society, St. Paul, MN.

An Unusual Pseudomonad Isolated From Diseased Parsley Roots in Serbia

A. OBRADOVIC, M. ARSENIJEVIC#, J.B. JONES, and G.V. MINSAVAGE
Plant Pathology Department, University of Florida, PO Box 110680, Gainesville 32611, Florida, USA; #Institute for Plant Protection, Faculty of Agriculture, Novi Sad, Yugoslavia.

Abstract: In autumn of 1998 and 1999, parsley root rot was observed either in the fields of northern Serbia or during storage. Diseased plants showed total destruction of the leaf rosette and root base resembling soft rot. However, when the remaining part of the root was removed from the soil, reddish brown areas were noticed on the root surface. The tissue was sunken and firm with no cracks visible on the surface. On the root, longitudinal brown discoloration spread from the upper part toward the root tip affecting the vascular cylinder and the surrounding tissue. Thirty-two bacterial isolates were obtained from inner tissue of many roots. All isolates caused a hypersensitive reaction in tobacco leaves within 24 h, were Gram negative, rod-shaped, motile, and produced shiny, greyish-white colonies on King's medium B (KB) without the typical green fluorescent pigment on KB medium in 24 h. They were levan positive but oxidase and arginine dihydrolase negative, non pectolytic, aerobic, utilised sucrose and aesculin, did not reduce nitrates, and did not grow at 37°C. Syringomycin and coronatine production was negative and ice nucleation was positive. Based on fatty acid profiles the strains had greatest similarity to *Pseudomonas cichorii* and *P. syringae*. Prick-inoculated roots of two-month-old parsley plants caused tissue depression around the point of inoculation and yellowing of older leaves followed by wilting of the foliage one week after inoculation. A week later the roots were pulled out and root rot resembling the symptoms of natural infection was observed. Considering their pathogenic and bacteriological characteristics, investigated bacterial isolates associated with parsley root rot in Serbia, belong to an unusual, non fluorescent *Pseudomonas* sp.

Key words: *Pseudomonas syringae*, *P. cichorii*, non fluorescent pseudomonad.

Author for correspondence: A. Obradovic; e-mail: oaleksa@verat.net.

N.S. Iacobellis *et. al.* (eds.), *Pseudomonas syringae* and Related Pathogens, 631-634.

1. INTRODUCTION

In Serbia, parsley (*Petroselinum sativum* L.) is grown for leaf and root production. Unlike parsley leaves sold in the fresh vegetable market or processed immediately after harvesting, most of the roots are kept in storage for later consumption. Increased market demand for this vegetable species encouraged farmers in northern Serbia to expand the area for parsley production. However, it is still considered a minor crop in Serbian agriculture and therefore, phytopathological problems have not been studied intensively.

Parsley root decay, occurring frequently every year under inadequate storage conditions, decreases the amount of marketable yield. However, in the autumn of 1998 and 1999, an unusual parsley root rot was observed in the fields of northern Serbia late in the season and following harvest during storage. Initially, diseased plants showed destruction of the leaf rosette and root base resembling soft rot symptoms caused by pectolytic erwinias (Gavrilovic *et al.*, 2001). However, when the remaining part of the root was removed from the soil, reddish-brown areas were noticed on the root surface. The tissue was sunken and firm with no cracks visible on the surface. On the root, longitudinal reddish-brown discoloration spread from the upper part toward the root tip affecting the vascular cylinder and then surrounding tissue. The root decay continued during the storage period and resulted in losses reaching up to 30% of the total yield harvested in particular fields. A high percentage of roots was affected. A survey of parsley fields and harvested roots resulted in many bacterial isolates, which were used for laboratory analysis to determine the causal agent of the disease.

2. MATERIALS AND METHODS

2.1 Isolation and Characterisation of the Pathogen

Several diseased parsley roots were inspected. Thirty-two bacterial isolates were obtained from inner tissue of roots. Standard bacteriological tests including toxin production, ice nucleation assay, carbon substrate utilisation (Biolog, Inc., Hayward, CA, USA) and membrane fatty acid composition (Microbial Identification System, MIDI, Newark, DE, USA) were performed as described previously (Arsenijevic, 1997; Schaad *et al.*, 2001) on 12 representative strains.

2.2 Pathogenicity Assay

Isolates were screened for pathogenicity by streaking a loopful of bacteria from pure culture onto the surface of parsley, parsnip, and carrot root slices and for hypersensitivity by infiltration of bacterial suspensions (10^8 CFU ml^{-1}) into tobacco leaves. Twelve isolates were chosen for further investigation. Later, two-month-old parsley plants cv. Eagle (*Petroselinum crispum*, Bejo Zaden, NL), grown in 10-cm pots in the greenhouse, were inoculated either by spraying the foliage with the bacterial suspensions (10^8 CFU ml^{-1}) or by pricking the root base with a toothpick previously dipped into the bacterial cultures. After inoculation, the plants were kept under high-humidity conditions at 28°C for 48 h, followed by transfer of the plants to a greenhouse for symptom development.

3. RESULTS AND DISCUSSION

A high incidence of unmarketable parsley roots, harvested in the seasons of 1998 and 1999 in Northern Serbia, indicated a serious disease problem. Intensity of the disease and unusual symptomatology turned our attention to this problem and contributed to detailed investigation of the disease aetiology.

3.1 Pathogen Identification

Based on the analysis of fatty acid profiles, the isolates possess greatest similarity either to *Pseudomonas cichorii* or *P. syringae*. *P. cichorii* appeared as a first choice for eight of twelve tested strains with similarity indices ranging from 0.852 to 0.901 while four strains were identified as *P. s.* pv. *tomato* (0.855÷0.899). The strains were Gram negative, rod-shaped, motile, and produced shiny, greyish-white colonies on King's medium B (KB). However, the strains did not produce the typical green fluorescent pigment on KB medium after 24 h. They were levan positive but oxidase and arginine dihydrolase negative, nonpectolytic, aerobic, utilised sucrose and aesculin, produced catalase, did not reduce nitrates, and did not grow at 37°C. Syringomycin and coronatine production were negative and ice nucleation was positive. Metabolism of the 95 carbon compounds available on the Biolog GN MicroPlate was tested for three of the twelve representative strains. When the data were analysed by using MicroLog 4.2 database, no species identification was possible. The strains were identified as members of the genus *Pseudomonas*.

3.2 Pathogenic Characteristics

Tissue discoloration and drops of an opaque liquid oozing from the vascular ring were noticed on parsley, parsnip and carrot root slices 48 h after inoculation. Three days later brown rot spread from the vascular ring toward the central part covering most of the inoculated tissue slice. The isolates produced water-soaked spots on the leaves of inoculated plants three days after inoculation, and rot of prick-inoculated roots after two weeks. The leaf spots coalesced causing necrosis of a larger leaf area. The first symptom on prick-inoculated roots was a tissue depression around the point of inoculation and yellowing of older leaves followed by wilting of the foliage one week after inoculation. A week later the roots were pulled out and root rot resembling the symptoms of natural infection was observed. No symptoms were observed on uninoculated root slices, parsley plants sprayed with water or roots pricked with a sterile toothpick. All strains caused a hypersensitive reaction in tobacco leaves within 24 h.

Most of the bacteriological characteristics indicate close relatedness of the isolates obtained from parsley with either *Pseudomonas cichorii* or *P. syringae*. However, the parsley isolates do not match completely with any of them. Different carbon source metabolism and lack of fluorescence on KB medium separate them from the group of fluorescent pseudomonads. Considering their bacteriological and pathogenic characteristics, investigated strains associated with parsley root rot in Serbia belong to an unusual, non fluorescent *Pseudomonas* sp.

REFERENCES

Arsenijevic, M., 1997, *Bacterial Plant Diseases*. S Print, Novi Sad, Yugoslavia.

Gavrilovic, V., Obradovic, A., and Arsenijevic, M., 2001, Bacterial soft rot of carrot, parsley and celery. In: *Plant Pathogenic Bacteria*, S.H. De Boer (ed.), Kluwer Academic Publishers, The Netherlands, pp. 269-271.

Schaad, N.W., Jones, J.B., and Chun, W., 2001, *Laboratory Guide for Identification of Plant Pathogenic Bacteria*. The American Phytopathological Society, St. Paul, MN.

Occurrence of Apical Shoot Chlorosis and Whitening of Pea Caused by Strains of *Pseudomonas syringae* pv. *pisi*

A. SUZUKI and Y. TAKIKAWA
Laboratory of Plant Pathology, Faculty of Agriculture, Shizuoka University, 836 Ohya, Shizuoka 422-8529, Japan.

Abstract: A new bacterial disease has been observed on pea in Shizuoka prefecture, Japan, since 1981. The disease occurs in early autumn when pea plants grow vigorously. The disease is characterised by chlorosis and whitening of apical shoot including leaflets, stipules and young pods. Usually, the Apical Shoot Chlorosis and Whitening (ASCW) symptoms are associated with extensive water soaking lesions on stems and on leaflets at the basal part of the diseased plants. The latter are the common symptoms of bacterial blight caused by *Pseudomonas syringae* pv. *pisi*. Fifty-one bacterial isolates, including 35 strains from ASCW and 16 from common bacterial blight of pea, were characterised and identified as *P. s.* pv. *pisi*. The bacteria were Gram-negative rods, having 1 to 3 polar flagella. The results of LOPAT tests were + – – – +, showing that they belong to *P. syringae*. In stab inoculation, the isolates from ASCW reproduced ASCW symptoms with water soaking spots 14 days after inoculation. *P. s.* pv. *pisi* strains never induced ASCW symptoms and, on the contrary, caused the typical bacterial blight ones. ASCW strains were not pathogenic on any other plants tested. *P. s.* pv. *pisi* strains were divided into 2 groups (A and B) according to their phenotypic properties. These were identical among ASCW strains and *P. s.* pv. *pisi* group A. Two distinct fingerprint profiles were identified by rep-PCR. ASCW strains and *P. s.* pv. *pisi* group A belonged to the same fingerprint type in rep-PCR, whereas a distinct fingerprint was shown by strains of the *P. s.* pv. *pisi* group B. We concluded that the ASCW strains should be included in *P. s.* pv. *pisi* as a distinct variant group distinguishable for symptom expression.

Key words: identification, rep-PCR, symptom.

Author for correspondence: Y. Takikawa; e-mail: abytaki@agr.shizuoka.ac.jp.

N.S. Iacobellis *et. al.* (eds.), *Pseudomonas syringae* and Related Pathogens, 635-641.

1. INTRODUCTION

In 1981, a disease of unknown cause was first observed on pea in Izu district of Shizuoka prefecture, one of the main pea-production area in Japan. The disease occurs in early spring and early autumn when pea plants grow vigorously, whereas in summer and winter, symptoms are rarely observed. The symptoms were observed on apical shoots including leaflets, stipules and young pods. The symptoms initially appeared as slight yellowing at the margin of expanding leaflet while veins and tips of leaflets remain light green. Later on the whole apical shoots become light yellow and then turned white. The apparently green healthy tissues and the shoot portions of affected white ones are sharply delineated. Usually, the apical shoot chlorosis and whitening symptoms are associated with extensive water soaking lesions on stems and on leaflets at the basal part of the infected plants. The latter are the common symptoms of bacterial blight of pea. Water soaking spots are also observed on chlorotic leaflets and pods. As the infected plant continues to grow, newly developed lateral shoots turn yellow to white. On the same affected plant, mature shoots never turn white. Earliest symptoms can be observed on plantlets two to three weeks after germination. The flowers formed on chlorotic shoots have whitish calyces and the formed pods are papery and unmarketable. Severely affected plants stop growing and are stunted. From the above lesions bacteria tentatively identified as strains of *Pseudomonas syringae* have been consistently isolated. No bacteria were isolated from chlorotic tissues

Objective of this work was to identify and characterise the pathogen isolated from ASCW lesions.

2. MATERIALS AND METHODS

2.1 Isolation of Pathogen

The pathogen was isolated by excising small tissue pieces from water soaking lesions as well as from boundary part of green and white tissue without water soaking. Samples were macerated in sterile water and the resulting suspensions were streaked onto Yeast-Peptone Agar.

2.2 Bacteriological Properties

Bacterial isolates were tested for Gram reaction by the non staining method. Fluorescent pigment production was determined on medium B of King *et al.* (1954). The methods of Lelliott *et al.* (1966) were used for LOPAT tests, gelatin liquefaction and esculin hydrolysis. Reaction in purple milk was tested by methods described by Dye (1968). Utilisation of organic compounds as sole source of carbon was determined by using the modified medium of Ayers *et al.* as described in the Manual of Microbiological Methods (1957). All tests were performed at 25°C.

2.3 Pathogenicity Tests

Pathogenicity tests were performed on 2 weeks old seedling of pea in greenhouse. Stems, stipules and leaflets were stabbed with a needle, and a 10 μl droplet of bacterial suspension (ca. 10^8 CFU ml^{-1}) was placed over each wounded site. Control plants received the same treatment but using distilled sterile water. The host range of pathogens was tested by stab inoculation on the following plants: *Actinidia chinensis* (kiwi), *Arachis hypogaea* (peanut), *Brassica oleracea* var. *capitata* (cabbage), *Capsicum annuum* (pepper), *Citrullus lanatus* (watermelon), *Citrus natsudaidai* (natsudaidai), *Cucumis sativus* (cucumber), *Eriobotrya japonica* (loquat), *Glycine max* (soybean), *Iris germanica* (german iris), *Lycopersicon esculentum* (tomato), *Morus bombycis* (mulberry), *Myrica rubra* (Chinese bayberry), *Nicotiana tabacum* (tobacco), *Oryza sativa* (rice), *Phaseolus vulgaris* (kidney bean), *Photinia glabra* (Japanese photinia), *Prunus mume* (Japanese apricot), *Prunus persica* (peach), *Syringa vulgaris* (lilac), *Tagetis erecta* (marigold), *Thea sinensis* (tea), *Trifolium repens* (white clover), *Triticum aestivum* (wheat), *Vicia faba* (broad bean) and *Zinnia elegans* (zinnia). Plants were evaluated for symptom appearance up to one month after inoculation.

2.4 Rep-PCR

Total bacterial genomic DNA was isolated by the CTAB purification method in Current Protocols in Molecular biology (Ausubel *et al.*, 1989). The rep-PCR reactions were carried out as described by Rademaker *et al.* (1998) with some modifications. PCR products were separated by electrophoresis in 1.5% agarose gels at 70V for 40 min in 1 x TAE buffer. Gels were stained with ethidium bromide and differences in DNA fingerprint patterns were visually assessed.

3. RESULTS

3.1 Bacterial Isolation and Characterisation

Thirty-five isolates were obtained from plants showing ASCW symptoms. They were compared with 16 strains of *Pseudomonas syringae* pv. *pisi* obtained from typical lesions of common bacterial blight of pea. All bacterial isolates were Gram- negative rods having 1 to 3 flagella. The results of LOPAT tests were + – – – +. According to their biological characters and symptom expression they were divided into 3 groups (ASCW strain, *P. s.* pv. *pisi* group A and B). ASCW and *P. s.* pv. *pisi* group A strains did not produce fluorescent pigment on King's B medium on the contrary of *P. s.* pv. *pisi* group B ones. Th latter strains showed negative reactions for esculin hydrolysis, gelatin liquefaction, alkali production and digestion in purple milk reaction, and did not utilise erythritol, DL-lactate and betain. On the contrary, ASCW and *P. s.* pv. *pisi* group A strains were positive for the above characters. All strains utilised homoserine. The results of these tests are summarised in Table 1.

Table 1. Bacteriological properties of *Pseudomonas syringae* strains isolated from pea in comparison to related pathovars.

Tests	ASCW	Ppi (A)	Ppi (B)	Pss	Pph	Pmc	Psc
Fluorescent pigment	–	–	+	+	+	–	+
Levan	+	+	+	+	+	+	+
Oxidase	–	–	–	–	–	–	–
Potato soft rot	–	–	–	–	–	–	–
Arginine dihydrolase	–	–	–	–	–	–	–
Tobacco HR	+	+	+	+	+	+	+
Gelatin liquefaction	+	+	–	+	+	–	–
Esculin hydrolysis	+	+	–	+	–	–	+
Purple milk	K D	K D	–	K D	–	–	K D
Utilisation of:							
Sucrose	+	+	+	+	+	+	+
Xylose	+	+	+	+	+	–	+
Erythritol	+	+	–	+	–	–	–
Mannitol	+	+	+	+	–	+	+
Inositol	+	+	+	+	–	+	+
Sorbitol	+	+	+	+	–	+	+
D-Tartrate	–	–	–	V	–	+	–
DL-Lactate	+	+	–	+	–	+	–
Betain	+	+	–	+	–	–	–
Homoserine	+	+	+	–	–	–	–

ASCW = ASCW strains; Ppi (A) and (B) = *Pseudomonas syringae* pv. *pisi* group A and B; Pss = *P. s.* pv. *syringae*; Pph = *P. s.* pv. *phaseolicola*; Pmc = *P. s.* pv. *maculicola*; Psc = *P. s.* pv. *coronafaciens*; + = positive reaction; – = negative reaction; V = variable reaction; K = alkali production; D = digestion.

3.2 Pathogenicity Tests

In stab inoculation on stem of pea, all strains caused, about four days after inoculation, dark water soaking elliptical area around the wounds. *Pseudomonas syringae* pv. *pisi* group A and B strains never induced ASCW symptoms and the inoculated plants showed typical bacterial blight symptoms. On the contrary, only ASCW strains, 14 days after inoculation, induced the typical ASCW symptoms. All strains were not pathogenic on any other host plants tested.

3.3 Rep-PCR

Genomic fingerprints were generated for *Pseudomonas syringae* pv. *pisi* group A, B and ASCW strains. Data from three primer sets were scored visually, taking into account differences for the presence or absence of main bands. More than 20 DNA fragments ranging from 0.1 to 2.0 kb were amplified with ERIC-, BOX-, and REP-PCR primers. Based on genomic fingerprints using ERIC primers, two distinct groups were identified for strains of *P. s.* pv. *pisi*. They were arbitrarily designated as PCR type I and II. Type I contained ASCW and *P. s.* pv. *pisi* group A strains (Fig. 1., lane 1-11). Type II consisted of *P. s.* pv. *pisi* group B (Fig. 1., lane 12 and 13). Type I fingerprints showed unique bands of about 500 bp and 200 bp in ERIC-PCR (Fig. 1). On the contrary, type II fingerprints showed typical bands of about 600 bp and 100 bp, respectively. ASCW and *P. s.* pv. *pisi* group A strains produced the same fingerprints in ERIC-, BOX- and REP-PCR (data not shown). Strains of *P. s.* pv. *syringae*, *P. s.* pv. *phaseolicola*, *P. s.* pv. *maculicola* and *P. s.* pv. *coronafaciens,* used for comparison (Fig. 1., lane 14-17), showed fingerprints different from both type I or II fingerprints in all primer sets.

4. DISCUSSION

The causal bacterium of the ASCW disease is a Gram-negative rod with polar flagella which on the bases of results of LOPAT tests belong to *Pseudomonas syringae*. The pathogen utilised homoserine and its bacteriological properties almost fit with those of *P. s.* pv. *pisi*.
The ASCW strains, four days after the inoculation in pea plants, caused first water-soaking lesions, the typical symptom of bacterial blight, and then, after 14 days, they caused the typical ASCW symptoms. On the contrary, typical *P. s.* pv. *pisi* strains never developed ASCW symptoms.
Bacterial blight of pea is a common disease in many countries but ASCW symptoms were never reported (Lawyer and Chun, 2001).

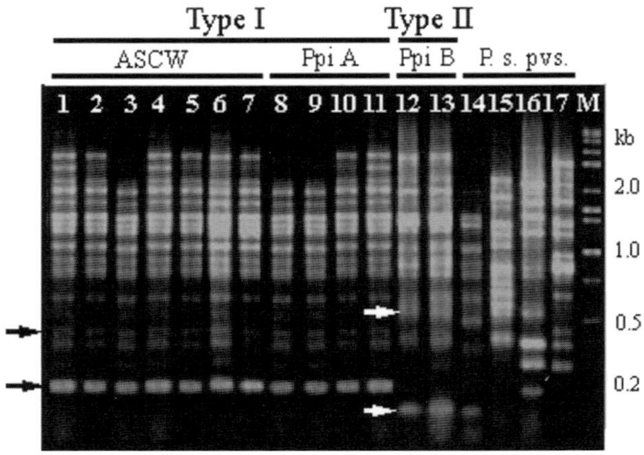

Figure 1. Rep-PCR fingerprint obtained from *Pseudomonas syringae* pv. *pisi* strains and closely related pathovars with the ERIC primers. Arrows indicated specific bands in each polymorphism types. M = DNA size marker; Lanes 1 to 7 were ASCW strains. Lanes 8 to 11 were *P. s.* pv. *pisi* group A. Lanes 12 to 13 were *P. s.* pv. *pisi* group B. Lane 14 to 17 were *P. s.* pv. *syringae,* pv. *phaseolicola*, pv. *maculicola*, pv. *coronafaciens* strains, respectively.

Genomic fingerprints generated by rep-PCR have been used to differentiate many plant pathogenic bacteria (Hollaway *et al*., 1997; Rademaker *et al*., 1998; Zhao *et al*., 2000). Several studies have shown that *P. s.* pv. *pisi* is heterogeneous in terms of phenotypic and genetic properties (Grondeau *et al.*, 1992; Arnold *et al.,* 1996). Our results revealed that Japanese isolates of *P. s.* pv. *pisi* consist of two groups (group A and B), confirming the heterogeneity this pathogen, and showed that ASCW and *P. s.* pv. *pisi* group A strains have the same fingerprints in all primer sets used. Thus, the causal agent of ASCW is identical with *P. s.* pv. *pisi* except for the symptoms. Therefore, we concluded that the above strains should be included in *P. s.* pv. *pisi* and considered as a distinct variant. This evidence has been reported for other phytopathogenic bacteria such as *P. s.* pv. *tabaci* and its non-toxigenic variant *P. s.* pv. *angulata* (Bradbury, 1967). The bacteria did not show differences in the bacteriologic characteristics and host range but *P. s.* pv. *angulata* did not produce tabtoxin and, on the contrary of *P. s.* pv. *tabaci*, it did not cause the typical chlorotic symptoms on tobacco. Therefore, *P. s.* pv. *angulata* was included in *P. s.* pv. *tabaci* (Bradbury, 1986).

ASCW symptoms on pea resemble the apical chlorosis of marigold incited by *P. s.* pv. *tagetis*, however, this pathogen is different from *P. s.* pv. *pisi* in bacteriological properties, host range and, furthermore, it belong to a different genomic group (Gardan *et al*., 1999). There still remains some possibility that *P. s.* pv. *pisi* ASCW strains and strains of *P. s.* pv. *tagetis* share similar mechanism in the pathogenetic mechanism. Studies on toxin production and its genetic background in *P. s.* pv. *pisi* ASCW strains are in progress.

REFERENCES

Arnold, D.L., Athey-Pollard, A., Gibbon, M.J., Taylor, J.D. and Vivian, A., 1996, Specific oligonucleotide primers for the identification of *Pseudomonas syringae* pv. *pisi* yield one of two possible DNA fragments by PCR amplification: evidence for phylogenetic divergence. *Physiol. Mol. Plant Pathol.* **49**: 233-245.

Ausubel, F.M., Brent, R., Kingston, R.E., Moore, D.D. Seidman, J.G., Smith, J.A. and Struhl, K., 1989, *Current protocols in molecular microbiology*. Greene Publishing Associate, New York.

Bradbury, J.F., 1967, *Pseudomonas tabaci*. C.M.I. Descriptions of Pathogenic Fungi and Bacteria. No. 129. CAB International, Wallingford, UK.

Bradbury, J.F. 1986, *Pseudomonas syringae* pv. *tabaci*. pp. 177-178. In *Guide to Plant Pathogenic Bacteria*. CAB International, Wallingford, UK.

Dye, D.W., 1962, The inadequacy of the usual determinative tests for the identification of *Xanthomonas* spp. *New Zealand J. Sci.* **5**: 393-416.

Gardan, L., Shafik, H., Belouin, R., Grimont, F., and Grimont, P.A.D., 1999, DNA relatedness among the pathovars of *Pseudomonas syringae* and description of *Pseudomonas tremae* sp. nov. and *Pseudomonas cannabina* sp. nov. (ex Sutic and Dowson 1959). *Int. J. Syst. Bacteriol.* **49**: 469-478.

Grondeau, C., Saunier, M., Poutier, F., and Samson, R., 1992, Evaluation of physiological and serological profiles of *Pseudomonas syringae* pv. *pisi* for pea blight identification. *Plant Pathol.* **41**: 495-505.

Hollaway, G.J., Gillings, M.R., and Fahy, P.C., 1997, Use of fatty acid profiles and repetitive element polymerase chain reaction (PCR) to assess the genetic diversity of *Pseudomonas syringae* pv *pisi* and *Pseudomonas syringe* pv. *syringae* isolated from field pea in Australia. *Australasian Plant Pathol.* **26**: 98-108.

Lawyer, A.S. and Chun, W., 2001, Bacterial Blight. pp. 22-23. In *Compendium of Pea Diseases and Pests*. 2nd. J. M. Kraft and F. L. Pfleger eds. APS press, Minnesota.

Lelliott, R.A., Billing, E., and Hayward, A.C., 1966, A determinative scheme for the fluorescent plant pathogenic pseudomoands. *J. Appl. Bacteriol.* **29**: 470-489.

Rademaker, J.L.W., Louws, F.J., and de Bruijn, F.J., 1998, Characterisation of the diversity of ecologically important microbes by rep-PCR genomic fingerprinting. *Mol. Micro. Ecol. Man*. 3. 4. 3: 1-27.

Society of American Bacteriologist, 1957, Manual of Microbiological Methods. McGraw-Hill Co., New York.

Zhao, Y., Damicone, J.P., Demezas, D.H., Rangaswamy, V., and Bender, C.L., 2000, Bacterial leaf spot of leafy crucifers in Oklahoma caused by *Pseudomonas syringae* pv. *maculicola*. *Plant Dis.* **84**: 1015-1020.

Taxonomy

Taxonomy of *Pseudomonas syringae* Pathovars: Classification and Nomenclature

Y. TAKIKAWA and Y. INOUE
Faculty of Agriculture, Shizuoka University, 836 Ohya, Shizuoka 422-8529, Japan.

Abstract: Recent developments in the taxonomy and nomenclature of *Pseudomonas syringae* pathovars are reviewed and a consensus classification within the species is outlined. Proposals for some conservative nomenclatural changes are discussed.

Key words: DNA-DNA hybridisation, sequencing, pathogenicity genes.

Author for correspondence: Y. Takikawa; e-mail: abytaki@agr.shizuoka.ac.jp.

1. INTRODUCTION

Since the amalgamation of many of fluorescent pseudomonad plant pathogens into a single species as *Pseudomonas syringae* in 1980, most of the former species names have been reduced to infrasubspecific rank (without taxonomic standing) as pathovars. Prior to the pathovar system, the heterogeneity of the fluorescent pseudomonads had been recognised in terms of their phytopathogenic, phenotypic and genetic characteristics. Until recently, those characters were not sufficiently different enough to discriminate subspecies or species within *P. syringae*. However, recent genetic data has allowed the differentiation of many of these pathovars. We present a brief historical review of the taxonomy of *P. syringae* and recommend several nomenclatural changes.

N.S. Iacobellis *et. al.* (eds.), *Pseudomonas syringae* and Related Pathogens, 645-652.

2. PHENOTYPIC ANALYSES

2.1 Phenotypic Properties

Many studies have reported the phenotypic properties of *Pseudomonas syringae* in detail (Lelliott *et al.*, 1966; Misaghi and Grogan, 1969; Sands *et al.*, 1970). Close resemblance has been observed among several pathovars, indicative of existence of several groups, though, many pathovars remained undefined. Recent reports also look like suggesting inability to distinguish genomic grouping only on phenotypic characters (Gardan *et al.*, 1999). Through these studies, the existence of several taxons whithin *P. syringae* has been confirmed. Determinative tests such as LOPAT, negative NO_3 reduction, utilization of trehalose, negative PHB-accumulation, fluorescence (not all) and so on are useful for differentiating some *P. syringae* pathovars from other *Pseudomonas* species. Recent data suggests species rank for some *P. syringae* members such as *P. amygdali*, *P. ficuserectae* and *P. meliae*.

2.2 Chemotaxonomic Analyses

Several new phenotypic analyses have been developed in the last 20 years including quinone analysis, cell wall component analysis and amine analysis. Such techniques are useful at the genus or higher level, but not effective enough for speciation of plant pathogenic bacteria such as *Pseudomonas syringae*. Fatty acid analysis and whole cell protein analysis have strong discriminating abilities. However, the technique is not applicable for species or subspecies level identification (Stead, 1997; Takikawa, unpublished observations).

2.3 Serological Analysis

Many attempts have been made to discriminate *Pseudomonas syringae* pathovars on serological methods. Most of those studies included only a limited number of pathovars. A recent report (Malandrin and Samson, 1999) on the serological analysis of flagellins of *P. syringae* pathovars showed that two serotypes, H1 and H2, could be distinguished. They correspond to the genetic groupings mentioned later.

3. GENETIC ANALYSES

3.1 DNA-DNA Hybridisation Studies

DNA homology studies on *Pseudomonas* including *P. syringae* were reported by Pecknold and Grogan (1973) and expanded by Hildebrand *et al.* (1982). The results suggested non-random variations in *P. syringae* and several species were proposed. In 1999, Gardan *et al.* (1999) reported a comprehensive study on several pathovars of *P. syringae* and proposed 9 genomospecies, including *P. virdiflava*.

Our analysis of *hrp* genes of *P. syringae* and related bacteria by DNA hybridisation revealed at least 4 major groups (Inoue and Takikawa, 1999, 2000). Our grouping coincided with the genomospecies differentiation of Gardan *et al.* (1999) and united some genomospecies into a *hrp*-homology group. This suggests that *hrp* genes are not likely to have moved horizontally among *P. syringae* pathovars, rather they have evolved together with whole genome. That they might be used as a marker for phylogenetic analysis of the organisms was confirmed by subsequent sequencing analyses described below.

3.2 Sequencing of Various Genes

Sequence data of various genes in *Pseudomonas syringae* pathovars have been rapidly accumulating. Most useful data for phylogenetic analysis was the work of Sawada *et al.* (1999) on two house-keeping genes, *rpoD* and *gyrB*, and two pathogenicity related genes, *hrpS* and *hrpL*. Their analysis suggests the presence of three groups. Our sequencing analysis on *hrpA* and *hrpZ* (Inoue and Takikawa, 2002) and hybridisation analysis (Inoue and Takikawa, 1999, 2000) also agree with the results of Sawada *et al.*(1999) pointing out 4 major groups. Our results from *hrpA* and *hrpZ* suggest that group I can split into two subgroups, however, other data do not support this subgrouping. Although considerable 16SrDNA data are available for *P. syringae* (Anzai *et al.*, 2000), the resolution is not high enough for speciation.

3.3 Distribution of Various Pathogenicity Related Genes

The distribution of several pathogenicity related genes was investigated in our laboratory using hybridisation and PCR (Table 1; Takikawa *et al.*, unpublished data). As expected, genes such as *iaaM* and *iaaH* and virulence genes such as tabtoxin, phaseolotoxin, coronatine and syringomycin were present in one or more pathovars and indicated their recent horizontal transfer. The *inaZ* gene or its homologue was detected in all *Pseudomonas syringae* pathovars except for group II pathovars (corresponding genomospecies 3 and 8). We suggest that the *inaZ* gene (and its neighboring region) might have been deleted from the genome of group II pathovars in the early stages of their evolution.

4. UNITED VIEW ON CLASSIFICATION

From these results, several of the *Pseudomonas syringae* pathovars can be classified into at least four major groups based on genetics. Among them, two groups (*hrpA,Z* group III and IV, corresponding genomospecies 1 and 4, respectively) are relatively well differentiated by phenotypic characteristics (Takikawa, unpublished data). One group (*hrpA,Z* group III or genomospecies 1) includes *P. s.* pv. *syringae* and its allies, many of which harbours genes for production of syringomycin or related toxins, having relatively wide and overlapped host ranges. The other group (*hrpA,Z* group IV or genomospecies 4) includes *P. s.* pv. *coronafaciens* and its allies, having monocotyledonous plants as hosts. These two groups might have evolved relatively recently.

In contrast, the other two groups are heterogeneous. *HrpA,Z* group I or genomospecies 2 includes many pathovars of various origins. Some are tightly grouped in terms of phenotypic characteristics (e.g. *P. s.* pv. *phaseolicola* and *P. s.* pv. *glycinea*), though, others have significant diversity. As Gardan *et al.* (1999) also noted, it is almost impossible to delineate this group by means of phenotypic characterisation. This group may be split into subgroups. However, subgroup division is not stable among the various grouping methods. Further DNA-DNA hybridisation studies may support creating new genomospecies. Most of *P. syringae* pathovars producing overgrowths (knots) belong in this group, including *P. s.* pv. *savastanoi*, *P. s.* pv. *myricae*, *P. s.* pv. *dendropanacis*, *P. meliae* and so on. Presence of *iaa* genes is not necessary for them. They might share some common genes for overgrowths formation. Further studies are awaited. Our preliminary PCR analysis suggested that another overgrowths former *P. s.* pv. *tremae* (genomospecies 5) may have close relationship with this group despite their distinct phenotype.

As for *hrpA, Z* group II or genomospecies 3 or 8, a situation similar to that of group I was observed. When comparing our results with those of Sawada *et al.* (1999), topologies of phylogenetic tree are slightly different even among the results based on the genes in the same *hrp* gene cluster. Only the above mentioned four major groups can consistently be identified. For example, *P.*

s. pv. *actinidiae* and *P. s.* pv. *theae* are suggested to be grouped together with *P. s.* pv. *avellanae* as genomospecies 8 (Gardan *et al.*, 1999, Scortichini *et al.*, 2002). They are positioned together in group II of our division. Their independence from other group II pathovars such as *P. s.* pv. *tomato* and *P. s.* pv. *maculicola* is not consistently supported by the sequence analyses. Therefore, we believe more data on more strains is necessary before making final decision for retaining these genomospecies species rank.

Higher grouping of these four major groups is not confirmed. It is likely that group I and III, and group II and IV are more closely related from the sequence data. This is contrast to the flagellin serology data (Table 1).

Table 1. Comparison of various groupings of *Pseudomonas syringae* pathovars.

Pseudomonas syringae pathovars	*hrp* Hybrid.	*hrp A, Z* sequences	groupI spec. PCR	*rpoD, gyrB, hrpS,L*	Genomo-species	Distribution of pathogenicity genes						Flagella serotype
						iaaM, H	*inaZ*	*tab*	*cor*	*phas*	*syr*	
References	a)	b)	c)	d)	e)	c)						f)
phaseolicola	I	Ia	+	3	2	-	+	-	-	+	-	H2
glycinea	I	Ia	+	3	2	+		-	+	-	-	H2
tabaci	I	Ia	+	3	2	-	+	+	-	-	-	H2
eriobotryae	I	Ia	+	3	2	-	+	-	-	-	-	
myricae	I	Ia	+	3	2	+	+	-	-	-	-	
dendropanacis	I	Ia	+		2	-	+	-	-	-	-	
morsprunorum	U	Ia	+	3	2	-	+	-	+	-	-	H2
? (kiwi)	I	Ia	+			-	+	-	-	-	+	
photiniae	I		+	3	2	-	+	-	-	-	-	
lachrymans	I	Ib	+	3	2	-	+	-	-	-	-	H2
mori	I	Ib	+	3	2	-	+	-	-	-	-	H2
sesami	I	Ib	+		2		+	-	-	-	-	
P. ficuserectae	I	Ib	+		2	+					-	
maculicola (IV)*	U	Ib	+			-	+	-	+	-	-	
P. meliae			+		2	-						
tremae			+		5	-						
maculicola(I,II, III)*	II	II		1	3	+	-	-	+	-	-	H2
tomato	II	II		1	3	-	-	-	+	-	-	H2
theae	II	II		1	8	-	-	-	-	-	-	H2
delphinii	II	II			3	-	-	-	-	-	-	
? (magnolia)	II	II				-	-	-	-	-	-	
actinidiae	II	II		1	8 ?	-	-	-	-	+	-	
spinaciae	II					-	+	-	+	-	-	
syringae	III	III		2	1	+	+	-	-	-	+	H1
lapsa	III	III			1	-	+	-	-	-	+	
japonica	III	III		2	1	-	+	-	-	-	+	
aptata	III	III		2	1		+	-	-	-	+	H1
aceris	III	III		2	1		+	-	-	-	+	
pisi	III	III		2	1	-	+	-	-	-	+/-	H1
coronafaciens	IV	IV			4		+	+	-	-	-	H1
striafaciens	IV	IV			4	-	+	-	+	-	-	
? (onion)	IV	IV				-	+	-	+	-	-	
atropurprea	IV				4		+	-	+	-	-	
oryzae	IV	IV			4		+	+	-	-	-	H1
porri	IV				4	-	+	-	+	-	-	H1

References: a) Inoue and Takikawa (1999, 2000); b) Inoue and Takikawa (2002); c) Takikawa *et al.* (unpublished); d) Sawada *et al.* (1999); e) Gardan *et al.* (1999); f) Malandrin and Samson (1999).

*: grouping of pv. *maculicola* by Takikawa *et al.* (1992).

5. NOMENCLATURAL PROBLEMS

As mentioned above, pathovars of *Pseudomonas syringae* can be divided into four major lineage and some can be subdivided into smaller subgroups, some of which correspond to genomospecies designations. This classification, which is taking phylogenetic view points into account, will not radically change for the future. However, naming of these groupings requires cautious treatment.

One problem is the lack of distinguishing phenotypic characters. Not to mention the proposal of Wayne *et al.* (1987), without appropriate determinative scheme, many people cannot identify the species. This defect may be overcome in the future by use of some genetic markers as distinguishing criteria. These markers should not be selected arbitrarily. To ensure this, a new methods such as multilocus sequence typing (Stackebrandt *et al.*, 2002) may be useful.

Another problem is the selection of species names for each groups. As Gardan *et al.* (1999) pointed, genomospecies 2 (our group I) include several valid species names, *P. amygdali* as the oldest. Acceptance of this species name as the representative may lead to creation of an atypical type strain which we encountered in the case of *Xanthomonas axonopodis* (Young *et al.*, 2001).

Under such circumstances, we believe that the pathovars should be retained under *P. syringae* and that rather some related species such as *P. amygdali* should be once included in *P. syringae* as pathovars, until sufficient accumulation of data can clearly delineate the grouping within *P. syringae* pathovars.

It certainly will be necessary to investigate more field strains to understand the diversity of each pathovar. For example, we have demonstrated that some radish isolates of *P. s.* pv. *maculicola* (*maculicola* group IV by Takikawa *et al.*, 1992) are more closely related to group I or genomospecies 2 than to group II or genomospecies 3. Several pathovars are also heterogeneous, indicating possible subdivisions. Some pathotype strains are known as not being typical because they may have been misidentified or contaminated. Therefore, future works should be concentrated on clarifying heterogeneity of each pathovars in terms of polyphasic approaches. Also, pathogenicity test must be conducted when receiving cultures from collections to be sure it is the correct organism.

6. CONCLUDING REMARKS

Progress on the taxonomy of *Pseudomonas* will enable us to understanding diversity of plant pathogenic bacteria. Despite detailed studies, naming of each taxa sometimes cause serious trouble. *P. syringae* group of bacteria is deduced to have evolved rapidly and diversely comparing to other species in the genus *Pseudomonas*. We believe that keeping the unity of *P. syringae* group as a single species still has merits on understanding phytopathogenic nature of this group. Degree of independence of the genomospecies in *P. syringae* group is not similar to that of *P. cichorii* or *P. aeruginosa*. It will be better to handle nomenclature problems in a rather conservative way than to keep changing the names.

REFERENCES

Anzai, Y., Kim, H., Park, J.-Y., Wakabayashi, H., and Oyaizu, H., 2000, Phylogenetic affiliation of the pseudomonads based on 16S rRNA sequence. *Int. J. Syst. Evol. Microbiol.* **50**: 1563-1589.

Gardan, L., Shafik, H., Belouin, S., Broch, R., Grimont, F., and Grimont, P.A.D., 1999, DNA relatedness among the pathovars of *Pseudomonas syringae* and description of *Pseudomonas tremae* sp. nov. and *Pseudomonas cannabina* sp. nov. (ex Sutic and Dowson 1959). *Int. J. Syst. Bacteriol.* **49**: 469-478.

Hildebrand, D.C., Schroth, M.N., and Huisman, O.C., 1982, The DNA homology matrix and non-random variation concepts as the basis for the taxonomic treatment of plant pathogenic and other bacteria. *Ann. Rev. Phytopathol.* **20**: 235-256.

Inoue, Y., and Takikawa, Y., 2000, *Pseudomonas syringae* strains are classified into five groups by comparing DNA homology at the *hrp* neighboring regions. *J. Gen. Plant Pathol.* **66**: 238-241.

Inoue, Y., and Takikawa, Y., 1999, Grouping of *Pseudomonas syringae* strains by comparing DNA homology at the *hrp* gene cluster and its neighboring regions. *Ann. Phytopathol. Soc. Japan* **65**: 32-41.

Inoue, Y., and Takikawa, Y., 2002, Phylogenic analysis of DNA sequences around the *hrpL* and *hrpZ* regions of *Pseudomonas syringae* group bacteria. In *6th International Conference of* Pseudomonas syringae *Pathovars and Related Pathogens*, Maratea, Italy, Book of Abstracts, p. 101.

Lelliott, R.A., Billing, E., and Hayward, A.C., 1966, A determinative scheme for the fluorescent plant pathogenic pseudomonads. *J. Appl. Bacteriol.* **29**: 470-489.

Malandrin, L., and Samson, R., 1999, Serological and molecular size characterisation of flagellins of *Pseudomonas syringae* pathovars and related bacteria. *Syst. Appl. Microbiol.* **22**: 534-545.

Misaghi, I., and Grogan, R.G., 1969, Nutritional and biochemical comparisons of plant-pathogenic and saprophytic fluorescent pseudomonads. *Phytopathology* **59**: 1436-1450.

Pecknold, P.C., and Grogan, R.G., 1973, Deoxyribonucleic acid homology groups among phytopathogenic *Pseudomonas* species. *Inter. J. Syst. Bacteriol.* **23**: 111-121.

Sands, D.C., Schroth, M.N. and Hildebrand, D.C., 1970, Taxonomy of phytopathogenic pseudomonads. *J. Bacteriol.* **101**: 9-23.

Sawada, H., Suzuki, F., Matsuda, I., and Saitou, N., 1999, Phylogenetic analysis of *Pseudomonas syringae* pathovars suggests the horizontal gene transfer of *argK* and the evolutionary stability of *hrp* gene cluster. *J. Mol. Evol.* **49**: 627-644.

Scortichini, M., Marchesi, U., and Di Prospero, P., 2002, Genetic relatedness among *Pseudomonas avellanae*, *P. syringae* pv. *theae* and *P.s.* pv. *actinidiae*, and their identification. *Euro. J. Plant Pathol.* **108**: 269-278.

Stackebrandt, E., Frederiksen, W., Garrity, G.M., Grimont, P.A.D., Kampfer, P., Maiden, M. C. J., Nesme, X., Rossello-Mora, R., Swings, J., Truper, H., Vauterin, L., Ward, A.C. and Whitman, W.B., 2002, Report of the ad hoc committee for the re-evaluation of the species definition in bacteriology. *Int. J. Syst. Evol. Microbiol.* **52**: 1043-1047.

Stead, D.E., Hennessy, J., Elphinstone, J.G., and Judith, K., 1997, Modern methods for classification of plant pathogenic bacteria including *Pseudomonas syringae*. In Pseudomonas syringae *pathovars and related Pathogens* (Rudolph, K., Burr, T.J., Mansfield, J.W., Stead, D., Vivian, A. and von Kietzell, J. eds.) Kluyver Academic, Dordrecht. pp. 427-434.

Takikawa, Y., Nishiyama, N., Ohba, K., Tsuyumu, S., and Goto, M., 1992, Synonymy of *Pseudomonas syringae* pv. *maculicola* and *Pseudomonas syringae* pv. *tomato*. In *Proceedings of 8th International Conference of Plant Pathogenic Bacteria*, INRA, Paris. pp. 199-204.

Wayne, L.G., Brenner, D.J., Colwel, R.R., Grimont, P.A.D., Kandler, O., Krichevsky, M.I., Moore, L.H., Moore, W.E.C., Murray, R.G.E., Stackebrandt, E., Starr, M.P. and Truper, H.G., 1987, Repot of ad hoc committee on reconciliation of approaches to bacterial systematics. *Inter. J. Syst. Bacteriol.* **37**: 463-464.

Young, J.M., Bull, C.T., De Boer, S.H., Firrao, G., Gardan, L., Saddler, G.E., Stead, D.E. and Takikawa, Y., 2001, Classification, nomenclature, and plant pathogenic bacteria - a clarification. *Phytopathology* **91**: 617-620.

Phytopathogenic *"Pseudomonas"* Species: a Taxonomic Overview

A. WILLEMS and P. VANDAMME
Laboratory of Microbiology, Dept. Biochemistry, Physiology and Microbiology, Faculty of Sciences, Ghent University, Ledeganckstraat 35, Gent, Belgium.

Abstract: Over the past two decades, the original genus *Pseudomonas* has been thoroughly revised and reorganised following the discovery of its heterogeneity on the basis of DNA:rRNA hybridisation. Only the nucleus of species around *P. fluorescens* and *P. aeruginosa* (rRNA group I) in the gamma-Proteobacteria has been retained in *Pseudomonas sensu stricto*. Other species have been reallocated to other genera in the alfa-, beta- and gamma-Proteobacteria.
This taxonomic overview will cover the phytopathogenic ex-*Pseudomonas* species of rRNA groups II and III. These organisms have been reallocated to *Burkholderia*, *Ralstonia*, *Herbaspirillum* and *Acidovorax*, all in the beta-Proteobacteria. They include: *P. caryophylli*, *P. cepacia*, *P. gladioli*, *P. glumae*, *P. plantarii*, *P. andropogonis*, and *P. stizolobii*, all now in *Burkholderia*; *P. solanacearum*, now in *Ralstonia*; *P. rubrisubalbicans*, now in *Herbaspirillum*; and *P. avenae*, *P. rubrilineans*, *P. cattleyae*, *P. pseudoalcaligenes* subsp. *citrulli* and subsp. *konjaci*, now in *Acidovorax*. *Acidovorax anthurii* has been described more recently. In addition to these phytopathogenic species, these four genera contain non-phytopathogenic species as well as clinical isolates. Recent findings reported in literature are summarised and discussed.

Key words: *Pseudomonas*, *Burkholderia*, *Ralstonia*, *Acidovorax*.

Author for correspondence: A. Willems; e-mail: Anne.Willems@UGent.be.

1. INTRODUCTION

Pseudomonas was created in 1894 for rod-shaped, Gram negative, aerobic bacteria that were motile by means of polar flagella and had a strictly aerobic

N.S. Iacobellis *et. al.* (eds.), *Pseudomonas syringae* and Related Pathogens, 653-665.

respiratory metabolism. This broad description, like many genus descriptions at the time, allowed a rather wide variety of species to be placed in this genus and the number of species therefore quickly rose. The eight edition of Bergey's Manual of Determinative Bacteriology listed an impressive 235 *Pseudomonas* species (Doudoroff and Palleroni, 1974). Some of these were incompletely described or had no representative strains and some were synonyms. Consequently, in the Approved lists of bacterial names (Skerman *et al.*, 1980) the number of *Pseudomonas* species was drastically reduced to 87, still a respectable size for a genus. Meanwhile, the broad genus definition had not changed and therefore the number of species started rising again steadily. In Bergey's Manual of Systematic Bacteriology, there were 92 *Pseudomonas* species (Palleroni, 1984). However, it was also made clear, mostly on the basis of DNA:rRNA hybridisation, that the genus in fact consisted of 5 large groups that were only remotely related to each other (Palleroni, 1984; De Vos and De Ley, 1983; De Vos *et al.*, 1985).

Pseudomonas rRNA group I contained the type species *P. aeruginosa* and was therefore regarded as the true genus *Pseudomonas*. It also contained *P. syringae* and *P. fluorescens*. Group II included *P. cepacia, P. gladioli* and *P. solanacearum*, species that would later be placed in *Burkholderia* and *Ralstonia* (Yabuuchi *et al.*, 1992; 1995). Group III, the *P. acidovorans* group, included some phytopathogenic species: *P. avenae, P. rubrilineans, P. konjaci, P. pseudoalcaligenes* subsp. *citrulli* and subsp. *cattleyae*. These have since been transferred to *Acidovorax* (Willems *et al.*, 1992). Group IV contained *P. diminuta* and *P. vesicularis* [now both in *Brevundimonas* (Segers *et al.*, 1994)] and group V included *P. maltophilia* [now *Stenotrophomonas maltophilia* (Palleroni and Bradbury, 1993)] together with the plant pathogenic genus *Xanthomonas*.

These groups were later confirmed by 16S rDNA sequencing, which again demonstrated that they were in fact only remotely related and should not be regarded as a single genus (reviewed by Kersters *et al.*, 1996; Anzai *et al.*, 2000). Over several years, the *Pseudomonas* species of groups II to V have therefore been removed from *Pseudomonas* and placed in other genera, some existing and others newly created. As a result the number of true *Pseudomonas* species had dropped to 78 by the year 2000 (-34) and now stands at 84, a slow rise as new species are being discovered and named.

2. POLYPHASIC TAXONOMY

The transfer of many of the old *Pseudomonas* species to new or other genera was mostly the result of polyphasic taxonomic studies. The polyphasic approach (Vandamme *et al.*, 1996) is now the most widely used method to characterise and classify new bacteria. In this approach different types of data (genetic, phenotypic, chemotaxonomic, phylogenetic) are compiled to produce a comprehensive picture of the relationships of the bacteria and to propose a suitable consensus classification (Vandamme *et al.*, 1996). This can be seen as a composite classification, because not all the techniques provide information at the same taxonomic level (Fig. 1). Each method should be used to draw conclusions at the taxonomic level that it provides information at (in a particular groups of organisms). The comparison of 16S rDNA sequences has proved particularly influential in bacterial taxonomy because for the first time it permitted the comparison of any bacterium with any other bacterium.

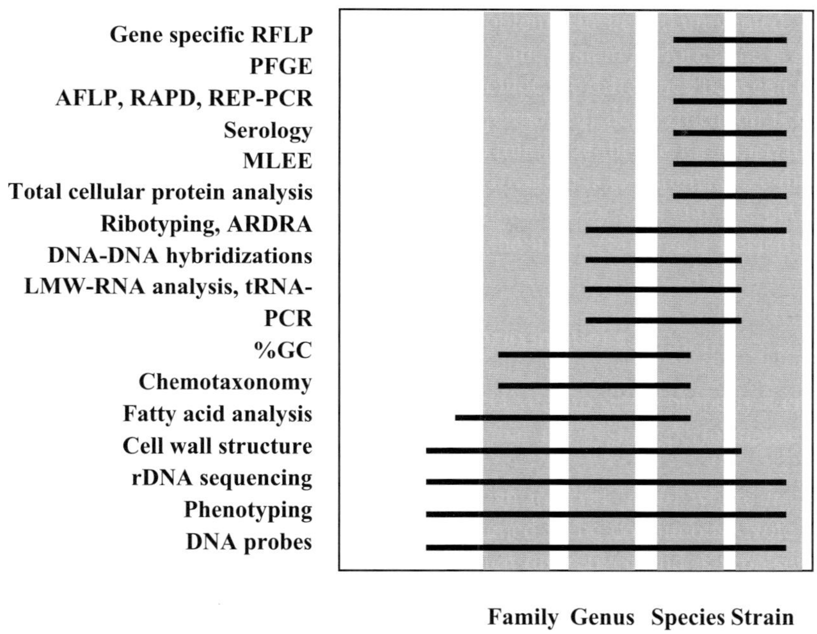

Figure 1. Taxonomic resolution of some techniques regularly used in polyphasic studies. Abbreviations: RFLP = restriction fragment length polymorphism; PFGE = pulsed field gel electrophoresis; AFLP = amplified fragment length polymorphism; RAPD = random amplified polymorphic DNA; REP-PCR = repetitive element PCR; MLEE = multilocus enzyme electrophoresis; ARDRA = amplified ribosomal DNA restriction analysis; LMW-RNA = low molecular weight RNA.

3. OVERVIEW OF THE PHYTOPATHOGENIC "PSEUDOMONADS"

In the 16S rDNA tree of the *Proteobacteria*, the old *Pseudomonas* groups are dispersed over the alpha-, beta- and gamma-*Proteobacteria* (Kersters *et al.*, 1996). The phytopathogenic "*Pseudomonas*" species outside the genus *Pseudomonas sensu stricto* (i.e. outside rRNA group I sensu Palleroni, 1984) are in four groups that all belong to the beta-*Proteobacteria* (Kersters *et al.*, 1996).

3.1 *Burkholderia* and *Ralstonia*

Burkholderia was created in 1992 for 7 species of *Pseudomonas* group II on the basis of a polyphasic study using 16S rRNA sequences, DNA-DNA homology values, cellular fatty acid contents and phenotypic characters. It contained *B. solanacearum, B. pickettii, B. caryophylli, B. pseudomallei, B. cepacia, B. gladioli* and *B. mallei* (Yabuuchi *et al.*, 1992). Later, in 1995, it became clear that the group was still quite diverse and a separate genus, *Ralstonia*, was created for *R. solanacearum, R. pickettii* and *R. eutropha* (originally *Alcaligenes eutropha*) (Yabuuchi *et al.*, 1995). Both genera have been steadily growing as new diversity is discovered and described. At present *Burkholderia* contains 24 valid species and *Ralstonia* 11. Among these, quite a few are phytopathogenic, others are associated with plants or the rhizospere, some are biocontrol strains and others still are from clinical sources (Table 1).

Of the plant pathogenic species, *B. caryophylli* is responsible for bacterial wilt and stem cracking on *Dianthus* (Palleroni, 1984); *B. cepacia* genomovars I and III cause soft rot on onions; *B. gladioli* pv. *alliicola* causes root rot in onions and pv. *gladioli* causes rot of stem bases and corms in gladiolus, crocus and other *Iridaceae*; *B. gladioli* is also responsible for sheath necrosis on rice (Palleroni, 1984; Cottyn *et al.*, 2001); *B. glumae* is responsible for bacterial grain rot and sheath necrosis of rice; *B. plantarii* causes bacterial blight of rice seedlings (Azegami *et al.*, 1987); *B. andropogonis* causes leaf stripe on grasses, leaf spot on legumes and it is also a pathogen of carnations (Coenye *et al.* 2001a); *R. solanacearum* causes wilting and brown rot on many plants [e.g. *Solanaceae*, banana, peanut, pelargonium (Palleroni, 1984)].

As an example, we will discuss in some more detail *B. cepacia.* In this species several genomovars have been described. *B. cepacia* is ubiquitous in diverse environments, including soil, rhizosphere, plant roots and shoots, water, the hospital environment and it is implicated in human infections. It has long been known as a plant-associated / plant pathogenic organism. It is also a soil organism with antifungal and bioremediation applications. Since the early 80s *B. cepacia* is also reported in infections in cystic fibrosis patients (Isles *et al.*, 1984).

Table 1. Chronological overview of validly published *Burkholderia* and *Ralstonia* species.

Species[1]	Year	Original name[1]	Current name[1] if changed (year of transfer)	*B. cepacia* genomovar	Plant pathogenic[2]	Plant associated[2]	Clinical[2]	Environmental[2]	Biocontrol[2]
B. caryophylli	1993	*P. caryophylli*			pp				
B. cepacia	1993	*P. cepacia*		I	pp		c	e	
				III	pp	pa	c	e	bc
				VI		pa	c		
B. gladioli	1993	*P. gladioli*			pp		c	e	bc
B. mallei	1993	*P. mallei*					c		
B. pickettii	1993	*P. pickettii*	*R. pickettii* (1996)				c	e	
B. pseudomallei	1993	*P. pseudomallei*					c	e	
B. solanacearum	1993	*P. solanacearum*	*R. solanacearum* (1996)		pp			e	
B. glumae	1994	*P. glumae*			pp				
B. plantarii	1994	*P. plantarii*			pp			e	
B. vandii	1994	*sp. nov.*	*B. plantarii* (1999)			pa		e	
B. andropogonis	1995	*P. andropogonis, P. stizolobii, P. woodsii*			pp				
B. cocovenenans	1995	*P. cocovenenans*	*B. gladioli* (1999)					e	
B. vietnamiensis	1995	*sp. nov.*		V	pp	pa	c	e	bc
B. glathei	1997	*P. glathei*						e	
B. multivorans	1997	*P. cepacia*		II			c	e	
B. pyrrocinia	1997	*P. pyrrocinia*		IX			c	e	
B. graminis	1998	*sp. nov.*				pa		e	
B. norimbergenis	1998	*sp. nov.*	*Pan. norimbergensis* (2000)					e	
B. phenazinium	1998	*P. phenazinium*						e	
B. thailandensis	1998	*sp. nov.*						e	
B. caribensis	1999	*sp. nov.*						e	
B. kururiensis	2000	*sp. nov.*						e	
B. stabilis	2000	*sp. nov.*		IV		pa	c	e	bc
B. ubonensis	2000	*sp. nov.*					c	e	
B. ambifaria	2001	*sp. nov.*		VII		pa	c	e	bc
B. caledonica	2001	*sp. nov.*						e	
B. fungorum	2001	*sp. nov.*					c	e	
B. sacchari	2001	*sp. nov.*						e	
B. anthina	2002	*sp. nov.*		VIII		pa	c	e	
Ralstonia pickettii	1996	*P. pickettii*					c	e	
R. solanacearum	1996	*P. solanacearum*			pp				
R. eutropha	1996	*A. eutrophus*						e	
R. basilensis	1999	*sp. nov.*						e	
R. gilardi	1999	*sp. nov.*					c	e	
R. paucula	1999	CDC group IVc-2					c	e	
R. oxalatica	2000	*P. oxalaticus*						e	
R. campinensis	2001	*sp. nov.*						e	
R. mannitolilytica	2001	*sp. nov.*					c	e	
R. metallidurans	2001	*sp. nov.*						e	
R. taiwanensis	2001	*sp. nov.*				pa	c		

[1]*A. = Alcaligenes; B. =Burkholderia; P. = Pseudomonas; Pan. = Pandoraea, R. = Ralstonia.*
[2]bc = some strains used for biocontrol; c = clinical strains; e = environmental strains; pa = plant associated strains; pp = plant pathogenic strains.

Since 1992 a polyphasic taxonomic study of over 4000 strains has been ongoing in our lab. This international collaborative study includes strains from various geographic origins, from various ecological niches (environmental, plant pathogenic, hospital, clinical and cystic fibrosis strains) and reference strains from *Burkholderia, Ralstonia* and *P. aeruginosa*. Techniques used include: SDS-PAGE protein analysis, fatty acid analysis, *recA* RFLP, AFLP and tRNA PCR for screening and DNA-DNA and DNA-rRNA hybridisations, 16S rDNA and *recA* sequence analyses and biochemical characterisation for further characterisation. With this polyphasic approach 9 genomic species were defined with a high degree of 16S rDNA (98-100%) and *recA* (94-95%) sequence similarity and moderate levels of DNA-DNA hybridisation (30-50%). These groups are referred to as genomovars. Genomovar I contains the type strain of *B. cepacia* and it is therefore retained as *B. cepacia*. The other genomovars are gradually being named as phenotypic and other data allow their differentiation.

Table 2. Overview of *Burkholderia cepacia* genomovars

Genomovar	Name	Reference
I	*Burkholderia cepacia*	Vandamme *et al.,* 1997
II	*B. multivorans*	Vandamme *et al.,* 1997
III		Vandamme *et al.,* 1997
IV	*B. stabilis*	Vandamme *et al.,* 2000
V	*B. vietnamiensis*	Gillis *et al.,* 1995
VI		Coenye *et al.*, 2001c
VII	*B. ambifaria*	Coenye *et al.,* 2001b
VIII	*B. anthina*	Vandamme *et al.,* 2002
IX	*B. pyrrocinia*	Vandamme *et al.,* 2002

The occurrence together of plant pathogenic, or plant-associated, clinical and soil strains is found in several *Burkholderia* species (Table 1). Among the soil and rhizosphere strains, some are used as biocontrol organisms that provide protection for plants against various fungi. For example, several *B. gladioli* strains provide protection against grey mould in lilies (Chiou and Wu, 2001); *B. cepacia* complex strains protect against carrot black rot (Chen and Wu, 1999), against *Fusarium* on bananas (Pan *et al.*, 1997) and maize (Bevivino *et al.*, 1998) and against oomycetes on peas (Heungens and Parke, 2001). The possible use of biocontrol strains from taxa that also harbour opportunistic or dangerous clinical strains underlines the absolute importance of correct identification and careful assessment of potential benefits and dangers (Parke and Gurian-Sherman, 2001). At present a combination of phenotypic and molecular tests such as *recA* PCR and 16S

rRNA RFLP is recommended to differentiate genomovars within the *B. cepacia* complex (Henry *et al.*, 2001).
The fact that such diverse ecological habitats are occupied by species of a single genus and regularly strains of a single species, indicates that there are probably common mechanisms that allow these bacteria to penetrate eukaryote defences. The enormous versatility of *Burkholderia* species has been linked with their complex genomes: strains can have up to 3 or 4 chromosomes and harbour several plasmids. They use a large number of substrates, produce many antibiotics and are resistant to several antibiotics and noxious substances. The total genome sequences of four Burkholderia species have been or are being determined and several additional sequencing projects are planned. This sequence information is expected to provide important new insights in the biology of these organisms.

3.2 *Herbaspirillum*

The genus *Herbaspirillum* contains one former *Pseudomonas* species, *P. rubrisubalbicans*, which was renamed *H. rubrisubalbicans* in 1996 (Baldani *et al.*, 1996). The genus *Herbaspirillum* had been created 10 years earlier for diazotrophic, endophytic bacteria associated with the roots, stems and leaves of *Poaceae* (Baldani *et al.*, 1996). There was just one species originally, *H. seropedicae* and this was not pathogenic (Baldani *et al.*, 1986). *H. rubrisubalbicans* is the causal agent of mottled stripe of sugarcane. Together with the transfer of *P. rubrilineans*, a third species was described but not named. It consisted of strains from clinical sources that were previously known as EF group 1. Thus, again we encounter the close relatedness of clinical and plant pathogenic or plant associated strains. Recently then, a fourth species was described, *H. frisingense*, for endophytic strains from roots, stems and leaves of C4-grasses (Kirchhof *et al.*, 2001).

3.3 *Acidovorax*

Acidovorax was created in 1990 for 2 *Pseudomonas* species from rRNA group III (the *P. acidovorans* complex): *P. facilis* was renamed *A. facilis* and *P. delafieldii* became *A. delafieldii*. In addition a third species was created, *A. temperans* (Willems *et al.*, 1990). None of these first three species is phytopathogenic. In a subsequent taxonomic revision (Willems *et al.*, 1992), several phytopathogenic species were added: *A. avenae* subsp. *avenae* (formerly *P. avenae* and *P. rubrilineans*), *A. avenae* subsp. *citruli* (formerly *P. avenae* subsp. *citrulli* and originally *P. pseudoalcaligenes* subsp. *citrulli*), *A. avenae* subsp. *cattleyae* (formerly *P. cattleyae*) and *A. konjaci* (formerly *P. avenae* subsp. *konjaci* and originally *P. pseudoalcaligenes* subsp. *konjaci*). In a polyphasic study using DNA:rRNA hybridisation, SDS-PAGE protein analysis, phenotypic analysis and DNA:DNA hybridisation, all these species were shown to be related (Willems *et al.*, 1992). More recently, two more species have been described: *A. defluvii* (Schulze *et al.*, 1999) from soil and *A. anthurii* from leaf spots of infected anthuriums (Gardan *et al.*, 2000). Also in this genus, plant pathogenic strains and clinical strains are closely related, although here they are not in the same species (Table 3).

Table 3. Acidovorax species.

Name	Plant pathogenic	Environmental	Clinical	Source
Acidovorax avenae subsp. *avenae*	+			Leaf stripe and top rot in grasses
A. avenae subsp. *cattleyae*	+			Leaf spot on *Cattleya, Phalaenopsis*
A. avenae subsp. *citrulli*[a]	+			Leaf spot on *Citrullus*
A. konjaci	+			Leaf spot and blight on konjac
A. facilis		+		Soil
A. delafieldii		+	+	Soil and clinical samples
A. temperans		+	+	Soil and clinical samples
A. defluvii		+		Activated sludge
A. anthurii	+			Leaf spot on *Anthurium*

[a] In *Acidovorax avenae* subsp. *citrulli* two subgroups were recently recognised by means of pulsed field gel electrophoresis of *Spe*I-digested DNA and fatty acid analysis. Group 1 includes strains from cantaloupe and pumpkin, plus the type strain and group 2 includes strains that cause watermelon fruit blotch (Walcott *et al.*, 2000).

4. CONCLUDING REMARKS

A point of criticism raised against some of the taxonomic changes is that polyphasic classification relies heavily on phylogenetic data (Young, 2000). In the past these were in particular 16S rDNA data. More recently, also other genes have been considered. However, phylogenetically close organisms are not necessarily phenotypically similar (Young, 2001; examples in Willems *et al.*, 1991 and Yurkov *et al.*, 1997). Therefore, phylogenetically delineated groups may be less coherent phenotypically than groups described mainly on phenotypic criteria. Hence, taxa delineated along phylogenetic lines may be more difficult to identify using phenotypic tests (Young, 2001). This is indeed a serious problem that has arisen from technological evolution in taxonomic research. It may seem regrettable, but on the other hand, it is almost unavoidable now that we are beginning – thanks to this ongoing technological revolution – to explore the full extent of bacterial diversity. Clearly, we are only at the beginning of this process.

At present about 5,200 species of *Bacteria* and *Archaea* have been described. Molecular ecological studies of different ecosystems demonstrate that this represents probably just a minor fraction of the actual bacterial diversity (Amann *et al.*, 1995). Predictions of up to 10^9 bacterial species have been made (Dykhuizen, 1998). It is most unlikely that we can develop a practical phenotypic system to classify and identify such diversity.

For the phytopathogenic bacteria that are discussed in this chapter, a picture of growing diversity emerges. The four genera also have a rising number of species (Fig. 2). Not all of these are phytopathogenic, but the identification of the phytopathogens should not be treated separately from the non-phytopathogenic species because that would make it impossible to recognise new phytopathogenic species in these genera. Therefore, also for the phytopathogenic species, the phenotype alone is not sufficient for identification or classification.

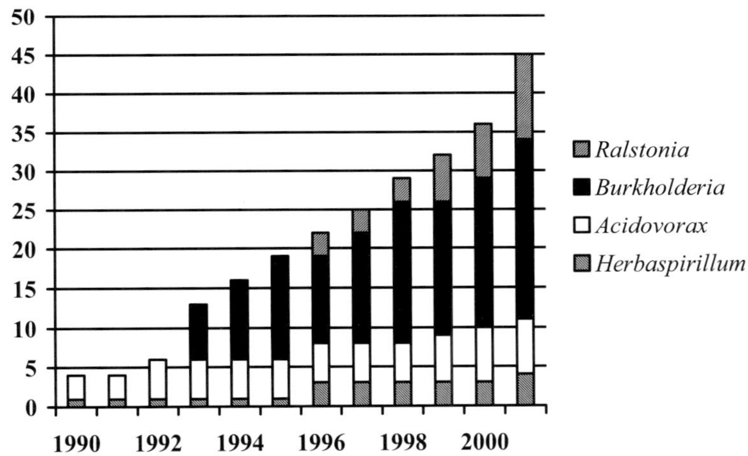

Figure 2. Evolution of the number of species.

PCR-based genotypic fingerprinting techniques such as REP PCR, tRNA PCR, specific PCR and PCR-RFLP tests, as well as multi-locus sequence typing (MLST) are increasingly important. Over the last 20 years, the genus *Pseudomonas* has been thoroughly revised. The framework for this revision was provided by ribosomal RNA sequence information, supplemented by additional data from other techniques. We have come from a largely phenotypic species concept to a polyphasic species concept, relying on different genotypic, chemotaxonomic and phenotypic characters. As total genome sequences accumulate, more extensive genetic comparisons will become possible. Especially in the genera discussed here, that contain phytopathogenic species or strains that are closely related to non-phytopathogenic species or strains, such whole genome comparisons may prove an important source for valuable insights into the similarities between ecologically different bacteria that are phylogenetically related.

ACKNOWLEDGEMENTS

The authors are grateful to the Fund for Scientific Research – Flanders for a Postdoctoral Fellowship (AW) and financial support (PV).

REFERENCES

Amann, R.I., Ludwig, W., and Schleifer, K.-H., 1995, Phylogenetic identification and in situ detection of individual microbial cells without cultivation. *Microbiol. Rev.* **59**: 143-169.

Anzai, Y., Kim, H., Park, J.-Y., Wakabayashi, H., and Oyaizu, H., 2000, Phylogenetic affiliation of the pseudomonads based on 16S rRNA sequences. *Int. J. Syst. Evol. Microbiol.* **50**: 1563-1589.

Azegami, K., Nishiyama, K., Watanabe, Y., Kadota, I., Ohuchi, A., and Fukazawa, C., 1987, *Pseudomonas plantarii* sp. nov., the causal agent of rice seedling blight. *Int. J. Syst. Bacteriol.* **37**: 144-152.

Baldani, J.I., Baldani, V.L.D., Olivares, L.F., Kirchhof, G., Hartmann, A., Pot, B., Hoste, B., Falsen, E., Kersters, K., Gillis, M., and Döbereiner, J., 1996, Emended description of *Herbaspirillum*, inclusion of [*Pseudomonas*] *rubrisubalbicans*, a mild plant pathogen as *Herbaspirillum rubrisubalbicans* comb.nov., and classification of a group of clinicl isolates (EF group 1) as *Herbaspirillum* species 3. *Int. J. Syst. Bacteriol.* **46**: 802-810.

Baldani, J.I., Baldani, V.L.D., Seldin, L., and Döbereiner, J., 1986, Characterisation of *Herbaspirillum seropedicae* gen. nov., sp. nov., a root-associated nitrogen-fixing bacterium. *Int. J. Syst. Bacteriol.* **36**: 86-93.

Bevivino, A., Sarrocco, S., Dalmastri, C., Tabacchioni, S., Cantale, C., and Chiarini, L., 1998, Characterization of a free-living maize-rhizosphere population of *Burkholderia cepacia*: effect of seed treatment on disease suppression and growth promotion of maize. *FEMS Microbiol. Ecol.* **27**: 225-237.

Chen, T.W., and Wu, W.S., 1999, Biological control of carrot black rot. *J. Phytopathol.* **147**: 99-104.

Chiou, A.L., and Wu, W.S., 2001, Isolation, identification and evaluation of bacterial antagonists against *Botrytis elliptica* on lily. *J. Phytopathol.* **149**: 319-324.

Coenye, T., Laevens, S., Gillis, M., and Vandamme, P., 2001a, Genotypic and chemotaxonomic evidence for the reclassification of *Pseudomonas woodsii* (Smith 1911) Stevens 1925 as *Burkholderia andropogonis* (Smith 1911) Gillis *et al.*, 1995. *Int J. Syst. Evol. Microbiol.* **51**: 183-185.

Coenye, T., Mahenthiralingam, E.,Henry, D., LiPuma, J.J., Laevens, S., Gillis, M., Speert, D.P., and Vandamme, P., 2001b, *Burkholderia ambifaria* sp. nov., a novel member of the *Burkholderia cepacia* complex including biocontrol and cystic fibrosis-related isolates. *Int. J. Syst. Evol. Microbiol.* **51**: 1481-1490.

Coenye, T., LiPuma, J.J., Henry, D., Hoste, B., Vandemeulebroucke, K., Gillis, M., Speert, D.P., and Vandamme, P., 2001c, *Burkholderia cepacia* genomovar VI, a new member of the *Burkholderia cepacia* complex isolated from cystiic fibrosis patients. *Int. J. Syst. Evol. Microbiol.* **51**: 271-279.

Coenye, T., Vandamme, P., Govan, J.R., and LiPuma, J.J., 2001d, Taxonomy and identification of the *Burkholderia cepacia* complex. *J. Clin. Microbiol.* **39**: 3427-3436.

Cottyn, B., Regalado, E., Lanoot, B, De Cleene, M., Mew, T.W., and Swings, J., 2001, Bacterial populations associated with rice seed in tropical environments. *Phytopathology* **91**: 282-292.

De Vos, P., Goor, M., and De Ley, J., 1985, Ribosomal ribonucleic acid cistron similarities of phytopathogenic *Pseudomonas* species. *Int. J. Syst. Bacteriol.* **35**: 169-184.

De Vos, P., and De Ley, J., 1983, Intra- and intergeneric similarities of *Pseudomonas* and *Xanthomonas* ribosomal ribonucleic acid cistrons. *Int. J. Syst. Bacteriol.* **33**: 487-509.

Dykhuizen, D.E., 1998, Santa Rosalia revisited: why are there so many species of bacteria? *Antonie van Leeuwenhoek* **73**: 25-33.

Doudoroff, M., and Palleroni, N.J., 1974, Genus I. *Pseudomonas* Migula 1894. In *Bergey's manual of determinative bacteriology*, 8th ed. (R. E. Buchanan and N. E. Gibbons, eds.), The Williams and Wilkins Co., Baltimore MD, pp. 217-243.

Gardan, L., Dauga, C., Prior, P., Gillis, M., and Saddler, G.S., 2000, *Acidovorax anthurii* sp. nov., a new phytopathogenic bacterium which causes bacterial leaf-spot of *Anthurium*. *Int. J. Syst. Evol. Microbiol.* **50**: 235-246.

Gillis, M., Van Van, T., Bardin, R., Goor, M., Hebbar, P., Willems, A., Segers, P., Kersters, K., Heulin, T., and Fernandez, M.P., 1995, Polyphasic taxonomy in the genus *Burkholderia* leading to an emended description of the genus and proposition of *Burkholderia vietnamiensis* sp. nov. for N_2-fixing isolates of rice from Vietnam. *Int. J. Syst. Bacteriol.* **45**: 274-289.

Henry, D.A., Mahenthiralingam, E., Vandamme, P., Coenye, T., and Speert, D.P., 2001, Phenotypic methods for determining genomovar status of the *Burkholderia cepacia* complex. *J. Clin. Microbiol.* **39**: 1073-1078.

Heungens, K., and Parke, J.L., 2001, Postinfection biological control of oomycetes pathogens of pea by *Burkholderia cepacia* AMMDR1. *Phytopathology* **91**: 383-391.

Isles, A., Maclusky, I., Corey, M., Gold, R., Prober, C., Fleming, P., and Levison, H., 1984, *Pseudomonas cepacia* infection in cystic fibrosis: an emerging problem. *J. Pediatr.* **104**: 206-210.

Kersters, K., Ludwig, W., Vancanneyt, M., De Vos, P., Gillis, M., and Schleifer, K.-H., 1996, Recent changes in the classification of the pseudomonads: an overview. *Syst. Appl. Microbiol.* **19**: 465-477.

Kirchhof, G., Eckert, B., Stoffels, M., Baldani, J.I., Reis, V.M., and Hartmann, A., 2001, *Herbaspirillum frisingense* sp. nov., a new nitrogen-fixing bacterial species that occurs in C4-fibre plants. *Int. J. Syst. Bacteriol.* **51**: 157-168.

Palleroni, N.J., 1984, Genus *Pseudomonas* Migula 1894, In *Bergey's Manual of Systematic Bacteriology*, vol. 1. (N. Krieg and R.G. Holt, eds.), The Williams and Wilkins Co., Baltimore MD, pp. 141-199.

Palleroni, N.J., and Bradbury, J.F., 1993, *Stenotrophomonas*, a new bacterial genus for *Xanthomonas maltophilia* (Hugh 1980) Swings et al. 1983. *Int. J. Syst. Bacteriol.* **43**: 606-609.

Pan, M.J., Rademan, S., Kunert, K., and Hastings, J.W., 1997, Ultrastructural studies on the colonization of banana tissue and *Fusarium oxysporum* f. sp. *cubense* race 4 by the endophytic bacterium *Burkholderia cepacia*. *J. Phytopathol.* **145**: 479-486.

Parke, J.L., and Gurian-Sherman, D., 2001, Diversity of the *Burkholderia cepacia* complex and implications for risk assessment of biological control agents. *Ann. Rev. Phytopathol.* **39**: 225-258.

Schulze, R., Spring, S., Amann, R., Huber, I., Ludwig, W., Schleifer, K.-H., and Kämpfer, P., 1999, Genotypic diversity of *Acidovorax* strains isolated from activated sludge and description of *Acidovorax defluvii* sp. nov. *Syst. Appl. Microbiol.* **22**: 205-214.

Segers, P., Vancanneyt, M., Pot, B., Torck, U., Hoste, B., Dewettinck, D., Falsen, E., Kersters, K., and De Vos, P., 1994, Classification of *Pseudomonas diminuta* Leifson and Hugh 1954 and *Pseudomonas vesicularis* Büsing, Döll, and Freytag 1953 in *Brevundimonas* gen. nov. as *Brevundimonas diminuta* comb. nov. and *Brevundimonas vesicularis* comb. nov., respectively. *Int. J. Syst. Bacteriol.* **44:** 499-510.

Skerman, V.B.D., McGowan, V., and Sneath, P.H.A. (eds.), 1980, Approved lists of bacterial names. *Int. J. Syst. Bacteriol.* **30**: 225-420.

Vandamme, P., Henry, D., Coenye, T., Nzula, S., Vancanneyt, M., LiPuma, J.J., Speert, D.P., Govan, J.R.W. and Mahenthiralingam, E., 2002, *Burkholderia anthina* sp. nov. and *Burkholderia pyrrocinia*, two additional *Burkholderia cepacia* complex bacteria, may confound results of new molecular diagnostic tools. *FEMS Immunol. Med. Microbiol.* **33**: 143-149.

Vandamme, P., Holmes, B., Vancanneyt, M., Coenye, T., Hoste,B., Coopman, R., Revets, H., Lauwers, S., Gillis, M., Kersters, K., and Govan, J.R.W., 1997, Occurrence of multiple genomovars of *Burkholderia cepacia* in cystic fibrosis patients and proposal of *Burkholderia multivorans* sp. nov. *Int. J. Syst. Bacteriol.* **47**: 1188-1200.

Vandamme, P., Mahenthiralingam, E., Holmes, B., Coenye, T., Hoste, B., De Vos, P., Henry, D., and Speert D.P., 2000, Identification and population structure of *Burkholderia stabilis* sp. nov. (formerly *Burkholderia cepacia* genomovar IV). *J. Clin. Microbiol.* **38**: 1042-1047.

Vandamme, P., Pot, B., Gillis, M., De Vos, P., Kersters, K., and Swings, J., 1996, Polyphasic taxonomy, a consensus approach to bacterial systematics. *Microbiol. Rev.* **60**: 407-438.

Walcott, R.R., Langston, D.B., Sanders, F.H., and Gitaitis, R.D., 2000, Investigating intraspecific variation of *Acidovorax avenae* subsp. *citrullii* using DNA fingerprinting and whole cell fatty acid analysis. *Phytopathology* **90**: 191-196.

Willems, A., De Ley, J., Gillis, M., and Kersters, K., 1991, *Comamonadaceae*, a new family encompassing the acidovorans rRNA complex, including *Variovorax paradoxus* gen. nov., comb. nov., for *Alcaligenes paradoxus* (Davis 1969). *Int. J. Syst. Bacteriol.* **41**: 445-450.

Willems, A., Falsen, E., Pot, B., Jantzen, E., Hoste, B., Vandamme, P., Gillis, M., Kersters, K., and De Ley, J., 1990, *Acidovorax*, a new genus for *Pseudomonas facilis*, *Pseudomonas delafieldii*, E. Falsen (EF) group 13, EF group 16, and several clinical isolates, with the species *Acidovorax facilis* comb. nov., *Acidovorax delafieldii* comb. nov., and *Acidovorax temperans* sp. nov. *Int. J. Syst. Bacteriol.* **40**: 384-398.

Willems, A., Goor, M., Thielemans, S., Gillis, M., Kersters, K., and De Ley, J., 1992, Transfer of several phytopathogenic *Pseudomonas* species to *Acidovorax* as *Acidovorax avenae* subsp. *avenae* subsp. nov., comb. nov., *Acidovorax avenae* subsp. *citrulli*, *Acidovorax avenae* subsp. *cattleyae*, and *Acidovorax konjaci*. *Int. J. Syst. Bacteriol.* **42**: 107-119.

Yabuuchi, E., Kosako, Y., Oyaizu, H., Yano, I., Hotta, H., Hashimoto, Y., Ezaki, T., and Arakawa, M., 1992, Proposal of *Burkholderia* gen. nov.; and transfer of seven species of the *Pseudomonas* homology group II to the new genus, with the type species *Burkholderia cepacia* (Palleroni and Holmes 1981)comb. nov. *Microbiol. Immunol.* **36**: 1251-1275.

Yabuuchi, E., Kosako, Y., Yano, I., Hotta, H., and Nishiuchi, Y., 1995, Transfer of two *Burkholderia* and an *Alcaligenes* species to *Ralstonia* gen. nov.: proposal of *Ralstonia picketti* (Ralston, Palleroni and Doudoroff 1973) comb. nov., *Ralstonia solanacearum* (Smith 1896) comb. nov. and *Ralstonia eutropha* (Davis 1969) comb. nov. *Microbiol. Immunol.* **39**: 897-904.

Young, J.M., 2000, Chapter 6. Recent developments in systematics and their implications for plant pathogenic bacteria, In *Applied Microbial Systematics* (F.G. Priest and M. Goodfellow, eds.), Kluwer Academic Publishers, pp. 135-163.

Young, J.M., 2001, Implications of alternative classifications and horizontal gene transfer for bacterial taxonomy. *Int. J. Syst. Bacteriol.* 51: 945-953.

Yurkov, V., Stackebrandt, E., Buss, O., Vermeglio, A., Gorlenko, V., and Beatty, J.T., 1997, Reorganisation of the genus *Erythromicrobium*: description of "*Erythromicrobium sibiricum*" as *Sandaracinobacter sibiricus* gen. nov., sp. nov., and of "*Erythromicrobium ursincola*" as *Erythromonas ursincola* gen. nov., sp. nov. *Int. J. Syst. Bacteriol.* **47**: 1172-1178.

Characterisation of an *Acidovorax* sp. Associated with Geranium and Petunia

J.B. JONES, R.E. STALL, G.V. MINSAVAGE, E.R. DICKSTEIN, H. BOUZAR[#], P.D. ROBERTS[§], P.M. STRICKLER[ϕ] and A.W. ENGELHARD[ϕ]

Plant Pathology Department, University of Florida, Gainesville, FL, USA; [#]Sakata Seed America, Inc. 105 Boronda Rd., Salinas, CA 93907; [§]Southwest Florida Research and Education Center, University of Florida, Immokalee, FL, USA; [ϕ]Gulf Coast Research and Education Center, University of Florida, Bradenton, FL, USA.

Abstract: Since the late 1980s, a bacterium has been routinely isolated from leafspots on geranium (*Pelargonium* X *hortorum* L. H. Bailey) and more infrequently from petunia (*Petunia hybrida*). When inoculated on healthy geranium and petunia leaves, these bacteria caused leafspots identical to those observed on naturally infected geranium plants. Carbon substrate utilisation (BIOLOG) and fatty acid analysis (FAME) identified these isolates to the genus *Acidovorax*. According to BIOLOG, FAME, and pulsed-field gel electrophoresis (PFGE) data, the geranium and petunia strains formed a very tight phenotypic group that was distinct from other *Acidovorax* strains tested. In DNA:DNA hybridisation analyses, the geranium strains had greater than 90% homology to each other and to the petunia strain, but less than 50% homology to other *Acidovorax* species with the exception of *A. konjaci* (52%). Sequence analysis of the 16S rRNA revealed that the geranium and petunia strains had 99% homology with *A. konjaci*. Based on PFGE, BIOLOG, FAME, and DNA:DNA hybridisation results, the geranium petunia strains are very closely related to each other and are most closely related to *A. konjaci*.

Key words: taxonomy, geranium, *Pelargonium*, *Acidovorax*.

Author for correspondence: J.B. Jones; e-mail: JBJones@ifas.ufl.edu.

N.S. Iacobellis *et. al.* (eds.), *Pseudomonas syringae* and Related Pathogens, 667-673.

1. INTRODUCTION

Simone *et al.* (1996) identified an *Acidovorax* sp. associated with a leaf spot of geranium (*Pelargonium* X *hortorum* L. H. Bailey). More recently a similar disease was observed on petunia (*Petunia hybrida*) seedlings. The symptoms consisted of irregularly shaped tan to brown lesions with slight yellowing around the margins. In humid conditions, the spots sometimes have been found to enlarge and coalesce, and leaf margins or entire leaves have become yellow, necrotic and dry. New symptom development ceased when the geraniums reach flowering stage. Acidovorax leaf spot is similar to bacterial blight of geranium incited by *Xanthomonas campestris* pv. *pelargonii.* However, *Acidovorax* does not cause leaves to wilt. Although first reported in 1996 by Simone *et al.*, the disease has been present in Florida since the late 1980s. In the report by Simone *et al.* (1996), the bacterium was identified as the genus *Acidovorax* based on fatty acid analysis. An extensive analysis of strains was undertaken using phenotypic and genotypic tests in order to identify an existing species for placement of these strains or to determine that this is in fact a new species. In this paper we describe the polyphasic approach used for characterising these strains.

2. MATERIALS AND METHODS

2.1 Bacterial Strains and Storage

Acidovorax strains isolated from geranium and various other crops were selected for characterisation. The origin of the strains is described in Table 1. The strains were stored in 15% glycerol at -70°C for long-term preservation.

2.2 Pathogenicity Test

Single colonies were transferred to nutrient agar and grown for 24 to 48 hours at 28°C. The resulting growth was suspended in sterile 0.01 M $MgSO_4*7H_2O$, and the suspensions of cells were standardised to an optical density at 600 nm of 0.3 (concentration equal to 5 x 10^8 CFU ml^{-1}). Geranium (*Pelargonium* X *hortorum*) and petunia (*Petunia hybrida*) seedlings were transplanted into 10-cm pots containing a soil-less mix and fertilised with a slow release fertiliser (Osmocote, Scotts/Sierra, USA) and grown in a greenhouse (23 to 35°C). Plants were inoculated by misting the leaves with a bacterial suspension. All strains were tested on geranium, whereas only the type strains, three geranium strains (UK32, UK39 and UK39), and the petunia strains were inoculated onto petunia. The inoculated plants were placed in clear polyethylene bags and incubated in a growth room at 28°C. The bags were removed after 36 h and the geranium plants remained in the growth-room for the remainder of the experiment, whereas the petunia plants were returned to the greenhouse. After inoculation, symptom development was observed periodically for 3 weeks.

Table 1. Bacterial strains.

Name	Strain	Original Host	Original designation	Source
Acidovorax sp.	UK32	*Pelargonium hortorum*	UK32	Jones[1]
Acidovorax sp.	UK37	*P. hortorum*	UK37	Jones
Acidovorax sp.	UK39	*P. hortorum*	UK39	Jones
Acidovorax sp.	AF9	*Petunia hybrida*	3795	DPI
Acidovorax sp.	AF10	*P. hybrida*	4424	DPI
A. avenae subsp. *avenae*	AAA 1	*Zea mays*	ATCC 19860	ATCC
A. a. subsp. *citrulli*	PPC1	*Citrullus vulgares*	ATCC 29625	ATCC
A. konjaci	AK2	*Amorphophallus rivieri* cv. Konjac	ATCC 33996	ATCC
A. anthurii	CFBP 3232	*Anthurium*	CFBP 3232	Gardan

[1]J.B. Jones, University of Florida, Department of Plant Pathology, Gainesville. DPI=Division of Plant Industry, Gainesville, FL, USA; ATCC=American Type Culture Collection, Manassas, VA, USA.

2.3 Carbon Source Utilisation

The strains were tested for utilisation of 95 carbon sources with GN MicroPlates (Biolog, Inc., Hayward, CA, USA) using standard protocols (Jones *et al.*, 1995). The plates were incubated at 28°C and then the A_{550} of each plate was determined with a plate reader (model EAR 400; SLT Lab Instruments USA, Inc., Ronkonkoma, NY, USA).

2.4 Fatty Acid Analyses

A single colony of each strain was transferred from a nutrient agar culture to trypticase soy broth agar, and after 24 h of growth at 28°C approximately 40 mg of cells was collected for analysis. The bacterial fatty acids were derivatized to their methyl esters, separated by gas chromatography, and identified using the Microbial Identification System software (version 3.6; MIDI, Newark, DE, USA).

2.5 Restriction Endonuclease Analysis

Genomic DNAs of the bacterial strains (Table 1) were prepared and digested with restriction enzymes *Xba*I, *Vsp*I and *Spe*I as described by Egel *et al.* (1991). The resulting large DNA fragments were separated by pulsed-field gel electrophoresis. The gels were stained with ethidium bromide and photographed with type 55 Polaroid film. All strains were run on the same gel, and similarity coefficients were determined for strains by comparing fragments greater than 100 kb (Egel *et al.*, 1991). The KITSCH program from the PHYLIP computer package was used to infer a rooted phylogenetic tree by using the Fitch-Margoliash method (Felsenstein, 1993). The input data consisted of a matrix of pairwise genetic distances determined as estimates of the numbers of nucleotide substitutions per site on the basis of similarity coefficients calculated for the combined digestion data of the three enzymes.

2.6 Sequence Determination

Isolation of total genomic DNA, PCR and electrophoresis conditions were as described (Ausubel *et al.*, 1987). Amplification of *rrs* (16SrRNA gene) and the intergenic spacer (IGS) located between the *rrs* and *rrl* (i.e., 23S rRNA gene) was performed with standard PCR conditions (Innes *et al.*, 1995) using primers FGPS6-63 (GGA GAG TTA GAT CTT GGC TCA G) and FGPL132'-38 (CCG GGT TTC CCC ATT CGG) as previously described (Jones *et al.*, 2000; Ponsonnet and Nesme, 1994). Samples from each strain were processed and sequenced as described by Jones *et al.* (2000).

2.7 DNA:DNA Hybridisation

DNA-DNA relatedness was performed in a microplate and determined by the fluorometric assay of Ezaki *et al.* (1989) with minor modifications. Large molecular-weight DNA was extracted from strains of strains listed in Table 1. DNA extraction and purification were performed following Marmur's procedure as described by Johnson (1994). The DNA was fragmented by three passages through a French pressure cell under 16,000 lb/in^2 , resulting in DNA fragments of ca. 0.5 KB in size. Hybridisation procedures and resulting DNA reassociation values were as described previously (Jones *et al.*, 2000).

3. RESULTS

3.1 Pathogenicity Tests

All strains isolated from geranium and petunia were pathogenic on geranium, whereas none of the other strains caused a disease reaction on geranium. Inoculations of petunia resulted in a strong disease reaction by the petunia strain and the three geranium strains (UK32, UK39 and UK39) tested, but a weak reaction with *Acidovorax avenae* subsp. *avenae* and no disease reaction with *A. konjaci* and *A. anthurii*.

3.2 Carbon Source Utilisation

All geranium and petunia strains were the genus *Acidovorax* based on carbon utilisation patterns. Cluster analysis revealed that this group formed a very tight cluster distinct from all other *Acidovorax* species (data not shown).

3.3 Fatty Acid Analysis

The bacterial species isolated from geranium were consistently identified to the genus, *Acidovorax*. The bacterium also contained 3-hydroxydecanoic acid, the only hydroxylated fatty acid, which is typical of members within the genus *Acidovorax* (Gardan *et al*., 2000; Willems *et al*., 1990). Cluster analysis revealed that this group formed a very tight cluster distinct from all other *Acidovorax* species (data not shown).

3.4 Restriction Endonuclease Analysis

The geranium and petunia strains had very similar banding patterns following digestion with *Spe*I compared to the other strains used for comparison (Fig. 1). Similar results were observed with the other two enzymes. Based on phylogenetic analysis using the Phylip software (Felsenstein, 1993), the geranium strains were very closely related (data not shown). The other *Acidovorax* species were distantly related based on banding patterns.

3.5 RNA Sequence Comparisons

A fragment of 1493 bases was resolved in UK37. Sequence comparisons of the 16S rRNA from UK37 resulted in 1493 bases. Sequence comparisons revealed that similarity of the geranium strain was 99% with *Acidovorax konjaci* and approximately 97% with *A. anthurii*.

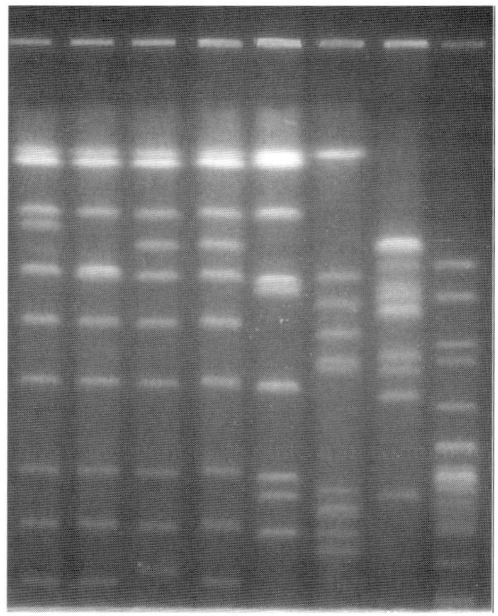

A B C D E F G H

Figure 1. Bands of DNA from *Acidovorax* strains separated by pulsed-field gel electrophoresis after genomic DNAs were digested with restriction enzyme *Spe*I. Lanes A-C contained DNAs from geranium strains UK37, UK39 and AF9. Lane D contained DNA from petunia strain AF10. Lane E-H contained DNAs from *A. avenae* subsp. *citrulli*, *A. a.* subsp. *avenae*, (AAA1), and *A. facilis* (AF15), *A. konjaci* (AK2), respectively.

3.6 DNA:DNA Hybridisation why this here?

UK37, which was used as the probe, had greater than 70% hybridisation with the geranium (UK32, UK37, and UK39) and the petunia strain (AF9 and AF10). It had less than 50% hybridisation with all other strains except for *Acidovorax konjaci* (AK2) with which it had 52.8%.

4. DISCUSSION

The geranium and petunia strains were determined to be very closely related based on phenotypic and genotypic analyses. Furthermore, pathogenicity tests also confirmed significant host specialisation by plant pathogenic members within this genus. Fatty acid analysis, carbon substrate utilisation patterns and sequence analysis of the 16S rRNA consistently placed this organism within the genus, *Acidovorax*. DNA:DNA hybridisation results were useful in demonstrating that these strains are significantly distinct from all other *Acidovorax* species tested. Although *A. anthurii* was not used in the hybridisation studies, comparison of the 16S rRNA revealed lower homology with this organism than with *A. konjaci.* Based on these findings, it is clear that the geranium strains represent a new plant pathogenic species within *Acidovorax* (Gardan *et al.*, 2000; Willems *et al.*, 1992).

REFERENCES

Ausubel, F.M., Brent, R., Kingston, R.E., Moore, D.D., and Seidman, K., 1987, Current protocols in molecular biology. John Wiley and Sons, New York.

Egel, D., Graham, J., and Stall, R.E., 1991, Genomic relatedness of *Xanthomonas campestris* strains causing diseases of citrus. *Appl. Env. Microbiol.* **57**: 2724-2730.

Ezaki, T., Hashimoto, Y., and Yabuuchi, E., 1989, Fluorometric deoxyribonucleic acid-deoxyribonucleic acid hybridisation in microdilution wells as an alternative to membrane filter hybridisation in which radioisotopes are used to determine genetic relatedness among bacterial strains. *Inst. J. Syst. Bacteriol.* **39**: 224-229.

Felsenstein, J., 1993, PHYLIP (Phylogeny Inference Package). Version 3.57. Computer program and manual distributed by the author.

Gardan, L, Dauga, C., Prior, P., Gillis, M., and Saddler, G.S., 2000, *Acidovorax anthurii* sp. nov., a new phytopathogenic bacterium which causes bacterial leaf-spot of anthurium. *Int. J. Syst. Evol. Microbiol.* **50**: 235-246.

Innes, M.A., Gelfand D.H., and Sninsky, J.J., 1995. PCR Strategies. Academic Press, New York, NY.

Johnson, J.L., 1994, Similarity analysis of DNAs. Pp. 655-682. In: *Methods for general and molecular bacteriology*. Gerhardt, P., Murray, T.G.E., and Krieg, N.R. (ed.). American Society for Microbiology, Washington, DC.

Jones, J.B., Bouzar, H., Stall, R.E., Almira, E.C., Roberts, P., Bowen, B.W., Sudberry, J., Strickler, P. Chun, J., 2000, Systematic analysis of xanthomonads (*Xanthomonas* spp.) asscoiated with pepper and tomato lesions. *Int. J. Syst. Bacteriol.* **50**: 1211-1219.

Jones, J.B., Stall, R.E., Somodi, G.C., Bouzar, H., and Hodge, N.C., 1995, A third tomato race of *Xanthomonas campestris* pv. *vesicatoria. Plant Dis.* **79**: 395-398.

Ponsonnet, C., and Nesme, X., 1994, Identification of *Agrobacterium* strains by PCR-RFLP analysis of pTi and chromosomal regions. *Arch. Microbiol.* **161**: 300-309.

Simone, G.W., Cullen, R.E., and Hodge, N.C., 1996, A new leaf spot disease of geranium caused by *Acidovorax* sp. *Phytopathology* **86**: S50.

Willems, A., Falsen, E., Pot, B., Jantzen, E., Hoste, B., Vandamme, P., Gillis, M., Kersters, K., and De Ley, J., 1990, *Acidovorax*, a new genus for *Pseudomonas facilis*, *Pseudomonas delafieldii*, E. Falsen (EF) group 13, EF group 16, and several clinical isolates, with the species Acidovorax *facilis* comb. nov., *Acidovorax delafieldii* comb. nov., and A*cidovorax temperans* sp. Nov. *Int. J. Syst. Bacteriol.* **40**: 384-398.

Willems, A., Goor, M., Thielemans, S., Gillis, M., Kersters, K., and De Ley J., 1992, Transfer of several phytopathogenic *Pseudomonas* species to *Acidovorax* as *Acidovorax avenae* subsp. *avenae* subsp. nov., comb. nov., *Acidovorax avenae* subsp. *citrulli*, *Acidovorax avenae* subsp. *cattleyae*, and *Acidovorax konjaci. Int. J. Syst. Bacteriol.* **42**: 107-119.

Bacterial Canker of Wild Cherry Tree in France Caused by a new Pathovar of *Pseudomonas syringae* pv. *avii* (pv. nov.)

L. SUTRA*, M. MENARD, J. LUISETTI#, J.P. PRUNIER§, and L. GARDAN

UMR de Pathologie Végétale, INRA, 42 rue Georges Morel, BP 57, 49071 Beaucouzé cedex, France; #INRA-CIRAD, Pôle de Protection des Plantes, 7 Chemin de l'IRAT, 97410 Saint-Pierre, La Réunion, France; §INRA, Direction Scientifique Productions Végétales, 147 rue de l'Université, 75338 Paris Cedex 07, France.

Abstract. This paper described a novel bacterial pathogen that causes canker on wild cherry tree. This bacterium belongs to the *Pseudomonas syringae* group (LOPAT Ia). Based on pathogenicity tests, phenotypic traits and molecular characteristics, it is proposed as a new *P. syringae* pathovar, pv. *avii*.

Key words: Etiology, pathogenicity, DNA-DNA hybridisation, rep-PCR.

Author for correspondence: M. Menard; e-mail: menard@angers.inra.fr.

1. INTRODUCTION

In 1989, a disease was observed on wild cherry (WC) trees in the north of France. Symptoms consisted of longitudinal bark cankers on the trunk and branches (Fig. 1), and often gum exudation around cankers. The infection can entirely girdle the trunk or branch causing the upper part dieback (Fig. 1).

Gram negative aerobic and non fluorescent bacterial strains that belong to *Pseudomonas syringae* LOPAT group Ia (+ - - - +) of Lelliott *et al.* (1966) were isolated from canker and necrotic lesions since 1991. Nine strains were

* deceased on December the 16th, 2002.

N.S. Iacobellis *et. al.* (eds.), *Pseudomonas syringae* and Related Pathogens, 675-679.

studied using a polyphasic approach based on pathogenicity tests, phenotypic characteristics, DNA-DNA hybridisation and rep-PCR profiles. The causal agent of bacterial canker on WC tree is proposed as a new pathovar of *P. syringae*, pv. *avii*.

A B

Figure 1. Major symptoms observed on wild cherry (WC) trees Stem canker (arrow) (A); Upper part dieback of a tree (B) (photographs: Département de la Santé des Forêts, Ministère de Agriculture, France).

2. MATERIAL AND METHODS

Nine strains isolated from diseased WC trees were selected and were compared to sixty eight type and pathotype strains belonging to the *Pseudomonas syringae* group.

Twenty biochemical tests were performed as described by Sutra *et al.* (1997). Assimilation of 99 carbon sources was studied using Biotype 100 galleries (bioMérieux, France). The 119 characteristics were analysed by numerical taxonomy as previously described (Sutra *et al.*, 1997).

For pathogenicity tests, five strains from WC trees were compared with strains of *P. s.* pv. *morsprunorum* race 1 CFBP 3840, *P. s.* pv. *morsprunorum* race 2 CFBP 3800 and *P. s.* pv. *persicae* CFBP 1573. Two year old wild cherry trees were inoculated by transverse incisions through the bark. A bacterial suspension (25 µl at about 10^8 CFU ml^{-1}) was introduced in each wound. Five trees were used for each strain. Sterilised distilled water was used as control. Canker severity was estimated using a girdling index (from 0, no symptom, to 5, stem entirely girdled) (Ridé and Ridé, 1978).

DNA-DNA hybridisation (S1 nuclease procedure) and ΔTm measurement were performed as previously reported (Gardan *et al.*, 1999). Experiments

were done with ^{3}H-labelled DNA of WC strain CFBP 3846 at 70°C. The G+C content of strain CFBP 3846 was determined by thermal denaturation of DNA (Mamur and Doty, 1962).
Strains were analysed by rep-PCR with REP-, ERIC- and BOX- primer sets (Versalovic *et al*., 1991) using the PCR conditions described by Louws *et al.* (1994). For each strain, the presence and the absence of a band were encoded as 1 and 0, respectively, and data were analysed by numerical taxonomy as above.

3. RESULTS AND DISCUSSION

3.1 Phenotypic Traits

Strains isolated from WC trees have the general characteristics of *Pseudomonas syringae* (Gram-negative rods, creamy colonies on YPGA agar, no production of fluorescent pigment on King's B medium, obligate aerobic and oxidisation of glucose) and belong to LOPAT Ia group (+ - - - +) of Lelliott *et al.* (1966). Numerical analysis of phenotypic characteristics showed that strains from WC trees are different from other *P. syringae* pathovars and related species. They can be differentiated from *P. s.* pv. *persicae* and *P. amygdali*, the two bacteria with the most similar phenotypic characteristics, by the hydrolysis of gelatine (+ for WC strains), assimilation of D-galacturonate, D-ketoglutarate (+ for WC strains) and D-xylose (- for WC strains).

3.2 Pathogenicity on WC Trees

Typical symptoms of canker were obtained 6 months after inoculation of WC strains on wild cherry trees. Values of girdling index was higher for WC strains (average value for 5 strains, 2.5; range, 1.3÷3.6) than for strains of *Pseudomonas syringae* pv. *persicae* CFBP 1573 (0.2), *P. s.* pv. *morsprunorum* race 1 CFBP 3840 (1.2) and race 2 CFBP 3800 (0.3).

3.3 DNA Relatedness and G+C Content

The WC strains tested by DNA-DNA hybridisation were 87 to 100 % related to WC strain CFBP 3846 (Table 2). Strain CFBP 3846 was 63 to 84% related to pathovars of genomospecies 3 of Gardan *et al.* (1999) such as *Pseudomonas syringae* pv. *persicae*, pv. *tomato* and pv. *morsprunorum* race 2. For the lowest DNA reassociation values (63 to 67%), ΔTm values were ranged from 1.3 to 1.7°C. Values of DNA relatedness with the pathovars of genomospecies 1, 2 and 4 to 9 ranged from 26 to 51%. The G+C content of strain CFBP 3846 was 59.0 mol%. This is in the range of values of *P. syringae* pathovars and related species (De Vos *et al.,* 1985).

3.4 Analysis of rep-PCR Fingerprints

Fingerprinting based on rep-PCR has previously been used to differentiate *Pseudomonas syringae* pathovars (Louws *et al.*, 1994; Weingart and Völksh, 1997; Marques *et al.*, 2000). In this study, band patterns generated by PCR using ERIC-, BOX- and REP-primers (Versalovic *et al.*, 1994) were analysed together. The inferred dendrogram based on Jaccard coefficient and UPGMA showed that strains isolated from WC trees are grouped in a unique cluster (data not shown). This cluster is clearly separated from other *P. syringae* pathovars, particularly from pathovars pathogenic to WC tree, such as *P. s.* pv. *morsprunorum* races 1 and 2. However, WC strains were closely related, but different from, *P. s.* pv. *persicae*.

4. CONCLUSION

The bacterium that causes bacterial canker on WC tree belongs to the *Pseudomonas syringae* group and to genomospecies 3 of Gardan *et al.* (1999). It can be differentiated from other *P. syringae* pathovars by phenotypic tests, molecular characteristics and pathogenicity. Most of the genomospecies within the *P. syringae* group have not been converted into species, because of the lack of discriminative phenotypic tests (Gardan *et al.*, 1999). Thus, WC strains should be included *P. syringae* sensu lato as a new pathovar according to the international standards for naming new phytopathogenic bacteria (Young *et al.*, 1996). The name *P. s.* pv. *avii* is proposed (pathotype strain CFBP 3846; G+C = 59,0 mol%) (Ménard *et al.*, 2003).

REFERENCES

De Vos, P., Goor, M., Gillis, M., De Ley, J., 1985, Ribosomal ribonucleic acid cistron similarities of phytopathogenic *Pseudomonas* species. *Int. J. Syst. Bacteriol.* **35**: 169-184.

Marques, A.S., Corbière, R., Gardan, L., Tourte, C., Manceau, C., Taylor, J.D., Samson, R., 2000, Multiphasic approach for the identification of the different classification levels of *Pseudomonas savastanoi* pv. *phaseolicola*. *Eur. J. Plant Pathol.* **106**: 715-734.

Gardan, L., Shafik, H., Belouin, S., Broch, H., Grimont, F., Grimont P.A.D., 1999, DNA relatedness among the pathovars of *Pseudomonas syringae* and description of

Pseudomonas tremae sp. nov. and *Pseudomonas cannabina* sp. nov. (ex Sutic and Dowson 1959). *Int. J. Syst. Bacteriol.* **49**: 469-478.

Lelliott, R.A., Billing, E., Haywart, A.C., 1966, A determinative scheme for the fluorescent plant pathogenic *Pseudomonas*. *J. Appl. Bacteriol.* **29**: 470-489.

Louws, F.J., Fulbright, D.W., Stephens, C.T., De Bruijn, F.J., 1994, Specific genomic fingerprints of phytopathogenic *Xanthomonas* and *Pseudomonas* pathovars and strains generated with repetitive sequence and PCR. *Appl. Environ. Microbiol.* **60**: 2286-2295.

Marnur, J., and Doty, P., 1962, Determination of the base composition of dexoyribo-nucleic acid from its thermal denaturation temperature. *J. Mol. Biol.* **5**: 109-118.

Ménard, M., Sutra, L., Prunier, J. P., Luisetti, J., Gardan, L., 2003 *Pseudomonas syringae* pv. *avii* (pv. nov.) the causal agent of bacterial canker of wild cherries (*Prunus avium)* in France. *Eur. J. Plant Pathol.* (in press).

Ridé, M., Ridé, S., 1978, Factors affecting inoculation success in woody plants. Proceedings of 4th International Conference of Plant Pathogenic Bacteria **2**: 957-968, INRA, Angers France.

Sutra, L., Siverio, F., Lopez, M.M., Hunault, G., Bollet, C., Gardan, L., 1997, Taxonomy of *Pseudomonas* strains isolated from *tomato* pith necrosis: emended description of *Pseudomonas corrugata* and proposal of three unnamed fluorescent *Pseudomonas* genomospecies. *Int. J. Syst. Bacteriol.* **47**: 1020-1033.

Versalovic, J., Schneider, M., De Bruijn, F. J., Lupski J. R., 1994, Genomic fingerprinting of bacteria using repetitive sequence based (rep-PCR). *Meth. Mol. Cell. Biol.* **5**: 25-40.

Weingart, H., Völksh, B., 1997, Genetic fingerprinting of *Pseudomonas syringae* pathovars using ERIC-, REP- and IS50-PCR. *J. Phytopathol.* **145**: 339-345.

Young, J. M., Saddler, G. S., Takikawa, Y., De Boer, S. H., Vauterin, L., Gardan, L., Gvodzdyak, R. I., Stead, D.E., 1996, Names of plant pathogenic bacteria 1864-1995. *Rev. Plant Pathol.* **75**: 721-763.

Phenotypic Characteristics of *Pseudomonas savastanoi* Strains from Various Hosts

P. BELLA, V. CATARA, L. SUTRA[#*], C. GUARINO, G. CIRVILLERI and L. GARDAN[#]

Dipartimento di Scienze e Tecnologie Fitosanitarie, Sezione Patologia Vegetale, Università degli Studi di Catania, Italy; [#]UMR de Pathologie Végétale INRA-INH-Universitè, Beaucouzé, France.

Abstract: Assimilation of 99 carbon sources by eighty-eight *Pseudomonas savastanoi* strains isolated from oleander (41), olive (32), and various other host plants was performed. The results of numerical analysis showed that the majority of the olive and oleander strains were clustered in one phenon. Strains from other hosts such as *Myrtus* sp., *Ligustrum* sp., *Phillyrea* sp. and *Retama sphaerocarpa*, were grouped into a different phenon together with five oleander strains.

Key words: oleander, olive, oleander knot disease.

Author for correspondence: P. Bella; e-mail: patrizia.bella@tiscali.it.

1. INTRODUCTION

Knot-inducing *Pseudomonas savastanoi* strains from olive, various genera of *Oleaceae* and oleander were assigned to *P. s.* pv. *savastanoi* (Gardan *et al.*, 1992). *P. savastanoi* strains were also isolated from tissue overgrowth on Spanish broom (*Retama sphaerocarpa*), *Fraxinus* sp., *Ligustrum* sp., *Phillyrea* sp., and *Jasminum* sp. In the "Names of Plant Pathogenic Bacteria" (www.isppweb.org), four pathovars are reported according to the host from which the respective strains were isolated: *P. s.* pv. *savastanoi*, *P. s.* pv. *nerii*, *P. s.* pv. *fraxini* and *P. s.* pv. *retacarpa*. Severe infections caused by *P.*

[*] deceased on December the 16th, 2002.

N.S. Iacobellis *et. al.* (eds.), *Pseudomonas syringae* and Related Pathogens, 681-686.

savastanoi on oleander and olive have been observed in Sicily. In this study, a subset of strains from a collection set up from oleander and olive in this region was compared to *P. savastanoi* strains isolated from different geographic areas from various hosts (*Fraxinus* sp., *Myrtus* sp., *Ligustrum* sp., *Phillyrea* sp., *Jasminum* sp. and *R. sphaerocarpa*).

2. MATERIALS AND METHODS

2.1 Bacterial Strains

Eighty-eight *Pseudomonas savastanoi* strains were included in this study. Three sets of strains were used. The first set comprised 15 *P. savastanoi* strains isolated from olive, the second set included 22 *P. savastanoi* strains isolated from oleander. All 37 strains were isolated in Sicily. The third set contained strains of *P. savastanoi* isolated from olive, oleander, *Myrtus* sp., *Fraxinus* sp., *Ligustrum* sp., *Phillyrea* sp., *Jasminum* sp. and *Retama sphaerocarpa* from different geographic origin.

2.2 Biochemical Characters and Data Analysis

Assimilation of 99 biochemical substrates (carbohydrates, amino acids and organic acids) was studied using Biotype 100 strips (BioMérieux, Italia). For Biotype 100 tests, strains were grown on KB medium at 28°C for 48h. Preparation of bacterial suspensions and inoculation were carried out as described by the manufacturer. The strips were incubated at 28°C and examined for growth daily up to 6 days. The results were included in the numerical taxonomy analysis. Distance matrix was calculated using the Jaccard coefficient and cluster analysis was done using the unweighted pair group method with averages (UPGMA) (Sneath and Sokal, 1973).

3. RESULTS

Distance relationships between the strains were displayed by a dendrogram (Fig. 1). At a distance of 0.200, six phena and 7 isolated strains were observed. The main phenon included 57 (65%) out of 88 strains. Among these 57 strains, 27 were isolated from oleander, 26 from olive and four from other hosts (*Ligustrum* sp. and *Retama sphaerocarpa*). Most Sicilian oleander (17/29) and olive (14/15) strains were grouped in this phenon. Olive and oleander strains from other geographic origin included in this phenon were isolated from Portugal, Algeria, Spain, Morocco, France,

Tunisia and other regions of Italy. Minor phena included two or three strains each. Distribution of strains in the six major phena is shown in Table 1.

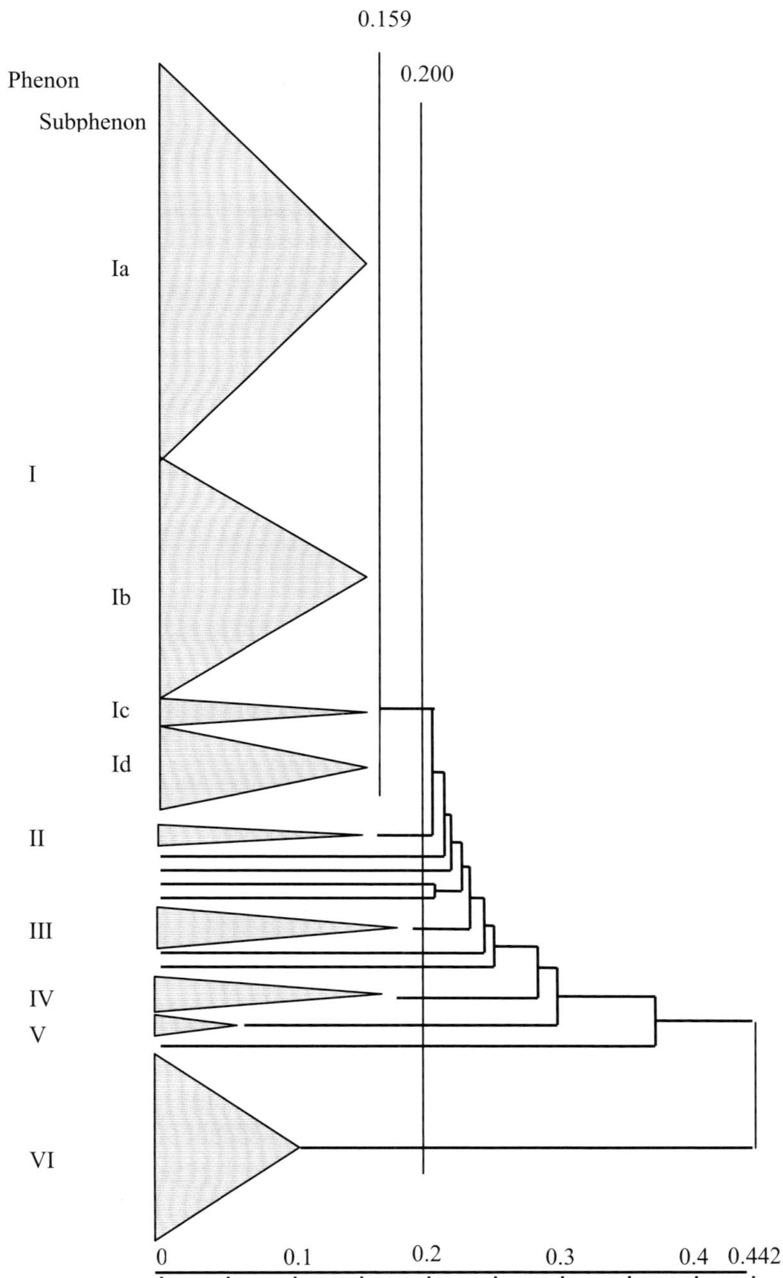

Figure 1. Dendrogram of the distances among 88 strains of *Pseudomonas savastanoi*.

The main phenon I was subdivided into four subphena at a distance of 0.159. Assimilation of malonate was a differential characteristic between these subphena. All strain of subphena Ia and Id and 33% of strains in subphena Ic could utilise malonate as carbon source whereas strains of subphenon Ib could not.

Table 1. Strains of *Pseudomonas savastanoi* included in the six phena and isolated phenotypes at a distance of 0.200.

Phenon	Subphenon	No. Isolates	Host	Geographic origin
I	Ia	31	Olive (12); Oleander (15); *Ligustrum* sp. (2); Spanish broom (2)	Portugal (1); Spain (3); Tunisia (1); Morocco (2); Algeria (2); France (1); Italy (21)
	Ib	17	Olive (14); Oleander (3)	Ex Yugoslavia (1); France (1); Italy (15)
	Ic	3	Oleander	Sicily
	Id	6	Oleander	Sicily
II		2	Olive (1); Oleander (1)	Ex Yugoslavia (1); France (1)
III		3	Olive (1); Oleander (2)	France (1); Greece (1); Algeria (1)
IV		3	Oleander	France (1); Italy (1)
V		2	Oleander	Sicily
VI		14	Oleander (5); *Phillyrea* sp. (3); *Jasminum* sp. (2); *Myrtus* sp. (2); *Ligustrum* sp. (1); Spanish broom (1)	Algeria (3); Greece (2); Syria (2); Portugal (2); Spain (1); Italy (4)
Isolated phenotypes		7	Olive (4); Oleander (1); Ash (2)	Algeria (1); Ex Yugoslavia (1); France (1); Syria (1); Italy (3)

It is noteworthy that strains from *Myrtus* sp. (2), *Ligustrum* sp. (1), *Jasminum* sp. (2), *Phillyrea* sp. (3) and *R. sphaerocarpa* (1) were clustered in a different phenon with five oleander strains. No olive strains were grouped in this phenon. Several biochemical characters differentiated phenon I from phenon VI. Strains in phenon I could not utilize malonate, quinate, D-(-)-tartrate, maltose, esculin, DL-lactate, D-(+)-trehalose, D-glucosammine, propionate, D-(+)-raffinose, α-lactose, lactulose, D-(+)-cellobiose, 1-O-methyl-β-D-glucopyranoside, maltotriose, α-L-rhamnose, 1-O-methyl-β-D-galactopyranoside, α-L-(-)-fucose, N-acetyl-D-glucosamide, 2-keto-D-gluconate. All these carbon sources were utilised by strains of the phenon VI.

4. DISCUSSION

DNA-DNA hybridisation analysis showed that *Pseudomonas. savastanoi* strains isolated from olive, oleander, ash, privet, jasmine and *Phillyrea* sp. were members of a single species, together with *P. s.* pv. *phaseolicola* and *P. s.* pv. *glycinea* (Gardan *et al.*, 1992). In this study, we investigated the phenotypic characteristics of a large collection of strains of *P. savastanoi* from various hosts and from 10 different Mediterranean Countries. Our results showed no correlation between hosts, geographical origin and phenotypic traits. This finding is in accordance with previous studies that showed that biochemical characteristics could not differentiate *P. savastanoi* strains according to the hosts (Janse *et al.*, 1982, Gardan *et al.*, 1992). However, strains from olive, oleander and ash, the three major hosts of *P. savastanoi*, could be distinguished on the basis of location of genes for IAA synthesis (Comai and Kosuge, 1980; Comai *et al.*, 1982; Caponero *et al.*, 1995), production of bacteriocins (Iacobellis *et al.*, 1995), fatty acids profiles (Janse, 1991), DNA fingerprints (Mugnai *et al.*, 1994; Sisto *et al.*, 2002). In pathogenicity tests, strains of *P. savastanoi* showed host specificity. Strains from olive, ash and Spanish broom induced the disease symptoms only on the host plant they were isolated from (Gardan *et al.*, 1992; Caponero *et al.*, 1995; Alvarez *et al.*, 1998; Garcia de los Rios, 1999). Strains from oleander caused characteristics symptoms of the disease both on oleander and olive plants (Surico *et al.*, 1985). Atypical responses were also obtained. For example, Alvarez *et al.* (1998) showed that olive strains produced a slight swelling but not a knot on oleander; an ash strain caused necrosis at inoculation point when inoculated on Spanish broom (Garcia de los Rios, 1999). In this study, strains from *Myrtus* sp. (2), *Ligustrum* sp. (1), *Jasminum* sp. (2), *Phillyrea* sp. (3) and *Retama sphaerocarpa* (1) were clustered in a different phenon with five oleander strains, 3 from Sicily and 2 from Portugal. Strains belonging to this phenon are currently under investigation for cross-pathogenicity tests using different hosts, molecular characterisation based on whole-genome analysis (rep-PCR, AFLP) and ITS sequence polymorphism.

ACKNOWLEDGEMENTS

We thank M. Scortichini, Istituto Sperimentale per la Frutticoltura, Roma, Italy, for supplying strains ISF020 and ISF021,and M. Zaccardelli, Istituto Sperimentale per le Colture Industriali, Mi.P.A.F., Battipaglia (SA) Italy, for providing some olive samples.

REFERENCES

Alvarez, F., Garcia de los Rios, J.E., Jimenez, P., Rojas, A., Reche, P., and Troya, M.T., 1998, Phenotypic variability in different strains of *Pseudomonas syringae* subsp. *savastanoi* isolated from different hosts. *Eur. J. Plant Pathol.* **104**: 603-609.

Caponero, A., Contesini, A.M., and Iacobellis, N.S., 1995, Population diversity of *Pseudomonas syringae* subsp. *savastanoi* on olive and oleander. *Plant Pathol.* **44**: 848-855.

Comai, L., and Kosuge, T., 1980, Involvement of plasmid deoxyribonucleic acid in indoleacetic acid synthesis in *Pseudomonas savastanoi*. *J. Bacteriol.* **143**: 950-957.

Comai, L., Surico, G., and Kosuge, T., 1982, Relation of plasmid DNA to indolacetic acid production in different strains of *Pseudomonas syringae* pv. *savastanoi*. *J. Gen. Microbiol.* **128**: 2157-2163.

Garcia de los Rios, J.E., 1999, *Retama sphaerocarpa* (L.) Boiss., a new host of *Pseudomonas savastanoi*. *Phytopathol. Medit.* **38**: 54-60.

Gardan, L., Bollet, C., Abu Ghorrah, M., Grimont, F., and Grimont, P.A.D., 1992, DNA relatedness among the pathovar strains of *Pseudomonas syringae* subsp. *savastanoi* Janse (1982) and proposal of *Pseudomonas savastanoi* sp. nov. *Int. J. Syst. Bacteriol.* **42**: 606-612.

Iacobellis, N.S., Contesini, A.M., and Surico, G., 1995, Bacteriocin production by *Pseudomonas syringae* subsp. *savastanoi*. *Phytopathol. Medit.* **34**: 15-22.

Janse, J.D., 1982, *Pseudomonas syringae* subsp. *savastanoi* (ex Smith) subsp. nov., nom., rev., the bacterium causing excrescences on *Oleaceae* and *Nerium oleander* L. *Int. J. Syst. Bacteriol.* **32**: 166-169.

Janse, J.D., 1991, Pathovars discrimination within *Pseudomonas syringae* subsp. *savastanoi* using whole-cell fatty acids and pathogenicity as criteria. *Syst. Appl. Microbiol.* **13**: 79-84.

Mugnai, L., Giovanetti, L., Ventura, S., and Surico, G., 1994, The grouping of strains of *Pseudomonas syringae* subsp. *savastanoi* by DNA restriction fingerprinting. *J. Phytopathol.* **142**: 209-218.

Sisto, A., Morea, M., Baruzzi, F., and Palumbo, G., 2002, Differentiation of *Pseudomonas syringae* subsp. *savastanoi* strains isolated from various host plants by restriction fragment length polymorphism. *Phytopathol. Medit.* **41**: 63-71.

Sneath, P.H.A., and Sokal, R.R., 1973, Numerical taxonomy. The principles and practice of numerical classification. Freeman & Co., San Francisco.

Surico, G., Iacobellis, N.S., and Sisto, A., 1985, Studies on the role of indole-3-acetic acid and cytokinins in the formation of knots on olive and oleander plants by *Pseudomonas syringae* pv. *savastanoi*. *Physiol. Plant Pathol.* **26**: 309-320.

Phylogenic Analysis of DNA Sequences Around the *hrpL* and *hrpZ* Regions of *Pseudomonas syringae* Group Bacteria

Y. INOUE and Y. TAKIKAWA[#]
National Institute of Agrobiological Sciences, Tsukuba, Ibaraki, Japan; [#]Faculty of Agriculture, Shizuoka University, 836 Ohya, Shizuoka 422-8529, Japan

Abstract: Based on DNA homology of the *hrp* cluster and its neighboring regions, *Pseudomonas syringae* strains were classified into groups I, II, III and IV, represented by *P. s.* pv. *phaseolicola*, *P. s.* pv. *maculicola*, *P. s.* pv. *syringae* and *P. s.* pv. *coronafaciens*, respectively. Based on these results, we compared the sequences of neighbouring region of the *hrpZ* (n-*hrpZ*) and that of *hrpL* (n-*hrpL*) for grouping *P. syringae* pathovars. The structure of the n-*hrpL* was quite different among the groups, and much more different among members of each group. Most strains of Group I had the *avrPphE* next to the *hrpY* and another common DNA sequence (MP) next to the *avrPphE*. The *avrPphE* and its adjoining sequence were distributed widely irrespective of the groups in *P. syringae*. Group II strains carried the gene area with a region specific to Group II (AM) in n-*hrpL*. In other regions, homology area was arranged in splinters by each strain. These results showed that outside the *hrpL* is a variable region and a lot of mutations have occurred.Therefore, some fragments could be used as probes to differentiates groups of *P. syringae*. In contrast, the structure of the n-*hrpZ* was the same from the *hrpS* gene to the *hrpB* gene. Deletion in the *hrpZ* gene was detected only in *P. s.* pv. *tabaci*. Based on comparison of DNA sequence of the *hrpZ*, strains of *P. syringae* divided into five groups without contradicting previous grouping. Group I was divided into two categories based on this study, and the unknown strains belong to this group. These results suggest that the n-*hrpZ* region changed with the evolution of *P. syringae*

Key words: classification, based on DNA sequence, the outside of the *hrpL*, the *hrpZ* neighbouring region.

Author for correspondence: Y. Inoue; e-mail: yasinoue@affrc.go.jp.

N.S. Iacobellis *et. al.* (eds.), *Pseudomonas syringae* and Related Pathogens, 687-695.

1. INTRODUCTION

Pseudomonas syringae has a number of pathovars that are distinct in their host plants and/or disease symptoms. *P. syringae* is identified with several characteristics, though, each pathovar and strain has considerable phenotypic variations in its nutritional and physiological characteristics. In recent years, DNA-DNA hybridisation and comparison of DNA sequences have been used to distinguish the strains or pathovars of the *P. syringae* group bacteria and to classify *P. syringae* into several groups (Gardan *et al.* 1999, Sawada *et al.* 1999).

Causing disease on host plants and eliciting the hypersensitive response (HR) on non-host plants are fundamental features controlled by the *hrp* genes (Peet *et al.* 1986). The *hrp* genes are present in some pathovars of *P. syringae* and several gram-negative plant pathogenic bacteria. These *hrp* genes from different strains, species or genera hybridised to each other to varying degrees. Interestingly, amino acid sequences of several *hrp* gene loci have high homology to secretion proteins of animal pathogens (Hueck 1998). Sequences similar to *hrp* genes are widely distributed in several bacteria and have differentiated into various functional genes through evolution. On the other hand, a cluster of *avrE* genes from *P. s.* pv. *tomato* (Lorang *et al.* 1995), *avrPphE* gene from *P. s.* pv. *phaseolicola* (Mansfield *et al.* 1995) and *hrmA* gene from *P. s.* pv. *syringae* (Heu *et al.* 1993), which are adjacent to a cluster of *hrp* genes, are shown to participate in the manifestation of pathogenicity. Pathogenicity is one of the most important characters in the classification of *P. syringae*. This region including *hrp* genes and their neighbours might be expected to be useful for characterising *P. syringae*. Therefore, we studied the utility of comparing large genomic regions including *hrp* and its neighbouring regions for grouping of *P. syringae* pathovars.

We have shown that *P. syringae* strains contain *hrp* and *avrE* genes in the same order and that they can be grouped into at least five types by comparing DNA homology at the *hrp* cluster and its neighbouring regions (Inoue and Takikawa 1999a; 2000). The first group contained strains of *P. s.* pv. *phaseolicola*, *P. s.* pv. *glycinea*, *P. s.* pv. *tabaci*, *P. s.* pv. *lachrymans*, *P. s.* pv. *mori*, *P. s.* pv. *dendropanacis*, *P. s.* pv. *eriobotryae*, *P. s.* pv. *myricae*, *P. s.* pv. *photiniae*, *P. s.* pv. *sesami*, an undetermined pathovar causing kiwifruit blossom rot disease and *P. ficuserectae*. This group (group I) showed high homology to probes from *P. s.* pv. *phaseolicola*. The second group contained strains of *P. s.* pv. *maculicola*, *P. s.* pv. *tomato*, *P. s.* pv. *theae*, *P. s.* pv. *delphinii*, *P. s.* pv. *actinidiae*, *P. s.* pv. *spinaciae* and *P. s.* pv. "magnoliae". This group (group II) showed high homology to the probes from *P. s.* pv. *maculicola*. The third group contained strains of *P. s.* pv.

syringae, *P. s.* pv. *aceris*, *P. s.* pv. *aptata*, *P. s.* pv. *japonica*, *P. s.* pv. *lapsa* and *P. s.* pv. *pisi*. This group (group III) showed high homology to the probes from *P. s.* pv. *syringae*. The fourth group contained strains of *P. s.* pv. *coronafaciens*, *P. s.* pv. *atropurpurea*, *P. s.* pv. *oryzae*, *P. s.* pv. *porri*, *P. s.* pv. *striafaciens* and an undetermined pathovar causing onion disease. This group (group IV) showed high homology to the probes from *P. s.* pv. *coronafaciens*. The fifth group (Group U) contained strains of *P. s.* pv. *morsprunorum* and *P. s.* pv. *maculicola* strain R1, and is defined as strains showing moderate or low homology to any of the four probes. Results of comparing DNA homology suggested that it might be useful to compare the *hrpL* and *hrpZ* neighbouring regions for grouping *P. syringae*. Therefore, in this study, we compared the structure and DNA sequence of these regions for genotyping *P. syringae* pathovars.

2. MATERIALS AND METHODS

2.1 Bacterial Strains and Culture Conditions

35 strains of 32 pathovars of *Pseudomonas syringae* and one strain of *P. ficuserectae* were used in this study (Inoue and Takikawa 1999a). *Pseudomonas* strains were grown in YP medium (yeast extract 5 g, peptone 10 g, water 1 liter, pH 6.8) with or without agar at 27°C. *Escherichia coli* strains were grown in Luria-Bertani (LB) medium (Maniatis *et al.* 1982) with or without agar at 37°C. Antibiotics were used at 50 μg ml^{-1} for ampicillin and 30 μg ml^{-1} for tetracycline.

2.2 DNA Sequence Analysis of the *hrpL* Neighbouring regions

The cosmid clones, pM32F5 delivered from *Pseudomonas syringae* pv. *maculicula* strain PMC8301, pAK8C6 delivered from *P. s* pv. *actinidiae* strain KW11 and pPIC6 delivered from *P. s.* pv. *phaseolicola* strain NPS3121 were digested with appropriate restriction endonucleases and subcloned into pGEM7Zf(+). These subclones were used as template DNA for sequencing. The single-stranded DNAs for sequencing were provided by PCR method used with fluorescent labeled primers and Sequetherm DNA sequencing kit (Epicenter Technology, Inc., Madison, WI, USA) according to the manufacturer's instructions. Determination of DNA sequences were performed by DNA sequencer LI-COR model 4000 and analysed by Base imager program (LI-COR, inc., Lincoln, NE, USA).

2.3 DNA Sequence Analysis of the *hrpZ* Neighbouring Regions

A DNA fragment was amplified by PCR using degenerate primers designed from published data of *hrpS* and *hrpB* DNA sequences. The amplified DNA fragment was cloned into pCR2.1 and the DNA sequence was analysed by primer walking method. The single-stranded DNAs for sequencing were provided by PCR using non-labelled primers and BigDye terminator kit v.2 (Applied Biosystems, Foster City, CA, USA) according to the manufacturer's instructions. Determination of DNA sequences were performed by ABI Prism 377 DNA sequencer and analysed by Sequencing Analysis V.3.4 software (Applied Biosystems, Foster City, CA, USA). For classification of the DNA sequences the clustalX 1.8 program, obtained from National Center for Biotechnology Information, was used.

3. RESULTS AND DISCUSSION

3.1 Comparison of the neighbouring DNA sequence of the *hrpL*

A comparison of the neighbouring regions of the *hrpL* (n-*hrpL*) among five strains is shown in Fig. 1. All tested strains contained the *hrpY* sequence but

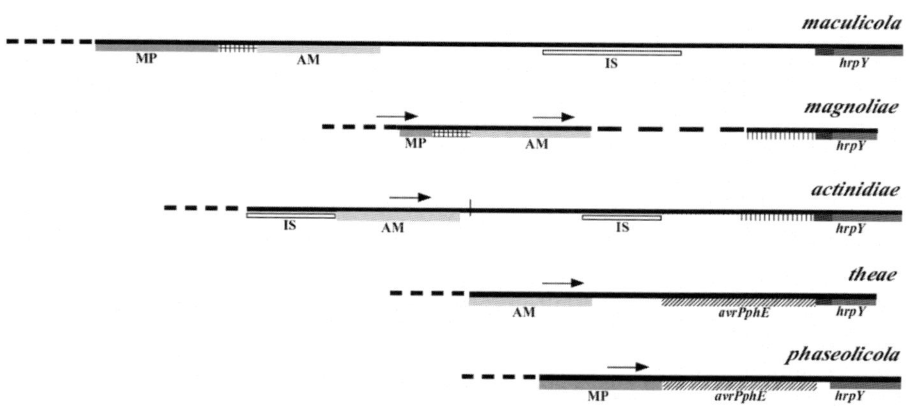

Figure 1. Comparison maps of regions of *hrpL* of *Pseudomonas syringae* pathovars. Filled patterns and coloured lines show homology area. Insertion sequences (IS) did not have any significant homology. Dotted lines show the area which has not been analysed.

had different sequences downstream the *hrpY* stop codon. *Pseudomonas syringae* pv. *phaseolicola* had the *avrPphE* next to the *hrpY* (Mansfield *et al.* 1995). In *P. s.* pv. *maculicola* and *P. s.* pv. *actinidiae*, a conserved region was detected, provisionally named AM. AM is 900 bp length region, located next to the *hrpY* and specific regions. *P. s.* pv. *phaseolicola* did not have any significant homology to AM. At the downstream of AM in *P. s.* pv. *maculicola*, another sequence was identified to have homology with the sequence downstream of the *avrPphE* gene of *P. s.* pv. *phaseolicola*. This region was provisionally named MP. MP did not have any significant homology against the sequence of n-*hrpL* of *P. s.* pv. *actinidiae*. The sequence downstream of AM in *P. s.* pv. *actinidiae* did not show homology to the sequence of n-*hrpL* of any other strains. We investigated distribution of the *hrpY*, *avrPphE*, AM and MP region in other strains of *P. syringae* by the genomic Southern hybridisation. A probe corresponding to *hrpY* of *P. s.* pv. *phaseolicola* hybridised to all *P. syringae* strains with relatively homogeneous strength. A probe corresponding to AM hybridised only to group II strains. In contrast, a probe corresponding to MP or the *avrPphE* hybridised to many strains, though, the distribution of these region did not correspond to our previous grouping (Table 1). PCRs were performed using various primer sets, which were designed based on the *hrpY*, *avrPphE*, MP, and AM sequence data. Those group I strains which had homology to MP region amplified a PCR product using the *hrpY* and MP primers set or the *hrpY* and the *avrPphE* primers set. Partial DNA sequence of PCR fragments showed that group I strains which possessed homology to MP region had the same structure in this region. In contrast, a PCR product was amplified only from the strains of group II when the *hrpY* and AM primers set was used. The size of the DNA fragment varied according to the strain. Partial DNA sequence of PCR fragments showed that the homology area was arranged by DNA splinters in the strains of *P. s.* pv. *magnoliae* and *P. s.* pv. *theae* (Fig. 1). These data suggested that genetic modifications have accumulated in the n-*hrpL* in many strains of *P. syringae*. We previously reported that some insertion sequences were detected in this region (Inoue and Takikawa 1999b). Alfano *et al.* (2000) reported that this region was an unique exchangeable effector locus (EEL). It might be a border of pathogenicity island containing *hrp* genes and some avirulence genes. The AM region existed only in group II strains, therefore, it might be useful for detection and classification of group II strains.

Table 1. Summary of hybridisation results of *Pseudomonas* species and *P. syringae* pathovars. Parts of the downstreem of *hrpL* were used as probes.

Species or pathovar	**A**	**B**	**C**	**D**	**Species or pathovar**	**A**	**B**	**C**	**D**
P.s pv. *phaseolicola* a)	I	+	+	-	pv. *syringae*	III	+	-	-
pv. *glycinea*	I	+	+	-	pv. *lapsa*	III	+	+	-
pv. *tabaci*	I	+	+	-	pv. *japonica*	III	+	+	-
pv. *lachrymans*	I	+	-	-	pv. *aptata*	III	+	+	-
pv. *mori*	I	+	+	-	pv. *aceris*	III	+	-	-
pv. *dendropanacis*	I	+	+	-	pv. *pisi*	III	+	-	-
pv. *eriobotryae*	I	+	-	-					
pv. *myricae*	I	+	-	-	pv. *coronafaciens*	IV	+	+	-
pv. *photiniae*	I	+	+	-	pv. *striafaciens*	IV	+	+	-
pv. *sesami*	I	+	+	-	pv. ?(onion)	IV	+	+	-
pv. ? (kiwi)	I	+	-	-	pv. *atropurprea*	IV	+	+	-
P. ficuserectae	I	+	+	-	pv. *oryzae*	IV	+	+	-
					pv. *porri*	IV	+	+	-
pv. *maculicola* b)	II	+	+	+					
pv. *tomato*	II	+	+	+	pv. *morsprunorum*	U	+	+	-
pv. *theae*	II	+	+	+	pv. *maculicola*(R1)	U	+	-	-
pv. *delphinii*	II	+	+	+					
pv. *magnoliae*	II	+	+	+					
pv. *actinidiae*	II	+	+	+					
pv. *spinaciae*	II	+	-	+					

A: Result of grouping by comparing DNA Homology at the *hrp* Neighbouring Regions (Inoue and Takikawa 1999, 2000); B: Result of genomic southern hybridisation using a probe corresponding to *hrpY*; C: Result of genomic southern hybridisation using a probe corresponding to *avrPphE*; D: Result of genomic southern hybridisation using a probe corresponding to AM.

+: indicates strong hybridisation signal; -: indicates no hybridisation signal; a): containing two strains; b): containing three strains.

3.2 Comparison of DNA Sequence at the Neighbouring Region of *hrpZ*

PCR reactions were carried out using degenerate primers designed from published data of *hrpS* and *hrpB* DNA sequences. 30 of the 34 strains tested were successfully amplified. Comparison of DNA sequences showed that the sequence of the neighbouring region of the *hrpZ* (n-*hrpZ*) was the same from the *hrpS* gene to the *hrpB* gene in each of the 30 strains. A partial deletion in the *hrpZ* gene was detected only in *Pseudomonas syringae* pv. *tabaci*. Based on a comparison of DNA sequences of the *hrpZ* and *hrpA* gene, strains of *P. syringae* were divided into five groups without contradiction to previous grouping (Fig. 2 and Table 2). Group I was sub-divided into two categories based on this study, and the undetermined strains belong to either one of these group (hereafter referred to as group I-A and I-B). Sawada *et al.* (1999) reported 3 groups by comparing sequences of two house-keeping genes (*rpoD*, *gyrB*) and two *hrp* genes (*hrpS*, *L*). Gardan *et al.* (1999) reported the comprehensive grouping of *P. syringae* pathovars through total DNA homology experiments. Our results are in very close agreement with them. Using a small portion of genomic DNA as probes, we demonstrated significant diversity that will lead to the easy and rapid identification of *P. syringae* pathovars. On the other hand, a structural comparison of the large genomic region might give us information on the structural similarity of the genomes of these organisms. This information is never obtained through single gene analysis and may reflect more precisely the history of evolution of organisms than that based on single gene analysis.

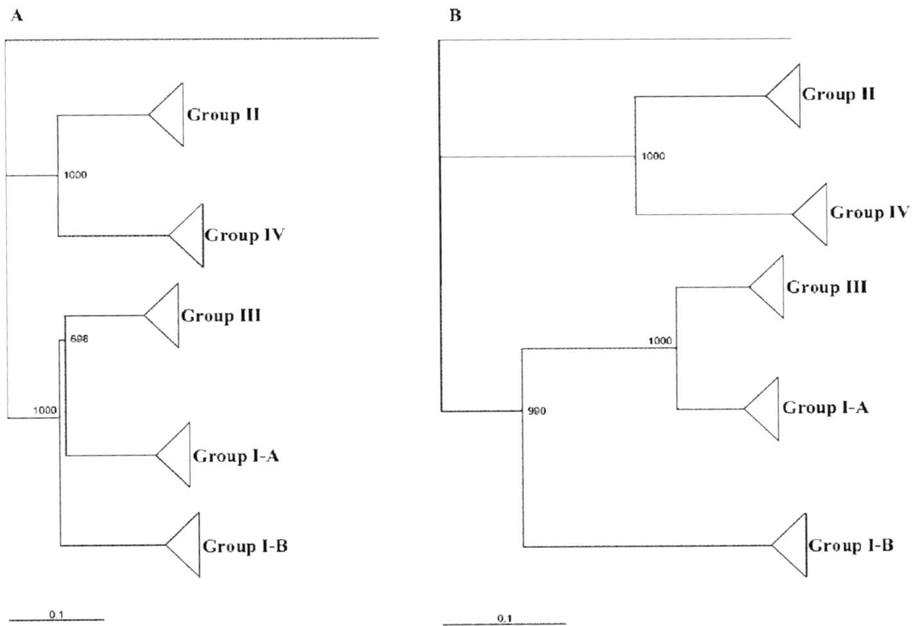

Figure 2. Cluster analysis of the groups of *Pseudomonas syringae* strains based on the *hrpZ* (A) and *hrpA* (B) DNA sequence, using ClustalX 1.8 program (National Center for Biotechnology Information).

Table 2. Results of grouping of *Pseudomonas syringae* pathovars (pv.) based on various criteria.

Species or pathovar	A	B	C	D	E	Species or pathovar	A	B	C	D	E
P.s. pv. *phaseolicola*	I	I-A	I-A	3	2	pv. *syringae*	III	III	III	2	1
pv. *glycinea*	I	I-A	I-A	3	2	pv. *lapsa*	III	III	III	ND	1
pv. *tabaci*	I	I-A	ND	3	2	pv. *japonica*	III	III	III	2	1
pv. *dendropanacis*	I	I-A	I-A	ND	2	pv. *aptata*	III	III	III	2	1
pv. *eriobotryae*	I	I-A	I-A	3	2	pv. *aceris*	III	III	III	2	1
pv. *myricae*	I	I-A	I-A	3	2	pv. *pisi*	III	III	III	2	1
pv. ? (kiwi)	I	I-A	I-A	ND	ND						
pv. *morsprunorum*	U	I-A	I-A	3	2	pv. *coronafaciens*	IV	IV	IV	ND	4
						pv. *striafaciens*	IV	IV	IV	ND	4
pv. *mori*	I	I-B	I-B	3	2	pv. ?(onion)	IV	IV	IV	ND	ND
pv. *lachrymans*	I	I-B	I-B	3	2	pv. *atropurprea*	IV	ND	ND	ND	4
pv. *sesami*	I	I-B	I-B	ND	2	pv. *oryzae*	IV	IV	IV	ND	4
P. ficuserectae	I	I-B	I-B	ND	2	pv. *porri*	IV	ND	ND	ND	4
pv. *maculicola* (R1)	U	I-B	I-B	ND	ND						
pv. *maculicola*	II	II	II	1	3						
pv. *tomato*	II	II	II	1	3						
pv. *theae*	II	II	II	1	8						
pv. *delphinii*	II	II	II	ND	3						
pv. *magnoliae*	II	II	II	ND	ND						
pv. *actinidiae*	II	II	II	1	ND						
pv. *spinaciae*	II	ND	ND	ND	ND						

A: Result of Comparing DNA Homology at the *hrp* Neighboring Regions (Inoue and Takikawa 1999; 2000); B: Result of classification based DNA sequence of *hrpA*; C: Result of classification based DNA sequence of *hrpZ*; D: Genomospecies assigned by Sawada *et al.* (1999); E: Genomospecies assigned by Gardan *et al.* (1999).

REFERENCES

Alfano, J.R., Charkowski, A.O,, Deng, W.L., Badel, J.L., Petnicki-Ocwieja, T., van Dijk K., and Collmer, A., 2000, The *Pseudomonas syringae* Hrp pathogenicity island has a tripartite mosaic structure composed of a cluster of type III secretion genes bounded by exchangeable effector and conserved effector loci that contribute to parasitic fitness and pathogenicity in plants. *Proc. Natl. Acad. Sci. USA*. **97**(9): 4856-4861.

Ausubel, F.M., Brent, R., Kingston, R.E., Moore, D.D., Seidman, J.G., Smith, J.A., and Struhl, K., 1992, *Current Protocols in Molecular Biology*. John Wiley and Sons, New York.

Gardan, L., Shafik, H., Belouin, S., Broch, R., Grimont, F., and Grimont, P.A.D. 1999, DNA relatedness among the pathovars of *Pseudomonas syringae* and description of *Pseudomonas tremae* sp. nov. and *Pseudomonas cannabina* sp. nov. (ex Sutic and Dowson 1959). *Int. J. Syst. Bacteriol.* **49**: 469-478.

Heu, S., and Hutcheson, S.W., 1993, Nucleotide Sequence and Properties of the *hrmA* locus associated with the *Pseudomonas syringae* pv. *syringae* 61 *hrp* gene cluster. *Mol. Plant-Microbe Interact.* **6**: 553-564.

Hueck, C.J., 1998, TypeIII protein secretion systems in bacterial pathogens of animals and plants. *Microbiol. Mol. Biol. Rev.* **62**: 379-433.

Inoue, Y., and Takikawa, Y., 1999a, Grouping *Pseudomonas syringae* strains by comparing DNA homology at the *hrp* gene cluster and its neighboring regions. *Ann. Phytopathol. Soc. Japan*. **65**: 32-41.

Inoue, Y., and Takikawa, Y., 1999b, Investigation of repeating sequences in *hrpL* neighboring region of *Pseudomonas syringae* strains. *Ann. Phytopathol. Soc. Japan*. 65: 100-109.

Inoue, Y., and Takikawa, Y., 2000, *Pseudomonas syringae* strains are classified into five groups by comparing DNA homology at the *hrp* neighboring regions. *J. Gen. Plant Pathol.* 66: 238-241.

Lorang, J.M., and Keen, N.T., 1995, Characterisation of *avrE* from *Pseudomonas syringae* pv. *tomato*: a *hrp*-linked avirulence locus consisting of at least two transcriptional units. *Mol. Plant-Microbe Interact*. **8**: 49-57.

Maniatis, T., Fritsch, E.F., and Sambrook, J., 1982, *Molecular cloning: A Laboratory Manual.* Cold Spring Harbor Laboratory Press, Cold Spring Harbor, New York.

Mansfield, J., Jenner, C., Hockenhull, R., Bennett, M.A., and Stewart, R., 1995, Characterisation of *avrPphE*, a gene for cultivar-specific avirulence from *Pseudomonas syringae* pv. *phaseolicola* which is physically linked to *hrpY*, a new *hrp* gene identified in the halo-blight bacterium. *Mol. Plant-Microbe Interact*. **7**: 726-739.

Peet, R.C., Lindgren, P.B., Willis, D.K., and Panopoulos, N.J., 1986, Identification and cloning of genes involved in phaseolotoxin production by *Pseudomonas syringae* pv. "*phaseolicola*". *J. Bacteriol*. **166**: 1096-1105.

Sawada, H., Suzuki, F., Matsuda, I. and Saitou, N., 1999, Phylogenetic analysis of *Pseudomonas syringae* pathovars suggests the horizontal gene transfer of *argK* and the evolutionary stability of *hrp* gene cluster. *J. Mol. Evolution*. **49**: 627-644

Molecular Characterisation of Spanish *Pseudomonas syringae* pv. *phaseolicola* Isolates

J.A. OGUIZA, A. RICO, L. SUTRA[#*], A. VIVIAN[§] and J. MURILLO
Departamento de Producción Agraria, Universidad Pública de Navarra, 31006 Pamplona, Spain; [#]UMR de Pathologie Vegétale INRA-INH-Université, Beaucouzé, France; [§]Centre for Research in Plant Science, University of West of England, Bristol, UK.

Abstract: In this work, we focused on the molecular characterisation of Spanish *Pseudomonas syringae* pv. *phaseolicola* isolates. PCR amplification profiles of repetitive DNA were similar among the Spanish *P. s.* pv. *phaseolicola* isolates and the reference *P. s.* pv. *phaseolicola* strains, but partially different from the patterns shown by *P. s.* pv. *glycinea*. Taking these profiles into consideration, two *P. s.* pv. *phaseolicola* genomic groups have been distinguished. Additionally, using a DNA heteroduplex mobility assay (HMA) performed on the internal transcribed spacer (ITS) region between 16S and 23S rRNA genes, it was possible to differentiate Group 1 isolates, which clearly present a unique HMA band, from Group 2 isolates, which show multiple HMA bands.

Key words: Genomic groups, rep-PCR, phaseolotoxin, heteroduplex mobility assay.

Author for correspondence: J. Murillo; e-mail: jesus@unavarra.es.

1. INTRODUCTION

Pseudomonas syringae pv. *phaseolicola* is an economically important seedborne pathogen of beans (*Phaseolus vulgaris*) world-wide, causing halo blight disease (Smith *et al.*, 1988). Additionally, *P. s.* pv. *phaseolicola* is intensively used as a model pathogen for the study of the molecular basis of pathogenicity. Recently, the ability of *P. s.* pv. *phaseolicola* to produce disease in its plant host bean has been shown to be based on the possession of a

[*] Deceased on December the 16th, 2002.

N.S. Iacobellis *et. al.* (eds.), *Pseudomonas syringae* and Related Pathogens, 697-703.

pathogenicity island (PAI) localised to an approx. 150 kb plasmid (Jackson *et al.*, 1999). The main pathogenicity genes identified in this PAI were *virPphA* and *avrPphF* (Jackson *et al.*, 1999; Tsiamis *et al.*, 2000). In addition to the PAI, *P. s.* pv. *phaseolicola* strains are defined by possession of the *argK-tox* gene cluster region, which directs phaseolotoxin biosynthesis and appears to increase virulence (Patil *et al.*, 1974; Mitchell, 1978). It is generally believed that only isolates containing this cluster (Tox+ isolates) are of epidemiological significance and, hence, this DNA region is commonly used as a target for PCR detection and identification of *P. s.* pv. *phaseolicola* (Rudolph, 1995; Schaad *et al.*, 1995).

P. s. pv. *phaseolicola* can easily be distinguished from other common pathovars of *P. syringae* pathogenic to beans, such as pathovars *syringae*, *glycinea* and *tabaci*, by nutritional characteristics and because only *P. s.* pv. *phaseolicola* isolates produce water soaked lesions on bean pods (Völksch and Weingart, 1997; dos A. Marques *et al.*, 2000). In general, *P. s.* pv. *phaseolicola* appears to be a more or less homogeneous pathovar, although it displays a degree of genetic and phenotypic variation that overlaps with isolates from pathovars *glycinea* and *tabaci* (dos A. Marques *et al.*, 2000). Additionally, intrapathovar variation in *P. s.* pv. *phaseolicola* can be linked in some cases to the host of isolation. Isolates that produce natural infections on kudzu vine (*Pueraria lobata*) are distinguished for carrying a plasmid-borne *efe* gene (Nagahama *et al.*, 1994) and, similar to those isolates from *Vigna radiata*, by their rep-PCR profile with ERIC primers (Völksch and Weingart, 1997; dos A. Marques *et al.*, 2000).

We have shown that, in sharp contrast with the expected prevalence of Tox+ strains, more than 67% of the Spanish *P. s.* pv. *phaseolicola* natural isolates lack the phaseolotoxin gene cluster region (Rico *et al.*, 2003). In the present study, we present preliminary results that suggest that these Tox– isolates could constitute a distinct genomic line that may have evolved separately from that of the Tox+ strains.

2. MATERIALS AND METHODS

2.1 Bacterial Strains

The Tox+ (CYL207, CYL215, CYL285 and CYL286) and Tox– Spanish *Pseudomonas syringae* pv. *phaseolicola* isolates (CYL233, CYL275, CYL309, CYL314, CYL315 and CYL325) were obtained from naturally diseased beans and have been previously characterised (Rico *et al.*, 2003). As reference strains we used strains 1449B (race 7) and 2709A (race 9) of *P.*

s. pv. *phaseolicola* (Taylor *et al.*, 1996), and strain 49a/90 of *P. s.* pv. *glycinea* (Ullrich *et al.*, 1993).

2.2 Rep-PCR Analysis

The genetic relationship among the different isolates was examined by PCR using primers for enterobacterial repetitive intergenic consensus (ERIC), repetitive extragenic palindromic (REP), and the arbitrarily primed PCR (AP-PCR) techniques. For ERIC and REP analyses, primers were as described (McManus and Jones, 1995). AP-PCR was carried out with a single 20 bp oligonucleotide primer complementary to the IS*50* portion of Tn*5*, as described (Sundin and Murillo, 1999), or using the universal M13 reverse primer. Cell suspensions of the bacterial isolates prepared in sterile water were used as PCR templates, and PCR reactions were carried out as described by McManus and Jones (1995).

2.3 DNA Heteroduplex Mobility Assay (HMA)

The sequence polymorphism of the internal transcribed spacer (ITS) was analysed using a DNA heteroduplex mobility assay (HMA) performed as described by Delwart *et al.* (1993). The ITS was amplified using primers D21 and D22 as described by Manceau and Horvais (1997) and PCR products were migrated in 5% polyacrylamide gels (Delwart *et al.*, 1993).

3. RESULTS AND DISCUSSION

Genetic fingerprinting of repetitive DNA sequences present in bacterial genomes (rep-PCR) has been used as a diagnostic tool (Versalovic *et al.*, 1991) to assess genetic diversity and differentiation of plant-associated bacteria (de Bruijn, 1992; Louws *et al.*, 1994; Sundin *et al.*, 1994). Therefore, we applied this technique for the characterisation of the Spanish population of the halo blight pathogen. The rep-PCR amplification profiles were similar among the Spanish *Pseudomonas syringae* pv. *phaseolicola* isolates and the reference *P. s.* pv. *phaseolicola* strains (1449B and 2709A), but could be easily differentiated from the patterns shown by strains of *P. s.* pv. *glycinea*. An example of the different fingerprints observed is shown in Fig. 1.

From the differences observed between the rep-PCR profiles of the Spanish *P. s.* pv. *phaseolicola* isolates (Fig. 1 and data not shown), two groups of strains could be distinguished. Group 1 contained the Tox+ Spanish isolates CYL215, CYL285, CYL207, CYL286, as well as the *P. s.* pv. *phaseolicola* strains used as reference (1449B and 2709A; also Tox+ strains). In turn,

Group 2 contained all the Tox– Spanish isolates (CYL233, CYL275, CYL325, CYL309, CYL314 and CYL315).

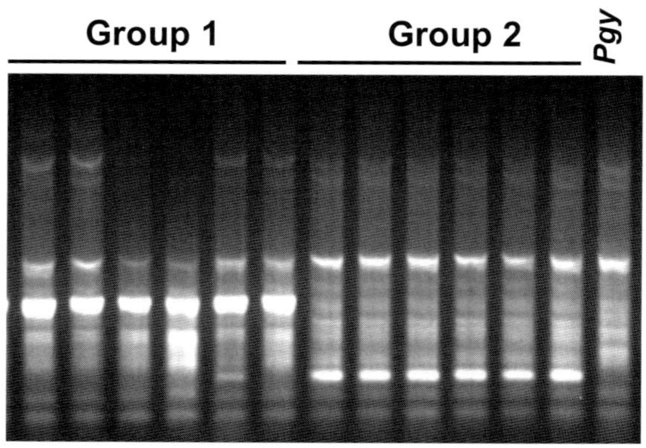

Figure 1. Agarose gel electrophoresis of AP-PCR (IS*50* primer) fingerprinting patterns from strains of *Pseudomonas syringae* pv. *phaseolicola* and *P. s.* pv. *glycinea* (*Pgy*).

Other authors have reported the generation of distinct ERIC profiles for strains of *P. s.* pv. *phaseolicola* isolated from kudzu and for those isolated from *Vigna radiata* (Völksch and Weingart, 1997; dos A. Marques *et al.*, 2000). On the basis of these and other differences, these authors have proposed that kudzu strains could represent a separate group or even a different pathovar. Kudzu strains, unlike bean isolates of *P. s.* pv. *phaseolicola* (including the Spanish Tox–; Rico *et al.*, this volume), have the capacity to produce ethylene (Nagahama *et al.*, 1991), although this capacity is due to the possession of a plasmid-borne *efe* gene (Nagahama *et al.*, 1994). Since plasmids carrying *efe* were shown to be mobilizable, this character is probably of little value for the differentiation of groups of isolates. However, kudzu strains are also differentiated by their esterase zymotypes and O-serogroup (dos A. Marques *et al.*, 2000). These results point to the existence of significant genotypic differences between kudzu and bean isolates, and suggests the validity of the rep-PCR technique for the separation of groups of strains. Nevertheless, we decided to evaluate the diversity among Tox+ and Tox– isolates by using other techniques.

Heteroduplex mobility assay (HMA) is a powerful technique for screening sequence dissimilarities between annealed strands of DNA (Delwart *et al.*, 1993), and has been applied to determine the 16S rDNA similarity of bacterial isolates (Espejo *et al.*, 1998). This technique was therefore used to evaluate possible differences in the sequence of the ITS regions of the Spanish *P. s.* pv. *phaseolicola* isolates.

Group 1 isolates had a profile with a unique band (homoduplex), while Group 2 isolates had a profile with two additional bands of reduced mobility that correspond to heteroduplexes (Fig. 2). These results indicate that the different copies of ITS are identical within each isolate of Group 1, but that they are different within each isolate of Group 2, so that heteroduplexes are formed (Delwart *et al.,* 1993).

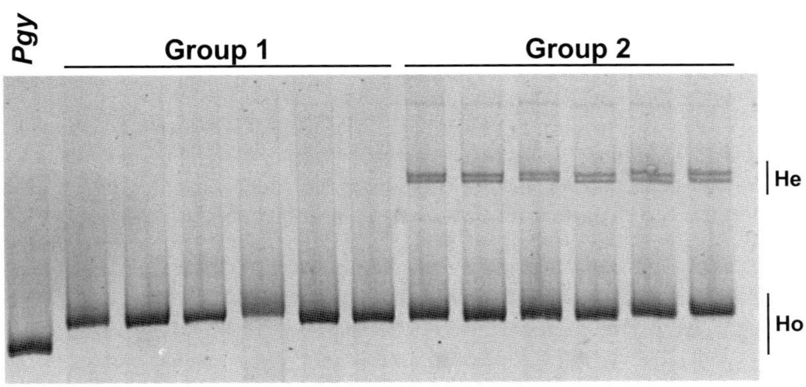

Figure 2. Electrophoretic HMA profiles from *Pseudomonas syringae* pv. *phaseolicola* and *P. s.* pv. *glycinea* (*Pgy*) strains obtained by direct migration of PCR-amplified ITS. Ho: homoduplex; He: heteroduplex.

Our results, therefore, imply that the differences observed between Group 1 and Group 2 isolates by rep-PCR fingerprinting reflect the existence of important differences at a genomic level, and support the existence of two genomic groups among the Spanish *P. s.* pv. *phaseolicola* isolates. It would be of interest to determine if isolates from kudzu also represent another genomic line.

ACKNOWLEDGEMENTS

We wish to thank Sophie Bonneau for technical assistance and Carmen Asensio (Servicio de Investigación y Tecnología Agraria, Junta de Castilla y León, Valladolid) for the Spanish *Pseudomonas syringae* pv. *phaseolicola* isolates.
JAO was supported by the Ramón y Cajal program of the Spanish Ministerio de Ciencia y Tecnología. This work was supported by grants AGL2001-1948-CO2-01 from CICYT to JM and the British Council Acciones Integradas program and BBSRC to AV.

REFERENCES

de Bruijn, F.J., 1992, Use of repetitive (repetitive extragenic palindromic and enterobacterial repetitive intergenic consensus) sequences and the polymerase chain reaction to fingerprint the genomes of Rhizobium meliloti isolates and other soil bacteria. *Appl. Environ. Microbiol.* **58**: 2180-2187.

Delwart, E.L., Shpaer, E.G., Louwagie, J., Mc Cutchan, E., Grez, M., Rübsamen-Waigmann, H., and Mullins, J.I., 1993, Genetic relationships determined by a DNA heteroduplex mobility assay: analysis of HIV-1 env genes. *Science* **262**: 1257-1261.

Espejo, R.T., Feijóo, C.G., Romero, J., and Vásquez, M., 1998, PAGE analysis of the heteroduplexes formed between PCR-amplified 16S rRNA genes: estimation of sequence similarity and rDNA complexity. *Microbiology* **144**: 1611-1617.

Jackson, R.W., Athanassopoulos, E., Tsiamis, G., Mansfield, J.W., Sesma, A., Arnold, D.L., Gibbon, M.J., Murillo, J., Taylor, J.D., and Vivian, A., 1999, Identification of a pathogenicity island, which contains genes for virulence and avirulence, on a large native plasmid in the bean pathogen *Pseudomonas syringae* pathovar phaseolicola. *Proc. Natl. Acad. Sci. USA* **96**: 10875-10880.

Louws, F.J., Fullbright, D.W., Stephens, C.T., and de Bruijn, F.J., 1994, Specific genomic fingerprints of phytopathogenic *Xanthomonas* and *Pseudomonas* pathovars and strains generated with repetitive sequences and PCR. *Appl. Environ. Microbiol.* **60**: 2286-2295.

Manceau, C., and Horvais, A., 1997, Assessment of genetic diversity among strains of *Pseudomonas syringae* by PCR-restriction fragment length polymorphism analysis of rRNA operons with special emphasis on *P. syringae* pv. tomato. *Appl. Environ. Microbiol.* **63**: 498-505.

Marques, A.S., Corbiere, R., Gardan, L., Tourte, C., Manceau, C., Taylor, J.D., and Samson, R., 2000, Multiphasic approach for the identification of the different classification levels of *Pseudomonas savastanoi* pv. *phaseolicola*. *Eur. J. Plant Pathol.* **106**: 715-734.

McManus, P.S., and Jones, A.L., 1995, Genetic fingerprinting of *Erwinia amylovora* strains isolated from tree-fruit crops and *Rubus* spp. *Phytopathology* **85**: 1547-1553.

Mitchell, R.E., 1978. Halo blight of beans: Toxin production by several *Pseudomonas phaseolicola* isolates. *Physiol. Plant Pathol.* **13**: 37-49.

Nagahama, K., Ogawa, T., Fujii, T., Tazaki, M., Tanase, S., Morino, Y., and Fukuda, H., 1991, Purification and properties of an ethylene-forming enzyme from *Pseudomonas syringae* pv *phaseolicola* PK2. *J. Gen. Microbiol.* **137**: 2281-2286.

Nagahama, K., Yoshino, K., Matsuloa, M., Sato, M., Tanase, S., Ogawa, T., and Fukuda, H., 1994, Ethylene production by strains of the plant-pathogenic bacterium *Pseudomonas syringae* depends upon the presence of indigenous plasmids carrying homologous genes for the ethylene-forming enzyme. *Microbiology* **140**: 2309-2313.

Patil, S.S., Hayward, A.C., and Emmons, R., 1974, An ultraviolet-induced non-toxigenic mutant of *Pseudomonas phaseolicola* of altered pathogenicity. *Phytopathology* **64**: 590-595.

Rico, A., López, R., Aizpún, M.T., Asensio, C., and Murillo, J., 2003, Limitations of the Current Methods for the Detection of Spanish Strains of *Pseudomonas syringae* pv. *phaseolicola*, 545-552, (this volume).

Rudolph, K.W.E., 1995, *Pseudomonas syringae* pathovars. In *Pathogenesis and host specificity in plant diseases* (U.S. Singh, R.P. Singh and K. Kohmoto, eds.), Elsevier Science Ltd., Oxford, U.K., Vol. 1, pp. 47-138.

Schaad, N.W., Cheong, S.S., Tamaki, S., Hatziloukas, E., and Panopoulos, N.J., 1995, A combined biological and enzymatic amplification (BIO-PCR) technique to detect

Pseudomonas syringae pv. *phaseolicola* in bean seed extracts. *Phytopathology* **85**: 243-248.

Smith, I.M., Dunez, J., Lelliott, R.A., Phillips, D.H., and Archer, S.A., 1988, *European handbook of plant diseases*. Blackwell Scientific Publications, London.

Sundin, G.W., and Murillo, J., 1999, Functional analysis of the *Pseudomonas syringae rulAB* determinant in tolerance to ultraviolet B (290-320 nm) radiation and distribution of *rulAB* among *P. syringae* pathovars. *Environ. Microbiol.* **1**: 75-88.

Sundin, G.W., Demezas, D.H., and Bender, C.L., 1994, Genetic and plasmid diversity within natural populations of *Pseudomonas syringae* with various exposures to copper and streptomycin bactericides. *Appl. Environ. Microbiol.* **60**: 4421-4431.

Taylor, J.D., Teverson, D.M., Allen, D.J., and Pastor-Corrales, M.A., 1996, Identification and origin of races of *Pseudomonas syringae* pv. *phaseolicola* from Africa and other bean growing areas. *Plant Pathol.* **45**: 469-478.

Tsiamis, G., Mansfield, J.W., Hockenhull, R., Jackson, R.W., Sesma, A., Athanassopoulos, E., Bennett, M.A., Stevens, C., Vivian, A., Taylor, J.D., and Murillo, J., 2000, Cultivar specific avirulence and virulence functions assigned to *avrPphF* in *Pseudomonas syringae* pv. *phaseolicola*, the cause of bean halo-blight disease. *EMBO J.* **19**: 3204-3214.

Ullrich, M., Bereswill, S., Volksch, B., Fritsche, W., and Geider, K., 1993, Molecular characterization of field isolates of *Pseudomonas syringae* pv. *glycinea* differing in coronatine production. *J. Gen. Microbiol.* **139**: 1927-1937.

Versalovic, J., Koeuth, T., and Lupski, J.R., 1991, Distribution of repetitive DNA sequences in eubacteria and application to fingerprinting of bacterial genomes. *Nucl. Acids Res.* **19**: 6823-6831.

Völksch, B., and Weingart, H., 1997, Comparison of ethylene-producing *Pseudomonas syringae* strains isolated from kudzu (*Pueraria lobata*) with *Pseudomonas syringae* pv. *phaseolicola* and *Pseudomonas syringae* pv. *glycinea*. *Eur. J. Plant Pathol.* **103**: 795-802.

Author Index